HANDBOOK OF CLINICAL NUTRITION AND AGING

NUTRITION ◊ AND ◊ HEALTH
Adrianne Bendich, Series Editor

Handbook of Clinical Nutrition and Aging, *Second Edition*, edited by **Connie Watkins Bales and Christine Seel Ritchie**, 2009

Handbook of Nutrition and Pregnancy, edited by **Carol J. Lammi-Keefe, Sarah Collins Couch, and Elliot H. Philipson**, 2008

Nutrition and Health in Developing Countries, *Second Edition*, edited by **Richard D. Semba and Martin W. Bloem**, 2008

Nutrition and Rheumatic Disease, edited by **Laura A. Coleman**, 2008

Nutrition in Kidney Disease, edited by **Laura D. Byham-Gray, Jerrilynn D. Burrowes, and Glenn M. Chertow**, 2008

Handbook of Nutrition and Ophthalmology, edited by **Richard D. Semba**, 2007

Adipose Tissue and Adipokines in Health and Disease, edited by **Giamila Fantuzzi and Theodore Mazzone**, 2007

Nutritional Health: Strategies for Disease Prevention, Second Edition, edited by **Norman J. Temple, Ted Wilson, and David R. Jacobs, Jr.**, 2006

Nutrients, Stress, and Medical Disorders, edited by **Shlomo Yehuda and David I. Mostofsky**, 2006

Calcium in Human Health, edited by **Connie M. Weaver and Robert P. Heaney**, 2006

Preventive Nutrition: The Comprehensive Guide for Health Professionals, Third Edition, edited by **Adrianne Bendich and Richard J. Deckelbaum**, 2005

The Management of Eating Disorders and Obesity, Second Edition, edited by **David J. Goldstein**, 2005

Nutrition and Oral Medicine, edited by **Riva Touger-Decker, David A. Sirois, and Connie C. Mobley**, 2005

IGF and Nutrition in Health and Disease, edited by **M. Sue Houston, Jeffrey M. P. Holly, and Eva L. Feldman**, 2005

Epilepsy and the Ketogenic Diet, edited by **Carl E. Stafstrom and Jong M. Rho**, 2004

Handbook of DrugNutrient Interactions, edited by **Joseph I. Boullata and Vincent T. Armenti**, 2004

Nutrition and Bone Health, edited by **Michael F. Holick and Bess Dawson-Hughes**, 2004

Diet and Human Immune Function, edited by **David A. Hughes, L. Gail Darlington, and Adrianne Bendich**, 2004

Beverages in Nutrition and Health, edited by **Ted Wilson and Norman J. Temple**, 2004

HANDBOOK OF CLINICAL NUTRITION AND AGING

Second Edition

Edited by

CONNIE WATKINS BALES, PhD, RD, FACN

Durham VA Medical Center and Duke University Medical Center, Durham, NC

and

CHRISTINE SEEL RITCHIE, MD, MSPH

Birmingham VA Medical Center, University of Alabama at Birmingham, Birmingham AL

Foreword by

NANCY S. WELLMAN, PhD, RD, FADA

Former Director, National Resource Center on Nutrition, Physical Activity and Aging, Florida International University, Miami, FL Past President, The American Dietetic Association

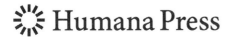 Humana Press

Editors

Connie Watkins Bales
Durham VA Medical Center and Duke
University Medical Center
Durham, NC

Christine Seel Ritchie
Birmingham VA Medical Center
University of Alabama at Birmingham
Birmingham, AL

Series Editor

Adrianne Bendich
GlaxoSmithKline Consumer Healthcare
Parsippany, NJ

ISBN 978-1-60327-384-8   e-ISBN 978-1-60327-385-5
DOI 10.1007/978-1-60327-385-5

Library of Congress Control Number: 2009920207

springer.com

Dedications

Connie Watkins Bales dedicates this volume to her children, Audrey Ashburn Bales Britton and William Brittain Bales, in appreciation of all the ways they have enriched her life and with enthusiastic anticipation of all they are becoming.

Christine Seel Ritchie dedicates this volume to the memory of her father, David John Seel, MD, FACS, a man of compassion who was devoted to life-long learning.

Acknowledgements

CWB would like to recognize and thank Tien Thi Ho who, as a Duke student assistant, worked tirelessly for two academic years on the management and copy editing of this text, contributing substantially to its quality. Thanks also to Justin (Cody) Maxwell and Caroline Friedman for their contributions to this project. CWB and CSR thank our series editor, Dr. Adrianne Bendich, for her encouragement to begin what has become an ongoing and exciting set of encounters with critical clinical issues in geriatric nutrition and the gifted and dedicated scientists who study them. Without the creative contributions of these scientist-authors, this book would not have been possible.

Series Introduction

The Nutrition and Health series of books have, as an overriding mission, to provide health professionals with texts that are considered essential because each includes (1) a synthesis of the state of the science, (2) timely, in-depth reviews by the leading researchers in their respective fields, (3) extensive, up-to-date fully annotated reference lists, (4) a detailed index, (5) relevant tables and figures, (6) identification of paradigm shifts and the consequences, (7) virtually no overlap of information between chapters, but targeted, inter-chapter referrals, (8) suggestions of areas for future research and (9) balanced, data-driven answers to patient/health professionals questions that are based upon the totality of evidence rather than the findings of any single study.

The series volumes are not the outcome of a symposium. Rather, each editor has the potential to examine a chosen area with a broad perspective, both in subject matter as well as in the choice of chapter authors. The international perspective, especially with regard to public health initiatives, is emphasized where appropriate. The editors, whose trainings are both research and practice oriented, have the opportunity to develop a primary objective for their book; define the scope and focus; and then invite the leading authorities from around the world to be part of their initiative. The authors are encouraged to provide an overview of the field, discuss their own research and relate the research findings to potential human health consequences. Because each book is developed de novo, the chapters are coordinated so that the resulting volume imparts greater knowledge than the sum of the information contained in the individual chapters.

"Handbook of Clinical Nutrition and Aging, Second Edition" edited by Connie Watkins Bales and Christine Seel Ritchie fully exemplifies the Nutrition and Health Series' goals. The first volume of the handbook, published in 2004, was acknowledged by reviewers as the most comprehensive volume available concerning the role of clinical nutrition in preserving the health of older adults – especially those suffering from established chronic disease. The second edition is very timely as the fastest growing population in the US as well as globally is those over 60 years of age and especially the oldest-old, those over 80 years of age. This important text provides practical, data-driven options to enhance this at-risk population's potential for optimal health and disease prevention with special emphasis on secondary disease prevention and therapeutic nutritional interventions. The overarching goal of the editors is to provide fully referenced information to health professionals, so that they may enhance the nutritional welfare and overall health of their older adult clients and family members. This excellent, up-to-date volume will add great value to the practicing health professional as well as those professionals and students who have an interest in the latest information on the science behind the aging process, and the potential for nutrition to modulate the effects of chronic diseases and conditions that are widely seen in the geriatric population.

Drs. Bales and Ritchie, who have edited the first and second editions, are internationally recognized leaders in the field of clinical nutrition and aging. Both editors are excellent communicators and they have worked tirelessly to develop a book that continues to be the benchmark in the field because of its extensive, in-depth chapters covering the most important aspects of the complex interactions between cellular functions, diet and nutrient requirements and their impact on the chronic diseases as well as the acute conditions that can adversely affect the quality of life and health of older individuals. The editors have chosen 40 of the most well-recognized and respected authors, internationally distinguished researchers, clinicians and epidemiologists, who provide a comprehensive foundation for understanding the role of nutrients and other dietary factors in the clinical aspects of nutritional management of the elderly.

Hallmarks of all the 29 chapters include complete explanations of terms, with the abbreviations fully defined for the reader, and consistent use of terminology between chapters. Key features of this comprehensive volume include the informative bulleted summary points and key words that are at the beginning of each chapter and appendices that include a detailed list of relevant nutrition resources, including lists of books, journals and websites. Glossaries of terms and abbreviations are provided as needed and recommendations for clinicians are included at the end of relevant chapters. The volume contains more than 45 detailed tables and informative figures, an extensive, detailed index and more than 1100 up-to-date references that provide the reader with excellent sources of worthwhile information about nutrition options to help maintain the health of seniors.

The first section of the volume contains three chapters that examine overarching issues for nutritional well-being in later life. The first chapter examines the complex factors that affect food choices. As one ages, the social interactions at mealtimes greatly affect food choices and intake. Also relevant is where the meals are consumed – in the home, in a hospital or nursing home or other type of institution, as examples. National feeding programs available in the US are described and relevant details about how these affect the access to food for the elderly are reviewed. The second chapter reviews the role of behavior modification in assuring the benefits of therapeutic nutritional changes. Two major determinants of success in adherence to dietary compliance are enhancement of patient knowledge and understanding of the value of the change for their own health and secondly, enhancement of patient confidence that they can make the changes and maintain them over the long term. Six behavioral theories are discussed in detail and helpful educational materials are also provided in this informative chapter. The third chapter highlights changes in population demographics in both the developed and developing world, the so-called "global graying" attributed to the combination of lower birth rates and increased longevity. A detailed discussion of demographics, diet and disease trends in China serves as an example of the potential effects of the Westernized diet on causes of death as they shift from infectious to chronic diseases associated with obesity. As in 36 other developing countries, in China overweight exceeds underweight as a nutritional problem. Although population growth has been curtailed due to the one child/family policy, lifespan has increased dramatically in the past 40 years. There are currently more than 100 million Chinese who are 65 years or older, and that number is increasing annually (from 8% now to 24% of the population by 2050). At present, China has more people 65 and older than all European countries combined. Family care of elderly parents remains the norm in China and may be a major factor that differentiates elder care in China from that seen in the US and other Western cultures. However, urbanization and smaller living spaces may lead to changes in the care of older family members in future generations.

The second section deals with the fundamentals of nutrition and geriatric syndromes in 10 chapters. The first chapter in this section reviews the majority of nutrition screening tools available for dietary intake assessment geared to seniors and examples are included in the nine tables. Tools for assessment of frailty are also discussed. The most critical information for assessment of overall nutritional status remains body mass index and recent weight loss. Sensory signals, including taste and smell, are key factors affecting the nutritional status of seniors and we are reminded in Chapter 5 that many of the medications that are commonly taken as we age affect these senses negatively. Visual and auditory losses also affect responses to food and eating experiences. There are somatosensory changes with aging that result in lowered oral, touch and other temperature-related sensations. A separate chapter reviews the role of certain environmental factors, such as smoking and sunlight exposure, in increasing the risk of vision loss. The latest data on the potential for essential nutrients to prevent cataracts and age-related macular degeneration – the two major causes of blindness in the elderly – are included in detailed tables. Nutrients reviewed include vitamins C and E, carotenoids including lutein and zeaxanthin, zinc and omega-3 fatty acids. The recommendation is to consume diets that are rich in these micronutrients. To this end, extensive tables listing foods that contain these nutrients are included.

The important changes that occur throughout the gastrointestinal tract, beginning in the mouth, are outlined in the seventh comprehensive chapter. Topics such as dysphagia, gastroesophageal reflux disease (GERD), gastritis, ulcers, diarrhea, fecal incontinence, constipation, colitis, inflammatory bowel disease, lactose intolerance, GI bleeding, anemia and hepatitis are all discussed and clinical recommendations are provided. There is an important chapter on the changes in the stimulus for thirst and potential for dehydration in the elderly. Deficiencies in sodium and certain trace minerals and electrolyte imbalances that may be drug, illness or age induced are reviewed.

Nutritional frailty, which is characterized by the loss of both muscle and fat, is often the consequence of unintentional progressive decreases in food intake in the elderly. Nutritional frailty differs from sarcopenia and cachexia, and, thus, each of these conditions that significantly affect health in the aging population is given its own in-depth chapter. In contrast to the loss of weight in the overweight or obese adult <65 years that is associated with reduced mortality risk, even a small loss of weight over age 65 is associated with an increased risk of death. The difference may be due to the change in body composition in older adults with the replacement of muscle with fat and the loss of bone. There may also be a loss of appetite and hormonal changes may also increase the potential for unintended weight loss. Information is given about the interactions between physiological, psychological and socioeconomic factors that may increase the risk of weight loss. Guidance is also provided on the introduction of nutritional supplements and drugs that may enhance appetite, and enteral and parenteral nutrition options in older adults who continue to lose weight. Sarcopenia, which is defined as age-related loss of skeletal muscle mass, is most prevalent in individuals who consume low protein diets and who are sedentary, but occurs almost universally as adults grow older. Lower body exercises that include resistance activities and protein-rich diets may help to avert the loss of muscle, functional impairments and loss of mobility seen in those with muscle loss. Cachexia includes sarcopenia and, because of its relevance to aging, is discussed in Chapter 11. Cachexia, the wasting of skeletal muscle and loss of protein and energy stores resulting from disease, is directly related to inflammatory states such as seen in immune-related diseases and cancer. In contrast to starvation, which can be reversed with increased intake, cachexia is driven by inflammatory cytokines that reduce hunger that is not abated with provision of food. The use of anti-inflammatory agents is discussed. An often seen consequence of cachexia in bed-ridden elderly is pressure sores. The chapter on pressure ulcers documents the strong association between nutritional status and incidence, progression and severity of these sores. The review of macronutrient and micronutrient interventions to prevent and/or treat pressure sores concludes that general nutritional support can help to prevent diet deficiencies and this may or may not affect the progression of pressure sores. The final chapter in this section addresses the sensitive issue of provision of nutrients at the end of life. Careful consideration must be given by family members in consultation with health providers concerning the legal and ethical distinction between acts of omission and acts of commission with regard to terminal nutrition and hydration. Religious considerations may also affect decisions about initiating artificial nutrition and hydration. Nutritional support for end-stage cancer patients with cachexia has not yet been shown to improve survival. Some studies have found that terminally ill patients are neither hungry nor thirsty and small amounts of food and liquid satisfy their needs. However, decisions about tube feeding for patients with terminal stages of Alzheimer's or other dementias may have more emotional than objective considerations. The chapter provides valuable guidance to attending physicians as well as caregivers of the terminally ill.

The third section of the volume relates to common clinical conditions seen in the geriatric population. The first chapter in this section looks at the importance of dental health to the overall nutritional status in the elderly. The major issues are dental and root caries, periodontal disease and tooth loss; loss of saliva (xerostomia) impacts these factors as well as affecting the ability to swallow food. Survey data confirm that about 1/3 of adults 75 years and older have no teeth (edentulous). The incidence of oral cancers and consequent mortality increases above age 65. Diet is implicated in all aspects of oral health and diet-related diseases such as diabetes increase the risk of tooth loss whereas lifestyle habits, such as smoking, increase the risk of head and neck cancers and are linked to lowered diet quality; oral cancer therapies also can further decrease nutritional status.

Obesity is a common global clinical condition and is also seen in the elderly. Obesity is associated with increased risk of mortality and morbidity, including decreased mobility and decreases in other activities of daily living. Gradual, modest weight loss is recommended in Chapter 15; however, this should include an exercise program to preserve muscle mass and sufficient calcium and vitamin D to help counteract any attendant bone loss. Along with the increased prevalence of obesity in the elderly, we see increased prevalence of diabetes; almost half of individuals with self-reported diabetes are 65 years or older. Over 40% of US adults over 70 years have been diagnosed with metabolic syndrome. Diabetes, and its co-morbidities, such as decreased vision and depression, can adversely affect diet quality. Both type 1 and type 2 diabetes diagnoses, treatments, and dietary and lifestyle approaches are well described in Chapter 16. Detailed information is also provided about the metabolic syndrome as well as identification and treatment of hypoglycemia and co-morbidities in the aged.

Cardiovascular disease remains the leading cause of death in older adults, and two important chapters review the nutritional aspects of heart disease in particular. In the chapter on cardiac rehabilitation, emphasis is placed on the multidisciplinary team that includes nutrition counseling with emphasis on lipids, antioxidants, salt reduction and increased whole grains, fruits and vegetables. The chapter on heart failure documents its effects on nutritional requirements, which are often dependent upon the types and dosages of medications given to treat the disease. Two of the major causes of heart failure are hypertension and coronary heart disease. Heart failure is the number one cause of hospitalization in the Medicare population, and hospitalization affects food intake and nutritional status, usually adversely. Activity levels are greatly reduced in heart failure and, in the end stage, cachexia is common. Nutritional interventions are complex and described in detail and recommendations are included.

Cancer is a disease of aging and the development of cancer as well as its treatment greatly impacts the nutritional status of the senior patient. The type of cancer and its stage are relevant factors in the development of malnutrition in the cancer patient regardless of age, but aging adds to the potential severity of the nutritional deficits. Specifically, by using assessment tools such as the comprehensive geriatric assessment, often there is the finding that protein intake, vitamins D, B_6, B_{12}, calcium and iron status may be reduced and further decreased with cancer treatments.

The next four chapters deal individually with chronic conditions and diseases including chronic obstructive pulmonary disease (COPD), chronic kidney disease (CKD), osteoporosis and osteoarthritis. Each of these chronic conditions is characterized by a decrease in mobility, significant changes in lifestyles and frequent pain. Chapter 20 describes the consequences of COPD that include a loss of weight and increase in basal metabolic rate. Use of multiple drugs is common and adverse drug/nutrient interactions can be found in all of these complex conditions. CKD is often a consequence of long-term hypertension, obesity and diabetes. At end stage, dialysis requires careful monitoring of mineral and protein intakes. A useful summary table of diet recommendations through the stages of CKD is provided. Osteoporosis is defined by the WHO as a loss of bone mineral density (BMD) greater than 2 standard deviations below the mean compared to the BMD seen in young adults. Lower than optimal intakes of calcium and vitamin D, as well as several other key essential nutrients, over the lifetime significantly increases the risk of low BMD and fractures. Of importance, and not well known, even during treatment with drugs to treat osteoporosis, there is a continued need for optimal intake of calcium, vitamin D and protein to help maintain bone strength and density. Osteoarthritis is the most common arthritis seen in seniors and is often the reason behind joint replacement operations. Osteoarthritis is associated with damage to the cartilage at joints. Anti-inflammatory drugs are commonly used to treat the pain associated with osteoarthritis. The 23rd chapter contains an extensive review of the clinical studies with dietary supplements including glucosamine, chondroitin, omega-3 fatty acids, avocado and soybean unsaponifiables, iodine and antioxidants including selenium, vitamins C and E and the bone-related nutrients, vitamins D and K.

The last three chapters in this section emphasize the role of nutrition in brain function. The separate chapters emphasize the effects of stroke, Alzheimer's disease, Parkinson's disease and other neurodegenerative disorders, and late-life depression on nutritional status and provide relevant dietary recommendations. Specifically, in stroke patients, changes in brain function often include dysphagia with consequent decreased consumption of foods that require chewing and, in some cases, swallowing. Stroke-related neurological deficits may include an inability to feed oneself, shop for food, carry food packages, cook, etc.

Malnutrition is frequently seen following stroke and is related to the level of impairment. Swallowing assessment tools are described in detail and have been shown to be helpful in patient evaluations. A major risk factor in stroke patients is aspiration of food into the lungs and subsequent development of respiratory tract infections. The use of enteral and parenteral nutrition options is also discussed.

In contrast with many stroke patients, those suffering from Alzheimer's disease have significant mental deterioration and those with Parkinson's disease have progressive loss of voluntary movements, but neither may include dysphagia. However, all of these patients may become malnourished over time for different reasons. Dietary factors that have been associated with decreased risk of developing the neurological diseases include higher intakes of omega-3 fatty acids, B vitamins, antioxidants and lowered intakes of saturated fats, total calories and sugar. Once the neurological disease is documented, weight loss is often seen. In fact, retrospective data suggest that weight loss precedes diagnosis of Alzheimer's as well as Parkinson's diseases. All of these conditions and diseases that are seen in the elderly can easily result in depressing thoughts about the future for the aging person. When the depressive mood overtakes activities of daily living and the individual becomes vegetative and withdrawn, a clinical mental condition may have developed. Factors that may result in late-life depression are reviewed and the potential for dietary factors to reduce the risk of depression is included. Dietary constituents, such as omega-3 fatty acids and folic acid, associated with reduced risk of neurological diseases, are also linked to reduced risk of depression. Vitamin B_{12} deficiency may also result in symptoms of depression.

The final section of this volume looks at new frontiers in preventive nutrition and includes separate chapters in the areas of long-term living arrangements for older adults, an in-depth examination of dietary supplements and the effects of complex health emergencies on nutritional status in the geriatric population. As the number of older adults increases exponentially over the next decades, planning by insurance and government agencies for elder care has included greater emphasis on home and community care rather than nursing facilities. However, it is critical to assure that disease condition needs as well as dietary needs are met. Education is important for the family caregiver so that dietary requirements through foods and/or supplements are met and drug–nutrient interactions are avoided. The importance of dieticians and nutritionists will be even greater as the level of care provided by non-specialists increases. Another option for preventing essential nutrient deficiencies is the use of dietary supplements. More than half of US adults over age 50 take a dietary supplement daily. Dietary supplements include those containing vitamins and minerals as well as herbal supplements. The in-depth chapter on dietary supplements includes a discussion of the regulatory environment as well as the scientific data supporting the use of certain supplements for chronic disease prevention and detailed tables that contain critical information about the most widely used ingredients in non-essential nutrient-containing dietary supplements. The final chapter in this comprehensive volume deals with responses to complex emergencies such as environmental or man-made disasters that can acutely affect the elderly, especially those who are infirmed. Even without the occurrence of an emergency, many elderly who are home bound are malnourished and are on waiting lists to receive Federally funded meals. Nevertheless, this chapter reviews examples of the state and local plans that are being made to cope with emergencies that include help for the elderly. Basic recommendations, such as always having on hand a 2-week supply of water, food and medicines as well as sources of power, are suggested for each senior whether at the individual, community or state-wide level.

Understanding the complexities of the aging process, drug use, physical debilities and mental changes that also affect nutrient status is not simple and the technologies used can often seem daunting. However, the volume's editors and authors have focused on assisting those who are unfamiliar with this field in understanding the critical issues and important new research findings that can impact the field of senior nutrition. The editors have taken special care to use the same terms and abbreviations between chapters, and provide guidance on the location of relevant material between chapters. Moreover, the Foreword by the well-acknowledged leader in the field, Dr. Nancy S. Wellman provides a clear overview of the value of this volume for increasing the understanding of the importance of clinical nutrition to the health of the aging population.

In conclusion, "Handbook of Clinical Nutrition and Aging, Second Edition", edited by Connie Watkins Bales and Christine Seel Ritchie provides health professionals in many areas of research and practice with the most up-to-date, well-referenced volume on the importance of nutrition in determining the potential for chronic diseases to affect overall health of the aging population. This volume will serve the reader as the benchmark in this complex area of interrelationships between the senses, immune function, heart, lungs, kidney, muscle, bone, cartilage, brain and other relevant organ systems in the human body and the substances that we consume. Moreover, the interactions between genetic and environmental factors and the numerous co-morbidities seen as the aging process progresses are clearly delineated so that students as well as practitioners can better understand the complexities of these interactions. Drs. Bales and Ritchie are applauded for their efforts to develop the most authoritative resource in the field to date and this excellent text is a very welcome addition to the Nutrition and Health Series.

Adrianne Bendich, PhD, FACN

Foreword

Aging, a multifaceted natural phenomenon, is dramatically changing the landscape of our country. We have not only the opportunity but also the obligation to broaden the nutrition services available to older persons. This book will help make that happen. It substantiates the connections between nutrition and successful aging. While comprehensively and convincingly focusing on nutrition's vital role in preventing, delaying onset, and managing costly and debilitating chronic diseases, the book explains the nutrition services and interventions that evidence shows work to keep older Americans more independent with a good quality of life.

As our nation addresses not only its obesity epidemic, but its impending age wave, alarms are sounding as Medicare and Medicaid costs for the poor and the old explode. Our skyrocketing health-care costs have resulted in a greater emphasis on the importance of healthy diets. Nutrition has become part of or has received increased emphasis in all major health promotion and risk reduction initiatives. The Dietary Guidelines for Americans now recognize people over age 50 as one of the "Specific Population Groups" that need special consideration. Steps to a HealthierUS, a US Department of Health and Human Services initiative, encourages Americans to live longer, better, and healthier lives by eating a nutritious diet as one of its four focal points. The Older Americans Update 2006 and 2008: Key Indicators of Well-Being list dietary quality as one of the 7 modifiable "Health Risks and Behaviors." The most recent White House Conference on Aging included a "Healthy Nutrition" recommendation for the first time in decades. Among its suggested strategies is greater access to nutrition therapy and education, as well as healthy diets, in any and all aging-related settings. As nutrition services for older adults move out of hospitals and institutions and into homes and communities, the new and updated chapters in this book are key to understanding cost containment trends where nutrition should play an essential role but does not yet – for example, nursing home diversion efforts dictated by the federal Deficit Reduction Act. Appropriately, the information in this book can be used to justify the need for greater availability of bona fide nutrition expertise in all programs and settings that serve older adults.

Clinicians, policymakers, faculty, and graduate and undergraduate students will find that this book fills practice and education gaps. As the most youth-obsessed, death-denying nation in the world, our culture's negative attitude qualifies as "ageist." It is based primarily on myths, stereotypes, and misinformation. It is therefore not surprising that some of our colleagues and many of our students have little interest in nutrition and aging. Geriatricians and gerontologists alike will find that the evidence in this book obligates them to include the nutritional status of older persons in comprehensive care management. It will help dietitians and other health professionals value the importance of healthy diets. For those new to aging, the book includes the many important approaches to improving the nutritional status of those they serve – from setting up screening programs, to recognizing when to make referrals to dietitians for individualized assessments and chronic disease management, to connecting older persons with community nutrition assistance "safety net" programs.

This book can help rectify longstanding educational gaps in nutrition and aging. Our national research found that knowledge about aging was lower in nutrition curricula than in some other disciplines and more than half of nutrition students had negative views about older adults. Students ranked working with them as their least preferred choice. Our review of curricular content nationally found relatively few undergraduate and graduate courses in aging compared to maternal and child courses. Rightfully, more than half of the program directors were not satisfied with the aging content in their curriculum, citing "curriculum already full" and "lack of faculty expertise in aging" as common obstacles. Our national review of nutrition textbooks identified problems that other disciplines had also found in their textbooks. Overall, nutrition textbooks generally fail to present aging comprehensively, across topics, or positively.

In contrast, this book is a standout. It provides an ideal structure for designing a course syllabus; it is a rich resource for faculty interested in strengthening components in their other courses such as aging in nutrition therapy, nutrition in geriatrics, and diet and health in gerontology; and it is useful for special topics or contemporary issues courses. It can be used in internships to amplify students' understanding of nutrition in aging. As such, this book will help today's students overcome their aging apathy. It tunes them into today's aging reality: most older Americans are living longer, healthier, and more actively. Older persons' determination to live independently makes them the most receptive and attentive to our guidance. They want to lessen their potential for illness, speed their recovery, shorten their hospital stays, and stay out of nursing homes. Students, both graduate and undergraduate, will value this book not only in their courses but as a "keeper" resource for their professional library. It may be just the antidote against deterring students and others from wanting to work with older adults. The rewards of working with older persons are real indeed!

Older people want to hear the good news about nutrition—that it is indeed never too late and that even small steps can make a difference at any age. The new aging reality says both the quantity and quality of life count. A healthy lifestyle and being active leads to greater longevity, adds more years of independence, and compresses morbidity in later years. However, people do not want to live longer to have more years of illness and unhappiness; the added years must be healthier ones. Good nutrition not only adds years to life, but life to years. This book has all the information needed to make universal access to quality nutrition services a reality for older Americans today. Doing so will positively improve their longevity and quality of life, prolong their independence in later years, and conserve the health care resources of our nation. Nutrition as depicted in this book fits a life-affirming view of aging—one that is long overdue in America.

Nancy S. Wellman PhD, RD, FADA

Preface

We opened the first edition of this handbook with a preface highlighting the unique challenges of the new millennium resulting from "successful aging" and reduced birth rates of the twentieth century. We emphasized, along with the "graying globe", the diversities of aging with regard to the influence of geographic location, gender, economic status, and even age (younger versus older old age). During the intervening years, we have come to understand that the majority of the health-related challenges faced by older adults, be they physical, social, or economic, are globally relevant. This is thanks in large part to incredible advances in telecommunication, along with a growing understanding of the nature of the finite resources of the planet.

In this context, it is clear that geriatric health issues and the behaviors that shape our responses to them (Chapters 1 and 2) are universal concerns, shared by the majority of older adults and their health-care providers. The same often applies to concerns about life threatening chronic diseases (Chapters 14–26), complex emergencies (Chapter 29), and related challenges for older citizens of the world (Chapter 3). With the inevitable graying of the globe, there will be exponential increases in expenditures for health care, increasing needs for long-term care services, and a demand for more focused health-care services for older adults living at home (Chapter 27). Concurrently, we expect an unfortunate shortage of geriatricians, especially in the US. Thus, understanding the unique interactions of geriatric syndromes (Chapters 4–13) with nutritional factors will be increasingly important for all health care givers attending to older patients.

We are indebted to many individuals who contributed as we put together this edition of this *Handbook*. Our sincere thanks and congratulations on a job well done go to Thuytien Thi Ho, a Duke senior who has worked tirelessly for most of the past two academic years on the management and copy editing of this text. Dr. Bales also thanks Justin (Cody) Maxwell and Caroline Friedman for their contributions to this project. We also thank Paul Dolgert, Richard Hruska, and the rest of the Humana staff for their support and offer a special tribute to the late Tom Lanigan and Julia Lanigan, whose vision and creativity gave us, through Humana, a unique opportunity to publish on a topic very dear to us. Finally, we offer warm wishes and sincere gratitude to our series editor, Dr. Adrianne Bendich, for her encouragement to begin what has become an ongoing and exciting interaction with some of the most critical clinical issues in geriatric nutrition and the gifted and dedicated scientists who study them. Without the creative contributions of these scientist-authors, this book would not have been possible.

A Pulitzer Prize winning author in the field of aging, Dr. Robert Butler predicts that what he terms "the Longevity Revolution" will become a worldwide geopolitical issue in the 21st century, with the need for global adaptations to accommodate the extension of human lifespan. The challenge is a formidable one but we propose that state-of-the-art nutritional interventions can help to meet it. This *Handbook* was written to assist health care givers for older adults by providing strategies for effective secondary interventions for established diseases and conditions amenable to dietary modulation. We believe it is a uniquely comprehensive resource and hope that it will be a valuable guide to all (including physicians, nurses, dietitians, and speech language and occupational therapists) who provide care for this high-risk population. It is our sincere intention that the nutritional welfare and overall health of older adults be enhanced at a global level through the application of the information contained here.

Connie Watkins Bales, PhD, RD, FACN
Christine Seel Ritchie, MD, MSPH

Contents

Acknowledgements vii

Series Introduction ix

Foreword xvii

Preface xix

Contributors xxv

Part I: Over-Arching Issues For Nutritional Well-Being in Late Life

 1 An Ecological Perspective on Older Adult Eating Behavior 3
   Julie L. Locher and Joseph R. Sharkey

 2 Behavioral Theories Applied to Nutritional Therapies for Chronic Diseases
   in Older Adults 19
   James M. Shikany, Charlotte S. Bragg, and Christine Seel Ritchie

 3 Global Graying, Nutrition, and Disease Prevention: An Update on China
   and Future Priorities 33
   Yanfang Wang and Connie Watkins Bales

Part II: Fundamentals of Nutrition and Geriatric Syndromes

 4 Update on Nutritional Assessment Strategies 65
   John E. Morley

 5 Sensory Impairment: Taste and Smell Impairments with Aging 77
   Susan Schiffman

 6 Nutrition and the Aging Eye 99
   Elizabeth J. Johnson

 7 Common Gastrointestinal Complaints in Older Adults 121
   Stephen A. McClave

 8 Hydration, Electrolyte, and Mineral Needs 137
   Robert D. Lindeman

 9 Redefining Nutritional Frailty: Interventions for Weight Loss Due
   to Undernutrition 157
   Connie Watkins Bales and Christine Seel Ritchie

10 Sarcopenia 183
   Ian Janssen

11 Cachexia: Diagnosis and Treatment 207
       David R. Thomas

12 The Relationship of Nutrition and Pressure Ulcers 219
       David R. Thomas

13 Nutrition at the End of Life: Ethical Issues. 235
       Christine Seel Ritchie and Elizabeth Kvale

Part III: Common Clinical Conditions

14 Nutrition and Oral Health: A Two-Way Relationship 247
       Kaumudi Joshipura and Thomas Dietrich

15 Obesity in Older Adults – A Growing Problem. 263
       Dennis T. Villareal and Krupa Shah

16 Nutrition and Lifestyle Change in Older Adults with Diabetes Mellitus
       and Metabolic Syndrome. 279
       Barbara Stetson and Sri Prakash Mokshagundam

17 Cardiac Rehabilitation: The Nutrition Counseling Component 319
       William E. Kraus and Julie D. Pruitt

18 Chronic Heart Failure 333
       Christopher Holley and Michael W. Rich

19 Nutrition Support in Cancer 355
       Elizabeth Kvale, Christine Seel Ritchie, and Lodovico Balducci

20 Nutrition and Chronic Obstructive Pulmonary Disease 373
       Danielle St-Arnaud McKenzie and Katherine Gray-Donald

21 Nutrition and Chronic Kidney Disease 403
       Srinivasan Beddhu

22 Nutritional and Pharmacological Aspects of Osteoporosis. 417
       David A. Ontjes and John J.B. Anderson

23 Osteoarthritis 439
       Paola de Pablo and Timothy E. McAlindon

24 Post-stroke Malnutrition and Dysphagia 479
       *Candice Hudson Scharver, Carol Smith Hammond, and
       Larry B. Goldstein*

25 Alzheimer's Disease and Other Neurodegenerative Disorders 499
       Ling Li and Terry L. Lewis

26 Nutrition and Late-Life Depression 523
       Martha E. Payne

Part IV: New Frontiers in Preventive Nutrition

27 Providing Food and Nutrition Choices for Home and Community
       Long-Term Living 539
       Dian O. Weddle and Nancy S. Wellman

28 Dietary Supplements: Current Knowledge and Future Frontiers............. 553
        Rebecca B. Costello, Maureen Leser, and Paul M. Coates

29 Minimizing the Impact of Complex Emergencies on Nutrition and Geriatric
        Health: Planning for Prevention is Key............................. 635
        Connie Watkins Bales and Nina Tumosa

Index 655

Contributors

JOHN J.B. ANDERSON, PHD • *Departments of Medicine and Nutrition, School of Medicine and School of Public Health, University of North Carolina, Chapel Hill, NC*

LODOVICO BALDUCCI, MD • *Moffitt Cancer Center, Tampa, FL*

CONNIE WATKINS BALES, PHD, RD, FACN • *Department of Medicine, Geriatrics Research, Education, and Clinical Center, Durham VA Medical Center, Duke University Medical Center, Durham, NC*

SRINIVASAN BEDDHU, MD • *Salt Lake Veterans Affairs Healthcare System; Division of Nephrology & Hypertension, University of Utah School of Medicine, Salt Lake City, UT*

CHARLOTTE S. BRAGG, MS, RD • *Division of Preventive Medicine, School of Medicine, University of Alabama at Birmingham, Birmingham, AL*

PAUL M. COATES, PHD • *Office of Dietary Supplements, National Institutes of Health, Bethesda, MD*

REBECCA B. COSTELLO, PHD • *Office of Dietary Supplements, National Institutes of Health, Bethesda, MD*

PAOLA DE PABLO, MD, MPH • *Division of Rheumatology, Tufts-New England Medical Center, Boston, MA*

THOMAS DIETRICH, DMD, MD, MPH • *School of Dentistry, University of Birmingham, Birmingham, UK; Health Policy and Health Services Research, Boston University Goldman School of Dental Medicine, Boston, MA*

LARRY B. GOLDSTEIN, MD, FAAN, FAHA • *Duke University Medical Center, Durham, NC*

KATHERINE GRAY-DONALD, PHD • *School of Dietetics and Human Nutrition, McGill University, Quebec, Canada*

CAROL SEEL HAMMOND, PHD • *Audiology, and Speech Pathology, Nutrition, and Radiology Services, Durham VA Medical Center; Department of Medicine, Duke Medical Center, Durham, NC*

CHRISTOPHER HOLLEY, MD, PHD • *Washington University School of Medicine, St. Louis, MO*

IAN JANSSEN, PHD • *Department of Community Health and Epidemiology, School of Kinesiology and Health Studies, Queen's University, Kingston, Ontario, Canada*

ELIZABETH J. JOHNSON, PHD • *Jean Mayer USDA Human Nutrition Research Center on Aging at Tufts University, Boston, MA*

KAUMUDI JOSHIPURA, BDS, SCD • *School of Dental Medicine, Medical Sciences Campus, Center for Clinical Research and Health Promotion, University of Puerto Rico, San Juan, Puerto Rico*

WILLIAM E. KRAUS, MD, FACC, FAHA, FACSM • *Duke University Medical Center, Durham, NC*

ELIZABETH KVALE, MD • *Division of Gerontology, University of Alabama at Birmingham, Geriatrics and Palliative Medicine and Birmingham VA Medical Center, Birmingham, AL*

MAUREEN LESER, MS, RD • *NIH Clinical Center Nutrition Department, National Institutes of Health, Bethesda, MD*

TERRY L. LEWIS, BS • *University of Alabama at Birmingham, Birmingham, AL*

LING LI, DVM, PHD • *University of Alabama at Birmingham, Birmingham, AL*

ROBERT D. LINDEMAN, MD • *University of New Mexico School of Medicine, Albuquerque, NM*

JULIE L. LOCHER, PHD, MSPH • *Division of Gerontology, Geriatrics, and Palliative Care, Department of Health Care Organization and Policy, Center for Aging and Lister Hill Center for Health Policy, University of Alabama at Birmingham, Birmingham, AL*

TIMOTHY E. MCALINDON, MD, MPH • *Division of Rheumatology, Tufts-New England Medical Center, Boston, MA*

STEPHEN A. MCCLAVE, MD • *Division of Gastroenterology/Hepatology, Department of Medicine, University of Louisville School of Medicine, Louisville, KY*

SRI PRAKASH MOKSHAGUNDAM, MD • *Division of Endocrinology, Department of Medicine, University of Louisville, Louisville, KY*

JOHN E. MORLEY, MB, BCH • *Division of Geriatric Medicine, GRECC, VA Medical Center, Saint Louis University School of Medicine, St. Louis, MO*

DAVID A. ONTJES, MD • *Departments of Medicine, School of Medicine, University of North Carolina, Chapel Hill, NC*

MARTHA E. PAYNE, PHD, MPH, RD • *Department of Psychiatry and Behavioral Sciences, The Neuropsychiatric Imaging Research Laboratory, Duke University, Durham, NC*

JULIE D. PRUITT, MS, RD, LDN • *Duke University Medical Center, Durham, NC*

MICHAEL W. RICH, MD • *Cardiovascular Division, Washington University School of Medicine, St. Louis, MO*

CHRISTINE SEEL RITCHIE, MD, MSPH • *University of Alabama at Birmingham; Birmingham-Atlanta VA Geriatric Research, Education, and Clinical Center (GRECC), Birmingham, AL*

CANDICE H. SCHARVER, MA • *Audiology and Speech Pathology, Durham VA Medical Center, Durham, NC*

SUSAN SCHIFFMAN, PHD • *Department of Psychiatry and Behavioral Sciences, Duke University Medical Center, Durham, NC*

KRUPA SHAH, MD • *Division of Geriatrics and Nutritional Science, Center for Human Nutrition, Washington University in St. Louis, St. Louis, MO*

JOSEPH R. SHARKEY, PHD, MPH, RD • *Department of Social and Behavioral Health, School of Rural Public Health, Texas A&M Health Science Center, College Station, TX*

JAMES M. SHIKANY, DRPH • *Division of Preventive Medicine, School of Medicine, University of Alabama at Birmingham, Birmingham, AL*

DANIELLE ST-ARNAUD MCKENZIE, PHD • *Département des sciences de la santé communautaire, Université de Sherbrooke, Sherbrooke, Quebec, Canada*

BARBARA STETSON, PHD • *Department of Psychological and Brain Sciences, University of Louisville, Louisville, KY*

DAVID R. THOMAS, MD, FACP, AGSF, GSAF • *Division of IM-Geriatric Medicine, Saint Louis University Health Sciences Center, St. Louis, MO*

NINA TUMOSA, PHD, GRECC • *Department of Internal Medicine, St. Louis VA Medical Center, Saint Louis University, St. Louis, MO*

DENNIS T. VILLAREAL, MD, FACP, FACE • *Division of Geriatrics and Nutritional Science, Center for Human Nutrition, Washington University in St. Louis, St. Louis, MO*

YANFANG WANG, MD, PHD, MHS • *Academy of Health and Development, Health and Development Foundation, Beijing, China*

DIAN O. WEDDLE, PHD, RD, FADA • *National Policy and Resource Center on Nutrition, Physical Activity and Aging, Florida International University, Miami, FL*

NANCY S. WELLMAN, PHD, RD, FADA • *National Policy and Resource Center on Nutrition, Physical Activity and Aging, Florida International University, Miami, FL*

I   OVER-ARCHING ISSUES FOR NUTRITIONAL WELL-BEING IN LATE LIFE

1 An Ecological Perspective on Older Adult Eating Behavior

Julie L. Locher and Joseph R. Sharkey

Key Points

- An ecological perspective is a useful conceptual framework that takes into account multiple levels of influence that affect eating behavior of older adults.
- Eating behavior of older adults is influenced simultaneously by intrapersonal (i.e., individual characteristics), interpersonal (i.e., interpersonal processes and primary groups), institutional (i.e., norms and structures), community (i.e., social networks and norms), and public policy factors (local, state, and federal policies and laws).
- Addressing these multiple factors is key to effectively implementing nutritional interventions.

Key Words: Ecological perspective; social factors; social interventions

1.1 INTRODUCTION

The achievement and maintenance of good nutritional health are essential to physical and cognitive functions, the prevention or delay of chronic disease and disease-related complications, and overall quality of life *(1–3)*. Poor nutritional health represents nutrient deficiency, undernutrition, nutritional imbalances, and excesses, such as obesity, and is influenced by a number of factors that relate to food acquisition and food intake. These factors include food insecurity/food insufficiency, inadequacy of personal and community resources, functional impairments (e.g., inability to acquire, prepare, and eat food that is available), social isolation, multi-morbidity, oral problems, limited nutritional knowledge, and regular use of multiple medications *(4–9)*.

We know that personal, structural, and neighborhood characteristics influence differential access to health care or serve as barriers or enhancements to lifestyle behaviors such as physical activity or healthy eating *(10,11)*. A social ecological model, adapted from the work of McLeroy (1988) and Booth (2001) of healthy eating

From: *Nutrition and Health: Handbook of Clinical Nutrition and Aging, Second Edition*
Edited by: C. W. Bales and C. S. Ritchie, DOI 10.1007/978-1-60327-385-5_1,
© Humana Press, a part of Springer Science+Business Media, LLC 2009

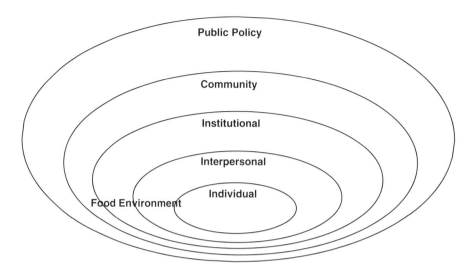

Fig. 1.1. Social ecological model of healthy eating.

is depicted in Fig. 1.1 *(10,12)*. The ecological model incorporates concepts from many theoretical perspectives that are key to specifying conditions that need to be changed to effect desired health actions *(13)*. This model emphasizes the influence that various levels of the ecology may exert on individual behavior, and thus has the potential to improve food choice, diet quality, and nutritional and health outcomes.

The eating behavior of older adults is influenced simultaneously by many different factors that may contribute to either healthy or unhealthy outcomes. A growing body of literature has focused attention on the important role that social factors play in the eating behavior of older adults. As efforts to promote dietary change or to attain and maintain optimal weight (either through weight loss or weight gain) are increasingly being targeted at older adults for the prevention or management of chronic disease and in the enhancement of function and independence, it is crucial that behavioral nutritional interventions take into account these social factors. Such an approach is consistent with a larger public health agenda focused on health promotion and disease prevention that routinely includes social factors ranging from individual-level attitudes and beliefs to environmental level governmental regulations in various interventions *(12,14,15)*.

As shown in Fig. 1.1, an ecological perspective is a conceptual framework that takes into account multiple levels of influence that affect behavior, including intrapersonal (i.e., individual characteristics), interpersonal (i.e., interpersonal processes and primary groups), institutional (i.e., norms and structures), community (i.e., social networks and norms), and public policy factors (local, state, and federal policies and laws). The ecological perspective emphasizes the mutually reinforcing relationship that exists between individual behavior and the multiple levels of influence whereby each is shaped by the other. The "reciprocal causation" outlined in the ecological perspective also emphasizes that each level of influence is interdependent with the others. Eating behaviors among older adults have been shown to be related to all of these levels of influence.

Using this conceptual framework, this chapter will outline social factors that work to support or maintain healthy or unhealthy eating behaviors in older adults. For each factor, health promotion strategies will be offered that may produce desired changes in both individuals and the social environment to enhance the nutritional well-being of older adults. The concluding section will briefly discuss the interplay between social factors and other factors that may influence eating behaviors among older adults, and suggest opportunities for developing evidence-based interventions to improve eating behaviors.

1.2 INTRAPERSONAL LEVEL

Intrapersonal factors that influence behavior include characteristics of the individual such as knowledge, attitudes, beliefs, self-concept, and skills that are acquired throughout the life-course of the individual *(15)*. Individuals make food choices based upon notions and capabilities they have about food, eating, and meals that are acquired through their lifelong interactions with others in consumption-related activities. These choices may be made either deliberately or "mindlessly." Brian Wansink's research lab has demonstrated that an average person makes over 200 decisions about food every day—most of these without much conscious consideration *(16)*. Furthermore, the behavioral choices individuals make regarding food are influenced simultaneously by many and varied factors.

In an effort to systematically identify both health- and non-health-related motives associated with food choices, Andrew Steptoe and his colleagues developed The Food Choice Questionnaire using factor analysis based on responses from a sample of 358 adults ranging in age from 18 to 87 years *(17)*. The 68-item multidimensional scale uncovered nine factors that were associated with individuals' food choices related to attributes of the food. These included health (food is perceived as healthy), mood (food helps to cope with stress or help relax), convenience (food preparation and shopping is easy), sensory appeal (food looks, smells, and tastes good), natural content (food contains natural ingredients), price (food is economical), weight control (food has low calories), familiarity (food is what person usually eats and/or ate as a child), and ethical concern (food associations are environmentally and politically acceptable). Differences in motives were observed for age, gender, and income. Of most relevance for this chapter, for both women and men, higher age was associated with food choices being made based on familiarity, natural content, and ethical concerns. Additionally, for women, older age was associated with food choices being based on health and sensory appeal; and for men, with mood and weight control. Using Steptoe's questionnaire, Locher and her colleagues found that the most salient motives underlying homebound older adults' food choices were sensory appeal, convenience, and price *(18)*.

Focusing more directly on qualities of individuals, other groups of researchers have found that beliefs motivating food choices emerge out of particular social contexts and serve to define and reinforce individuals' sense of self identity, including both social and personal identities *(19–22)*. Social identities refer to the group categories that individuals belong to or the social roles that individuals occupy. Group memberships that are important in influencing food choices are those involving ethnicity, region, social class, age cohort, and various significant reference groups. So, for example, persons in the southern United States might regularly consume grits for breakfast, while persons in the northern United States might more frequently consume oatmeal. Social roles that individuals occupy may be married versus un-married or widowed person or healthy versus sick role. Persons who are healthy may eat a full course meal with varied food items, while persons who define themselves as sick may consume smaller meals or only particular food items that they associate with being sick, such as chicken soup.

Intrapersonal factors related to one's attitudes, feelings, and behaviors toward food and eating that influence food choices may or may not be factually correct. This may be especially true in regard to individuals' perceptions regarding what constitutes healthy food choices or the right amount of food to consume. Regardless of whether individuals' beliefs are factually based, those choices are real in their consequences as individuals make decisions based upon what they believe to be true and that are consistent with their sense of self. These choices often conflict with health providers' recommendations. Researchers studying older adults' beliefs about consumption of salt and its effect on health in the rural, southern United States found that older adults' narratives of the role of traditional foods in maintaining their identities as Southerners, including consumption of salt, had to be reconciled with their attempts to meet medical recommendations for healthy eating (23). These efforts were complicated by the older adults' lay interpretations of illnesses and their declining changes in taste perceptions (which increased their consumption of salt).

1.2.1 Intervention Strategies at the Intrapersonal Level

Over two decades ago, Mary Douglas pointed out that it is no wonder that efforts designed to improve nutrition have so often failed because these efforts rarely consider that attempts to change a persons' food habits are perceived as assaults on a fundamental component of one's moral and social order (24,25). Whether it involves individual counseling or broader societal-level interventions, the perspective of individuals must be taken into account. Unfortunately, most efforts aimed at modifying or improving individual and societal eating habits have focused on the health meanings of food from particular nutritional disciplinary perspectives. They have not adequately accounted for the social meanings attached to food- and eating-related activities that are extremely important to individuals. In the case of older adults, familiarity of foods seems to be particularly salient.

Health promotion interventions targeted at changing older adults' food and eating behaviors must consider the many motivations that are involved in food choices beyond those related to health (17). Additionally, traditional models of educating individuals to make changes in food behaviors that places the clinician in the expert role and the older adult in the role of recipient of information does not sufficiently capitalize on individuals' knowledge of their own food and eating preferences and concerns, and how individuals' identities are expressed through their consumption of particular foods (26). In her review of nutrition education interventions for older adults, Sahyoun and her colleagues found that among those that were most effective were the ones that included personalized messages and hands-on activities, incentives, and cues *(27)*. At a minimum, this should include incorporating familiar and favorite foods that are central to individuals' sense of identity into any proposed dietary changes and to actively engage and encourage individuals' participation in those proposed changes. Last, to the extent that special diets interfere with adequate food consumption, moderation of those diets may be recommended.

1.3 INTERPERSONAL LEVEL

Interpersonal factors include formal and informal social network and social support systems comprising family, friends, neighbors, peers, and paid providers of services who provide support to individuals and serve to define and reinforce individuals' roles and senses of self and purpose *(15)*. There is strong and compelling theoretical and empirical evidence covering an extended time period that links social support and networks with positive health outcomes *(28,29)*. A large body of research consistently shows that older adults with better social support systems experience better health; and a number of social network and social support factors have been found to correlate specifically with better nutrition in older adults.

Most notably, the positive benefits conferred on those who live with others or who are married, particularly men, have been repeatedly demonstrated in regard to nutritional health in older adults *(30–34)*. Persons who are married are less likely to skip meals and better able to afford them. Older men who are not married, particularly those who are widowed, are vulnerable to experiencing poor nutritional health because they have not been socialized to be feeders and often do not know how to shop or cook for themselves. Women, especially those who are widowed, are also vulnerable to poor nutritional health because they may not be able to afford an adequate diet. Additionally, women traditionally cook for others; and it is one of the primary ways they express their care for others *(35)*. When older women no longer have anyone to cook for, they may be less inclined to cook for themselves.

In work examining the presence of others during meals, John de Castro and his colleagues have consistently found that persons, including older adults, who eat in the presence of others consume more than those who eat alone *(36–39)*. Additionally, they find a power function associated with the presence of others such that the number of calories consumed increases as a function of the number of persons present and that the effect is strongest when those present are family members or friends. De Castro's work supports the theory of social facilitation in that food intake is increased because the duration of the meal is extended. Alternative explanations for why food intake might be increased or decreased because of the presence of others is supported by research conducted in younger samples using either modeling or impression management theory. Findings indicate that individuals' food intake is influenced by social cues; persons will eat similarly as those with whom they are dining, or persons may control their eating behavior in an attempt to project a desirable image of the self *(40)*.

Work by Alex McIntosh and his colleagues and Locher and colleagues has attempted to disentangle the relationship between having someone present in the household versus having someone present during meals *(41–43)*. Both research teams have found that it is companionship at mealtime or help with cooking, and not either being married or having someone present in the household, that affect nutritional intake. Having someone present in the household may increase the natural opportunities for others to be present during meals. The presence of others during meals may result in better nutritional intake for a number of reasons, including: (1) duration of the meal may have been increased; (2) persons with whom the older person is eating consumed more; (3) persons may have received help or encouragement to eat; (4) and persons may have wanted to please caregivers/others by eating more.

The mechanisms by which social networks and social support contribute to better nutritional health may extend beyond the presence of others at meals. For example, individuals may be encouraged to participate in healthy behaviors and discouraged to participate in unhealthy behaviors or vice versa depending upon the social network and support system. Additionally, and especially relevant for older adults in need of caregiving, receipt of social support may directly or indirectly enhance one's personal competence and enable one to access needed resources or services. Another aspect of interpersonal influence on food intake is in the situation where reliance on others for help with food-related activities, such as grocery shopping and meal preparation, can influence the type and amount of food consumed (Keller, 2005) *(44)*.

1.3.1 *Intervention Strategies at the Interpersonal Level*

At a minimum, interventions targeted at changing dietary behaviors among older adults must consider individuals' social network and support system and the role these may play in successfully implementing a strategy for change. As noted by Rimer and Glanz, the opinion, thoughts, behavior, advice, and support of those surrounding an individual may be very influential; and this may be especially true for older adults who are dependent on caregivers *(15)*. For these individuals in particular, it may be especially beneficial to involve caregivers in any proposed dietary interventions, particularly if they may affect caregiver responsibilities. This may include providing nutritional counseling to caregivers, as well as vocalizing individuals' food preferences to caregivers. Last, for all older adults, and in particular for those who are not consuming enough calories, recommendations to eat with family or friends or to eat at the table with others present may be useful in promoting better dietary intake.

1.4 INSTITUTIONAL LEVEL

Institutional factors refer to elements of social structure within an institution or organization such as formal and informal rules and regulations that affect individuals' behavior *(15)*. Institutional factors are related mostly to the environment within which older adults reside. The majority of older adults reside in the community in a variety of settings; such as in their own homes, in someone else's home in the case of frailty, or in continuing care retirement communities, assisted living facilities, or some other communal type arrangement. A small percentage of older adults reside in nursing homes. For all older adults, there is the possibility that they may be hospitalized for some period of time. Very little research has examined institutional or structural level factors that may affect older individuals' food and eating behavior.

Regardless of living arrangements, older adults are exposed to implicit and explicit rules regarding accessibility and availability of food and the patterned activities surrounding the structure of food consumption *(24,45–47)*. Individuals make choices regarding what to eat, how much to eat, how often to eat, with whom to eat, when to eat, where to eat, and the manner in which food is to be eaten. There are norms, traditions, and rituals related to each of these, and they vary according to living arrangement and stage in the life-course. For example, as described in the previous section, older adults who live alone may skip meals compared with those who live with others. Previous structures (e.g., rules and policies) of consuming breakfast, lunch, and dinner that were part of one's earlier life with spouses who worked and children who went to school may no longer be relevant if living alone and unemployed. However, if older adults are living with adult children, they may follow the structure of that household and consume three meals a day.

The nature of mealtimes in hospitals has been examined by Xia and McCutheon (2006) and in nursing homes by Kayser-Jones (1997) and Simmons and her collea-gues *(48–50)*. These researchers report that many older adults do not receive needed assistance during meals, disruptions occur frequently, social interaction and verbal cuing are neglected, and the eating environment and food are either unpleasant or unfamiliar. Delays may result in the loss of appeal for the meal as a result of change in meal temperature or texture. Social disruption of routine and "normal" eating habits that occurs during hospitalization may result in a failure to resume normal eating patterns upon return to home. Previous research has demon-strated the effect that aging has on body energy regulation, such that older adults are especially vulnerable to periods of under-eating—regardless of the circum-stance(s) for that under-eating *(51)*. One influential experimental study of healthy men found that older participants, in contrast to younger ones, did not resume prior "normal" eating habits and did not regain weight they lost following a period of voluntary under-eating *(52)*.

The manner in which food is to be prepared and consumed and the ways in which this affects older adults' eating behaviors have been inadequately studied. Select manual tasks, which may require muscle strength, are necessary to prepare and consume meals. Although these tasks—opening cans, milk or juice carton, lifting a cup, and opening plastic, frozen, and single-serving packages—are important to independent living, older adults, especially persons exposed to increasing presence of single-serving packages through receipt of nutrition programs (e.g., home-delivered meals) are experiencing difficulty in these nutrition-related tasks *(6)*. Additionally, older adults with physical and functional limitations may not be able to consume foods in a way that is considered civilized or culturally appropriate. This is especially true if they cannot chew or see well or are unable to use eating utensils in a conven-tional manner. This may cause embarrassment and lead them to not consume foods in the presence of others or to alter the types of foods they eat. Locher and colleagues have observed this in cases where persons resided either in their own or someone else's home or in assisted living facilities *(53)*. Older adults also report not wanting to consume meals in the presence of others if the environment is too hectic or busy, if conflict is taking place, or if they are the focus of attention to eat in a particular way. Older adults perceive themselves as the focus of attention if, for example, others draw attention to them not eating enough or taking too long to eat.

1.4.1 Intervention Strategies at the Institutional Level

Interventions at the level of the institution require an understanding of institutional-, interpersonal-, and individual-level processes. Possible interventions at the institutional level include changing the type of food containers used; making recommendations to create the dining area as calm an environment as possible by reducing noise levels; minimizing distractions and interruptions; and avoiding negative social interaction during meal time. Seemingly simple changes such as placing food within easy physical access (or having older adults in need use a bell to indicate hunger or thirst), placing food within visual range (including by having good lighting), eating food in small portions, and eating finger foods may involve multiple strategies. For example, one may need to convince the food manager in an institution to provide chicken nuggets or tenders instead of whole chicken breasts so that older adults with disabilities may continue to eat. At the same time, it may be necessary to enable older adults to overcome embarrassment associated with eating chicken with one's fingers instead of a fork.

Additionally, formal feeding assistance may be required by some older adults. Use of feeding assistants have been shown to be effective in improving nutritional care in nursing homes, and may be effective in the community setting as well *(54)*. Simmons and her colleagues have also demonstrated that older nursing home residents who ate their meals in the dining room as opposed to privately in their rooms consumed better quality meals explained by the additional feeding assistance that they received in the formal dining area *(55)*. Thus, changing normative expectations for where food ought to be consumed may be encouraged.

1.5 COMMUNITY LEVEL

Community factors refer to organizations, institutions, and informal networks and the relationships that exist among these within defined boundaries that individuals may access *(15)*. To the extent that these public resources are open to individuals through their engagement in various community and social structures and they produce some beneficial outcome, community factors might also be thought of as social capital. Kawachi and Berkman (2000) maintain that social capital within communities affects health by promoting healthy behaviors and discouraging unhealthy ones, by increasing access to health services and amenities, and by enhancing psychosocial process through the provision of emotional support in trusting social environments *(56)*. Very little research has been performed at the community level that examines the association between social capital and nutritional well-being among older adults.

Locher and colleagues have examined social isolation, support, and capital and nutritional risk in an older sample, measures of social capital were associated with nutritional risk only for African American men *(57)*. Specifically, African American men who did not participate regularly in a religious community, who limited their activity for fear they would be attacked, and who experienced recent discrimination were more likely to experience nutritional risk. For all groups except black men, not having reliable transportation was associated with a greater likelihood of experiencing nutritional risk.

Community influences food choice and dietary intake through the spatial access to food stores and food service places that comprise the food environment *(58)*. Certain characteristics of the food environment influence potential spatial access, including the number and type of food stores and food service places, organization (chain or independent), size, distribution/location, and distance to the neighborhoods where people live *(59–61)*. This is especially problematic for older adults who reside in rural areas, and face greater risk to poor nutritional health *(62)*.

Research has shown that populations with limited or difficult access to a supermarket tend to live in rural areas or urban neighborhoods with greater disadvantages and/or with a higher proportion of minorities *(63–65)*. Without easy geographic access to supermarkets or full grocery stores, older individuals either have to pay higher travel costs to reach a supermarket/grocery store or are only able to shop at convenience or small grocery stores and pay higher prices for limited selections of food products *(63,66,67)*.

1.5.1 Intervention Strategies at the Community Level

Interventions targeted at improving nutritional well-being at this level involve community level as well as individual- and interpersonal-level involvement. At the community level, advocacy for greater community resources are required. At the individual and interpersonal levels it requires making older adults and their caregivers aware of community resources that already exist. It may also involve encouragement of utilization of those resources that exist. Support may be available from non-governmental community providers including neighborhood and religious organizations.

Interventions targeted at this level also involve educating service providers, including especially registered dieticians, case managers, and social workers, about appropriate programs in the community that may target food insufficiency. These may include programs that are not specific to food, such as utility, pharmacy, housing, and transportation assistance, where the needs are related.

This may additionally mean greater collaboration between care providers to ensure that older adults receive appropriate nutrition interventions, including nutrition assessment, nutrition care planning, referrals to food assistance or other support services, meals, nutrition counseling, medical nutrition therapy, and caregiver counseling.

1.6 POLICY LEVEL

Public policy factors refer to national, state, and local laws, policies, and services *(15)*. There is strong evidence documenting the association of nutrition with health among older adults with data indicating also that poor nutrition leads to poor health outcomes and ultimately increased healthcare utilization and costs. Despite this evidence, there is no coherent policy regarding nutrition for older adults that might improve their health and well-being and curtail rising healthcare costs.

Further, policies and services that do exist are fragmented and reflect the specific goals and objectives of the program *(68)*. We briefly review some of these here as they have been outlined by Wellman and Johnson in an earlier article *(68)*. More detail is provided in the chapter by Weddle and Wellman.

At the national level, nutrition policies and services emanate from the Department of Health and Human Services (DHHS) and the United States Department of Agriculture (USDA). Within DHHS, nutrition services offered are administered by the Centers for Medicare and Medicaid Services (CMS), the Administration on Aging (AoA), and the Indian Health Services (HIS). Within USDA, nutrition services include the Food Stamps Program, the Food Stamps Nutrition Education Program, the Commodity Supplemental Food Program, the Child and Adult Care Food Program, The Emergency Food Assistance Program, and the Food Distribution Program on Indian Reservations.

CMS programs offer nutrition services through the auspices of both Medicare and Medicaid. Medicare policies and services focus on acute health-related aspects of nutrition. Specifically, Medicare Part B enrollees may receive nutrition counseling for diabetes or renal disease with a physician's referral. Additionally, while Medicare (under Part A Services) requires that hospitals and home health agencies have nutritional services available, use of prospective payment systems frequently means that many patients who are in need of support either do not receive them or receive inadequate services *(1)*. Despite the potential benefit, traditional Medicare does not provide reimbursement for preventive nutrition counseling. Enrollees who choose one of Medicare's managed care plans over traditional Medicare may be eligible for nutritional services, depending upon the details of their particular plan.

The focus of Medicaid nutrition services is to keep older adults out of nursing homes. Through Medicaid Home- and Community-Based Care Service Waiver Programs (HCBC), which are federal–state partnerships, approved nutrition services may be offered. These services include home-delivered meals, nutrition counseling, and supplements. Only 38 states offer nutrition services, and these vary greatly across states and time depending upon state budgets and priorities *(68)*.

The focus of Administration on Aging Older Americans Act (OAA) nutrition services is on food insecurity, hunger, and poor diets among low-income older adults. Services provided include those supported by Title IIIC of the OAA, including especially congregate meal and home-delivered meal services. The US government spends only about $1 billion on food programs for older adults and reaches between 6 and 7% of those in need *(68)*. OAA services can also include Title IIIB supportive services (which may include meals provided in adult daycares) and information and assistance (which may include nutritional consultation). Many persons who are in need of services may not receive them because of limited budgets or lack of service (either because of location or lack of volunteers).

The foci of USDA programs are also on food insecurity, hunger, and poor diets for all low-income Americans, including older adults. The Food Stamp Program provides electronic benefit cards or coupons to low-income people. Poor older adults are less likely to receive food stamps than poor younger persons. It is estimated that only one-third of eligible older adults participate in the program.

Reasons why older adults are less likely to participate include lack of information, perceived lack of need, low expected and actual benefits, difficulty applying, and stigma of receiving benefits *(69)*.

In contrast, older adults are more likely than younger persons to participate in the USDA Commodity Supplemental Food Program *(68)*. Thirty-three states participate in the commodity food program which targets older adults with income less than 130% of the poverty level. The USDA also serves meals in adult day care centers through the Child and Adult Care Food Program. The Emergency Food Assistance Program is another service offered through the USDA. It is a commodity food distribution program wherein food is allotted to states based upon need, and distributed through local agencies such as food banks, pantries, and soup kitchens. Another USDA program is the Senior Farmers' Market Nutrition Program. This program provides coupons for low-income older adults to buy fresh, unprepared food at markets, roadside stands, etc. The program exists in 40 states, and older adults receive an average annual benefit of $25. The Food Distribution Program on Indian Reservations is another commodity program used instead of food stamps at particular times. All of the USDA services provide limited benefits and are not available in all locations.

1.6.1 *Interventions Targeted at the Policy Level*

Strategies to improve nutritional well-being of older adults at the policy level involve empowering older adults to utilize services that may be available to them. This first entails making older adults aware of those services. It also involves identification of older adults' preference for use of particular services and barriers to use. This may additionally entail increasing older adults' knowledge of the benefits of any nutritional programs that may be available. These strategies are contingent on care providers being fully aware of services that are available. This involves care providers gathering up-to-date information on nutrition programs and services that are available from different sources. It also involves care providers identifying ways in which to address gaps in services.

1.7 IMPLICATIONS OF AN ECOLOGICAL APPROACH AND OPPORTUNITIES FOR INTERVENTION

Eating is a complex biological and psychosocial phenomenon involving many factors. Taking a social ecological approach allows one to systematically explain contextual aspects of older adult eating behavior in order to design interventions that may lead to improved nutritional health. The use of a social ecological framework is especially useful in understanding how change at one level may be influenced by changes at another level.

For example, a low-income older adult may be eligible for food stamps, but not receiving them. At the intrapersonal level, this person may believe that he or she will receive limited benefits. At the interpersonal level, the person's caregiver may believe that it is embarrassing to receive food stamps, and discourage the older adult from applying for them and, further, not provide transportation to either the food stamp office or social security administration office where it is necessary to apply. At the community level, the older adult may not have access to adequate public transportation to take him or her to the office to apply or the hours of operation may not be convenient. At the policy level, rules that require persons to apply for food stamps in-person discourage use of the program by making it difficult for many older adults to benefit from the program. An intervention strategy that draws from the ecological model would correct misconceptions about the actual benefits that would be provided (intrapersonal level), provide information regarding use and benefit of food stamps by older adults to remove stigma for the caregiver (interpersonal level), enable older adult to either identify alternative means of transportation or apply for food stamps from home, where available (community and policy level).

In addition to the social factors, there are many other factors that affect eating behaviors among older adults. These include those which will be discussed in subsequent chapters, namely, medical, functional, economic, oral health, and psychological factors. All of these must be taken into account in applying a social ecological perspective to older adult eating behavior.

The overwhelming majority of work related to nutrition in older adults has been observational, cross-sectional, and included small sample sizes. Much of this work has focused on the intrapersonal levels of influence, particularly the influence of social support on dietary health. More observational research is needed in identifying especially institutional, community, and policy level influences on nutritional behavior, particularly in more vulnerable populations such as those residing in rural areas and those who are homebound. Additionally, further research is required that evaluates the efficacy of targeted social and behavioral nutritional interventions that are non-invasive and cost-effective. There is tremendous opportunity to increase the scientific evidence upon which practice can be based in the area of nutrition and older adults.

1.8 RECOMMENDATIONS

1. At the intrapersonal level, actively engage and encourage individuals' participation in proposed dietary changes.
2. At the interpersonal level, consider individuals' social network and support system and the role these may play in successfully implementing a strategy for change.
3. At the institutional level, take into account the social and environmental context within which food and eating activities take place.
4. At the community level, service providers first must become aware of community resources that are available and, then, make older adults and their caregivers aware of these and encourage utilization of resources.
5. At the policy level, make older adults aware of and empower them to utilize government programs and services that may be available to them.

REFERENCES

1. Institute of Medicine, Committee on Nutritional Services for Medicare Beneficiaries. The Role of Nutrition in Maintaining Health in the Nation's Elderly: Interventions and Assessments Can Help Beneficiaries. Washington, DC: National Academy Press, 2001.

2. Sharkey JR. Nutrition and public health. In Markides KS, ed. Encyclopedia of health and aging. Sage Publications, 2007, 425–28.
3. Diet, nutrition and the prevention of chronic disease. Report of a Joint WHO/FAO Expert Consultation. World Health Organization. Geneva, Switzerland; Technical Report Series 916, 2003.
4. Sharkey JR. Risk and presence of food insufficiency are associated with low nutrient intakes and multimorbidity among homebound older women who receive home-delivered meals. J Nutr 2003;133:3485–91.
5. Sharkey JR. Longitudinal examination of homebound older adults who experience heightened food insufficiency: effect of diabetes status and implications for service provision. Gerontologist 2005;45:773–82.
6. Sharkey JR, Branch LG, Zohoori N, Giuliani C, Busby-Whitehead J, Haines PS. Inadequate nutrient intake among homebound older persons in the community and its correlation with individual characteristics and health-related factors. Am J Clin Nutr 2002;76:1435–45.
7. Edington J. Problems of nutritional assessment in the community. Proc Nutr Soc 1999;58:47–51.
8. Ponza M, Ohls JC, Millen BE. Serving elders at risk: The Older Americans Act nutrition programs, national evaluation of the elderly nutrition program, 1993–1995. Princeton, NJ: Mathematica Policy Research, Inc., 1996.
9. White J. Risk factors for poor nutritional status in older Americans. Am Family Phys 1991;446: 2087–97.
10. Booth SL, Sallis JF, Ritenbaugh C, et al. Environmental and societal factors affect food choice and physical activity: rationale, influences, and leverage points. Nut Rev 2001;59:S21–39.
11. Fisher KJ, Li F, Michael Y, Cleveland M. Neighborhood-level influences on physical activity among older adults: A multilevel analysis. J Aging Phys Activ 2004;11:45–63.
12. McLeroy KR, Bibeau D, Steckler A, Glanz K. An ecological perspective on health promotion programs. Health Educ Quart 1988;15:351–77.
13. Grzywacz JG Fuqua J. The social ecology of health: Leverage points and linkages. Behav Med 2000;26:101–16.
14. Hovell MF, Wahlgren DR, Gehrman C. The Behavioral Ecological Model: Integrating public health and behavioral science. In DiClemente RJ, Crosby R, Kegler M, eds. New and Emerging Theories in Health Promotion Practice & Research. Jossey-Bass Inc., San Francisco, California, 2002:347–85.
15. Rimer B Glanz K. Theory at a glance: A guide for health promotion practice, 2nd Ed. US Department of Health and Human Services, National Institutes of Health, National Cancer Institute, 2005. (Accessed August 31, 2007, at cancer.gov/aboutnci/oc/theory-at-a-glance/print)
16. Wansink B. Mindless eating: Why we eat more than we think. New York: Bantam Books, 2006.
17. Steptoe A, Pollard TM, Wardle J. Development of a measure of the motives underlying the selection of food: the food choice questionnaire. Appetite 1995; 25:267–84.
18. Locher JL, et al. Food choice among homebound older adults: Motives and perceived barriers. J Nutr Health Aging 2009; In press.
19. Falk LW, Bisogni CA, Sobal J. Food choice processes of older adults. J Nutr Educ 1996;28:257–65.
20. Bisogni CA, Connors M, Devine CM, Sobal J. Who we are and how we eat: a qualitative study of identities in food choice. J Nutr Educ Behav. 2002;34:128–39.
21. Fischler C. Food, self and identity. Soc Sci Inform 1988;27:275–92.
22. Lupton D. Food, the body, and the self. London: Sage, 1996.
23. Smith SL, Quandt SA, Arcury TA, Wetmore LK, Bell RA, Vitolins MZ. Aging and eating in the rural, southern United States: beliefs about salt and its effect on health. Soc Sci Med 2006;62:189–98.
24. Douglas MT. Standard social uses of food. In: Douglas MT, ed. Food In the social order: Studies of food and festivities in three American communities. New York: Russell Sage Foundation, 1984:1–37.
25. Douglas MT. Fundamental issues in food problems. Curr Anthropol 1984;25:408–9.
26. Vickers K. Personal communication. 2005.

27. Sahyoun NR Pratt CA, Anderson A. Evaluation of nutrition education interventions for older adults: A proposed framework. J Am Diet Assoc 2004;104:58–69.

28. Thoits PA. Stress, coping, and social support processes: Where are we? What Next? J Health Soc Behav (Extra Issue) 1995:53–79.

29. House JS, Umberson D, Landis K. Social relationships and health. Science 1988;241:540–5.

30. Davis MA, Murphy SP, Neuhaus JM, Gee L, Quiroga SS. Living arrangements affect dietary quality for US adults aged 50 years and older: NHANES III 1988–1994. J Nutr 2000;130: 2256–64.

31. Frongillo EA, Rauschebach BS, Roe DA, Williamson DF. Characteristics related to elderly person's not eating for 1 or more days: implications for meal programs. Am J Public Health 1992;82:600–2.

32. Quandt SA, McDonald J, Arcury TA, Bell RA, Vitolins MZ. Nutritional self-management of elderly widows in rural communities. Gerontologist 2000;40: 86–96.

33. Torres CC, McIntosh WA, Kubena KS. Social network and social background characteristics of elderly who live and eat alone. J Aging Health 1992;4: 564–78.

34. McDonald J, Quandt SA, Arcury TA, Bell RA, Vitolins MZ. Nutritional self-management strategies of rural widowers. Gerontologist 2000;40:480–91.

35. DeVault M. Feeding the family: The social organization of caring as gendered work. Chicago: University of Chicago Press, 1991.

36. de Castro JM Stroebele N. Food intake in the real world: implications for nutrition and aging. In: Thomas D, ed. Undernutrition in older adults: Clinics in geriatric medicine 18. Philadelphia, PA: WB Saunders Company, 2002:685–97.

37. De Castro JM. Age-related changes in the social, psychological, and temporal influences on food intake in free-living, healthy, adult humans. J Gerontol Med Sci 2002;57A:368–77.

38. De Castro JM. Family and friends produce greater social facilitation of food intake than other companions. Phys Behav 1994;56:445–55.

39. De Castro JM Brewer EM. The amount eaten in meals by humans is a power function of the number of people present. Physiol Behav 1992;51:121–5.

40. Herman PC, Roth DA, Polivy J. Effect of the presence of others on food intake: a normative interpretation. Psychol Bull 2003;129:873–86.

41. McIntosh WA, Shifflett PA, Picou SJ. Social support, stressful events, strain, dietary intake, and the elderly. Med Care 1989;27:140–53.

42. McIntosh WA, Shifflett PA. Influence of social support systems on dietary intake of the elderly. J Nutr Elder 1989;4:5–18.

43. Locher JL, Robinson CO, Roth DL, Ritchie CS, Burgio KL. The effect of the presence of others on caloric intake in homebound older adults. J Gerontol Med Sci 2005;60A:1475–8.

44. Keller HH. Reliance on others for food-related activities of daily living. J Nutr Elder 2005;25:43–59.

45. Douglas MT. Deciphering a meal. Daedulus 1972;101:61–81.

46. Simmel G. The sociology of the meal (Translated by Michael Symons). Food and Foodways 1910/ 1994;5:345–51.

47. Locher JL, Burgio KL, Yoels WC, Ritchie CS. The social significance of food and eating in the lives of older adult recipients of meals on wheels. J Nutr Elder 1997;17:15–33.

48. Xia C, McCutcheon H. Mealtime in hospital—who does what? J Clin Nurs 2006;15:1221–7.

49. Kayser-Jones J. Inadequate staffing at mealtime: implications for nursing and health policy. J Gerontol Nurs 1997;23:14–21.

50. Simmons SF, Babinou S, Garcia E, Schnelle JF. Quality assessment in nursing homes by systematic direct observations: Feeding assistance. J Gerontol Med Sci 2002;57A:M665–71.

51. MacIntosh C, Morley JE, Chapman IM. The anorexia of aging. Nutrition 2000; 16:983–95.

52. Roberts SB, Fuss P, Heyman MB, et al. Control of food intake in older men. J Am Med Assoc 1994;272:1601–6.

53. Locher JL, Robinson CO, Bailey FA, et al. Social Factors Contribute to Under-Eating in Older Adults with Cancer. (Under review, Journal of Supportive Care).

54. Simmons SF, Bertrand R, Shier V, Sweetland R, Moore TJ, Hurd DT, Schnelle JF. A preliminary evaluation of the paid feeding assistant regulation: impact on feeding assistance care process quality in nursing homes. Gerontologist 2007;47:184–92.

55. Simmons SF, Levy-Storms L. The effect of dining location on nutritional care quality in nursing homes. J Nutr Health Aging 2005;9:434–9.

56. Kawachi I Berkman LF. Social cohesion, social capital, and health. In: Berkman LF, Kawachi I, eds. Social epidemiology. New York: Oxford University Press, 2002.

57. Locher JL, Ritchie CS, Roth DL, Baker PS, Bodner EV, Allman RM. Social isolation, support, and capital and nutritional risk in an older sample: Ethnic and gender differences. Soc Sci Med 2005;60:747–61.

58. Sharkey JR, Horel S. Neighborhood socioeconomic deprivation and minority composition are associated with better potential spatial access to the food environment in a large rural area, 2007, under review.

59. Furst T, Connors M, Bisogni CA, Sobal J, Falk LW. Food Choice: A conceptual Model of the process. Appetite 1996;26:247–66.

60. Mela DJ. Food choice and intake: the human factor. Proc Nutr Soc 1999;58:513–21.

61. Zenk SN, Schulz AJ, Hollis-Neely T, et al. Fruit and vegetable intake in African Americans: Income and store characteristics. Am J Prev Med 2005;29:1–9.

62. Sharkey JR, Bolin JN. Health and nutrition in rural areas. In: Goins RT, Krout JA, eds. Service delivery to rural older adults. New York: Springer Publishing Company, 2006:79–101.

63. Liese AD, Weis KE, Pluto D. Food store types, availability and cost of foods in a rural environment. J Am Diet Assoc In press 2007.

64. Zenk SN, Schulz AJ, Israel BA, James SA, Bao S, Wilson ML. Neighborhood racial composition, neighborhood poverty, and the spatial accessibility of supermarkets in metropolitan Detroit. Am J Public Health 2005;95:660–7.

65. Morland K, Wing S, Roux AD, Poole C. Neighborhood characteristics associated with the location of food stores and food service places. Am J Prev Med 2002;22:23–9.

66. Clifton KJ. Mobility strategies and food shopping for low-income families. J Plan Educ Res 2004;23:402–13.

67. Blanchard T, Lyson T. Food availability & food deserts in the non-metropolitan south. Southern Rural Development Center, Mississippi State, MS; April 2006. Policy Report No.: 12.

68. Wellman NS, Johnson MA. Federal food and nutrition assistance programs for older people. Generat J Am Soc Aging 2004;33:78–85.

69. US General Accounting Office. Food assistance: Options for improving nutrition for older Americans. (GAO/RCED-00-238). Washington, DC: US GAO, 2000. (Accessed August 30, 2007, at www.gao.gov/archive/2000/rc00238.pdf.).

2 Behavioral Theories Applied to Nutritional Therapies for Chronic Diseases in Older Adults

James M. Shikany, Charlotte S. Bragg, and Christine Seel Ritchie

Key Points

- Behavioral theories can guide clinicians in developing the best strategies for promoting a therapeutic nutritional change.
- Commonly utilized behavioral models in nutrition interventions in geriatric populations include the Ecological Perspective, the Health Belief Model, the Stages of Change Model, the Theory of Meaningful Learning, the Information Processing Model, and Social Cognitive Theory.
- Integration of behavioral theories in clinical care can facilitate improved chronic disease self-management by supporting adoption and maintenance of healthful nutritional practices.

Key Words: Behavioral intervention; behavioral theories; nutrition education; lifestyle modification; stages of change

2.1 INTRODUCTION

With the aging of our population and with the successes of public health and technology, an increasing number of older adults are living longer but many are affected by chronic conditions. Many of these conditions, such as atherosclerosis, hypertension, congestive heart failure, and diabetes, are responsive to changes in diet and nutrient intake. However, adopting more health-promoting dietary practices often requires significant changes in behaviors that are deeply ingrained. Health care professionals can facilitate enhanced self-management of chronic conditions by understanding behavioral theories and their role in behavior change.

From: *Nutrition and Health: Handbook of Clinical Nutrition and Aging, Second Edition*
Edited by: C. W. Bales and C. S. Ritchie, DOI 10.1007/978-1-60327-385-5_2,
© Humana Press, a part of Springer Science+Business Media, LLC 2009

2.2 BENEFITS OF BEHAVIORAL THEORIES IN PRACTICE

Behavioral theories benefit health care professionals in a number of ways. They assist clinicians in studying a clinical problem, in developing and implementing the appropriate interventions, and evaluating their progress. Because they help explain both the dynamics of health behaviors and the influences of factors affecting health behaviors, including social and physical environments, they can guide clinicians in developing the best strategies for addressing a particular nutritional issue.

2.3 OVERVIEW OF COMMON BEHAVIORAL THEORIES/MODELS USED IN NUTRITION INTERVENTIONS FOR OLDER ADULTS

Many behavioral models have been used to guide nutrition interventions in older adults. For the purpose of this chapter, we will focus on those that have been most commonly utilized in nutrition interventions in geriatric populations.

2.3.1 Social Ecologic Theory or the Ecological Perspective

The *Ecological Perspective* highlights the interaction between, and interrelationship between, factors within and across all levels of a health problem. It is discussed in greater detail in Chapter 1. The key tenet of the Ecological Perspective is that behavior both affects and is affected by multiple levels of influence. McLeroy and colleagues *(1)* identified five levels of influence: (1) intrapersonal or individual factors, (2) interpersonal factors, (3) institutional or organizational factors, (4) community factors, and (5) public policy factors. At the individual level, characteristics such as knowledge, attitudes, beliefs, and personality traits all influence, for example, an older person's eating patterns and preferences. At the interpersonal level, family, friends, and peers may have an equally important impact on dietary intake, especially if the older adult depends on others for food preparation or procurement. Institutional or organizational factors may include rules, regulations, policies, and informal structures that support or impede adequate or health-promoting dietary intake. At the community level, social norms or standards often influence an older adult's ability to adhere to a particular dietary strategy, especially if that strategy runs counter to prevailing social norms. Many public policy factors at the local, state, and federal level affect nutritional issues in older adults. For example, state and federal pressures to prevent weight loss in nursing home settings have both increased positive attention to nutritional issues in this setting and, at times, contributed to potentially excessively aggressive interventions (such as enteral nutrition in residents with advanced illness).

Many health behavior theories focus on intrapersonal (individual) and interpersonal factors in behavior change. Examples of theories that focus primarily on these intrapersonal factors include the *Health Belief Model,* the *Stages of Change (Transtheoretical) Model,* the *Theory of Planned Behavior (TPB),* the *Theory of Meaningful Learning,* and the *Information Processing Model.*

2.3.2 The Health Belief Model

The Health Belief Model (HBM) focuses on perceptions individuals have of the threat posed by a health problem (susceptibility, severity), the potential benefits of avoiding the threat, and factors influencing the decision to act (barriers, cues to action, and self-efficacy). The tenet of this model is that for individuals to adopt a new health behavior or change their current health behavior, they have to (1) believe they are susceptible to the condition, (2) believe the condition will have serious consequences, (3) believe that changing their behavior will reduce their susceptibility to the condition or its severity, and (4) believe costs of taking action (perceived barriers) are outweighed by the benefits. Health behavior change in this model is also facilitated by specific factors that prompt action such as a reminder from one's provider (also called a "cue to action") or when the individual is confident in their ability to successfully perform an action (also called "self-efficacy") *(2)*.

2.3.3 Stages of Change

The Stages of Change Model *(3)* posits that behavior change is a process, not an event. This model asserts that as people attempt to change their behavior, they move through five stages: precontemplation, contemplation, preparation, action, and maintenance. In the precontemplation stage, the individual has no intention of taking action (some definitions include a time period; e.g., no intention to take action within the next 6 months). In the contemplation stage, the individual intends to take action in foreseeable future. In the preparation stage, the individual plans to take action within the next 30 days and is taking some steps in this direction. In the action stage, the individual has successfully changed behavior for a short period of time, whereas in the maintenance stage, the individual has changed behavior for a longer period of time or at least 6 months. The Stages of Change Model, in addition to emphasizing the process of behavior change, recommends stage-specific interventional strategies tailored to where the person is in their transition from one behavior to another more health-promoting behavior (see Table 2.1).

2.3.4 The Theory of Meaningful Learning

The Theory of Meaningful Learning posits that each individual must construct his or her own understanding of concepts and relationships. While health care providers and others can assist an older adult in learning, the construction of meanings and understandings, and ultimately learning and behavior change, is a unique process that only each person can achieve on their own *(4)*.

2.3.5 The Information Processing Model

The Consumer Information Processing Model states that individuals must be exposed to, comprehend, retain, and retrieve pertinent information in order to make a decision and engage in behavior change *(5)*. In another words, health information is important but not sufficient for people to adopt healthful behaviors. Central assumptions of this model are that (1) individuals have limitations in how

Table 2.1
Stage-based dietary counseling strategies

Stage	Patient needs	Counseling messages
Not interested in dietary change (*Precontemplation*)	Motivation to engage in dietary change	Ask what patient/caregiver likes/dislikes about recommended dietary changes
		Discuss pros and cons along with perceived barriers
		Reinforce and build on patient/caregiver's personal reasons for making dietary change
		Discuss positive effects of dietary change on health, lifestyle, and quality of life
		Restate desire to support the patient and assist with change
		Follow up with patient and let them know you will
Interested in dietary change in the next 6 months but not in the next 30 days (*Contemplation*)	Motivation to engage in dietary change sooner than later	Strengthen the benefits for dietary change and weaken the cons
Interested in making a dietary change in the next 30 days (*Preparation*)	Skill building Support specific planning strategies	Encourage patient/caregiver to make a specific plan using small, achievable steps Address expected obstacles
In process of making a dietary change – has consistently made this change in the past 6 months (*Action*)	Relapse prevention	Congratulate on success Review concerns If brief relapse (e.g., consumption of too much sodium during a holiday), encourage to cycle back to recommended diet right away and use experience as an opportunity for learning rather than discouragement
Has engaged in dietary change for more than 6 months (*Maintenance*)	Relapse prevention	Support, encourage, and review plans for relapse prevention

much information they can process at one time and (2) information is more useable if combined into manageable "chunks." Individuals are more likely to use information if it is perceived as relevant to their situation, useful, new, and easy to use *(6)*.

2.3.6 *Social Cognitive Theory*

Social Cognitive Theory (SCT) posits that whether a person will change a health behavior depends on (1) self-efficacy, (2) goals, and (3) outcome expectancies. If individuals have a high level of confidence, they can change even when they are faced with many obstacles. If they are not confident about the behavior in question, they will be less motivated to act or to persevere through obstacles or challenges as they arise. Important elements of SCT include reciprocal determinism (the interaction of the person, behavior, and the environment), behavioral capability (knowledge and skills needed to perform a particular behavior), expectations (the individual's anticipated outcome of the behavior), self-efficacy (confidence in one's ability to overcome the barriers encountered during behavior change), observational learning (watching the actions and outcomes of others' behavior), and reinforcements (factors that increase or decrease the likelihood of the desired behavior) *(7)*.

2.4 EXAMPLES OF NUTRITION INTERVENTIONS GROUNDED IN BEHAVIORAL MODELS

A variety of behavioral models and theories have been used to formulate interventions to reduce the impact of common chronic diseases such as diabetes mellitus, hypertension, heart failure, hyperlipidemia, atherosclerosis, and to improve overall dietary quality. Examples of effective dietary interventions based on theories and models of health-related behavior are provided below and summarized in Table 2.2.

2.4.1 *Diabetes Mellitus*

Behavioral interventions have been applied to the prevention of complications in patients with impaired glucose tolerance and diabetes mellitus (see also Sections 13 and 14 in Chapter 16). Oldroyd and colleagues conducted a randomized, controlled trial evaluating the effectiveness of a behavioral intervention to modify cardiovascular risk in men and women (mean age 58 years) with impaired glucose tolerance *(8)*. The intervention consisted of regular diet and physical activity counseling from a dietitian and physiotherapist using the Stages of Change Model of behavior change *(3,9,10)*. After 6 months, subjects randomized to the intensive lifestyle intervention experienced an increase in physical activity, positive changes in selected dietary variables (such as a reduction in total fat), a reduction in body weight, and a reduction in insulin resistance.

The impact of a nutrition education intervention on blood glucose and lipoprotein levels in adults 65 years of age and older with diabetes was assessed by Miller and colleagues *(11)*. The theoretical framework for the intervention integrated concepts from the Theory of Meaningful Learning *(12)*, the Information Processing Model *(5)*, and Social Cognitive Theory *(7)*. Following 10 weeks of intervention, the intervention group had greater improvements in fasting plasma glucose and glycosylated hemoglobin than the control group. In addition, a significantly greater proportion of subjects in the intervention group than the control group met

Table 2.2
Effective behavioral theory-based dietary interventions in older adults

Author	Behavioral theory/model	Disease/Dietary indicator	Mean age of subjects (years)	Men/women
Oldroyd, 2001	SCM	Diabetes mellitus	57.9	38/29
Miller, 2002	SCT, TML, IPM	Diabetes mellitus	72.2 ± 4.3 (I) 73.0 ± 4.2 (C)	43/50
Chapman-Novakofski, 2005	SCM, SCT	Diabetes mellitus	63 ± 10	65/174
Toobert, 2002	SCT, SET	Diabetes mellitus	? (postmenopausal)	0/279
Miura, 2004	SCT	Hypertension	62 ± 10	29/28
Rankins, 2005	SCT	Hypertension	55.2 ± 6.1	?
Sethares, 2004	HBM	Heart failure	75.7 ± 12.3 (I) 76.8 ± 10.5 (C)	33/37
Nasser, 2006 (25)	SCM	Hyperlipidemia	50 ± 11	73/68
van der Veen, 2002 (26)	SCM	Cardiovascular disease	58.5 ± 7.1 (I) 58.2 ± 6.9 (C)	38/105
Manios, 2007 (29)	SCT, HBM	Diet quality	60.0 ± 4.8	0/75

SCM = Stages of Change Model; SCT = Social Cognitive Theory; TML = Theory of Meaningful Learning; IPM = Information Processing Model; SEM = Social Ecologic Theory; HBM = Health Belief Model; I = intervention; C = control.

the treatment goal for serum total cholesterol. In a second report from this study analyzing the effect of this intervention on changes in knowledge and skills necessary for diabetes management, the intervention group had greater improvement in total knowledge scores, positive outcome expectations, promoters of diabetes management, and decision-making skills than the control group and greater reduction in barriers to diabetes management *(13)*.

A community-based diabetes education program incorporating Social Cognitive Theory and Stages of Change Theory was conducted in patients with diabetes (mean age 63 years) by Chapman-Novakofski and colleagues *(14)*. The program included three group sessions focused on meal planning and cooking demonstrations, with pre- and post-intervention evaluation. At posttest, significantly more participants reported positive dietary changes (such as using herbs in the place of salt, cooking with olive or canola oils, using artificial sweeteners in baking) and having more confidence in changing one's diet, in preparing healthful meals, in using the Nutrition Facts label, and in overcoming the degree of difficulty in meal preparation than pretest.

The Mediterranean Lifestyle Trial evaluated the effects of a comprehensive lifestyle management intervention in postmenopausal women with type 2 diabetes *(15)*. The intervention, based on a synthesis of Social Cognitive Theory and Social Ecologic Theory *(16,1,17)*, was focused on dietary factors (i.e., adoption of a Mediterranean diet), physical activity, social support, and stress management. Participants randomized to the intervention adhered to all aspects of the Mediterranean diet more days per week than participants randomized to the usual care condition *(18)*. Significantly greater reductions in total fat and saturated fat and significantly greater increases in daily fruit and vegetable consumption occurred in intervention participants. Intervention participants also demonstrated significantly greater improvement in behavioral patterns related to low-fat eating. Dietary improvements in the intervention group were sustained at 24 months *(19)*.

2.4.2 Hypertension

A multicomponent lifestyle modification intervention study was conducted in 57 subjects with hypertension (mean age 62 years) in Japan by Miura and colleagues *(20)*. The intervention included patient-centered assessment and exercise and nutrition counseling based on Social Cognitive Theory. Following 24 weeks of intervention, systolic blood pressure was decreased to a significantly greater degree in the intervention group compared to the control group, which received no counseling. The intervention group also experienced significantly greater reductions in urinary sodium excretion, total energy intake, and fat intake (as percent of total energy intake) and a significantly greater increase in exercise energy expenditure compared to the control group.

DASH – Dinner with Your Nutritionist, a university neighborhood health care center intervention to promote the Dietary Approaches to Stop Hypertension (DASH) diet, was tested in low-income African-American adults (mean age 55 years) with poorly controlled blood pressure *(21)*. This weekly program followed a structured syllabus of objectives that included concepts based on Social Cognitive Theory as described by Baranowski et al. *(22)* and featured dinners based on the

DASH diet plan. The objective of the study was blood pressure reduction, to be accomplished through the following behavioral objectives: building behavioral capability to identify DASH foods; instruction in planning, shopping for, and preparing DASH foods; instruction in using the Nutrition Facts label to select foods to meet DASH dietary goals; and building self-control by providing options for study participants to limit their intake of fat and sodium. Systolic and diastolic blood pressure was significantly lowered among participants who missed no more than two of the eight weekly sessions, while there was no change in blood pressure in participants who missed three or more sessions.

2.4.3 Heart Failure

An evaluation of a tailored message intervention addressing heart failure readmission rates, quality of life, and health beliefs in patients (mean age 76 years) with heart failure was reported by Sethares and colleagues (23). This randomized trial compared patients hospitalized with heart failure who received a tailored message intervention based on the Health Belief Model (24) to patients hospitalized with heart failure who did not receive the intervention. Heart failure readmission rates at 3 months; quality of life at 1 month after hospital discharge; and the effect of the intervention on the perceived benefit and barrier beliefs of the treatment at 1 week and 1 month were compared. A sample benefit question related to diet was "Salty food is not good for me," while a sample barrier question was "Food does not taste good on a low-salt diet." Although hospital readmission rates and quality of life did not differ significantly between the treatment and control groups following the intervention period, beliefs regarding a sodium-restricted diet were positively impacted, including an increase in the perceived benefits and a reduction in the perceived barriers to following this diet.

2.4.4 Hyperlipidemia

An educational approach based on the Stages of Change Model (25,26) was compared with usual care in reducing dietary fat intake and serum lipids in individuals (mean age 51 years) with hyperlipidemia (27). This randomized, controlled study included 40 stage-based tailored dietary activities for fat reduction in the Stages of Change intervention and different activities that did not incorporate the Stages of Change approach in the usual care intervention. Total cholesterol and low-density lipoprotein cholesterol decreased significantly in both groups at 4 weeks. These reductions were sustained over 40 weeks, with no differences noted between the groups.

2.4.5 Cardiovascular Disease Prevention

A randomized intervention study conducted by van der Veen and colleagues evaluated the effects of nutrition counseling delivered by a dietitian in men and women (mean age 58 years) at elevated risk of cardiovascular disease (28). Patients randomized to the intervention received nutrition information based on the Stages of Change Model (29,30). Compared to patients randomized to usual care, those receiving the intervention had significantly greater reductions in dietary total fat, saturated fat, and cholesterol both 6 and 12 months after the initiation of the intervention.

2.4.6 Diet Quality

Manios and colleagues evaluated the effectiveness of a nutrition education program on postmenopausal women (mean age 60 years) in a randomized clinical trial*(31)*. Women were randomized to either a dietary intervention group and attended regular nutrition education sessions for 5 months based on a combined application of the Health Belief Model *(32)* and the Social Cognitive Theory or to a control group. The goal of the intervention group was to increase nutritional knowledge and self-efficacy of the subjects to adopt and maintain healthy dietary choices as assessed by the Healthy Eating Index (HEI). Subjects in the intervention group increased selected individual HEI scores to a significantly greater degree compared with the control group at the end of the intervention period; however, there were no significant changes in total HEI scores between the groups.

2.5 PRACTICAL APPLICATIONS OF BEHAVIORAL THEORY IN DIETARY INTERVENTIONS FOR OLDER ADULTS

The behavioral theory-based dietary interventions in older adults and the reported impact of the interventions to reduce common chronic disease described earlier in this chapter were most often studied in free-living populations with several contacts over a significant period of time. Dietary adherence for the treatment and/or prevention of disease is often viewed as the most difficult part of a therapy. The key to increasing the adherence in dietary compliance is to improve the patient's knowledge and perceived confidence in making necessary lifestyle changes. To improve those clinical outcomes, education should be delivered by health professionals with appropriate training, knowledge, and skills on the topic. Being sensitive to individual differences and tailoring the instructions not only to the educational level/need but addressing the environmental influence in which a patient lives impacts adherence to the dietary therapy.

Effective nutritional counseling is rarely achieved in a single contact. Quality follow-up through face-to-face, phone, mail, internet, or a combination of any of these over a sufficient duration improves outcomes and provides accountability between the patient (or caregiver) and the health professional. It also offers opportunities for continued assessment of barriers/challenges to behavior change. Unfortunately, educational resources and reimbursement for continued counseling is often inadequate in the current health care environment.

Table 2.3 identifies the six behavioral theories in this chapter and offers scripts of questions and talking points representing each theory. Which theories and which questions are used will vary depending on whether the aging patient is free living, in an assisted living setting, or long-term care facility.

Table 2.3
Practical applications of behavioral theory for nutrition interventions in older adults

Social Ecologic Theory or Ecologic Perspective	Prior to the delivery of any effective dietary counseling, it is necessary to have an appreciation of the patient's environment. Some areas to probe might include • *The people, places, things that affect the foods consumed.* Where does the patient live and with whom? It is helpful to have the ages and relationships of the people living and eating with the patient as dietary choices may reflect others' preferences • *The dietary preferences and customs.* Completing a simple 24-hour dietary recall or asking what a typical breakfast, lunch, dinner, and snack the past week included can provide a wealth of information. It may be the patient's diet needs only a simple tweaking to comply with the dietary recommendation and not a major overall. Dietary restrictions and current nutritional therapy being employed may or may not be apparent from the recall. Ask what dietary regime has been prescribed and adherence to those recommendations the past 7 days • *Primary person(s) responsible for the purchase and preparation of the food.* If there is a person, other than or in addition to the patient, responsible for meals, the counseling session should include them • *Where meals are eaten?* If a person eats all meals at fast food places, restaurants, or community centers, counseling instructions to purchase and prepare meals for home is unlikely to be accomplished. How can the dietary therapy be tailored to fit into the patient's current lifestyle?
Health Belief Model	Have the patient verbalize what "he believes" his health status to be and listen for words and phrases that would indicate he feels that making dietary changes would be of benefit to his health and quality of life. Ask "If you made no changes in your dietary habits and you continue eating with no regard for improving your health, what will life be like for you this time next year/in 5 years?" After getting responses, flip the question to say "If you made the dietary changes recommended as a part of the therapy for your disease, how do you see life 3 months from now?" The discussion can be directed to repeating to the patient those benefits that are more immediately measured, i.e., lab values, blood pressure, weight, blood sugar

(continued)

Table 2.3
(continued)

Stages of Change Model	Assess the patient's "readiness" to engage in the dietary behaviors to improve health. Counseling patients and/or their caregivers who are at a precontemplation or contemplation phase is very different than counseling those who are in the action and maintenance phase (see Table 2.1). There may be other factors that are of higher priority to the patient than making dietary change, i.e., financial concerns, care of a family member, divorce, recent death of a loved one, loss of job. A question to ask might be "On a scale of 1–10, how confident are you that in the next week you will be able to prepare five of your seven dinner meals at home?" One (1) being you are not at all confident you will be able to do this and 10 being I know for sure I can prepare five of my seven dinner meals at home. Exploring the number given will give the counselor information on how best to address the nutritional therapy and plans for future counseling sessions
Social Cognitive Theory	Incorporating principles such as setting goals, breaking larger goal down to smaller more manageable steps, and building on the success of each step promotes self-efficacy. The patient will feel successful and good about his ability to accomplish goals. Information overload can and will stop the learning and behavioral change process, decreasing the motivation to move forward
Theory of Meaningful Learning	Information from the educational materials and counseling session must have meaning to the patient. The patient should understand what is being asked of her since she is ultimately the person in charge of herself (in most cases). Use terminology familiar to the patient and utilize information obtained from the 24-hour dietary recall and environmental survey to individualize the dietary goals. Know that the patient may feel uncomfortable asking questions or admitting a lack of understanding
Information Processing Model	Look and listen for understanding. Stop to ask open-ended questions of the patient on how specific dietary change(s) might be accomplished. "To eat fruit in place of the candy bar for afternoon snack will require that you have fruit readily available." What do you need to do to have that fruit available? Walking the patient through making a grocery list, buying the fruit, and eating it for the snack is helpful. Visualize the process with the patient

2.6 RECOMMENDATIONS

The following recommendations reflect perspectives of well-established behavioral theories:

1. Consider the patient's environment when designing the intervention strategy. Keep in mind the patient's environment (who lives with the patient and who procures the patient's food), cultural background, and dietary preferences (Ecological Perspective).
2. Assess the patient's health beliefs and determine their level of motivation for a dietary change (Health Belief Model).
3. Assess the patient's "readiness" to engage in specific dietary behaviors to improve their health (Stages of Change Model).
4. Agree on "SMART" (specific, measureable, attainable, realistic, time-limited) goals that patients/caregivers feels confident they can achieve (Social Cognitive Theory).
5. After providing patients and caregivers with nutritional information, close the loop by assessing the patient/caregiver's understanding (Information Processing).

REFERENCES

1. McElroy KR, Bibeau D, Steckler A, Glanz K. An ecological perspective on health promotion programs. Health Educ Q 1988;15:351–77.
2. Rosentock IM, Strecher VJ, Becker MH. Social learning theory and the health belief model. Health Educ Q 1988;15:175–83.
3. Prochaska JO, DiClemente CC. Transtheoretical therapy: toward a more integrative model of change. Psychother Theory Res Pract 2002;19:276–88.
4. Novak JD. The Promise of New Ideas and New Technology for Improving Teaching and Learning. Cell Biol Educ 2003;2:122–32.
5. McGuire WJ. Some internal psychological factors influencing consumer choice. J Consumer Res 1976;2:302–19.
6. Theory at a Glance: A Guide for Health Promotion Practice (NIH Publication No. 97-3896), by the National Cancer Institute, 1995. Bethesda, MD.
7. Bandura, A. (1986). Social foundations of thought and action: A social cognitive theory. Englewood Cliffs, NJ: Prentice-Hall.
8. Oldroyd JC, Unwin NC, White M, Imrie K, Mathers JC, Alberti KG. Randomised controlled trial evaluating the effectiveness of behavioural interventions to modify cardiovascular risk factors in men and women with impaired glucose tolerance outcomes at 6 moths. Diabetes Res Clin Pract 2001;52:29–43.
9. Rollnick S, Kinnersley P, Stott N. Methods of helping patients with behaviour change. Br Med J 1993;307:188–90.
10. Lamb R, Joshi MS. The stage model and process of change in dietary fat reduction. J Hum Nutr Diet 1996;9:43–53.
11. Miller CK, Edwards L, Kissling G, Sanville L. Nutrition education improves metabolic outcomes among older adults with diabetes mellitus: results from a randomized controlled trial. Prev Med 2002;34:252–9.
12. Novak JD. Learning, creating, and using knowledge: concept maps as facilitative tools in schools and corporations. Mahwah (NJ): Erlbaum, 1998.
13. Miller CK, Edwards L, Kissling G, Sanville L. Evaluation of a theory-based nutrition intervention for older adults with diabetes mellitus. J Am Diet Assoc 2002;102:1069–74.
14. Chapman-Novakofski K, Karduck J. Improvement in knowledge, Social Cognitive Theory variables, and movement through Stages of Change after a community-based diabetes education program. J Am Diet Assoc 2005;105:1613–6.

15. Toobert DJ, Strycker LA, Glasgow RE, Barrera M, Bagdade JD. Enhancing support for health behavior change among women at risk for heart disease: the Mediterranean Lifestyle Trial. Health Educ Res 2002;17:574–85.

16. Moos RH. Social ecological perspectives on health. In: Stone GC, Cohen F, Alder NE, editors. Health psychology: a handbook. San Francisco: Jossey Bass; 1979. p. 523–47.

17. Stokols D. Translating social ecological theory into guidelines for community health promotion. Am J Health Prom 1996;10:282–98.

18. Toobert DJ, Strycker LA, Glasgow RE, Barrera M, Angell K. Effects of the Mediterranean Lifestyle Program on multiple risk behaviors and psychosocial outcomes among women at risk for heart disease. Ann Behav Med 2005;29:128–37.

19. Toobert DJ, Glasgow RE, Strycker LA, Barrera M, Ritzwoller DP, Weidner G. Long-term effects of the Mediterranean Lifestyle Program: a randomized clinical trial for postmenopausal women with type 2 diabetes. Int J Behav Nutr Phys Activ 2007;4:1.

20. Miura SI, Yamaguchi Y, Urata H, Himeshima Y, Otsuka N, Tomita S, Yamatsu K, Nishida S, Saku K. Efficacy of a multicomponent program (patient-centered assessment and counseling for exercise plus nutrition [PACE + Japan] for lifestyle modification in patients with essential hypertension. Hypertens Res 2004;27:859–64.

21. Rankins J, Sampson W, Brown B, Jenkins-Salley T. Dietary Approaches to Stop Hypertension (DASH) intervention reduces blood pressure among hypertensive African American patients in a neighborhood health care center. J Nutr Educ Behav 2005;37:259–64.

22. Baranowski T, Perry CL, Parcel GS. How individuals, environments, and health interact. In: Glanz K, Lewis FM, Rimer BK, editors. Health behavior and health education: theory, research and practice. 2nd ed. San Francisco: Jossey Bass; 1997. p. 153–78.

23. Sethares KA, Elliott K. The effect of a tailored message intervention on heart failure readmission rates, quality of life, and benefit and barrier beliefs in persons with heart failure. Heart Lung 2004;33:249–60.

24. Maiman LA, Becker MH. The Heath Belief Model: origins and correlates in psychological theory. In: Becker MH, editor. The Health Belief Model: personal health behavior. Thorofare (NJ): Charles Slack; 1974. p. 9–26.

25. Greene GW, Rossi SR, Reed GR, Willey C, Prochaska JO. Stages of change for reducing dietary fat to 30% of energy or less. J Am Diet Assoc 1994;94:1105–10.

26. Curry SJ, Kristal AR, Bowen DJ. An application of the stage model of behavior change to dietary fat reduction. Health Educ Res 1992;7:97–105.

27. Nasser R, Cook SL, Dorsch KD, Haennel. Comparison of two nutrition education approaches to reduce dietary fat intake and serum lipids reveals registered dietitians are effective at disseminating information regardless of the educational approach. J Am Diet Assoc 2006;106:850–9.

28. van der Veen J, Bakx C, van den Hoogen H, Verheijden M, van Denbosch W, van Weel C, van Staveren W. Stage-matched nutrition guidance for patients at elevated risk for cardiovascular disease: a randomized intervention study in family practice. J Fam Pract 2002;51:751–8.

29. Horwath CC. Applying the transtheoretical model to eating behaviour change: challenges and opportunities. Nutr Res Rev 1999;12:281–317.

30. Prochaska JO, DiClemente CC. In search of how people change: applications to addictive behaviors. Am Psychol 1992;47:1102–14.

31. Manios Y, Moschonis G, Katsaroli I, Grammatikaki E, and Tanagra S. Changes in diet quality score, macro and micronutrients intake following a nutrition education intervention in postmenopausal women. J Hum Nutr Diet 2007;20:126–31.

32. Janz NK, Becker MH. The Health Belief Model: a decade later. Health Educ Q 1984;11:1–47.

3 Global Graying, Nutrition, and Disease Prevention: An Update on China and Future Priorities

Yanfang Wang and Connie Watkins Bales

Key Points

- In accompaniment with a rapidly increasing proportion of older adults (global graying), the worldwide trend is toward illness and death due to chronic diseases rather than infectious ones.
- China has one of the most rapidly "graying" populations of any country in the world and is, at the same time, progressing toward a wealthier economy.
- With its growing affluence, the Chinese population has also begun to experience dramatic increases in the incidence and prevalence of diet-related chronic diseases, highlighting the critical need for behavioral interventions for the diseases of old age that may be modulated by nutrition.
- One of the most effective strategies for promoting global health would be one that guides countries that are emerging to a more affluent status toward healthy diet and exercise patterns.

Key Words: Global health; preventive nutrition; China; chronic disease

3.1 INTRODUCTION TO THE CHALLENGE

The challenge of the "graying globe" is unique for each country and yet common to all. As illustrated in Fig. 3.1 and Table 3.1, dramatic increases in the number and proportion of older adults are already occurring in both developed and developing countries; a continuation of these trends is expected for several more decades. Over the past 50 years, the (absolute) number of older persons has tripled and will more than triple again over the next half century *(1)*. Figure 3.1 also shows the overall trend downward in total population growth due to declining fertility rates in many parts of the world. The fertility

From: *Nutrition and Health: Handbook of Clinical Nutrition and Aging, Second Edition*
Edited by: C. W. Bales and C. S. Ritchie, DOI 10.1007/978-1-60327-385-5_3,
© Humana Press, a part of Springer Science+Business Media, LLC 2009

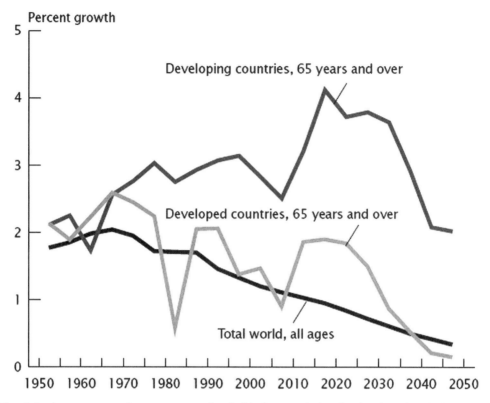

Fig. 3.1. Average annual percent growth of elderly population in developed and developing countries.
Source: United Nations, 1999, as published in An Aging World 2001: International Population Report. Issued November 2001. http://www.census.gov/prod/2001pubs/p95-01-1.pdf accessed 7.24.08.

Table 3.1

Number and percent of world and regional populations aged ≥ 60 years in 1950, 2000, and 2005

Region		1950	2000	2050
World	n	205,475	605,785	1,963,767
	percent	8	10	21
More developed regions	n	95,473	231,442	395,106
	percent	12	19	34
Less developed regions	n	110,003	374,343	1,568,660
	percent	6	8	19
Least developed regions	n	10,733	32,167	173,222
	percent	5	5	10

Source: United Nations (2001). *World Population Prospects: The 2000 Revision*, vol. 1.

rate is below replacement level in almost all industrialized nations. Coupled with the continued increases in life expectancy (see next section), this accounts for the rapid increase in the proportion of the world's population made up by older cohorts, i.e., the "graying" of the globe. The shift toward a predominantly older population is taking place along with the phenomenon of the world "becoming smaller". Advances in technology that provide worldwide real time communication are helping citizens of the world to understand that our future economies, ecologies, and health-related qualities of life will be increasingly intertwined. Thus the status of public health at the global level will be a major determinant of quality of life for all older adults in the decades to come.

Because international trade of food commodities, produce, and manufactured food products—as well as global proliferation of restaurant chains—is increasingly "blending" our food supplies and food habits, positive dietary interventions have the potential to produce marked improvements in health scenarios for older adults on a global scale. Such interventions could reduce health care needs and costs and enhance the future quality of life for the elder citizens of the world. The flip side of this coin could be catastrophically problematic and expensive. If societies do not learn from past mistakes with regards to the effects of detrimental dietary patterns on health, the countries now emerging to a more economically advantaged status (including China) are at risk of a disappointing and costly repetition of the diet-related health problems seen in many developed countries during the 20th century.

3.2 GLOBAL LIFE EXPECTANCY TRENDS

As noted in Table 3.1, by 2050 the proportion of the population that will be aged >60 years will be 21% globally, with regional proportions ranging from 10 to 34%. One important component of the global demographic transition is the continued increase in human life expectancy. Expected survival rates obviously vary depending upon the region of the world but all are generally increasing; Fig. 3.2 illustrates the years of life expectancy in the past and future for developed versus developing regions of the world.

During the first half of the last century, the increases in life expectancy were mainly attributable to improvements in the standard of living, public health initiatives, and immunization efforts that reduced the number of deaths due to infectious disease. These kinds of improvements may still benefit life expectancy in the developing world. But in developed and emerging countries, increases in life expectancy will mainly be due to changes in rates and severity of the major chronic diseases. In addition to impacting the years of survival, these factors will also strongly influence the health-related quality of life experienced by older adults. Thus, an important component of total life expectancy is the "healthy life expectancy", which is defined by the World Health Organization (WHO) as

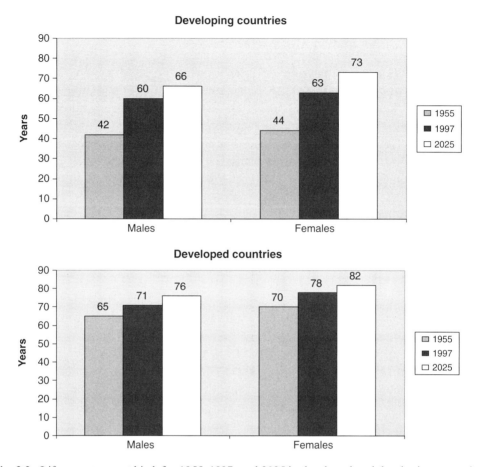

Fig. 3.2. Life expectancy at birth for 1955, 1997, and 2025 in developed and developing countries. *Source*: United Nations, 1995 and U.S. Bureau of the Census, International Programs Center.

the "average number of years that a person can expect to live in full health" *(2)*. One of the main goals of nutritional intervention is to maximize the years of healthy life expectancy.

Overall, women experience a considerable survival advantage over men. Thus women do and will continue to make up a substantial majority of the older population. Figure 3.3 illustrates the predicted increases in life expectancy for women in some representative countries. It should be noted, however, that this is not a universal pattern with regards to healthy years. Figure 3.4 illustrates the gender differentials in healthy life expectancy in a sampling of 23 countries of the world. At the two ends of the extremes of gender survival differentials are Russia, where men are affected by major illness or injury 11 years before women, and Qatar, where men have an average of 2.9 healthy years more than women.

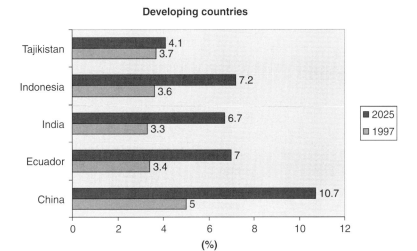

Fig. 3.3. Percent older women in the populations of developed and developing countries for 1997 and 2025.
Source: U.S. Bureau of the Census, International Programs Center.

3.3 PROFILE OF GLOBAL MORTALITY AND MORBIDITY CAUSES

Other than for deaths associated with HIV/AIDS, the causes for mortality on a global scale are increasingly chronic diseases rather than infectious ones (see Fig. 3.5), although differences do remain as determined by economic resources. The leading three causes of death are: for high-income countries—coronary heart disease (CHD), stroke, and cancers of the lung; for middle-income nations—CHD, stroke, and chronic obstructive pulmonary disease; and for low-income countries—CHD, lower respiratory infections and HIV/AIDS *(3)*. CHD is the most prevalent cause of morbidity and mortality overall, accounting for one-third of deaths globally. It is followed closely by stroke and cerebrovascular disease in high- and middle-income countries.

Healthy Life Expectancy Differential

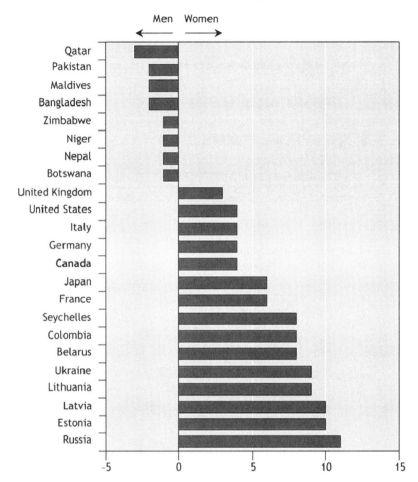

Fig. 3.4. Differences in healthy life expectancy by gender. Used with permission *(2)*.

Worldwide, close to 25 million individuals are living with cancer, with the eight most common being cancers of the lung, breast, colon and rectum, stomach, prostate, liver, cervix, and esophagus *(4)*. In association with rapidly increasing rates of obesity, the burden of type 2 diabetes (T2D) is also a heavy one, especially in older adults. The number of cases of T2D has increased so rapidly during the past two decades that the term "diabesity" has been coined to describe the concomitant epidemic of T2D and obesity *(5)*. A summary of findings from 13 European countries showed T2D prevalence rates of 16, 23, and 19% in men aged 60–69 years, 70–79 years, and 80–89 years, and 16, 27, and 43% in women of similar age groups, respectively. Between the survey years of 1994–1995 and 2003–2004, the annual incidence of T2D in the US population aged > 65 years increased by 23%, with an increase in prevalence of 62% *(6)*.

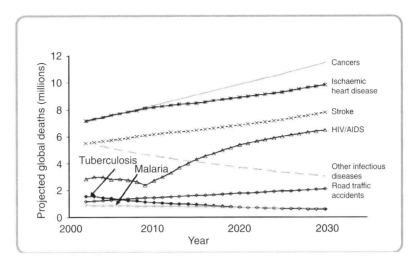

Fig. 3.5. Projected global deaths for selected causes of death for 2002 through 2030.
Source: http://www.who.int/whosis/whostat2007_10highlights.pdf

3.4 INTERACTIONS OF HEALTH BEHAVIORS WITH EXPECTED MORBIDITY AND MORTALITY

Some risk factors for chronic disease are non-modifiable, including genetic propensities and environmental exposures in earlier life. But, as illustrated in Fig. 3.6, lifestyle influences (especially diet and physical activity) play a critical role as intermediate determinants of risks for most of the leading causes of illness and death. This emphasizes the potential for prudent dietary patterns to influence the incidence of chronic disease and thus the length of healthy life expectancy.

When developing countries experience rapid economic growth, the associated changes in nutritional status of the population are not always fully positive. While

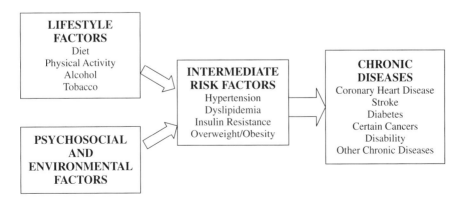

Fig. 3.6. Conceptual diagram of modifiable etiologies of chronic diseases.
Source: Adapted in part from Darnton-Hill et al. *(70)*.

the incidence of frank malnutrition is definitely lessened, there is often an increase in intake of nutrient-poor, energy-rich foods. Industries that produce, process, and manufacture foods, as well as the culinary/food service industries, "respond to mankind's inherent demand for sugary, salty and fatty foods" (7) by making access to them easy and affordable. In addition to dietary changes, life styles tend to transition to be less active and more sedentary as societies become more urban and wealthier.

Foods of high energy density can serve a valuable purpose when food supplies are short and demand great (e.g., famine, disaster relief rations). But in times of relative plenty, predominant consumption of such foods is strongly disadvantageous to health, especially when coupled with a sedentary activity pattern. This unhealthy lifestyle transition is being observed in emerging countries on a worldwide scale. An example of how Western influences might have a detrimental effect on global dietary patterns is found in a new approach known as "glocalization". This approach is being test marketed by Yum Brands (a company based in Louisville, KY, that also owns Kentucky Fried Chicken, Pizza Hut, and Taco Bell). Via the restaurant franchise "East Dawning", the company is successfully marketing "fast Chinese food" to Chinese consumers. The company is also testing the same approach with "fast Mexican food" in Mexico (7). Such marketing of high calorie, inexpensive and tasty foods—even when the food served mimics local cuisines—can lead to a dramatic change in diet and, potentially, a burgeoning problem of chronic disease associated with excess consumption of calories, fat, and sodium.

3.5 CROSS-CULTURAL ISSUES: FOCUS ON CHINA

The unique characteristics of the living environment, including socioeconomic factors, lifestyle and dietary habits, are known determinants of "successful aging" in all societies. Cross-cultural examinations can provide unique perspectives on population aging by demonstrating how functional impairment, morbidity, and mortality are variably influenced by the cultural, socioeconomic, and physical environments. In some cases, differences in health and nutritional status may be due more to environmental and lifestyle factors that occur over the entire life span of the individual than to variations in the underlying physiological mechanisms associated with morbid and functional change (8). Therefore, the impact of environmental determinants including socioeconomic factors, lifestyle, and dietary habits is an important consideration.

The following sections describe the dramatic speed of population aging in China, the transition in dietary habits and nutritional status that has developed during its economic strengthening, the potential interaction of these changes with aging, the impact of population aging on the public health and social system, and the response of the country to its future problems. New study results, including those from the 2002 National Health and Nutrition Survey, have become available, enabling us to update information on the general and older Chinese populations and to extend the implications for future aging concerns across cultures.

3.6 CHINA: UPDATE ON TRANSITIONS IN DIET AND DISEASE PATTERNS

3.6.1 Changes in Dietary Patterns

As previously noted, emerging nations experience a dramatic transition in diet and other lifestyle behaviors due to increased industrialization, urbanization, economic development, and market globalization. These processes of modernization and economic transition ultimately result in an improved standard of living. However, the shift from the traditional to a more Westernized diet may also have produced significant negative consequences in China. Age-group-specific dietary information from the 2002 China National Nutrition Survey is not available at the time of this writing; however, we can examine the overall population trends of dietary patterns among adults (which have recently become available) to help provide some update on changes relevant for older Chinese (see Table 3.2). Figure 3.7 demonstrates changes in the intake of meat, poultry, and seafood from 1982 to 2002, as observed from three national nutrition surveys. The average intake of meat and poultry increased 68% and more than 200% for urban and rural areas, respectively, over the past 30 years. The intake of seafood doubled in urban areas and tripled in rural areas during the same period. In contrast, total energy intake from grain sources decreased steadily in both urban and rural areas, from 70 and 80% of total daily energy intake in 1982 to 47% and 61% in 2002, respectively (Fig. 3.8a). Meanwhile, energy consumption from fat increased 32% in both urban and rural areas of China (Fig. 3.8b). Specifically, in 2002, fat accounted for more than 35% of total calories in urban areas, reaching up to 38% of total kcal in large cities like Beijing and Shanghai.

Fifteen years ago, Popkin et al. *(9)* reported strong evidence that the dietary pattern of the Chinese population was rapidly changing toward the typical high-fat,

Table 3.2
Intakes (g/day) by food group in China for the years 1982, 1992, and 2002

	1982	1992	2002
Grains	509.7	439.9	402.1
Dry bean	8.9	3.3	4.2
Bean products	4.5	7.9	11.8
Vegetables	316.1	310.3	276.2
Fruits	37.4	49.2	45.0
Nuts	2.2	3.1	3.8
Meat	34.2	58.9	78.6
Milk and its products	8.1	14.9	45.0
Egg and its products	7.3	16.0	23.7
Fish and other seafood	11.1	27.5	29.6
Cooking oil	12.9	22.4	32.9
Animal fat	5.3	7.1	8.7
Salt	12.7	13.9	12.0

Source: 1982, 1992, and 2002 China National Nutrition Survey.

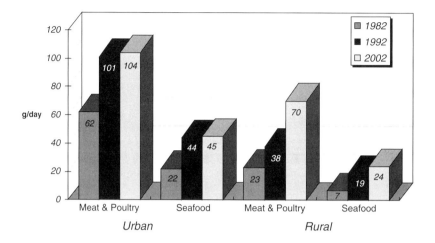

Fig. 3.7. Meat, poultry, and seafood intake in urban and rural areas.
Source:
For 2002 data: from 2002 national nutrition survey: Yang XG, Zhei F. Diet and Nutrition intake (Chinese). Beijing, China: People's Publishing House; 2005 *(30)*.
For 1992 data: from 1992 national nutrition survey: Ge K. Editor, The Dietary and Nutritional Status of Chinese Population: 1992 National Nutritional Survey. People's Medical Publishing House, Beijing China, 1996.
For 1982 data: Report on 1982 National Nutritional Survey. Institute of Nutrition and Food Hygiene, Chinese Academy of Preventive Medicine. Institute document, 1986.

high-sugar diet of the West. The authors also indicated that higher income levels, particularly in urban areas, were associated with consumption of such a diet and the ensuing problem of obesity. Even today, there remains a large differential in animal food/fat intake between urban and rural populations and across regions of varying levels of economic development. However, as economic improvements proliferate and reach more rural areas, the predictions by Popkin et al. *(9)* can increasingly be applied to rural populations. Indeed, rural regions have begun following the trend of their urban peers in the transition of dietary and disease patterns.

3.6.2 Changes in Disease Patterns and Prevalences

As already emphasized, diet-related diseases (e.g., obesity, T2D, CHD, hypertension, stroke, and certain cancers) are significant causes of disability and premature death in both developing and newly developed countries *(10)*. In China, as in many other emerging countries, such diseases are replacing more traditional public health concerns such as malnutrition and infectious disease. This phenomenon has been widely demonstrated in Western societies and is likely to occur in many emerging countries, including China, as they continue to gain economic strength *(11)*. Although nutritional deficiency and infectious diseases have not been eradicated, these conditions are now largely confined to certain economic and age groups within particular regions of the country.

The overall mortality in China has declined from 20 per thousand in the early 1950s to 6.8 per thousand in 2006 *(12)*. Communicable diseases that caused about

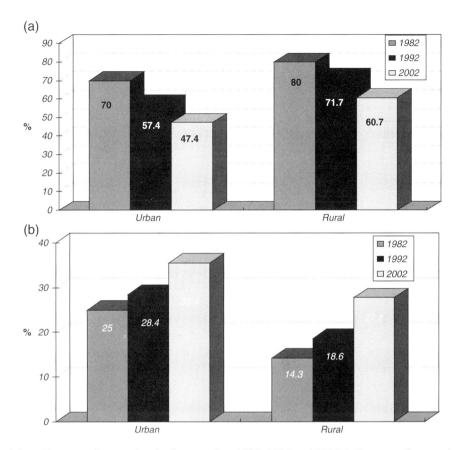

Fig. 3.8. a: Percent of energy intake from grains: 1982, 1992 and 2002. **b:** Percent of energy intake from fat: 1982, 1992 and 2002.

Source:

For 2002 data: from 2002 national nutrition survey: Yang XG, Zhei F. Diet and Nutrition intake (Chinese). Beijing, China: People's Publishing House; 2005 *(30)*.

For 1992 data: from 1992 national nutrition survey: Ge K. Editor, The Dietary and Nutritional Status of Chinese Population: 1992 National Nutritional Survey. People's Medical Publishing House, Beijing China, 1996.

For 1982 data: Report on 1982 National Nutritional Survey. Institute of Nutrition and Food Hygiene, Chinese Academy of Preventive Medicine. Institute document, 1986.

8% of deaths in 1957 have been largely reduced, while chronic diseases are now considered a major cause of death. A large, international collaborative study found heart disease (23%), cancer (22%), and cerebrovascular (21%) diseases to top the list of the leading causes of death for the total population of China. Pneumonia and influenza (3.2%), infectious disease (3.1%), accidents (2.8%), COPD (1.8%), chronic liver disease (1.5%), diabetes (1.5%), and kidney disease (1.4%) rounded out the list of causes of mortality *(13)*.

3.6.2.1 Obesity

Dietary energy and fat intakes are known to be positively and significantly associated with body mass index (BMI: note that Asian BMI criteria are overweight = OW : BMI

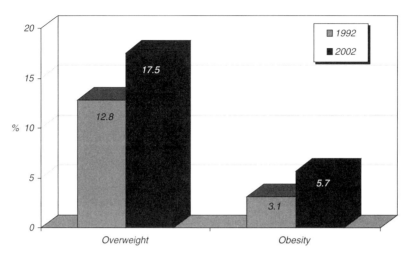

Fig. 3.9. Percentage of overweight and obese Chinese: 1992 and 2002.
Source: China National Nutrition and Health Survey: 1992 and 2002.

24.0–27.9 and obesity = OB : BMI > 28.0) in the Chinese population. Although the prevalence is still much lower in China than in Western societies and other developing countries, a 10 year increase of 84% and 38%, for OW and OB, respectively (Fig. 3.9), is a potential forerunner of OB-associated chronic diseases and a subsequent public health burden *(14,15)*. Rates of OW/OB are generally higher in urban than rural areas, but Zhang et al. *(16)* recently reported an 18.6% prevalence of OW in a rural Chinese population. Zhao et al. used the third (2002) National Health Services Survey to assess direct medical costs attributable to OW/OB in Mainland China and estimated the figure to be about 21.11 billion Yuan (approximately US $2.74 billion) *(17)*.

3.6.2.2 Hypertension

Hypertension, a major risk factor for stroke, accounts for 11.7% of the total mortality in the Chinese population and does not vary substantially by gender, extent of urbanization, or geographic region *(13)*. The prevalence of hypertension has been increasing in China in recent decades, whereas rates of awareness, treatment, and control remain unacceptably low *(18)*. Figure 3.10 shows that the prevalence of hypertension has increased 3.5 times from 1959 to 2002 in a steady fashion. Considering the large total population of the country, this increase has already created a huge burden for public health and with especially immense effects in the older population.

3.6.2.3 Type 2 Diabetes (T2D)

In recent years, the prevalence of T2D, which is closely associated with high-risk dietary behaviors and OW/OB issues, has spread nationwide; this is particularly true in the urban population. In the last review of T2D prevalence, we reported an increase by fivefold between 1995 and 1982 *(19)*. The overall prevalence of T2D in China is now at an all time high, with increases of 39 and 15% reported for large and middle-sized Chinese urban cities, respectively, between 1996 and 2002 (Fig. 3.11) *(20)*.

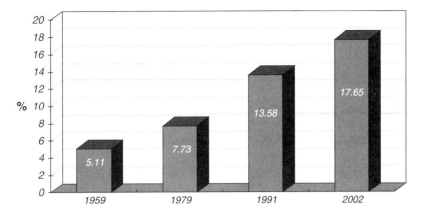

Fig. 3.10. Prevalence of hypertension (age 15 +): 1959 through 2002.
Source: Adapted from Wang et al. *(46)*.

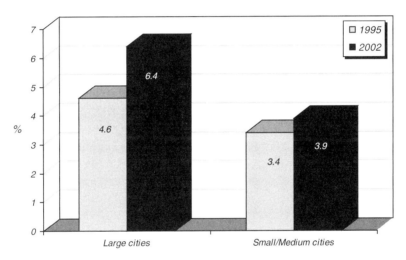

Fig. 3.11. Changes in prevalence of type 2 diabetes in China between 1995 and 2002. By definition, large cities have a non-agricultural population $\geq$ 500,000. Small/medium cities are all other cities.
Source: This figure was based on published information from the 1995 National Diabetes Survey and the 2002 National Nutrition and Health Survey.

3.7 RAPID GROWTH OF THE ELDERLY POPULATION IN CHINA

3.7.1 Prolonged Life and Changes of Population Structure

China has been experiencing an extraordinarily rapid age structure transition since the 1980s. The dramatic fertility decline and improved longevity over the past four decades are causing China's population to age at one of the fastest rates ever recorded *(21)*. Increases in average life expectancy contribute to this, as well as the implementation of family planning. The one-child policy has markedly slowed population growth; the crude birth rate declined from 33.43 per thousand in 1970 to 22.28 per thousand in

1982 (18), then continuously decreased to 15.23 per thousand in 1999 (19), and to 12.3 per thousand by 2004. The annual growth rate of the population has also consistently declined from 25.83 per thousand in the 1970s to 5.87 per thousand in 2004 *(22)*.

Just 30 years ago the country was concerned that it had too many children to support, but today the country is facing the opposite problem—there are now too few young people to provide for such a rapidly aging population. Figure 3.12 presents the most recent population pyramids for China, which show the projected change of population structure between 2000 and 2050. As reported by the State Bureau of Statistics *(22)*, the population aged > 65 years exceeded 100 million in China in 2005, accounting for 7.7% of the total population and an increase of 3% from 2000 (Table 3.3). The growth of the elderly population in China is faster than the overall growth of the total population. The average annual increase of total population for those aged 65+ and 80+ was 2.68 and 4.67% from 2000 to 2005, compared to 0.63% for total population growth during the same span *(23)*.

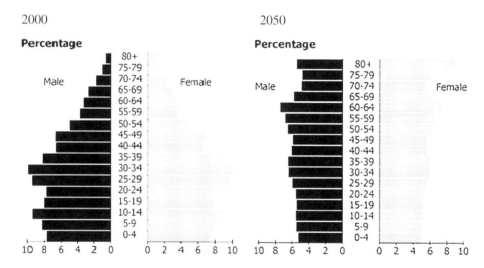

Fig. 3.12. Population pyramids for China: 2000 and 2050.
Source: *World Population Prospects: The 2004 Revision* (2005).

Table 3.3
Size and proportion of the elderly population in China by year 2000–2005

	Year-end figure in millions		
Year	Total population	Elderly population (65+)	Proportion of elderly population (%)
2000	1267.43	88.11	7
2001	1276.27	90.62	7.1
2002	1284.53	93.77	7.3
2003	1292.27	96.92	7.5
2004	1299.88	98.57	7.6
2005	1307.56	100.55	7.7

Source: Feng N, Xiao N. *23rd Population Census Conference*, Christchurch, New Zealand *(23)*.

China is considered by Western standards to be a youthful country, with the elderly constituting only a moderate percentage of population. However, with a population of 1.3 billion, of which 7.8% is aged > 65 years, China has a larger total number of people aged > 65 years than in all European countries combined. In addition, the trend of population aging in large urban cities is much faster than in small cities and rural areas. For example, in 2004, the percent of people in Shanghai and Beijing aged > 65 years had already exceeded 15 and 10%, respectively *(22)*.

3.7.2 *Nutritional Status and Health Behaviors of Chinese Older Adults*

The previously reviewed changes in Chinese dietary patterns have important implications for health and the prevalence of chronic disease for those currently aged and those entering this demographic group in the near future. With an increase in prevalence of OW and OB of the middle-aged population, it is predicted that related chronic health conditions, including CHD, T2D, and cancer, will affect an unprecedented number of older people. On a more positive note, the prevalence of malnutrition has consistently declined in all populations of China, particularly within the category of those aged > 60 years, although there is still a disparity between urban and rural residents. The prevalence of malnutrition decreased from 9.0 and 20.3% in 1992 to 5.4 and 14.9% in 2002 for urban and rural populations, respectively.

3.7.2.1 CIGARETTE SMOKING

Because health behaviors often cluster, we examined other important lifestyle factors to help gauge the current trends of adherence to prudent versus poor choices, looking at rates of smoking and alcohol use and physical activity patterns overall and in older adults. China has the largest number of cigarette smokers of any country in the world. As reported in the 2002 China National Health and Nutrition Survey, more than 50% of men aged 15 and above currently smoke, while fewer than 3% women of the same age group are smokers. About 62.5% men aged 50–54 are smokers (Fig. 3.13a), and 40% of those aged 75 and older report smoking. The overall smoking rate for people 15 years and older was higher in the rural (37.8%) than in the urban population (29.5%) in 2002, while the percentages were 39.2 and 34.5%, respectively, in 1996. Thus the smoking rate has slightly decreased in recent years, but most of the reduction was in urban rather than in rural areas *(24)*. A 2005 report indicated that cigarette smoking was responsible for 7.9% of the total mortality in China; not surprisingly, the estimated risk was higher among men than women *(13)*. With the high prevalence of smoking in China, lung cancer has consistently increased as a cause of mortality, moving from the 4th leading cause of death (in the 1970s) to the No. 1 leading cause of death among all cancers in the Chinese population in 2000. From 2000 to 2005, the incidence of lung cancer continued to increase, from 43.0 to 49.0 per 100,000 in males, and from 19.1 to 22.9 per 100,000 in females, increases of 14 and 20%, respectively *(25)*. Elevated mortality risks from all causes were observed for current smokers of both sexes in a

a

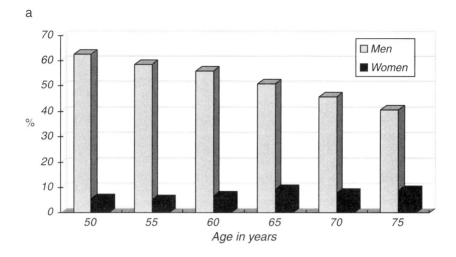

b

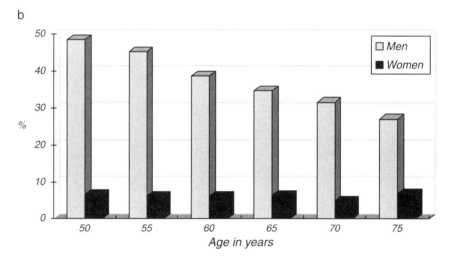

Fig. 3.13. a: Percentage of smokers in the population by age and gender. **b:** Percentage of alcohol drinkers in the population by age and gender.
Source: Ma G, Kung L. *Behavior and Lifestyle* People's Publishing House: Beijing, China, 2006 *(27)*.

3-year longitudinal study of 2,030 Hong Kong Chinese subjects aged >70 years, indicating that the effect of smoking on health is still apparent at older ages *(26)*, and thus smoking cessation would be beneficial even at advanced ages.

3.7.2.2 ALCOHOL CONSUMPTION

Figure 3.13 b shows the proportion of Chinese who regularly consume alcohol (mainly liquor with 40–60% alcohol content). About 40% of men and 4.5% of women reported regularly drinking alcohol; men aged 50–54 showed the highest percent of alcohol use (48.4%). Alcohol use also appeared to decrease with age, and

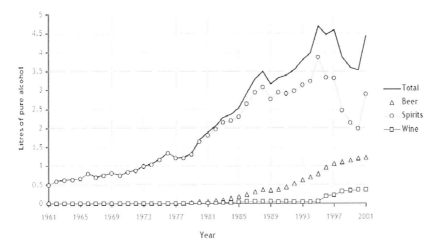

Fig. 3.14. Adult consumption of alcohol in China (per capita) from 1961 to 2001.
Source: FAO (Food and Agriculture Organization of the United Nations), World Drink Trends 2003.

urban residents had slightly lower consumption rates than rural residents *(27)*. In recent decades, increase in alcohol consumption and related problems in China have become significant. While alcohol is a traditional part of Chinese life, commercial alcohol production in China has increased more than 50-fold per capita since 1952 (Fig. 3.14). Evidence suggests that people living in Northern China have higher levels of alcohol consumption than those in the south, that urban residents drink lower-strength beverages than do rural residents, and that some minority ethnic groups, such as those of Tibetan and Mongolian background, drink more than other ethnic groups *(28)*. Evidence also indicates a marked increase in the prevalence of alcohol dependence, which has become the third most prevalent mental illness in China *(29)*.

3.7.2.3 Physical Activity Behaviors

He et al. *(13)* linked physical inactivity with 6.8% of the total mortality in China, with the estimated risk of death being slightly higher among men than women and among urban residents than rural residents (especially among women). However, Chinese elderly are living quite actively in general, as most people over the age of 60 years do not own a car or drive. People who still work must walk to take a bus/subway/train or ride a bike to go to the work place, while many elderly people take care of their grandchildren, the housework, and daily grocery shopping and exercise every day after retiring from work. Moreover, as illustrated in Fig. 3.15, participation in organized programs of physical activity is quite common in the older Chinese community. Findings from the 2002 China National Health and Nutrition Survey indicated that those who participate in regular exercise were mostly those aged 50 and older (Table 3.4). About 9% men and 10% women aged 50–54 reported participating in regular exercise, with the most active group (21%) being men aged 70–74 (older age groups showed a decrease in activity). For women, the highest

Fig. 3.15. Organized physical activity for older adults is common in Chinese communities. Here, in the "Temple of Heaven Park" in Beijing, older adults (aged around 50 years and up, the typical age for retirement in China) gather daily to participate in a variety of group activities.

Table 3.4
Percent of Chinese adults (aged > 50 years) who exercise regularly

Age group	Men (%)	Women (%)
50–	9.2	10.0
55–	13.2	13.9
60–	17.6	17.3
65–	20.5	17.3
70–	21.2	13.9
75–	17.4	9.5

Source: 2002 China National Nutrition Survey.

percent (17%) participating exercise was observed for those aged 65–69. A very large proportion of elderly people reported participating in exercise more than 3 times each week. Table 3.5 shows self-reported physical activity levels for men and women aged 50 and over who exercise regularly. There were more than 50% of men and 43% of women aged 50–54 who reported heavy physical activity levels, although the number declined with age *(27)*.

Table 3.5
Physical activity levels of Chinese adults (aged > 50 years) who exercise regularly

	Men (%)			Women (%)		
Age group	Light	Moderate	Heavy	Light	Moderate	Heavy
50–	27.7	16.3	56.0	41.2	16.0	42.8
55–	35.0	14.8	50.0	50.5	14.0	35.5
60–	45.6	12.4	42.0	58.5	15.2	26.3
65–	51.7	13.9	34.4	65.8	14.7	19.6
70–	64.6	12.0	23.4	76.3	12.2	11.6
75–	76.9	8.4	14.7	86.4	7.4	6.5

Source: 2002 China National Nutrition Survey.

In summary, lifestyle choices in China continue to be shaped more by traditional parameters rather than any newly instituted efforts to pursue healthier behaviors. Thus, while the negative risk factors of smoking and heavy alcohol use are continuing or increasing, the tradition of physical activity also continues as an important positive determinant of health.

3.7.3 Causes of Death, Illness, and Disability in Older Chinese Adults

3.7.3.1 LEADING CAUSES OF DEATH

Age is an important risk factor for all degenerative diseases, along with factors of changing lifestyles and dietary patterns. Similar to the general population, the leading causes of death in older Chinese are also dominated by chronic diseases. CHD, stroke, and cancer are shown to be the top three causes of death for people aged 65+, ranking 8.1, 6.5, and 2.9 times higher than in those aged <65 (Fig. 3.16) *(13)*.

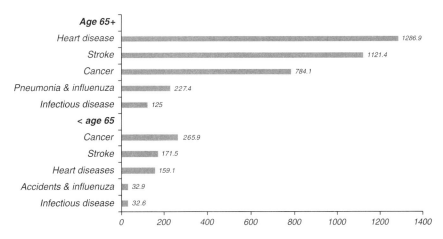

Fig. 3.16. Cause of death in China (100,000 person/yr) for those ≥65 and <65 years of age. *Source*: He et al. *(13)*.

Table 3.6
Percentage of Chinese adults with chronic health problems by age cohort

	Total	Men	Women	Urban	Rural
Age 45–59 years					
Overweight	29.0	26.3	31.4	37.4	25.8
Obesity	10.2	7.2	12.9	15.1	8.4
Hypertension	29.3	28.6	30.0	32.8	28.0
Diabetes	4.3	3.4	4.6	7.8	3.0
High cholesterol	4.7	4.0	5.4	7.0	3.9
High triglyceride	15.7	16.1	15.5	20.4	13.9
Age > 60 years					
Overweight	24.3	23.5	25.2	37.2	19.5
Obesity	8.9	6.6	11.2	16.0	6.2
Hypertension	49.1	48.1	52.0	54.4	47.2
Diabetes	6.8	6.5	7.1	13.1	4.41
High cholesterol	6.1	4.0	8.3	10.6	4.5
High triglyceride	14.8	11.8	17.7	20.5	12.6

Source: 2002 China National Health and Nutrition Survey.

The prevalence of certain chronic health problems is shown in Table 3.6, for both urban and rural male and female populations. T2D, obesity, and high total cholesterol are 1.9-, 1.5-, and 1.3-fold higher in urban compared to rural areas for the older age group, with a similar regional pattern observed in the younger age group. Women have a higher occurrence of all diseases than men in both age groups *(30)*.

3.7.3.2 SPECIAL CONCERNS ABOUT ALZHEIMER'S DISEASE AND QUALITY OF LIFE

For a country with an already large and rapidly growing proportion of elderly, the increasing prevalence of Alzheimer's disease (AD) is a major health issue affecting nutritional and overall quality of life. While AD is unfamiliar to most people in China, a recent survey by the Peking Union Medical College Hospital of 34,807 people aged > 55 years from Beijing, Xian, Shanghai, and Chengdu revealed that China had 3.1 million Alzheimer's patients, accounting for 5.9% of the population above age 55 years. The death rate due to Alzheimer's disease was 14.4 out of 100 affected persons per year in China, similar to findings for Japan, England, and the United States *(31)*. AD was more common in northern than southern China and in women than men.

Due to poor understanding of this disease, AD has been largely ignored by the Chinese public. The Peking Union Medical College Hospital study also found that only 23.3% of China's AD patients sought medical advice, with 21.3% ultimately receiving medical treatment *(31)*. About 48.8% of the study participants believed that the disease was a normal part of aging and nearly 96% of the people who took care of AD patients had never received any form of standard training.

3.7.3.3 OTHER RISK FACTORS FOR POOR HEALTH-RELATED QUALITY OF LIFE

Other health problems such as arthritis, hearing loss, dental disease, gastrointestinal conditions, liver disease, and various disabilities may also interact with the need for dietary and other long-term care services for elderly Chinese individuals. Psychological changes, especially depression, may also influence the nutritional and health status of some Chinese elderly; unfortunately, these changes have yet to be adequately studied in China.

3.8 TRADITIONAL VIEWS OF AGING IN CHINA: A POSITIVE MODEL BUT CONCERNS FOR THE FUTURE

The health- and nutrition-related issues of concern in China's aging population must be considered within the context of the traditional position held by the elderly in Chinese society *(32)*. In comparison to Western culture, Chinese culture values the benefits of old age to a much greater extent. Respect for older people is a generally ingrained and pervasive value of the Chinese. In the traditional Chinese family structure, age is considered to be a key determinant of authority, and therefore, elderly family members hold a particularly high status *(33)*. While increases in educational opportunities and modern technology in urban areas have given younger Chinese greater status in today's society, older citizens continue to command a high degree of respect. In addition, the traditional patterns of interdependence between generations have largely been maintained *(34)*. The position of aged adults is defined by the Constitution of 1982: "Children who have come of age have the duty to support and assist their parents" *(35)*. The newly updated Marriage Law of 2001 also embraces this commitment and further stipulates that "when children fail to perform the duty of supporting their parents, parents who have lost the ability to work or have difficulties in providing for themselves have the right to demand that their children pay for their support" *(36)*. The Family Law essentially works to make the tradition of family care obligatory for all. Those who do not care for their older parents might thus be criticized or even penalized *(34)*.

In the present day Chinese family, the elderly have very close relationships with their children; most older people do not live alone, especially in rural areas. Thus many elderly Chinese adults depend on support from spouses and children for financial, emotional, and physical aid *(37)*. The level of respect for and close family relations of older people in China very likely has a positive impact on their health and nutritional status. Foreman et al. *(38)* studied the relationship between "social support" (including marriage, family size, proximity, and relationship) and the use of health services in a group of Chinese elderly in Beijing. This study indicated that the "structural" social support of marriage and children in the home results in a variegated pattern of increased physician use (due to increased access) but diminished hospitalization. Feelings of self-worth and respect were thought to increase access to and use of physician services. "Functional" social support, such as respect and harmony (filial piety) also contributed to an increased usage of health care services, while household harmony was related to lower use. In this case, care provided by family members probably substituted for hospital care.

While the traditional model of family care has greatly benefited the health of older adults, it will undoubtedly soon be altered due to the recent changes in socioeconomic, educational, and cultural aspects of Chinese society. Changes within family context and structure that inevitably accompany fertility reduction and the increasing life expectancy, along with an increasing working pressure on only children and the fact that many youth are choosing to seek a brighter future outside their impoverished hometowns will all contribute to this change. At the same time, the living arrangements of older people in China remain greatly dependent on the family. There is a relatively high pension coverage rate of 35.3% for urban workers. However, China's pension system provides low coverage rate for rural farmers *(39)*. According to the "China Urban and Rural Elderly Survey", in 2000 about 7% of rural people aged > 60 years received pension benefits or social old age insurance, whereas 85% relied on family support. Other reported estimates of rural pension coverage are only slightly higher (9–11%). *(40)* Overall, only 22% of elderly Chinese receive retirement pension, while 52% still rely on filial relations for economic support. In 2005, 45.5% of urban elderly versus 4.6% of rural elderly lived on retirement pensions. Accordingly, for those living in rural areas family dependency for economic support was 65.2% (Table 3.7) *(23)*. Such huge regional disparities in pension coverage dictate that rural elders are more dependent on their children or relatives, and in turn, their health in later life is not protected by any government program and likely more problematic for themselves and their families.

China is currently in the process of developing the largest pension system in the world to accommodate a rapidly aging society within a rapidly growing, but still under-developed economy. However, in the transition from a planned toward a "socialist market" economy, the complex interactions between numerous challenges such as the urban–rural divide, growing economic inequality, and the ongoing reform of formerly state-owned enterprises make an incremental approach increasingly difficult *(41)*. Thus in the near future for China, while family care continues to be viewed as the "best option," there may be dramatically challenging burdens felt by the child or children in a family. This is an almost inevitable consequence of a strict (one-child) population policy and increasing longevity in China. Concerns about this issue were poignantly emphasized when the earthquake of May, 2008, devastated a large segment of rural China. Many victims were children who died in the collapse of their classrooms. Thus many families lost their only child and future caregiver in this tragedy *(42)*. The country now faces the challenge of how to balance the traditional provision of care to the elderly and the available social support of a modern society.

Table 3.7
Major sources of income for urban and rural Chinese elderly

Order	Urban	Percent	Order	Rural	Percent
1	Pensions	45.5	1	Family dependency	65.0
2	Family dependency	41.3	2	Income from work	26.0
3	Income from work	8.2	3	Pensions	4.6

Source: Feng N, Xiao N. *23rd Population Census Conference*, Christchurch, New Zealand *(23)*.

3.9 COMING CHALLENGES FOR SOCIAL SERVICES AND THE HEALTH CARE SYSTEM IN CHINA

3.9.1 Medical and Economic Impact of a Graying Population

As we have seen in the preceding sections, the percentage of elderly persons in China is rapidly expanding (will triple from 8 to 24 percent between 2006 and 2050), there has been a steady increase in chronic disease associated with modifiable risk factors such as smoking and high-fat/high-calorie diets, and age-related health problems are increasingly prevalent. Chronic diseases (heart disease, cancer, and stroke) already account for as much as 60% of all deaths in China *(13)*. As China exceeds the ability of the traditional family approach to meet the health and long-term care needs of this growing elderly population, soaring health care costs will undoubtedly result. Unfortunately, the size of the working-age population (who bear much of the health care cost burden) is simultaneously shrinking. The *elderly support ratio*, defined as the number of working-age adults (ages 15–64 years) per number of elderly (age 65 years and above), is projected to decline drastically, from 9 persons to 2.5 persons by 2050 *(43)*. This is in contrast to the growth of the elderly populations in more developed countries, which has not been as drastic, allowing those nations more time to adjust to this structural change. For example, the United States experienced a doubling of the aged population over a span of 45 years (1930–1975), while that of China is predicted to double (from 7 to 14%) within 26 years (2000–2026) *(44,45)*. China is already feeling the economic impact of her demographic transitions, with a per capita GDP of less than $1,000 in 1999 and $1,700 in 2005. This phenomenon of "getting older before becoming richer" poses a serious challenge for a health care system that already faces a number of adversities, the most important of which is the rapid increase in overall costs in private health care spending since shifting to a market-oriented system in the early 1980s *(14)*. Thus, rising out-of-pocket costs prevent many Chinese from seeking early (preventive) medical care. This has resulted in wide disparities in health care access, particularly between urban and rural areas. These trends are of special concern to the elderly, who likely have more extensive health care needs, yet fewer resources for affording such care *(21)*.

3.9.2 Public Health Responses

The Chinese government has recently acknowledged and begun to address the consequences of rapid population aging. Programs have been established that are targeted towards preventing and managing specific age-related diseases. One example is the community-based intervention for management of hypertension and diabetes conducted in Beijing, Shanghai, and Changsha between 1991 and 2000. Similarly, the formation of the Program of Cancer Prevention and Control in China and ratification of the WHO Framework Convention of Tobacco Control are helping to address important health issues affecting older adults in China, despite the limited funding that exists *(46)*. At the same time, new opportunities for entrepreneurship in the health service industry have been established as a result of China's social-welfare reform of the 1990s, which decentralized government-funded welfare institutions and significantly reduced their government financing *(47)*. Today, an increasing number of private elder homes as well as the country's former government-sponsored elder homes (which were once reserved exclusively for elderly without children or other means of support) are providing an alternative to familial elder care *(47)*. However, these facilities are insufficient in number, of varying standards, and often too expensive for many older adults and their families. Community-based long-term care services for elderly persons in China—both informally and locally supported by the government—have also begun to emerge, especially in urban areas *(48)*. These efforts are serving a variety of needs of older clients and their family caregivers, including daily care, home maintenance, and referral services.

3.10 IMPLEMENTATION OF PREVENTIVE NUTRITION: GLOBAL IMPLICATIONS

Lessons learned from China and its shift toward an elderly population highlight both the benefits and the risks conveyed by lifestyle determinants of late life health trajectories. Effective implementation of preventive nutrition strategies could play a key role in diminishing the future chronic disease burden. However, an intensive, on-going, and incredibly flexible approach will be necessary to accomplish meaningful dietary change. Many challenges exist, including the need for culturally and ethnically appropriate nutrition education, the difficulty of changing dietary behaviors (see Chapter 2), and, most recently, a dramatic increase in global food costs.

3.10.1 Achieving Global Behavior Change in Older Adults

The Global Strategy on Diet, Physical Activity and Health of the WHO *(49,50)* urges prudent lifestyle behaviors to promote better health, yet available survey data suggest that these goals are not being fully met *(51)*. For example, Pomerleau et al. *(52)* studied worldwide efforts to help reduce the burden of chronic diseases by increasing intakes of fruit and vegetables. They found positive effects with face-to-face education or counseling, as well as interventions using telephone contacts or computer-based information; however, the amount of improvement achieved was modest, with an approximate increase of only 0.1–1.4 fruit/vegetable servings per day. The investigators noted that more research is needed on approaches for promoting healthy behaviors, especially in the developing world.

Acknowledging that behavioral change is always difficult to achieve, there is a common notion that it is especially difficult for older individuals to change their lifestyles and follow recommended healthy eating plans. However, there is evidence that older adults can both implement and benefit from health promotion programs that emphasize dietary changes *(53,54)*. One major challenge for any behavioral intervention is to bridge the gap between having the knowledge of the dietary changes that are needed and actually implementing these changes. Investigators in the Study to Help Improve Early evaluation and management of risk factors Leading to Diabetes (SHIELD) examined the knowledge, attitudes, and behaviors of subjects (mean age ∼ 60 years) with T2D (n = 3, 867) and at high risk for developing T2D (n = 5,419). In agreement with other reports *(55)*, they found that knowledge alone did not predict appropriate behavioral modifications. Despite reporting healthy attitudes and knowledge conducive to good health, the majority of subjects did not translate these positive traits into healthy behaviors with respect to diet, exercise, and weight loss *(56)*. Suggestions for addressing this problem include counseling to assist the individual in establishing values, motivations and goals, and to guide them in coping with real and perceived barriers to behavior change. Estabrooks et al. *(57)* conducted behavioral assessments in a group of randomized controlled trial participants with T2D (n = 422) and found that when they personally set appropriate goals, there were significant corresponding behavioral changes over the 6-month study period. With regards to barriers, Folta et al. *(58)* studied commonly reported barriers to achieving a heart-healthy diet in middle- and older-aged women and reported that time constraints and concerns about "wasting" food topped the list. In addition to the usual challenges, for some high-risk older persons, the barriers to good nutrition listed elsewhere in this text, for example, social isolation and a limited ability to shop for and prepare meals, also clearly come into play *(59)*.

3.10.2 Soaring Global Food Costs and the Dual Challenge of Under- and Over-Nutrition

One of the most important barriers to achieving dietary behavior change in any culture is the cost and availability of foods that are rich in health-promoting nutrients (e.g., protein, vitamins and minerals, fiber) and are not excessive in terms of calories, sodium, certain fats (saturated, trans), and simple sugars. Generally, foods associated with a healthier nutrient profile and lower rates of obesity are more expensive than foods of poor nutrient density *(60,61)*. These foods are also more perishable (thus more likely to be wasted after purchase) and more difficult to shop for than highly processed foods, which tend to be readily available, have long shelf lives, and be high in calories, fat, sugar, and/or sodium. Recent increases in food prices are exacerbating concerns about the global affordability of nutritious food. In the time period from March 2007 to March 2008, the global food price index showed a sharp increase (see Fig. 3.17), with escalations in prices for almost all food categories *(62)*. There were multiple causes for this food cost crisis, including poor recent harvests, restrictive trade policies, the increasing price of oil, diversion of crops for bio-fuels, and increasing world demand for food in fast-growing economies of countries with large populations, including China and India *(63)*. The dramatic rise in the proportion of income that must go for food hits hardest in the poorest countries and civil unrest has resulted in at least 20 countries *(64)*. Leaders at the United Nations called for emergency aid to help avoid widespread starvation and the World Bank plans to offer emergency financing to boost agricultural productivity, projecting that food prices will remain elevated for at least another year *(63,64)*.

While the increase in food costs is leading to critical concerns about nutritional adequacy for those with subsistence-level incomes, it also has implications for middle-income groups. As previously noted, in addition to being appealing and convenient, refined grains, sugars, and added fats are among the most affordable sources of dietary energy and yet often are of poor nutritional value *(61)*. As rising food costs put increasing pressure on family food budgets, these foods are likely out of necessity to replace "healthier" choices such as fresh fruits and vegetables. In fact, in developed countries like the United States, obesity has been linked with the price disparity between

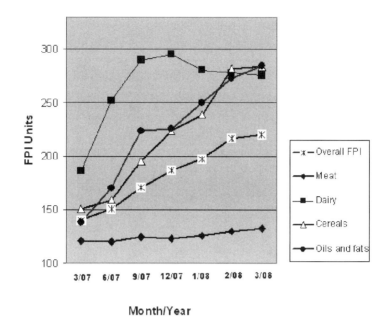

Fig. 3.17. Overall and specific food price indices by quarter from March 2007 to March 2008. The food price index (a monthly measure of price changes in major food commodities traded internationally) has averaged as much as 80 points (57%) higher over this time period. The increase was driven by rising prices for almost all food commodities. The price of staple foods (grains, oils sugar) has increased by 50%, including an increase of 90% for rice. *Sources*: World Bank; Food and Agriculture Organization of the United Nations.

"healthy" and "unhealthy" foods. Likewise, in emerging countries where there are substantial income and health disparities, over-nutrition is becoming a major nutritional concern, creating a "dual burden" to health of under-nutrition and obesity. Mendez et al. *(65)* studied patterns of under- and over-nutrition in women of 36 developing countries (in Africa, Latin America, the Caribbean, Asia, and the Middle East) and found that in almost all countries overweight exceeded underweight as a nutritional problem. This helps to explain the ironic observation that increasing income does not necessarily predict better health outcomes *(66,67)*. As previously noted in the findings from China, growth of disposable income in emerging economies is often associated with an acceleration of the kinds of health problems associated with dietary excess. It is clear that one size does not fit all with regards to needed nutritional improvements—some groups mainly need more nutrients, some need fewer calories, and some need both modifications to achieve a health-promoting diet.

3.10.3 Summary

The United Nations has identified promotion of healthy aging as one of the major emerging nutritional challenges that will dominate the global agenda in the coming years *(68)*. In 2004, the World Health Assembly endorsed the Global Strategy on Diet, Physical Activity and Health of the WHO, which recognizes the shift in the balance of major causes of death and disease toward non-communicable diseases. The nutritional guidelines outlined in this plan are as follows:

- Achieve energy balance and a healthy weight
- Limit energy intake from total fats and shift fat consumption away from saturated fats to unsaturated fats and toward the elimination of transfatty acids
- Increase consumption of fruits and vegetables, and legumes, whole grains, and nuts
- Limit the intake of free sugars
- Limit sodium consumption from all sources and ensure that salt is iodized

Clearly, an array of resources will need to be applied in an integrated, trans-disciplinary, international approach if nutritional interventions are to be successful in reducing the incidence of chronic disease *(5,69)*. Regional political, epidemiological, environmental, infrastructural, and genetic determinants of health must all be taken into account *(69)* and health-promoting behaviors need to be integrated into the normal daily life if they are to be sustained *(70)*. Innovative thinking and use of technology could hold promise—in the future it may be possible to create and integrate into the food supply unique new foods that are enhanced in flavor and texture, enriched with nutrients, and yet low in undesirable attributes *(71)*. We recommend that both traditional and novel avenues be very actively pursued—the future health and well-being of the entire world may very well be at stake.

3.11 RECOMMENDATIONS

1. Nutritional and other lifestyle interventions have the potential to improve quality of health scenarios for the elderly and help control medical care costs.
2. An intensive, integrated, trans-disciplinary, international approach is needed to achieve success in reducing the incidence of chronic disease on a global level.
3. We applaud the efforts of the WHO and other international organizations that have recognized the need for making preventive nutrition for populations of all ages a global priority. We also encourage "outside the box" approaches to address urgent global nutrition issues.

REFERENCES

1. Fund UNP. Population aging and development-social health and gender issues: Population and development strategies issue. 2002.
2. Baerlocher MO. Differences in healthy life expectancy among men and women. CMAJ 2007;177(10):1174.
3. The ten leading causes of death by broad income group 2002. http://www.who.int/mediacentre/factsheets/fs310.pdf)
4. Kamangar F, Dores GM, Anderson WF. Patterns of cancer incidence, mortality, and prevalence across five continents: Defining priorities to reduce cancer disparities in different geographic regions of the world. J Clin Oncol 2006;24(14):2137–50.
5. Zimmet P, Alberti KG, Shaw J. Global and societal implications of the diabetes epidemic. Nature 2001;414(6865):782–7.
6. Sloan FA, Bethel MA, Ruiz D, Jr., Shea AH, Feinglos MN. The growing burden of diabetes mellitus in the US elderly population. Arch Intern Med 2008;168(2):192–9; discussion 9.
7. Cullen LT. When Eat Meets West. Time 2008; 171(4):44–6.

8. Fried LP. Epidemiology of aging. Epidemiol Rev 2000;22(1):95–106.
9. Popkin BM, Keyou G, Zhai F, Guo X, Ma H, Zohoori N. The nutrition transition in China: a cross-sectional analysis. Eur J Clin Nutr 1993;47(5):333–46.
10. World Health Organization. Diet, nutrition and the prevention of chronic diseases : report of a joint WHO/FAO expert consultation. Geneva: WHO Technical report Series 916; 2003.
11. Zimmet ZP. The pathogenesis and prevention of diabetes in adults: genes, autoimmunity, and demography. 1995;18(7):1050–64.
12. State Bureau of Statistics. China Statistics – 2006. Beijing, China: China Statistical Press; 2007.
13. He J, Gu D, Wu X, et al. Major causes of death among men and women in China. N Engl J Med 2005;353(11):1124–34.
14. Popkin B, Paeratakul S, Ge K. Zhai F. Body weight patterns among the Chinese: Results from the 1989 and 1991 China Health and Nutrition Surveys. Am J Public Health 1995;85(5):690–4.
15. Gu X, Chen M. Health services in Shanghai County; Vital statistics. Am J Public Health 1982;72(9 (Suppl)):19–23.
16. Zhang X, Sun Z, Zhang X, et al. Prevalence and associated factors of overweight and obesity in a Chinese rural population. Obesity (Silver Spring) 2008;16(1):168–71.
17. Zhao W, Zhai Y, Hu J, et al. Economic burden of obesity-related chronic diseases in Mainland China. Obes Rev 2008;9(Suppl 1):62–7.
18. Gu D, Reynolds K, Wu X, et al. Prevalence, awareness, treatment, and control of hypertension in China. Hypertension 2002; 40(6):920–7.
19. Wang K, Li T, Xiang H. Study on the epidemiological characteristics of diabetes mellitus and IGT in China. Zhonghua Liu Xing Bing Xue Za Zhi 1998;19(5):282–5.
20. Wang L. Overall Report 2002 National Health and Nutrition Survey. Beijing, China: People's Publishing House; 2005.
21. Kaneda T. China's concern over population aging and health. In: Population Reference Bureau; 2007.
22. State Bureau of Statistics. China Statistics – 2005. Beijing, China: China Statistical Press; 2006.
23. Feng N, Xiao N. Population aging in China as reflected by the results of the 2005 Population Sample Survey. In: 23rd Population Census Conference, Utilization of the 2000 and 2005 Rounds of Asia-Pacific Censuses. Christchurch, New Zealand; 2007.
24. Yang G. Deaths and their risk factors among Chinese population. In: Publishing house of Peking Union Medical College, Beijing China 2004.
25. China Cancer Foundation. Report on Smoking and Health in China. Beijing, China 2006.
26. Woo J, Ho SC, Yu AL. Lifestyle factors and health outcomes in elderly Hong Kong Chinese aged 70 years and over. Gerontology 2002;48(4):234–40.
27. Ma G, Kung L. *Behavior and Lifestyle* 2002 National Health and Nutrition Survey. Beijing, China: People's Publishing House; 2006.
28. Hao W, Su Z, Liu B, et al. Drinking and drinking patterns and health status in the general population of five areas of China. Alcohol 2004;39(1):43–52.
29. Cochrane J, Chen H, Conigrave KM, Hao W. Alcohol use in China. Alcohol 2003; 38(6):537–42.
30. Yang X, Zhei F. Diet and nutrition intake 2002 National Health and Nutrition Survey. Beijing, China: People's Publishing House; 2005.
31. Zhang ZX, Zahner GE, Roman GC, et al. Dementia subtypes in China: Prevalence in Beijing, Xian, Shanghai, and Chengdu. Arch Neurol 2005;62(3):447–53.
32. Flaherty JH, Liu ML, Ding L, et al. China: The aging giant. J Am Geriatr Soc 2007;55(8):1295–300.
33. Yank C. Chinese Communist Society: the family and the village. Cambridge, MA: MIT Press; 1965.
34. Davis-Friedmann D. *Long lives: Chinese Elderly and the Communist Revolution*. Stanford, California: Stanford University Press; 1991.
35. China National People's Congress. Marriage Law. In. Beijing, China; 1983.
36. China National People's Congress. Marriage Law. In. Beijing, China; 2001.
37. Zeng Y, George L. Extremely rapid ageing and the living arrangements of older persons: the case of China. In: *Technical Meeting on Population Aging and Living Arrangements of Older Persons: Critical Issues and Policy Responses*. New York: Population Division, Department of Economic and Social Affairs, United Nations Secretariat; 2000.

38. Foreman S, Earl J, Lu L. Use of health serves by Chinese elderly in Beijing. Med Care 1998;36:1265–82.
39. Wang D. China's Urban and Rural Old Age Security System: Challenges and Options. China & World Economy 2006;14(1):102–16.
40. Hu Y-W. Pension reform in China. A Case Study. In. London: Brunel University, London; 2006.
41. Salditt F WP, Adema W. Pension Reform in China: Progress and Prospects. 2007.
42. Bodeen. China's One-Child Policy Causes Extra Pain. Associated Press 2008 May 16, 2008.
43. United Nations, Department of Economic and Social Affairs, Population Division. World Population Prospects. The 2004 Revision. New York; 2005.
44. Kinsella K, Gist Y. Older Workers, Retirement, and Pensions: A Comparative. International Chartbook. In: U.S. Department of Commerce, Economics and Statistics Administration, Bureau of the Census, eds.; 1995.
45. Kinsella K, Phillips D. Global aging: The challenge of success. Population Bulletin 2005; 60:1–40.
46. Wang L, Kong L, Wu F, Bai Y, Burton R. Preventing chronic diseases in China. Lancet 2005;366(9499):1821–4.
47. Zhan H. Recent developments in institutional elder care in China: Changing concepts and attitudes. J Aging Soc Policy 2006; 18(2):85–108.
48. Wu B, Carter M, Goins R. Emerging services for community-based long-term care in urban China: A systematic analysis of Shanghai's community-based agencies. J Aging Soc Policy 2006;17(4).
49. Global Strategy on Diet, Physical Activity and Health. 2004. (Accessed May 21, 2008, at http://www.who.int/dietphysicalactivity/strategy/eb11344/en/index.html.)
50. Waxman A. Prevention of chronic diseases: WHO global strategy on diet, physical, activity and health. Food Nutr Bull 2003;24(3).
51. Pomerleau J, Lock K, McKee M, Altmann DR. The challenge of measuring global fruit and vegetable intake. J Nutr 2004;134(5):1175–80.
52. Pomerleau J, Lock K, Knai C, McKee M. Interventions designed to increase adult fruit and vegetable intake can be effective: a systematic review of the literature. J Nutr 2005;135(10):2486–95.
53. Masley SC, Weaver W, Peri G, Phillips SE. Efficacy of lifestyle changes in modifying practical markers of wellness and aging. Altern Ther Health Med 2008;14(2):24–9.
54. Chernoff R. Nutrition and health promotion in older adults. J Gerontol A Biol Sci Med Sci 2001;56(Spec No 2):47–53.
55. Kim S, Love F, Quistberg DA, Shea JA. Association of health literacy with self-management behavior in patients with diabetes. Diabetes Care 2004;27(12):2980–2.
56. Green AJ, Bazata DD, Fox KM, Grandy S. Health-related behaviours of people with diabetes and those with cardiometabolic risk factors: results from SHIELD. Int J Clin Pract 2007;61(11):1791–7.
57. Estabrooks PA, Nelson CC, Xu S, et al. The frequency and behavioral outcomes of goal choices in the self-management of diabetes. Diabetes Educ 2005;31(3):391–400.
58. Folta SC, Goldberg JP, Lichtenstein AH, Seguin R, Reed PN, Nelson ME. Factors related to cardiovascular disease risk reduction in midlife and older women: a qualitative study. Prev Chronic Dis 2008;5(1):A06.
59. Old and alone: barriers to healthy eating in older men living on their own. Appetite 2004;43(3):269–76.
60. High monetary costs of dietary patterns associated with lower body mass index: a population-based study. Int J Obes (Lond) 2006;30(10):1574–9.
61. Drewnowski A, Darmon N. The economics of obesity: dietary energy density and energy cost. Am J Clin Nutr 2005;82(1 Suppl):265S–73S.
62. World Food Situation: Food Prices Index. 2008. (Accessed May 14, 2008, at http://www.fao.org/worldfoodsituation/FoodPricesIndex.)
63. Dykman J. Why the World Can't Afford Food-And why higher prices are here to stay. Time 2008 May 19, 2008:35–7.
64. Amid food riots and shaken governments IFI scramble to develop a coherent response. 2008. (Accessed May 20, 2008, at http://www.bicusa.org/en/Article.3763.aspx.)

65. Mendez MA, Monteiro CA, Popkin BM. Overweight exceeds underweight among women in most developing countries. Am J Clin Nutr 2005;81(3):714–21.
66. Wider income gaps, wider waistbands? An ecological study of obesity and income inequality. J Epidemiol Commun Health 2005;59(8):670–4.
67. Overweight exceeds underweight among women in most developing countries. Am J Clin Nutr 2005;81(3):714–21.
68. Dangour AD, Uauy R. Nutrition challenges for the twenty-first century. Br J Nutr 2006;96(Suppl 1):S2–7.
69. Schwarz PE, Reimann M, Li J, et al. The Metabolic Syndrome – A global challenge for prevention. Horm Metab Res 2007;39(11):777–80.
70. Darnton-Hill I, Nishida C, James WP. A life course approach to diet, nutrition and the prevention of chronic diseases. Public Health Nutr 2004;7(1A):101–21.
71. Hsieh YH, Ofori JA. Innovations in food technology for health. Asia Pac J Clin Nutr 2007;16(Suppl 1):65–73.

II

Fundamentals of Nutrition and Geriatric Syndromes

4 Update on Nutritional Assessment Strategies

John E. Morley

Key Points

- Assessing the true nutritional status of an older person is a task not easily performed, being frequently complicated by the presence of excess cytokine production (which leads to many of the same effects as protein undernutrition) as well as other factors.
- Assessment of recent weight status is thus generally the best indicator of nutritional deficit.
- This chapter reviews a number of recently developed nutritional screening tools and laboratory tests used for nutritional assessment.
- While the most practical nutrition assessment tools are simple ones, the interpretation is still fraught with difficulty.

Key Words: Simplified nutrition assessment questionnaire; mini nutritional assessment; anorexia; weight loss; albumin; anemia; midarm circumference; MUST; screen III

4.1 INTRODUCTION

Assessing the true nutritional status of an older person is a complex process that is not easily performed. Precise measurements of energy expenditure, such as direct calorimetry or doubly labeled water, are expensive and difficult to obtain. Even the availability of a "metabolic cart" to do indirect calorimetry would represent an unusual occurrence in a longer term care setting. Measurements of proteins classically associated with nutritional status are more effected by cytokines than nutritional status, making them extremely poor nutritional markers. Measurements of micronutrients are difficult, subject to error, and often expensive and problematic to interpret. For these reasons the most practical nutrition assessment tools are simple, but the interpretation is fraught with difficulty.

From: *Nutrition and Health: Handbook of Clinical Nutrition and Aging, Second Edition*
Edited by: C. W. Bales and C. S. Ritchie, DOI 10.1007/978-1-60327-385-5_4,
© Humana Press, a part of Springer Science+Business Media, LLC 2009

4.2 BODY MASS AS AN INDICATOR OF NUTRITIONAL STATE

As a person ages perhaps the single best measure of a nutritional problem is a loss of weight. Accurate weights remain a problem to obtain. Scales are often poorly calibrated. Persons are weighed with different clothing at different seasons. Little attention is paid to the time of the day the person is weighed, despite a well-recognized circadian rhythm. Despite these problems it is recognized that a weight loss of 5% of body weight in any period up to a year clearly is indicative of a problem. The cause of the loss of weight is often less clear and can include poor nutrient intake or absorption, age-related loss of muscle mass (sarcopenia), severe osteoporosis, loss of fat and muscle mass (cachexia) and dehydration. While weight loss is the nutritional gold standard in older persons, it is unclear what level of weight loss less than 5% should represent an early warning sign of nutritional problems. Persons who are losing weight should have the possible causes evaluated using the MEALS-ON-WHEELS mnemonic (Table 4.1 Similarly, weight gain in older persons can be indicative of future problems leading to functional decline.

Body mass index (BMI) [weight (kg)/height^2(cm)] is a simple way to adjust body mass for height. It is independent of sex. It correlates well with body fat and can be used as a surrogate for fat mass. In general, in older persons low BMI's (<22) are associated with increased mortality (1). A problem with BMI in older persons is shrinkage due to loss of bone and changes in posture. For this reason, some authorities use arm length or knee height as a proxy for height (2).

Waist circumference is a useful measure to allow estimation of visceral fat. A waist circumference greater than 120 cm (40 in.) in men or 88 cm (35 in.) in women is a good predictor of risk for the metabolic syndrome. Lower waist circumferences should be used in persons of Asian origin (3,4).

Table 4.1
MEALS-ON-WHEELS mnemonic for treatable causes of weight loss

*M*edications (e.g., digoxin, theophylline, cimetidine)
*E*motional (e.g., depression)
*A*lcoholism, elder abuse, anorexia tardive
*L*ate-life paranoia
*S*wallowing problems

*O*ral factors
*N*osocomial infections (e.g., tuberculosis)

*W*andering and other dementia-related factors
*H*yperthyroidism, hypercalcemia, hypoadrenalism
*E*nteral problems (e.g., gluten enteropathy)
*E*ating problems
*L*ow salt, low cholesterol and other therapeutic diets
*S*tones (cholecystitis)

4.3 NUTRITIONAL QUESTIONNAIRES

A number of questionnaires have been developed and validated to determine nutritional risk (Table 4.2). The Simplified Nutrition Assessment Questionnaire (SNAQ) is a set of four questions that have high sensitivity and specificity to determine older persons at subsequent risk for weight loss *(5)* (Table 4.3). Keller and her colleagues *(6–8)* have developed SCREEN II (Seniors in the Community: Risk Evaluation for Eating and Nutrition). This index determines nutritional risk using four factors, viz. food intake, physiological, adaptive and functional. It has been validated and has good inter-rater and test–retest reliability as well as excellent sensitivity and specificity. It is widely used in the community in Canada.

The "Malnutrition Universal Screening Tool" (MUST) is a simple screening tool for use in both elderly inpatients and outpatients *(9,10)*. It does not require a measured weight or height but can utilize recalled height and weight. It has become a commonly used screening test in the United Kingdom. Its components are BMI, weight loss in 3–6 months and anorexia for 5 days due to disease. It appears to be particularly sensitive to recognize protein energy undernutrition in hospitalized patients.

The Appetite Hunger and Sensory Perception (AHSP) questionnaire was developed in Dutch nursing homes *(11,12)*. This questionnaire focused on loss of appetite and alterations in taste and smell. While relatively useful in healthy community-dwelling older persons, it performed poorly in frail elderly.

DETERMINE is a checklist using 10 items to increase awareness of nutritional risk *(13–15)*. It was developed by the US Nutrition Screening Initiative. It tends to identify frail, ill persons rather than those in whom nutritional risk is the primary problem. SCALES is a simple tool that identifies older persons who are likely to have cachexia *(16)* (Table 4.4).

The Subjective Global Assessment is an in-hospital tool to assess the level of nutrition in hospitalized patients *(17)*. It involves a medical history of weight change, dietary intake, gastrointestinal symptoms and functional impairment as well as a brief physical examination looking for loss of subcutaneous fat, muscle wasting, edema and ascites. It was originally validated for younger persons with gastrointestinal disorders and appears less useful in older persons.

Table 4.2
Commonly used nutritional screening tools in older persons

1.	Simplified Nutritional Assessment Questionnaire (SNAQ) (Table 4.3)
2.	SCREEN II (Seniors in the Community: Risk Evaluation for Eating and Nutrition) *(6–8)*.
3.	Malnutrition Universal Screening Tool (MUST) *(9,10)*.
4.	Appetite Hunger and Sensory Perception Questionnaire (AHSP) *(11,12)*.
5.	DETERMINE *(13–15)*.
6.	SCALES (Table 4.4)
7.	Subjective Global Assessment *(17)*.
8.	Mini Nutritional Assessment (MNA) (Table 4.5)

Table 4.3

Appetite questionnaire to predict weight loss in older persons—
SNAQ (Simplified Nutritional Appetite Questionnaire)

1. My appetite is
   A. very poor
   B. poor
   C. average
   D. good
   E. very good
2. When I eat
   A. I feel full after eating only a few mouthfuls
   B. I feel full after eating about a third of a meal
   C. I feel full after eating over half a meal
   D. I feel full after eating most of the meal
   E. I hardly ever feel full
3. Food tastes
   A. very bad
   B. bad
   C. average
   D. good
   E. very good
4. Normally I eat
   A. less than one meal a day
   B. one meal a day
   C. two meals a day
   D. three meals a day
   E. more than three meals a day

Instructions: Complete the questionnaire by circling the correct answers and then tally the results based upon the following numerical scale: A = 1, B = 2, C = 3, D = 4, E = 5.

Scoring: If the mini-CNAQ is less than 14, there is a significant risk of weight loss.

Table 4.4
SCALES

Sadness (Geriatric Depression Scale)
Cholesterol <160 mg/dL
Albumin <3.5 mg/dL
Loss of 5% body weight
Eating Problems (physical or cognitive)
Shopping/food preparation

The Mini Nutritional Assessment (MNA) (Table 4.5) consists of global assessment questions, specific dietary questions, questions on subjective perception of health and a series of anthropomorphic measurements (18–22). It has been widely validated. It correlates with more sophisticated measurements. It is highly predictive

Table 4.5
The Mini Nutritional Assessment (MNA) scale

Last name: _____ First name: _____ Sex: _____ Date: _____

Age: _____ Weight, kg: _____ Height, cm: _____ I.D. Number: _____

Complete the screen by filling in the boxes with the appropriate numbers.
Add the numbers for the screen. If score is 11 or less, continue with the assessment to gain a Malnutrition Indicator Score.

SCREENING

A Has food intake declined over the past 3 months
due to loss of appetite, digestive problems,
chewing or swallowing difficulties?
0 = severe loss of appetite
1 = moderate loss of appetite
2 = no loss of appetite □

B Weight loss during last months
0 = weight loss greater than 3 kg (6.6 lbs)
1 = does not know
2 = weight loss between 1 and 3 kg (2.2 and
6.6 lbs)
3 = no weight loss □

C Mobility
0 = bed or chair bound
1 = able to get out of bed/chair but does not
go out
2 = goes out □

D Has suffered psychological stress or acute
disease in the past 3 months
0 = yes   2 = no □

E Neuropsychological problems
0 = severe dementia or depression
1 = mild dementia
2 = no psychological problems □

F Body Mass Index (BMI) (weight in kg) / (height in m)²
0 = BMI less than 19
1 = BMI 19 to less than 21
2 = BMI 21 to less than 23
3 = BMI 23 or greater □

Screening Score (subtotal max. 14 points) □□

12 points or greater   Normal—not at risk—
no need to complete assessment

11 points or below   Possible malnutrition—
continue assessment

ASSESSMENT

G Lives independently (not in a nursing home or
hospital)
0 = no   1 = yes □

H Takes more than 3 prescription drugs per day
0 = yes   1 = no □

I Pressure sores or skin ulcers
0 = yes   1 = no □

J How many full meals does the patient eat daily?
0 = 1 meal
1 = 2 meals
2 = 3 meals □

K Selected consumption markers for protein intake
• At least one serving of dairy products
(milk, cheese, yogurt) per day? yes □ no □
• Two or more servings of legumes
or eggs per week? yes □ no □
• Meat, fish or poultry every day yes □ no □
0.0 = if 0 or 1 yes
0.5 = if 2 yes
1.0 = if 3 yes □.□

L Consumes two or more servings of fruits or
vegetables per day?
0 = no   1 = yes □

M How much fluid (water, juice, coffee, tea, milk. . .)
is consumed per day?
0.0 = less than 3 cups
0.5 = 3 to 5 cups
1.0 = more than 5 cups □.□

N Mode of feeding
0 = unable to eat without assistance
1 = self-fed with some difficulty
2 = self-fed without any problem □

O Self view of nutritional status
0 = views self as being malnourished
1 = is uncertain of nutritional state
2 = views self as having no nutritional problem □

P In comparison with other people of the same age,
how do they consider their health status?
0.0 = not as good
0.5 = does not know
1.0 = as good
2.0 = better □.□

Q Mid-arm circumference (MAC) in cm
0.0 = MAC less than 21
0.5 = MAC 21 to 22
1.0 = MAC 22 or greater □.□

R Calf circumference (CC) in cm
0 = CC less than 31   1 = CC 31 or greater □

Assessment (max. 16 points) □□.□
Screening score □□
Total Assessment (max. 30 points) □□.□

Malnutrition Indicator Score
17 to 23.5 points   at risk of malnutrition □
Less than 17 points   malnourished □

of poor outcomes. A six-question short-form screening needs to be undertaken before completion of the whole MNA *(23)*.

The tools discussed above are the most used and/or best validated tools. Many tools that are available have "an inadequate assessment of their effectiveness" *(24)*. Green and Watson *(25)* reviewed 21 published instruments available for nutritional screening and assessment of older adults. They concluded that because of inadequate levels of validation, the clinician should carefully consider whether or not a particular tool is appropriate for their situation.

4.4 ASSESSMENT OF DIETARY INTAKE

A number of methods exist to assess dietary intake. Reported energy intakes tend to be lower than measured intakes. The level of underreporting of food intake ranges from 10 to 45%. Recently a picture-sort food-frequency questionnaire for older persons of low socioeconomic status has been developed *(26,27)*. While this tends to perform better, nevertheless approximately 40% of older persons still underreport food intake. Overall, food-frequency recall has limited use in older persons. Food diaries may be slightly better.

4.5 ANTHROPOMORPHIC MEASURES

Skinfold thickness can be measured with a caliper in older persons. Commonly measured areas are biceps, triceps, subscapular and suprailiac. In older persons the subscapular may be the most accurate *(28)*.

Lean tissue mass (fat free mass) can be calculated by measuring the midpoint of the upper arm muscle or the midpoint of the calf muscle. Upper limbs lose lean mass more rapidly than lower limbs. Limb muscle area can be more accurately estimated by subtracting out the amount of skinfold thickness using appropriate equations. These approaches overestimate muscle mass. Accuracy is less in older persons. However, in persons with excess water retention these can be useful *(29,30)*.

Lean tissue mass and fat mass can be measured by bioelectrical impedance *(31,32)*. Specific age-adjusted equations need to be used. Body fluid disturbances alter the accuracy of bioelectrical impedance.

Free fat mass can be measured by DEXA. To define the amount of sarcopenia, the appendicular skeletal mass divided by height squared is a useful equation. CT scan or MRI is useful to detect fat infiltration into muscle, i.e., myosteatosis.

4.6 UTILITY OF SERUM PROTEIN, CHOLESTEROL AND HEMOGLOBIN

Albumin and a number of shorter lived proteins, viz. transferrin, prealbumin (transthyretin) and retinol-binding protein, have been used as nutritional markers *(33)*. Their levels are markedly decreased by circulating cytokines, such as tumor necrosis factor-alpha. In addition, liver disease results in decreased production. For this reason they are good markers for cachexia, but poor markers for general nutrition screening. Similarly, low cholesterol levels are markers of poor outcomes, but are highly sensitive to cytokines.

Anemia can occur because of a variety of nutritional factors. These include low protein intake, iron, vitamin B_{12}, folate and pyridoxine. Cytokine excess can also result in anemia.

4.7 DEHYDRATION

The clinical diagnosis of dehydration is difficult in older persons because of the poor specificity of signs such as poor skin turgor (tenting), sunken eyes, dry mucous membranes and dry axilla (see also Chapter 8). Postural hypotension is a useful sign but only with relatively severe dehydration. In addition, there are numerous other causes of orthostasis in older persons. Delirium is a common presentation of dehydration.

A ratio of blood urea nitrogen to creatinine of greater than 20 to 1 is often used to detect dehydration. Unfortunately, bleeding and renal, heart and liver failure can all produce this, giving it little specificity. A serum sodium greater than 145 mmol/L or a serum osmolality greater than 300 mmol/kg is diagnostic of hypertonic intravascular volume depletion. Hypotonic dehydration occurs with excess diuretic use. Both the serum sodium (<135 mmol/L) and osmolality (<280 mmol/kg) are low in this condition.

4.8 IMMUNE ASSESSMENT AND NUTRITION

Delayed-type cutaneous hypersensitivity has been classically used to suggest nutritional deficiencies. It occurs in response to low protein, vitamin deficiencies and deficiency of essential omega-3 fatty acids. While older persons tend to have a decreased response to antigens introduced under the skin, anergy to common antigens in well-nourished older persons is extremely rare. In older persons with protein malnutrition or micronutrient deficiencies there is a decrease in T cell subsets (CD3+, CD8+ and CD4+). In severely malnourished older persons, CD4+ levels can fall to less than $400/mm^3$ (34). This is similar to levels seen in persons with acquired immune deficiency.

Evidence of cytokine excess can be useful in helping to determine whether or not the person has cachexia. C-reactive protein can be used as a nonspecific measure of cytokine excess. Measurement of tumor necrosis factor-alpha, interleukin-2 or interleukin-6 or their soluble receptors gives a more accurate determination of cytokine excess. However, these measurements are usually limited to research at present.

4.9 ASSESSING FRAILTY, STRENGTH AND MOBILITY

Frailty, a condition of pre-disability, is often associated with malnutrition and weight loss. It can be objectively measured using the Fried criteria (35) (Table 4.6). Frailty can also be identified if three of the following criteria are identified:

*F*atigue (self-report)
*R*esistance (inability to climb one flight of stairs)
*A*mbulation (inability to walk one block)
*I*llness (more than four diseases)
*L*oss of weight (greater than 5%)

Table 4.6
Objective definition of frailty* as proposed by Fried et al. *(35).*

Parameter		Frailty indicator
1.	Weight loss	>10 lbs. weight loss in 1 year
2.	Grip strength	Lowest 20% of population
3.	Exhaustion	Self-report
4.	Walking speed	Lowest 20% of population
5.	Low activity	Males <383 kcals/week
		Females <270 kcals/week

*Frailty is defined as having three or more of the above criteria.

Table 4.7
Objective measurements of physical function

Test	Normative values
Get-up-and-go	<20 sec—adequate mobility
	>30 sec—high risk for falls and functional dependence
6-m walk	<5.8 sec
Gait speed	>6.0 sec
6-minute walk	<300 m predicts mortality
	<400 m predicts functional impairment

Strength is best measured by using a hand-held dynamometer. It requires the average of best of three measurements. It cannot be done in persons with severe arthritis.

Mobility can be measured by using a 6-m or 6-minute walk. In addition, the "get-up-and-go" test is a reasonable test of lower limb strength and balance. Normal values for these tests are given in Table 4.7.

4.10 ASSESSING DISABILITY

Disability can be due to poor nutritional status or it can aggravate nutritional deficits. The Katz basic Activities of Daily Living are highly predictive of mortality and need for nursing home placement (Table 4.8). Instrumental Activities of Daily Living identify a slightly less-impaired individual (Table 4.9).

4.11 SCREENING FOR OSTEOPOROSIS

While loss of height can be due to changes in posture, significant height loss almost always suggests osteoporosis. All women by the age of 50 years and all men by 75 years should be screened for osteoporosis utilizing dual X-ray absorptiometry (DXA). Rate of bone loss can be determined by repeating the DXA 2 years later at the same season. In nursing homes heel ultrasound has proven to be a useful screening test for osteoporosis.

Table 4.8 Katz ADLs	Table 4.9 Lawton's IADLs
Basic ADLS	Instrument ADLs
Bathing	Using the telephone
Dressing	Shopping
Toileting	Food preparation
Transfers	Housekeeping
Continence	Laundry
Feeding	Transportation
ADL score: ____/16	Taking medicine
	Managing money
	IADL score _____/8

4.12 MEASUREMENTS OF VITAMINS AND TRACE ELEMENTS

While mild degrees of vitamin and trace element deficiency are common in older persons, these levels are generally not measured. In hospitalized patients, low levels of B vitamins are associated with delirium. In persons with dementia or macrocytic anemia both folate and vitamin B_{12} should be measured. If the levels of vitamin B_{12} are borderline, a serum methylmalonic acid should be obtained. Measurements of serum homocysteine are generally not useful. In persons where it is suspected that weight loss may be due to malabsorption, measurement of vitamin A and beta-carotene may be useful.

It is now recognized that 25(OH) vitamin D levels below 30 ng/mL are associated with hip fracture, sarcopenia and falls. Many older persons who are ingesting 800 IU of vitamin D have levels below this. For this reason, 25(OH) vitamin D levels should be measured in all older persons *(36)*.

The difficulty in obtaining interpretable levels of trace elements and the lack of controlled trials make it difficult to recommend their measurement. In areas where the soil is known to be poor in certain trace elements it may be useful to measure them. A reasonable argument could be made for measuring zinc (serum, leukocyte or hair levels) in persons with pressure ulcers. This, however, is not routinely done.

4.13 RECOMMENDATIONS

1. Nutritional assessment includes history taking, nutritional screening (with a focus on risk stratification) and a nutritional evaluation, including anthropometry and laboratory evaluations.
2. Body weight (mass) measurement is the single most practical and dependable assessment tool and should be carefully measured over time and compared to previous weights.
3. A well-designed nutritional surveillance program should recognize threats to nutritional status before full-blown malnutrition is detected using biochemical indices.
4. Quick screening for anorexia can be done with the SNAQ questionnaire.

5. Persons losing weight should be investigated for treatable causes of weight loss using the "MEALS-ON-WHEELS" mnemonic.

6. 25(OH) vitamin D levels should be measured in persons over 70 years of age.

REFERENCES

1. Thomas DR, Kamel H, Azharrudin M, Ali AS, Khan A, Javaid U, Morley JE. The relationship of functional status, nutritional assessment, and severity of illness to in-hospital mortality. J Nutr Health Aging 2005;9(3):169–75.

2. Van Lier AM, Roy MA, Payette H. Knee height to predict stature in North American Caucasian frail free-living elderly receiving community services. J Nutr Health Aging 2007;11(4):372–9.

3. Lear SA, Humphries KH, Kohli S, Birmingham CL. The use of BMI and waist circumference as surrogates of body fat differs by ethnicity. Obesity (Silver Spring) 2007;15(11):2817–24.

4. Chinuki D, Amano Y, Ishihara S, Moriyama N, Ishimura N, Kazumori H, Kadowaki Y, Takasawa S, Okamoto H, Kinoshita Y. REG Ialpha protein expression in Barrett's esophagus. J Gastroenterol Hepatol 2008;23:296–302.

5. Wilson MM, Thomas DR. Rubenstein LZ, Chibnall JT, Anderson S, Baxi A, Diebold MR, Morley JE. Appetite assessment: simple appetite questionnaire predicts weight loss in community-dwelling adults and nursing home residents. Am J Clin Nutr 2005;82(5):1074–81.

6. Keller HH. The SCREEN (Seniors in the Community: Risk Evaluation for Eating and Nutrition) index adequately represents nutritional risk. J Clin Epidemiol 2006;59(8):836–41.

7. Keller HH, Goy R, Kane SL. Validity and reliability of SCREEN II (Seniors in the community: risk evaluation for eating and nutrition, Version II). Eur J Clin Nutr 2005;59(10):1149–57.

8. Keller HH, McKenzie JD, Goy RE. Construct validation and test-retest reliability of the seniors in the community: risk evaluation for eating and nutrition questionnaire. J Gerontol A Biol Sci Med Sci 2001;56:M552–8.

9. Stratton RJ, Hackston A, Longmore D, Dixon R, Price S, Stroud M, King C, Elia M. Malnutrition in hospital outpatients and inpatients: prevalence, concurrent validity and ease of use of the 'malnutrition universal screening tool' ('MUST') for adults. Br J Nutr 2004;92(5):799–808.

10. Stratton RJ, King CL, Stroud MA, Jackson AA, Elia M. Malnutrition Universal Screening Tool' predicts mortality and length of hospital stay in acutely ill elderly. Br J Nutr 2006;95(2):325–30.

11. Mathey MF. Assessing appetite in Dutch elderly with the Appetite, Hunger and Sensory Perception (AHSP) questionnaire. J Nutr Health Aging 2001;5(1):22–8.

12. Savina C, Donini LM, Anzivino R, De Felice MR, De Bernardini L, Cannella C. Administering the "AHSP Questionnaire" (appetite, hunger, sensory perception) in a geriatric rehabilitation care. J Nutr Health Aging 2003;7(6):385–9.

13. Posner BM, Jette AM, Smith KW, Miller DR. Nutrition and health risks in the elderly: the nutrition screening initiative. Am J Publ Hlth 1993;83(7):972–8.

14. De Groot LC, Beck AM, Schroll M, van Staveren WA. Evaluating the DETERMINE Your Nutritional Health Checklist and the Mini Nutritional Assessment as tools to identify nutritional problems in elderly Europeans. Eur J Clin Nutr 1998;52:877–83.

15. Miller DK, Carter ME, Sigmund RH, Smith JW, Miller JP, Bentley JA, McDonald K, Coe RM, Morley JE. Nutritional risk in inner-city-dwelling older black Americans. J Am Geriatr Soc 1996;44(8):959–62.

16. Morley JE. Death by starvation: a modern American problem? J Am Geriatr Soc 1989;37(2):184–5.

17. McCann L. Using subjective global assessment to identify malnutrition in the ESRD patient. Nephrol News Issues 1999;13(2):18–9.

18. Sieber CC. Nutritional screening tools—How does the MNA compare? Proceedings of the session held in Chicago May 2–3, 2006 (15 Years of Mini Nutritional Assessment). J Nutr Health Aging 2006;10(6):488–92.

19. Langkamp-Henken B. Usefulness of the MNA in the long-term and acute-care settings within the United States. J Nutr Health Aging 2006;10(6):502–6.

20. Vellas B, Villars H, Abellan G, Soto ME, Rolland Y, Guigoz Y, Morley JE, Chumlea W, Salva A, Rubenstein LZ, Garry P. Overview of the MNA—its history and challenges. J Nutr Health Aging 2006;10(6):456–63.

21. Guigoz Y, Vellas B. The Mini Nutritional Assessment (MNA) for grading the nutritional state of elderly patients: presentation of the MNA, history and validation. Nestle Nutr Workshop Ser Clin Perform Programme 1999;1:3–11.

22. Vellas B, Guigoz Y, Garry PJ, Nourhashemi F, Bennahum D, Lauque S, Albarede JL. The MiniNutritional Assessment (MNA) and its use in grading the nutritional state of elderly patients. Nutrition 1999;15(2):159–61.

23. Rubenstein LZ, Harker JO, Salva A, Guigoz Y, Vellas B. Screening for undernutrition in geriatric practice: developing the short-form mini-nutritional assessment (MNA-SF). J Gerontol A Biol Sci Med Sci 2001;56(6):M366–72.

24. Jones JM. The methodology of nutritional screening and assessment tools. J Human Nutr Diet 2002;15:59–71.

25. Green SM, Watson R. Nutritional Screening and assessment tools for older adults: literature review. J Adv Nurs 2006;54(4):477–90.

26. Tooze JA, Vitolins MZ, Smith SL, Arcury TA, Davis CC, Bell RA, DeVellis RF, Quandt SA. High levels of low energy reporting on 24-hour recalls and three questionnaire in an elderly low-socioeconomic status population. J Nutr 2007;137(5):1286–93.

27. Quandt SA, Vitolins MZ, Smith SL, Tooze JA, Bell RA, Davis CC, DeVellis RF, Arcury TA. Comparative validation of standard, picture-sort and meal-based food-frequency questionnaires adapted for an elderly population of low socio-economic status. Public Health Nutr 2007;10(5):524–32.

28. The Merck Manual Online. In: Robert S Porter, ed.: Merck Research Laboratories; 2005. (Accessed at www.merck.com/mmpe/sec01/ch006/ch006a.html.)

29. Omran ML, Morley JE. Assessment of protein energy malnutrition in older persons, part I: History, examination, body composition, and screening tools. Nutrition 2000;16(1):50–63.

30. Morley JE, Silver AJ. Nutritional issues in nursing home care. Ann Intern Med 1995;123:850–9.

31. Silver AJ, Guillen CP, Kahl MJ, Morley JE. Effect of aging on body fat. J Am Geriatr Soc 1993;41:211–3.

32. Davidson J, Getz M. Nutritional risk and body composition in free-living elderly participating in congregate meal-site programs. J Nutr Elder 2004;24(1):53–68.

33. Omran ML, Morley JE. Assessment of protein energy malnutrition in older persons. Part II: Laboratory evaluation. Nutrition 2000;16:131–40.

34. Kaiser FE, Morley JE. Idiopathic CD4+ T lymphopenia in older persons. J Am Geriatr Soc 1994;42:1291–94.

35. Fried LP, Tangen CM, Walston J, Newman AB, Hirsch C, Gottdiener J, Seeman T, Tracy R, Kop WJ, Burke G, McBurnie MA; Cardiovascular Health Study Collaborative Research Group. Frailty in older adults: evidence for a phenotype. J Gerontol A Biol Med Sci 2001;56(3):M134–5.

36. Morley JE. Should all long-term care residents receive vitamin D? J Am med Dir Assoc 2007;8:69–70.

5 Sensory Impairment: Taste and Smell Impairments with Aging

Susan Schiffman

Key Points

- The sensory properties of foods influence food choices and provide cues about a food's nutritional value.
- Sensory signals from food elicit salivary, gastric acid, and pancreatic secretions associated with digestion and ultimately absorption of nutrients. Taste and smell are especially important in activating these digestive secretions because the taste and olfactory systems have closer anatomical connections to the neural pathways involved in digestion than the other senses.
- Deficits in taste and smell perception as well as the other senses occur during the course of normal aging and are exacerbated by medical conditions and treatments including medications.
- When sensory signals are compromised, food selection and intake, absorption of nutrients, and ultimately nutritional status are negatively impacted.

Key Words: Taste; sensory; smell; olfaction; vision; auditory function; somatosensory system

5.1 INTRODUCTION

This chapter provides an overview of sensory impairments of all five senses (taste, smell, vision, hearing, and touch) in older persons with special emphasis on the chemical senses of taste and smell. Sensory losses that interfere with the ability to procure and appreciate food can lead to inadequate intake of calories and nutrients, weight loss, and ultimately increased risk of morbidity and mortality *(1)*. Thus, awareness of sensory alterations in older individuals is crucial for making decisions regarding appropriate medical care and nutritional support. Impairments of the chemical senses of taste and smell are among the primary reasons for deficiencies, excesses, or imbalances in the dietary intake among older adults *(2–9)*. Losses and

From: *Nutrition and Health: Handbook of Clinical Nutrition and Aging, Second Edition*
Edited by: C. W. Bales and C. S. Ritchie, DOI 10.1007/978-1-60327-385-5_5,
© Humana Press, a part of Springer Science+Business Media, LLC 2009

distortions of the senses of taste and smell can reduce the motivation to eat as well as interfere with the ability to select appropriate foods and to modulate intake as nutritional requirements vary over time. Furthermore, without the simple pleasures of taste and smell sensations, overall quality of life is greatly reduced, especially for those in later life whose other senses (vision, hearing, and touch) have also declined during the aging process. The prevalence of chemosensory losses and distortions begins to increase around 60 years of age and becomes more pronounced in subsequent decades of life. Currently, there are no known medical treatments for taste or smell losses.

The purpose of this chapter is to review the literature on the chemical senses of taste and smell changes with age. Age-related impairments of vision, hearing, and touch that impact nutritional status of older persons are also delineated. The magnitude of sensory losses tends to be greatest for those in later life who have a history of critical medical conditions (e.g., coronary artery bypass surgery). Experimental evidence is also presented that indicates multiple testing rather than a single test is required to accurately assess sensory perception in the older persons due to large intra-individual fluctuations in performance in this age cohort with repeated testing over a short time period.

5.2 TASTE

The sense of taste is an oropharyngeal chemical sense that plays a critical role in food selection and food safety. Taste signals, along with odor signals, trigger cephalic phase secretions (e.g., salivary, gastric, pancreatic) that prepare for the digestion of food before it reaches the stomach. In addition, learned association of taste sensations with the metabolic consequences of food enables meal size and food choices to be modulated in anticipation of nutritional needs. Taste sensations are initiated when chemical stimuli interact with receptors and ion channels located on taste cells that are clustered into buds on the tongue surface and other discrete areas of the oral cavity (2,5). Taste signals from taste buds are carried by the seventh, ninth, and tenth cranial nerves to the nucleus of the solitary tract (NST) in the medulla of the brainstem that projects to the ventroposteromedial nucleus of the thalamus and finally the insular-opercular cortex. The NST receives information not only from the taste system but also from visceral sensory fibers that originate in the esophagus, stomach, intestines, and liver. Information from the olfactory nerve (cranial nerve I) that transmits information about smell also converges in the NST. This convergence of neural input in the NST enables taste and odor signals to impact ingestive and digestive activity by producing gastric and pancreatic secretions. The qualitative range of taste includes sweet, sour, salty, and bitter as well as other less familiar sensations that are also carried by taste nerves including umami (the taste of glutamate salts, brothy, savory, or meat-like), fatty, metallic, starchy/polysaccharide, chalky, and astringent (9).

The sense of taste gradually declines with aging, with differential losses depending on the chemical structure of individual taste stimuli. The decrements in taste sensitivity that occur during normal aging are exacerbated by certain disease states, pharmacologic and surgical interventions, radiation, and environmental exposure

(2,5,7). The cause of taste losses in normal aging independent of disease or medications is not known; some researchers have found losses in the number of taste buds in older individuals while others have not. Whatever the cause, taste losses reduce the motivation to eat, interfere with the ability to modulate appetite and food choices, impair quality of life, and can lead to inadequate nutritional status especially in the sick or malnourished older persons. When taste and smell losses no longer play a major role in initiating, sustaining, and terminating ingestion, the quantity of food that is eaten and the size of meals can be affected. Cephalic phase responses including salivary, gastric, pancreatic, and intestinal secretions can be blunted which can affect digestion of food and absorption of nutrients.

5.2.1 Taste Losses at Threshold Levels

Older adults have losses in the ability to detect and recognize all taste qualities as well as other oral stimuli *(4,8,9)*. The detection thresholds (DT) for tastes are elevated in older persons, and hence they require the presence of more molecules (or ions) for a sensation to be perceived compared to a younger cohort. The recognition thresholds (RT) are also elevated so a greater concentration of a tastant is required to correctly recognize its quality. Table 5.1 compares mean DTs and RTs for older persons with those for the young persons for a broad range of compounds including sodium salts with different anions, bitter compounds, sweeteners, acids, astringent compounds, amino acids including glutamate salts, metallic compounds, fats, gums, and astringent compounds *(4,9)*. The older adult subjects in these studies took an average of 3.4 medications but otherwise led active, normal lives. For detection thresholds (DTs), the ratio of DT (older)/DT (young) revealed that DTs in older persons were higher by the following amounts: 11.6 times higher for sodium salts; 7.0 times higher for bitter compounds; 2.7 times higher for sweeteners; 4.3 times higher for acids; 2.8 for astringent compounds; 2.5 times higher for amino acids; 5.0 times higher for glutamate salts; 3.1 times higher for fats/oils; 3.7 times higher for polysaccharides/gums; and 2.2 times higher for metallic compounds. For recognition thresholds (RTs), the ratio of RT (older)/RT (young) revealed that RTs in older persons were higher by the following amounts: 5.8 times higher for sodium salts; 7.5 times higher for bitter compounds; 2.1 times higher for sweeteners; 6.8 times higher for acids with sour tastes; 3.0 for astringent compounds; 3.0 times higher for polysaccharides/gums; and 2.0 times higher for metallic compounds.

Examination of Table 5.1 reveals that age-related decrements as determined by the ratio DT (older)/DT (young) varied widely over the different compounds tested. For sodium salts, the losses at the threshold level (i.e., elevated thresholds) were greatest for anions with the largest molar conductivity (Na sulfate, Na tartrate, Na citrate, and Na succinate). Molar conductivity (λ) is a measure of the electrical charge carried by the anion per unit time. That is, age-related losses in sensitivity to sodium salts were greatest for anions with the highest charge mobility. For bitter compounds, the greatest losses for older adults at the threshold level were for the least lipophilic compounds, i.e., $MgNO_3$, $MgSO_4$, and KNO_3. For sour acids, the greatest loss in sensitivity in older adults was for HCl, the acid with the lowest

Table 5.1
Comparison of taste detection and recognition thresholds for older and young subjects for a broad range of stimuli

Sodium salts	Detection thresholds			Recognition thresholds for saltiness		
	Older (O)	Young (Y)	O/Y	Older (O)	Young (Y)	O/Y
MSG (monosodium glutamate)	0.00638 M	0.00126 M	5.06	0.0091 M	0.00207 M	4.41
Na acetate	0.0190 M	0.00242 M	7.84	0.0229 M	0.00952 M	2.41
Na ascorbate	0.0250 M	0.00404 M	6.19	0.0265 M	0.00809 M	3.28
Na carbonate	0.00829 M	0.00218 M	3.79	0.0234 M	0.00425 M	5.51
Na chloride	0.01850 M	0.00238 M	7.76	0.0227 M	0.00815 M	2.79
Na citrate	0.0130 M	0.000531 M	24.5	0.0187 M	0.00190 M	9.84
Na phosphate monobasic	0.0160 M	0.00307 M	5.21	0.0253 M	0.01140 M	2.22
Na succinate	0.0138 M	0.000854 M	16.2	0.0167 M	0.00217 M	7.71
Na sulfate	0.0283 M	0.000981 M	28.8	0.0349 M	0.00322 M	10.86
Na tartrate	0.0159 M	0.00151 M	10.5	0.0277 M	0.00295 M	9.39

Bitter compounds	Detection thresholds			Recognition thresholds for bitterness		
	Older (O)	Young (Y)	O/Y	Older (O)	Young (Y)	O/Y
Caffeine	1.99 mM	1.30 mM	1.53	6.74 mM	1.87 mM	3.60
Denatonium benzoate	0.0323 μM	0.0115 μM	2.81	0.0387 μM	0.0123 μM	3.14
KNO_3	32.7 mM	1.91 mM	17.1	271 mM	5.97 mM	45.4
$MgCl_2$	5.20 mM	1.02 mM	5.10	21.8 mM	20.3 mM	1.07
$MgNO_3$	33.3 mM	1.40 mM	23.8	191 mM	14.8 mM	12.9
$MgSO_4$	6.08 mM	0.323 mM	18.8	14.8 mM	2.59 mM	5.71
Naringin	0.138 mM	0.0427 mM	3.23	0.195 mM	0.0561 mM	3.48
Phenylthiocarbamide	1.26 mM	0.591 mM	2.13	1.74 mM	1.21 mM	1.44
Quinine HCl	8.07 μM	3.99 μM	2.02	12.3 μM	4.75 μM	2.59

(continued)

Table 5.1
(continued)

	Detection thresholds			Recognition thresholds		
	Older (O)	Young (Y)	O/Y	Older (O)	Young (Y)	O/Y
Quinine sulfate	8.75 µM	2.04 µM	4.29	12.3 µM	2.53 µM	4.86
Sucrose octaacetate	5.32 µM	3.89 µM	1.37	22.8 µM	5.30 µM	4.30
Urea	0.116 M	0.103 M	1.12	0.245 M	0.134 M	1.83

Sweeteners

	Detection thresholds			Recognition thresholds for sweetness		
	Older (O)	Young (Y)	O/Y	Older (O)	Young (Y)	O/Y
Acesulfame-K	74.7 µM	44.4 µM	1.68	239 µM	161 µM	1.48
Aspartame	91.3 µM	22.4 µM	4.07	124 µM	44.9 µM	2.76
Calcium cyclamate	0.412 mM	0.266 mM	1.55	1.69 mM	1.33 mM	1.27
Fructose	10.1 mM	4.39 mM	2.30	26.4 mM	16.6 mM	1.59
Monellin	0.0913 µM	0.0195 µM	4.67		0.0676 µM	
Neohesperidin dihydrochalcone	4.60 µM	2.20 µM	2.09		5.28 µM	
Rebaudioside	13.0 µM	4.61 µM	2.82		13.6 µM	
Sodium saccharin	42.4 µM	14.7 µM	2.88	137 µM	49.7 µM	2.76
Stevioside	16.0 µM	5.31 µM	3.02		23.7 µM	
Thaumatin	0.133 µM	0.0716 µM	1.86		0.201 µM	
D-tryptophan	0.322 mM	0.109 mM	2.95	1.45 mM	0.546 mM	2.66

Sour compounds

	Detection thresholds			Recognition thresholds for sourness		
	Older (O)	Young (Y)	O/Y	Older (O)	Young (Y)	O/Y
Acetic acid	0.273 mM	0.106 mM	2.58	0.819 mM	0.294 mM	2.79
Ascorbic acid	0.725 mM	0.281 mM	2.58	2.190 mM	0.396 mM	5.53
Citric acid	0.375 mM	0.0498 mM	7.53	0.816 mM	0.131 mM	6.23
Glutamic acid	0.463 mM	0.0920 mM	5.03	1.500 mM	0.309 mM	4.85
Hydrochloric acid	0.200 mM	0.0179 mM	11.17	0.477 mM	0.0226 mM	21.11

(continued)

Table 5.1
(continued)

	Detection thresholds			Recognition thresholds for astringency		
	Older (O)	Young (Y)	O/Y	Older (O)	Young (Y)	O/Y
Succinic acid	0.188 mM	0.132 mM	1.42	1.330 mM	0.174 mM	7.64
Sulfuric acid	0.100 mM	0.0468 mM	2.14	0.170 mM	0.0468 mM	3.63
Tartaric acid	0.163 mM	0.0864 mM	1.89	0.297 mM	0.131 mM	2.27
Astringent compounds						
Gallic acid	0.780 mM	0.250 mM	3.12	2.07 mM	1.10 mM	1.88
Tartaric acid	0.220 mM	0.0549 mM	4.01	0.324 mM	0.0689 mM	4.70
Tannic acid	0.072 mM	0.0271 mM	2.66	0.295 mM	0.0528 mM	5.59
Catechin	1.48 mM	1.18 mM	1.25	2.500 mM	1.56 mM	1.60
Ammonium alum	0.172 mM	0.0780 mM	2.21	0.487 mM	0.244 mM	2.00
Potassium alum	0.454 mM	0.120 mM	3.78	1.380 mM	0.723 mM	1.91

Amino acids

Detection thresholds			
	Older (O)	Young (Y)	O/Y
L-alanine	19.5 mM	16.2 mM	1.20
L-arginine	1.12 mM	1.20 mM	0.93
L-arginine HCl	2.39 mM	1.23 mM	1.94
L-asparagine	9.33 mM	1.62 mM	5.75
L-aspartic acid	0.501 mM	0.182 mM	2.75
L-cysteine	0.390 mM	0.0630 mM	6.19
L-cysteine HCl	20.0 μM	16.0 μM	1.25
L-glutamic acid	0.100 mM	0.0630 mM	1.59
L-glutamine	26.9 mM	9.77 mM	2.75
L-glycine	0.0617 M	0.0309 M	2.00
L-histidine	6.45 mM	1.23 mM	5.24
L-histidine HCl	0.389 mM	0.0794 mM	4.90

(continued)

Table 5.1
(continued)

L-isoleucine	12.0 mM	7.41 mM	1.62
L-leucine	12.9 mM	6.45 mM	2.00
L-lysine	2.24 mM	0.708 mM	3.16
L-lysine HCl	2.09 mM	0.447 mM	4.68
L-methionine	2.63 mM	3.72 mM	0.71
L-phenylalanine	19.1 mM	6.61 mM	2.89
L-proline	0.0372 M	0.0151 M	2.46
L-serine	0.0263 M	0.0209 M	1.26
L-threonine	0.020 M	0.0257 M	0.78
L-tryptophan	2.88 mM	2.29 mM	1.26
L-valine	0.0115 M	0.00416 M	2.76

Glutamate salts (with and without the enhancer IMP[1])

	Detection thresholds			Recognition thresholds for umami		
	Older (O)	Young (Y)	O/Y	Older (O)	Young (Y)	O/Y
Sodium glutamate	2.83 mM	0.902 mM	3.14	5.24 mM	2.55 mM	2.05
Sodium glutamate with 0.1 mM IMP[1]	0.888 mM	0.113 mM	7.86	1.82 mM	0.183 mM	9.94
Sodium glutamate with 1 mM IMP[1]	0.145 mM	0.0480 mM	3.02	0.328 mM	0.0964 mM	3.40
Potassium glutamate	7.69 mM	0.902 mM	8.53	10.1 mM	5.13 mM	1.97
Potassium glutamate with 0.1 mM IMP	0.549 mM	0.106 mM	5.18	0.205 mM	0.189 mM	1.08
Potassium glutamate with 1 mM IMP	0.0928 mM	0.0108 mM	8.59	0.231 mM	0.0378 mM	6.11
Ammonium glutamate	4.26 mM	1.08 mM	3.94	8.70 mM	2.75 mM	3.16
Ammonium glutamate with 0.1 mM IMP	0.458 mM	0.139 mM	3.29	0.581 mM	0.252 mM	2.30

(continued)

Table 5.1 (continued)

	Older (O)	Young (Y)	O/Y			
Ammonium glutamate with 1 mM IMP	0.129 mM	0.0343 mM	3.76	0.274 mM	0.065 mM	4.22
Calcium diglutamate	1.09 mM	0.292 mM	3.73	1.06 mM	1.29 mM	0.82
Calcium diglutamate with 0.1 mM IMP	0.327 mM	0.0606 mM	5.40	0.409 mM	0.0848 mM	4.82
Calcium diglutamate with 1 mM IMP	0.0692 mM	0.0190 mM	3.64	0.0692 mM	0.033 mM	2.09
Magnesium diglutamate	1.86 mM	0.253 mM	7.35	3.15 mM	0.854 mM	3.69
Magnesium diglutamate with 0.1 mM IMP	0.289 mM	0.0421 mM	6.86	0.795 mM	0.0674 mM	11.8
Magnesium diglutamate with 1 mM IMP	0.0452 mM	0.0257 mM	1.76	0.109 mM	0.0524 mM	2.08
IMP (inosine 5'-monophosphate)	1.99 mM	0.430 mM	4.63	2.12 mM	1.07 mM	1.98

Oils in four different emulsifiers[2]

	Detection thresholds		
	Older (O)	Young (Y)	O/Y
MCT[2] (in acacia)	10.1%	2.85%	3.54
Soybean (in acacia)	12.9%	4.02%	3.20
Mineral (in acacia)	9.77%	4.43%	2.20
MCT (in Emplex)	25.0%	3.93%	6.37
Soybean (in Emplex)	14.9%	6.52%	2.28
Mineral (in Emplex)	20.0%	8.85%	2.26
MCT (in Tween-80)	19.3%	5.35%	3.60
Soybean (in Tween-80)	17.7%	5.85%	3.02
Mineral (in Tween-80)	19.9%	5.77%	3.44

(continued)

Table 5.1
(continued)

	Older (O)	Young (Y)	O/Y
MCT (in Na caseinate)	13.6%	6.18%	2.20
Soybean (in Na caseinate)	13.0%	5.35%	2.43
Mineral (in Na caseinate)	13.4%	4.27%	3.13

Polysaccharides/gums

	Detection thresholds			Recognition thresholds for thickness		
	Older (O)	Young (Y)	O/Y	Older (O)	Young (Y)	O/Y
Acacia gum	1.02%	0.644%	1.58	3.12%	1.44%	2.17
Guar gum	0.116%	0.057%	2.04	0.42%	0.22%	1.91
Locust bean gum	0.43%	0.061%	7.05	0.74%	0.22%	3.36
Xanthan gum	0.238%	0.0396%	6.01	0.41%	0.071%	5.77
Algin	0.115%	0.0605%	1.90	0.26%	0.15%	1.73

Metallic compounds

	Detection thresholds			Recognition thresholds for metallic taste		
	Older (O)	Young (Y)	O/Y	Older (O)	Young (Y)	O/Y
$FeSO_4$	0.343 mM	0.143 mM	2.4	2.14 mM	1.07 mM	2.0
$FeCl_2$	1.663 mM	0.875 mM	1.9	1.76 mM	0.924 mM	1.9
$ZnSO_4$	1.050 mM	0.456 mM	2.3	1.46 mM	0.695 mM	2.1
$ZnCl_2$	0.774 mM	0.387 mM	2.0	0.890 mM	0.445 mM	2.0

[1]Thresholds were obtained for five glutamate salts (sodium glutamate, potassium glutamate, ammonium glutamate, calcium diglutamate, and magnesium diglutamate). The effect of inosine 5'-monophosphate (IMP), a taste enhancer, at two concentrations (0.1 mM IMP and 1.0 mM IMP) on taste thresholds of these glutamate compounds was also investigated. While 0.1 mM IMP lowered thresholds for all five salts for young but not older subjects, 1 mM IMP lowered thresholds in both young and older groups.

[2]The oils were refined, bleached, deodorized soybean oil (LCT; long-chain triglyceride) oil; MCT (medium chain triglyceride) oil; and light mineral oil. These fats are long-chain triglycerides, medium-chain triglycerides, and a mixture of liquid hydrocarbons from petroleum, respectively. Oil-in-water emulsions were made with each oil using one of four different emulsifiers: Polysorbate-80 (Tween-80), sodium stearoyl lactylate (Emplex), sodium caseinate, and acacia gum.

molecular weight. For amino acids, age-related losses tended to be higher for two amino acids with side chains containing basic groups (L-histidine and L-lysine) and their monohydrochloride derivations.

Table 5.1 reveals two very important points about taste perception. First, the relative differences in loss for individual compounds with age contribute to the distortions of taste (called dysgeusia) experienced by many persons in later life. Foods are a mixture of many different compounds, and that mixture will taste different to an older person than a younger one because the relative sensory salience of the individual compounds will differ based on age. Second, compounds with high caloric or nutritional value such as sugars, fats, and amino acids tend to have higher detection thresholds than certain noxious bitter compounds that can be detected in minute amounts. A possible explanation for higher concentrations required to detect sugars, amino acids, and fats is that too much taste at low concentrations could inhibit intake of adequate calories.

5.2.2 Suprathreshold Taste Perception

Suprathreshold taste studies that relate perceived intensity to concentration indicate that tastes are less intense for older persons compared with the young. Like threshold measurements, the degree of loss is not uniform across compounds but rather depends on the chemical structure of the tastant (2–9); the lack of uniform loss plays a role in taste distortions (dysgeusia) experienced by older persons for foods comprised of mixtures of numerous compounds. For 23 amino acids, there was loss in perceived intensity with age but the losses for L-aspartic acid and L-glutamic acid were far greater than for L-lysine and L-proline. For sweeteners, the suprathreshold losses were greater for large sweetener molecules such as thaumatin, rebaudioside, and neohesperidin dihydrochalcone than for sweeteners with lower molecular weights. Furthermore, the ability to discriminate between different suprathreshold intensities of the same stimulus is also impaired with age. For example, while young subjects required a 34% difference in concentration to perceive a perceptible difference in the bitterness of caffeine, older subjects required an increment of 74%. Decrements in the ability to perceive suprathreshold concentrations of NaCl (salty) or sucrose (sweet) can tempt older individuals with hypertension (who must comply with sodium-restricted diets) or those with diabetes (who must monitor their carbohydrate intake) to use too much salt or sugar to improve the taste of their food. Reduced ability to perceive the oral component of fats (taste and mouth-feel) makes it difficult for older adults to comply with a low-fat diet and thus can potentially increase the risk from medical conditions such as cardiovascular disease, diabetes, and hypertension in which high-fat intake is contraindicated. Many older individuals unknowingly consume large amounts of fat without being able to really perceive it.

5.2.3 Medications and Medical Conditions Associated with Taste Alterations

While the actual incidence and prevalence of drug-induced taste disorders are not well documented, several community and longitudinal studies of older persons suggest that between 11 and 33% experience medication-related alterations in taste *(10)*. Medicated older individuals are more likely to complain of "loss of taste", "altered taste", and "metallic taste" than either age-matched non-medicated controls or younger medicated and non-medicated controls *(5–9)*. Hundreds of medications, including most major drug classes, have been associated clinically with taste complaints *(11)*. However, there is individual variability in predisposition to adverse taste side effects experienced from use of medications. Genetic testing for variability in genes such as the cytochrome P450 2D6 (CYP2D6) that determine the rate of drug metabolism may ultimately be helpful in predicting whether an individual is vulnerable to taste disorders from drugs. Genetic tests for variations in genes (genotypes) associated with drug metabolism are available from medical providers as well as directly to consumers *(12)*.

Neither the sites of action nor cascade of cellular events by which medications induce taste complaints is well understood. Medications can potentially alter taste perception by affecting the peripheral receptors, neural pathways, and/or the brainstem and brain. At the periphery, drugs in the saliva can generate a taste of their own or modify transduction mechanisms in taste receptor cells. For some drugs, the plasma concentrations are high enough to stimulate taste receptors on the basolateral side of taste cells (called intravascular taste). Even when the salivary or plasma concentrations of medications are lower than the taste threshold values, drugs or their metabolites can accumulate in taste buds over time to reach (especially lipophilic bitter-tasting drugs) concentrations that are greater than taste detection thresholds. Drugs can also alter neurotransmitter levels along neural pathways or interfere with taste signals in the brainstem and brain. Drugs that cause a substantial percentage of taste disorders (such as the antifungal agent terbinafine) tend to be highly lipophilic and are thus readily distributed into the brain and brainstem.

The metabolism and absorption of drugs can be modified by dietary constituents, and thus food choices can play a role in potential taste disorders. When lipophilic medications are ingested with a fatty meal, absorption of these drugs can increase. Ingestion of dietary protein with methyl-dopa and L-dopa (used to treat Parkinson's disease) can affect the metabolism and absorption of these medications because they are amino acid derivatives. Intestinal transit time, which affects the absorption of drugs, can be modified by ingestion of spicy foods, dietary fiber, and other food components *(13)*.

A vast range of medical conditions have been reported to alter the sense of taste including infectious and parasitic diseases; cancer; endocrine, nutritional, and metabolic diseases; as well as diseases of the nervous, circulatory, digestive, respiratory, and musculoskeletal systems. Cancer is an example of a medical condition in which patients are especially vulnerable to taste disorders. Taste alterations occur in both untreated cancer patients and those receiving radiation therapy or chemotherapy. Some complaints, such as taste aversions in cancer patients, are not due to altered sensory physiology per se but to learned aversions in which the taste of foods is associated with the noxious effects of treatment. Clinical observations indicate that inflammatory conditions and wasting also predispose sick older individuals to taste disorders.

5.3 SMELL

The sense of smell is a nasal chemical sense that plays a critical role in the motivation to eat, selection of edible foods, avoidance of spoiled food, and ultimately nutritional status (8,9). Odor sensations, like taste sensations, trigger cephalic phase secretions that prepare the body for the digestion of food. Odor sensations are generated by volatile compounds (called odorants) that vary widely in chemical structure and include many molecular classes including organic acids, alcohols, aldehydes, amides, amines, aromatics, esters, ethers, fixed gases, halogenated hydrocarbons, hydrocarbons, ketones, nitriles, phenols, nitrogen-containing compounds, and sulfur-containing compounds (such as mercaptans). The sense of smell is exquisitely sensitive to some of these chemical classes, and it is estimated that as few as 40 molecules of some chemical types such as mercaptans are sufficient to perceive an odor (14). Odor sensations from foods and beverages are produced by mixtures of hundreds of odorant types rather than by a single compound. Odorant mixtures tend to harmonize or blend together leading to perceptual fusion that is labeled "banana" or "orange". The range of distinctive odor sensations is enormous, and a skilled odor expert, such as perfume chemist, can recognize and distinguish 8,000–10,000 different substances on the basis of their odor quality (15).

Odor sensations are initiated when hydrophobic volatile compounds (i.e., the odorants) activate olfactory receptors (ORs) located on the cilia of bipolar olfactory receptor neurons (ORNs) situated high in the nasal vault in the olfactory epithelium (8,9). Humans have several hundred distinct genes that encode for a broad range of ORs. This extensive range of receptor types permits the detection of odor sources comprised of unpredictable mixtures of molecular species and even allows for detection of newly synthesized compounds with no known function. Odorants can reach the ORs by two different routes-via the nostrils during nasal inhalation while sniffing food or from the nasal pharynx (up the back of the throat) which occurs during mastication of food. The latter route is called retronasal olfaction. Much of what is often termed "taste" is actually smell because odorous molecules placed in the mouth reach the olfactory receptors by the retronasal route.

The signals induced by activation of ORs are transmitted along the axons of the ORNs that coarse through small holes in a bone called the cribriform plate to synapse in neural masses called glomeruli in the olfactory bulb. Olfactory signals from the olfactory bulb are then transmitted via the olfactory tract to brain structures including the anterior olfactory nucleus, the olfactory tubercle, the prepyriform cortex, the amygdala, and ultimately to higher brain centers that process the olfactory signals. The prepyriform cortex and the amygdala are part of the limbic system that processes emotions and memories in addition to olfactory signals. Olfactory signals are ultimately transmitted to the hypothalamus (which mediates food intake) and to the neocortex.

At elevated concentrations, odorants can also stimulate free nerve endings of the trigeminal nerve in the nasal cavity. Trigeminal activation by odorous chemicals induces sensations such as irritation, tickling, burning, stinging, scratching, prickling, and itching. Sensory signals transmitted by the trigeminal nerve are not considered an "odor" but rather involve a different sense called chemesthesis which is related to nociception (e.g., pain).

5.3.1 Perceptual Olfactory Losses in Older Persons

Perceptual olfactory losses in older adults occur in five domains: threshold detection, recognition, discrimination, identification, and olfactory memory *(5,8,9)*. These losses occur during normal aging and are exacerbated by certain disease states, pharmacologic and surgical interventions, radiation, and environmental exposure. The magnitude of impairment of odor perception in the older adults is generally greater than for taste. Losses and distortions of the sense of smell can reduce the motivation to initiate and sustain ingestion as well as interfere with the ability to select appropriate foods and to modulate intake as nutritional requirements vary over time. Furthermore, decrements in odor perception put older adults at risk for food-borne illness, fires, and blunting of digestive secretions (from reduced cephalic phase responses).

An overview of odor studies indicates that losses in sensitivity for a broad range of individual odorants and odor mixtures tend to begin around 60 years of age with progressively greater losses with each decade of life. Older individuals also lose the ability to make discriminations among different odors. The number of odor sensations that can be differentiated is greatly reduced in the older persons who first lose the ability to make fine discriminations between odors with similar qualities (e.g., different types of nuts), and ultimately, in more extensive loss, between odors with different qualities (e.g., orange versus lamb). The prevalence of impaired ability to identify suprathreshold concentrations of odors increases from 17.3% for persons 60–69 years of age to 29.2% for persons aged 70–79 years and to 62.5% for the 80–97-year-olds according to a population-based, cross-sectional study in the United States (USA) *(16)*. Given these US prevalence figures along with United Nations projections of global population growth, the number of individuals projected to have impaired smell function worldwide by 2050 is at least 230 million.

5.3.2 Causes of Olfactory Losses in Older Persons

Many alterations in the anatomy and physiology of the olfactory system occur with age. The olfactory epithelium undergoes numerous changes including increased ORN apoptotsis (programmed cell death), decreased basal cell proliferation, decreased thickness of the olfactory epithelium, decreased number of cilia and supporting microvilli, and increased accumulation of electron-dense granules in supporting cells *(2–9)*. Neurons in the olfactory bulb begin to degenerate, and the bulb takes on a moth-eaten appearance as glomeruli atrophy and fibers disappear. Central olfactory projection areas are especially impacted by the aging process, with changes occurring in the hippocampus, amygdaloid complex, and hypothalamus including reductions in cell number, damage to cells, and diminished levels of neurotransmitters.

The anatomical and physiological losses that occur during normal aging are compounded by a wide range of medical conditions (e.g., endocrine, neurological, nutritional, psychiatric) as well as environmental exposures. Olfactory losses are especially profound in neurodegenerative disorders such as Alzheimer's disease and Parkinson's disease that are prevalent in an older population *(7,17)*. Loss of smell also occurs during the course of cancer and its treatment *(6)*. Many medications such as antianginal drugs (diltiazem, nifedipine), antimicrobial agents (allicin, streptomycin), antithyroid agents (carbimazole, methimazole, methylthiouracil, propylthiouracil), as well as radiation therapy and chemotherapy have been reported clinically to impair the sense of smell *(2–9,11)*.

5.4 OTHER AGE-RELATED SENSORY LOSSES

Sensory losses of visual, auditory, and somatosensory systems with age further exacerbate the impact of taste and smell losses on nutritional status in older adults.

5.4.1 Vision

Many adverse physiological changes in the structure and function of the eye occur during the aging process that lead to sensory losses *(18–20)*. By 45 years of age, there is a loss in the ability to focus on near objects such as a book due to reduced elasticity of the lens (called presbyopia). By 65 years of age, many older individuals need brighter illumination due to decreases in pupil size (called "senile miosis") and reduced transparency of the lens. Problems with glare in bright light become more frequent due to opaque particles in the lens. With increased age, more time is required to adapt from light to darkness; impairments in the rate and maximum level of dark adaptation can have profound effects on activities such as driving at night or reading a menu in a restaurant. Far visual acuity (resolution of spatial detail) and contrast sensitivity (ability to recognize subtle differences in shading and color between an object and its background) also begin to decline around 65 years of age. Losses in peripheral vision occur as well so that adults in later life do not recognize movement or objects at the periphery of the field. Reduced tear production (e.g., "dry eye") in older individuals can impair the clarity of an image.

Progression of degenerative changes in the eye beyond those of normal aging leads to a variety of age-related ocular diseases including macular degeneration, glaucoma, cataracts, and diabetic retinopathy. Age-related macular degeneration (AMD) is a condition that affects the macula (i.e., that part of the retina analogous to the bull's-eye of a target) that is responsible for sharp central vision. Persons with AMD are unable to recognize faces, read, or drive as the condition advances. When the macula deteriorates due to growth as well as leakage of tiny new blood vessels (neovascularization) under the retina, straight lines begin to appear wavy and ultimately there can be complete loss of central vision. Glaucoma is also a condition that afflicts many older persons. It results from a rise in intraocular pressure that damages the optic disk and optic nerve in the eye and results in gradual loss of outer (peripheral) vision which is opposite of that found in AMD. The peripheral visual field loss in end-stage glaucoma can be compared to looking down the barrel of a gun. Genetic factors play an integral role in predisposition to eye diseases. Genetic testing is now available in medical clinics and directly to consumers *(12)* to screen for genes associated with eye diseases. Knowledge of the genetic factors for eye disease in an older individual can make it easier to identify non-genetic factors.

With the exception of cataracts, none of these eye diseases is curable, although some treatments are available. Visual loss from cataracts can be effectively cured by removal of the clouded natural lens by a process called phacoemulsification and replacement with an intraocular lens (e.g., silicone or acrylic) that can be mono- or multifocal. For AMD caused by growth of blood vessels, laser therapies are used to destroy leaking blood vessels and thus reduce the risk of advancing vision loss. Surgical macular rotation can also be performed so that the macula is positioned over an area free of leaking blood vessels. For glaucoma, methods for lowering intraocular pressure (IOP) including medications and surgery (e.g., trabeculectomy or glaucoma drainage devices) are used. Low-vision devices such as magnification systems (ranging from simple lenses to complex electronic visual enhancement systems) can also be employed to compensate for visual loss and thus improve quality of life.

Many medications *(21)* and medical conditions (other than eye diseases) also affect the visual system. Victims of stroke, for example, with damage to the striate cortex experience blindness in the visual fields that correspond to the cortical area that is affected. However, stroke victims may have some residual ability to respond to a visual target placed within the blind visual field without acknowledged awareness of the presence of the stimulus (termed "blindsight") *(22)*. An example of blindsight is pointing in the direction of a visual target, such as a plate of food or a glass of water, without conscious awareness of its presence. Blindsight appears to be mediated by subcortical neural structures such as the superior colliculus that were not affected by the stroke *(23)*. An important new finding regarding "blindsight" is that daily detection training in cortically blind patients can improve visual sensitivities in the very depths of the field defect *(22)*.

Impaired vision can profoundly affect nutritional status because it interferes with mobility (including driving), activities of daily living, food preparation, and use of utensils. Ocular diseases such as AMD make it difficult to monitor food quality and safety by appearance and to identify food and eating utensils by visual cues. Loss in the ability to discriminate the color and arrangement of foods at the table greatly reduces the enjoyment of the eating experience and motivation to eat.

5.4.2 Auditory System

The function of the auditory system is to transmit information about sound waves (pressure variations) in the air. The auditory system consists of (1) the external ear or pinna which captures sound waves, (2) the middle ear which transmits sound vibrations from the ear drum to the inner ear via three small bones called the malleus, incus, and stapes, (3) the inner ear (the cochlea along with hair cells), and (4) the auditory nerve and pathways to various central neural structures. Approximately 80% of all hearing loss is sensorineural; i.e., the hair cells in the inner ear or the auditory nerve are damaged due to normal aging, illness, ototoxic drugs (e.g., streptomycin), injury (e.g., noise), or a hereditary condition. Aging-related hearing loss (presbycusis), which initially tends to impair perception of high frequencies, afflicts between 40 and 50% of people aged 75 years and older *(24)* and up to 90% of older persons in nursing homes *(25)*. A small percentage of hearing loss is due to conductive problems in which sound waves are not passed to the inner ear due to the presence of excessive earwax, infection, fluid in the middle ear from ear infection, or a punctured eardrum. While conductive hearing loss can sometimes be medically or surgically corrected, sensorineural hearing loss is currently not reversible nor is it amenable to medical or surgical treatment. A variety of hearing devices are currently available to compensate for hearing loss including external hearing aids (to amplify sound waves), bone-anchored hearing aids (in which sound is conducted through the bone rather than via the middle ear), cochlear implants (to compensate for damaged parts of the inner ear with an electronic device), and auditory brainstem implants (to compensate for dysfunction of the auditory nerve).

Hearing loss can adversely affect nutritional status as well as lead to social isolation and depression. Furthermore, impairment of hearing results in reduced sensory input from textural cues from food such as crispiness (high-frequency sounds), and crunchiness (low-frequency sounds).

5.4.3 *Somatosensory System with Age*

The somatosensory system is comprised of a variety of receptors in the skin, including free nerve endings and encapsulated nerve endings, that transduce sensations of touch, temperature, pressure, and pain. The number of somatosensory receptors in the skin is reduced in older individuals, and these physiological losses result in perceptual decreases in touch sensitivity especially at the fingertips *(26)* as well as thermal sensitivity *(27)*. Peripheral neuropathies from a variety of medical conditions including diabetes, thyroid disorders, rheumatoid arthritis, alcoholism, vitamin B_{12} deficiency, and peripheral vascular disease exacerbate age-related somatosensory impairments. Certain drugs and medical treatments including statins and chemotherapy have somatosensory side effects such as numbness and tingling *(11)*.

Somatosensory perception can have an adverse effect on nutritional status because reduced oral sensitivity impairs the ability to discriminate different textures of food. Reduced tactile sensitivity can make food preparation and manipulation of utensils more difficult. Decrements in the ability to perceive heat can reduce pleasure from food and become a safety issue if foods or beverages are served at boiling temperatures (e.g., with a potential for oral burns). Reduced temperature perception in the fingers is also risk factor for burns during cooking.

5.5 CHALLENGES FOR ASSESSING SENSORY FUNCTIONING IN OLDER PERSONS: COMPARISON OF TASTE AND SMELL WITH OTHER SENSES

In current clinical practice, physicians and other medical personnel generally make diagnoses, treatment decisions, and clinical prognoses based on tests at a single point in time. Clinical research, however, suggests that tests obtained at a single time point are often inadequate to make a definitive diagnosis for a variety of medical conditions. Sensory, sensorimotor, and psychological assessments *(28)* can be highly variable for a given individual over short time periods. This temporal variation for an individual relative to his/her own mean is called intra-individual variation.

A recent study emphasizes the difficulty of relying on a single assessment of sensory function at one time point in older individuals. Schiffman *(9)* evaluated sensory performance (all five sensory modalities as well as cognition) concomitantly in three groups of non-demented older subjects including coronary artery bypass surgery patients, patients with cardiovascular conditions but with no history of surgery,

Table 5.2

Sensory (and cognitive) tests ranked by percent of non-medicated older persons who performed best on the first test of a series of four repeated tests

Test modality	*Specific test type*	*Percentage of persons who performed best on the first of four repetitions*
Cognitive	Immediate Recall[1]	11.3
Cognitive	Delayed Recall[2]	18.3
Touch	Tactile special sensitivity thresholds[3]	19.0
Cognitive	Symbol Digit Modalities Test[4]	19.7
Taste	Sucrose detection threshold[5]	26.1
Taste	Sucrose (sweet) recognition threshold[5]	26.8
Taste	Quinine HCl detection threshold[5]	26.8
Taste	NaCl detection threshold[5]	28.2
Hearing	Thresholds (in decibels) at 6000 Hz[6]	29.6
Taste	NaCl (salty) recognition threshold[5]	31.7
Taste	Quinine HCl (bitter) recognition threshold[5]	31.7
Hearing	Thresholds (in decibels) at 8000 Hz[6]	33.1
Vision	Contrast Sensitivity Row D (somewhat difficult)[7]	38.7
Cognitive	Mini Mental Status Examination (MMSE)[8]	39.4
Smell	Smell identification[9]	40.8
Smell	Smell Memory[10]	40.8
Vision	Contrast Sensitivity Row E (most difficult)[7]	40.8
Smell	Butanol detection threshold[11]	41.5
Hearing	Thresholds (in decibels) at 4000 Hz[6]	42.3
Vision	Contrast Sensitivity Row A (easiest)[7]	43.0
Hearing	Thresholds (in decibels) at 500 Hz[6]	43.7
Hearing	Thresholds (in decibels) at 2000 Hz[6]	50.0
Vision	Contrast Sensitivity Row C (moderate difficulty)[7]	50.7

(continued)

Table 5.2
(continued)

Hearing	Thresholds (in decibels) at 1000 Hz[6]	50.7
Vision	Near Vision[12]	52.1
	Contrast Sensitivity Row B (relatively	53.5
Vision	easy)[7]	

[1]Immediate recall: Subjects read aloud a list of 13 word pairs, with each pair presented at 3 s intervals. After a second presentation of these same word pairs, subjects were given the cue words (first words of each pair) on a page, but the target words (second words of each pair) were absent (blank line). Subjects were asked to write on the blank lines as many of the pair-associated words as they could remember. The maximum correct score attainable was 13.

[2]Delayed recall: Same as immediate recall but after a 5–10 min delay.

[3]Tactile spatial sensitivity thresholds: Cutaneous spatial resolution on the fingertip of the index finger was measured by assessing orientation of spatial gratings (JVP Domes, Stoelting Company, Wood Dale, IL) (29). The score is the narrowest spatial resolution that the subject can detect in millimeters (mm).

[4]The Symbol Digit Modalities Test (SDMT) evaluates processing speed (30). Subjects were presented with a key that pairs each digit from 1 to 9 with a specific geometric symbol. The subject was then instructed to fill in rows of blank boxes with the digits that correspond to the symbols presented directly above the boxes. Subjects were allotted 90 s to write the digits in as many possible consecutive boxes in a row without skipping any items.

[5]The detection threshold is the concentration at which the subject correctly distinguished the tastant as stronger than a water control. The recognition threshold (recognition threshold) is the concentration at which the subject correctly identified the taste as salty (NaCl), sweet (sucrose), or bitter (quinine HCl).

[6]Thresholds (in decibels) for six pure tones (500, 1000, 2000, 4000, 6000, and 8000 Hz) were determined for each ear using the Maico 25 portable air conduction audiometer and an audiocup headset (Eden Prairie, MN). A total score was computed by adding the decibels required to reach threshold at each of the six pure tones and dividing by six. Lower total scores indicated better performances.

[7]Contrast sensitivity refers to the visual ability to distinguish between an object and its background and is considered a more sensitive measure of visual status than standard visual acuity measures. The Functional Acuity Contrast Test (FACT) Chart (Stereo Optical Co., Chicago, IL) was used to measure contrast sensitivity. Subjects reported the orientation (up, left, or right) of gray sine-wave grating lines at progressively decreasing contrast. The FACT chart tests five spatial frequencies in rows labeled A (easiest), B, C, D, and E (most difficult).

[8]The MMSE is a brief screening instrument that assesses orientation and cognitive status in adults (31).

[9]Subjects were asked to identify 12 odors with the use of a list of 27 names of odorous substances. The test score was the total number correct out of 12 target substances.

[10]Smell memory was assessed for four odorants using procedures described by Schiffman et al. (17). Subjects were instructed to sniff one bottle containing the target odor and then try to remember it. Subjects then counted backward by threes for 7 s from a number given by the examiner. Subjects were sequentially presented with a set of the four odorants and were asked to identify which of the four had the odor that they were asked to remember. Each odor was presented twice, for a total of eight trials, and scoring was based on the total number of correctly remembered smells, with eight as the maximum score.

[11]The butanol detection threshold was the concentration at which the subject correctly distinguished the odor as stronger than a water control. The threshold is expressed as concentration (% v/v) of 1-butanol in deionized water.

[12]Near visual acuity for each eye was assessed using a Rosenbaum Pocket Vision Screening Chart (Western Ophthalmics Corporation, Lynnwood, WA), with the subject wearing his/her personal visual correction when needed. The near visual acuity score was determined to be the lowest line at which a subject could correctly identify at least half the numbers or symbols. Mean score is the denominator of the distance equivalent, e.g., 20/20, 20/30, i.e., 20/36.9 is worse than 20/33.1.

and healthy non-medicated age-matched controls. All subjects were tested three times over a 2-month period in order to investigate and compare short-term intra-individual fluctuation or lability of the five senses and cognition over a short time period when little change in perception was expected. The non-medicated control group was also tested a fourth time four months after the 2-month testing in triplicate.

The main findings of the study were as follows. First, the intra-individual variability was extensive for all three groups although it was significantly greater for the coronary artery bypass surgery patients than for the other two groups that were statistically equivalent. Second, the greatest individual variability (as well as magnitude of sensory losses for the coronary bypass patients) was for taste and smell thresholds. Third, correlations among the tests of different modalities revealed that performance in one sensory modality does not necessarily correlate with performance in another modality. Fourth, the initial or first sensory (and cognitive) assessment of the repetitions seldom provided the best performance for the majority of the tests (see Table 5.2 for data on the non-medicated cohort). The variability for the medicated (but not surgical) patients was similar to the non-medicated cohort, but the surgical patients showed significantly more fluctuation (9). This finding is important because it emphasizes that sensory sensitivity varies from day to day, and multiple assessments of the senses are necessary to establish the range (maximum to minimum) of sensory performance.

5.6 FINAL COMMENT

Currently there are no proven pharmacological methods to treat age-related impairment of taste or smell, and the prognosis for recovery of chemosensory sensations is poor. Use of stem cells that are capable of differentiating into taste, smell, or other sensory receptors may be valuable for therapy in the future. Hyposmia (reduced perception of smell) but not anosmia (total loss of smell) can be "treated" by adding simulated food flavors to nutritious foods such as meat and vegetables to amplify the odor intensity (5,6). Simulated flavors (e.g., cheese, butter, bacon) can be obtained commercially and are mixtures of odorous molecules that are synthesized after chemical analysis of the target food or, in some cases, extracted from natural products. Flavors are essentially concentrated odors that are comparable to frozen concentrated orange juice or extract of vanilla. Because flavors are not spices, they do not irritate the stomach or cause gastric intolerance which is a common complaint in older individuals. Amplification of the flavor levels to compensate for taste and smell losses can improve food enjoyment, have a positive effect on food intake, and foster appropriate nutritional intake.

5.7 RECOMMENDATIONS

1. For patients with nutritional disorders, clinicians should assess whether a patient has sensory deficits that interfere with food intake. Importantly, clinicians should be familiar with prescription medications that impair sensory functioning because substitution of the offending drug (i.e., one causing sensory impairment) with one lacking sensory side effects may improve food intake.

2. For persons with diminished chemosensory acuity, the addition of simulated food flavors (flavor amplification) to meat, vegetables, and other nutritious foods may improve intake and compensate for chemosensory loss. Other helpful approaches include varying the texture, color, and temperature of foods and instructing the patient to chew foods thoroughly and to rotate among different foods on the plate to reduce sensory fatigue.
3. Awareness of safety issues when temperature or tactile sensations are blunted is important to ensure that patients do not burn the inside of the oral cavity or their fingers with overheated food and beverages.

Acknowledgment Research reported in this paper was supported by a grant to Dr. Susan Schiffman and Duke University from the National Institute on Aging AG00443.

REFERENCES

1. Toth MJ, Poehlman ET. Energetic adaptation to chronic disease in the elderly. Nutr Rev 2000; 58(3 Pt 1): 61–6.
2. Schiffman SS. Taste and smell in disease. N Engl J Med 1983; 308: 1275–9, 1337–43.
3. Schiffman SS. Sensory properties of food: their role in nutrition. In: Schlierf G, ed. Recent advances in clinical nutrition. III. London: Smith-Gordon, 1993; 67–81.
4. Schiffman SS. Perception of taste and smell in elderly persons. Crit Rev Food Sci Nutr 1993; 33: 17–26.
5. Schiffman SS. Taste and smell losses in normal aging and disease. JAMA 1997; 278: 1357–62.
6. Schiffman SS, Graham BG. Taste and smell perception affect appetite and immunity in the elderly. Eur J Clin Nutr 2000; 54 Suppl 3: S54–63.
7. Schiffman SS, Zervakis J. Taste and smell perception in the elderly: effect of medications and disease. Adv Food Nutr Res 2002; 44: 247–346.
8. Schiffman SS. Smell and taste. In: Birren JE, ed. Encyclopedia of Gerontology (Second Edition): Age, Aging, and the Aged. Elsevier; 2007: 515–25.
9. Schiffman SS. Critical illness and changes in sensory perception. Proc Nutr Soc 2007; 66: 331–45.
10. Shinkai RS, Hatch JP, Schmidt CB, Sartori EA. Exposure to the oral side effects of medication in a community-based sample. Spec Care Dentist 2006; 26(3):116–20.
11. Physicians' Desk Reference 59th edition. DesMoines: Medical Economics, 2005.
12. Genetic tests. "DNAdirect" tests for cytochrome P450 2D6 (CYP2D6); "CyGene" for AMD and glaucoma.
13. Knapp HR. Nutrient-drug interactions. In: Ziegler EE, Filer Jr. LJ, eds. Present knowledge in Nutrition. Seventh Edition. Washington, DC: ILSI Press. 1996: 540–46.
14. DeVries H, Stuiver M. The absolute sensitivity of the human sense of smell. In: Rosenblith WA, ed. Sensory Communication. New York: John Wiley and Sons. 1961: 159–67.
15. Axel R. The molecular logic of smell. Sci Am 1995; 273: 154–59.
16. Murphy C, Schubert CR, Cruickshanks KJ, Klein BE, Klein R, Nondahl DM. Prevalence of olfactory impairment in older adults. JAMA 2002; 288: 2307–12.
17. Schiffman SS, Graham BG, Sattely-Miller EA, Zervakis J, Welsh-Bohmer K. Taste, smell and neuropsychological performance of individuals at familial risk for Alzheimer's disease. Neurobiol Aging 2002; 23: 397–404.
18. Sekuler R, Sekuler AB. Visual perception and cognition. In: Evans JG, Williams TF, Beattie BL, Michel J-P, Wilcock GK, eds. Oxford Textbook of Geriatric Medicine. Oxford: Oxford University Press. 2000: 874–80.
19. Faye EE. Poor vision. In: Evans JG, Williams TF, Beattie BL, Michel J-P, Wilcock GK, eds. Oxford Textbook of Geriatric Medicine. Oxford: Oxford University Press. 2000: 881–93.
20. Weale RA. The eye and senescence. In: Evans JG, Williams TF, Beattie BL, Michel J-P, Wilcock GK, eds. Oxford Textbook of Geriatric Medicine. Oxford: Oxford University Press. 2000: 863–73.

21. Fraunfelder FT, Fraunfelder FW. Drug-Induced Ocular Side Effects. Boston: Butterworth-Heinemann. 2001: 1–824.

22. Sahraie A, Trevethan CT, MacLeod MJ, Murray AD, Olson JA, Weiskrantz L. Increased sensitivity after repeated stimulation of residual spatial channels in blindsight. Proc Natl Acad Sci USA 2006; 103(40):14971–6.

23. Danckert J, Rossetti Y. Blindsight in action: what can the different sub-types of blindsight tell us about the control of visually guided actions?. Neurosci Biobehav Rev 2005; 29(7):1035–46

24. NIDCD: National Institute on Deafness and Other Communication Disorders (2005): http://www.nidcd.nih.gov/health/statistics/hearing.asp http://www.nidcd.nih.gov/health/hearing/coch.asp as of Nov. 2005.

25. Cohen-Mansfield J, Taylor JW. Hearing aid use in nursing homes. Part 1: Prevalence rates of hearing impairment and hearing aid use. J Am Med Dir Assoc. 2004; 5(5): 283–88.

26. Stevens JC, Patterson MQ. Dimensions of spatial acuity in the touch sense: Changes over the life span. Somatosens Mot Res 1995; 12: 19–47.

27. Stevens JC, Choo KK. Temperature sensitivity of the body surface over the life span. Somatosens Mot Res 1998; 15: 13–28.

28. Li S-C, Aggen SH, Nesselroade JR, Baltes PB. Short-term fluctuations in elderly people's sensorimotor functioning predict text and spatial memory performance: the MacArthur successful aging studies. Gerontology 2001; 47: 100–16.

29. Van Boven RW, Johnson KO. The limit of tactile spatial resolution in humans: grating orientation discrimination at the lip, tongue, and finger. Neurology 1994; 44: 2361–6.

30. Smith A. Symbol Digit Modalities Test (SDMT) manual, revised. Los Angeles: Western Psychological Services, 1982.

31. Folstein MF, Folstein SE, McHugh PR. "Mini-mental state." A practical method for grading the cognitive state of patients for the clinician. J Psychiatr Res 1975; 12: 189–98.

6 Nutrition and the Aging Eye

Elizabeth J. Johnson

Key Points

- There is growing interest in the role that nutrition plays in modifying the development and/or progression of vision disorders in older persons, including age-related cataract and macular degeneration.
- Available evidence to date supports a possible protective role of several nutrients, including vitamins C and E and the carotenoids lutein and zeaxanthin.
- Due to inconsistencies among the findings of currently available studies regarding doses and combinations of nutrients, it may be most practical to recommend specific natural diet choices rich in vitamins C and E, lutein and zeaxanthin, omega-3 fatty acids and zinc, which would also provide potential benefits from other components of these natural food sources.

Key Words: Cataract; macular degeneration; eye disease; retina

6.1 INTRODUCTION

Vision loss among the elderly is an important health problem. Approximately one person in three has some form of vision-reducing eye disease by the age of 65 *(1)*. Age-related cataract and age-related macular degeneration (AMD) are the major causes of visual impairment and blindness in the aging US population. Approximately 50% of the 30–50 million cases of blindness worldwide result from unoperated cataract *(2,3)*. A clinically significant cataract is present in about 5% of Caucasian Americans aged 52–64 yrs and rises to 46% in those aged 75–85 yrs *(4)*. In the United States, cataract extraction accompanied by ocular lens implant is currently the most common surgical procedure done in Medicare beneficiaries *(5)*. Lens implantation is highly successful in restoring vision; however, the procedure is costly, accounting for 12% of the Medicare budget and accounts for more than $3 billion in annual health expenditures *(5,6)*. For these reasons, there is much interest in the prevention of cataract as an alternative to surgery. The prevalence of AMD

From: *Nutrition and Health: Handbook of Clinical Nutrition and Aging, Second Edition*
Edited by: C. W. Bales and C. S. Ritchie, DOI 10.1007/978-1-60327-385-5_6,
© Humana Press, a part of Springer Science+Business Media, LLC 2009

also increases dramatically with age. Nearly 30% of Americans over the age of 75 have early signs AMD and 7% have late stage disease, whereas the respective prevalence among people 43–54 yrs are 8 and 0.1% *(4)*. AMD is the leading cause of blindness among the elderly in industrialized countries. Because there are currently no effective treatment strategies for most patients with AMD, attention has focused on efforts to stop the progression of the disease or to prevent the damage leading to this condition *(7)*.

Cataract and AMD share common modifiable risk factors, such as light exposure and smoking *(7,8)*. Of particular interest is the possibility that nutritional counseling or intervention might reduce the incidence or retard the progression of these diseases. The components of the diet that may be important in the prevention of cataract and AMD are vitamins C and E and the carotenoids, lutein and zeaxanthin. Given that the lens and retina suffer oxidative damage, these nutrients are thought to be protective through their role as antioxidants. Additionally, lutein and zeaxanthin may provide protection as filters against light damage, i.e., absorbers of blue light.

6.2 PHYSIOLOGICAL BASIS OF CATARACTS AND AMD

The role of the lens is to transmit and focus light on the retina. Therefore, for optimal performance the lens must be transparent. The lens is an encapsulated organ without blood vessels or nerves (*see* Fig. 6.1). The anterior hemisphere is covered by a single layer of epithelial cells containing subcellular organelles. At the lens equator the epithelial cells begin to elongate and differentiate to become fiber cells. Fully differentiated fiber cells have no organelles but are filled with proteins called crystallins, organized in a repeating lattice. The high density and repetitive spatial arrangement of crystallins produce a medium of nearly uniform refractive index with dimensions similar to light wavelengths *(9)*. Cataracts result when certain events, e.g., light exposure, cause a loss of order and result in abrupt fluctuations in refractive index causing increased light scattering and loss in

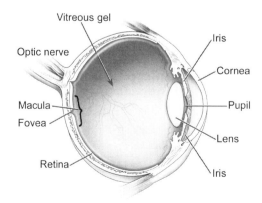

Fig. 6.1. The human eye showing typical organization and terminology (from National Eye Institute, National Institute of Health).

transparency in the lens. It is proposed that lens opacity results from damage to lens enzymes, proteins, and membranes by activated oxygen species, e.g., hydrogen peroxide, superoxide anion, and hydroxyl free radicals, which are the results of exposure to light and other types of radiation. For these reasons dietary antioxidants may be important in the prevention of cataract.

AMD is a disease affecting the central area of the retina (macula) (Fig. 6.1) resulting in loss of central vision. In the early stages of the disease, lipid material accumulates in deposits underneath the retinal pigment epithelium (RPE). This is believed to arise after failure of the RPE to perform its digestive function adequately. These lipid deposits are known as drusen, and can be seen as pale yellow spots on the retina. The pigment of the RPE may become disturbed with areas of hyperpigmentation and hypopigmentation. In the later stages of the disease, the RPE may atrophy completely. This loss can occur in small focal areas or can be widespread. In some cases, new blood vessels grow under the RPE and occasionally into the subretinal space (exudative or neovascular AMD). Hemorrhage can occur which often results in increased scarring of the retina. The early stages of AMD are in general asymptomatic. In the later stages there may be considerable distortion of vision and complete loss of visual function, particularly in the central area of vision *(7)*. Although the specific pathogenesis of AMD is still unknown, chemical and light-induced oxidative damage to the photoreceptors is thought to be important in the dysfunction of the RPE. The retina is particularly susceptible to oxidative stress because of its high consumption of oxygen, its high proportion of polyunsaturated fatty acids, and its exposure to visible light. Currently, there is no treatment which can restore vision in AMD. Therefore, efforts have focussed on its prevention. As with cataract, dietary antioxidants have been suggested to play an important role in the prevention of AMD.

The antioxidants, vitamins C and E, lutein and zeaxanthin are common components of our diet that are most often implicated as protective against eye disease. These antioxidants may prevent damage in the lens by reacting with free radicals produced in the process of light absorption. Photoreceptors in the retina are subject to oxidative stress throughout life due to combined exposures to light and oxygen.

Vitamin E and carotenoids are lipid-soluble oxidant scavangers that protect biomembranes. Vitamin C is an important water-soluble antioxidant and also promotes the regeneration of vitamin E. Both vitamins C and E are found in the lens *(10–12)*. Of the 20–30 carotenoids found in human blood and tissues *(13)* only lutein and zeaxanthin are found in the lens and retina *(10,14)*. Lutein and zeaxanthin are concentrated in the macula or central region of the retina and are referred to as macular pigment. In addition to their role as antioxidants, lutein and zeaxanthin are believed to limit retinal oxidative damage by absorbing incoming blue light and/or quenching reactive oxygen species. Many putative risk factors for AMD have been linked to a lack of macular pigment, including female gender, lens density, smoking, light iris color, and reduced visual sensitivity *(7)*. Omega-3 fatty acids and zinc are highly concentrated in the eye and have been implicated in eye health.

6.3 HUMAN STUDIES ON DIETARY INTAKE AND BLOOD LEVELS OF ANTIOXIDANTS AND EYE DISEASE

Studies with human subjects provide information on the strength of associations between nutritional factors and the frequency of a disease. Such studies can be a valuable means of identifying and evaluating risk factors. Although there are limitations to such studies, consistency of findings among studies lends to the credibility of nutritional and disease associations.

6.3.1 Cataract

6.3.1.1 VITAMIN C

Several studies have found a relationship between increased dietary vitamin C and decreased risk of cataract (*see* Table 6.1) *(15–17)*. For example, it was observed that the prevalence of nuclear cataract was lower for men with total vitamin C intakes in the highest quintile category relative to the lowest intake quintile *(18)*. It has also been observed that the prevalence of cataract was about 75% lower in persons with vitamin C intakes >490 mg/d than in those with intakes <125 mg/d *(16)*. However, such a relationship was not always observed *(15,19–22)* (Table 6.1).

Table 6.1
Summary of epidemiologic studies of dietary vitamin C* and cataract

Data analysis method	*Result*	*Reference*
Positive outcome		
Highest vs. lowest quintile (men, 104 vs. 33 mg/d)	Lower prevalence of cataract in highest quintile	Mares-Perlman, J.A., et al., 1995 *(15)*
>490 mg/d vs. <125 mg/d	Lower prevalence of cataract with high intake	Jacques, P.F., et al., 1991 *(16)*
Highest vs. lowest quintile	Lower prevalence of cataract in highest quintile	Leske, M.C., et al., 1991 *(17)*
Null outcome		
Highest vs. lowest quintile (women, 171 vs. 34 mg/d)	No difference in prevalence of cataract between groups	Mares-Perlman, J.A.,et al.,1995 *(15)*
Highest vs. lowest quintile (705 vs. 70 mg/d) (women)	No difference in prevalence of cataract between groups	Hankinson, S.E., et al., 1992 *(19)*
Highest vs. lowest quintile	No difference in prevalence of cataract extraction between groups	Tavani, A., et al., 1996 *(20)*
Highest vs. lowest quartile (261.1 vs. 114.4 mg/d)	No difference in prevalence or nuclear or cortical cataract between groups	Vitale, S., et al., 1993 *(21)*
Multiple logistic regression	No association of dietary vit C with nuclear or cortical cataract	Italia- American Cataract Study Group, 1991 *(22)*

*RDA for adults: 60 mg/d.

Table 6.2
Summary of epidemiologic studies of plasma vitamin C* and cataract

Data analysis method	Result	Reference
Positive outcome		
Multiple logistic regression	Serum vit C inversely associated with prevalence of cataract ($p = 0.03$)	Simon, J.A., et al., 1999 (23)
>90 μmol/L vs. <40 μmol/L	Lower prevalence of cataract with high plasma levels	Jacques, P.F., et al., 1991 (16)
Null outcome		
Highest vs. lowest quartile	Plasma vit C levels were not associated with risk of cortical or nuclear cataract	Vitale, S., et al., 1993 (21)
Negative outcome		
	Higher prevalence of cataract with increased plasma vit C	Mohan, M., et al., 1989 (24) cataract with increased plasma vit C

*Reference range: 23–125 μmol/L (Nutrition Evaluation Laboratory, Tufts University, 2001.

Compared to diet, serum concentrations of a nutrient are considered to be a better measure of nutrient status. Therefore, an evaluation of serum vitamin C relationships with cataract may be useful. Serum ascorbic acid level has been reported to be inversely associated with prevalence of cataract (*see* Table 6.2) (*16,23*). However, Vitale et al. (*21*) observed that plasma vitamin C concentrations were not associated with risk of nuclear or cortical cataract. In contrast to these studies, one study found an increased prevalence of cataract with increased plasma vitamin C (*24*) (Table 6.2).

6.3.1.2 VITAMIN E

A protective effect of dietary vitamin E has been observed in several studies (*see* Table 6.3). Persons in the highest quintile for vitamin E intake were reported to be 50% less likely to undergo cataract extraction compared to those in the lowest quintile for vitamin E intake (*20*). Mares-Perlman et al. (*15*) observed a lower prevalence of nuclear cataract in men in the highest quintile category of total vitamin E intake relative to those in the lowest vitamin E intake. Similarly, Leske et al. (*17*) reported that persons with vitamin E intakes in the highest quintile category had an approximately 40% lower prevalence of cataract relative to persons with intakes in the lowest quintile category. Jacques and Chylack (*16*) found that although persons with vitamin E intake greater than 35.7 mg/d had a 55% lower prevalence of cataract than did persons with intakes less than 8.4 mg/d, a significant difference was not found. Two other studies also reported no difference in cataract prevalence between persons with high and low vitamin E intake (*15,19*) (Table 6.3). In the first of these two studies, the null relationship was in women only (in men, dietary vitamin E was protective). In the second study only women were studied.

Table 6.3
Summary of epidemiologic studies of dietary vitamin E* and cataract

Data analysis method	Result	Reference
Positive outcome		
Highest vs. lowest quintile	Lower cataract extraction in highest quintile	Tavani, A., et al., 1996 *(20)*
Highest vs. lowest quintile (men, 12.8 vs. 4.0 mg/d)	Lower prevalence of nuclear cataract in highest quintile	Mares-Perlman, J.A. et al., 1995 *(15)*
Highest vs. lowest quintile	40% Lower prevalence of cataract in highest quintile	Leske, M.C., et al., 1991 *(17)*
Null outcome		
>35.7 mg/d vs. <8.4 mg/d	No difference between groups	Jacques, P.F., et al., 1991 *(16)*
Highest vs. lowest quintile (women, 19.9 vs. 5 mg/d)	No difference between groups	Mares-Perlman, J.A., et al., 1995 *(15)*
	No relationship between dietary vit E and cataract extraction in women	Hankinson, S.E., et al., 1992 *(19)*

*RDA: 8 and 10 mg/d, women and men, respectively.

As with dietary vitamin E, results from studies reporting relationships between plasma vitamin E and cataract have been mixed (*see* Table 6.4). In five of eight studies examining this issue, increased plasma vitamin E was observed to be protective against the risk of cataract *(21,25–28)* (Table 6.4). However, one study observed that the prevalence of cataract was not related to plasma vitamin E concentrations *(22)* and one study observed that the prevalence of cortical cataract did not differ between those with high and low plasma vitamin E concentrations *(21)*. In contrast, one study found increased levels of plasma vitamin E to be a risk factor for cataract *(18)*. In this study there was a significantly increased prevalence of nuclear cataract among women and men in the highest serum vitamin E quintile relative to those in the lowest quintile (Table 6.4).

6.3.1.3 LUTEIN AND ZEAXANTHIN

Few studies have specifically examined the relationship between lutein and zeaxanthin with cataract risk. In a recent report, Chasen-Taber et al. *(29)* observed in women that those with the highest intake of lutein and zeaxanthin had a 22% decreased risk of cataract extraction compared with those in the lowest quintile. Brown et al. *(30)* also observed that there was a lower risk of cataract extraction in men with higher intakes of lutein and zeaxanthin but not other carotenoids. Men in the highest fifth of lutein and zeaxanthin intake had a 19% lower risk of cataract relative to men in the lowest fifth. Mares-Perlman et al. *(31,32)* observed in women a significant inverse trend across quintiles of lutein intake. Women in the highest quintile of lutein intake (median 0.95 mg/d) had a 27% lower prevalence of nuclear

Table 6.4
Summary of epidemiologic studies of plasma vitamin E* and cataract

Data analysis method	Result	Reference
Positive outcome		
Highest vs. lowest quartile	Decrease in cortical cataract progression in highest quintile	Rouhiainen, P., et al., 1996 (20)
>30 µmol/L vs. <19 µmol/L	Less nuclear cataract in high plasma vit E group	Vitale, S., et al., 1993 (21)
>20 µmol/L vs. <20 µmol/L	Higher plasma levels of vit E had 1/2 the amount of cataract surgery	Knekt, P., et al., 1992 (26)
Highest vs. lowest quintile	Lower prevalence of nuclear cataract in highest quintile	Leske, M.C., et al., 1995 (27)
Regression model	High plasma vit E related to decreased prevalence of nuclear cataract	Leske, M.C., et al., 1998 (28)
Null outcome		
>30 µmol/L vs. <19 µmol/L	No difference in cortical cataract progression	Vitale, S., et al., 1993 (21)
Multivariate logistic regression	No relationship between plasma vit E and cataract prevalence	Italian-American Cataract Study Group, 1991 (22)
Negative outcome		
Highest vs. quintile men, 37.8 vs. 16.9 µmol/L women, 46.5 vs. 18.2 µmol/L	Increased prevalence of lowest nuclear cataract in highest	Mares-Perlman, J.A., et al., 1995 (31)

*Reference range: 12.0–43.2 µmol/L (Nutrition Evaluation Laboratory, Tufts University, 2001.

cataract than women in the lowest lutein intake quintile (median 0.28 g/d). The trend was in the same direction in men, but did not reach significance. Hankinson et al. (19) reported that the rate of cataract surgery was associated with lower intakes of lutein-rich foods such as spinach and other green vegetables. Although the data are few, the studies suggest that dietary lutein and zeaxanthin play a role in cataract prevention. Table 6.5 summarizes the epidemiologic studies that evaluated lutein status and cataract risk.

6.3.1.4 OMEGA-3 FATTY ACIDS

Cataract formation is associated with perturbations of lens membrane composition, structure, and function (33–35) as well as changes in fatty acid composition (36). Studies in rats found that high intake of polyunsaturated fatty acids delays the cataract formation (37,38). In a prospective study examining the relationship between dietary fat and cataract extraction in women (n = 71,083, 16-yr follow-up), women in the highest quintile of long-chain omega-3 fatty acids (0.21%

Table 6.5
Summary of epidemiologic studies of lutein status and cataract

Data analysis method	Result	Reference
Highest vs. lowest quintile lutein/ zeaxanthin intake(women)	22% Decrease risk of cataract extraction	Chasen-Taber, L., et al., 1999 (29)
Highest vs. lowest quintile lutein/ zeaxanthin intake (men)	19% Decrease risk of cataract extraction	Brown, L., et al., 1999 (30)
Highest vs. lowest quintile lutein/ zeaxanthin intake (women)	27% Lower prevalence of nuclear cataract	Mares-Perlman, J.A., et al., 1995 (15)
Highest vs. lowest quintile lutein intake (men)	No difference in prevalence of nuclear cataract	Mares-Perlman, J.A., et al., 1995 (15)
Consumption >5×/wk vs. <1×/ mo spinach (women)	29% Decrease risk of cataract extraction	Hankinson, S.E., et al., 1992 (19)

of energy) had a 12% lower risk of cataract extraction compared to those in the lowest quintile (0.03% of energy) (39) (relative risk = 0.88, 95% CI: 0.79–0.98, P for trend = 0.02).

6.3.2 AMD

6.3.2.1 VITAMIN C

Only one epidemiologic study has evaluated the role of dietary vitamin C and AMD risk (see Table 6.6). Seddon et al. (40) observed that persons in the highest and lowest intake quintiles for vitamin C had the same prevalence of advanced AMD. However, results examining relationships between plasma levels of vitamin C and AMD suggest that increased plasma vitamin C may decrease the risk of AMD. West et al. (41) reported that individuals with plasma vitamin C

Table 6.6
Summary of epidemiologic studies of dietary and plasma vitamin C* with AMD

Data analysis method	Result	Reference
Dietary vitamin C		
Highest vs. lowest quintile for vit C intake (1039 vs. 65 mg/d)	No difference in prevalence of advanced AMD between groups	Seddon, J.M., et al., 1994 (44)
Plasma vitamin C		
>80 μmol/L vs. <60 μmol/L	Lower prevalence of AMD in high plasma vit C group	West, S., et al.,1994 (41)
>91 μmol/L <40 μmol/L	Lower prevalence of AMD in high plasma vit C group	EDCCSG, 1993 (42)

*RDA for adults: 60 mg/d; reference range 23–125 μmol/L (Nutrition Evaluation Laboratory, Tufts University, 2001).

concentrations >80 µmol/L had a 45% lower prevalence of AMD compared with individuals who had concentrations <60 µmol/L. Others have reported that individuals with serum vitamin C concentrations ≤91 µmol/L had a 30% lower prevalence of AMD compared with those who had concentrations <40 µmol/L [42] (Table 6.6).

6.3.2.2 VITAMIN E

The one study that has evaluated the role of dietary vitamin E and AMD risk reported no difference in prevalence of advanced AMD between individuals in the highest and lowest vitamin E intake quintiles (see Table 6.7) (40). A protective effect of increased plasma vitamin E against AMD has been found in some studies (41,42), but not in others (31,43) (Table 6.7).

6.3.2.3 LUTEIN AND ZEAXANTHIN

Results of a multicenter case–control study suggest that high intakes of carotenoid, particularly lutein and zeaxanthin, are related to lower risk of advanced neovascular AMD (44). This is consistent with earlier findings from the First Health and Nutrition Examination Survey, in which low intakes of fruits and vegetables providing vitamin A were related to higher rates of all types of advanced AMD (45). The Eye Disease Case–Control Study (42) found after adjusting for other risk factors, people in the highest fifth of carotenoid intake had a 43% lower risk for neovascular AMD compared to those in the lowest fifth. Among the specific carotenoids, lutein and zeaxanthin, which are primarily obtained from dark green, leafy vegetables, were most strongly correlated with a reduced risk for age-related

Table 6.7
Summary of epidemiologic studies of dietary and plasma vitamin E* with AMD

Data analysis method	Result	Reference
Dietary vitamin E		
Highest vs.1994 (40) lowest quintile for vit E intake (405 vs. 3.4 mg/d)	No difference in prevalence of advanced AMD between groups	Seddon, J.M., et al.,
Plasma vitamin E		
>30 µmol/L vs.1994 (41) <19 µmol/L	Lower prevalence of AMD with high plasma vit E	West, S., et al.,
>43 µmol/L vs. <25 µmol/L	Lower prevalence of AMD with high plasma vit E	EDCCSG, 1993 (42).
>23 µmol/L vs.et al., 1995 (31) <23 µmol/L	No difference between groups in prevalence of AMD	Mares-Perlman, J.A.,
Patients with AMD vs.1993 (43) age-, sex-matched controls	No difference in plasma concentration of vit E between groups	Sanders, T.A.B, et al.,

*RDA: 8 and 10 mg/d, women and men, respectively; reference range: 12.0–43.2 µmol/L (Nutrition Evaluation Laboratory, Tufts University).

macular degeneration. However, a nested case–control study as part of the Beaver Dam Eye Study found no association with serum levels of lutein and zeaxanthin in 167 cases of (largely) early age-related macular degeneration and age-, sex-, and smoking-matched controls *(18)*.

6.3.2.4 OMEGA-3 FATTY ACIDS

It has been suggested that atherosclerosis of the blood vessels that supply the retina contributes to the risk of AMD, analogous to the mechanism underlying coronary heart disease *(46)*. Long-chain omega-3 fatty acids may have a special role in the function of the retina in addition to their antithrombotic and hypolipidemic effects on the cardiovascular system. Docosahexaenoic acid (DHA) is the omega-3 fatty acid of key interest. DHA is a major fatty acid found in the retina *(47)*. Rod outer segments of vertebrate retina have a high DHA content *(47,48)*. Since photoreceptor outer segments are constantly being renewed, a constant supply of DHA may be required for proper retinal function and a marginal depletion may impair retinal function and influence the development of AMD.

Epidemiologic studies examining the relationship of DHA or fish intake with AMD suggest a trend toward a protective relationship. In a prospective follow-up study of the Nurses' Health Study and the Health Professionals Follow-up Study, men and women ($n = 72,489$) with no diagnosis of AMD were followed for 10–12 yrs. Odds of AMD decreased with increased DHA intake (top vs. bottom quintile of RR: 0.70; 95% CI: 0.52–0.93; *P* for trend = 0.05). However, the relationship of DHA does not remain (OR for highest vs. lowest of DHA intake = 0.8; 95% CI, 0.5–1.1) when modeled simultaneously with intake of other dietary lipids. These investigators also examined the association of fish intake (a major source of DHA) with AMD risk. Consumption of >4 servings of fish per week was associated with a 35% lower risk of AMD compared with ≤3 servings per month (RR: 0.65; 95% CI: 0.46–0.91; *P* for trend = 0.0009) in pooled multivariate analysis *(49)*. Of the individual fish types examined, a significant inverse association was found only with tuna intake. The pooled RR of participants who ate canned tuna more than once per week compared with those who consumed it less than once per month was 0.61 (95% CI: 0.45–0.83).

The Dietary Ancillary Study of the Eye Disease Case–Control Study *(42)* reported results for 349 participants with neovascular AMD and 504 control subjects without AMD *(50)*. In demographically adjusted analyses, increasing intake of linoleic acid was significantly associated with higher prevalence of AMD (*P* for trend, 0.004). This association remained in multivariate analyses, with an OR for the fifth vs. first quintile of 2.00 (95% CI, 1.19–3.37) (*P* for trend, 0.02). In contrast, intake of omega-3 fatty acids showed an inverse relationship with AMD in demographically adjusted analyses (*P* for trend, 0.01) but became non-significant after controlling for confounding variables, e.g., cigarette smoking. When the study population was stratified by linoleic acid intake (≤5.5 or ≥5.6 g/d), the risk for AMD was significantly reduced with high intake of omega-3 fatty acids among those with low linoleic acid intake (*P* for trend, 0.05; *P* for continuous variable, 0.03). In contrast, among individuals with high linoleic acid intake, no significant association was seen for omega-3 fatty acid intake after controlling for other confounding variables. The authors commented that these findings suggest a competition between *n*-3 and *n*-6 fatty acids and that both the levels of *n*-3 fatty acids and its ratio to the *n*-6 acids are important. These results are similar to a more recent report involving a prospective cohort study of 261 persons aged 60 yrs or older at baseline with an average follow-up of 4.6 yrs. In this study, 101 patients with AMD progressed to advanced AMD. It was reported that higher fish intake (>2 servings/wk vs. <1 serving/wk) was associated with a lower risk of progression to advanced AMD among subjects with lower linoleic acid intake (OR 0.36; 95% CI, 0.14–0.95) *(51)*.

A relationship between fish intake and late ARM (neovascular AMD) or geographic atrophy was not measured in the Beaver Dam Eye Study, a retrospective population-based study. However, fish intakes were low *(52)* and it is possible that the intake of omega-3 fatty acids in this population was not varied enough to detect a difference in risk for AMD.

Heuberger et al. evaluated the associations between fish intake and age-related maculopathy (ARM) in the Third National Health and Nutrition Examination Survey. Persons aged 40–79 yrs (*n* = 7405) were included in analyses for early ARM (*n* = 644); those 60 yrs or older (*n* = 4294) were included in analyses for late ARM (*n* = 53). Consuming fish more than once a week compared with once a month or less was associated with ORs of 1.0 for early ARM (95% CI, 0.7–1.4) and 0.4 for late ARM (95% CI, 0.2–1.2) after adjusting for age and race. Adjusting for other possible risk factors did not influence these relationships *(53)*. These investigators concluded that no associations were observed between fish intake and ARM in this population. However, in associations with late ARM, while not statistically significant, the ORs were consistent with observations of inverse association reported by others *(49,54)*.

The Blue Mountains Eye Study (BMES) was a population-based survey of vision, common eye diseases, and diet in an urban population of 3654 people aged 49 yrs and older *(54)*. In the 2915 subjects evaluated for fish intake, there were 240 cases of early ARM and 72 cases of late ARM identified. In this study, more frequent consumption of fish appeared to protect against late ARM, after adjusting for age, sex, and smoking. The protective effect of fish intake commenced at a relatively low frequency of consumption (1–3 times per month compared with intake <1 time per month; OR: 0.23; 95% CI: 0.08–0.63). The ratio of cases to controls in these intake groups was 6/777 and 17/380, respectively. The OR for intake >5 times per week compared with <1 time per month was 0.46 (95% CI: 0.12–1.68). The authors suggested that there may be a threshold protective effect at low levels of fish intake, with no increased protection from ARM at increased fish intake. In this study, there was little evidence of protection against early ARM. A summary of the studies that evaluated omega-3 fatty acid intake and risk of AMD is found in Table 6.8.

6.3.2.5 ZINC

The zinc concentration in the retina is one of the highest levels in the body, suggesting a special importance to the eye *(55)*. In fact, in a study by Newsome et al. *(56)*, zinc supplementation given to elderly people with early stages of AMD resulted in better maintenance of visual acuity than in those receiving placebo.

Table 6.8
Multivariable odds ratios for neovascular AMD and late ARM and omega-3 fatty acid intake (adapted from *(70)*.

Study	Sample design	Exposure (high vs. low)	Outcome	Cases	OR	95%CI	Ref.
NHS/HPFU	Prospective	LCPUFA	NV AMD	9	0.4	0.2–1.2	*(49)*
EDCCS	Case–control	LCPUFA	NV AMD	349	0.6	0.3–1.4	*(50)*
BDES	Population based	Fish	Late ARM	30	0.8	0.2–1.5	*(52)*
NHANES	National Survey	Fish	Late ARM	9	0.4	0.2–1.2	*(53)*
BMES	Population based	Fish	Late ARM	46	0.5	0.2–1.2	*(54)*
Seddon et al.	Prospective	Fish	Late AMD	261	0.4	0.1–0.95	*(51)*

BDES, Beaver Dam Eye Study; BMES, Blue Mountains Eye Study; NHS/HPFU, Nurses Health Study/Health Professionals Follow-Up; NHANES, National Health and Nutrition Survey; EDCCS, Eye Disease Case–Control Study; AREDS, Age-Related Eye Disease Study. Late ARM – biographic atrophy or neovascular age-related macular degeneration; NV AMD, neovascular age-related macular degeneration.

However, in a prospective study of zinc intake and risk of AMD, in which 66,572 women and 37,636 men ($\geq$50 yrs) were followed for 10 yrs it was found that after multivariate adjustment for potential risk factors, the pooled relative risk was 1.13 (0.82–1.56) among participants in the highest quintile of total zinc intake (25.5 mg/d for women, 40.1 mg/d for men) compared with those in the lowest quintile (8.5 mg/d for women, 9.9 mg/d for men) *(57)*. The relative risk of highest compared with lowest quintile was 1.04 (95% CI, 0.59–1.83) for zinc intake from food. Subjects who took zinc supplements had a pooled multivariate relative risk of 1.04 (95% CI, 0.75–1.45). It was concluded from this large prospective study that zinc intake either in food or supplements was not associated with a reduced risk of AMD.

In summary, the studies examining nutrient and eye disease relationships are not entirely consistent. Methodology differences among studies may, in part, explain the inconsistencies. Also, there are limitations to such studies that examine relationships between a nutrient and disease because calculations from dietary recall may not always accurately estimate nutrient intakes due to limitations of the database or recall abilities of the subjects. Furthermore, a single blood value for a nutrient may not always be an accurate indicator of long-term status. In addition, the high degree of correlation in intake among the various dietary micronutrients makes it difficult to determine which specific nutrient or nutrients are related to the observed relationships. Despite these drawbacks, a possible protective role of vitamins C and E and the carotenoids lutein and zeaxanthin cannot be dismissed given the number of studies that found a protective effect and the very few studies that found a negative effect. In some cases, it may be difficult to measure an outcome if nutrient intake levels are at those found in diet alone. That is, dietary and plasma levels may not be sufficiently high to see an effect. In this regard, review of studies that have examined the relationship between supplemental nutrient intake with cataract and AMD risk may be useful.

6.4 THE EFFECT OF NUTRIENT SUPPLEMENTS ON EYE DISEASE RISK

Supplemental vitamins C and E have been long available to the general public. Currently, there are a variety of supplement products available in health food stores that contain lutein in amounts of 6–25 mg/capsule. At this point, lutein can be found in a few multivitamin products. Centrum was the first multivitamin supplement with lutein and contains 250 µg/capsule. Alcon Laboratories and Bausch & Lomb have recently made available multivitamin supplements formulated for eye care. These products contain lutein in higher amounts.

6.4.1 Cataract

Jacques et al. [58] observed a >75% lower prevalence of early opacities in women who used vitamin C supplement for ≥10 yrs (see Table 6.9). None of the 26 women who used vitamin C supplements for ≥10 yrs had more advanced nuclear cataract. Hankinson et al. (19) observed that women who reported use of vitamin C supplement for ≥10 yrs had a 45% reduction in rate of cataract surgery. The study of Robertson et al. (59) observed that the prevalence of cataract in persons who consumed vitamin C supplement of >300 mg/d was approximately one-third the prevalence in persons who did not consume vitamin C supplements. However, Chasan-Tabar et al. (60) prospectively examined the association between vitamin supplement intake and the incidence of cataract extraction during 12 yrs of follow-up in a cohort of 73,956 female nurses. After adjusting for cataract risk factors,

Table 6.9
Summary of epidemiologic studies of supplemental vitamin C* and cataract

Data analysis method	Result	Reference
Positive outcome		
Usage ≥10 yrs vs. Usage <10 yrs	Decrease in cataract in long-term users of vit C supplements	Jacques, P.F., et al., 1997 (58)
Usage ≥10 yrs vs. Usage <10 yrs	Decrease in cataract surgery with increase usage of vit C supplement	Hankinson, S.E., et al., 1992 (19)
>300 mg/d vs. non-users of vit C supplements	Lower prevalence of cataract in vit C supplement users	Robertson, J.M., et al., 1989 (59)
Null outcome		
Usage ≥10 yrs vs. non users	No difference between groups	Chasen-Taber, L., et al., 1999 (29)

*RDA for adults: 60 mg/d.

Table 6.10

Summary of epidemiologic studies of supplemental vitamin E* and cataract

Data analysis method	Result	Reference
Positive outcome		
Vit E supplement users vs. non-users	Decrease in cortical cataract in users of vit E supplements;	Nadalin, G., et al., 1999 (61)
Vit E supplements users vs. non-users	Decrease in nuclear cataract in users	Leske, M.C., et al., 1995 (27)
Vit E supplements users vs. non-users	Decrease in cataract in users	Robertson, J.M., et al., 1989 (59)
Null outcome		
Vit E supplement users vs. non-users	No difference in nuclear cataract	Nadalin, G., et al., 1999 (61)
Vit E supplement users vs. non-users	No difference in prevalence of cataract between users and non-users	Hankinson, S.E., et al., 1992 (19)

*RDA: 8 and 10 mg/d, women and men, respectively.

including cigarette smoking, body mass index, and diabetes mellitus, there was no difference in the incidence of cataract between users of vitamin C supplements for 10 yrs or more and non-users (Table 6.9).

Nadalin et al. (61) cross-sectionally examined the association between prior supplementation with vitamin E and early cataract changes in volunteers (see Table 6.10). Of 1111 participants 26% reported prior supplementation with vitamin E. Only 8.8% of these participants took supplementation greater than the recommended daily intake (10 mg/d). A statistically significant association was found between prior supplementation and the absence of cortical opacity, after adjusting for age. However, the levels of nuclear opacity were not statistically different between those who reported intake and those with no prior vitamin E supplementation. Leske et al. (28) examined the association of antioxidant nutrients and risk of nuclear opacification in a longitudinal study. The risk of nuclear opacification at follow-up was decreased in regular users of multivitamin supplements, vitamin E supplements, and in persons with higher plasma levels of vitamin E. The investigators concluded that in regular users of multivitamin supplements, the risk of nuclear opacification was reduced by one-third. They also reported that in regular users of vitamin E supplement and persons with higher plasma levels of vitamin E, the risk was reduced by approximately half. These results are confirmed by Robertson et al. (59) who reported that the prevalence of cataract was 56% lower in persons who consumed vitamin E supplement than in persons not consuming supplements. One study observed no relation between risk of cataract and vitamin E supplements (19) (Table 6.10).

To date, there are few data from intervention trials of vitamins and cataract risk. In a recent study, it was reported that a high-dose combination of antioxidants (vitamins C and E, beta-carotene, and zinc) had no significant effect on the development or progression of cataract (62). The LINXIAN trial (63) examined the role of antioxidants in prevention of cataract, and the effect is not clear. The intervention was a combination dose of 14 vitamins and 12 minerals. Therefore, a specific role of any one nutrient could not be accurately evaluated. The multi-vitamin component demonstrated that nutrition can modify the risk of nuclear cataract, but specific nutrients were not evaluated. Also, the population examined had suboptimal nutritional intakes at the study start and the effect may have been due to a correction of certain nutrient deficiencies.

The Roche European-American Anticataract Trial (REACT) was carried out to examine if a mixture of oral antioxidant micronutrients (beta-carotene, 18 mg/d; vitamin C, 750 mg/d; vitamin E, 600 mg/d) would modify the progression of age-related cataract (64). This was a multicenter prospective double-masked randomized placebo-controlled 3-yr trial in 445 patients with early age-related cataract. REACT demonstrated a statistically significant positive treatment effect after 2 yrs for US patients and for both subgroups (US, UK) after 3 yrs, but no effect for the UK patients alone. The conclusion from this study was that daily supplementation with these nutrients for 3 yrs produced a small deceleration in progression of age-related cataract.

6.4.2 AMD

A recent study reported that a high level of antioxidants and zinc significantly reduces the risk of AMD and its associated vision loss (62). In the Age-Related Eye Disease Study (AREDS) it was found that people at high risk for developing advanced stages of AMD (people with intermediate AMD or advanced AMD in one eye but not the other eye) lowered their risk by about 25% when treated with a high-dose combination of vitamins C and E, beta-carotene, and zinc. In the same high-risk group, the nutrients reduced the risk of vision loss caused by advance AMD by about 19%. For those subjects who had either no AMD or early AMD, the nutrients did not provide a measured benefit. Because single nutrients were not evaluated, specific effects could not be determined. AREDS 2 has begun and aims to refine the findings of AREDS by including the xanthophylls as well as omega-3 fatty acids into the test formulation.

It has reported that the prevalence of AMD in persons who consumed vitamin C supplement for >2 yrs was similar to those who never took vitamin C supplements [38] (see Table 6.11). In a study conducted by Seddon et al. (40) the prevalence of AMD was also similar between those who took vitamin E supplement for >2 yrs

Table 6.11
Summary of epidemiologic studies of supplement use and AMD

Data analysis method	Result	Reference
Null outcome		
Vit C supplement users (>2 yrs) vs. non-users	No difference between groups in AMD prevalence	EDCCSG, 1993 (42).
Vit E supplement users (>2 yrs) vs. non-users	No difference between groups in AMD prevalence	Seddon, J.M., et al., 1994 (40).
Intervention trial vit E, beta-carotene, or both supplements vs. placebo	No association of treatment Alpha-Tocopherol Beta-group with any sign of Carotene Cancer Prevention maculopathy	Study Group, 1994 (65). Eye Disease Prevalence Research Group, 2004 (71).

and those who never took vitamin E supplements. One primary prevention trial has been published on age-related macular degeneration *(65)* (Table 6.11). This trial evaluated the effect of nutritional antioxidants on AMD. Overall there were 728 people randomized to any antioxidant and 213 to placebo. The results of this study found that there was no association of treatment group with any sign of maculopathy. There were 216 cases of the disease in the antioxidant groups and 53 in the placebo group. The majority of these cases were early age-related maculopathy. The findings are similar when each of the antioxidant groups—vitamin E, beta-carotene, vitamin E and beta-carotene—are compared with placebo. Although this was a large, high-quality study there were few cases of late AMD (14 cases in total) which means that the study had limited power to address the question as to whether supplementation prevents AMD. There was no association with the treatment group and development of early stages of the disease. This study was conducted in Finnish male smokers and caution must be taken when extrapolating the findings to other geographical areas, to people in other age-groups, to women, and to non-smokers. However, the incidence of AMD, particularly neovascular disease, is likely to be higher in smokers *(66)* which means that they provide a good population to demonstrate any potential protective effects of antioxidant supplementation.

Lutein and zeaxanthin supplements have only recently become available to the general public. Therefore, time has not allowed for the adequate study of the effect of these nutrient supplements for the prevalence of either cataract or AMD.

In summary, of the studies that have examined nutrient supplement use vs. the risk of eye disease, it is difficult to determine if supplements provide any added protection against eye disease. The number of studies reporting a positive outcome, i.e., a decreased risk, was about the same as the number of null outcomes. Further, in a recent meta-analysis it was concluded that there is insufficient evidence to support the role of dietary antioxidants including the use of dietary antioxidant supplements for the primary prevention of AMD *(67)*. However, it appears that nutrient supplementation does not cause an increased risk to the eye disease.

6.5 CLINICAL SUMMARY AND TREATMENT GUIDELINES

The inconsistencies among studies in terms of the amount of nutrient required for protection against eye disease make it difficult to make specific recommendations for dietary intakes of these antioxidants. Therefore, it may be more practical to recommend specific food choices rich in vitamins C and E, lutein and zeaxanthin, omega-3 fatty acids and zinc, thereby benefiting from possible effects of the components in food that may also be important. This necessitates an awareness of dietary sources of nutritional antioxidants for both the patient and clinician. Good sources of vitamin C include citrus fruit, berries, tomatoes, and broccoli (*see* Table 6.12). Good sources of vitamin E are vegetable oils, wheat germ, whole grain cereals, nuts, and legumes (*see* Table 6.13). The two foods that were found to have the highest amount of lutein and zeaxanthin are kale and spinach (*see* Table 6.14). Other major sources include broccoli, peas, and brussel sprouts. Fish oils are the primary source of omega-3 fatty acids (*see* Table 6.15).

Table 6.12
Vitamin C content of foods* (72)

Food	Amount	Milligrams
Orange juice	1 cup	12
Green peppers	1/2 cup	96
Grapefruit juice	1 cup	94
Papaya	1/2 med	94
Brussel sprouts	4 sprouts	73
Broccoli, raw	1/2 cup	70
Orange	1 medium	70
Cantaloupe	1/4 melon	70
Turnip greens, cooked	1/2 cup	50
Cauliflower	1/2 cup	45
Strawberries	1/2 cup	42
Grapefruit	1/2 medium	41
Tomato juice	1 cup	39
Potato, boiled with peel	2 1/2" diam.	19
Cabbage, raw, chopped	1/2 cup	15
Blackberries	1/2 cup	15
Spinach, raw, chopped	1/2 cup	14
Blueberries	1/2 cup	9

*Edible portion.

Table 6.13
Vitamin E content of foods* (72)

Food	Amount	Milligrams (α-tocopherol equivalents)
Wheat germ oil	1 tb	26.2
Sunflower seeds	1/4 cup	16.0
Almonds	1/4 cup	14.0
Safflower oil	1 tb	4.7
Peanuts	1/4 cup	4.2
Corn oil	1 tb	2.9
Peanut butter	2 tb	4.0
Soybean oil	1 tb	2.0
Pecan, halves	1/4 cup	2.0

*Edible portion.

A healthy diet including a variety of fresh fruit and vegetables, legumes, fish, and nuts, will have many benefits, will not do any harm, and will be a good source of the antioxidant vitamins and minerals implicated (but not proven) in the etiology of cataract and age-related macular degeneration. There is no evidence that nutrient-dense diets high in these foods, which provide known and unknown antioxidant

Table 6.14
Lutein/zeaxanthin content of foods *(73)*.

Food	Amount	Milligrams
Kale, cooked	1/2 cup	8.7
Spinach, raw	1/2 cup	6.6
Spinach, cooked	1/2 cup	6.3
Broccoli, cooked	1/2 cup	2.0
Corn, sweet, cooked	1/2 cup	1.5
Peas, green, cooked	1/2 cup	1.1
Brussels sprouts, cooked	1/2 cup	0.9
Lettuce, raw	1/2 cup	0.7

*Edible portion.

Table 6.15
Omega-3 fatty acid content of selected seafood *(74,75)*.

Seafood	Omega-3 fatty acids (% by wt)
Mackerel	1.8–5.3
Herring	1.2–3.1
Salmon	1.0–1.4
Tuna	1.0–1.4
Trout	0.5–1.6
Halibut	0.4–0.9
Shrimp	0.2–0.5
Cod	0.2–0.3
Plaice	~0.2
Flounder	~0.2
Haddock	0.1–0.2

components, are harmful. In fact, intake of fruits and vegetables is associated with reduced risk of death due to cancer, cardiovascular disease, and all causes. Thus, recommendations such as consuming a more nutrient-dense diet, i.e., lower in sweets and fats, and increasing levels of fruit and vegetable intake do not appear to be harmful and may have other benefits despite their unproven efficacy in prevention or slowing disease. Until the efficacy and safety of taking supplements containing nutrients can be determined, current dietary recommendations *(68)* are advised.

In addition to antioxidant vitamins, patients ask about a wide variety of unproven and often untested nutritional supplements. These include bilberries, shark cartilage, and Ginkgo biloba extract. Unfortunately, little is known about the effect of these products on cataract or AMD: no clinical trials have been conducted. Patients with eye disease who are offered these often expensive and sometimes risky treatments are given little information as to their benefit or risk. Patients should be advised to avoid unproven treatments.

6.6 CONCLUSION

The hypothesis that antioxidant nutrients may protect against the cataract and AMD is a plausable one given the role of oxidative damage in the etiology of these diseases. It is not known at what stage the protective effect may be important. The question that needs to be addressed is whether people who begin to consume antioxidant vitamins in their 60s and 70s alter their risk of age-related macular degeneration. Although data regarding the use of nutrient supplements suggest protection in cataract, the data are less convincing for AMD. The research to date has not sufficiently evaluated the effectiveness vs. safety of nutrient supplements. But advocating the use of nutrient supplementation must be done with a cautionary note given that there have been trials which have suggested that supplementation with beta-carotene may have an adverse effect on the incidence of lung cancer in smokers and workers exposed to asbestos *(65,69)*. Clearly further trials are warranted to address the usefulness nutrient supplementation in eye disease prevention.

It is likely that cataract and AMD develop over many years and the etiology of these diseases is due to many factors. There are likely to be differences in the potential protective effect of antioxidant supplementation depending on the stage of the disease. Future research needs to take into account the stage at which oxidative damage, and therefore antioxidant supplementation, may be important.

REFERENCES

1. Quinlan, D.A., Common causes of vision loss in elderly patient. Am Fam Physician 1999; 60: 99–108.
2. Thylefors, B., et al., Global data on blindness. Bull World Health Organ 1995; 69: 115–21.
3. World Health Organization, Use of intraocular lenses in cataract surgery in developing countries. Bull World Health Organ 1991; 69: 657–66.
4. Institute, N.E., Prevalence and Causes of Visual Impairment and Blindness Among Adults 40 Years and Older in the United States, in Prevent Blindness in America. 2004.
5. Javitt, J.C., Who does cataract surgery in the United States? Arch Ophthalmol 1993; 111: 1329.
6. Steinberg, E.P., et al., The content and cost of cataract surgery. Arch Ophthalmol 1993; 111: 1041–9.
7. Snodderly, D.M., Evidence for protection against age-related macular degeneration by carotenoids and antioxidant vitamins. Am J Clin Nutr 1995; 62: 1448S–61S.
8. Taylor, H.R., Epidemiology of age-related cataract. Eye 1999; 13: 445–8.
9. Benedek, G.B., Theory of transparency of the eye. Appl Opt 1971; 10: 459–73.
10. Yeum, K.J., et al., Measurement of carotenoids, retinoids, and tocopherols in human lenses. Invest. Ophthalmol Vis Sci 1995; 3: 2756–61.
11. Yeum, K.J., et al., Fat-soluble nutrient concentrations in different layers of human cataractous lens. Current Eye Res 1999; 19: 502–5.
12. Taylor, A., et al., Relationship in humans between ascorbic acid consumption and levels of total and reduced ascorbic acid in lens, aqueous humor and plasma. Current Eye Res 1991; 10: 751–9.
13. Parker, R.S., Bioavailability of carotenoids. Eur J Clin Nutr 1997; 51: S86–90.
14. Bone, R.A., J.T. Landrum, and S.E. Tarsis, Preliminary identification of the human macular pigment. Vision Res 1985; 25: 1531–5.
15. Mares-Perlman, J.A., et al., Diet and nuclear lens opacities. Am J Epidemiol 1995; 141: 322–34.
16. Jacques, P.F. and L.T. Chylack, Jr., Epidemiologic evidence of a role for the antioxidant vitamins and carotenoids in cataract prevention. Am J Clin Nutr 1991; 53: 353S–5S.

17. Leske, M.C., L.T. Chylack, Jr., and C. Wu, The lens opacities case–control study risks factors for cataract. Arch Ophthalmol 1991; 109: 244–51.
18. Mares-Perlman, J.A., et al., Serum carotenoids and tocopherols and severity of nuclear and cortical opacities. IOVS 1995; 36: 276–88.
19. Hankinson, S.E., et al., Nutrient intake and cataract extraction in women: a prospective study. BMJ 1992; 305: 244–51.
20. Tavani, A., E. Negri, and C. LaVeccia, Food and nutrient intake and risk of cataract. Ann Physiol 1996; 6: 41–6.
21. Vitale, S., et al., Plasma antioxidants and risk of cortical and nuclear cataract. . Epidemiology 1993; 4: 195–203.
22. The Italian-American Cataract Study Group., Risk factors for age-related cortical, nuclear, and posterior subcapsular cataracts. Am J Epidemiol 1991; 133: 541–53.
23. Simon, J.A. and E.S. Hudes, Serum ascorbic acid and other correlates of self-reported cataract among older Americans. J Clin Epid 1999; 52: 1207–11.
24. Mohan, M., et al., Indian–US case–control study of age-related cataracts. India–US case–control Study Group. Arch Ophthalmol 1989; 107: 670–6.
25. Rouhiainen, P., H. Rouhiainen, and J.T. Saloneen, Association between low plasma vitamin E concentrations and progression of early cortical lens opacities. Am J Epidemiol 1996; 114: 496–500.
26. Knekt, P., et al., Serum antioxidant vitamins and risk of cataract. BMJ 1992; 304: 1392–4.
27. Leske, M.C., et al., Biochemical factors in the Lens Opacities Case–Control Study. Arch Ophthalmol 1995; 113: 1113–9.
28. Leske, M.C., et al., Antioxidant vitamins and nuclear opacities: the longitudinal study of cataract. Ophthalmology 1998; 105: 831–6.
29. Chasen-Taber, L., et al., A prospective study of carotenoid and vitamin A intakes and risk of cataract extraction in US women. Am J Clin Nutr 1999; 70: 517–24.
30. Brown, L., et al., A prospective study of carotenoid intake and risk of cataract extraction in US men. Am J Clin Nutr 1999; 70: 517–24.
31. Mares-Perlman, J.A., et al., Serum antioxidants and age-related macular degeneration in a population-based case–control study. Arch Ophthalmol 1995; 113: 1518–23.
32. Mares-Perlman, J.A., et al., Relationship between lens opacities and vitamin and mineral use. Ophthalmology 1994; 101: 315–25.
33. Borchman, D., R.J. Cenedella, and O.P. Lamba, Role of cholesterol in the structural order of lens membrane lipids. Exp Eye Res 1996; 62: 191–7.
34. Kistler, J. and S. Bullivant, Structural and molecular biology of the eye lens membranes. Crit Rev Biochem Mol Biol 1989; 24: 151–81.
35. Simonelli, F., et al., Fatty acid composition of membrane phospholipids of cataractous human lenses. Ophthalmic Res 1996; 28: 101–4.
36. Rosenfeld, L. and A. Spector, Comparison of polyunsaturated fatty acid levels in normal and mature cataractous human lenses. Exp Eye Res 1982; 35: 69–75.
37. Hatcher, H. and J.S. Andrews, Changes in lens fatty acid composition during galactose cataract formation. Invest Ophthalmol 1970; 9: 801–6.
38. Hutton, J.C., et al., The effect of an unsaturated-fat diet on cataract formation in streptozotocin-induced diabetic rats. Br J Nutr 1976; 36: 161–7.
39. Lu, M., et al., Prospective study of dietary fat and risk of cataract extraction among US women. Am J Epidemiol 2005; 161: 948–59.
40. Seddon, J.M., et al., Dietary fat intake and age-related macular degeneration. Invest. Ophthalmol Vis Sci 1994; 35: 2003.
41. West, S., et al., Are antioxidants or supplements protective for age-related macular degeneration? Arch Ophthalmol 1994; 117: 1384–90.
42. Eye Disease Case–Control Study Group and (EDCCSG). Antioxidant status and neovascular age-related macular degeneration. Arch Ophthalmol 1993; 111: 104–9.
43. Sanders, T.A.B., et al., Essential fatty acids, plasma cholesterol, and fat-soluble vitamins in subjects with age-related maculopathy and matched control subjects. Am J Clin Nutr 1993; 57: 428–33.

44. Seddon, J.M., et al., Dietary carotenoids, vitamins A, C, and E, and advanced age-related macular degeneration. Eye Disease Case–Control Study Group. JAMA 1994; 272(18): 1413–20.

45. Goldberg, J., et al., Age-related degeneration and cataract: are dietary antioxidants protective? Am J Epidemiol 1998; 128: 904–5.

46. Sarks, S.H. and J.P. Sarks, Age-related macular degeneration: atrophic form., in Retina, S.J. Ryan, S.P. Schachat, and R.M. Murphy, Editors. 1994, Mosby, Inc.: St. Louis. 149–73.

47. Fliesler, S.J. and R.E. Anderson, Chemistry and metabolism of lipids in the vertebrate retina. Prog Lipid Res 1983; 22: 79–131.

48. Bazan, N.G., et al., Metabolism of arachidonic and docosahexaenoic acids in the retina. Prog Lipid Res 1986; 25: 595–606.

49. Cho, E., et al., Prospective study of dietary fat and the risk of age-related macular degeneration. Am J Clin Nutr 2001; 73: 209–18.

50. Seddon, J.M., et al., Dietary fat and risk for advanced age-related macular degeneration. Arch Ophthalmol 2001; 119: 1191–9.

51. Seddon, J.M., J. Cote, and B. Rosner, Progression of age-related macular degeneration. Association with dietary fat, trans unsaturated fat, nuts and fish intake. Arch Ophthalmol 2003; 121: 1728–37.

52. Mares-Perlman, J.A., et al., Dietary fat and age-related maculopathy. Arch Ophthalmol 1995; 113: 743–8.

53. Heuberger, R.A., et al., Relationship of dietary fat to age-related maculopathy in the Third National Health and Nutrition Examination Survey. Arch Ophthalmol 2001; 119: 1833–8.

54. Smith, W., Mitchell P., and S. Leeder, R., Dietary fish and fish intake and age-related maculopathy. Arch Ophthalmol 2000; 118: 401–4.

55. Newsome, D.A., et al., Zinc uptake by primate retinal pigment epithelium and choroid. Curr Eye Res 1992; 11: 213–7.

56. Newsome, D.A., et al., Oral zinc in macular degeneration. Arch Ophthalmol 1988; 106: 192–8.

57. Cho, E., et al., Prospective study of zinc intake and the risk of age-related macular degeneration. Ann Epidemiol 2001; 11: 328–36.

58. Jacques, P.F., et al., Long-term vitamin C supplement and prevalence of age-related opacities. Am J Clin Nutr 1997; 66: 911–6.

59. Robertson, J.M., A.P. Donner, and J.R. Trevithick, Vitamin E intake and risk for cataracts in humans. Ann NY Acad Sci 1989; 570: 373–82.

60. Chasen-Taber, L., et al., A prospective study on vitamin supplement intake and cataract extraction among US women. Epidemiology 1999; 10: 679–84.

61. Nadalin, G., et al., The role of past intake of vitamin E in early cataract changes. Ophthal Epidemiol 1999; 6: 105–112.

62. Age-Related Eye Disease Study Research Group, A randomized, placebo-controlled, clinical trial of high-dose supplementation with vitamins C and E, beta-carotene, and zinc for age-related macular degeneration and vision loss. Arch Ophthalmol 2001; 119: 1417–36.

63. Sperduto, R.D., et al., The Linzian cataract studies. Two nutrition intervention trials. Arch Ophthalmol 1993; 111: 1246–53.

64. Chylack, L.T.J., et al., The Roche European American Cataract Trial (REACT): a randomized clinical trial to investigate the efficiency of a antioxidant micronutrient mixture to slow progression of age-related cataract. Ophthal Epidemiol 2002; 9: 49–80.

65. The Alpha-Tocopherol Beta-Carotene Cancer Prevention Study Group, The effect of vitamin E and beta-carotene on the incidence of lung cancer and other cancers in male smokers. New Eng J Med 1994; 330: 1029–35.

66. Solberg, Y., M. Posner, and M. Belkin, The association between cigarette smoking and ocular diseases (review). Survey Ophthalmol 1998; 42: 535–47.

67. Chong, E.W.T., et al. Dietary antioxidants and primary prevention of age related macular degeneration: systematic review and meta-analysis. BMJ 2007 [cited 2007 8 October]; Available from: www.bmj.com/cgi/conent/full/bmj.39350.500428.47v1#BIBL.

68. USDA (2005) Dietary Guidelines for Americans 2005. Volume,

69. Omenn, G.S., et al., Risk factors for lung cancer and for intervention effects in CARET, the beta-carotene and retinol efficiency trial. J Natl Cancer Inst 1996; 88: 1550–9.

70. SanGiovanni, J.P. and E.Y. Chew, The role of omega-3 long-chain polyunsaturated fatty acids in health and disease of the retina. Prog Retina Eye Res 2005; 24: 87–138.
71. The Eye Disease Prevalence Research Group, Prevalence of age-related macular degeneration in the United States. Arch Ophthalmol 2004; 122: 564–74.
72. USDA and ARS (2006) National Nutrient Database for Standard Reference, Release 19. Nutrient Data Laboratory Home Page.
73. USDA. USDA-NCC Carotenoid Database for U.S. Foods-1998. 1998 [cited; Available from: www.nal.usdea.gov/fnic/foodcomp/Data/car98/car98.html.
74. Hepburn, F.N., J. Exler, and J.L. Weilharach, Provisional tables on the content of omega-3 fatty acids and other fat components of selected foods. J Am Diet Assoc 1986; 86: 788–93.
75. Ackman, R.G., Concerns for utilization of marine lipids and oils. Food Technol 1988; 42: 151–5.

7

Common Gastrointestinal Complaints in Older Adults

Stephen A. McClave

Key Points

- Medication effects and specific dietary factors, such as acid and foods lowering esophageal sphincter pressure, require special attention in older adults with esophageal and gastric disorders.
- The high prevalence of atrophic gastritis in older adults increases the risk for vitamin B_{12} malabsorption and deficiency.
- Evaluation of both diarrhea and constipation requires a thorough history (paying attention to diet, activities, and bowel habits) and a careful physical examination (including a digital rectal exam).
- Hepatitis is more common among older adults than originally appreciated and is often secondary to medications or hepatitis C exposure.
- Evaluation of anemia in older adults should include assessment of iron stores, copper, zinc, B_{12}, and folate levels.

Key Words: Esophageal function; atrophic gastritis; dysphagia; diarrhea; constipation; anemia; hepatitis; gut

7.1 INTRODUCTION

Many gastrointestinal (GI) complaints are voiced by the geriatric patient population. Surprisingly, few of these complaints are due to specific aging processes of the gut. With rare exception, the onset of symptoms reflects underlying pathophysiologic disease processes which warrant full evaluation and appropriate management. Often other factors, unique to the older age group, predispose the patient to added risk. Toleration of the insult is diminished, resulting in the potential for greater morbidity and mortality than for younger age groups. The diagnosis can be difficult to make due to underreporting by the patient, atypical symptom presentation, and the existence of comorbid disease processes.

From: *Nutrition and Health: Handbook of Clinical Nutrition and Aging, Second Edition*
Edited by: C. W. Bales and C. S. Ritchie, DOI 10.1007/978-1-60327-385-5_7,
© Humana Press, a part of Springer Science+Business Media, LLC 2009

This chapter will review the most common GI complaints, discuss the clinical issues and their relevance to nutritional status, and briefly highlight a basic management scheme for each problem.

7.2 DYSPHAGIA

Dysphagia in the geriatric population cannot be attributed to presbyesophagus ("old esophagus"), a term used inappropriately in the past to attribute symptoms to the natural aging process of the esophagus. This is because only minor, clinically insignificant, changes in the physiology of the esophagus occur with aging (1). Invariably, any significant complaints of dysphagia are related to an underlying disease process such as diabetes, Parkinson's disease, malignancy, stricture, scleroderma, gastroesophageal reflux, or neuropathies (2,3). The incidence of this complaint is surprisingly common, occurring in approximately 10% of the population aged 50 years or greater (3).

Certain minor changes are seen with esophageal motility that occur with aging. While these changes are clinically insignificant, they may accentuate or aggravate preexisting swallowing disorders (1,2,4). Changes related to the upper esophageal sphincter (UES) include increased sensory threshold for relaxation, decreases in pressure, and delayed relaxation. Other changes include decreased frequency of contractions in response to swallowing and reduced amplitude of contractions for esophageal peristalsis (5). Esophageal sensitivity is diminished, the myenteric neurons in the esophageal wall are decreased in number, and there may be a nonspecific increase in synchronous, non-peristaltic contractions. Anatomically, decreases in the density of striated muscle occur in the esophagus, as well as decreased strength of the smooth muscle (which may be present in up to 20–60% of patients in older age groups) (5). Surprisingly, these subtle changes are rarely responsible for producing symptoms (5,6).

True dysphagia in the older adult population is comprised of two main categories—motility disorders and structural abnormalities. Motility disorders are more common in older age groups than structural abnormalities (3). Oropharyngeal transfer dysphagia (difficulty in passage of the food bolus from the oropharynx to the proximal esophagus) is associated with cognitive or perception changes and neurologic deficits (3), especially when a cerebral vascular accident (CVA) involves the brain stem (5). Oropharyngeal dysphagia is also addressed in Chapter 24.

The incidence of dysphagia following a CVA ranges from 25 to 45% (7). Approximately 40% of these patients will recover full function and return to normal eating within 6 weeks. Several disorders predispose specifically to oropharyngeal transfer dysphagia: Parkinson's disease, CVA, sclerodema, Alzheimer's dementia, multiple sclerosis, drug-induced extrapyramidal symptoms, motor-neuron disease, polymyositis, myasthenia gravis, myopathy, and amyloidosis (5). Structurally, Zenker's diverticulum and cervical vertebral spurs may contribute to oropharyngeal transfer dysphagia. Clinical signs that may be associated with oropharyngeal transfer dysphagia include facial weakness, oral apraxia, reduced oral sensation, cough or throat clearing, reduced laryngeal elevation with swallowing, nasal regurgitation, repeat attempts to swallow, drooling, food spillage, and food sticking at the back of the throat *(5)*. The clinical consequences of transfer dysphagia include dysarthria, nasal tone to the voice, avoidance of social activities and dining, cough, aspiration, and ultimately pneumonia *(5)*.

Structural abnormalities leading to tertiary or esophageal dysphagia include esophageal webs or rings, peptic strictures, malignancy, strictures related to caustic medication, or rarely dysphagia aortica. Webs or rings are mucosal in origin, are thin transverse membranes, and may be increased in incidence with advancing age *(5,8)*. The malignancy increasing the most in incidence in the older adult population (particularly in the caucasian male) is adenocarcinoma of the gastroesophageal junction. Multiple factors may be contributing to the increased incidence of this specific cancer, such as eradication of *Helicobacter pylori* infestation of the stomach, increased acid production by the stomach, and exacerbation of gastro-esophageal reflux disease (GERD) with formation of Barrett's esophagus *(9,10)*. Corrosive medications can lead to stricture formation, a process which is especially common in the older adult *(5)*. Dysphagia aortica is a rare condition caused by compression of the esophagus by an aortic aneurism or heavily calcified aorta *(5)*.

For structural abnormalities, aggressive dilation alleviates the need for diet restriction. Diet therapy may be useful for the patient with a difficult stricture that requires frequent dilation or for the patient in whom there are short intervals between dilation sessions. Specific motility disorders, such as achalasia, respond to treatment with an oversized pneumatic balloon dilation. Botox injection with *Clostridium botulinum* toxin is more effective in older patients than youth, presumably because the muscle of the lower esophageal sphincter (LES) may be thicker and stronger in the younger patients. Certainly aggressive treatment of GERD is indicated with high levels of acid suppression being required. Dietary factors which should be avoided because they exacerbate GERD would include acidic foods (such as orange juice, tomato juice, or apple juice) or any foods which lower pressure in the LES (such as alcohol, chocolate, peppermint, or nicotine). Obviously, any corrosive medications should be avoided when possible (such as doxycycline, potassium chloride, tetracycline, Fosamax, quinidine, theophyline, or vitamin C). Medications such as ibuprofen or non-steroidal anti-inflammatory drugs (NSAIDs) should be taken in liquid form or by ingesting pill form with sufficient volume of water to encourage passage through the esophagus into the stomach.

7.3 GASTROESOPHAGEAL REFLUX DISEASE (GERD) AND HEARTBURN

While GERD occurs throughout the lifespan, there does appear to be an increased incidence of this disorder with aging *(2,3)*. Overall, there is a 20% prevalence of GERD in older adults, but numbers may range from one geographic area to the next. In Finland, reports indicated that 54% of men and 66% of women over the age of 65 years had GERD symptoms on at least a monthly basis *(9,11)*. In the United States, 44% of the population over the age of 65 years had at least monthly symptoms *(9)*. With an increasing incidence of GERD seen with advancing age, there is expectedly an increased occurrence of associated complications (such as severe erosive esophagitis, stricture formation, or Barrett's esophagus). The peak incidence of Barrett's formation is in the seventh decade *(2,3)*.

A number of contributing factors may compromise defense systems that would normally protect against gastroesophageal reflux *(2,3)*. Esophageal motility disorders, delayed gastric emptying, decreased salivary flow, presence of a hiatal hernia, spending more time in the recumbent position, reduced LES pressure from medication or concomitant disease, and ingestion of a greater number of potentially corrosive medications, all may contribute to an increasing incidence of GERD and compromise of normal defense systems *(2,3)*.

Often making the diagnosis of GERD can be difficult in the geriatric population because of decreased visceral pain sensation, underreporting by the patients because of reduced esophageal sensitivity, or tolerance of GERD symptoms in the face of other comorbid conditions *(5,9)*. Physicians may be paying more attention to the other comorbidities and paying less attention to complications from GERD. Also, the older patient may present with atypical symptoms, such as regurgitation, vomiting, and chest pain, more often than classic heartburn. Confusion with other comorbid disease processes such as COPD or coronary artery disease may lead to delays in making the diagnosis. Certainly, the presence of GERD may exacerbate the preexisting comorbidities. Reflux may precipitate asthma, chronic pancreatis, or chronic obstructive pulmonary disease. Chest pain or heartburn occurring as a result of reflux may lead to precipitation of angina from coronary artery disease *(5,9)*.

Pressure in the LES may be diminished by any number of drugs, such as calcium channel blockers (Norvasc, Procardia, and Cardizem), alpha antagonists (Cardura or Flomax), or beta antagonists with alpha blocking properties (Coreg) *(9)*. Theophyline, dopamine, and nitrates similarly may reduce LES pressure. Delayed esophageal clearance occurs in older age groups *(9,12)*, and may be related to the minor esophageal manometric changes seen with aging. Similarly, there may be age-related decreases in UES pressure, and factors such as a recent stroke or neuromuscular disease may lead to further compromises in tone of this particular sphincter. The presence of a hiatal hernia may be a bigger factor in precipitating GERD for older than for younger patients. The incidence of hiatal hernia increases with age *(9,13)*. The incidence between ages 14 and 24 is 10%, while that between 55 and 64 years of age increases to over 64% *(9,14)*. The size or length of the hiatal hernia also correlates with increased development of Barrett's disease *(9,15)*.

Xerostomia caused by reduced salivary flow and a dry mouth may contribute to GERD. While there is no specific increase in incidence of this disorder because of aging alone, many drugs act as xenogenic agents with dry mouth as a side effect. Such medicines include anti-depressants, anti-hypertensives, diuretics, anticholinergic agents, sedatives, and antihistamines *(9)*.

The management of GERD in the geriatric population is similar to the management and treatment used for younger patients *(9)*. Lifestyle changes include cessation of smoking, reduction of fat in the diet, decreasing meal size, and avoiding agents which are corrosive (such as citric acid) or which decrease LES pressure (such as chocolate and peppermint). Many of these strategies may be more difficult to implement in the geriatric population *(9)*. It may be difficult to avoid medications which decrease LES pressure, for example, or meds which are particularly corrosive to the esophageal mucosa. Patients should always be advised to swallow pills in the upright position with a sufficient bolus volume of water. Symptom relief from GERD does not guarantee relief from acid exposure and injury. Adequate acid suppression with proton pump inhibitor therapy may require divided doses or "priming the pump" by giving the drug postprandially to maximize binding to the hydrogen ion pump at a time of active secretion. Use of proton pump inhibitors in geriatrics may generate increased concern for side effects, such as lethargy, disorientation, agitation, or confusion. Use of histamine-2 blockers may inhibit the P450 system of the liver and alter the metabolism of drugs such as theophyline, warfarin, or dilantin *(9)*.

7.4 GASTRITIS AND PEPTIC ULCER DISEASE

Chronic gastritis is common in older adults, with an overall prevalence of 11–50% *(2,16)*. Gastritis may be divided by pathophysiologic mechanism into type A and type B gastritis. Type A is less common of the two, occurring in less than 5% of patients over the age of 60. Type A gastritis is an autosomal dominant condition involving the proximal stomach, which is painless and symptomless. It is associated with parietal cell antibodies, achlorhydria, and pernicious anemia. Patients with type A gastritis also tend to have antibodies to intrinsic factor and to thyroid tissue. Type B gastritis tends to be symptomatic (dyspepsia and abdominal pain) and is associated with *H. pylori* infestation. This type of gastritis is more common in the geriatric population, occurring in 24% of patients over the age of 60 and 37% over the age of 80 *(2,16)*. Type B gastritis involves the more distal stomach (body and antrum). Susceptibility to peptic ulcer disease (PUD) increases with age, and as a result both the incidence and prevalence are increased in the geriatric population *(2,17)*. Factors which promote development of PUD in the older population include decreased mucosal defenses, increased prevalence of *H. pylori*, and increased use of NSAIDs. The morbidity from PUD increases with age and is more likely to require hospital admission. Mortality from PUD may be greater as well compared to younger age groups *(2,17)*.

Management depends to some degree on the type of gastritis and the underlying mechanisms which generate the disease process. For type A, the clinical endpoint evolves as significant malabsorption of vitamin B_{12}. As the gastritis progresses, there is a point where the degree of reduced secretion of intrinsic factor prevents absorption of B_{12} from the ileum. Additionally, the presence of achlorhydria reduces absorption of protein-bound vitamin B_{12} from the diet. Because type B gastritis is localized to the distal stomach, levels of intrinsic factor are high enough that B_{12} deficiency is much less frequent *(2,16)*. Achlorhydria however has many consequences. Bacterial overgrowth due to reduced acid may cause increased bacterial binding of ingested dietary B_{12}. Achlorhydria slows gastric emptying for solids and may actually increase emptying for liquids. Achlorhydria interferes with the pH-dependent absorption of ferric iron, calcium, and folate. Paradoxically, bacterial overgrowth may actually produce folate in the gut. Increased homocystine levels in the blood may be due either to a folate or B_{12} deficiency, and may increase risk for arteriosclerotic cardiovascular disease.

Treatment of the B_{12} deficiency seen in type A gastritis may be more difficult and require larger doses of B_{12} than type B gastritis, particularly if the goal is to reduce homocystine levels *(2,16)*. Parenteral B_{12} injections in the range of 100–1000 mcg may be required intramuscularly on a monthly basis. Even with pernicious anemia, most patients with B_{12} deficiency can be treated with oral B_{12}. Likewise, for both type A and type B gastritis, oral supplements with unbound free B_{12} may be sufficient to treat the deficiency *(2,16)*. In PUD, dietary changes are rarely required *(2)*. The traditional "bland" diet used in the past has been shown to have no effect on symptom relief, healing, or recurrence of PUD. Food intolerances may necessitate avoidance or restriction of certain foods which precipitate symptoms. However, only upon re-challenge with recurrence of symptoms should these foods be eliminated from the diet.

7.5 DIARRHEA

Chronic diarrhea lasting >4–6 weeks occurs in 3–5% of the geriatric population *(18,19)*. Diarrhea is defined medically by the passage of >250 g of stool by weight per day. Among older adults, the complaint of diarrhea may vary from low volume incontinence, to any loose stool, to fecal impaction with passage of liquid stool underneath. Physiologic factors which contribute to diarrhea in older adults include achlorhydria, atrophic gastritis, immune dysfunction, and reductions in release of secretory IgA. Comorbid conditions such as bacterial overgrowth, chemotherapy, or ischemia may contribute risk. Environmental factors such as residence in a long-term care facility, community epidemics of infectious diarrhea, and adverse side effects from polypharmacy increase the likelihood for diarrhea *(18)*. Older adults may experience greater morbidity from diarrhea, due to dehydration, electrolyte abnormalities, and organ hypoperfusion *(18)*. Approximately 75% of deaths related to diarrhea occur in adults over the age of 55 *(18,20)*. Risk factors specifically linked to death from diarrhea include increasing age, caucasian race, female gender, and residence in a long-term care facility *(18)*.

7.5.1 Infectious Gastroenteritis

Infectious gastroenteritis may be the most common cause of diarrhea in the geriatric population. The more frequent etiologic organisms include *Campylobacter*, *Giardia*, *Rotavirus*, Norwalk virus, and *Salmonella*. In other parts of the world, an identified bacteria accounts for 16.3% of diagnosed infectious diarrheas, a virus 15.4%, and a parasite 8.3%, respectively *(18,21)*. In the United States, viral gastroenteritis is probably the most common cause of infectious diarrhea. Patients over the age of 65 years have a 2.6–3.4 times greater likelihood of requiring hospitalization for diarrhea *(18)*.

7.5.2 Drug-Induced Diarrhea

Of particular concern in the geriatric population is drug-induced diarrhea. Diarrhea accounts for 7% of all drug side effects (18,28). Older adults in particular are at risk because of polypharmacy, comorbid conditions, and their increased vulnerability to side effects (18). The mechanism of drug-induced diarrhea may be multifactorial, including such factors as an osmotic effect (sorbitol), altered motility (erythromycin or metoclopramide), a secretory effect (such as stimulant laxatives), an inflammatory exudative process (NSAIDs), or simply interference with absorption of macronutrients. The diagnosis of a drug-induced diarrhea is suggested by the timing of onset of symptoms with initiation of the drug, and is confirmed by withdrawal of the drug and re-challenge (with subsequent recurrence of symptoms).

The management of diarrhea in older adults involves a careful workup and evaluation. Stool cultures, endoscopy with appropriate biopsy, and review of all medications should be the first steps in the management strategy. Disorders which precipitate nutrient deficiencies and malabsorption may require vitamin supplements, oral or tube feeding with formula, and a combination of water-soluble and/or fat-soluble vitamins. Individual therapy should be directed toward a specific diagnosis. Antibiotics should be used as appropriate for infectious diarrhea, *Lactobacillus* for *lactose intolerance*, or avoidance of dairy products, 5-aminosalicylic acid therapy or corticosteroids for IBD, and anti-peristaltic agents such as Lomotil (diphenoxylate) or Imodium (loperamide) for symptomatic relief.

7.6 OTHER GI DISORDERS

7.6.1 Fecal Impaction

Fecal impaction, which involves blockage of the colon with a hard dry stool, can lead to spurius diarrhea or seepage around the impaction. This issue may be the most common cause of diarrhea, specifically in long-term nursing home facilities (18,22). Predisposing factors which occur in older adults that lead to fecal impaction include decreased colonic motility, pelvic floor abnormalities, inactivity, low fiber in the diet, medications, dehydration, depression, and other comorbidities (such as Parkinson's disease, dementia, or diabetes). Fecal impaction is usually determined by physical exam, as 70% occur in the rectum within reach on digital exam (18). The treatment includes manual disimpaction, enemas, and glycerin suppositories.

7.6.2 Ischemic Colitis

Ischemic colitis is associated with older age groups, as over 90% of cases occur in the geriatric patient population (18,23). The most common locations for ischemic colitis to occur are "the watershed" areas of the colon where there is poor collateral flow, such as the rectosigmoid junction and the splenic flexure. Factors which may predispose to ischemic colitis include atherosclerosis, hypotension, pelvic irradiation, congestive heart failure, aortoiliac surgery, infection, low-flow states, and medications (such as NSAIDs, progesterone, or digitalis) (18). Patients may present with bloody diarrhea, with or without abdominal pain. The specific treatment for ischemic colitis involves bowel rest, broad spectrum antibiotics, and restoration of hemodynamic stability.

7.6.3 Inflammatory Bowel Disease

Inflammatory bowel disease (IBD) is common in the geriatric population as evidenced by the bimodal peak of incidence. Inflammatory bowel disease most commonly occurs in the second to fourth decade, but has a second lesser peak incidence in the sixth to eighth decade (18,24). Approximately 12% of patients with *ulcerative colitis* will have onset of disease over the age of 60, while upwards of 16% of Crohn's disease patients will have onset of the disease over this age (18,25). One out of eight cases of IBD will be diagnosed in the geriatric population (18). Compared to their younger counterparts, older patients are more likely to have colonic involvement, less likely to have a family history of IBD, and are more likely to respond to therapy. However, older patients will experience higher mortality due to severity of disease and the existence of comorbidities (18,26). Similar to younger age groups, geriatric patients with Crohn's disease are more likely to have nutritional sequelae than those patients with ulcerative colitis. Risk factors which contribute to nutritional sequelae include extent of the bowel involved, length of bowel previously resected at surgery, and the type of medical therapy required to achieve remission.

7.6.4 Microscopic Colitis

Microscopic colitis is a syndrome associated with diarrhea in which the endoscopic or radiographic evaluation of the colon is normal, but there is inflammation of the colonic mucosa seen microscopically on biopsy. Microscopic colitis is more common in females, with a mean age on presentation in the range of 55–65 years (18,27). Microscopic colitis may be precipitated by the use of NSAIDs (18,27). Treatment is similar to that for routine IBD. Five-aminosalicylic acid agents, corticosteroids, and immunosuppressants should be used appropriately.

7.6.5 Small Bowel Bacterial Overgrowth

Small bowel bacterial overgrowth (SBBO) or blind-loop syndrome may be increased in the geriatric population. Symptoms of SBBO range from bloating, diarrhea, steatorrhea, and weight loss to specific nutritional deficiencies. Risk factors which predispose the older patient to SBBO include achlorhydria or hypochlorhydria from chronic gastritis, motility abnormalities from diabetes and scleroderma, and disruption of intestinal continuity because of formation of blind-loops postoperatively (2). These factors usually lead to over-colonization of the small bowel with coliforms and anaerobes. Infestation of these organisms in turn leads to abnormalities associated with binding of vitamins, de-conjugation of bile salts, and fat malabsorption with subsequent diarrhea/steatorrhea.

7.6.6 Lactose Intolerance

Lactose intolerance may be increased in older adults, as lactase levels decrease with age *(2,16)*. Across all ethnic populations, the incidence of clinical lactase deficiency can be up to 75%. Overall, 25% of adults in the United States demonstrate lactase deficiency *(2,16)*. Symptoms are variable and include bloating, gas, and diarrhea.

7.6.7 Incontinence

Incontinence is defined by the *involuntary passage* of liquid or solid stool. Older patients in particular often misinterpret these symptoms. They assume that they are experiencing diarrhea because they cannot make it to the bathroom in time and end up soiling their clothes. A careful history may sometimes show, however, that the complaint really involves accidents where the patient simply loses continence and passes small amounts of solid or semi-liquid stool. These symptoms may be truly interspersed between otherwise normal stools. The prevalence of incontinence ranges widely in the nursing home population from 2.2% up to as high as 50% *(29)*. Across a general population, the overall incidence of incontinence is 4–5%, but in the geriatric population the incidence may be nearly twice that rate (up to 11–12%) *(29)*. Clearly the prevalence of incontinence increases with age *(29)*. In the past, incontinence has affected women more than men, but these differences may not be as great as previously thought *(29)*. Incontinence has been known to result in perianal dermatitis, pressure sores around the perineum, and urologic infections. For the individual patient, these symptoms may result in social isolation, anxiety, depression, or loss of self-esteem. A number of predisposing factors have been identified in the literature *(29)*. Age-related changes which predispose to incontinence include pelvic floor descent, stretch-induced pudendal nerve damage, increase in the anorectal angle, decreased rectal sensitivity, decreased anal squeeze pressure, reduced rectal reserve volume, and a lower threshold for internal anal sphincter relaxation. Comorbid conditions which predispose to incontinence include stroke, dementia, immobility, constipation, previous hemorrhoid surgery, or injury to the perineum (such as an episiotomy at the time of childbirth at a younger age) *(29)*.

Evaluation of the patient complaining of incontinence should include a thorough physical exam of the anus and perineum with a careful digital exam. Endoscopy may be required to rule out proctitis. Anorectal manometry may be needed to evaluate the integrity of the sphincters and whether or not there is appropriate relaxation of the internal anal sphincter with distention (indicating that the rectoanal inhibitory reflex is intact). Endoscopic ultrasound is a newer modality which affords the opportunity to see whether there is a structural abnormality of the anal sphincters. A defecating proctogram may be required to determine whether there is an increase in the anorectal angle.

Management of incontinence includes anti-peristaltic agents such as Lomotil or Imodium. Imodium may be the superior of the two drugs, as it helps to tighten anal sphincters in addition to slowing peristalsis or motility *(29)*. Lomotil only slows transit without affecting sphincter tone. Fiber supplementation may be important to decrease the liquidity of stools. Foods which promote diarrhea should be avoided, such as caffeine, alcohol, fruit juices, beans, or broccoli. Biofeedback training may help strengthen and improve control of the external anal sphincter. In severe cases, surgery may be required to repair disruption of the anal sphincters.

7.6.8 Constipation

Constipation may be the most common complaint in the geriatric population. While some reports indicate that the incidence of constipation is 4–8 times more common in the older adult population than in younger controls, the incidence may not be as high in healthy ambulatory geriatric citizens (2,17). In western countries, the incidence of constipation ranges between 25 and 50%, but may range as high as 80% in nursing home residents (30,31). Constipation is more common in women than men. Symptoms associated with constipation range from decreased stool frequency and difficult passage of hard stool, to bloating, cramping, sensation of inadequate evacuation, and painful defecation (30).

There are a number of physiologic age-related changes which may predispose to constipation (2,32). Aging is associated with decreased colonic wall elasticity, impaired rectal sensation, and reduced colonic propulsion (30). There appears to be little change in colonic transit time with age however. In the majority of cases, constipation is idiopathic with no obvious underlying cause (30). However, a number of specific predisposing conditions known to occur in older adults have been identified (2,32). Functional causes for constipation include confusion, depression, immobility, and pain from anorectal fissures. Drugs such as iron, opioid narcotics, calcium antagonists, anti-depressants, and diuretics may predispose to constipation, as does longstanding laxative abuse. Endocrine etiologies which promote constipation include hypothyroidism, hypercalcemia, diabetes, and hyperparathyroidism. Diets low in fiber, poor fluid intake, and excessive caffeine intake (causing a diuretic effect) all predispose to constipation (2,32). Mechanical issues such as diverticulosis and cancer, or gut ischemia and IBD (due to stricture formation) may actually promote constipation. Clinical conditions leading to predendal nerve damage (such as diabetes, child birth, or chronic straining) may also precipitate constipation (2,32).

The evaluation and management of constipation again starts with a thorough history, paying attention to diet, activities, and bowel habits. It is important to evaluate comorbidities and to identify any drugs which may contribute to constipation. The clinician should question the patient on what treatment modalities have already been attempted and ascertain what degree of compliance was maintained. The physical exam should include a digital rectal exam and visual inspection of the anus and perineum. Laboratory tests should focus on electrolyte abnormalities, iron stores, calcium metabolism, guaiac testing of the stool, glucose homeostasis, and parathyroid hormone levels. Colonoscopy may be required to rule out structural abnormalities. Treatment should include fiber supplement, increased water intake, increased activity, and avoidance of caffeine or other diuretic agents. Bowel training, where the patient sits on the commode at the same time every day, helps to promote better contractility and increases the chance of having a bowel movement. Reviewing and revising medications may be needed as well.

7.7 GASTROINTESTINAL BLEEDING

The incidence of GI bleeding increases with age, and geriatric patients are at increased risk for morbidity and mortality arising from a GI bleed compared to younger controls *(33)*. The risk for increased morbidity and mortality comes from the increase in comorbid illnesses (coronary artery disease, chronic obstructive pulmonary disease, and renal insufficiency) as well as the increased use of ulcero-genic medications (such as aspirin or NSAIDs). Patients are 5 times more likely to experience an upper GI than a lower GI bleed. Upper GI bleeding is more common in men than women *(33)*. Of all patients who present with an upper GI bleed, 35–45% are over the age of 60 years *(33–35)*. The mortality rate associated with any GI bleed is increased in the face of organ failure (liver, heart, kidney, or lung) and ranges between 25 and 65% *(33,34)*.

Of those patients who require hospital admission for an upper GI bleed, 90% are due to PUD, mucosal erosions, or portal hypertension and varices. PUD is more common in older adults than in the young, while Mallory-Weise tears are less common. Varices are less common among older adults, due to the fact that alcohol abuse may be less common *(33,36)*. Approximately 50% of patients who present with upper GI bleeding give a history of ingestion of aspirin or NSAIDs *(33,37)*. Cyclooxygenase-2 inhibitors have helped to decrease the incidence of PUD and bleeding in the geriatric population.

Symptoms of upper GI bleeding include melena, hematemesis, hematochezia, and abdominal pain. Risk from the upper GI bleeding is increased in the absence of pain due to decreased visceral sensitivity or masking of the pain symptoms by NSAIDs *(33,38)*. The increased risk occurs because of delays in the diagnosis and the increased chance for complications to develop. Risk is further increased by advancing age, concomitant anticoagulation therapy, and the presence of comorbid disorders.

Lower GI bleeding in general is less common than upper GI bleeding. Like its counterpart, the incidence of lower GI bleeding does increase with age (there is a 200-fold increase from the third to the ninth decade of life) *(33,39)*, and occurrence is more likely in men than women. The reason for the increase in lower GI bleeding with age is due to the increased incidence of diverticular disease, colonic neoplasms, angiodysplasias, and ischemic colitis (CDT). These four diagnoses make up approximately 60% of all cases of lower GI bleeding *(33,40,41)*. Of note is the fact that from 10 to 15% of cases which appear to present as a lower GI bleed in fact are found to have an upper GI source *(33,40)*.

Evaluation of GI bleeding in the geriatric population may be difficult because the history may be complicated by poor cognition. On physical exam, the presence of beta blockers may block the rise in pulse with orthostasis. Key issues in the evaluation of the patient with lower GI bleeding focus on whether or not the patient is taking NSAIDs which may mask pain, placement of an NG tube to rule out an upper GI bleed, and the fact that a certain percentage (at least 5%) will have a negative upper and lower endoscopy despite overt symptoms or occult detection by guaiac testing *(33)*.

The management of the GI bleeder, whether it is found ultimately to be an upper or lower tract source, involves adequate volume resuscitation and appropriate transfusion of blood products. Particularly for upper GI bleeders, starting an intravenous infusion of octreotide to reduce splanchnic pressure (in case the etiology turns out to be varices) and continuous intravenous infusion of a proton pump inhibitor to reduce acid output (in case the etiology is related to gastritis, erosions, or ulcerations) are appropriate for initial management. Obviously, endoscopy is needed to determine the source of the bleeding.

Dietary management of the GI bleeder depends on the results of the endoscopic evaluation. In PUD, patients found to have ulcers which have a clean base or a flat-pigmented spot may be fed immediately and can be discharged from the hospital fairly quickly. If instead, endoscopy finds that the ulcer has a visible vessel, clot, ooze, or an arterial spurter, then electrocoagulation therapy is applied. For these latter patients, it is important to hold further feeds for at least 48 h to allow stabilization of the blood clot following therapy. Patients with Mallory-Weise tears rarely require endoscopic therapy and can usually be fed immediately. For patients bleeding from esophageal varices, particularly if they undergo esophageal band ligation, it is important to wait 48 h to feed (because feeding may increase blood flow and subsequent splanchnic pressure, which can precipitate rebleeding). For patients with lower GI bleeds, the most important aspect is to rule out an upper GI bleed. Appropriate endoscopic procedures should include both upper and lower tract evaluation.

7.8 HEPATITIS

Consensus opinion in the past was that there were no specific age-related disease processes involving the liver (2). However, a number of important clinical trends have recently emerged in the literature to suggest the opposite. It is anticipated that there will be an increase in the incidence of viral hepatitis C over the next few decades. The prevalence of anti-hepatitis C antibodies is higher in the geriatric population than younger controls (2,3), a disease process which predisposes to an increased incidence of hepatocellular carcinoma later in life. Primary biliary cirrhosis is more common in older adults, with a mean onset at age 60 (2,42). Alcoholic liver disease may be more common in the geriatric population, as a result of accumulated years of alcohol abuse (although alcohol abuse per se is not more common). Increasing age correlates with increased severity of symptoms and increased mortality from alcoholic liver disease. Drug-induced liver disease is a common occurrence in the geriatric population due to the likelihood for polypharmacy. In fact, over the age of 50 years, 40% of cases of liver injury are found to be related to drugs (43,44). In the geriatric population, each patient is usually on an average of seven drugs at any given time (both prescription and over the counter) (43). Factors which predispose older adults to drug-induced liver injury include an increasing number of drugs prescribed, presence of malnutrition, poor compliance and mixing of drugs, comorbidities, and age-related changes in pharmacokinetics of the drug (43). Decreased clearance resulting in an increased half-life for medications may result from age-related decreases in liver blood flow, liver volume, and microsomal enzyme activity *(43)*. The most common potentially hepatotoxic drugs used in the geriatric population include NSAIDs, aspirin, and acetaminophen, tacrine for Alzheimer's dementia, dilantin and valproic acid for seizures, antibiotics (amoxicillin, augmentin, bactrim, and isoniazid), and oral hypoglycemic agents *(43)*.

Older patients presenting with abnormal liver enzymes should undergo a full investigation similar to any other age group. Initially it is important to rule out underlying liver disease such as hepatitis C or primary biliary cirrhosis. A pattern of enzyme elevation of AST>ALT may alert the clinician to the presence of alcohol abuse. Evaluation of abnormal liver enzymes should include iron and copper studies, viral hepatitis markers, and markers for autoimmune disease. The evaluation may ultimately require liver biopsy. The clinician should scrutinize both prescription and over-the-counter medications. The timing of new medications with the increase in liver enzymes may help identify the particular agent. Management may require simplifying or eliminating certain medications or substituting other drugs within a similar class.

7.9 ANEMIA

Anemia has been shown to increase in incidence with advancing age *(45)*. Anemia affects males more than females, and is more common in lower social economic groups and under-developed countries *(45)*. The most common micronutrient deficiencies which lead to anemia include iron, B_{12}, folate, and copper. Of these, iron is clearly the most common deficiency, occurring in anywhere from 60 to 80% of certain populations (with or without development of anemia) *(45)*.

There are a number of age-related factors which may contribute to micronutrient deficiencies. Atrophic gastritis from *H. pylori* infestation causes achlorhydria and reduced production of intrinsic factor, which in turn leads to impaired absorption of folate and food-bound vitamin B_{12}. Reduced acid output leads to decreased conversion of ferric to ferrous iron, which reduces overall iron absorption *(16,45)*. Increased incidence of SBBO occurs due to decreased contractility and reduced acid production, which in turn adversely effects absorption of fat-soluble vitamins and vitamin B_{12}. Deficiencies of folate are less frequent because of production of folate by bacteria within the lumen of the gut. Medications in older adults may interfere with availability of micronutrients *(45)*. Agents such as cholestyramine may directly bind folate. Other drugs such as alcohol and metformin may competitively inhibit the absorption of folate and B_{12}. Proton pump inhibitors (by altering pH) may interfere with absorption of iron, folate, and B_{12}. Increased blood loss from the GI tract resulting in iron deficiency may occur as a result of NSAIDs, colonic neoplasms, or use of anticoagulant therapy.

Evaluation and management of anemia in the geriatric population involves checking iron stores (ferritin, serum iron, iron saturation, and total iron-binding capacity). Of all the markers for iron, ferritin is the best marker for overall iron stores. It is important to check copper, zinc, B_{12}, and folate levels. A 24-h urine for homocystine and methylmalonic acid may be more sensitive and specific for B_{12} deficiency than serum B_{12} levels. If there is evidence of B_{12} deficiency, it is important to rule out atrophic gastritis, Crohn's disease of the terminal ileum, and SBBO. With evidence of iron-deficiency anemia, a full upper and lower endoscopic evaluation should be done. If endoscopy is negative, antibodies for celiac sprue (a cause for reduced small bowel absorption of iron) and an endocapsule evaluation (to rule out a small bowel source for blood loss) should be performed. It is important to reassess any medications which would interfere with absorption or availability of micronutrients.

7.10 RECOMMENDATIONS

A number of predictable GI complaints and problems may be anticipated in the geriatric population. Rarely, age-related changes in GI function contribute to the malady. More often, associated comorbidities, the prevalence of polypharmacy, and underlying disease processes generate the GI complaints. Atypical presentations, misinterpretation of the presenting signs and symptoms, and failure to recognize predisposing age-related factors contribute to delays in making the diagnosis. Unfortunately diminished reserve, concomitant disorders, alterations in hepatic and GI tract function may worsen severity of the condition and increase associated morbidity. Any new GI complaint warrants a full investigation and aggressive management to minimize the clinical sequelae.

Acknowledgement An excellent series of articles on geriatric gastroenterology, edited by T.S. Dharmarajan, MD and C.S. Pitchumoni, MD, was published in the journal *Practical Gastroenterology* in 2001–2006. These articles were a valuable source of information for this chapter, and the reader is advised to refer to them if further information is needed.

REFERENCES

1. Dharmarajan TS, Pitchumoni CS, Kokkat, AJ: The aging gut. Pract Gastroenterol 2001 Jan; 25:15–27.
2. Dryden GW, McClave SA: Gastrointestinal senescence and digestive diseases of the elderly, in Bales CW and Ritchie CS (eds): Handbook of Clinical Nutrition and Aging. Humana Press, 2004, 569–81.
3. Tack J, Vantrappen G: The aging oesophagus. Gut 1997 Oct; 41(4):422–4.
4. Ergun GA, Miskovitz PF: Aging and the esophagus: common pathologic conditions and their effect upon swallowing in the geriatric population. Dysphagia. 1992;7(2):58–63.
5. Jayadevan R, Pitchumoni CS, Dharmarajan TS: Dysphagia in the elderly. Geriatric Gastroenterology, Series #4, Practical Gastroenterology 2001, 75–88.
6. Gorman RC, Morris JB, Kaiser LR: Esophageal disease in the elderly patient. Surg Clin North Am 1994 Feb; 74(1):93–112.
7. O'Neil KH, Purdy M, Falk J, Gallo L: The Dysphagia Outcome and Severity Scale. Dysphagia 1999 Summer; 14(3):139–45.
8. Sallout H, Mayoral W, Benjamin SB: The aging esophagus. Clin Geriatr Med 1999 Aug; 15(3):439–56.
9. Pitchumoni CS, Lin M, Dharmarajan TS: Gastroesophageal reflux disease in the elderly. Geriatric Gastroenterology, Series #7, Practical Gastroenterology 2002. 13–29.
10. Fallone CA, Barkun AN, Friedman G, Mayrand S, Loo V, Beech R, Best L, Joseph L: Is *Helicobacter pylori* eradication associated with gastroesophageal reflux disease? Am J Gastroenterol 2000 Apr; 95(4):914–20.
11. Räihä IJ, Impivaara O, Seppälä M, Sourander LB: Prevalence and characteristics of symptomatic gastroesophageal reflux disease in the elderly. J Am Geriatr Soc 1992 Dec; 40(12):1209–11.

12. Castell DO: Esophageal disorders in the elderly. Gastroenterol Clin North Am 1990 Jun; 19(2):235–54.
13. Stilson WL, Sanders I, Gardiner GA, Gorman HC, Lodge DF: Hiatal hernia and gastroesophageal reflux. A clinicoradiological analysis of more than 1,000 cases. Radiology 1969 Dec; 93(6):1323–37.
14. Flora-Filho R, Zilberstein B: The importance of age as determining factor in hiatus hernia and gastroesophageal reflux. Cross-sectional study. Arq Gastroenterol 1999 Jan–Mar; 36(1):10–7.
15. Cameron AJ: Barrett's esophagus: Prevalence and size of hiatal hernia. Amer J Gastroent 1999; 94:2054–9.
16. Saltzman JR, Russell RM: The aging gut. Nutritional issues. Gastroenterol Clin North Am 1998 Jun; 27(2):309–24.
17. Holt PR: Are gastrointestinal disorders in the elderly important? J Clin Gastroenterol 1993 Apr; 16(3):186–8.
18. Adiga GU, Dharmarajan TS and Pitchumoni CS: Diarrhea in older adults. Geriatric Gastroenterology, Series #14, Practical Gastroenterology 2005, 63–82.
19. American Gastroenterological Association medical position statement: guidelines for the evaluation and management of chronic diarrhea. Gastroenterology 1999 Jun; 116(6):1461–3.
20. Lew JF, Glass RI, Gangarosa RE, Cohen IP, Bern C, Moe CL: Diarrheal deaths in the United States, 1979 through 1987. A special problem for the elderly. JAMA 1991 Jun 26; 265(24):3280–4.
21. de Wit MA, Koopmans MP, Kortbeek LM, van Leeuwen NJ, Bartelds AI, van Duynhoven YT: Gastroenteritis in sentinel general practices,The Netherlands. Emerg Infect Dis 2001 Jan–Feb; 7(1):82–91.
22. Kinnunen O, Jauhonen P, Salokannel J, Kivelä SL: Diarrhea and fecal impaction in elderly long-stay patients. Z Gerontol 1989 Nov–Dec; 22(6):321–3.
23. Powell DW: Approach to the patient with diarrhea. In: *Text Book of Gastroenterology*. Vol I. Eds. Yamada, Alpers, Lorine, Owyang and Powell. 3rd Ed. Lippincott Williams and Wilkins, Philadelphia 1999; 858–909.
24. Harper PC, McAuliffe TL, Beeken WL: Crohn's disease in the elderly. A statistical comparison with younger patients matched for sex and duration of disease. Arch Intern Med 1986 Apr; 146(4):753–5.
25. Softley A, Myren J, Clamp SE, Bouchier IA, Watkinson G, de Dombal FT: Inflammatory bowel disease in the elderly patient. Scand J Gastroenterol Suppl 1988; 144:27–30.
26. Camilleri M: Non malignant gastrointestinal syndromes in the elderly: Part II Midgut and hindgut diseases. Clin Geriatrics 2004. 12:24–33.
27. Loftus EV: Microscopic colitis: epidemiology and treatment. Am J Gastroenterol 2003; 98(12 Suppl):S31–6.
28. Chassany O, Michaux A, Bergmann JF: Drug-induced diarrhoea. Drug Saf 2000; 22(1):53–72.
29. Houghton, JP, Patankar, S: Fecal incontinence: a common, but under recognized problem in the elderly. Geriatric Gastroenterology, Series #11, Practical Gastroenterology 2004, 70–84.
30. Dharmarajan TS, Rao VSR, Pitchumoni CS: Constipation in older adults: An ancient malady, remains a management enigma. Geriatric Gastroenterology, Series #13, Practical Gastroenterology 2004. 40–64.
31. Locke GR 3rd, Pemberton JH, Phillips SF: AGA technical review on constipation. American Gastroenterological Association. Gastroenterology 2000 Dec; 119(6):1766–78.
32. Balson R, Gibson PR: Lower gastrointestinal tract. Med J Aust 1995 Feb 6; 162(3):155–7.
33. Trivedi, C.T., Pitchumoni, C.S: Gastrointestinal Bleeding in Older Adults. Pract Gastroenterol 2006. 30(3): 15–42.
34. Silverstein FE, Gilbert DA, Tedesco FJ, Buenger NK, Persing J: The national ASGE survey on upper gastrointestinal bleeding. Parts I–III. Study design and baseline data. Gastrointest Endosc 1981 May; 27(2):73–9.
35. Cooper BT, Weston CF, Neumann CS: Acute upper gastrointestinal haemorrhage in patients aged 80 years or more. Q J Med 1988 Oct; 68(258):765–74.
36. Antler AS, Pitchumoni CS, Thomas E, Orangio G, Scanlan BC: Gastrointestinal bleeding in the elderly. Morbidity, mortality and cause. Am J Surg 1981 Aug; 142(2):271–3.

37. Thomopoulos KC, Vagenas KA, Vagianos CE, Margaritis VG, Blikas AP, Katsakoulis EC, Nikolopoulou VN: Changes in aetiology and clinical outcome of acute upper gastrointestinal bleeding during the last 15 years. Eur J Gastroenterol Hepatol 2004 Feb; 16(2):177–82.
38. Hilton D, Iman N, Burke GJ, Moore A, O'Mara G, Signorini D, Lyons D, Banerjee AK, Clinch D: Absence of abdominal pain in older persons with endoscopic ulcers: a prospective study. Am J Gastroenterol 2001 Feb; 96(2):380–4.
39. Longstreth GF: Epidemiology and outcome of patients hospitalized with acute lower gastrointestinal hemorrhage: a population-based study. Am J Gastroenterol 1997 Mar; 92(3):419–24.
40. Vernava AM 3rd, Moore BA, Longo WE, Johnson FE: Lower gastrointestinal bleeding. Dis Colon Rectum 1997 Jul; 40(7):846–58.
41. Jensen DM, Machicado GA: Colonoscopy for diagnosis and treatment of severe lower gastrointestinal bleeding. Routine outcomes and cost analysis. Gastrointest Endosc Clin N Am 1997 Jul; 7(3):477–98.
42. James OF: Parenchymal liver disease in the elderly. Gut 1997 Oct; 41(4):430–2
43. Dharmarajan TS, Pitchumoni CS, Kumar KS: Drug-induced liver disease in older adults. Pract Gastroenterol 2001 Mar; 25:43–60.
44. Benhamou JP: Drug-induced hepatitis: Clinical aspects. In Fillastre JP (Ed): Hepatotoxicity of drugs. *Rouen*, University de Rouen, 1986, 23–30.
45. Dharmarajan, TS, Bullecer, MLF, Pitchumoni, CS: Anemia of gastrointestinal origin in the elderly. Pract Gastroenterol 2002; 26(9): 22–24, 29–31, 35–36.

8 Hydration, Electrolyte, and Mineral Needs

Robert D. Lindeman

Key Points

- Water is an essential dietary component; dehydration due to inadequate replacement of body fluids can become life-threatening in a matter of a few days.
- The stimulus to thirst is impaired in older persons, as is the ability to conserve sodium and concentrate the urine. This makes the older individual more susceptible to dehydration, especially in association with acute and chronic illnesses.
- A variety of acute and chronic illnesses make older persons more susceptible to electrolyte imbalances as often evidenced by high or low concentrations of the mineral in the serum, e.g., hyponatremia, hyperkalemia, hypercalcemia, and hypomagnesemia. Again, prompt medical interventions are necessary to prevent life-threatening developments.
- Older persons are also more vulnerable to deficiencies of the 10 essential trace minerals due to various physiologic, pathologic, and psychosocial factors. Since the onset of deficiency states is often insidious in onset affecting quality of life, it is important to recognize early signs and symptoms, and when persons are at risk.

Key Words: Hydration (water); sodium; potassium; calcium; magnesium; trace minerals

8.1 INTRODUCTION

Water is a component of the diet that warrants close attention, especially in older persons. Intracellular water (lean body mass) decreases with age so that total body water also decreases. Whereas a person can go weeks without solid foods and not starve to death, dehydration due to inadequate replacement of body fluids can become life-threatening in a matter of a few days.

Sodium, potassium, calcium, and magnesium are considered the bulk minerals in the human body. The approximate quantities of the four bulk minerals in a 70 kg (154 pound) man are 1000 g of calcium (mostly in bone), 140 g of potassium, 110 g of

From: *Nutrition and Health: Handbook of Clinical Nutrition and Aging, Second Edition*
Edited by: C. W. Bales and C. S. Ritchie, DOI 10.1007/978-1-60327-385-5_8,
© Humana Press, a part of Springer Science+Business Media, LLC 2009

sodium, and 20 g of magnesium. While we now have specific recommendations for daily intakes of calcium and magnesium, i.e., Recommended Dietary Allowances (RDAs), Dietary Reference Intakes (RDIs), or Adequate Intakes (AIs) to avoid deficiencies, no similar recommendations exist for sodium or potassium as deficiencies of these minerals generally do not develop on normal diets unless certain pathological conditions exist. More often, problems develop that require dietary restrictions of these latter two minerals.

There are also now at least 10 elements (minerals or metals) that have been shown to be essential for health in animals and/or man, specifically zinc, copper, manganese, chromium, molybdenum, nickel, vanadium, boron, tin, and silicon. Since induced deficiencies cause reproducible symptom complexes in animals, and, in many instances, in man, especially in older persons, and when certain disease states exist, they are referred to as essential trace elements. Most vitamin–mineral combinations contain small amounts of these trace elements, generally in sufficient amounts to prevent deficiencies from developing, but not enough to produce toxicity.

Table 8.1
Food sources and recommended intakes for selected minerals *(21)*

Mineral	*Generous food sources*	*RDA/AI for Individuals ≥50 years of Age*
Calcium	Dairy products, Chinese cabbage, kale, broccoli, spinach, sardines, salmon[**]	1200 mg/day[*]
Magnesium	Green vegetables (spinach), some legumes (beans and peas), nuts and seeds, whole/unrefined grains, yogurt, milk[**]	Men: 420 mg/day Women: 320 mg/day
Iron	Red meats, fish, poultry, cereal, oatmeal, beans, spinach[#]	8 mg/day
Zinc	Oysters, beans, nuts, whole grains, dairy products[**]	Men: 11 mg/day Women: 8 mg/day
Copper	Beef, seafood (oysters, lobster, crab, clams), mushrooms, tomato products, barley, beans, nuts, seeds[**]	900 µg/day
Selenium	Brazil nuts, walnuts, poultry, beef, tuna, egg, cheddar cheese[**]	55 µg/day
Chromium	Meat, whole-grain products, broccoli, grape juice, orange juice, red wine, apple, banana, green beans[**]	Men: 30 µg/day[*] Women: 20 µg/day[*]
Manganese	Oat bran, whole grains, pineapple, nuts, spinach, raspberries, beans, peas[**]	Men: 2.3 mg/day Women: 1.8 mg/day

*Adequate Intakes (AIs) represent the recommended average daily intake level based on observed or experimentally determined approximations. AIs are used when there is insufficient information to determine an RDA.

** http://dietary-supplements.info.nih.gov/index.aspx. Accessed 12/07.

http://www.ars.usda.gov/Services/docs.htm?docid = 9673 Accessed 12/07.

Table 8.1 lists the minerals for which there are RDA/RDI/AIs for individuals 50 years of age and older. When there is insufficient information to determine an RDA or RDI, AIs are proposed and recommended based on available observed or experimentally determined approximations. Also listed are some of the better food sources for each of these minerals.

8.2 DEHYDRATION AND HYPERNATREMIA DUE TO PRIMARY WATER LOSS

Dehydration is, by definition, a decrease in total body water. Total body water, which is normally about 60% of body weight, is made up of about two-thirds water inside the cells (intracellular water) and one-third water outside the cells (extracellular water or ECFV). A portion of this is the serum or plasma water that is necessary to maintain an adequate circulation of blood by the heart. Dehydration can be caused by either primary loss of sodium with its osmotically obligated water or by primary water loss.

A loss of body water without proportionate loss of salt (primary water loss) results in an increase in serum sodium concentration. As a result, water is moved from inside the cell out into the extracellular fluid in an attempt to reach an osmotic equilibrium between the inside and outside of the cell. This also helps to maintain the serum and ECFV, so that the circulation has sufficient blood to maintain the blood pressure. This is why one can estimate the amount of water loss in a dehydrated person, assuming there is no significant sodium loss, by measuring the percentage increase in the patient's serum sodium compared against a normal serum sodium concentration (mean of 140 mEq/L with a range of 136–144 mEq/L) by total body water (60% of body weight in kg or L) rather than just by ECFV (20% of body weight). For example, a 70 kg (154 pound) man has 42 kg or L of total body water. If his serum sodium concentration is found to be 154 mEq/L, the 14 mEq/L increase above the normal (140 mEq/L) represents a 10% increase. Ten percent of 42 L is 4.2 L, the amount of water the dehydrated person has had to lose to reach this level of dehydration and therefore the amount that needs to be replaced. Brain cells produce idiogenic molecules during hypertonic states to maintain cell volume. If the hypernatremia is corrected too rapidly, this can result in a hazardous shift of water intracellularly producing brain swelling.

Insuring adequate intake of fluids in older adults becomes a greater challenge than in younger adults. The ability to concentrate the urine and thereby conserve fluids decreases with age, as does the ability to conserve sodium. The main reason, however, for dehydration in older persons is impairment in the thirst mechanism. The primary stimulus to thirst is an increase in serum osmolality. Even normal older individuals when deprived of fluids over a period of time will not rehydrate to baseline levels after ad lib access to water is allowed like younger persons will *(1)*. This is true even though their serum osmolality increases more than the younger persons. Impairment of thirst is even greater in persons with central nervous system disorders, e.g., stroke.

Dehydration most often occurs in acutely ill, febrile persons, especially where nausea, vomiting, diarrhea, and excessive sweating complicate adequate replacement of fluids. It may also occur in bedfast or physically handicapped persons when they are not provided enough water to satisfy an already impaired thirst mechanism. Loss of fluids also can be due to a water diuresis (pituitary or nephrogenic diabetes insipidus) or an osmotic diuresis (glycosuria) in diabetics that can lead to a condition known as hyperosmolar non-ketotic diabetic acidosis.

Most of the symptoms of dehydration are due to the shift of fluid out of brain cells shrinking them so that intracranial damage to blood vessels with venous thrombosis, infarction, and/or hemorrhage can occur. The earliest manifestation of dehydration may or may not be thirst, followed by confusion and lethargy, and ultimately delirium, stupor, and coma. Because intravascular volume is preserved at the expense of cell water, a drop in blood pressure, and increase in heart rate tend not to be prominent features of hypernatremic dehydration.

What the optimum fluid intake is for older adults is not really known. The USDA Food Pyramid (2) for older adults sets a goal of eight 8-ounce glasses of water or fluids per day. However, a literature review provides little justification for such a high intake. The only benefit is that it might lower the risk for bladder cancer (3). A recent article studying a random sample of community-based elders showed no evidence of dehydration or related symptoms, e.g., constipation or chronic fatigue, in individuals drinking much less than this per day (4). On the other hand, reasons for not forcing such a high fluid intake might include decreasing the risk of incontinence, decreasing awakening at night creating insomnia, decreasing risk of falls at night associated with nocturia, and not creating a washout of the urine concentrating mechanism in the kidney so that one would lose the ability to concentrate the urine if conditions developed that would make dehydration imminent.

While dehydration may not be a problem in healthy, community-living older persons, it may be a problem in the acutely or chronically ill, or functionally dependent individuals, who often do not drink voluntarily, and have to be encouraged to drink fluids. Fluids should be pushed orally if tolerated, but if not tolerated, intravenous 5% glucose and water can be given as needed. As previously described, one can estimate the deficit knowing the serum sodium concentration. If salt depletion is also present, normal or half normal saline can also be given. To avoid a recurrence in individuals with demonstrated impaired thirst, a fluid prescription establishing the quantity of fluid to be ingested daily may become an important part of preventive management.

8.3 SODIUM

Essentially all of the body's sodium is located in the extracellular fluid volume (ECFV) or space, so that total body sodium can be closely estimated just by multiplying the serum sodium concentration times the ECFV. However, while the serum sodium concentration is easy to measure, ECFV is difficult to estimate. Sodium is the primary cation in the ECFV and is hydrophilic attracting osmotically obligated water that, in turn, is responsible for the maintenance of plasma and extracellular volume.

While there are no recommended dietary intakes for sodium, current US dietary guidelines recommend that persons greater than 2 years of age consume less than

2300 mg of sodium per day (5.8 g of salt) as high dietary sodium intakes increase the development of hypertension and the atherosclerotic vascular diseases *(5)*. Persons with normal renal function, when placed on severely restricted salt intakes, can very effectively conserve salt; however, older persons do so less effectively than younger persons *(6)*.

When primary salt depletion occurs (the causes will be discussed below), there is with it a loss of osmotically obligated water so that plasma volume and ECFV initially decrease at a rate proportionate to the sodium loss. If this sodium loss continues unabated, forces come into play, e.g., stimulation of release of antidiuretic hormone (ADH) in response to volume depletion that increases water reabsorption by the kidneys and stimulation of thirst that increases the ingestion of water, so that the concentration of sodium in the ECFV begins to drop (hyponatremia). This causes a decrease in the osmolality (osmotic pressure) of the serum and ECFV. Since sodium and other ions cannot move freely out of the cells to reach an equilibrium with concentrations of ions outside the cell, water instead shifts (is osmotically pulled) from the ECFV back into the cells producing swelling of the cells, further depleting the ECFV. As water shifts into the cell, however, it creates an osmotic equilibrium with the lower serum osmolality outside the cell. With depletion of serum and ECFV, subjective signs and symptoms of dehydration begin to develop (postural hypotension, tenting of the forearm skin, dryness of the mucous membranes).

A low serum sodium concentration may result from a loss of sodium in excess of osmotically obligated water (primary salt depletion), a retention of water in excess of sodium (dilutional hyponatremia), or a combination of both, as seen in the syndrome of inappropriate antidiuretic hormone (SIADH). Older persons studied in both acute and chronic care facilities have a higher prevalence of hyponatremia (serum sodium concentration less than 135 mEq/L) *(7,8)*. Even in an ambulatory geriatric clinic, 11% of the patients had evidence of hyponatremia *(9)*. SIADH appeared to be the etiology in over half the cases with no apparent underlying cause other than age found in seven patients (idiopathic). Reasons why older subjects are more prone to the development of hyponatremia, more specifically SIADH, are discussed elsewhere *(10)*.

A reduction in serum sodium concentration to below 120 mEq/L, regardless of etiology, produces symptoms ranging from mild non-specific complaints, such as malaise, apathy, irritability, muscle weakness, and change in personality, to marked central nervous system (CNS) impairment, often ending with seizures. The CNS symptoms result from shifts of fluid along osmotic gradients from the hypotonic ECFV into the isotonic brain cells, thereby increasing brain volume and intracranial pressure. Although this fluid shift occurs throughout the body, e.g., producing "finger printing" (pressure over the sternum leaves finger prints due to the increase in intracellular water), those tissues not entrapped in a bony structure (cranium) are little affected.

8.3.1 Hyponatremia with Contracted ECFV

The causes of hyponatremia are listed in Table 8.2. When primary salt depletion with contraction of ECFV is suspected, a measure of urinary sodium concentration is most helpful. A concentration less than 10 mEq/L suggests inadequate intake of

Table 8.2

Hyponatremic syndromes

1. Hyponatremia with contracted extracellular volume (ECFV)

    A. Urinary sodium <10 mEq/L (inadequate intake, excessive sweating, excessive gastrointestinal loss, e.g., diarrhea, bowel, and biliary fistulas)

    B. Urinary sodium >10 mEq/L

        1. Severe metabolic alkalosis due to vomiting

        2. Excessive urinary losses (salt wasting)

            (a) Adrenal insufficiency (Addison's disease, hypoaldosteronism)

            (b) Renal disease (renal tubular acidosis, interstitial nephritis)

            (c) Diuretic induced

2. Hyponatremia with normal ECFV

    A. Displacement syndromes (hyperglycemia, hyperlipidemia, hyperproteinemia)

    B. Syndrome of inappropriate antidiuretic hormone (SIADH)

        1. Malignancies, especially small cell carcinoma of lung

        2. Pulmonary diseases, including positive pressure breathing

        3. Cerebral conditions (trauma, infection, tumor, stroke)

        4. Drugs (sulfonylureas, thiazides, antitumor agents, psychotropics, antidepressants)

        5. Other (myxedema, porphyria, idiopathic)

    C. Water intoxication (schizophrenia)

3. Hyponatremia with expanded ECFV (dilutional hyponatremia) (congestive heart failure, cirrhosis, nephrotic syndrome, hypoalbuminemia, renal failure)

salt, excessive sweating, or, most often, excessive gastrointestinal losses (diarrhea, bowel, or biliary fistulas). If urinary sodium excretion exceeds this level, one must think of excessive use of diuretics, adrenal or pituitary insufficiency, or intrinsic renal causes, e.g., renal tubular acidosis, salt-losing nephritis, or renal insufficiency. Severe vomiting with resultant metabolic alkalosis can result in loss of sodium bound to the increased amounts of bicarbonate filtered and not reabsorbed in the proximal tubules, despite hyponatremia and hypovolemia.

Treatment is generally best accomplished by giving isotonic (0.9%) saline as the replacement; however, in individuals with symptomatic conditions, small amounts of hypertonic (5%) saline can be utilized. Caution needs to be exercised to avoid correcting the hyponatremia too rapidly as devastating demyelinating syndromes (cerebropontine myelinosis) can occur. The differential diagnosis between primary salt loss syndromes and SIADH can be problematic as edema is generally absent and urinary sodium excretions are high in both situations. The classic signs of dehydration may be unreliable, e.g., skin turgor in the normally hydrated older persons may appear poor due to loss of subcutaneous fat allowing "tenting" of skin over the backs of hands rather than being evidence of dehydration. An important clue in making the correct diagnosis can be serum urea nitrogen (SUN or BUN); it tends to be high in the former, and low, even sub-normal, in patients with SIADH, unless pre-existing renal impairment is present.

8.3.2 Hyponatremia with Expanded ECFV (Dilutional Hyponatremia)

Impairment of water excretion commonly occurs in conditions where salt excretion also is severely impaired. Patients with advanced cardiac, hepatic, and renal failure, and with generalized edema often are placed on diets that sharply curtail salt intake with little attention placed on putting limitations on fluid intake. When dilutional hyponatremia develops, considerations need to be given to also restricting fluid intake. Although total ECFV is increased, the blood flow returning to the heart and delivered to the arterial system is decreased, stimulating baroreceptors in the right atrium and arterial system to help the kidneys retain sodium, and baroreceptors in the left atrium and arterial system to stimulate ADH release and retain water. A marked increase in sodium and water reabsorption in the proximal tubule further limits water excretion by limiting the delivery of water and sodium to the distal nephron where dilution of urine below isotonic levels can take place. Although potent diuretics, e.g., furosemide, are available to increase sodium excretion, vasopressin antagonists that will block ADH effect on the distal nephron and increase water excretion are still in the investigative stages. One can use furosemide to increase salt excretion (this also creates a hypotonic urine), and give a hypertonic salt solution back to increase serum sodium concentration. As mentioned earlier, one must exercise caution in attempting to correct the hyponatremia too rapidly.

8.3.3 Hyponatremia with Normal ECFV (Syndrome of Inappropriate ADH)

Persistence of high levels of circulating ADH is considered inappropriate when neither hyperosmolality of serum nor volume depletion is present. The diagnosis of SIADH is usually made after other causes of hyponatremia are excluded. This means the urine remains concentrated and a high urinary sodium excretion persists after adrenal, renal, cardiac, and hepatic functions are determined to be normal. The inability to excrete water because of high levels of circulating ADH causes volume expansion that, in turn, increases sodium loss with its osmotically obligated water, thereby returning ECFV to near normal. The causes of SIADH are shown in Table 8.2, and include patients with different neoplasms, most notably oat cell carcinoma of the lung, where an ADH-like molecule is secreted by the tumor. Conditions limiting blood flow through the pulmonary circulation, including positive pressure breathing, limit filling of the left atrium thereby stimulating ADH release in the presence of normal or increased volumes elsewhere. A number of drugs also have been implicated in causing this syndrome.

Normovolemic hyponatremia also can result from addition to the serum of an uncharged solute, e.g., glucose or mannitol. This will increase serum osmotic pressure drawing water from inside the cell to expand and dilute the ECFV. This increases urinary sodium loss, thereby lowering serum and body stores of sodium. A pseudohyponatremia also can develop from displacement of water in serum/plasma with abnormal amounts of high molecular weight solute, e.g., proteins or lipids. While the concentration of sodium in the serum water is normal, the volume of water is less in the sample so that when further diluted prior to the assay, it gives a low serum sodium value.

8.4 POTASSIUM

Potassium is the primary intracellular cation, with less than 2% of total body potassium in the extracellular fluid compartment. Consequently, the serum potassium concentration may not accurately reflect total body potassium stores. Potassium enters cells during cell growth, with intracellular glucose and nitrogen deposition, and with increases in extracellular pH (respiratory and metabolic alkalosis); it leaves cells with cell destruction, glucose utilization, and decreases in extracellular pH (respiratory and metabolic acidosis). Normally, a steep concentration gradient (greater than 20:1) is maintained between intracellular stores and extracellular concentrations of potassium, so that one needs to appreciate that a small shift of potassium into or out of the cell can greatly affect extracellular potassium concentration.

As an example, the patient with diabetic ketoacidosis is likely to be admitted to the hospital with a high serum potassium concentration. With rehydration (dilution), correction of the acidosis with sodium bicarbonate, and treatment of the hyperglycemia with insulin, one can anticipate a dramatic fall in serum potassium concentration that can become life-threatening if not treated with adequate potassium replacements. Age alone does not appear to affect the ability to maintain this steep concentration gradient; however, isotopic dilution studies and muscle biopsies have shown that certain clinical conditions common to chronically ill older persons, e.g., respiratory and metabolic acidosis, congestive heart failure, and liver and renal failure cause them to develop depleted intracellular concentrations of potassium without affecting their ability to maintain normal circulating concentrations of potassium.

8.4.1 Hypokalemia

Although the normal kidney is not as efficient in conserving potassium as it is sodium, it still takes 2–3 weeks on a severely restricted potassium diet to lower serum potassium below normal. Again, there are no minimum dietary intakes identified for potassium intake. A reasonable criterion for establishing a diagnosis of urinary potassium wasting when the serum concentration falls below normal (3.5 mEq/L) would be a daily potassium excretion of more than 20 mEq/L per day.

The causes of hypokalemia are listed in Table 8.3. Often there are multiple contributors to the development of this electrolyte disorder. As an example, a patient with pernicious vomiting not only reduces potassium intake and loses small amount of potassium in the vomitus, but he/she loses hydrogen ions creating a metabolic alkalosis. This increases urinary potassium losses and shifts potassium intracellularly. A contracted ECFV then increases proximal tubular sodium and bicarbonate reabsorption potentiating the metabolic alkalosis, and stimulates increased aldosterone secretion further increasing potassium losses in the urine. The causes of excessive urinary potassium losses can be divided into four categories as shown in Table 8.3, specifically (1) pituitary–adrenal excesses, (2) renal defects, (3) drug-induced losses, and (4) other (idiopathic and miscellaneous).

Manifestations of hypokalemia include EKG changes, some related to focal myocardial necrosis (depressed ST segments, inversion of T waves, accentuated U

Table 8.3

Causes of hypokalemia

1. Inadequate intake, excessive sweating, dilution of extracellular fluid volume
2. Shift of potassium intracellularly (increase in blood pH (alkalosis), glucose and insulin, familial hypokalemic periodic paralysis)
3. Excessive gastrointestinal losses
   A. Vomiting
   B. Biliary, pancreatic, and intestinal drainage from fistulas and ostomies
   C. Chronic diarrhea (chronic infections and inflammatory lesions, malabsorption, villous adenomas of colon, excessive use of enemas and purgatives)
4. Increased urinary losses (potassium wasting)
   A. Pituitary–adrenal disturbances (primary aldosteronism, secondary aldosteronism due to renal artery stenosis, Cushing syndrome due to adrenal adenomas, carcinomas and/or hyperplasia, pituitary corticotrophin hypersecretion, ectopic corticotropin secretion secondary to tumor)
   B. Renal disorders (distal or proximal renal tubular acidosis, salt-losing nephritis, diuretic phase of acute tubular necrosis, post-obstructive diuresis)
   C. Drug induced (thiazide and loop diuretics, large non-absorbable anions, e.g., carbenicillin, cistplatin, aminoglycosides, amphotericin B, respiratory alkalosis due to acetylsalicylic acid, adrenergic agonists used to treat bronchospasm)
   D. Idiopathic, familial, and other pathologies (Liddle's and Gitelman's syndromes, hypomagnesemia)

waves) with ventricular arrhythmias, muscle weakness, pain, and tenderness, again with focal necrosis, decreased intestinal motility (paralytic ileus at the extreme), and polyuria.

Because an alkalosis (chloride depletion) generally accompanies hypokalemia, replacement therapy should be started with potassium chloride rather than the alkaline salts of potassium (the exception would be the patient with renal tubular acidosis). Foods high in potassium content (citrus and tomato juices, bananas, red meats, and some vegetables) provide the safest way to supplement potassium intake. When additional oral replacement therapy is needed, commercial preparations come in liquid, tablet, and powder forms, usually in 20 mEq dosages that can be given up to three times daily. Intravenous potassium can be administered if necessary, but becomes hazardous if given at rates exceeding 20 mEq/h, or in concentrations exceeding 40 mEq/L. EKG monitoring should be mandatory under such conditions.

Patients receiving loop (furosemide) or thiazide diuretics may develop hypokalemia. Concern that such individuals might be at increased risk for the development of potentially lethal, ventricular arrhythmias has led to the common practice of giving routine potassium supplements when receiving these diuretics. Patients with edematous and acidotic conditions are likely to have depleted intracellular potassium stores; hypertensive patients with comparable serum potassium concentrations undergoing diuretic therapy tend to maintain normal intracellular potassium levels. Routine replacement therapy for patients receiving the thiazide diuretics as treatment for hypertension is unnecessary. Rather it is better to monitor serum potassium levels for the occasional patient who will develop hypokalemia. A significant incidence of life-threatening hyperkalemia develops in patients receiving routine supplements meaning that the potential for benefit is out-weighed by the risk *(11)*.

8.4.2 Hyperkalemia

Hyperkalemia is most commonly observed in patients with impaired renal function; however, those with chronic kidney disease (CKD) do not develop significant hyperkalemia until the azotemia becomes life-threatening, unless there is some significant source of additional exogenous/endogenous potassium. Because the distal nephron of the kidney has such a large capacity for secreting potassium, hyperkalemia develops only when some associated factor is present, e.g., oliguria (acute renal failure), an excessive exogenous/endogenous potassium load (supplements, medications, protein catabolism, rhabdomyolysis), a metabolic or respiratory acidosis, a deficiency of endogenous steroid (aldosterone, corticosteroids), or administration of a drug that inhibits potassium secretion in the distal nephron, e.g., spironolactone, triamterene, an angiotensin-converting-enzyme (ACE) inhibitor, a beta-adrenergic antagonist (blocker), or a non-steroidal anti-inflammatory drug (NSAID).

Clinical manifestations of hyperkalemia often are subtle and non-specific, and occur only shortly before death from cardiac arrhythmia. Unexplained anxiety, apprehension, weakness, stupor, and hyporeflexia should alert the health-care professional to the potential existence of this condition. Characteristic EKG changes start with peaking of the T waves followed by widening and loss of p waves, and ultimately by widening of the QRS complex.

Dietary restriction of potassium-rich foods can be started in anticipation of a risk of hyperkalemia. More aggressive therapy should be started when the serum potassium concentration exceeds 5.5 mEq/L; a true medical emergency exist when it reaches 7.0 mEq/L. Sodium polystyrene sulfonate (Kayexalate) resins are used to remove excess potassium from the body, and can be given either orally or in enema forms. To avoid constipation and fecal impaction with oral administration, sorbitol solutions can be given titrating the dose to achieve loose stools. An acute decrease in serum potassium levels can be achieved with infusions of hypertonic glucose with insulin, along with calcium and sodium salts that act as physiologic antagonists. This will move potassium intracellularly, but the potassium will move back into the extracellular fluids once the infusion stops. When hyperkalemia is due to a mineralocorticoid or aldosterone deficiency, 9-fluorohydrocortisone (Florinef) can be given. Dialysis rarely is used as a last resort.

8.5 CALCIUM

Calcium differs from most other minerals in that the serum calcium concentration is not an adequate guide to the state of the individual's calcium nutrition. A finely tuned endocrine system exists to maintain serum-ionized calcium levels within a narrow normal range by controlling intestinal absorption, bone exchange, and renal excretion of calcium. When serum-ionized calcium concentration decreases, parathyroid hormone (PTH) secretion increases causing mobilization of calcium from bone, a decrease in renal tubular phosphate reabsorption, and a decrease in serum phosphate concentration. The change in serum phosphate concentration facilitates bone resorption of calcium, increases renal tubular calcium reabsorption (conserves calcium), and increases intestinal calcium absorption, either directly or by enhancing the effects of vitamin D. Vitamin D is converted in the liver to the carrier metabolite, 25-OH cholecalciferol, and by the kidney to the active metabolite $1,25\text{-}(OH)_2$ cholecalciferol . The active form of vitamin D acts primarily to increase calcium absorption in the intestine, but it also increases bone resorption of calcium, and decreases urinary calcium and phosphate excretion. PTH produces its effect on the intestine by accelerating conversion of the carrier form of vitamin D to its active form. Because there are over 1000 g of calcium in the skeleton and only 1 g in the extracellular fluid, it requires little shift of calcium out of the skeleton to replenish serum calcium concentrations within the normal range.

Serum calcium exists in both the ionized and bound states, but only the ionized fraction is physiologically active. The latter fraction is bound either to protein (albumin) or complexed with various anions, e.g., citrate. Binding of calcium is dependent upon the concentration of serum proteins (albumin), and the blood pH, as calcium binding increases as blood pH increases. Because most clinical laboratories report only total serum calcium concentration, these effects need to be recognized in evaluation of any specific calcium concentration.

Both hypocalcemia and hypercalcemia occur with increasing frequency in older persons, primarily because specific disease entities causing these abnormalities are more common in older individuals. Of greater importance clinically is the age-related loss of bone calcium (osteoporosis) that occurs in both genders, but more severely in women. This is discussed in more detail in Chapter 22 along with nutritional requirements for calcium and vitamin D. The Food and Nutrition Board of the Institute of Medicine published Dietary Reference Intakes (DRIs) for calcium of 1200 mg/day, recognizing that the majority of adults do not consume these levels of calcium without taking supplements *(12–14)*. One study of older persons showed that 62% ingested less than 500 mg/day and 21% less than 300 mg/day in their diet *(15)*. They also recommended 10 μg of vitamin D (400 IU) per day for persons aged 50–70 years, and 15 μg (600 IU) for those over the age of 70 years. They recognized that not only do older persons fail to ingest enough calcium, but they do not absorb it as well, and they frequently have insufficient blood levels of vitamin D.

8.5.1 Hypocalcemia

Most of the causes of hypocalcemia (serum total calcium <8.6 mg/dL) are due to disturbances in PTH metabolism, e.g., decreased production or end-organ unresponsiveness, or to deficiencies in vitamin D intake or metabolism, e.g., a deficiency or malabsorption of vitamin D, decreased production of 25-OH cholecalciferol due to liver disease, decreased conversion of 25-OH to $1,25\text{-}(OH)_2$ cholecalciferol in renal disease, accelerated vitamin D catabolism due to anticonvulsant therapy, or end-organ resistance to vitamin D (vitamin D-resistant rickets). Other causes of hypocalcemia include a low serum albumin concentration (total serum calcium is low, but ionized calcium is normal), acute pancreatitis, hyperphosphatemia (renal insufficiency), and calcitonin-producing tumors (medullary carcinoma of the thyroid). In unexplained hypocalcemia/hypokalemia, a low serum magnesium (hypomagnesemia) should be looked for as it produces a peripheral resistance to PTH and decreases the release of PTH in response to the hypocalcemic stimulus.

The symptoms associated with hypocalcemia are primarily related to increased neuromuscular excitability, i.e., tetany, parasthesias, muscle weakness and spasms, and movement disorders. Chronic manifestations include cataracts, abnormalities of nails, skin, and teeth, and mental retardation.

Correction of acute symptomatic hypocalcemia can be accomplished with parenteral calcium gluconate. Mild or latent hypocalcemia can be treated with oral calcium salts (carbonate, gluconate, lactate). Vitamin D or its active metabolite (Calcitriol) can be used to increase calcium absorption.

8.5.2 Hypercalcemia

Primary hyperparathyroidism and malignancy together are responsible for 80–90% of all cases of hypercalcemia. Malignancy-associated hypercalcemia is the most common paraneoplastic syndrome, occurring in up to 10% of all patients with breast, renal cell, and lung cancers, squamous cell cancers of the head and neck, and multiple myeloma. These tumors secrete substances with PTH-like activities. These and other types of tumors metastasizing to bone also can release bone-resorbing cytokines and growth factors resulting in osteolysis thereby raising serum calcium levels. Other less common causes of hypercalcemia include endocrine disorders (hyperthyroidism, adrenal insufficiency, pheochromocytoma), granulomatous diseases (sarcoidosis), and vitamin D intoxication.

The early signs of hypercalcemia are vague and non-specific, e.g., fatigue, anorexia, nausea, vomiting, constipation, confusion, somnolence, weakness, apathy, depression, and personality change. Hypercalcemia often results in dehydration as it decreases the ability of the kidney to reabsorb salt and water in the tubular portions of the kidney. The initial therapy is rehydration with normal saline and administration of a loop diuretic (furosemide) as this will greatly increase urinary calcium excretion. Thiazide diuretics, in contrast, decrease urinary calcium excretion and potentiate hypercalcemia. The use of drugs that inhibit bone resorption, namely the biphosphonates, e.g., etidronate, pamidronate, and zoledronic acid, are usually used first in the chronic management of malignancy-associated hypercalcemia.

8.6 MAGNESIUM

Magnesium is the second most important intracellular cation with about 60% of body magnesium located in bone, about 40% in intracellular spaces (primarily muscle), and only 1% located extracellularly. The normal serum magnesium concentration ranges from 1.4 to 2.2 mEq/L. About 30% of serum magnesium is protein bound (albumin) with most of the rest ionized, physiologically active, and ultrafilterable through the kidney. The magnesium inside the cells primarily is bound to protein- and energy-rich phosphates being indispensable to the metabolism of adenosine triphosphate (ATP). Magnesium is the metal component that activates more than 300 different enzyme systems affecting glucose metabolism, synthesis of fat, protein, and nucleic acids, muscle contraction, and several membrane transport systems.

The Dietary Reference Intakes (DRIs) recommended by the Food and Nutrition Board are 420 mg/day for men over the age of 50 years and 320 mg/day for women of that age *(12–14)*. Studies on 37,000 healthy adults showed mean intakes of magnesium in their diets for men and women were 266 and 228 mg/day, respectively *(16)*. The magnesium content of a typical senior multivitamin tablet to be taken once daily is 100 mg. One can assume that for older persons, especially for those with chronic conditions that might affect food intake, gastrointestinal absorption, or urinary excretion of magnesium, that a deficiency of magnesium might be a real concern.

8.6.1 Magnesium Deficiency (Hypomagnesemia)

Magnesium deficiency is most commonly seen in gastrointestinal disorders causing malabsorption/chronic diarrhea, and in chronic alcoholism. It can also be seen associated with protein–calorie malnutrition, acute pancreatitis, and in a variety of conditions associated with wasting of magnesium in the urine, e.g., diuretic therapy (furosemide), endocrine disorders (hyperthyroidism, hyperparathyroidism), and after nephrotoxic drug therapy (aminoglycoside antibiotics, cyclosporine, cisplatin).

Symptoms associated with hypomagnesemia include neuromuscular manifestations (muscular hyperirritability, tetany, hyperacousis, seizures, vertigo, and tremors) and mental psychiatric changes (irritability, aggressiveness). Other manifestations include development of polyuria, hypocalcemia, hypokalemia, and hypophosphatemia.

Significant depletion of intracellular magnesium can occur before the serum magnesium concentration falls below the normal range *(17)*. Because symptoms appear more related to intracellular concentrations of magnesium than to serum levels, the question then arises how to best measure intracellular magnesium. There are no clinically practical ways short of muscle biopsy, so it is best if one suspects a deficiency exists to treat empirically with supplemental magnesium. This is safe unless some degree of renal insufficiency is present.

8.6.2 Hypermagnesemia

Hypermagnesemia normally occurs only in patients with some degree of renal insufficiency and also taking magnesium containing antacids or cathartics. Somnolence and hypotension are the earliest manifestations, and EKG changes may develop, i.e., prolongation of the PR interval and QRS duration with peaking of T waves. Later, respiratory depression (respiratory acidosis) or paralysis may develop terminating with a fatal cardiac arrhythmia.

8.7 TRACE MINERALS

Studies of dietary intake and nutritional status of the trace minerals indicate that persons in later life are more vulnerable to deficiencies than are younger persons as a result of multiple physiological, psychological, and socioeconomic factors. Reasons for this include an increased prevalence of chronic diseases and medication use, e.g., diuretics, that may affect nutrient turnover in the former, physiologic changes with age that may influence absorption, distribution, and metabolism of the minerals, and psychological, e.g., loneliness, and socioeconomic conditions that may affect dietary choices and eating patterns. Brief discussions of the biologic importance of each of the minerals will be followed by presenting features and functional consequences of each trace mineral deficiency, along with how best to establish a diagnosis.

When available, Recommended Daily Allowances (RDAs) are provided for each of the trace minerals (Table 8.1). The RDA is defined as the intake that meets the nutrient need of 97–98% of the individuals in that group. Food sources providing the highest content of each mineral are listed. The quantity of each mineral in a typical once-daily multivitamin preparation for seniors is shown along with additional sources of single element supplements. Also provided when available is a tolerable upper intake level defined as the maximum intake that is unlikely to produce adverse health effects in most (97–98%) individuals. More comprehensive reviews are available for those desiring additional information on the vast literature describing evidence for trace mineral deficiency and ranges of dietary intake of these minerals in population studies *(18–20)*.

8.7.1 Zinc

Zinc is the intrinsic metal component or activating cofactor for over 70 important enzyme systems, including carbonic anhydrase, the alkaline phosphatases, the dehydrogenases, and the carboxypeptidases. It governs the rate of synthesis of the nucleoproteins (DNA and RNA polymerases are zinc-dependent enzymes) and other proteins, thereby influencing growth and the reparative processes. It is also an important regulator of the activity of various inflammatory cells (macrophages, polymorphonuclear leukocytes) thereby affecting the immune system, plays a role in certain endocrine functions (growth, sexual maturation), affects carbohydrate tolerance by decreasing the rate of insulin secretion in response to glucose administration, and is an important regulator of bone repair.

Human zinc deficiency was first described in adolescent males (Iran, Egypt) receiving inadequate intakes of zinc and having excessive binding of ingested zinc to phytates (unleavened bread), fiber, and clay. These youths developed severe growth retardation (dwarfism), anemia, rough skin, apathy (general lethargy), and delayed sexual maturation that responded dramatically to zinc supplementation. Similar manifestations of severe zinc deficiency have been seen in patients started on TPN and given no mineral supplements, and in certain disease states, especially those affecting GI absorption, e.g., regional enteritis (Crohn's disease). Potential manifestations of zinc deficiency include impaired taste and smell, anorexia, alopecia, dermatitis starting in the nasolabial folds and extending to the scrotum, perineum, and extensor aspects of the elbows, impaired wound healing, hypogonadism and sexual dysfunction, and immune dysfunction *(19,20)*.

The Recommended Dietary Allowances (RDA) for adult males is 11 mg/day and for females 8 mg/day *(21)*. While normal healthy elderly should be ingesting quantities near these amounts daily, only about two-thirds of the zinc ingested is normally absorbed through the GI tract. Zinc from meats is absorbed better than that from vegetables. Diets high in phytates (breads) interfere with absorption.

The normal plasma zinc concentration ranges from 80 to130 µg/mL. Low plasma zinc concentrations can result either from redistribution of zinc due to acute infectious or inflammatory conditions (release of cytokines into the circulation), or from a true zinc deficiency state *(19)*. Large population studies (NHANES II) have shown that 12% of men and women had plasma zinc levels below 70 µg/mL, and were strikingly higher in the older populations, those in poor health, and/or those with low socioeconomic status *(20)*.

A typical once-daily multivitamin preparation for seniors contains 15 mg of elemental zinc; however, zinc sulfate tablets containing 25 and 50 mg of elemental zinc are also available. The tolerable upper intake level has been established at 40 mg/day *(19)*.

8.7.2 Copper

Copper is the metal component of a number of enzymes important in human metabolism. It is involved in such diverse enzymatic and metabolic activities as hemoglobin synthesis, bone and elastic tissue development, and the normal function of the central nervous system. The usual first manifestations of a copper deficiency in man are anemia unresponsive to iron supplements, leucopenia, and neutropenia. Much less commonly seen are hypopigmentation, immune dysfunction, and skeletal abnormalities. Animals made severely deficient will develop central nervous system deficiencies (ataxia, mental deficiency), skeletal defects (osteoporosis with fractures), aortic rupture, and skin and hair (red) changes. Significant copper deficiency in adults does not appear to be common, unless the individual is on TPN without mineral supplements, or the patient has intestinal or malabsorptive disease. It is seen primarily in infants with GI disorders (diarrhea) on cow's milk, and in children with a hereditary disorder (Mencke's kinky hair syndrome) in which copper absorption from the GI tract is defective.

Normal plasma copper concentrations range from 80 to 140 µg/mL, but may be higher in acute infectious or inflammatory states. Wilson disease (hepatolenticular degeneration) is a rare, inherited inborn error of copper metabolism characterized by a low serum ceruloplasmin concentration (the major copper-binding protein in blood), usually low serum copper levels, but increased urinary copper and deposition of copper in vital organs (brain, liver, kidney, eyes) producing dysfunction in these systems.

The Recommended Dietary Allowance for older men and women is 900 µg of elemental copper per day *(19,21)*. Most once-daily multivitamin preparations for seniors contain 2 mg of elemental copper. Copper gluconate is also available in tablet form in dosages containing 3 mg of elemental copper. The tolerable upper intake level is 10 mg/day *(19)*.

8.7.3 Iron

About 70% of bodily stores of iron are found in hemoglobin and myoglobin, with most of the rest stored in a labile form (ferritin, hemosiderin) in the liver, spleen, and bone marrow to be used in the regeneration of hemoglobin in case of blood loss. A hypochromic, microcytic anemia (hemoglobin <12 g/dL) with decreased serum iron, normal to increased iron-binding capacity, and decreased serum transferrin levels, is the primary manifestation of iron deficiency. Other manifestations of iron deficiency (Plummer–Vinson syndrome) include listlessness, fatigue, palpitations, glossitis, stomatitis, and dysphagia caused by a post-cricoid stricture.

The range of normal serum iron concentration is 60–175 μg/dL in men and 50–170 μg/dL in women, the iron-binding capacity is 250–450 μg/dL in both genders, and serum transferrin is 200–400 mg/dL in both genders.

While many feel that post-menopausal females and males get enough iron in their diets to avoid a deficiency state, and that excess iron may actually have detrimental effects, RDAs for elemental iron have been established for both of these groups at 8 mg/day. While some once-daily multivitamin preparations for seniors include this amount of elemental iron in their preparations, many no longer contain any iron. Oral ferrous sulfate tablets and capsules, both enteric-coated and extended release (commonly 325 mg of ferrous sulfate equals 65 mg elemental iron), are available for treatment of documented iron-deficiency anemia.

8.7.4 Selenium

Selenium is concerned with growth, muscle function, the integrity of the liver, and fertility in ways that are poorly understood. A number of animal disease models with deficiencies of both vitamin E and selenium can be cured by administration of one or the other showing the interacting biochemical roles of these two agents. Selenium is the metal component of glutathione peroxidase; both this and vitamin E act as antioxidants in the detoxification of peroxides and free radicals that have their most damaging effects on cell membranes in blood, liver, and other tissues. Selenium-deficient animals develop hair loss, growth retardation, and reproductive failure. Selenium deficiency has been implicated in Keshan disease, an endemic cardiomyopathy found in selenium-deficient areas of China. While selenium supplements will prevent the disease, they will not cure it once established. Selenium deficiency is not likely to occur in free-living, healthy individuals living in the United States; however, individuals in poor health may be at risk. A number of animal and human studies suggest that selenium may play a role in protecting against the development of cancer (20).

Serum/plasma selenium levels can be quantified (normal 100–180 ng/mL).

The best sources of selenium are grains (bread) (depending on the selenium content of soil), seafood, liver, and meats. The RDAs for selenium have been established for both men and women at 55 μg/day (19,21). The typical once-daily multivitamin preparations for seniors contain 20 μg of elemental selenium. Elemental selenium is also available in tablet form at dosages of 50, 100, and 200 μg. The tolerable upper intake level for both men and women is 400 μg/day. Symptoms associated with selenium toxicity include nausea and vomiting, hair loss, nail changes, irritability, fatigue, peripheral neuropathy, and changes in breath odor (garlic or sour milk).

8.7.5 Chromium

Chromium is an integral and active part of the glucose tolerance factor (GTF), a dinicotinic acid glutathione complex that increases the influx of glucose into cells in response to insulin. Because it is difficult to quantify levels of chromium in tissue and blood, chromium deficiency is best demonstrated by showing a response to chromium supplementation. A deficiency of chromium in animals produces an abnormal glucose tolerance (insulin resistance). Chromium deficiency has been reported in humans on long-term TPN and is characterized by impaired glucose tolerance (insulin resistance) not responsive to added exogenous insulin, decreased HDL lipoprotein cholesterol, increased LDL lipoprotein cholesterol, and peripheral neuropathy or encephalopathy. All these are reversible with chromium supplementation. Diabetics tend to have low tissue stores of chromium and lose increased amounts in the urine. Investigators have speculated that there may be a subset of diabetics with glucose intolerance that will respond to chromium supplementation, but the evidence for this remains inconclusive. Chromium supplements have also been reported to be performance enhancing, but the data to support this claim is also very limited.

Processed meats, whole grains, and certain vegetables have high chromium concentrations. There is no RDA for chromium, but intakes adequate (AIs) to sustain nutritional status are felt to be 30 μg/day in adult males and 20 μg/day in adult females *(19,21)*. Once-daily multivitamin supplements for seniors commonly contain 150 μg of elemental chromium. Chromium capsules are available that contain 200 μg of elemental chromium. Toxicity has not been reported.

8.7.6 Manganese

Manifestations of manganese deficiency in animals include impaired growth, skeletal abnormalities, reproductive dysfunction, ataxia, and alterations in lipid and carbohydrate metabolism. A single case study describes manganese deficiency in a human volunteer inadvertently deprived of this mineral during an experimental study, who developed hypocholesterolemia, weight loss, transient dermatitis, and reddening and slow growth of normally black hair and beard. All returned to normal with repletion.

Manganese is ubiquitous being more available in vegetarian than in non-vegetarian diets. Although no RDAs have been published, the Food and Nutrition Board identified Adequate Intakes as 2.3 mg/day for men and 1.8 mg/day for women *(19,21)* the levels easily obtained with the normal diet. The typical once-daily multivitamin for seniors contains 2 mg of manganese. Parenteral manganese is available for use in TPN solutions. The tolerable upper limit for manganese is 11 mg/day.

8.7.7 Molybdenum

Molybdenum is a cofactor of xanthine, sulfite, and aldehyde oxidases, and is a copper antagonist. Molybdenum deficiency in animals impairs the conversion of hypoxanthine and xanthine to uric acid causing xanthine renal calculi. Apparent molybdenum deficiency developed in a patient on TPN characterized by an intolerance to the sulfur-containing amino acids. She/he developed tachycardia, tachypnea, central scotomas, night blindness, irritability, and finally coma that cleared when the sulfur-containing amino acids were stopped. There was biochemical evidence of impaired conversion of sulfite and thiosulfite to sulfates by sulfite oxidase, and xanthine and hypoxanthine to uric acid by xanthine oxidase, both reactions normally facilitated by molybdenum-containing enzymes. Treatment with molybdenum reversed the sulfur-handling defect.

The RDAs for molybdenum are 45 µg/day for men and 34 µg/day for women, both easily achievable with a normal diet. The typical once-daily multivitamin preparation for seniors contains 75 µg. The tolerable upper intake level is 2 mg/day.

8.7.8 Other Trace Minerals

Although studies in animals depleting and replacing single trace minerals have convincingly shown that vanadium, nickel, silicon, boron, and tin are essential for normal growth and maturation, the evidence that clinically apparent deficiencies develop in man is non-existent (18). Once-daily multivitamin–mineral supplements for seniors, however, commonly contain 150 µg of boron, 5 µg of nickel, 2 mg of silicon, and 10 µg of vanadium. Cobalt is essential as the metal cofactor in vitamin B_{12}, but because it is such a commonly available nutrient and such small amounts are needed to meet demands, cobalt deficiency has not been reported in animals or man. Finally, there are five non-essential minerals (cadmium, lead, mercury, arsenic, and aluminum) that can accumulate acutely or chronically to produce toxicity (18).

8.8 RECOMMENDATIONS

1. While maintenance of adequate hydration is important in older persons, especially during periods of acute/chronic illness, overemphasis on maintaining a high fluid intake in healthy, community-dwelling older individuals is unnecessary, and can have a negative impact on quality of life by increasing incontinence, nocturia, and the incidence of related falls.
2. Diets consumed by older persons contain sufficient amounts of sodium and potassium, so that deficiencies of these minerals (electrolytes) do not occur unless pathological conditions exist, or excessive medications, e.g., diuretics, are in use. Hence, there is no need for an RDA/AI. More often, the clinician needs to limit intake or increase loss, e.g., increase urine excretion, to prevent excessive accumulation of these two minerals in the body.
3. Dietary calcium and magnesium intakes in most older persons fall well short of RDA/AI recommended levels, thereby increasing the risk for development of osteoporosis, especially in women. The use of dietary supplements is the easiest way to insure that these recommended standards are met.
4. The development of trace mineral deficiencies can be very insidious in onset, and occur primarily in persons with acute/chronic illnesses. Balanced diets generally meet the needs of healthy older persons. Clinicians should be aware of the "failure to thrive" symptom complexes resulting from deficiencies of these trace minerals.

REFERENCES

1. Phillips PA, Rolls BJ, Ledingham JG, et al. Reduced thirst after water deprivation in healthy elderly men. N Engl J Med 1984;311:753–9.
2. Russell RM, Rasmussen H, Lichtenstein, AH. Modified food guide pyramid for people over the age of 70 years. J Nutr 1999;129:751–3.
3. Michaud DS, Spiegelman D, Clinton SK, et al. Fluid intake and the risk of bladder cancer in men. N Engl J Med 1999;340:1390–7.
4. Lindeman RD, Romero LJ, Liang HC, et al. Do elderly persons need to be encouraged to drink more fluids? J Gerontol 2000;55A:M361–5.
5. US Department of Health and Human Services and US Department of Agriculture. Dietary Guidelines for Americans, 2005, 6th edition. Washington, D.C. US Government Printing Office, 2005.
6. Epstein M, Hollenberg N. Age as a determinant of renal sodium conservation in normal man. J Lab Clin Med 1976;87:411–7.
7. Kleinfeld M, Casimir M, Borra S. Hyponatremia as observed in a chronic disease facility. J Am Geriatr Soc 1979;27:156–61.
8. Miller M, Morley JE, Rubenstein LZ. Hyponatremia in a nursing home population. J Am Geriatr Soc 1995;43:1410–3.
9. Miller M, Hecker MS, Friedlander DA, et al. Apparent idiopathic hyponatremia in an ambulatory geriatric population. J Am Geriatr Soc 1996;44:404–8.
10. Hirshberg R, Ben-Yehuda A. The syndrome of inappropriate antidiuretic hormone secretion in the elderly. Am J Med 1997;103:270–3.
11. Lawson DH. Adverse reactions to potassium chloride. Q J Med 1974;171:433–40.
12. Institute of Medicine, Food and Nutrition Board. Dietary Reference Intakes for Calcium, Phosphorus, Magnesium, Vitamin D and Fluoride. Washington, D.C.: National Academy Press, 1997.
13. Yates AA, Schlicker SA, Suitor CW. Dietary Reference Intakes: the new basis for recommendations for calcium and related nutrients, B vitamins, and choline. J Am Diet Assoc 1998;98:699–706.
14. Bryant RJ, Cadogan J, Weaver CM. The new dietary reference intakes for calcium: implications for osteoporosis. J Am Coll Nutr 1999;18(Suppl):406S–12S.
15. Chapuy MC, Chapuy P, Meunier PJ. Calcium and vitamin D supplements: effects of calcium metabolism in elderly people. Am J Clin Nutr 1987;46:324–8.
16. Pao EM, Mickle SJ. Problem nutrients in the United States. Food Technol. 1981;35:58–69.
17. Reinhart RA. Magnesium metabolism: a review with special reference to the relationship between intracellular content and serum levels. Arch Intern Med 1988;148:2415–20.
18. Lindeman R. Chapter 131. Trace minerals: Hormonal and metabolic interrelationships. In: Principles and Practice of Endocrinology and Metabolism, 3rd edition. Becker KL, ed., Lipincott, Williams and Wilkins, Philadelphia, 2001, 1277–86.
19. McClain CJ, McClain M, Barve S, Boosalis MG. Trace metals and the elderly. Clin Geriatr Med 2002;18:801–18.
20. Fosmire G. Chapter 6. Trace metal requirements. In: Geriatric Nutrition, 3rd edition. Chernoff R, editor. Jones and Bartlett Publishers, Inc., Sudbury, MA, 2006, 95–114.
21. Food and Nutrition Board, Institute of Medicine. Dietary Reference Intakes: The Essential Guide to Nutrient Requirements. The National Academies Press, 2006.

9 Redefining Nutritional Frailty: Interventions for Weight Loss Due to Undernutrition

Connie Watkins Bales and Christine Seel Ritchie

Key Points

- Nutritional frailty is an unintentional, precipitous loss of both lean and fat mass resulting almost entirely from a reduction in food intake.
- Unintentional weight loss is associated with functional decline and increased mortality.
- Efforts to minimize nutritional frailty should emphasize early interventions to maintain and enhance oral food intake; these efforts can include improved esthetics, social support at mealtime, assistance with eating/feeding, and treatment with orexigenic agents.
- When oral food intake is insufficient, protein/calorie supplements or enteral/parenteral feedings by tube may be utilized. However, the extent of positive impact on medical outcomes is generally modest and complications are common with tube feedings.

Key Words: Undernutrition; cachexia; sarcopenia; nutritional supplements enteral nutrition; end-of-life nutrition

9.1 NUTRITIONAL FRAILTY: A REVISED DEFINITION BASED ON ETIOLOGY

During the time since the first edition of this book addressed the topic of nutritional frailty, substantial progress has been made toward the characterization and understanding of the causes of unintentional weight loss in older adults. Thanks to an on-going international effort to develop a consensus definition of cachexia *(1)* and comprehensive work characterizing the natural course of sarcopenia *(2–5)*, there is now a much better understanding of the unique contributions that loss of

From: *Nutrition and Health: Handbook of Clinical Nutrition and Aging, Second Edition*
Edited by: C. W. Bales and C. S. Ritchie, DOI 10.1007/978-1-60327-385-5_9,
© Humana Press, a part of Springer Science+Business Media, LLC 2009

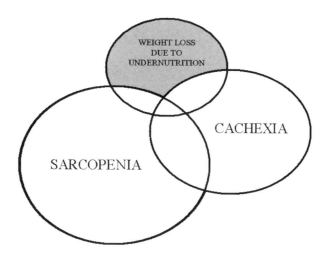

Fig. 9.1. The Unhappy Triad. Three distinct causes of weight loss and physical frailty have different prevalences and overlap to varying degrees. While sarcopenia occurs very commonly with aging, cachexia occurs mainly in association with acute or chronic disease. Weight loss due strictly to undereating is the least common of the three but has obvious overlap with the other causes of frailty. This conceptual presentation was originally proposed by Dr. Thomas at the Third Cachexia Consensus Conference in Rome, Italy, in December, 2005. It was derived from a paper published by Dr. Thomas on "Distinguishing Starvation from Cachexia" *(81)*. Dr. C.C. Seiber assigned the figure the name of "The Unhappy Triad".

appetite and poor food intake make to the overall phenomenon of weight loss and tissue wasting in older patients. Thus, while acknowledging that often more than one condition exists in a given older individual (see Fig. 9.1), this edition of the *Handbook* includes separate discussions of undernutrition-related weight loss (this chapter), sarcopenia (Chapter 10), and cachexia (Chapter 11). This chapter explores the development of nutritional frailty as specifically defined, its contribution to overall frailty *(6)*, and potential interventions to lessen its detrimental effects.

9.1.1. Definitions and Risk Factors

Table 9.1 lists definitions for a number of salient terms; the one most integral to this chapter is the definition of nutritional frailty, which is an unintentional, precipitous loss of both lean and fat mass resulting almost entirely from a reduction in food intake. In contrast, sarcopenia is an age-related loss of muscle and cachexia is a complex metabolic syndrome associated with underlying illness and (often) inflammation in which there is loss of muscle with or without loss of fat mass. In particular, it should be emphasized that nutritional frailty as a sole cause of weight loss is less common than sarcopenia (although the two can overlap) and that both nutritional frailty and sarcopenia can occur in the absence or presence of cachexia (again, see Fig. 9.1). It should be noted that the consensus definition for cachexia shown in Table 9.1 was recently developed by an International Cachexia Consensus group *(1)*.

Table 9.1
Glossary of weight-loss-related definitions

Nutritional frailty (weight loss due to undernutrition): An unintentional, precipitous decrease in body mass (both adipose and lean) that produces detrimental effects on function and other health outcomes and results almost entirely from a reduction in the intake of energy yielding nutrients. Nutritional frailty is distinct from age-related muscle loss (sarcopenia) and can occur in the absence or presence of cachexia.

Frailty: The presence of three or more of the following: unintentional weight loss, self-reported exhaustion, weakness, slow gait speed, and low physical activity *(83)*.

Sarcopenia: The process of age-related loss of muscle mass and strength that occurs with aging; often the term sarcopenia is used to refer to older persons with skeletal muscle values in an unhealthy range, which has traditionally been defined as a height-adjusted muscle mass of ≥ 2 standard deviations below the mean of a young and healthy population; see Chapter 10.

Cachexia: "Cachexia is a complex metabolic syndrome associated with underlying illness and characterized by loss of muscle with or without loss of fat mass. The prominent clinical feature of cachexia is weight loss in adults (corrected for fluid retention) or growth failure in children (excluding endocrine disorders). Anorexia, inflammation, insulin resistance and increased muscle protein breakdown are frequently associated with wasting disease. Wasting disease is distinct from starvation, age-related loss of muscle mass, primary depression, malabsorption and hyperthyroidism and is associated with increased morbidity." (Source: International Consensus Conference *(1)*); also see Chapter 11.

Sarcopenic obesity: Individuals with sarcopenia (defined above) who also have a percent body fat greater than sex-specific cutoff values and an approximate BMI of ≥ 30 kg/m^2 *(19)*.

9.1.2. The Mortality Impact of Body Weight and Weight Loss

It might seem very obvious that decrements in food intake that result in undernutrition and weight loss would adversely impact health. In the case of pronounced weight loss and subnormal body weights (expressed as a low Body Mass Index, calculated as weight (kg)/[height (m)]^2; BMI), this is surely known to be the case. Less well understood, however, are the complex interactions that occur between health and body mass when BMIs are within the range of normal to overweight (BMI 25–29.9 kg/m^2). Recent concerns about the effects of the obesity epidemic on the health of adults of all ages *(7,8)* have raised the consciousness of both older patients and their caregivers regarding the potential detrimental effects of excess body weight. It is true that obesity in old age contributes to disability and frailty *(9)*; it can also intensify metabolic disorders like insulin resistance and dyslipidemias. However, the ideal BMI for older adults remains elusive. Epidemiological studies of the relationship of BMI with mortality reveal that the extent of the negative impact of being overweight tends to decline with aging *(10,11)*. In fact, most studies show a beneficial or neutral, rather than a detrimental, effect of a BMI in the overweight range on length of life after the age of 65 years. Additionally, there is a well-established link between unintentional weight loss and poor health out-comes in later life. A number of studies have associated recent weight loss with shortened survival. Newman et al. *(12)* found that even a modest decline in body weight was an important, independent marker of risk for mortality in older adults. Locher et al. *(13)* demonstrated a 1.7-fold increase in mortality risk with uninten-tional weight loss in community-dwelling older persons. A prospective study of weight loss and non-cancer-related mortality in 5,722 overweight/obese but other-wise healthy Swedish men showed that those who lost weight had higher mortality rates than weight stable men in the same weight range *(14)*. Lee and Paffenberger *(15)* reviewed 17 studies of weight loss and subsequent all-cause mortality and concluded that those who "faired best" were those who remained weight stable. The observation of increased survival in heavier adults with wasting diseases like end-stage renal disease, heart failure, and COPD (the so-called "reverse epide-miology" of obesity and wasting diseases) is another factor that contributes to concerns about the best approach for managing body weight issues in late life *(16)*. Excess adiposity in late life can serve as an energy reserve in times of food deprivation or illness; additionally, the proportion of energy expenditure derived from protein oxidation is lower and lean tissue is better preserved in persons with large fat stores *(17)*.

Body composition changes that occur with weight loss may also explain the negative effect of weight loss on function in older adults *(18)*. Changes in body composition (particularly muscle and bone) with aging can lead to undesirable effects that would not be apparent from a change in body weight alone. In fact, due to hormonal and other changes discussed in more detail in Chapter 10, even adults who remain weight stable have a shift in body composition in later life characterized by more fat and less muscle and bone. When the loss of lean mass is dramatic, an individual may be frail even with a normal or elevated BMI. Sarco-penic obesity is associated with functional decline, including the development of disabilities for Instrumental Activities of Daily Living (IADLs) *(19)*.

9.2 CAUSES OF UNDERNUTRITION

It is well established that food intake diminishes with age, with a gradual down-ward trend over the later decades of life *(20–23)*. This almost universally observed phenomenon is due to lower energy demands as a result of reduced physical activity, decreases in resting energy expenditure, and/or loss of lean body mass (sarcopenia). Over time, these changes produce a decrease in demand for calories and thus for food intake *(24,25)*.

Other common causes of reduced food intake are listed in Table 9.2 and include physiologic changes such as loss of appetite, alterations in taste and smell, poor oral health, gastrointestinal changes, and a reduced ability to regulate appetite in response to acute weight changes. In addition, psychological and socioeconomic factors may lead to undernutrition and a variety of pathologic factors, including illness and medication effects, may contribute to a chronic, progressive reduction in food intake.

Table 9.2
Causes of undereating-related weight loss

Physiologic/pathologic	Psychologic	Socioeconomic
Appetite/food intake regulation *(84)*	Depression	Social and/or geographical isolation, loneliness *(31,85)*
Oral health problems *(86,87)*	Dementia *(88–90)*	Lack of caregiver support
Sensory impairments	Other cognitive impairments	Food insecurity
Altered absorption or digestion	Longstanding emotional or mental illness	Institutionalization and decreased access to food *(91,92)*
Acute and chronic diseases, associated therapies		
Difficulty with self-feeding (physical or neurological disability)		
Dependence on enteral or parenteral feeding *(93,94)*		

9.3 EARLY INTERVENTIONS TO INCREASE FOOD AND FLUID INTAKES

Body weight, although sometimes difficult to obtain in older patients, should be routinely monitored. A chair or bed scale can be used for those unable to stand. Height measures can also be problematic. When ascertaining height for calculation of BMI, it will be best to use the patient's height before age 50 as the reference height in order to avoid the effects of kyphosis due to osteoporosis *(26)*. If weight loss and/ or low BMI are confirmed, the frequency of monitoring weight for that patient should be increased to weekly intervals *(27)*. In the nursing home setting, interventions should be considered for patients who have lost 5% of usual body weight in 30 days or 10% in 6 months *(28)*, using the calculation of weight change percent as usual weight—current weight/usual weight × 100. The algorithm in Fig. 9.2 illustrates many of the recommended approaches for correcting problems with nutritional adequacy when unintentional weight loss is a concern. Strategies to utilize in patients for whom improving oral intake is a viable option are discussed in this section.

9.3.1. Therapeutic Interventions

Clearly, correction of any underlying medical cause for weight loss should be the first step employed to help counteract low food intake. Unfortunately, it is the rare situation when this is possible, either because no treatment is able to resolve the condition or because the primary cause of the anorexia is unknown. Other causes

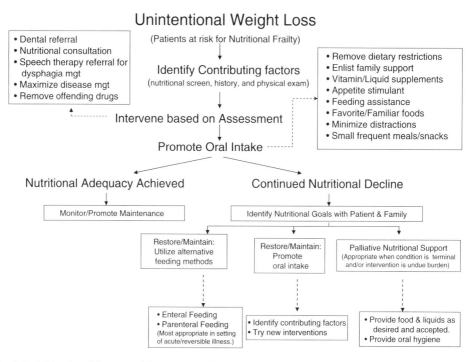

Fig. 9.2. This algorithm to guide care providers as they identify, assess, and treat weight loss and nutritional frailty in older adults living in the community was previously published by Bales and White *(82)* and is reprinted with permission.

such as reduced access to food, depression, social isolation, poor dentition, and over-medication are more amenable to correction and every effort should be made to remove these barriers when they are impeding food intake.

9.3.2. Improving the Dining Environment and the Nutritional Value of Food Choices

There is growing evidence of the measurable impact of improving the aesthetics of the food and dining environment on the food intake of frail patients. Mathey et al. *(29)* instituted improvements in meal ambiance focused on the physical environment and atmosphere of the dining room, quality of food service, and nursing staff assistance for 38 elderly residents of a Dutch nursing home. After 1 year, in contrast to the control group, the 22 residents who completed the study had increased body weights ($+3.3$ kg, $P < 0.05$) and stable health status assessments. The importance of socialization and support at mealtime should not be under-estimated *(30)*. Locher et al. *(31)* demonstrated the value of the presence of others at mealtime, reporting that persons who had others present during meals consumed an average of 114.0 calories more per meal than those who ate alone ($p = 0.009$). Nijs et al. *(32)* conducted a randomized controlled trial of providing family-style meals to older residents in five Dutch nursing homes and observed significant improvements in daily energy intakes. There was a decrease in the proportion of residents in the intervention group scored as malnourished by the Mini Nutritional

Assessment from 17 to 4%, whereas the percentage malnourished increased from 11 to 23% in the control group. Altus et al. *(33)* found that serving meals family style led to modest increases in mealtime participation and communication even in residents with dementia and very low pre-study rates of appropriate communication. These and other findings provide strong justification for emphasizing improvements in the taste, presentation, and social setting of meals for the older patient with clinically important unintentional weight loss.

Along with improving the esthetic appeal of the food and the presence of others at meals, it is also important to optimize the nutrients provided in the meal and snacks that are offered. Lorefalt et al. *(34)* enriched the energy and protein content of small meals offered to 10 patients on a geriatric rehabilitation ward and increased the daily energy intake by 37%, accompanied by an increase in the intakes of protein, fat, carbohydrate, certain vitamins, and minerals. Odlund Olin et al. *(35)* were able to demonstrate similar findings, along with preservation of Activities of Daily Living (ADLs), by providing energy-dense meals to 35 nursing home residents (median age = 83 years). Zizza et al. *(36)* found that older adults who ate between-meal snacks had significantly higher daily intakes of energy, protein, carbohydrate, and total fat, with snacking contributing 14% of their daily protein intakes. Table 9.3 provides a list of easy-to-eat foods and snacks along with their caloric and protein density per gram of food.

Bernstein et al. *(37)* found measurable improvements in dietary quality when they provided nutrition education to community-dwelling elders with functional impairments, demonstrating the value of including older adults in the effort to improve their nutrient intakes whenever this is feasible. Included in the nutrition education should be an emphasis on choosing a wide variety of nutritious foods. Roberts et al. *(38)* reported that older adults with low BMIs ($<22 \text{ kg/m}^2$) consumed a lower variety of energy-dense foods compared with older adults with higher BMIs ($p < 0.05$) and Bernstein et al. *(39)* found a highly varied diet to be linked with better nutritional status in nursing home residents as assessed by nutrient intake, biochemical measures, and body composition measures.

Another important step in enhancing food intake and nutritional status is the lifting of dietary restrictions whenever possible so as to offer a wider selection of food choices. A therapeutic diet prescription may be unnecessary for frail, undernourished patients and can often be bypassed even in diabetic patients, provided they are being regularly monitored *(40)*. The American Dietetic Association supports removing restrictions as a way of enhancing food intakes *(41)*.

9.3.3. Providing Feeding Support and Assistance

When the process of eating and swallowing are physically affected by age- or disease-related changes, the loss of the ability to adequately self-feed can place the older patient at high nutritional risk. An approach called functional feeding suggested by Van Ort et al. *(42)* emphasizes a behavioral approach that involves the patient and encourages interaction between the patient and feeder throughout the meal. Feeding approaches that emphasize touch and/or verbal cueing have been used successfully for older patients who were severely cognitively impaired *(43)*. However, it should be

Table 9.3
Calorie and protein content of familiar foods and snacks of high caloric density

Food	Serving size, typical and (g)	Calories	Energy density (kcal/g)	Protein (g)	Protein density (g protein/g)
Main meal components and meal replacements					
Bagel with cashew butter	½ Bagel (28)	72	2.6	2.5	0.089
	2 T. cashew butter (30)	180	6.0	2.8	0.093
	Total (58)	252	4.3	4.5	0.078
Whole-wheat eggo waffle	1 Waffle (35)	90	2.6	2.5	0.071
With butter and syrup	2 T. butter (9.5)	68	7.2	0.08	0.008
	1 T. syrup (60)	158	2.6	0	0
	Total (104)	316	3.0	2.6	0.025
Carnation instant breakfast drink	8 Fluid ounces (281)	250	0.9	13.0	0.046
Peanut butter sandwich	2 Slices bread, wheat (50)	130	2.6	5.0	0.10
	2 T. peanut butter (32)	190	5.9	8.2	0.256
	Total (82)	320	8.5	13.2	0.16
Cream soup, chicken, made with whole milk	6 oz. (186)	109	0.6	5.5	0.029
Macaroni and cheese					
–Home recipe	½ cup (100)	215	2.2	8.4	0.084
–Box type	½ cup (100)	160	1.6	7.0	0.07
Fried chicken	1 leg (62)	162	2.6	16.6	0.268
	½ Boneless chicken breast (98 g)	218	2.2	31.2	0.318
Mashed potatoes with gravy	½ Cup potatoes (105)	81	0.8	2.1	0.02
	2 T. gravy (30)	25	0.8	0.5	0.016
	Total (135)	106	1.6	2.6	0.019
Health choice frozen dinner –Grilled chicken	1 Meal (284)	270	0.95	22	0.077
Apple pie	½ Slice pie (89)	265	2.9	2.0	0.022
Ice cream, vanilla	1 Scoop ice cream (70)	150	2.2	3.0	0.043

(continued)

Table 9.3
(continued)

Food	Serving size, typical and (g)	Calories	Energy density (kcal/g)	Protein (g)	Protein density (g protein/g)
Pudding, chocolate, ready to eat	1 Snack cup (113)	110	1.0	2.0	0.017
Between-meal snacks					
Cheddar cheese on crackers	2 oz. Cheese (57)	223	3.9	14.2	0.249
	4 Saltine squares (11)	50	4.6	1.0	0.09
	Total (68)	273	8.5	15.2	0.223
Yogurt, full fat, flavored	8 oz. (227)	253	1.1	10.7	0.047
Crunchy granola bar	2 Bars (42)	180	4.3	5.0	0.119
Cliff energy bar	1 Bar (68)	230	3.4	10.0	0.147
Peanuts, dry roasted	2 T. (19)	107	5.8	4.3	0.226
Walnuts, pieces	¼ Cup (25)	180	7.2	3.5	0.14
Hershey chocolate kisses	6 (28)	145	5.1	2.0	0.071
Milkshake, chocolate					
–Fast food type	8 oz. (147)	174	1.2	6.3	0.043
Condiments and add-ons					
Olive oil*	2 T. (27)	239	8.8	0	0
Nonfat dry milk powder**	2 T. (15)	54	3.6	5.4	0.36
Soy sauce	2 T. (36)	22	0.61	4.0	0.11
Barbeque sauce	2 T. (30)	60	0.66	1.0	0.03
Mustard	2 tsp. (10)	6.6	0.66	0.4	0.04

*Can be added as source of calories to savory sauces and foods.
**Can be used to fortify beverages, soups, puddings, and sauces.

noted that not all patients benefit equally from feeding assistance *(44)*. Because of the intensive and time-consuming nature of this approach, Simmons et al. *(45)* recommend a trial of feeding assistance for the purpose of identifying those individuals who are likely to be responsive to intense feeding assistance.

9.3.4. *Use of Appetite Stimulants (Orexigenic Agents)*

As previously noted, in many cases of undernutrition the underlying cause of poor food intake is NOT inferior appetite. However, in situations when poor appetite is known to be an important contributor to undernutrition, pharmacologic appetite stimulants (orexigenic agents) may be considered if improved appetite would be an important contributor to the individual's quality of life. As illustrated in Table 9.4, the available armamentarium for improving appetite in older adults is rather limited, depending primarily on agents approved for use in patients with

Table 9.4
Medications and supplements with potential benefits for appetite and/or weight gain*

Medications/ supplements	Potential side effects	Recommended regimens
Megestrol acetate (48,50,57,95,96)	Edema, hypertension, deep vein thrombosis, adrenal suppression, blunting of muscle response to exercise (51)	400–800 mg/day; treatment course should be no longer than 8–12 weeks
Dronabinol (52,95,97)	Sedation, fatigue, and hallucinations. Not recommended for patients who are cognitively compromised or prone to falls	2.5 mg initially in the evening. Increase to 5 mg per day after 2–4 weeks
Anabolic agents	Masculinization, fluid retention, hepatic toxicity	Oxandrolone: 2.5 mg 2–4 times/day
Protein and amino acid supplements	Renal protein overload, gout and hyperuricemia-related complications	Total protein intake up to 1.6 g protein/kg/day Isoleucine RDA: 25 mg/g protein** Leucine RDA: 55 mg/g protein** Valine RDA: 32 mg/g protein** Creatine: 20 g/day for 5–6 days***
Omega-3 fatty acids and fish oil (57)	Gastrointestinal side effects, potential effects on red blood cell structure, function	Recommendation for healthy adults is 0.3–0.5 g/day EPA + DHA, with about 1 g/day for patients with coronary heart disease (98). Dose level for treatment of weight loss is unknown

*Table is based on part on the review by Yeh et al., 2007 (46).

** Rehabilitation-level doses of specific amino acids are not determined. Values shown are recommendations for healthy adults. From Dietary Reference Intakes: Macronutrients.

*** Rehabilitation-level doses of creatine are unknown. Values shown are from Greenhaff et al. (99).

cancer or HIV/AIDS *(46)*. All of the pharmacologic agents have a meaningful negative side effect profile. Currently, there are no orexigenic drugs approved by the FDA for the treatment of age-related anorexia.

Megestrol acetate (Megace) is a synthetic progesterone derivative that has anti-inflammatory and glucocorticoid activity. Several studies of megestrol acetate have shown moderately encouraging results, with modest positive effects on appetite, weight gain, and quality of life. Because megestrol is a strong catabolic hormone, it should only be given on a temporary basis (no longer than 8–12 weeks at a time). However, its effects can continue to be seen after discontinuation and the drug can be given again if needed after a 3- to 6-month rest period *(46)*. Recently, a nanocrystal formulation of the drug (Megace ES) has become available; it can be given at a lower dosage and is more soluble so that it may be better absorbed when given between meals *(46)*.

In one of the first studies of this medication in older adults, Yeh et al. *(47)* reported that 800 mg/day of megestrol acetate increased appetite, food intake, and (in 3 months) body weight in nursing home patients who had lost 5% or more of their body weight. Karcic et al. *(48)* reported improvement in nutritional parameters (including food intake and BMI) in 13 megestrol-treated nursing home patients for whom other methods of nutritional support had failed. Simmons et al. *(49)* noted little benefit using Megace OS alone but, when used in combination with optimal mealtime feeding assistance, there was a significant increase in oral intakes of frail nursing home patients at high risk for weight loss. Reuben et al. *(50)* randomized 47 older persons (mean age 83 years) who were post-hospital discharge with fair or poor appetite to a placebo or a megestrol acetate suspension providing 200, 400, or 800 mg daily for 9 weeks. The megestrol acetate doses of 400 and 800 mg were associated with increased levels of plasma prealbumin, but cortisol suppression was *common* at the higher doses. Other potential side effects of this orexigenic agent include edema, hypertension, adrenal suppression, and thromboembolism. Another potential concern about megestrol acetate is the observation by Sullivan et al. *(51)* that in subjects (mean age 79.4 ± 7.4 years) participating in a randomized trial of resistance training alone or with megestrol acetate, the addition of the drug appeared to blunt the beneficial effects of resistance training, lessening the anticipated gains in muscle strength and functional performance. One solution for this problem might be to institute the resistance training in post-megestrol-treated patients *(46)*.

Dronabinol (tetrahydrocannabinol, a cannabis derivative with FDA approval for use in HIV/AIDS and cancer) has been suggested as a possible agent to increase appetite and lead to weight gain in geriatric patients. It has been suggested as a promising choice when anorexia is secondary to nausea and is best given in the evenings due to its sedative effects. A retrospective observational study of residents with anorexia and weight loss in five long-term care facilities on a 12-week course of dronabinol showed that the drug was well tolerated; 15 of the 28 treated subjects gained weight *(52)*. In a placebo-controlled crossover study, with each treatment period lasting 6 weeks, 15 patients with a diagnosis of probable Alzheimer's disease who were refusing food were treated with dronabinol. In the patients who tolerated the treatment and could remain in the study (n = 11), there was an increase in body weight and a decrease in severity of disturbed behavior. The adverse reactions of euphoria, somnolence, and tiredness did not necessitate discontinuation of the treatment. Nonetheless, with the potential side effects of sedation and hallucinations, dronabinol may not be an appropriate choice for confused elderly patients or those prone to falls (46). A study comparing megestrol acetate and dronabinol treatment in advanced cancer patients showed superior effectiveness for megestrol; combination therapy with the two drugs provided no additional benefit (53).

Anabolic drugs such as oxandrolone and nandrolone are known to produce increases in lean mass in body builders and have been considered for their potential benefit in frail older patients with loss of lean body mass and in those with chronic wasting conditions like COPD (54). Schroeder et al. (55) found that oxandrolone (10 mg twice daily) produced substantial increases in lean body mass as well as decreases in total body and trunk fat mass in older men (72 ± 6 years). While such effects are hormonal, not nutritional, they could interact with nutritional parameters like metabolic rate and insulin sensitivity. Anabolic agents have a host of potential side effects such as hepatotoxicity, hypogonadism, testicular atrophy, gynecomastia, and psychiatric disturbances. There is insufficient study in older subjects at this time to enable us to make clinical recommendations and their use is confined to controlled studies (56).

Various amino acids and fatty acids are listed in Table 9.4 but the potential for these compounds to be used as agents to improve weight status have not been fully studied. To date, findings regarding their benefits are not particularly encouraging. A comparison of eicosapentaenoic acid (EPA) supplement with Megace treatment showed that subjects did not have better improvements for weight or appetite, whether the EPA was given alone or in combination with the Megace (57).

9.4 INTERVENTION WITH NUTRITIONAL (PROTEIN/CALORIE) SUPPLEMENTS

When intensive efforts to encourage adequate oral intake of a natural diet fail to produce the desired preservation of body weight and lean mass, individuals who can still take oral nutrition may be offered a commercially prepared supplemental food. Protein/calorie supplements (usually provided as a liquid drink or pudding) are used as an "add on" to the traditional diet to help improve nutrient intake, particularly in the institutional setting. Table 9.5 provides a current listing (as of December 2008) of many of the commonly utilized commercial supplements in the US, with their nutritional content, potential application (e.g. oral or tube, tube only), and manufacturer contact information. Many of these products are intensely marketed, with glossy color advertisements in medical/nutrition journals. The estimated cost of around $1.50 per day ($45.00/month) is not insignificant for older persons on a fixed budget. However, Arnaud-Battandier et al. (58) conducted a cost analysis with 311 patients (age >70 years) living at home or in institutions and reported that those who received oral nutritional supplements had lower medical care consumption, fewer hospitalization days, and reduced medical care costs. The cost of the supplement use was exceeded by the estimated savings in medical costs.

While the expense and the monotony of their use are definite concerns, there can be advantages for these products. Generally speaking, they (1) provide a ready source of nutrients, often including fortification with essential micronutrients, (2) help assure safety from food-borne illness and inadvertent contamination during preparation due to standardized manufacturing processes, and (3) provide easy access to nutrition, being ready to serve and suitable for long-term storage at room temperature.

The disadvantages of these supplements have been widely extolled. Concerns about relatively low protein content, lack of fiber and other beneficial components of natural foods, potential misuse as meal replacements, avoidance of liquid intake by elderly patients with urinary incontinence, and the chance of electrolyte and carbohydrate overload in diabetes and chronic renal insufficiency, respectively, have all been expressed (59–61). Fortunately, an evidence-based approach for decision-making about the effectiveness of these products is on the horizon as intervention trials continue to accumulate. The designs and findings of many of the trials published within the past 10 years are summarized in Table 9.6.

Two recent analyses have considered the overall trends for the effects of nutritional supplements on mortality outcomes. Milne et al. (62) reported a meta-analysis of 55 randomized trials evaluating the effect of nutritional supplements on 9,187 older participants. In long-term care subjects (receiving ≥35 days of intervention) who were undernourished at baseline, 75 years of age or older, and receiving greater than 1674 kJ in supplements, a reduced mortality was found. No such effect was observed for well-nourished patients receiving supplementation. In a systematic review, Koretz et al. (63) examined the effects of volitional nutritional support (VNS) on mortality. Based on the 16 trials included, there was an increased survival for malnourished and/or institutionalized geriatric patients receiving VNS. Of these, two trials were of high quality (Price et al. (64) and Potter et al. (65)) and, although a mortality benefit was favored in these trials, neither showed a significant difference in survival due to VNS treatment. The findings of eight of the 11 studies presented in Table 9.6 show positive outcomes for body weight, nutrient intake, and/or nutritional status.

While we see a clear trend of positive findings with regard to the effects of these nutritional supplements, often the overall impact on outcomes is fairly modest. Obviously, many factors that are not attributable to the formulation of the products affect clinical outcomes. For example, inconsistencies in the administration of nutritional supplements are commonly observed in the institutional setting. Simmons et al. (66) observed supplement provision in a nursing home setting where 88% of patients had an order to receive a supplement one to three times daily and 12% of patients had an order of four to six times daily and found supplements provided on average less than once per participant per day. Fewer than 10% of patients received the supplement consistent with their orders during the 2-day observation.

To summarize, the use of commercial protein/calorie supplements should be reserved for patients who have specific limitations in their oral food intake, such as food intolerances, inability or unwillingness to eat adequate amounts of nutritious foods, and in situations where having the patient do their own food

Table 9.5
Compositional profiles of selected commercially available nutritional supplements

Name of product/ manufacturer Nutritional description	Boost/Nestle	Ensure/Abbott	Jevity/Cal/Abbott	Osmolite/ Cal/Abbott
kcal/mL	1.01	1.06	1.06	1.06
Protein (g/L)	42	38	44.3	44.3
Protein source	Milk protein concentrate	N/A	N/A	N/A
Carbohydrate (g/L)	173	169	154.7	143.9
Carbohydrate source	Corn syrup solids, sugar	N/A	N/A	N/A
Fat (g/L) total (t), saturated (s)	17.8 (t), 2 (s)	25 (t), 4 (s)	34.7 (t), N/A (s)	34.7 (t), N/A (s)
Fat source	Canola, high oleic sunflower and corn oils	N/A	N/A	N/A
Fiber (g/mL)	0	0	0.0144	NA
Osmolality	610-670 mOsm/kg H_2O	590-600 mOsm/ kg H_2O	300 mOsm/kg H_2O	300 mOsm/kg H_2O
Unit amount/ flavors	8 fl oz serving; chocolate, vanilla, strawberry	8 fl oz serving; vanilla, chocolate, strawberries and cream, butter pecan, coffee latte	8 fl oz serving; unflavored, for tube feeding	8 fl oz serving; unflavored
Web site	http://www.boost. com/healthcare professional.htm	http://abbott nutrition.com/ products/ products. aspx?pid = 222	http://abbott nutrition.com/ products/index.aspx	http://abbott nutrition. com/ products /products. aspx? pid = 32# factlabel
Access date	11/12/2007	11/12/2007	11/12/2007	12/9/2008
Notes	Novartis has boost glucose contro			

N/A, information not available.

Promote/Abbott	Replete/Nestle	Slim Fast/Slim Fast Foods	Carnation Instant Breakfast/Nestle	Clinutren/Nestle
1	1	0.68	0.76–0.79	1.25
62.5	62.4	30.7	41	40
N/A	N/A	N/A	N/S	N/A
130	113	123	127	270
N/A	N/A	N/A	N/A	N/A
26 (t), N/A (s)	34 (t)	9.2 (t), 3.1 (s)	15.9 (t), 4.8 (s)	<2 (t), ? (s)
N/A	N/A	N/A	N/A	N/A
NA	NA	0.015	0.003	<0.002
340 mOsm/kg H_2O	300 mOsm/kg H_2O (unflavored), 350 mOsm/kg H_2O (vanilla)	N/A	N/A	N/A
8 fl oz serving; vanilla	8.4 fl oz serving; vanilla and unflavored	11 fl oz serving; French vanilla, strawberries n' cream, and cappuccino delight contain 180 cals; creamy milk chocolate and rich chocolate royale contain 190 cals	10.6 oz serving; creamy milk chocolate, French vanilla, strawberry crème	200 mL (6.8 oz) serving; orange, grapefruit, rasberry/blackcurrant, pear/cherry
http://abbottnutrition.com/products/products.aspx?pid=35	http://www.nestle-nutrition.com/product.aspx?objectID=7DCD5E99-7979-462C-99EA-590C28B009FE	http://www.slim-fast.com/products/products.asp	http://www.carnationinstantbreakfast.com/Products/Details.aspx?ProductId=258A92B0-A7BA-4509-B5EB-6C291AA4117B#	http://www.nestlenutrition.com/NR/rdonlyres/7019D37A-DBCB-42D4-BDF1-DD330D0A33A6/0/ClinutrenFruit.pdf
11/19/2007	11/19/2007	11/19/2007	11/19/2007	03/27/2008
				Clinutren also offers soups, desserts for those with chewing/swallowing problems, high fiber milky drinks, high calorie/protein milky drinks, high calorie, and regular milky drinks

Table 9.6
Studies* of nutritional supplements with mortality, weight, and/or nutritional status outcomes

Characteristics of geriatric study population	N** and mean ± SEM age, years	Study design	Protein/energy supply/day	Intervention duration	Mortality, BW, and nutritional outcomes in treatment group	Reference
Newly discharged from hospital in the UK	51/49, 77 ± 5.3 and 79 ± 8.0	RCT	600–1000 kcal as a supplemented drink, pudding, or bar	8-Week intervention, f/u 24 weeks	No difference in nutritional status between groups at 24 weeks	Edington et al., 2004 (100)
Hospitalized geriatric patients in France	39/41, 82 ± 7.6 and 79 ± 6.1	RCT	500 kcal and 21 g protein per day	2 Months	Increased energy intake, maintained BW, increased MNA scores ($p = 0.004$)	Gazzotti, et al., 2003 (101)
Nursing home residents who were not eating well	29/11, NA	Purposive sampling	Supplements ordered by own physician	3 Days	BW was not preserved, nearly half continued to lose weight	Kayser-Jones et al., 1998 (102)
Nursing home residents in France	13/22, 85 ± 5.5 and 85 ± 5.5	RCT	300–500 kcal and up to 37.5 g protein	2 Months	Increased daily protein and energy intake, increased BW (1.5±0.4 kg, 1.4±0.5 kg) and nutritional status for malnourished patients and patients at risk of malnutrition, respectively	Lauque, et al., 2000 (103)

(continued)

Table 9.6
(continued)

Characteristics of geriatric study population	N** and mean ± SEM age, years	Study design	Protein/energy supply/day	Intervention duration	Mortality, BW, and nutritional outcomes in treatment group	Reference
Geriatric wards and daycare centers in France	46/45, 80 ± 5.9 and 78 ± 4.8	RCT	300–500 kcal and up to 21 g protein	3 Months intervention, 6-month f/u	Increased **BW** (1.90±2.33 kg) and fat-free mass (0.78±1.20 kg) at 3 months, benefit maintained at 6 months (1.57±3.35 kg) and (0.63±1.60 kg), respectively	Lauque, et al., 2004 (104)
Elderly patients from nursing care units of a geriatric facility	$n = 143$, 60–103, 83 (median)	Prospective (uncontrolled) intervention	1020–2040 kcal and 42–84 g protein	1–6 Years	Increased **BW** (0.319 kg/ month for 0–23 months), (0.174 kg/ month for 24–60 months), and maintained protein status	Levinson et al., 2005 (105)
Elders receiving community at-home care service	42/41, 82 ± 7.5 and 79 ± 6.1	RCT	250 kcal	16 Weeks	Increased weight gain (1.62±1.77 kg), excessive weight loss stabilized or reversed	Payette et al., 2002 (106)
Emergency admission patients to elderly unit in Scotland hospital	165/162, > 60	RCT	540 kcal and 22.5 g protein	18 Months	Reduced mortality ($p<0.05$), BW gain ($p = 0.003$), and better energy intake ($p = 0.001$)	Potter et al., 2001 (65)

(continued)

**Table 9.6
(continued)**

Characteristics of geriatric study population	N** and mean ± SEM age, years	Study design	Protein/energy supply/day	Intervention duration	Mortality, BW, and nutritional outcomes in treatment group	Reference
Undernourished patients, post-hospital discharge	35/41, 85	RCT	600 kcal. 24 g protein	8 Weeks after hospital discharge	Insignificant ($p=0188$) BW increase, greater increase in handgrip strength	Price et al., 2005 (64)
Psychogeriatric nursing home residents in the Netherlands	19/16, 85 ± 84 and 79 ± 8.8	RCT	273 kcal and 8.5 g protein	3 Months	Significant improvement was observed for BW (1.4 ± 2.4 kg, $p=0.03$) and nutritional status	Wouters et al., 2002 (107)
Residence/ sheltered housing for older people in the Netherlands	34/34, 84 ± 63 and 81 ± 6.9	RCT	250 kcal, 8.8 g protein	6 Months	Increased BW gain (1.6 kg, $p=0.03$), increase in and positive influence on sleep	Wouters et al., 2003 (108)

*Only studies conducted since 1997 are included.

**N, intervention/control whenever this information is available.

RCT, randomized controlled trial; BW, body weight; MNA, Mini Nutritional Assessment.

preparation is prohibited or unsafe. Timing of supplements with meals must be considered, recognizing that older adults reach satiation more quickly and have slower gastric emptying (67). Liquid dietary supplements should be administered between meals in order to optimize net energy consumption for the day (68).

9.5 PARTIAL OR TOTAL NUTRITION SUPPORT

Enteral feedings are often considered if the patient is unable to ingest sufficient calories through other means and the patient's condition and patient/family wishes dictate such an approach (also see Chapter 13). Only a few randomized controlled trials have been performed to evaluate the effectiveness of enteral nutrition outside of the critical care setting. These trials have been in stroke patients, hip fracture patients, and patients with cancer (see Chapter 19 for studies in cancer patients). There have been no trials of enteral nutrition in patients with neurodegenerative disorders such as Parkinson's or Alzheimer's disease. As a result, information regarding the effectiveness of enteral nutrition in these patient populations is based on observational data and must be understood to be highly confounded.

Two of the FOOD (Feed Or Ordinary Diet) trials evaluated enteral nutrition in stroke patients with dysphagia (69). The first trial evaluated early versus late enteral nutrition. When early (less than a week) versus late (after a week) enteral feeding was the intervention, there was an absolute reduction in mortality of 5.8% (95% CI –0.8 to 12.5, $p = 0.09$) in those who received early enteral feeding. Not surprisingly, those patients, in part due to survivor effect, were more heavily dependent on others for activities of daily living. In the second trial, nasogastric (NG) feeding was tested against feeding by percutaneous gastric tube (PEG) (69). PEG feeding was associated with an increase in death and poor outcome at a just significant level ($p = 0.05$). The conclusions of these studies were that NG feeding should be used for dysphagic patients early and that PEG feeding should be reserved for patients either who do not tolerate NG feeding or who require long-term enteral nutrition.

Adults ≥65 years of age who underwent surgical repair of an acute hip fracture were randomized to nightly nasoenteral tube feeding versus standard care (70). Although during the first week postoperatively the intervention group received a greater amount of calories, by the second week, there was no difference in intake between the intervention group and the control group. There were also no differences in mortality or complications between the two groups. Additionally, the tube feeding was poorly tolerated in the intervention group.

Because the population being considered for enteral nutrition often has advanced illness, the impact of enteral nutrition on mortality has been difficult to ascertain. Reported 30-day mortality is high, with rates in the 10–25% range (71–73). In a large study from the Veterans Administration, 24% of patients died before leaving the hospital (74). However, a subset of patients receiving enteral nutrition live for a long time; in women less than 80 years of age referred from a nursing home, the median survival following PEG tube placement was over 2 years (780 days) (71). The variation in mortality rates reflects the great variation in underlying disease states of patients. Poor prognostic features include increased age (>80 years), chewing and swallowing disorders, diabetes, and the presence of underlying malignancy *(72,75)*.

In cases where enteral nutrition is planned, the clinical situation will dictate the use of either NG, nasointestinal, PEG, or percutaneous jejunal (PEJ) tubes for delivery of nutritional support. Because of patient discomfort and the risk for sinusitis with long-term use of NG and nasointestinal tubes, a PEG or PEJ should be considered if enteral feeding is expected to continue for more than 30 days. PEG tubes are often larger than PEJ tubes and thus facilitate easier administration of medications without causing clogging. Klor and Milianti *(76)* also found a favorable outcome when PEGs were used in patients with dysphagia—when provided an individualized treatment program, the oral intakes of all patients were upgraded and 10 of the 16 eventually has their PEGs removed. Loser et al. *(77)* followed 210 patients after PEG placement (not all were elderly) for an average of 133 days and found an increase in body weight at 1 year in the 34.4% of the population who survived. In a meta-analysis of nine studies, aspiration pneumonia was reduced to 24% for post-pyloric (e.g., PEJ) compared to gastric feeding *(78)*.

It is important that nutritional status continue to be monitored after the feeding regime is initiated as some patients may have higher nutritional requirements than provided for in the initial diet order. Nitrogen balance should also be assessed. Refeeding syndrome (characterized by hypophosphatemia and potentially fatal arrhythmia) is underrecognized and should be considered in patients who have had chronically poor intake prior to initiation of nutrition support.

Parenteral nutrition may be considered when the enteral route absolutely cannot be used. However, very little data exist regarding the efficacy of this form of nutrition support for the elderly. Complications are multiple and include increased risk of systemic bacterial and fungal infections, hyperglycemia, and intestinal atrophy. Delivery by larger central veins is preferable to peripheral delivery if the support is expected to be needed for more than 10–14 days. Careful monitoring for tolerance, nutritional adequacy, and complications by a dedicated multidisciplinary team (including a physician, nurse, nutritionist, and pharmacist) is required to reduce complications. Glucose levels should be carefully monitered and aggressively controlled. Exclusive parenteral nutrition is generally not recommended since there is a lack of evidence that long-term outcomes will be improved using this approach.

As already mentioned, artificial nutrition support cannot be guaranteed to provide a solution to nutritional problems for high-risk patients who cannot or would not eat. In addition, artificial nutrition support may increase the likelihood for physical (restraints) and psychosocial (sensory deprivation, diminished social contact) adverse effects, particularly for institutionalized or frail older adults. The discussion of alternatives to tube feeding should include information on aggressive hand feeding for demented and non-demented patients. Because of a lack of randomized controlled trials, providers cannot offer accurate information regarding survival benefit for PEG tube versus aggressive hand feeding in this population and decisions about the use of tube feedings and the potential benefits which might be expected are not straightforward (see Chapter 13). While one-third of nursing home patients say they would choose to be fed by tube if they could no longer eat because of cognitive deficits *(79)*, this preference may be based on the potentially erroneous conclusion that a prolonged survival would be the outcome *(80)*. A sensitive but honest presentation of the physiological aspects of nutrition support at the end of life and expectations about its impact need to be provided to the patient/family/caregiver by the health-care team so that decisions about nutrition support can be guided by the best interest of the patient.

9.6 RECOMMENDATIONS

1. Early interventions to maintain and enhance oral intake of natural foods are strongly encouraged. These interventions should emphasize improved taste and esthetics, as well as the presence of others at mealtime, and feeding assistance with verbal cueing for eating-dependent individuals.
2. The use of commercially prepared oral nutrient supplements can be a useful choice when the patient is unable to consume adequate amounts of nutrients from a natural foods diet. However, expected weight gain and extent of positive impact on medical outcomes are generally modest.
3. If oral consumption of calories is not possible, feedings may be delivered by tube. In the case of stroke, NG feeding should be used for dysphagic patients early, and PEG feeding should be reserved for patients either who do not tolerate NG feeding or who require long-term enteral nutrition.
4. Complication rates with parenteral nutrition support are high. Decisions to provide aggressive nutritional intervention should be weighed carefully and decisions about feeding tubes should be included in the patient's advance directives (see Chapter 13).

REFERENCES

1. Evans WJ, Morley JE, Argiles J, et al. Cachexia: A new definition. Clin Nutr 2008;27:793–9.
2. Janssen I, Heymsfield SB, Ross R. Low relative skeletal muscle mass (sarcopenia) in older persons is associated with functional impairment and physical disability. J Am Geriatr Soc 2002;50(5):889–96.
3. Janssen I, Baumgartner RN, Ross R, Rosenberg IH, Roubenoff R. Skeletal muscle cutpoints associated with elevated physical disability risk in older men and women. Am J Epidemiol 2004;159(4):413–21.
4. Newman AB, Kupelian V, Visser M, et al. Sarcopenia: alternative definitions and associations with lower extremity function. J Am Geriatr Soc 2003;51(11):1602–9.
5. Goodpaster BH, Park SW, Harris TB, et al. The loss of skeletal muscle strength, mass, and quality in older adults: the health, aging and body composition study. J Gerontol A Biol Sci Med Sci 2006;61(10):1059–64.
6. Bartali B, Frongillo EA, Bandinelli S, et al. Low nutrient intake is an essential component of frailty in older persons. J Gerontol A Biol Sci Med Sci 2006;61(6):589–93.
7. Wang YC, Colditz GA, Kuntz KM. Forecasting the obesity epidemic in the aging U.S. Population. Obesity (Silver Spring) 2007;15(11):2855–65.
8. Bales CW and Buhr G. Is obesity bad for older persons? A systematic review of the pros and cons of weight reduction in later life. J Am Med Dir Assoc 2008;9:302–12.
9. Blaum CS, Xue QL, Michelon E, Semba RD, Fried LP. The association between obesity and the frailty syndrome in older women: the Women's Health and Aging Studies. J Am Geriatr Soc 2005;53(6):927–34.
10. Horiuchi S, Finch CE, Mesle F, Vallin J. Differential patterns of age-related mortality increase in middle age and old age. J Gerontol A Biol Sci Med Sci 2003;58(6):495–507.

11. Janssen I, Katzmarzyk PT, Ross R. Body mass index is inversely related to mortality in older people after adjustment for waist circumference. J Am Geriatr Soc 2005;53(12):2112–8.

12. Newman AB, Yanez D, Harris T, Duxbury A, Enright PL, Fried LP. Weight change in old age and its association with mortality. J Am Geriatr Soc 2001;49(10):1309–18.

13. Locher JL, Roth DL, Ritchie CS, et al. Body mass index, weight loss, and mortality in community-dwelling older adults. J Gerontol A Biol Sci Med Sci 2007;62(12):1389–92.

14. Nilsson PM, Nilsson JA, Hedblad B, Berglund G, Lindgarde F. The enigma of increased non-cancer mortality after weight loss in healthy men who are overweight or obese. J Intern Med 2002;252(1):70–8.

15. Lee IM, Paffenbarger RS, Jr. Is weight loss hazardous? Nutr Rev 1996;54(4 Pt 2):S116–24.

16. Kalantar-Zadeh K, Horwich TB, Oreopoulos A, et al. Risk factor paradox in wasting diseases. Curr Opin Clin Nutr Metab Care 2007;10(4):433–42.

17. Elia M. Hunger disease. Clin Nutr 2000;19(6):379–86.

18. Ritchie CS, Locher JL, Roth DL, McVie T, Sawyer P, Allman R. Unintentional weight loss predicts decline in activities of daily living function and life-space mobility over 4 years among community-dwelling older adults. J Gerontol A Biol Sci Med Sci 2008;63(1):67–75.

19. Baumgartner RN, Wayne SJ, Waters DL, Janssen I, Gallagher D, Morley JE. Sarcopenic obesity predicts instrumental activities of daily living disability in the elderly. Obes Res 2004;12(12):1995–2004.

20. Morley JE. Decreased food intake with aging. J Gerontol A Biol Sci Med Sci 2001;56 Spec No 2:81–8.

21. Wakimoto P, Block G. Dietary intake, dietary patterns, and changes with age: an epidemiological perspective. J Gerontol A Biol Sci Med Sci 2001;56(2 Suppl):65–80.

22. Roberts SB. A review of age-related changes in energy regulation and suggested mechanisms. Mech Ageing Dev 2000;116(2–3):157–67.

23. Vellas BJ, Hunt WC, Romero LJ, Koehler KM, Baumgartner RN, Garry PJ. Changes in nutritional status and patterns of morbidity among free-living elderly persons: a 10-year longitudinal study. Nutrition 1997;13(6):515–9.

24. Hunter GR, Weinsier RL, Gower BA, Wetzstein C. Age-related decrease in resting energy expenditure in sedentary white women: effects of regional differences in lean and fat mass. Am J Clin Nutr 2001;73(2):333–7.

25. Klausen B, Toubro S, Astrup A. Age and sex effects on energy expenditure. Am J Clin Nutr 1997;65(4):895–907.

26. Thomas DR, Zdrowski CD, Wilson MM, et al. Malnutrition in subacute care. Am J Clin Nutr 2002;75(2):308–13.

27. Fabiny AR, Kiel DP. Assessing and treating weight loss in nursing home patients. Clin Geriatr Med 1997;13(4):737–51.

28. MDS Reference Manual, Appendix section A-4. In. Natick, MA: Eliot Press; 1993.

29. Mathey MF, Vanneste VG, de Graaf C, de Groot LC, van Staveren WA. Health effect of improved meal ambiance in a Dutch nursing home: a 1-year intervention study. Prev Med 2001;32(5):416–23.

30. de Castro JM. Age-related changes in the social, psychological, and temporal influences on food intake in free-living, healthy, adult humans. J Gerontol A Biol Sci Med Sci 2002;57(6):M368–77.

31. Locher JL, Robinson CO, Roth DL, Ritchie CS, Burgio KL. The effect of the presence of others on caloric intake in homebound older adults. J Gerontol A Biol Sci Med Sci 2005;60(11):1475–8.

32. Nijs KA, de Graaf C, Siebelink E, et al. Effect of family-style meals on energy intake and risk of malnutrition in Dutch nursing home residents: a randomized controlled trial. J Gerontol A Biol Sci Med Sci 2006;61(9):935–42.

33. Altus DE, Engelman KK, Mathews RM. Using family-style meals to increase participation and communication in persons with dementia. J Gerontol Nurs 2002;28(9):47–53.

34. Lorefalt B, Wissing U, Unosson M. Smaller but energy and protein-enriched meals improve energy and nutrient intakes in elderly patients. J Nutr Health Aging 2005;9(4):243–7.

35. Odlund Olin A, Armyr I, Soop M, et al. Energy-dense meals improve energy intake in elderly residents in a nursing home. Clin Nutr 2003;22(2):125–31.

36. Zizza CA, Tayie FA, Lino M. Benefits of snacking in older Americans. J Am Diet Assoc 2007;107(5):800–6.

37. Bernstein A, Nelson ME, Tucker KL, et al. A home-based nutrition intervention to increase consumption of fruits, vegetables, and calcium-rich foods in community dwelling elders. J Am Diet Assoc 2002;102(10):1421–7.

38. Roberts SB, Hajduk CL, Howarth NC, Russell R, McCrory MA. Dietary variety predicts low body mass index and inadequate macronutrient and micronutrient intakes in community-dwelling older adults. J Gerontol A Biol Sci Med Sci 2005;60(5):613–21.

39. Bernstein MA, Tucker KL, Ryan ND, et al. Higher dietary variety is associated with better nutritional status in frail elderly people. J Am Diet Assoc 2002;102(8):1096–104.

40. Tariq SH, Karcic E, Thomas DR, et al. The use of a no-concentrated-sweets diet in the management of type 2 diabetes in nursing homes. J Am Diet Assoc 2001;101:1463–6.

41. Niedert KC. Position of the American Dietetic Association: Liberalization of the diet prescription improves quality of life for older adults in long-term care. J Am Diet Assoc 2005; 105(12):1955–65.

42. Van Ort S, Phillips LR. Nursing intervention to promote functional feeding. J Gerontol Nurs 1995;21(10):6–14.

43. Lange-Alberts ME, Shott S. Nutritional intake. Use of touch and verbal cuing. J Gerontol Nurs 1994;20(2):36–40.

44. Taylor KA, Barr SI. Provision of small, frequent meals does not improve energy intake of elderly residents with dysphagia who live in an extended-care facility. J Am Diet Assoc 2006; 106(7):1115–8.

45. Simmons SF, Osterweil D, Schnelle JF. Improving food intake in nursing home residents with feeding assistance: a staffing analysis. J Gerontol A Biol Sci Med Sci 2001;56(12): M790–4.

46. Yeh SS, Lovitt S, Schuster MW. Pharmacological treatment of geriatric cachexia: evidence and safety in perspective. J Am Med Dir Assoc 2007;8(6):363–77.

47. Yeh SS, Wu SY, Lee TP, et al. Improvement in quality-of-life measures and stimulation of weight gain after treatment with megestrol acetate oral suspension in geriatric cachexia: results of a double-blind, placebo-controlled study. J Am Geriatr Soc 2000;48(5):485–92.

48. Karcic E, Philpot C, Morley JE. Treating malnutrition with megestrol acetate: literature review and review of our experience. J Nutr Health Aging 2002;6(3):191–200.

49. Simmons SF, Walker KA, Osterweil D. The effect of megestrol acetate on oral food and fluid intake in nursing home residents: a pilot study. J Am Med Dir Assoc 2005;6(3 Suppl):S5–11.

50. Reuben DB, Hirsch SH, Zhou K, Greendale GA. The effects of megestrol acetate suspension for elderly patients with reduced appetite after hospitalization: a phase II randomized clinical trial. J Am Geriatr Soc 2005;53(6):970–5.

51. Sullivan DH, Roberson PK, Smith ES, Price JA, Bopp MM. Effects of muscle strength training and megestrol acetate on strength, muscle mass, and function in frail older people. J Am Geriatr Soc 2007;55(1):20–8.

52. Wilson MM, Philpot C, Morley JE. Anorexia of aging in long term care: is dronabinol an effective appetite stimulant? – a pilot study. J Nutr Health Aging 2007;11(2):195–8.

53. Jatoi A, Windschitl HE, Loprinzi CL, et al. Dronabinol versus megestrol acetate versus combination therapy for cancer-associated anorexia: a North Central Cancer Treatment Group study. Journal of Clinical Oncology 2002;20(2):567–73.

54. Yeh SS, DeGuzman B, Kramer T. Reversal of COPD-associated weight loss using the anabolic agent oxandrolone. Chest 2002;122(2):421–8.

55. Schroeder ET, Vallejo AF, Zheng L, et al. Six-week improvements in muscle mass and strength during androgen therapy in older men. J Gerontol A Biol Sci Med Sci 2005;60(12):1586–92.

56. Blackman MR, Sorkin JD, Munzer T, et al. Growth hormone and sex steroid administration in healthy aged women and men: a randomized controlled trial. JAMA 2002;288(18):2282–92.

57. Jatoi A, Rowland K, Loprinzi CL, et al. An eicosapentaenoic acid supplement versus megestrol acetate versus both for patients with cancer-associated wasting: a North Central Cancer Treatment Group and National Cancer Institute of Canada collaborative effort. J Clin Oncol 2004;22(12):2469–76.

58. Arnaud-Battandier F, Malvy D, Jeandel C, et al. Use of oral supplements in malnourished elderly patients living in the community: a pharmaco-economic study. Clin Nutr 2004;23(5):1096–103.

59. Johnsen C, East JM, Glassman P. Management of malnutrition in the elderly and the appropriate use of commercially manufactured oral nutritional supplements. J Nutr Health Aging 2000;4(1):42–6.

60. Steigh C, Glassman PA, Fajardo F. Physician and dietitian prescribing of a commercially available oral nutritional supplement. Am J Manag Care 1998;4(4):567–72.

61. Fiatarone Singh MA, Bernstein MA, Ryan AD, O'Neill EF, Clements KM, Evans WJ. The effect of oral nutritional supplements on habitual dietary quality and quantity in frail elders. J Nutr Health Aging 2000;4(1):5–12.

62. Milne AC, Avenell A, Potter J. Meta-analysis: protein and energy supplementation in older people. Ann Intern Med 2006;144(1):37–48.

63. Koretz RL, Avenell A, Lipman TO, Braunschweig CL, Milne AC. Does enteral nutrition affect clinical outcome? A systematic review of the randomized trials. Am J Gastroenterol 2007;102(2):412–29.

64. Price R, Daly F, Pennington CR, McMurdo ME. Nutritional supplementation of very old people at hospital discharge increases muscle strength: a randomised controlled trial. Gerontology 2005;51(3):179–85.

65. Potter JM, Roberts MA, McColl JH, Reilly JJ. Protein energy supplements in unwell elderly patients – a randomized controlled trial. JPEN J Parenter Enteral Nutr 2001;25(6):323–9.

66. Simmons SF, Patel AV. Nursing home staff delivery of oral liquid nutritional supplements to residents at risk for unintentional weight loss. J Am Geriatr Soc 2006;54(9):1372–6.

67. Cook CG, Andrews JM, Jones KL, et al. Effects of small intestinal nutrient infusion on appetite and pyloric motility are modified by age. Am J Physiol 1997;273(2 Pt 2):R755–61.

68. Wilson MM, Purushothaman R, Morley JE. Effect of liquid dietary supplements on energy intake in the elderly. Am J Clin Nutr 2002;75(5):944–7.

69. Dennis M, Lewis S, Cranswick G, Forbes J. FOOD: a multicentre randomised trial evaluating feeding policies in patients admitted to hospital with a recent stroke. Health Technol Assess 2006;10(2):iii–iv, ix–x, 1–120.

70. Sullivan DH, Nelson CL, Klimberg VS, Bopp MM. Nightly enteral nutrition support of elderly hip fracture patients: a pilot study. J Am Coll Nutr 2004;23(6):683–91.

71. Dharmarajan TS, Unnikrishnan D, Pitchumoni CS. Percutaneous endoscopic gastrostomy and outcome in dementia. Am J Gastroenterol 2001;96(9):2556–63.

72. Rimon E, Kagansky N, Levy S. Percutaneous endoscopic gastrostomy; evidence of different prognosis in various patient subgroups. Age Ageing 2005;34(4):353–7.

73. Grant MD, Rudberg MA, Brody JA. Gastrostomy placement and mortality among hospitalized Medicare beneficiaries. Jama 1998;279(24):1973–6.

74. Rabeneck L, Wray NP, Petersen NJ. Long-term outcomes of patients receiving percutaneous endoscopic gastrostomy tubes. J Gen Intern Med 1996;11(5):287–93.

75. Mitchell SL, Tetroe JM. Survival after percutaneous endoscopic gastrostomy placement in older persons. J Gerontol A Biol Sci Med Sci 2000;55(12):M735–9.

76. Klor BM, Milianti FJ. Rehabilitation of neurogenic dysphagia with percutaneous endoscopic gastrostomy. Dysphagia 1999;14(3):162–4.

77. Loser C, Wolters S, Folsch UR. Enteral long-term nutrition via percutaneous endoscopic gastrostomy (PEG) in 210 patients: a four-year prospective study. Dig Dis Sci 1998;43(11):2549–57.

78. Heyland DK, Drover JW, Dhaliwal R, Greenwood J. Optimizing the benefits and minimizing the risks of enteral nutrition in the critically ill: role of small bowel feeding. JPEN J Parenter Enteral Nutr 2002;26(6 Suppl):S51–5.

79. O'Brien LA, Grisso JA, Maislin G, et al. Nursing home residents' preferences for life-sustaining treatments. Jama 1995;274(22):1775–9.

80. Ouslander JG, Tymchuk AJ, Krynski MD. Decisions about enteral tube feeding among the elderly. J Am Geriatr Soc 1993;41(1):70–7.

81. Thomas DR. Distinguishing starvation from cachexia. Clin Geriatr Med 2002;18(4):883–91.

82. Bales CW, White H.K. Nutrition for geriatric primary care. In: Deen D HL, ed. The Complete Guide to Nutrition in Primary Care. Malden, MA: Blackwell Publishing; 2007:209–31.

83. Fried LP, Tangen CM, Walston J, et al. Frailty in older adults: evidence for a phenotype. J Gerontol A Biol Sci Med Sci 2001;56(3):M146–56.

84. Wilson MM, Thomas DR, Rubenstein LZ, et al. Appetite assessment: simple appetite questionnaire predicts weight loss in community-dwelling adults and nursing home residents. Am J Clin Nutr 2005;82(5):1074–81.

85. Ferry M, Sidobre B, Lambertin A, Barberger-Gateau P. The SOLINUT study: analysis of the interaction between nutrition and loneliness in persons aged over 70 years. J Nutr Health Aging 2005;9(4):261–8.

86. Hebling E, Pereira AC. Oral health-related quality of life: a critical appraisal of assessment tools used in elderly people. Gerodontology 2007;24(3):151–61.

87. Semba RD, Blaum CS, Bartali B, et al. Denture use, malnutrition, frailty, and mortality among older women living in the community. J Nutr Health Aging 2006;10(2):161–7.

88. Amella EJ. Feeding and hydration issues for older adults with dementia. Nurs Clin North Am 2004;39(3):607–23.

89. Reed PS, Zimmerman S, Sloane PD, Williams CS, Boustani M. Characteristics associated with low food and fluid intake in long-term care residents with dementia. Gerontologist 2005;45 Spec No 1(1):74–80.

90. Zimmerman S, Sloane PD, Williams CS, et al. Dementia care and quality of life in assisted living and nursing homes. Gerontologist 2005;45 Spec No 1(1):133–46.

91. Pauly L, Stehle P, Volkert D. Nutritional situation of elderly nursing home residents. Z Gerontol Geriatr 2007;40(1):3–12.

92. Crogan NL, Pasvogel A. The influence of protein-calorie malnutrition on quality of life in nursing homes. J Gerontol A Biol Sci Med Sci 2003;58(2):159–64.

93. Mitchell SL, Teno JM, Roy J, Kabumoto G, Mor V. Clinical and organizational factors associated with feeding tube use among nursing home residents with advanced cognitive impairment. Jama 2003;290(1):73–80.

94. Shah PM, Sen S, Perlmuter LC, Feller A. Survival after percutaneous endoscopic gastrostomy: the role of dementia. J Nutr Health Aging 2005;9(4):255–9.

95. Jatoi A, Windschitl HE, Loprinzi CL, et al. Dronabinol versus megestrol acetate versus combination therapy for cancer-associated anorexia: A north central cancer treatment group study. 2002;20(2):567–73.

96. Simmons SF, Levy-Storms L. The effect of dining location on nutritional care quality in nursing homes. J Nutr Health Aging 2005;9(6):434–9.

97. Volicer L, Stelly M, Morris J, McLaughlin J, Volicer BJ. Effects of dronabinol on anorexia and disturbed behavior in patients with Alzheimer's disease. Int J Geriatr Psychiatry 1997;12(9):913–9.

98. Kris-Etherton PM HW, Appel LJ. Fish consumption, fish oil, omega-3 fatty acids, and cardiovascular disease. Circulation 2002;106(21):2747–57.

99. Greenhaff PL CA, Short AH, Harris R, Soderlund K, Hultman E. Influence of oral creatine supplementation of muscle torque during repeated bouts of maximal voluntary exercise in man. Clin Sci (Lond) 1993;84(5):565–71.

100. Edington J, Barnes R, Bryan F, et al. A prospective randomised controlled trial of nutritional supplementation in malnourished elderly in the community: clinical and health economic outcomes. Clin Nutr 2004;23(2):195–204.

101. Gazzotti C, Arnaud-Battandier F, Parello M, et al. Prevention of malnutrition in older people during and after hospitalisation: results from a randomised controlled clinical trial. Age Ageing 2003;32(3):321–5.

102. Kayser-Jones J, Schell ES, Porter C, et al. A prospective study of the use of liquid oral dietary supplements in nursing homes. J Am Geriatr Soc 1998;46(11):1378–86.

103. Lauque S, Arnaud-Battandier F, Mansourian R, et al. Protein-energy oral supplementation in malnourished nursing-home residents. A controlled trial. Age Ageing 2000;29(1):51–6.

104. Lauque S, Arnaud-Battandier F, Gillette S, et al. Improvement of weight and fat-free mass with oral nutritional supplementation in patients with Alzheimer's disease at risk of malnutrition: a prospective randomized study. J Am Geriatr Soc 2004;52(10):1702–7.
105. Levinson Y, Dwolatzky T, Epstein A, Adler B, Epstein L. Is it possible to increase weight and maintain the protein status of debilitated elderly residents of nursing homes? J Gerontol A Biol Sci Med Sci 2005;60(7):878–81.
106. Payette H, Boutier V, Coulombe C, Gray-Donald K. Benefits of nutritional supplementation in free-living, frail, undernourished elderly people: a prospective randomized community trial. J Am Diet Assoc 2002;102(8):1088–95.
107. Wouters-Wesseling W, Wouters AE, Kleijer CN, Bindels JG, de Groot CP, van Staveren WA. Study of the effect of a liquid nutrition supplement on the nutritional status of psycho-geriatric nursing home patients. Eur J Clin Nutr 2002;56(3):245–51.
108. Wouters-Wesseling W, Van Hooljdonk C, Wagenaar L, Bindels JG, De Groot LC, Van Staveren WA. The effect of a liquid nutrition supplement on body composition and physical functioning in elderly people. Clinical Nutrition 2003;22(4):371–7.

10 Sarcopenia

Ian Janssen

Key Points

- Sarcopenia refers to the process of age-related skeletal muscle loss. Although all humans lose muscle mass as they age, the term sarcopenia is often used to refer to older persons with skeletal muscle values in an unhealthy range. This has traditionally been defined as a height-adjusted muscle mass of two standard deviations or more below the mean of a young and healthy population.
- Several cross-sectional studies published in the past decade report that sarcopenia is strongly associated with functional impairment and physical disability. Newer findings based on longitudinal analyses of large cohort studies have reported much weaker effects of sarcopenia on physical disability risk, implying that the effects of sarcopenia on function and disability inferred from earlier cross-sectional studies were overestimated.
- An initial strategy that may retard the normal progression of sarcopenia in older persons is to ensure that they consume adequate protein in their diet. The average protein intake is lower in older adults even though they may have increased protein needs. Long-term protein-feeding studies are required to demonstrate whether the acute benefits of protein supplementation on muscle protein synthesis can translate into long-term improvements in muscle mass and strength in older persons.
- Physical activity, particularly resistance exercise, is the most promising approach for preventing and treating sarcopenia. To maximize strength development, a resistance should be used that allows 10–15 repetitions for each exercise. Older adults should perform at least one set of repetitions for 8–10 exercises that train the major muscle groups, and exercises for each of the major muscle groups should occur on two or three non-consecutive days of the week.

Key Words: Skeletal muscle; functional impairment; resistance exercise; protein intake

From: *Nutrition and Health: Handbook of Clinical Nutrition and Aging, Second Edition*
Edited by: C. W. Bales and C. S. Ritchie, DOI 10.1007/978-1-60327-385-5_10,
© Humana Press, a part of Springer Science+Business Media, LLC 2009

10.1 INTRODUCTION

The purpose of this chapter is to examine the impact of *sarcopenia* – a term used to describe the age-related loss in skeletal muscle mass – on the health and well-being of older adults. The chapter begins by describing the changes in skeletal muscle that occur during the aging process. This is followed by an examination of the effect sarcopenia has on physical disability, the development of cardiometabolic diseases, and mortality risk. The chapter concludes with a discussion of the effectiveness of strategies for preventing and treating sarcopenia.

10.2 DEFINITION OF SARCOPENIA

One of the most dramatic and clinically relevant age-related anatomical changes in humans is that which occurs to skeletal muscle. In 1989, Irwin Rosenberg coined the term "sarcopenia" to refer to the process of age-related skeletal muscle loss [1]. Sarcopenia comes from the Greek words s*arx* (flesh) and *penia* (loss), respectively.

10.3 PROCESS OF SARCOPENIA

It has been reported that skeletal muscle mass and/or size is lower in older adults by comparison to younger adults as measured by imaging techniques [2,3] total body potassium and/or nitrogen [4,5] creatinine excretion [3,6] muscle biopsy [7,8] and dual energy X-ray absorptiometry [9,10]. With few exceptions [6]. these cross-sectional studies observed that muscle size remained relatively constant in the third and fourth decades, and then started to decline noticeably at around the fifth decade. It is currently believed that the age-related reduction in skeletal muscle size is accounted for by a loss in the number of type I (slow twitch) and type II (fast twitch) muscle fibers and a reduction in the size of type II muscle fibers [7,8].

A number of longitudinal studies, with follow-up lengths of up to 12 years, have also examined changes in skeletal muscle mass and size with advancing age. Without exception these studies reported a loss of skeletal muscle with advancing age [11–15]. From these longitudinal studies it has been estimated that skeletal muscle mass decreases by approximately 6% per decade after mid-life [16]. Thus, by the time someone reaches 85 years of age, they will have lost, on an average, about one-quarter of the skeletal muscle mass they had when they were 45 years old.

The loss in skeletal muscle mass [2,17] and strength [18,19] occur to a greater extent in the lower extremities than in the upper extremities. The aging associated reduction in physical activity may be partially responsible for these regional differences in sarcopenia [20,21]. It is reasonable to assume that a reduction in physical activity would primarily be associated with a decreased use of lower body muscles as these are the muscles used for most common activities (i.e., walking, climbing stairs, etc.). This pattern of age-related muscle loss, while interesting from a biological standpoint, may also have important functional implications given the strong reliance on the musculature in the lower body to perform most activities of daily living.

10.4 CLASSIFICATION AND PREVALENCE OF SARCOPENIA

All humans lose skeletal muscle mass as they age. Even older adults who are healthy and free of disease are not immune to the sarcopenia process. Thus, by definition, the prevalence of sarcopenia in the older adult population is 100%. However, because there are individual differences in peak skeletal muscle mass and the rate at which muscle loss occurs, skeletal muscle mass varies widely in older adults. Some older persons have skeletal muscle mass values comparable to young healthy adults, whereas skeletal muscle mass is so low in some older adults that their ability to perform simple functional tasks of daily living is compromised. Thus, for clinical purposes and for comparisons in research studies, an alternative definition of sarcopenia is used to indicate which older persons have skeletal muscle values in the healthy (i.e., normal) and unhealthy (i.e., sarcopenic) ranges. In 1998, Baumgartner and colleagues *(22)* proposed a dichotomous process to determine which older persons have sarcopenia. They defined sarcopenia as a height-adjusted appendicular (arm + leg) skeletal muscle mass (muscle mass/height^2) of two standard deviations or more below the mean of a young and healthy reference population. Using this approach the prevalence of sarcopenia in participants in the New Mexico Elder Health Survey was 14% in those aged 65–69 years and over 50% in those aged 80 years or older *(22)*.

Since Baumgarner and colleagues' *(22)* landmark paper, many other researchers have employed a similar threshold approach for classifying sarcopenia *(22–25)*. Recently, Janssen and colleagues *(26)* have proposed thresholds for defining sarcopenia that are based on the relation between skeletal muscle mass and physical disability. Specifically, statistical techniques were applied to a large and representative data set of Americans aged 60 or older to determine which skeletal muscle values corresponded with a high and low likelihood of physical disability. Whole-body skeletal muscle thresholds of <5.75 kg/m^2 in women and <8.50 kg/m^2 in men corresponded to a *high* likelihood of disability and denoted severe levels of sarcopenia. Whole-body skeletal muscle thresholds of >6.75 kg/m^2 in women and >10.75 kg/m^2 in men corresponded to a *low* likelihood of disability and denoted a normal, healthy muscle mass. Skeletal muscle mass values between 5.75 and 6.74 kg/m^2 in women and 8.50 and 10.74 kg/m^2 in men were associated with a modest increased risk of disability and were denoted as moderate sarcopenia. Based on these thresholds, 9% of the American women and 11% of American men aged 60 or older were considered to have severe sarcopenia, while 22% of the older women and 53.1% of the older men were considered to have moderate sarcopenia *(26)*. These findings suggest that the prevalence of sarcopenia, defined here as having a muscle mass that increases physical disability risk, is extremely high in the older adult population.

10.5 INFLUENCE OF SARCOPENIA ON STRENGTH, FUNCTIONAL IMPAIRMENT, MORBIDITY, AND MORTALITY

The next section of this chapter will consider the influence that sarcopenia has on strength and physical function, illness and disease, and mortality.

10.5.1 Changes in Strength

What is more important to older persons and clinicians than the actual reduction in skeletal muscle mass is whether or not this reduction leads to a loss of strength, power, and ultimately functional impairment and physical disability (i.e., difficulty performing simple activities of daily living). It is clear that normal aging is associated with a reduction in skeletal muscle strength *(3,27)*. Frontera and colleagues *(3)* have proposed that the majority of the age-related loss in muscle strength is caused by the corresponding decrease in muscle mass. These authors found that strength decreased with advancing age in a cross-sectional study of more than 200 men and women aged 45–78 years. However, when corrected for whole-body skeletal muscle mass, the age-related strength differences were severely diminished *(3)*. Although the loss of skeletal muscle mass at least in part explains the age-related reduction in muscle strength, it is important to highlight that other factors also play a role as demonstrated by the fact that the force produced by skeletal muscle per unit area, as measured by computed tomography (CT) or magnetic resonance imaging (MRI), decreases with advancing age *(28,29)*. Even when adjusted for age-related differences in fiber size, the force produced by a single muscle fiber is smaller in older adults than young adults, which suggests that there are alterations at the level of cross-bridge formation and interaction *(30)*. Finally, there are age-related neuromuscular changes, including a decline in the number *(31,32)* and firing rates *(33)* of motor units, which likely in part explains the reduced strength and force produced per pound of skeletal muscle in older persons.

10.5.2 Functional Impairment and Physical Disability

The next question to address is whether the age-related reductions in muscle mass and size are responsible for the corresponding increase in functional impairment and physical disability. The sarcopenia literature is replete with studies that examine measures of physical function such as activities of daily living (ADLs) and instrumental activities of daily living (IADLs). ADLs relate to personal care tasks of bathing and washing, dressing, feeding, getting in and out of bed, and getting to and from the toilet *(34,35)*. IADLs relate to domestic tasks such as shopping, laundry, vacuuming, cooking a main meal, and handling personal affairs *(34,35)*. Difficulty in carrying out ADLs and IADLs denotes a critical physical activity limitation and a level of dependency. In this chapter difficulty or inability to perform ADLs and IADLs is referred to as functional impairment and physical disability.

The results of the published cross-sectional studies examining the relation between sarcopenia and functional impairment in older men and women are summarized in the beginning portion of Table 10.1. The results from a few key studies are further highlighted below. Using a cutoff value for stature-adjusted appendicular muscle mass of two standard deviations or more below the mean of young adults to define sarcopenia, Baumgartner and colleagues *(22)* report that the likelihood of having disability is approximately four times greater in sarcopenic older men and women by comparison to older persons with a normal muscle mass. In the Health Aging and Body Composition Study, older adults in the lowest height and fat mass-adjusted skeletal muscle quintile were 80–90% more likely to have mobility

Table 10.1
Summary of cross-sectional and longitudinal epidemiological studies examining the relation between sarcopenia and physical function in older adults

Study cohort	Disability measure	Sample size (n)	Follow-up length (years)	Age range (years)	Exclusion criteria	Risk factors controlled for	Sarcopenia classification	Risk estimate for impaired function [OR or RR (95% CI)]	Method of measuring muscle
Cross-sectional studies									
Cardiovascular Health Study (women) (39)	Difficulty with IADLs	2842	N/A	≥65	Institutionalized	Age, race, income, smoking, adiposity, cognitive function, prevalent disease	*Whole-body muscle (kg/m²)* Normal (≥6.76) Moderate sarcopenia (5.76–6.75) Severe sarcopenia (≤5.75)	1.00 1.03 (0.83–1.27) 1.77 (1.28–2.44)	Bioimpedance
Cardiovascular Health Study (men) (39)	Difficulty with IADLs	2194	N/A	≥65	Institutionalized	Age, race, income, smoking, adiposity, cognitive function, prevalent disease	*Whole-body muscle (kg/m²)* Normal (≥10.76) Moderate sarcopenia (8.51–10.75) Severe sarcopenia (≤8.50)	1.00 1.39 (0.94–2.09) 2.17 (1.35–3.55)	Bioimpedance
NHANES III (women) (96)	Difficulty with ADLs	2278	N/A	≥60		Age, race, BMI, health behaviors, and comorbidity	*Relative muscle (% body mass)* > –1 SD of young adult mean –1 to –2 SD of young adult mean < –2 SD of young adult mean	1.00 1.49 (1.12–2.00) 3.96 (2.03–7.70)	Bioimpedance
NHANES III (men) (96)	Difficulty with ADLs	2224	N/A	≥60		Age, race, BMI, health behaviors, comorbidity	*Relative muscle (% body mass)* > –1 SD of young adult mean	1.00 1.57 (1.17–2.09) 1.87 (1.17–2.99)	Bioimpedance

(continued)

Table 10.1 (continued)

Study cohort	Disability measure	Sample size (n)	Follow-up length (years)	Age range (years)	Exclusion criteria	Risk factors controlled for	Sarcopenia classification	Risk estimate for impaired function [OR or RR (95% CI)]	Method of measuring muscle
Health ABC Study (women) (36)	Impaired lower extremity function	1549	N/A	70–79	Difficulty performing ADLs, cancer	Race, age, obesity, smoking, alcohol, comorbidity, physical activity	−1 to −2 SD of young adult mean < −2 SD of young adult mean *Height-adjusted muscle (kg/m²)* Highest quintile Lowest quintile *Height- and fat-adjusted muscle mass* Highest quintile Lowest quintile	1.00 0.9 (0.7–1.2) 1.00 1.9 (1.4–2.5)	Dual energy X-ray absorptiometry
Health ABC Study (men) (36)	Impaired lower extremity function	1435	N/A	70–79	Difficulty performing ADLs, cancer	Race, age, obesity, smoking, alcohol, comorbidity, physical activity	*Height-adjusted muscle (kg/m²)* Highest quintile Lowest quintile *Height- and fat-adjusted muscle mass* Highest quintile Lowest quintile	1.00 1.5 (1.1–2.1) 1.00 1.8 (1.3–2.5)	Dual energy X-ray absorptiometry
Framingham Heart Study (women) (43)	Physical disability	478	N/A	72–95		Age, education, self-rated health, chronic illness, physical activity, estrogen use, alcohol, smoking, height	*Whole-body muscle mass (kg)* Tertile 3 Tertile 2 Tertile 1	1.00 0.80 (0.42–1.50) 0.55 (0.27–1.13)	Dual energy X-ray absorptiometry

(continued)

Table 10.1 (continued)

Study cohort	Disability measure	Sample size (n)	Follow-up length (years)	Age range (years)	Exclusion criteria	Risk factors controlled for	Sarcopenia classification	Risk estimate for impaired function [OR or RR (95% CI)]	Method of measuring muscle
Framingham Heart Study (men) (43)	Physical disability	275	N/A	72–95		Age, education, self-rated health, chronic illness, physical activity, alcohol, smoking, height	Whole-body muscle mass (kg) Tertile 3 Tertile 2 Tertile 1	1.00 0.53 (0.20–1.41) 1.06 (0.35–3.18)	Dual energy X-ray absorptiometry
Italian sample (women) (97)	ADL or IADL limitations	167	N/A	67–78	Weight change >5 kg in past 6 months, cognitive impairment, inability to walk 800 m without difficulty, regular exercise	Age, heart disease, hypertension, diabetes, arthritis	Appendicular muscle mass (kg/m²) >−1 SD of young adult mean −1 to −2 SD of young adult mean <−2 SD of young adult mean % Muscle (% body mass) >−1 SD of young adult mean −1 to −2 SD of young adult mean −1 to −2 SD of young adult mean	1.00 0.54 (0.26–1.11) 0.98 (0.35–2.78) 1.00 1.57 (0.39–6.33) 3.86 (1.01–14.87)	Dual energy X-ray absorptiometry
NHANES III (women) (26)	IADL limitations	2276	N/A	≥60	Races other than black and white	Age, race, smoking, alcohol, comorbidity, body fat	Whole-body muscle mass (kg/m²) Normal (≥6.76) Moderate sarcopenia (5.75–6.75) Severe sarcopenia (≤5.75)	1.00 1.41 (0.97–2.04) 3.31 (1.91–5.73)	Bioimpedance

(continued)

Table 10.1 (continued)

Study cohort	Disability measure	Sample size (n)	Follow-up length (years)	Age range (years)	Exclusion criteria	Risk factors controlled for	Sarcopenia classification	Risk estimate for impaired function [OR or RR (95% CI)]	Method of measuring muscle
NHANES III (men) (26)	IADL limitations	2223	N/A	≥60	Races other than black and white	Age, race, smoking, alcohol, comorbidity, body fat	Whole-body muscle mass (kg/m²) Normal (≥10.76) Moderate sarcopenia (8.51–10.75) Severe sarcopenia (≤8.50)	1.00 3.65 (1.92–6.94) 4.71 (2.28–9.74)	Bioimpedance
NHANES III (women) (98)	ADL limitations	1526	N/A	≥70	Ethnicity other than white, black, and Hispanic	Age, ethnicity, education, diabetes, hypertension, heart disease, stroke, hip fracture, arthritis	% Muscle (% body mass) Quintile 5 Quintile 4 Quintile 3 Quintile 2 Quintile 1	1.00 1.07 (0.75–1.51) 0.88 (0.50–1.54) 0.67 (0.33–1.36) 0.87 (0.43–1.77)	Bioimpedance
NHANES III (men) (98)	ADL limitations	1391	N/A	≥70	Ethnicity other than white, black, and Hispanic	Age, ethnicity, education, diabetes, hypertension, heart disease, stroke, hip fracture, arthritis	% Muscle (% body mass) Quintile 5 Quintile 4 Quintile 3 Quintile 2 Quintile 1	1.00 0.75 (0.38–1.40) 1.01 (0.52–1.97) 0.98 (0.58–1.65) 1.20 (0.61–2.35)	Bioimpedance
New Mexico Elder Health Survey (women) (22)	ADL and IADL limitations	382	N/A	70–79	Ethnicity other than non-Hispanic white or Hispanic white	Age, ethnicity, obesity, income, alcohol, physical activity, smoking, comorbidity	Appendicular muscle mass (kg/m²) >−2 SD of young adult mean ≤−2 SD of young adult mean	1.00 4.08 (1.52–11.31)	Anthropometry

(continued)

Table 10.1 (continued)

Study cohort	Disability measure	Sample size (n)	Follow-up length (years)	Age range (years)	Exclusion criteria	Risk factors controlled for	Sarcopenia classification	Risk estimate for impaired function [OR or RR (95% CI)]	Method of measuring muscle
New Mexico Elder Health Survey (men) (22)	ADL and IADL limitations	426		70–79	Ethnicity other than non-Hispanic white or Hispanic white	Age, ethnicity, obesity, income, alcohol, physical activity, smoking, comorbidity	Appendicular muscle mass (kg/m²) > −2 SD of young adult mean ≤ −2 SD of young adult mean	1.00 3.66 (1.42–10.02)	Anthropometry
Longitudinal studies									
Health ABC Study (women) (37)	Mobility difficulty	1345	2.5	70–79		Age, race, height, fat mass, education, alcohol, smoking, physical activity, prevalent disease, self-rated health, depression, cognitive status	Thigh muscle area (cm²) Quartile 4 Quartile 3 Quartile 2 Quartile 1	1.00 1.05 (0.79–1.38) 1.14 (0.84–1.54) 1.68 (1.23–2.31)	Computed tomography
Health ABC Study (men) (37)	Mobility difficulty	1286	2.5	70–79		Age, race, height, fat mass, education, alcohol, smoking, physical activity, prevalent disease, self-rated health, depression, cognitive status	Thigh muscle area (cm²) Quartile 4 Quartile 3 Quartile 2 Quartile 1	1.00 1.53 (1.06–2.21) 1.40 (0.96–2.05) 1.90 (1.27–2.84)	Computed tomography
Cardiovascular Health Study (women) (39)	IADL limitations	1964	8	≥65		Age, race, income, smoking, adiposity, cognitive function, prevalent and incident disease	Whole-body muscle mass (kg/m²) Normal (≥6.76) Moderate sarcopenia (5.76–6.75)	1.00 1.09 (0.94–1.25) 1.37 (1.10–1.72)	Bioimpedance

(continued)

Table 10.1 (continued)

Study cohort	*Disability measure*	*Sample size (n)*	*Follow-up length (years)*	*Age range (years)*	*Exclusion criteria*	*Risk factors controlled for*	*Sarcopenia classification*	*Risk estimate for impaired function [OR or RR (95% CI)]*	*Method of measuring muscle*
							Severe sarcopenia (≤5.75)		
Cardiovascular Health Study (men) *(39)*	IADL limitations	1730	8	≥65		Age, race, income, smoking, adiposity, cognitive function, prevalent and incident disease	*Whole-body muscle mass (kg/m²)* Normal (≥10.76) Moderate sarcopenia (8.51–10.75) Severe sarcopenia (≤8.50)	1.00 1.08 (0.86–1.34) 1.20 (0.90–1.61)	Bioimpedance

Abbreviations: OR, odds ratio; HR, hazards ratio; CI, confidence intervals; ADL, activities of daily living; IADL, instrumental activities of daily living; SES, socioeconomic status; BMI, body mass index; SD, standard deviation.

impairment by comparison to older adults in the highest quintile *(36)*. Finally, in the Third National Health and Nutrition Examination Survey (NHANES III) the likelihood of functional impairment and disability was two- to threefold higher in older adults with severe sarcopenia than in older adults with a normal muscle mass *(23)*.

The cross-sectional studies covered in Table 10.1 measured skeletal muscle mass using a variety of methods, defined sarcopenia using numerous different approaches, and studied population groups with various sociodemographic and physical characteristics. Despite these differences, sarcopenia has consistently, although not universally, been shown to be strongly associated with functional impairment. However, these cross-sectional studies cannot infer causation about the relationship between sarcopenia and disability, leaving open the possibility that functional impairment proceeds rather than follows sarcopenia.

More recent findings from prospective (longitudinal follow-up) cohort studies provide a stronger form of scientific evidence on the temporal relationship between sarcopenia and functional decline in older adults. The findings from these are summarized in the bottom of Table 10.1. Two published reports on the Health Aging Body Composition cohort indicate that low muscle is predictive of a loss in physical function over 2–3 years of follow-up *(37,38)*. In these studies, the effects of muscle size on the loss in function were attributable to muscle strength, implying that the association between low muscle mass and functional decline is a function of the underlying loss in muscle strength. In an 8-year follow-up of older adult participants in the Cardiovascular Health Study cohort, the risk of developing physical disability was 27% greater in those with severe sarcopenia than in those with a normal muscle mass *(39)*. In the same study, the likelihood of having disability at the start of the study in the baseline exam was 79% greater in those with severe sarcopenia than in those with normal muscle mass. Thus, the effect of sarcopenia on disability risk was considerably smaller in the longitudinal analysis than in the cross-sectional analysis, implying that the effects of sarcopenia on functional impairment and disability inferred from the cross-sectional studies published in the 1990s and early 2000s may have been overestimated.

10.5.3 Morbidity

Although the majority of research in the sarcopenia field has considered strength and physical function outcomes, attention has also been given to the potential impact of sarcopenia on metabolic function, illness, and disease.

The metabolic effects of sarcopenia include a decreased resting metabolic rate subsequent to the loss in skeletal muscle mass *(40)*. It has also been postulated that sarcopenia contributes to cardiometabolic diseases such as insulin resistance, type 2 diabetes, dyslipidemia, and hypertension *(41)*. However, a recent study in a group of obese postmenopausal women found that individuals with sarcopenia had a better lipid and lipoprotein profile than those without sarcopenia *(42)*. Additional studies are needed to determine what effect, if any, sarcopenia has on cardiometabolic risk factors and related diseases.

The strong correlation between skeletal muscle mass and bone mineral density in older adults has been well documented *(43,44)*. From a mechanistic standpoint this may be accounted for by the load that contracting skeletal muscle places on the skeleton. This observation suggests that sarcopenia may play a causal role in the development of osteoporosis. Indeed, older women with osteoporosis have a significantly lower muscle mass than older women with a healthy bone mineral density (45,46).

10.5.4 Mortality

A number of investigators have examined whether anthropometric estimates of muscle area in the upper arm are related to mortality risk. These studies were primarily based on the assumption that low muscle area in the arm is a reflection of nutritional deficiencies. In a series of studies, Friedman and colleagues have shown that a skeletal muscle cross-sectional area in the upper arm of less than about 21 cm^2 has a prognostic value in older persons (47). Other studies of community dwelling (48,49) and institutionalized (50) older adults have also shown that low muscle area in the upper arm (e.g., <21 cm^2), a crude measure of sarcopenia, is predictive of both short- and long-term mortality risk. For example, in an 8-year follow-up study of approximately 1400 older Australians, Miller and colleagues (48) found that mortality risk was increased by about twofold in those with an upper arm muscle area below ~21 cm^2.

Consquéric and colleagues (51) have recently reported that older sarcopenic patients were more likely to contract infection during a hospital stay than were older patients with a normal muscle mass. The apparent decreased immunity in sarcopenic individuals, which has also been observed in individuals with protein malnutrition (52), may provide a mechanistic link between sarcopenia and mortality risk. Although it would appear that greater degrees of muscle wasting increase the risk of mortality in older men and women, the extent of the risk caused by sarcopenia is unclear at the present time. Studies in which more specific and accurate measures of muscle mass are employed are needed to clarify the magnitude of risk induced by sarcopenia.

10.6 AGE-RELATED CHANGES IN SKELETAL MUSCLE COMPOSITION

As with muscle size, with advancing age there is a change in the composition of skeletal muscle. That is, a pound of muscle in an older adult is not the same as a pound of muscle in a young adult. One composition change that is particularly important is the increase in skeletal muscle lipid (fat) content. In skeletal muscle, lipids can be stored both intracellularly (within the muscle fibers) and extracellularly (outside of the muscle fibers). The intramyocellular lipids are primarily stored in large droplets close to the mitochondria whereas the extramyocellular lipids accumulate to form adipocytes between the muscle fibers and bundles (53,54). When large numbers of these adipocytes accumulate it is seen as marbling within the skeletal muscle. Aging is associated with an increase in both intramyocellular (55,56) and extramyocellular (57,58) lipids.

Recent evidence suggests that skeletal muscle lipid content influences strength and function. Goodpaster and colleagues (59) report that the attenuation of the mid-thigh muscles on CT images, which is a marker of skeletal muscle lipid content, was associated with age in the Health Aging Body Composition cohort. Further studies in this cohort report that lower skeletal muscle attenuation values in CT images, which is akin to higher skeletal muscle lipid content, is associated with lower specific strength values (e.g., force produced per unit size of skeletal muscle) *(59)*. Observations from this cohort also indicate that lower skeletal muscle attenuation values on CT images are independently associated with poor lower extremity performance scores on 6-m walk and chair-stand tests *(60)*. The authors of these studies hypothesized that the age-related increase in the proportion of type I muscle fibers may explain their findings. Type I fibers are known to have a greater intramyocellular lipid content *(61,62)* and smaller force production *(30)* by comparison to type II muscle fibers.

Visser and colleagues *(37)* recently examined whether greater fat infiltration in the muscle predicts the development of mobility limitation, again within the Health Aging Body Composition cohort. Their analyses included 3075 well-functioning black and white men and women aged 70–79 years. In this study the quartile with the highest muscle fat infiltration had a 91% greater chance of developing mobility limitation over the 2.5-year follow-up by comparison to the quartile with the lowest muscle fat infiltration. Also noteworthy is the fact that muscle fat infiltration predicted mobility limitation independent of muscle mass and muscle strength. These findings provide additional evidence that muscle quality, and not just muscle size, is important to consider when discussing the role of muscle on function and quality of life in older adults.

10.7 PREVENTION AND TREATMENT OF SARCOPENIA

Given the limited knowledge on the interaction of the many causal factors involved in the pathogenesis of sarcopenia, a comprehensive approach for the prevention and treatment of sarcopenia does not exist. Rather, current prevention and treatment strategies are targeted towards the individual factors involved in sarcopenia.

Although the long-term effect of high-protein diets on skeletal muscle mass in the older adults remains to be evaluated, an initial prevention strategy that may retard the normal progression of sarcopenia in older persons is to ensure that they consume adequate protein in their diet. Data suggests that the average protein intake is lower in older than in young adults *(63,64)*. Although a matter of debate, it has also been suggested that the recommended daily allowance (RDA) for protein (0.8 g/kg) may be slightly below requirements in a significant proportion of the older adult population *(65,66)*. Subsequently, the protein intake in a considerable proportion of older adults may be lower than their nutritional requirements. However, as findings from existing research studies are controversial and mixed, sufficient evidence does not exist to establish a new protein RDA for older adults, and it is currently unclear as to what percentage of the older adult population has an insufficient amount of protein in their diet. Nonetheless, ensuring that the protein content in an older individual is at or slightly above the RDA seems appropriate. This will likely require individual consultation with a dietitian to obtain an accurate diet history and, if necessary, provide an appropriate dietary prescription.

Protein supplements can be considered in addition to the individual's ad libitum diet. Protein-based nutritional supplements are attractive given their low cost, ease of administration, and tolerability in older adults (67). When used appropriately, protein/nutritional supplements may result in an increase in daily energy and protein intake in older adults (67). However, a protein/nutritional supplement may also replace ad libitum food consumption, thereby acting as a food/protein replacement rather than a dietary supplement (68).

An abundance of published studies have demonstrated that direct infusion of essential amino acids into the blood stream or oral ingestion of essential amino acids acutely increases muscle protein synthesis in both young and older individuals (69,70). Unfortunately, these protein sources have limited relevance for the "real world" in which older persons live and eat. From a practical standpoint the cost, availability, and palatability of proteins need to be considered. All of these factors favor the use of intact proteins over essential amino acids. Indeed, the vast majority of commercially available protein supplements are comprised of intact proteins such as whey. With this in mind, Paddon-Jones and colleagues (71) recently quantified muscle protein synthesis in a group of healthy older individuals following ingestion of 15 g of whey protein or 15 g of essential amino acids. The results of this study demonstrated that, while not as effective as essential amino acids, the oral ingestion of intact whey protein stimulates muscle protein synthesis in older adults. It is noteworthy that the 15 g whey protein supplement only provided 60 kcal of energy, an amount of energy that only represents a small portion of the total daily energy intake, even if the whey supplement was provided two or three times a day. It is possible that the small increase in energy from a supplement of this nature would not be enough to decrease ad libitum food and protein consumption, as shown in other studies which used protein/nutritional supplements with a considerably larger caloric content (68).

Although protein supplement studies have demonstrated promising results on short-term protein synthesis within skeletal muscle, the long-term effectiveness of protein supplements on muscle mass has yet to be demonstrated. Thus, long-term protein-feeding studies are still required to demonstrate whether the acute benefits of protein supplementation on muscle protein synthesis can translate into long-term improvements in muscle mass and strength.

Given that the pathogenesis of sarcopenia is multifactorial in nature, ensuring that older adults maintain an adequate protein intake is not the only nutritional strategy that may be effective at preventing and treating age-related muscle loss. Elevations in inflammatory cytokines, such as interleukin-6 (IL-6), have been associated with a reduced muscle size, poor strength, and physical disability (72). Dietary carotenoids, which are provided primarily in fruits and vegetables, comprise an important component of the antioxidant defense system that helps keep the inflammatory cytokines in check. As recently reviewed, there is evidence to suggest that older adults with low plasma carotenoid levels are more likely to have elevated IL-6 concentrations, poor muscle strength, and impaired functional performance

(73). Although long-term prospective studies and intervention studies are lacking, this early evidence suggests that adequate intake of carotenoids may also be important to consider when developing dietary strategies to combat sarcopenia.

At present, physical activity appears to be the most promising approach for preventing and treating sarcopenia. Although no prolonged longitudinal studies have examined the effect of chronic physical activity on skeletal muscle mass, it is well known that physically active older persons, particularly those who perform resistance exercise on a regular basis, have larger muscles than their sedentary counterparts *(74,75)*. Furthermore, it is well documented that strength training can increase skeletal muscle mass and strength in previously sedentary older men and women. Table 10.2 summarizes the results of two randomized and seven non-randomized trials that have examined the effects of resistance training per se on skeletal muscle mass, as measured by imaging modalities (CT or MRI), in older persons *(76–84)*. These studies ranged in duration from 8 to 16 weeks, and with the exception of one of the randomized trials, *(77)* all noted a significant increase in skeletal muscle size. From these studies it can be estimated (i.e., average of the group means obtained from all nine studies) that skeletal muscle size increases by approximately 1% for every week of resistance training performed in previously sedentary older persons. When combined with the observation that skeletal muscle mass decreases by approximately 6% per decade, this suggests that 6 weeks of resistance exercise, when appropriately prescribed and adhered to, can reverse a decade's worth of muscle wasting. Thus, resistance exercise can have a dramatic influence on sarcopenia.

Newly released physical activity guidelines from the American College of Sports Medicine recommend resistance exercise training of a moderate intensity for most older adults *(85)*. High-intensity resistance training is an option for older adults in supervised settings or those with sufficient fitness, experience, and knowledge of resistance exercise. Specific recommendations are: *To maximize strength development, a resistance should be used that allows 10–15 repetitions for each exercise. The level of effort for muscle-strengthening activities should be moderate to high. On a 10-point scale, where no movement is 0, and maximal effort of a muscle group is 10, moderate-intensity effort is a 5 or 6 and high-intensity effort is a 7 or 8. Muscle-strengthening activities include a progressive-weight training program, weight-bearing calisthenics, and similar resistance exercises that use the major muscle groups.* Older adults should perform at least one set of repetitions for 8–10 exercises that train the major muscle groups, and exercises for each of the major muscle groups should occur on two or three non-consecutive days of the week *(85)*.

Some controversy exists as to whether or not a protein supplement combined with resistance exercise training will have a greater effect on muscle hypertrophy than resistance exercise alone. Whereas two studies observed a beneficial effect of protein supplementation *(86,87)* another study did not *(77)*. The dose of protein required to increase skeletal muscle hypertrophy is unclear as one of the studies that demonstrated a beneficial effect *(86)* used a smaller dose of protein (10 g vs. 15 g) than the study showing no effect *(77)*. The timing of protein intake around the exercise may be even more relevant in determining whether or not additional muscle hypertrophy occurs. Esmarck and colleagues *(86)* report that a liquid protein

Table 10.2
Influence of resistance exercise on skeletal muscle size, as measured by imaging modalities, in older adults

Reference	Subject characteristics			Treatment	Resistance exercise protocol	Study duration (weeks)	Region of measurement	% Change in skeletal muscle
	N	Gender	Age (yr)*					
Randomized, controlled trials								
Hurley et al. (76)	12	Men	56 ± 4	Control	3 days/wk, 14 exercises for whole body, 1 set of 15 reps at maximal effort	16	Mid-thigh	−1.0
	23		60 ± 5	Resistance				+7.2 [†]
Fiatarone et al. (77)	26	Both	89 ± 4	Control	3 days/wk, leg extensors and flexors, 3 sets of 8 repetitions at 80% 1-RM	10	Mid-thigh	−0.9
	25		86 ± 5	Resistance				+2.0
Non-randomized trials								
Brown et al. (78)	14	Men	63 ± 3	Resistance	3 days/wk, elbow flexion, 4 sets of 10–20 repetitions at 70–90% 1-RM	12	Elbow flexors – CA TA	+7.2
								+17.4 [‡]
Grimby et al. (79)	9	Men	81 ± 2	Resistance	~3 days/wk, knee extension and flexion at maximal effort on Cybex II machine	~9	Mid-thigh	+2.8 [¶]
Roman et al. (80)	5	Men	68 ± 5	Resistance	2 days/wk, 4 exercises for elbow flexors, 4 sets of 8 repetitions to fatigue	12	Elbow flexors	+13.9 [¶]
Fiatarone et al. (81)	7	Both	90 ± 3	Resistance	3 days/wk, leg extensors and flexors, 3 sets of 8 repetitions at 80% 1-RM	8	Mid-thigh	+9.0 [¶]
Frontera et al. (82)	12	Men	60 to 72	Resistance	3 days/wk, knee extension and flexion, 3 sets of 8 repetitions at 80% 1-RM	12	Mid-thigh	+10.6 [¶]

(continued)

Table 10.2
(continued)

	Subject characteristics				Treatment	Resistance exercise protocol	Study duration (weeks)	Region of measurement	% Change in skeletal muscle
Reference	N	Gender	Age (yr)*						
Tracy et al. (83)	12	Men	69 ± 3		Resistance	3 days/wk, knee extension, 4 sets of 5–10 repetitions at maximal effort	9	Quadriceps	+11.5 ¶
	11	Women	68 ± 3		Resistance				+12.0 ¶
Treuth et al. (84)	14	Women	67 ± 4		Resistance	3 days/wk, 12 exercises for whole body, 2 sets of 12 repetitions at ~70% 1-RM	16	Mid-thigh	9.6

* Group means ± standard deviation.
† Significantly greater than change in control group ($P < 0.05$).
‡ Change in trained arm significantly greater than change in control arm ($P < 0.05$).
¶ Significant within group change ($P < 0.05$).
Abbreviations: N = number of subjects. 1-RM = one repetition maximum, CA = control arm, TA = trained arm, SM = skeletal muscle.

supplement (10 g protein) ingested immediately post-exercise in older men increased skeletal muscle hypertrophy whereas ingesting the protein supplement 2 hours post-exercise did not.

The age-related decrease in anabolic hormones such as growth hormone, estrogen, testosterone, and dehydroepiandosterone (DHEA) have been implicated in the development of sarcopenia *(88,89)*. It should therefore come as no surprise that hormone replacement therapy has also been targeted as a sarcopenia treatment strategy. It is important to note, however, that unlike physical activity and diet, these pharmacological-based treatment options are not widely prescribed in the clinical setting.

Pharmacological doses of testosterone increase skeletal muscle mass and strength in young and old hypogonadal men *(90,91)*. A recent meta-analysis of 11 randomized clinical trials reported that testosterone replacement therapy produced a moderate increase in muscle strength in older men, with significant increases in some studies and not in others *(92)*. Although growth hormone administration increases muscle mass and strength in individuals with hypopituitarism *(89)*. most of the initial studies of older persons found that growth hormone therapy did not increase muscle mass and strength *(88,89)*. However, more recently, alternative strategies for stimulating the growth hormone/insulin-like growth factor pathway have shown promising results and more clinical studies in this area are ongoing *(89)*. In general, replacement studies of estrogen *(93)* and DHEA *(94,95)* have shown poor results.

10.8 SUMMARY AND CONCLUSIONS

Sarcopenia is a commonly occurring body composition abnormality in older adults. While the effects of sarcopenia on functional impairment and disability were thought to be quite strong based on early cross-sectional studies, more recently completed longitudinal studies suggest that the effect of sarcopenia on functional impairment risk is modest. However, other scientific evidence indicates that sarcopenia increases osteoporosis and mortality risk, thereby confirming that maintaining an adequate skeletal muscle mass plays an important role in the health and well-being of older men and women. Treatment strategies for sarcopenia include nutritional counseling to ensure that the dietary protein content is sufficient, physical activity counseling to ensure that older adults participate regularly in resistance exercise, and pharmacological targeting of the anabolic hormones that tend to decline with advancing age. While all of these treatment options have their strengths and weaknesses, the available evidence clearly indicates that resistance exercise, when performed according to current physical activity guidelines, is the most effective prevention and treatment strategy for sarcopenia.

10.9 RECOMMENDATIONS

1. To help reduce the normal progression of sarcopenia in older persons, ensure that their protein intake is at or slightly above the RDA. This will likely require individual consultation with a dietitian to obtain an accurate diet history and, if necessary, provide an appropriate dietary prescription.

2. Protein supplements can be considered in addition to the individual's ad libitum diet as they have a low cost, are easy to administer, and are tolerable. When used appropriately, protein/nutritional supplements can increase daily energy and protein intake in older adults.

3. Physical activity is the most promising approach for preventing and treating sarcopenia. Resistance exercise training of a moderate intensity is appropriate for most older adults. To maximize strength development, a regimen should be used that allows 10–15 repetitions for each exercise. Muscle-strengthening activities include a progressive-weight training program, weight-bearing calisthenics, and similar resistance exercises that use the major muscle groups.

4. Older adults should perform at least one set of repetitions for 8–10 exercises that train the major muscle groups, and exercises for each of the major muscle groups should occur on two or three non-consecutive days of the week.

REFERENCES

1. Rosenberg IR. Summary comments. Am J Clin Nutr 1989;50:1231–3.
2. Janssen I, Heymsfield SB, Wang ZM, Ross R. Skeletal muscle mass and distribution in 468 men and women aged 18–88 yr. J Appl Physiol 2000;89(1):81–8.
3. Frontera WR, Hughes VA, Lutz KJ, Evans WJ. A cross-sectional study of muscle strength and mass in 45- to 78-yr-old men and women. J Appl Physiol 1991;71(2):644–50.
4. Ellis KJ, Shukla KK, Cohn SH, Pierson RN. A predictor for total body potassium in man based on height, weight, sex and age: applications in metabolic disorders. J Lab Clin Med 1974;83(5):716–27.
5. Novak LP. Aging, total body potassium, fat-free mass, and cell mass in males and females between ages 18 and 85 years. J Gerontol 1972;27(4):438–43.
6. Tzankoff SP, Norris AH. Effect of muscle mass decrease on age-related BMR changes. J Appl Physiol 1977;43(6):1001–6.
7. Larsson L, Sjodin B, Karlsson J. Histochemical and biochemical changes in human skeletal muscle with age in sedentary males, age 22–65 years. Acta Physiol Scand 1978;103(1):31–9.
8. Lexell J. Human aging, muscle mass, and fiber type composition. J Gerontol A Biol Sci Med Sci 1995;50 Spec No:11–6.
9. Lynch NA, Metter EJ, Lindle RS, et al. Muscle quality. I. Age-associated differences between arm and leg muscle groups. J Appl Physiol 1999;86(1):188–94.
10. Gallagher D, Visser M, De Meersman RE, et al. Appendicular skeletal muscle mass: effects of age, gender, and ethnicity. J Appl Physiol 1997;83(1):229–39.
11. Greig CA, Botella J, Young A. The quadriceps strength of healthy elderly people remeasured after eight years. Muscle Nerve 1993;16(1):6–10.
12. Frontera WR, Hughes VA, Fielding RA, Fiatarone MA, Evans WJ, Roubenoff R. Aging of skeletal muscle: a 12-yr longitudinal study. J Appl Physiol 2000;88(4):1321–6.
13. Aniansson A, Hedberg M, Henning GB, Grimby G. Muscle morphology, enzymatic activity, and muscle strength in elderly men: a follow-up study. Muscle Nerve 1986;9(7):585–91.
14. Forbes GB, Reina JC. Adult lean body mass declines with age: some longitudinal observations. Metabolism 1970;19(9):653–63.
15. Flynn MA, Nolph GB, Baker AS, Martin WM, Krause G. Total body potassium in aging humans: a longitudinal study. Am J Clin Nutr 1989;50(4):713–7.
16. Janssen I, Ross R. Linking age-related changes in skeletal muscle mass and composition with metabolism and disease. J Nutr Health Aging 2005;9:408–19.
17. Grimby G. Muscle performance and structure in the elderly as studied cross-sectionally and longitudinally. J Gerontol A Biol Sci Med Sci 1995;50 Spec No:17–22.

18. Bemben MG, Massey BH, Bemben DA, Misner JE, Boileau RA. Isometric muscle force production as a function of age in healthy 20- to 74-yr-old men. Med Sci Sports Exerc 1991;23(11):1302–10.

19. Bemben MG, Massey BH, Bemben DA, Misner JE, Boileau RA. Isometric intermittent endurance of four muscle groups in men aged 20–74 yr. Med Sci Sports Exerc 1996;28(1):145–54.

20. Bennett KM. Gender and longitudinal changes in physical activities in later life. Age Ageing 1998;27 Suppl 3:24–8.

21. Westerterp KR. Daily physical activity and ageing. Curr Opin Clin Nutr Metab Care 2000;3(6):485–8.

22. Baumgartner RN, Koehler KM, Gallagher D, et al. Epidemiology of sarcopenia among the elderly in New Mexico. Am J Epidemiol 1998;147(8):755–63.

23. Janssen I, Heymsfield SB, Ross R. Low relative skeletal muscle mass (sarcopenia) in older persons is associated with functional impairment and physical disability. J Am Geriatr Soc 2002;50(5):889–96.

24. Melton LJ, 3rd, Khosla S, Crowson CS, O'Connor MK, O'Fallon WM, Riggs BL. Epidemiology of sarcopenia. J Am Geriatr Soc 2000;48(6):625–30.

25. Tanko LB, Movsesyan L, Mouritzen U, Christiansen C, Svendsen OL. Appendicular lean tissue mass and the prevalence of sarcopenia among healthy women. Metabolism 2002;51(1): 69–74.

26. Janssen I, Baumgartner RN, Ross R, Rosenberg IR, Roubenoff R. Skeletal muscle cutpoints associated with elevated physical disability risk in older men and women. Am J Epidemiol 2004;159(4):413–21.

27. Harries UJ, Bassey EJ. Torque-velocity relationships for the knee extensors in women in their 3rd and 7th decades. Eur J Appl Physiol Occup Physiol 1990;60(3):187–90.

28. Brooks SV, Faulkner JA. Skeletal muscle weakness in old age: underlying mechanisms. Med Sci Sports Exerc 1994;26(4):432–9.

29. Metter EJ, Lynch N, Conwit R, Lindle R, Tobin J, Hurley B. Muscle quality and age: cross-sectional and longitudinal comparisons. J Gerontol A Biol Sci Med Sci 1999;54(5):B207–18.

30. Frontera WR, Suh D, Krivickas LS, Hughes VA, Goldstein R, Roubenoff R. Skeletal muscle fiber quality in older men and women. Am J Physiol Cell Physiol 2000;279(3):C611–8.

31. Doherty TJ, Vandervoort AA, Brown WF. Effects of ageing on the motor unit: a brief review. Can J Appl Physiol 1993;18(4):331–58.

32. Tomlinson BE, Irving D. The numbers of limb motor neurons in the human lumbosacral cord throughout life. J Neurol Sci 1977;34(2):213–9.

33. Roos MR, Rice CL, Connelly DM, Vandervoort AA. Quadriceps muscle strength, contractile properties, and motor unit firing rates in young and old men. Muscle Nerve 1999;22(8):1094–103.

34. Rosow I, Breslau N. A Guttman health scale for the aged. J Gerontol 1966;21(4):556–9.

35. Nagi SZ. An epidemiology of disability among adults in the United States. Milbank Mem Fund Q Health Soc 1976;54(4):439–67.

36. Newman AB, Kupelian V, Visser M, et al. Sarcopenia: alternative definitions and associations with lower extremity function. J Am Geriatr Soc 2003;51(11):1602–9.

37. Visser M, Goodpaster BH, Kritchevsky SB, et al. Muscle mass, muscle strength, and muscle fat infiltration as predictors of incident mobility limitations in well-functioning older persons. J Gerontol A Biol Sci Med Sci 2005;60(3):324–33.

38. Goodpaster BH, Park SW, Harris TB, et al. The loss of skeletal muscle strength, mass, and quality in older adults: the health, aging and body composition study. J Gerontol A Biol Sci Med Sci 2006;61(10):1059–64.

39. Janssen I. Influence of sarcopenia on the development of physical disability: the Cardiovascular Health Study. J Am Geriatr Soc 2006;54(1):56–62.

40. Lammes E, Akner G. Resting metabolic rate in elderly nursing home patients with multiple diagnoses. J Nutr Health Aging 2006;10(4):263–70.

41. Karakelides H, Sreekumaran Nair K. Sarcopenia of aging and its metabolic impact. Curr Top Dev Biol 2005;68:123–48.

42. Aubertin-Leheudre M, Lord C, Goulet ED, Khalil A, Dionne IJ. Effect of sarcopenia on cardiovascular disease risk factors in obese postmenopausal women. Obesity (Silver Spring) 2006;14(12):2277–83.

43. Visser M, Harris TB, Langlois J, et al. Body fat and skeletal muscle mass in relation to physical disability in very old men and women of the Framingham Heart Study. J Gerontol A Biol Sci Med Sci 1998;53(3):M214–21.

44. Blain H, Vuillemin A, Teissier A, Hanesse B, Guillemin F, Jeandel C. Influence of muscle strength and body weight and composition on regional bone mineral density in healthy women aged 60 years and over. Gerontology 2001;47(4):207–12.

45. Gillette-Guyonnet S, Nourhashemi F, Lauque S, Grandjean H, Vellas B. Body composition and osteoporosis in elderly women. Gerontology 2000;46(4):189–93.

46. Walsh MC, Hunter GR, Livingstone MB. Sarcopenia in premenopausal and postmenopausal women with osteopenia, osteoporosis and normal bone mineral density. Osteoporos Int 2006;17(1):61–7.

47. Friedman J, Campbell AJ, Caradoc-Davies TH. Prospective trial of new diagnostic criteria for severe wasting malnutrition in the elderly. Age Ageing 1985;14:149–54.

48. Miller MD, Crotty M, Giles LC, et al. Corrected arm muscle area: an independent predictor of long-term mortality in community-dwelling older adults? J Am Geriatr Soc 2002;50(7):1272–7.

49. Prothro JW, Rosenbloom CA. Body measurements of black and white elderly persons with emphasis on body composition. Gerontology 1995;41(1):22–38.

50. Muhlethaler R, Stuck AE, Minder CE, Frey BM. The prognostic significance of protein-energy malnutrition in geriatric patients. Age Ageing 1995;24(3):193–7.

51. Cosqueric G, Sebag A, Ducolombier C, Thomas C, Piette F, Weill-Engerer S. Sarcopenia is predictive of nosocomial infection in care of the elderly. Br J Nutr 2006;96(5):895–901.

52. Chandra RK. Protein-energy malnutrition and immunological responses. J Nutr 1992;122 (3 Suppl):597–600.

53. Levin K, Daa Schroeder H, Alford FP, Beck-Nielsen H. Morphometric documentation of abnormal intramyocellular fat storage and reduced glycogen in obese patients with Type II diabetes. Diabetologia 2001;44(7):824–33.

54. Vock R, Hoppeler H, Claassen H, et al. Design of the oxygen and substrate pathways. VI. structural basis of intracellular substrate supply to mitochondria in muscle cells. J Exp Biol 1996;199 (Pt 8):1689–97.

55. Tsubahara A, Chino N, Akaboshi K, Okajima Y, Takahashi H. Age-related changes of water and fat content in muscles estimated by magnetic resonance (MR) imaging. Disabil Rehabil 1995;17(6):298–304.

56. Forsberg AM, Nilsson E, Werneman J, Bergstrom J, Hultman E. Muscle composition in relation to age and sex. Clin Sci (Lond) 1991;81(2):249–56.

57. Rice CL, Cunningham DA, Paterson DH, Lefcoe MS. Arm and leg composition determined by computed tomography in young and elderly men. Clin Physiol 1989;9(3):207–20.

58. Overend TJ, Cunningham DA, Paterson DH, Lefcoe MS. Thigh composition in young and elderly men determined by computed tomography. Clin Physiol 1992;12(6):629–40.

59. Goodpaster BH, Kelley DE, Thaete FL, He J, Ross R. Skeletal muscle attenuation determined by computed tomography is associated with skeletal muscle lipid content. J Appl Physiol 2000;89(1):104–10.

60. Visser M, Kritchevsky SB, Goodpaster BH, et al. Leg muscle mass and composition in relation to lower extremity performance in men and women aged 70–79: the health, aging and body composition study. J Am Geriatr Soc 2002;50(5):897–904.

61. Askanas V, Engel WK. Distinct subtypes of type I fibers of human skeletal muscle. Neurology 1975;25(9):879–87.

62. Malenfant P, Joanisse DR, Theriault R, Goodpaster BH, Kelley DE, Simoneau JA. Fat content in individual muscle fibers of lean and obese subjects. Int J Obes Relat Metab Disord 2001;25(9):1316–21.

63. Rousset S, Patureau Mirand P, Brandolini M, Martin JF, Boirie Y. Daily protein intakes and eating patterns in young and elderly French. Br J Nutr 2003;90(6):1107–15.

64. Morley JE. Anorexia of aging: physiologic and pathologic. Am J Clin Nutr 1997;66(4):760–73.

65. Campbell WW, Trappe TA, Wolfe RR, Evans WJ. The recommended dietary allowance for protein may not be adequate for older people to maintain skeletal muscle. J Gerontol A Biol Sci Med Sci 2001;56(6):M373–80.

66. Morse MH, Haub MD, Evans WJ, Campbell WW. Protein requirement of elderly women: nitrogen balance responses to three levels of protein intake. J Gerontol A Biol Sci Med Sci 2001;56(11):M724–30.

67. Lauque S, Arnaud-Battandier F, Mansourian R, et al. Protein-energy oral supplementation in malnourished nursing-home residents. A controlled trial. Age Ageing 2000;29(1):51–6.

68. Fiatarone-Singh MA, Bernstein MA, Ryan AD, O'Neill EF, Clements KM, Evans WJ. The effect of oral nutritional supplements on habitual dietary quality and quantity in frail elders. J Nutr Health Aging 2000;4(1):5–12.

69. Walrand S, Boirie Y. Optimizing protein intake in aging. Curr Opin Clin Nutr Metab Care 2005;8(1):89–94.

70. Fujita S, Volpi E. Amino acids and muscle loss with aging. J Nutr 2006;136(1 Suppl):277S–80S.

71. Paddon-Jones D, Sheffield-Moore M, Katsanos CS, Zhang XJ, Wolfe RR. Differential stimulation of muscle protein synthesis in elderly humans following isocaloric ingestion of amino acids or whey protein. Exp Gerontol 2006;41(2):215–9.

72. Ferrucci L, Harris TB, Guralnik JM, et al. Serum IL-6 level and the development of disability in older persons. J Am Geriatr Soc 1999;47(6):639–46.

73. Semba RD, Lauretani F, Ferrucci L. Carotenoids as protection against sarcopenia in older adults. Arch Biochem Biophys 2007;458(2):141–5.

74. Klitgaard H, Mantoni M, Schiaffino S, et al. Function, morphology and protein expression of ageing skeletal muscle: a cross-sectional study of elderly men with different training backgrounds. Acta Physiol Scand 1990;140(1):41–54.

75. Melichna J, Zauner CW, Havlickova L, Novak J, Hill DW, Colman RJ. Morphologic differences in skeletal muscle with age in normally active human males and their well-trained counterparts. Hum Biol 1990;62(2):205–20.

76. Hurley BF, Redmond RA, Pratley RE, Treuth MS, Rogers MA, Goldberg AP. Effects of strength training on muscle hypertrophy and muscle cell disruption in older men. Int J Sports Med 1995;16(6):378–84.

77. Fiatarone MA, O'Neill EF, Ryan ND, et al. Exercise training and nutritional supplementation for physical frailty in very elderly people. N Engl J Med 1994;330(25):1769–75.

78. Brown AB, McCartney N, Sale DG. Positive adaptations to weight-lifting training in the elderly. J Appl Physiol 1990;69(5):1725–33.

79. Grimby G, Aniansson A, Hedberg M, Henning GB, Grangard U, Kvist H. Training can improve muscle strength and endurance in 78- to 84-yr-old men. J Appl Physiol 1992;73(6):2517–23.

80. Roman WJ, Fleckenstein J, Stray-Gundersen J, Alway SE, Peshock R, Gonyea WJ. Adaptations in the elbow flexors of elderly males after heavy-resistance training. J Appl Physiol 1993;74(2):750–4.

81. Fiatarone MA, Marks EC, Ryan ND, Meredith CN, Lipsitz LA, Evans WJ. High-intensity strength training in nonagenarians. Effects on skeletal muscle. JAMA 1990;263(22):3029–34.

82. Frontera WR, Meredith CN, O'Reilly KP, Knuttgen HG, Evans WJ. Strength conditioning in older men: skeletal muscle hypertrophy and improved function. J Appl Physiol 1988;64(3):1038–44.

83. Tracy BL, Ivey FM, Hurlbut D, et al. Muscle quality. II. Effects Of strength training in 65- to 75-yr-old men and women. J Appl Physiol 1999;86(1):195–201.

84. Treuth MS, Hunter GR, Kekes-Szabo T, Weinsier RL, Goran MI, Berland L. Reduction in intra-abdominal adipose tissue after strength training in older women. J Appl Physiol 1995;78(4):1425–31.

85. Nelson ME, Rejeski WJ, Blair SN, et al. Physical Activity and Public Health in Older Adults. Recommendation From the American College of Sports Medicine and the American Heart Association. Circulation 2007.

86. Esmarck B, Andersen JL, Olsen S, Richter EA, Mizuno M, Kjaer M. Timing of postexercise protein intake is important for muscle hypertrophy with resistance training in elderly humans. J Physiol 2001;535(Pt 1):301–11.

87. Meredith CN, Frontera WR, O'Reilly KP, Evans WJ. Body composition in elderly men: effect of dietary modification during strength training. J Am Geriatr Soc 1992;40(2):155–62.

88. Roubenoff R. Origins and clinical relevance of sarcopenia. Can J Appl Physiol 2001;26(1):78–89.

89. Borst SE. Interventions for sarcopenia and muscle weakness in older people. Age Ageing 2004;33(6):548–55.

90. Brodsky IG, Balagopal P, Nair KS. Effects of testosterone replacement on muscle mass and muscle protein synthesis in hypogonadal men – a clinical research center study. J Clin Endocrinol Metab 1996;81(10):3469–75.

91. Urban RJ, Bodenburg YH, Gilkison C, et al. Testosterone administration to elderly men increases skeletal muscle strength and protein synthesis. Am J Physiol 1995;269(5 Pt 1):E820–6.

92. Ottenbacher KJ, Ottenbacher ME, Ottenbacher AJ, Acha AA, Ostir GV. Androgen treatment and muscle strength in elderly men: A meta-analysis. J Am Geriatr Soc 2006;54(11):1666–73.

93. Dionne IJ, Kinaman KA, Poehlman ET. Sarcopenia and muscle function during menopause and hormone-replacement therapy. J Nutr Health Aging 2000;4(3):156–61.

94. Yen SS, Morales AJ, Khorram O. Replacement of DHEA in aging men and women. Potential remedial effects. Ann N Y Acad Sci 1995;774:128–42.

95. Balagopal P, Proctor D, Nair KS. Sarcopenia and hormonal changes. Endocrine 1997;7(1):57–60.

96. Williams MJ, Hunter GR, Kekes-Szabo T, et al. Intra-abdominal adipose tissue cut-points related to elevated cardiovascular risk in women. Int J Obes Relat Metab Disord 1996;20(7):613–7.

97. Zoico E, Di Francesco V, Guralnik JM, et al. Physical disability and muscular strength in relation to obesity and different body composition indexes in a sample of healthy elderly women. Int J Obes Relat Metab Disord 2004;28(2):234–41.

98. Davison KK, Ford ES, Cogswell ME, Dietz WH. Percentage of body fat and body mass index are associated with mobility limitations in people aged 70 and older from NHANES III. J Am Geriatr Soc 2002;50(11):1802–9.

11 Cachexia: Diagnosis and Treatment

David R. Thomas

"..for wasting which represents old age [sarcopenia] and wasting that is secondary to fever [cachexia] and wasting which is called doalgashi [starvation]"
—*Maimonides (1135–1204)*

Key Points

- Three primary categories of loss of body weight include starvation (including undernutrition), sarcopenia, and cachexia.
- Cachexia is the cytokine-associated wasting of protein and energy stores due to the effects of disease; there is a progressive, severe loss of skeletal muscle with relative preservation of visceral protein reserves.
- Cachexia is directly related to inflammatory states, such as cancer or acquired immunodeficiency syndrome, and also occurs in rheumatoid arthritis, chronic renal insufficiency, chronic obstructive pulmonary disease, ischemic cardiomyopathy, and infectious diseases.
- In developing appropriate interventions for weight loss, it is important to recognize the distinction between cachexia and starvation.
- While starvation due to protein-energy undernutrition is widely regarded as the primary cause of loss of fat and fat-free mass in older persons, a failure to improve with nutritional replacement should trigger a consideration of other causes.

Key Words: Wasting; sarcopenia; starvation; lean mass; cytokines

11.1 INTRODUCTION

As progress in the field continues, we are gaining a better understanding of the distinctions and commonalities of weight loss-related syndromes in geriatric patients. Loss of body weight in older persons can result from voluntary or involuntary causes (see Table 11.1). Three primary categories of involuntary loss of body weight include starvation, sarcopenia, and cachexia. Cachexia is best viewed as the cytokine-associated wasting of protein and energy stores due to the effects of disease *(1)*. Persons with cachexia generally lose roughly equal amounts of fat and fat-free mass, while

From: *Nutrition and Health: Handbook of Clinical Nutrition and Aging, Second Edition*
Edited by: C. W. Bales and C. S. Ritchie, DOI 10.1007/978-1-60327-385-5_11,
© Humana Press, a part of Springer Science+Business Media, LLC 2009

Table 11.1
Causes of weight loss in older persons

Voluntary	Food restriction
	Increased exercise
Involuntary	Starvation (also undernutrition)
	Cachexia/anorexia
	Sarcopenia

maintaining extracellular water and intracellular potassium. The loss of fat-free mass is mainly from the skeletal muscle. Systemic inflammation mediated through cell injury or activation of the immune system triggers the acute inflammatory response. Starvation (or weight loss due purely to low food intake, addressed in Chapter 9) results from a pure protein-energy deficiency, thus forcing a reduction in both fat and fat-free mass. Sarcopenia (addressed in Chapter 10) reflects an age-related decline in muscle mass. The term sarcopenia was coined from the Greek "sarx," or "flesh," and "penia," or "loss." (2) Cachexia results from severe wasting of both fat and fat-free mass; the term cachexia is derived from the Greek "kak" or "cac," meaning "bad," and "hexis," or "condition." Cachexia is widely recognized as severe wasting accompanying disease states such as cancer or immunodeficiency disease, but does not have a single widely accepted definition. Recently, an International Cachexia Consensus group proposed this definition: "A complex metabolic syndrome associated with underlying illness and characterized by loss of muscle with or without loss of fat mass. The prominent clinical feature of cachexia is weight loss (corrected for fluid retention). Anorexia, inflammation, insulin resistance and increased muscle protein breakdown are frequently associated with wasting disease. Wasting disease is distinct from starvation, age-related loss of muscle mass, primary depression, malabsorption and hyperthyroidism and is associated with increased morbidity" (3). As illustrated in Fig. 11.1, cachexia is a common cause of weight loss compared to pure starvation. Sarcopenia is the most common cause of weight loss and can occur in the absence or presence of cachexia and/or starvation.

Historically, weight loss and undernutrition have been classified as due to either a relative lack of dietary protein (kwashiorkor) or lack of both dietary protein and calories (marasmus) (1). This classification system focused clinical attention on a lack of adequate food. Simply put, weight loss in older persons was attributed to starvation. Since kwashiorkor is accompanied by hypoalbuminemia late in the course, measures of serum proteins were thought to be able to detect early starvation. Thus, low serum levels of albumin or prealbumin, or a decline in body mass index, suggested inadequate food intake. Remedial nutritional strategies were aimed at increasing voluntary food intake, prescribing hyper-caloric supplements, or by instituting parenteral or enteral feeding (4). In general, these interventions have had only a modest effect in reversing weight loss (5,6).

Recent data suggest that the major cause of weight loss in older adults is the anorexia/cachexia syndrome (7). Simply put, underlying cachexia may be the most common cause of weight loss in long-term care settings. This has resulted in a new paradigm for defining the mechanism of weight loss in older persons.

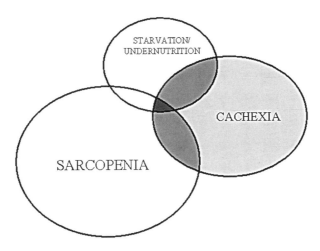

Fig. 11.1. The Unhappy Triad. Weight loss in older adults occurs for a variety of reasons. Cachexia occurs mainly in association with acute or chronic disease, while sarcopenia is commonly observed even in otherwise healthy individuals. Weight loss due purely to starvation is the least common. However, the conditions can overlap and when all three mechanisms come into play, the challenge of stabilizing body weight and lean mass is magnified. This conceptual presentation was originally proposed by D.R. Thomas at the 3rd Cachexia Consensus Conference in Rome, Italy, in December 2005. It derived from a paper published by Dr. D.R. Thomas on "Distinguishing Starvation from Cachexia" *(1)*. Dr. C.C. Seiber assigned the figure the name of "The Unhappy Triad."

Measures of serum proteins are increasingly seen as acute phase reactants reflecting underlying inflammation rather than measures of starvation. Physiological stress (such as surgical operations), cortisol excess, and hypermetabolic states reduce serum albumin even in the presence of adequate protein intake. Soluble IL-2 receptors (and other pro-inflammatory cytokines) are negatively associated with albumin, prealbumin, cholesterol, transferrin, and hemoglobin. Decreases in serum albumin may reflect the presence of inflammatory cytokine production or co-morbidity rather than nutritional status *(8)*, and the use of serum proteins as nutritional markers can frequently lead to overdiagnosis of undernutrition *(9)*.

11.2 CAUSES AND MECHANISMS OF CACHEXIA

Cachexia is directly related to inflammatory states, such as cancer or acquired immunodeficiency syndrome, and also occurs in other common conditions such as rheumatoid arthritis, chronic renal insufficiency, chronic obstructive pulmonary disease, ischemic cardiomyopathy, and infectious diseases *(10)*. Persons with cachexia due to cancer may deplete up to 80% of their muscle mass *(11)*. More than 80% of persons with upper gastrointestinal cancer have cachexia at diagnosis and more than 60% of lung cancer patients develop cachexia.

This same cachexia process is thought to occur in chronic infection, inflammatory myopathies, liver disease, malabsorptive syndromes, and perhaps in normal aging *(12)*. A list of conditions that have been associated with cachexia are shown in Table 11.2. Tumor necrosis factor (TNF)-α levels are elevated in patients with

Table 11.2
Conditions associated with cachexia

Infections, such as tuberculosis, AIDS
Cancer
Rheumatoid arthritis
Ischemic cardiomyopathy
End-stage renal disease
Chronic obstructive pulmonary disease
Cystic fibrosis
Crohn's disease
Alcoholic liver disease
Malabsorption diseases
Elderly persons without obvious cause

severe undernutrition and congestive heart failure, but not in patients with congestive heart failure who do not have severe undernutrition *(13)*. Severe undernutrition occurs in both chronic infections and neoplastic disorders, suggesting that severe undernutrition develops along a common pathway and is not dependent on a specific infection or a particular neoplasm.

11.2.1 Mechanisms of Cachexia

Patients with cachexia experience progressive severe loss of skeletal muscles with relative preservation of visceral protein reserves *(10)*. The loss of skeletal muscle mass is due to a combination of reduced protein synthesis and increased protein degradation. While reduced protein synthesis plays a role, protein degradation is the major cause of loss of skeletal muscle mass in cachexia.

Lysosomal protease cathepsin B probably plays a role in early protein breakdown, as it is elevated in skeletal muscle biopsies from patients with lung cancer and minimal weight loss *(14)*. In more established cachexia, the ubiquitin proteasome-dependent proteolytic pathway is upregulated and is the predominant pathway for protein degradation *(15)*. The underlying mechanism(s) appears to involve the induction of muscle-specific ubiquitin ligases by catabolic hormones, such as the glucocorticoids, but also the inhibition of anabolic pathways such as those controlled by insulin-like growth factor-1, phosphatidylinositol-3-kinase/Akt, and mammalian target of rapamycin (mTOR) *(16)*.

Several cytokines, including tumor necrosis factor-alpha, interleukin-6, interleukin-1-beta and gamma interferon, reproduce symptoms of cachexia in animal models, although individually they have not been shown to produce full-blown cachexia syndrome *(17)*. Direct infusion of interleukin-6 into a mouse muscle decreased myofibrillar protein by 17% at 14 days, suggesting a direct effect on muscle *(18)*.

In addition to the effect of cytokines on skeletal muscle, cytokines act in the hypothalamus to cause an imbalance between the orexigenic and anorexigenic regulatory pathways. In anorexia–cachexia syndrome, the peripheral signals of an energy deficit reach the hypothalamus but fail to produce a response, which propagates the cachectic process.

11.3 CONSEQUENCES OF CACHEXIA

A primary effect of pro-inflammatory cytokines is a decrease in nutrient intake. Cytokines directly result in feeding suppression and thus lower intake of nutrients; cachexia is nearly always accompanied by anorexia. Interleukin-1-beta and tumor necrosis factor act on the glucose-sensitive neurons in the ventromedial hypothalamic nucleus (a "satiety" site) and the lateral hypothalamic area (a "hunger" site) (19). This response is the most common cause of anorexia observed in the acute care setting (20). However, this effect appears to contribute to, but not directly cause, the loss of body mass.

Cytokines have a direct negative effect on muscle mass, and increased concentrations of inflammatory markers have been associated with a reduced lean mass (21–23). This direct effect also has been associated with a decline in muscle strength in older adults. A combination of elevated tumor necrosis factor and interleukin-6 was found in 31% of white males and 29% of black males and in 24% of white women and 22% of black women. For each standard deviation increase in tumor necrosis factor, a 1.2–1.3 kg decrease in grip strength was observed, after adjusting for age, clinic site, health status, medications, physical activity, smoking, height, and body fat. For each standard deviation of interleukin-6, a 1.1–2.4 kg decrease in grip strength was observed (21).

In women followed for 3 years, the baseline level of interleukin-6 predicts walking limitations and knee strength, diminished activities of daily living (24). In a study population of persons at high risk for cardiovascular disease, an inverse relationship was found between fat-adjusted appendicular lean mass and both C-reactive protein and interleukin-6 and also between appendicular lean mass and C-reactive protein (25). In a sample of older persons with a mean age of 71 years and no mobility or activities of daily living deficit at baseline, levels of interleukin-6 predicted mortality at 4 years (26).

11.4 AGE-RELATED CHANGES IN PRO-INFLAMMATORY CYTOKINES

Pro-inflammatory cytokines have been found in apparently healthy older persons as a function of age. Age greater than 70 years is associated with increased circulating plasma levels of interleukin-6 independent of disease states and disorders of aging (27). Increased levels of circulating inflammatory components including tumor necrosis factor-alpha, interleukin-6, interleukin-1 receptor antagonist, soluble tumor necrosis factor receptor, C-reactive protein, serum amyloid A, and high neutrophil counts have been observed in older adults. These age-related changes in immune function are associated with progressively increased levels of glucocorticoids and catecholamines and decreased growth and sex hormones, a pattern reminiscent of that seen in chronic stress. The difference in levels of interleukin-6 in randomly selected older persons compared with strictly selected healthy older persons suggests that inflammatory activity may be a marker of health status (28). Interleukin (IL)-1 concentrations are elevated in elderly patients with severe undernutrition of unknown etiology (29), and levels of IL-1β and IL-6 can be increased in elderly persons without evidence of infection or cancer (30).

However, the increase in circulating inflammatory parameters in healthy elderly humans is small and far less than levels seen during acute infections. Increased cytokine production with aging is inconsistent, resulting in uncertainty whether changes in cytokine levels are due to age itself or to underlying disease (31). The observed increase in levels of interleukin-6 with age may occur as the result of catecholamine hypersecretion and sex-steroid hyposecretion (32). In addition, numerous other conditions (visceral obesity, smoking, stress, etc.) also trigger IL-6 release (33). Subclinical infections such as *Chlamydia pneumoniae* or *Helicobacter pylori* or dental infections and asymptomatic bacteriuria have been postulated to play a role in the observed increase in pro-inflammatory cytokine levels (34).

11.5 DISTINGUISHING CACHEXIA FROM STARVATION

In developing a therapeutic approach to skeletal muscle loss and muscle strength in older persons, it is important to make the distinction between cachexia and starvation. In all persons, the first consideration should be an evaluation of nutritional intake. Starvation resulting from an inability to eat due to mechanical problems or a hypermetabolic state can directly lead to weight loss. In fact,

Table 11.3
Distinguishing starvation from cachexia

	Starvation (and undernutrition)	Cachexia
Appetite	Suppressed in late phase	Suppressed in early phase
Body mass index	Not predictive of mortality	Predictive of mortality
Serum albumin, transthyretin, transferrin, retinol-binding protein	Low in late phase	Low in early phase
Cholesterol	May remain normal	Low
Total lymphocyte count	Low, responds to refeeding	Low, unresponsive to refeeding
C-reactive protein	Little data	Elevated
Inflammatory disease	Usually not present	Present
Response to refeeding	Reversible	Resistant

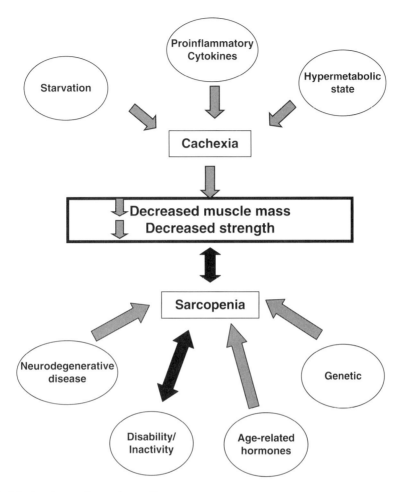

Fig. 11.2. Mechanisms of muscle wasting.

starvation, from whatever cause, will ultimately lead to a loss of muscle mass and strength indistinguishable from that produced by cachexia or sarcopenia. The key physiological sign of starvation is that it is reversed solely by the replenishment of nutrients *(35)*. Failure to improve nutritional status with adequate intake should trigger a concern for underlying cachexia.

In cachexia, pro-inflammatory cytokines have a direct effect on muscle mass, leading to a loss of muscle mass indistinguishable from sarcopenia. On the other hand, sarcopenia alone has not been shown to lead to a decrease in appetite or to loss of fat mass similar to that associated with cachexia. Whether aging itself in some persons, in the absence of any defined inflammatory disease, is associated with elevated pro-inflammatory cytokines that lead to sarcopenia is not clear.

Cachexia is often accompanied by anorexia. A helpful instrument is the simplified appetite nutritional questionnaire (SNAQ; see Chapter 4). Based on the assessment of appetite, this instrument accurately predicts weight gain in the ensuing 6 months *(36)*. Failure of appetite may also be related to depression or adverse drug effect. Deficits in intake should be corrected whenever possible.

The failure of appetite associated with cachexia may limit the success of the nutritional interventions. Cachexia defines a distinct clinical syndrome where the activation of pro-inflammatory cytokines has a direct effect on muscle metabolism and anorexia. The anorexia resulting from the effect of pro-inflammatory cytokines can initiate a vicious feedback loop leading to starvation. Distinguishing sarcopenia from cachexia can be difficult, since there can be an overlap between hormonal deficiency and disease activation causes. Some guidelines are suggested in Table 11.3. A summary of the proposed mechanisms for age-related decline in muscle mass and strength is shown in Fig. 11.2.

11.6 INTERVENTIONS FOR CACHEXIA

In contrast to starvation, cachexia is remarkably resistant to hyper-caloric feeding. Trials of both enteral and parenteral feeding in cancer cachexia have consistently failed to show any benefit in terms of weight gain, nutritional status, quality of life, or survival *(15)*. Pharmacological treatment of anorexia with agents that modulate cytokine production may produce weight gain in cachexia states *(37)*. Steroids and hormonal agents such as megestrol acetate are currently widely used in the treatment of cachexia and anorexia *(38)*. They act through multiple pathways, such as increasing neuropeptide-Y levels to increase appetite and downregulating pro-inflammatory cytokines. Thalidomide significantly attenuated both total weight loss and loss of lean body mass in patients with cancer and acquired immunodeficiency syndrome *(39)*. The action is linked to inhibition and degradation of tumor necrosis factor-alpha. Eicosapentaenoic acid can halt weight loss in cancer cachexia and may increase lean body mass at high doses *(40)*. The effect is postulated to result from its ability to downregulate pro-inflammatory cytokines and proteolysis-inducing factor. The results of these pharmacological trials suggest that improvement in cachexia results from a common effect of these agents on pro-inflammatory cytokines.

11.7 RECOMMENDATIONS

A therapeutic approach to the loss of skeletal muscle mass and strength in older persons depends on correct classification. The term sarcopenia should be reserved for age-related decline in muscle mass not attributable to the presence of pro-inflammatory cytokines. Cachexia may be a better term for a decline in muscle mass associated with known inflammatory disease states. While starvation due to protein-energy undernutrition is widely regarded as the primary cause of loss of fat and fat-free mass in older persons, a failure to improve with nutritional replacement should trigger a consideration of other causes.

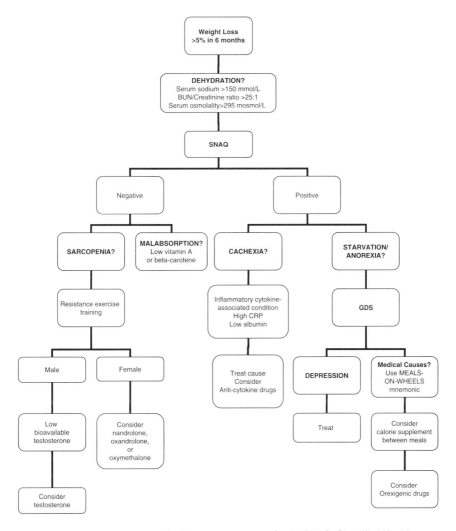

CRP: C-reactive protein; GDS: Geriatric Depression Scale; SNAQ: Simplified Nutrition Assessment Questionnaire

Fig. 11.3. Approach to the management of cachexia.

The complexity of cachexia requires a highly specialized approach depending upon the underlying cause and the presence/absence of concurrent starvation and/ or sarcopenia. A proposed approach to the management of starvation and cachexia is shown in Fig. 11.3.

REFERENCES

1. Thomas DR. Distinguishing starvation from cachexia. Geriatric Clinics of North America 2002;18:883–92.
2. Rosenberg IH. Summary comments. Am J Clin Nutr 1989;50:1231–33.
3. Morley JE, Thomas D. Cachexia: New Advances in the Management of Wasting Diseases. J Am Med Dir Assoc 2008;9(4):205–10.

4. Thomas DR. Weight loss in older adults. Rev Endocr Metab 2005; 6:129–36.

5. Haddad RY, Thomas DR. Enteral nutrition and tube feeding: A review of the evidence. Geriatr Clin North Am 2002;18:867–82.

6. Thomas DR. A prospective, randomized clinical study of adjunctive peripheral parenteral nutrition in adult subacute care patients. J Nutr Health Aging 2005;9:321–5.

7. Morley JE. Thomas DR. Wilson MM. Cachexia: pathophysiology and clinical relevance. Am J Clin Nutr 2006;83(4):735–43.

8. Friedman FJ, Campbell AJ, Caradoc-Davies. Hypoalbuminemia in the elderly is due to disease not malnutrition. Clin Exp Gerontol 1985;7:191–203.

9. Rosenthal AJ, Sanders KM, McMurtry CT, Jacobs MA, Thompson DD, Gheorghiu D, Little KL, Adler RA. Is malnutrition overdiagnosed in older hospitalized patients? Association between the soluble interleukin-2 receptor and serum markers of malnutrition. J Gerontol. Series A, Biological Sciences & Medical Sciences 1998;53:M81–6.

10. Thomas DR. Loss of skeletal muscle mass in aging: examining the relationship of starvation, sarcopenia and cachexia. Clin Nutr, 2007;26(4):389–99

11. Baracos VE. Management of muscle wasting in cancer-associated cachexia: understanding gained from experimental studies. Cancer 2001;92(6 Suppl):1669–77.

12. Westerblad, H, Allen DG. Recent advances in the understanding of skeletal muscle fatigue. Curr Opin Rheumatol 2002;14(6):648–52.

13. Ikeda U, Yamamoto K, Akazawa H, et al. Plasma cytokine levels in cardiac chambers of patients with mitral stenosis with congestive heart failure. Cardiology 1996;87:476–8.

14. Jagoe RT, Redfern CP, Roberts RG, Gibson GJ, Goodship TH. Skeletal muscle mRNA levels for cathepsin B, but not components of the ubiquitin-proteasome pathway, are increased in patients with lung cancer referred for thoracotomy. Clin Sci 2002;102:353–61.

15. Gordon JN, Green SR, Goggin PM. Cancer cachexia. Q J Med 2005;98(11):779–88.

16. Frost RA, Lang CH. Skeletal muscle cytokines: regulation by pathogen-associated molecules and catabolic hormones. Curr Opin Clin Nutr Metabolic Care 2005;8(3):255–63.

17. Murray S, Schell K, McCarthy DO, Albertini MR. Tumor growth, weight loss and cytokines in SCID mice. Cancer Lett 1997;111:111–15.

18. Haddad F, Zaldivar FP, Cooper DM, Adams GR. IL-6 induced skeletal muscle atrophy. J Appl Physiol 2005;98(3):911–7.

19. Espat NJ, Moldawer LL, Copeland EM 3rd. Cytokine-mediated alterations in host metabolism prevent nutritional repletion in cachectic cancer patients. J Surg Oncol 1995;58:77–82.

20. Rote NS. Inflammation. In: Pathophysiology: The biological basis for disease in adults and children. McCance KL, Huether SE, eds. St. Louis: Mosby, 1998:205–36.

21. Visser M, Pahor M, Taaffe DR, Goodpaster BH, Simonsick EM, Newman AB, Nevitt M, Harris TB. Relationship of interleukin-6 and tumor necrosis factor-alpha with muscle mass and muscle strength in elderly men and women: the Health ABC Study. J Gerontol Ser A-Biol Sci Med Sci 2002;57(5):M326–32.

22. Schols AM, Buurman WA, Staal van den Brekel AJ, Dentener MA, Wouters EF. Evidence for a relation between metabolic derangements and increased levels of inflammatory mediators in a subgroup of patients with chronic obstructive pulmonary disease. Thorax 1996;51:819–24.

23. Anker SD, Ponikowski PP, Clark AL, et al. Cytokines and neurohormones relating to body composition alterations in the wasting syndrome of chronic heart failure. Eur Heart J 1999;20: 683–93.

24. Ferrucci L, Penninx BW, Volpato S, Harris TB, Bandeen-Roche K, Balfour J, Leveille SG, Fried LP, Md JM. Change in muscle strength explains accelerated decline of physical function in older women with high interleukin-6 serum levels. J Am Geriatr Soc 2002;50(12):1947–54.

25. Cesari M, Kritchevsky SB, Baumgartner RN, Atkinson HH, Penninx BWHJ, Lenchik L, Palla SL, Ambrosius WT, Tracy RP, Pahor M. Sarcopenia, obesity, and inflammation—results from the Trial of Angiotensin Converting Enzyme Inhibition and Novel Cardiovascular Risk Factors study. Am J Clin N 2005;82:428–34.

26. Ferrucci L, Harris TB, Guralnik JM, Tracy RP, Corti MC, Cohen HJ, Penninx B, Pahor M, Wallace R, Havlik RJ. Serum IL-6 level and the development of disability in older persons. J Am Geriatr Soc 1999;47(6):639–46.

27. Cohen HJ, Pieper CF, Harris T, Rao KM, Currie MS. The association of plasma IL-6 levels with functional disability in community-dwelling elderly. J Gerontol Ser A-Biol Sci Med Sci 1997;52(4):M201–8.

28. Baggio G, Donazzan S, Monti D, Mari D, Martini S, Gabelli C, Dalla Vestra M, Previato L, Guido M, Pigozzo S, Cortella I, Crepaldi G, Franceschi C. Lipoprotein(a) and lipoprotein profile in healthy centenarians: a reappraisal of vascular risk factors. FASEB J 1998;12(6):433–7.

29. Liso Z, Tu JH, Small CB, Schnipper SM, Rosenstreich DL. Increased urine IL-1 levels in aging. Gerontology 1993;39:19–27.

30. Cederholm T, Whetline B, Hollstrom K, et al. Enhanced generation of Interleukin 1β and 6 may contribute to the cachexia of chronic disease. Am J Clin Nutr 1997;65:876–82.

31. Gardner EM, Murasko DM. Age-related changes in Type 1 and Type 2 cytokine production in humans. Biogerontology 2002;3(5):271–90.

32. Straub RH, Miller LE, Scholmerich J, Zietz B. Cytokines and hormones as possible links between endocrinosenescence and immunosenescence. J Neuroimmunol 2000;109(1):10–50.

33. Yudkin JS, Kumari M, Humphries SE, Mohamed-Ali V. Inflammation, obesity, stress and coronary heart disease: is interleukin-6 the link? Atherosclerosis 2000;148(2):209–14.

34. Crossley KB, Peterson PK. Infections in the elderly. Clin Infect Dis 1996;22(2):209–15.

35. American Society of Parenteral and Enteral Nutrition. The clinical guidelines task force, guidelines for the use of parenteral and enteral nutrition in adult and pediatric patients. J Parenter Enteral Nutr 2002:26:S1-? Single supplement issue.

36. Wilson MM. Thomas DR. Rubenstein LZ. Chibnall JT. Anderson S. Baxi A. Diebold MR. Morley JE. Appetite assessment: simple appetite questionnaire predicts weight loss in community-dwelling adults and nursing home residents. Am J Clin Nutr 2005;82(5):1074–81.

37. Thomas DR. Guidelines for the use of orexigenic drugs in long-term care. Nutr Clin Pract 2006;21(1):82–7.

38. Deans C, Wigmore SJ. Systemic inflammation, cachexia and prognosis in patients with cancer. Curr Opin Clin Nutr Metabolic Care 2005;8:265–9.

39. Gordon JN, Trebble TM, Ellis RD, Duncan HD, Johns T, Goggin PM. Thalidomide in the treatment of cancer cachexia: a randomised placebo controlled trial. Gut 2005; 54:540–5.

40. Fearon KC, Von Meyenfeldt MF, Moses AG, Van Geenen R, Roy A, Gouma DJ, Giacosa A, Van Gossum A, Bauer J, Barber MD, Aaronson NK, Voss AC, Tisdale MJ. Effect of a protein and energy dense N-3 fatty acid enriched oral supplement on loss of weight and lean tissue in cancer cachexia: a randomised double blind trial. Gut 2003;52:1479.

12 The Relationship of Nutrition and Pressure Ulcers

David R. Thomas

Key Points

- A strong epidemiological association exists between nutritional status and the incidence, progression, and severity of pressure sores.
- The results of trials of prevention and treatment of pressure ulcers with nutritional interventions to date have been disappointing. While nutrient deficiencies are linked with poor wound healing, providing supplements to patients who are not deficient has not been shown to be of benefit for pressure ulcers.
- This paradoxical finding could be explained by a mechanism of weight loss occurring in a cycle of anorexia and cachexia. Cytokine-induced cachexia is remarkably resistant to hypercaloric feeding.
- Acknowledging these ambivalent findings, it is still important that general nutritional support be provided to persons with pressure ulcers, consistent with medical goals and patient wishes.

Key Words: Wound healing; undernutrition; nutritional supplements; cytokines

12.1 INTRODUCTION AND BACKGROUND

Wound healing is intricately linked to nutrition. Severe protein-calorie undernutrition in humans alters tissue regeneration, the inflammatory reaction, and immune function *(1)*. After vascular surgery, hypoalbuminemia and low serum transferrin levels predict wound-healing complications *(2)*. Undernourished patients are more likely to have post-operative complications than well-nourished patients *(3)*. Although these markers do predict outcome, they do not correlate well with nutritional status *(4)*.

Experimental studies in animal models suggest a biologically plausible relationship between undernutrition and development of pressure ulcers. When pressure was applied for 4 h to the skin of well-nourished animals and malnourished animals,

From: *Nutrition and Health: Handbook of Clinical Nutrition and Aging, Second Edition*
Edited by: C. W. Bales and C. S. Ritchie, DOI 10.1007/978-1-60327-385-5_12,
© Humana Press, a part of Springer Science+Business Media, LLC 2009

pressure ulcers occurred equally in both groups. However, the degree of ischemic skin destruction was more severe in the malnourished animals. Epithelialization of the pressure lesions occurred in normal animals at 3 days post-injury, while necrosis of the epidermis was still present in the malnourished animals *(5)*. This data suggest that while pressure damage may occur independently of nutritional status, malnourished animals may have impaired healing after a pressure injury.

12.2 EPIDEMIOLOGICAL ASSOCIATIONS OF NUTRITION AND PRESSURE ULCERS

A strong epidemiological association exists between nutritional status and the incidence, progression, and severity of pressure sores. Hospitalized patients with undernutrition were twice as likely to develop pressure ulcers as non-undernourished patients *(6)*. In a long-term care setting, pressure ulcers developed in 65% of residents were diagnosed as severely undernourished on admission *(7)*. In another long-term care setting, the estimated percent intake of dietary protein, but not total caloric intake, predicted development of pressure ulcers *(8)*. Impaired nutritional intake, defined as a persistently poor appetite, meals held due to gastrointestinal disease, or a prescribed diet less than 1100 kcal or 50 g protein per day, predicted pressure ulcer development in an additional long-term care setting *(9)*. Table 12.1

Table 12.1
Epidemiological association of nutritional markers with development of a pressure ulcer

First author	Setting	Associated with presence of PU	Not associated with presence of PU
Allman *(81)*	AC	Albumin	Weight, hemoglobin, TLC, nutritional assessment
Gorse *(82)*	AC	Albumin	Nutritional assessment score
Inman *(83)*	AC, ICU	Albumin (measured at 3 days)	Serum protein, hemoglobin, weight
Allman *(82)*	AC	BMI, TLC	Albumin, TSF, arm circumference, weight loss, hemoglobin, nitrogen balance
Hartgrink *(17)*	AC, orthopedic		Nocturnal enteral feeding
Anthony *(84)*	AC	Albumin <32 g/l	
Moolten *(85)*	LTC	Albumin <35 g/l	
Pinchcofsky-Devin *(7)*	LTC	Severe malnutrition	Mild-to-moderate malnutrition or normal nutrition

(continued)

Table 12.1
(continued)

First author	Setting	Associated with presence of PU	Not associated with presence of PU
Berlowitz (9)	LTC	Impaired nutritional intake	Albumin, serum protein, hemoglobin, TLC, BMI/weight
Bennett (86)	LTC		Weight, BMI, weight gain
Brandeis (87)	LTC	Dependency in feeding	BMI/weight, TSF
Trumbore (88)	LTC	Albumin, cholesterol	
Breslow (89)	LTC	Albumin, hemoglobin	Serum protein, cholesterol, zinc, copper, transferrin, body weight, BMI, TLC
Bergstrom (8)	LTC	Dietary protein intake 93% of RDA vs. 119%, dietary iron	Serum protein, cholesterol, zinc, copper, transferrin, weight, BMI, TLC
Ferrell (90)	LTC		Albumin, serum protein, BMI, hematocrit
Bourdel-Marchasson (12)	LTC		Oral nutritional supplement (26 vs. 20% incidence)
Guralnik (91)	Community		Albumin, BMI, impaired nutrition, hemoglobin

AC = acute care; LTC = long-term care; BMI = body mass index; TLC = total lymphocyte count; TSF = triceps skinfold thickness; ICU = intensive care unit; RDA = recommend daily allowance.

demonstrates the association of serum albumin and other nutritional variables with the development of a pressure ulcer. Pressure ulcers appear to be associated with traditional markers of nutritional status in some, but not all studies.

12.3 NUTRITION IN THE PREVENTION OF PRESSURE ULCERS

This strong association of undernutrition with the development of pressure ulcers led to a hypothesis that providing hypercaloric feeding to persons at risk for undernutrition might lead to the prevention of pressure ulcers. Several trials have

Table 12.2
Nutritional interventions in the prevention of pressure ulcers

First author	Setting	Intervention	Outcome
Delmi 1990 (14)	Hospitalized with femoral neck fracture	One oral nutrition supplement per day in addition to hospital diet vs. standard hospital diet alone	All stage pressure ulcers 9% in the nutritional intervention group vs. 7% in the control group RR 0.79 (0.14–4.39, $p = 0.8$)
Hartgrink 1998 (17)	Hospitalized with hip fracture and increased pressure ulcer risk	Overnight nasogastric tube feeding vs. standard hospital diet	Stage 2 or greater pressure ulcers 52% in the nutritional intervention group vs. 56% in the control group RR = 0.92 (0.64–1.32, $p = 0.6$)
Bourdel-Marchasson 2000 (13)	Acute phase of a critical illness	Two oral supplements per day in addition to normal diet vs. standard hospital diet alone	All stage pressure ulcers 40% in the nutritional intervention group vs. 48% in the control group RR 0.83 (0.70–0.99)
Houwing 2003 (15)	Hip fracture patients	One supplement daily in addition to the standard hospital diet vs. non-caloric water-based placebo and standard hospital diet	Stages 1 and 2 pressure ulcers 55% in the nutritional intervention group vs. 59% in the placebo group RR 0.92 (0.65–1.3)

RR = relative risk (95% confidence intervals).

examined this hypothesis (Table 12.2). However, the combined results of trials of nutritional intervention for the prevention of pressure ulcers have been disappointing (10,11).

The effect of oral nutrition supplements was observed in a non-randomized group of hospitalized, severely ill patients. Nutritional supplements were given to 33% of subjects in one ward and to 87% of subjects in another hospital ward. There was no difference between the groups in pressure ulcer incidence (26.4 vs. 20.2%), pressure ulcer prevalence at discharge (14.7 vs. 10.3%), mortality (15.6 vs. 14.2%), length of stay (17.3 vs. 17.4 days), or nosocomial infections (26.4 vs. 19.0%) (12).

A multi-center, cluster randomized trial in persons older than 65 years in the acute phase of a critical illness examined the effect of two oral supplements per day in addition to the normal hospital diet. At the end of 15 days, the supplemented group had a reduced incidence of all stages of pressure ulcers (40% in the nutritional intervention group vs. 48% in the control group (relative risk 0.83, 95% confidence intervals 0.70–0.99) *(13)*.

In hospitalized persons with femoral neck fracture, one group was randomized to receive one oral nutrition supplement in addition to the hospital diet or the standard hospital diet alone. There was no difference in the incidence of pressure ulcers between the groups *(14)*.

In another population of persons with hip fracture, subjects were randomized to receive either 400 ml daily of a nutritional supplement enriched with protein, arginine, zinc, and antioxidants in addition to the standard hospital diet or the standard hospital diet and a non-caloric water-based placebo. There was no difference in the incidence of stage 2 or greater pressure ulcers between the groups after 28 days (55% in the supplement group vs. 59% in the placebo group) *(15)*.

Of the four trials of oral nutritional supplements for the prevention of pressure ulcers, only one trial suggests that a nutritional intervention may reduce the incidence of pressure ulcers. Similar results were found when another trial that used a group randomization was included in a meta-analysis *(16)*. The calculated number needed to treat in this analysis suggests that 20 patients would need to receive oral nutritional supplements to prevent one pressure ulcer.

In a trial of overnight enteral feeding in patients with hip fracture, no difference in pressure ulcer incidence, total serum protein, serum albumin, or the severity of pressure sores after 1 and 2 weeks was observed. After 2 weeks, 52% of subjects in the enterally fed group and 56% of the control group developed stage 2 or greater pressure ulcers ($p = 0.06$). Of the 62 patients randomized for enteral feeding, only 25 tolerated their tube for more than 1 week, and only 16 tolerated their tube for 2 weeks. Comparison of the actually tube-fed group ($n = 25$ at 1 week, $n = 16$ at 2 weeks) and the control group showed two to three times higher protein and energy intake ($p < 0.0001$), and a significantly higher total serum protein and serum albumin after 1 and 2 weeks in the actually tube-fed group (all $p < 0.001$). In an intention to treat analysis there was also no difference in the incidence of sores of grade 2 or above *(17)*. It is possible that the lack of effect on supplemental enteral feeding was due to poor tolerance of the feedings.

A study of enteral tube feedings in patients with a pressure ulcer in a long-term care setting observed 49 patients for 3 months *(18)*. Patients received 1.6 times basal energy expenditure daily, 1.4 g of protein per kilogram per day, and 85% or more of their total recommended daily allowance. At the end of 3 months, there was no difference in number or in healing of pressure ulcers.

In a study of survival among residents in long-term care with severe cognitive impairment, 135 residents were followed for 24 months *(19)*. The reasons for the placement of a feeding tube included the presence of a pressure ulcer. Having a feeding tube was not associated with increased survival; in fact the risk of death was slightly increased (OR 1.09). There was no apparent effect on the prevalence of pressure ulcers in this group of enterally fed persons.

All of these clinical trials suffer from methodological problems in factors including study design and statistical power. Large, prospective, randomized controlled trials will be required to define the effect of nutritional interventions in prevention of pressure ulcers.

12.4 NUTRITION IN THE HEALING OF PRESSURE ULCERS

Although correction of poor nutrition is part of total patient care and should be addressed in each patient, controversy exists about the ability of nutritional support to reduce wound complications or improve wound healing (20,21).

Randomized, controlled trials have evaluated the effect of increased protein, vitamin C, zinc, and oral supplements in the treatment of pressure ulcers. One trial randomized by group found no difference in pressure ulcer healing between the nutritionally supplemented and control goups (16). With the exception of a single trial of higher protein intake, no trial has demonstrated improved healing with the intervention. Table 12.3 summarizes the interventional trials for pressure ulcers.

Table 12.3
Nutritional interventions in the treatment of pressure ulcers

First author	Setting	Intervention	Outcome
Breslow (26)	Long-term care	24% protein vs. 14% protein enteral feeding	−4.2 vs. −2.1 cm2 decrease in surface area
Chernoff (25)	Long-term care	1.8 g/kg protein vs. 1.2 g/ kg protein enteral feeding	73 vs. 43% improvement in surface area
Henderson (18)	Long-term care	1.6 times basal energy expenditure, 1.4 g of protein per kilogram per day	65% PU at onset; 61% prevalence at 3 months
Mitchell (19)	Long-term care	Enteral feeding	RR of death 1.49 (1.2–1.8) vs. RR of death 1.06 (0.8–1.4) after 2 years
ter Riet (38)	Long-term care	Vitamin C 10 vs. 1000 mg	No difference in healing rate
Taylor (39)	Acute surgical patients	Vitamin C large dose vs. none	84 vs. 43% (control) reduction surface area at 30 days
Norris (47)	Acute hip fracture patients	Zinc	No difference
Hartgrink (17)	Acute hip fracture patients	Enteral feeding	No difference in incidence

RR = relative risk (95% confidence intervals).

12.5 GENERAL NUTRITIONAL SUPPORT FOR PERSONS WITH PRESSURE ULCERS

It is clear that nutritional deficiency in the form of starvation is associated with increased mortality and morbidity. Therefore, nutritional requirements must be addressed in every patient and corrected when possible, consistent with the wishes and plan of care for each individual.

12.5.1 Energy

Daily caloric requirements range from 25 kcal/kg/day for sedentary adults to 40 kcal/kg/day for stressed adults. Stress generally includes persons with burns, pressure ulcers, cancer, infections, and other similar conditions. In general, caloric requirements can be met at 30–35 kcal/kg/day for elderly patients under moderate stress. Various formulas, including the Harris–Benedict equation, can be used to predict caloric requirements, but controversy exists over accuracy in obese or severely undernourished individuals (22). Other formulas have been adjusted for severely stressed hospitalized subjects (23). Considerable debate exists over whether to use ideal body weight or an adjusted body weight in calculations. The best instrument for predicting nutritional requirements in older, undernourished individuals in whom ideal or usual body weight is often unknown is not clear. Prediction equations have been published (24).

12.5.2 Protein

Greater healing of pressure ulcers has been reported with a higher protein intake irrespective of positive nitrogen balance (25). Clinical trials have examined dietary interventions in the healing of pressure ulcers. In 48 patients with stages 2–4 pressure ulcers who were being fed enterally, undernutrition was defined as a serum albumin below 35 g/l or body weight more than 10% below the midpoint of the age-specific weight range. Total truncal pressure ulcer surface area showed more decrease (-4.2 vs. -2.1 cm^2) in surface area in patients fed the enteral formula containing 24% protein compared to a formula containing 14% protein. However, changes in body weight or in biochemical parameters of nutritional status did not occur between groups. The study was limited by a small sample size (only 28 patients completed the study), non-random assignment to treatment groups, confounding effects of air-fluidized beds, and the use of two different feeding routes (26).

In a small study of 12 enterally fed patients with pressure ulcers, the group who received 1.8 g/kg of protein had a 73% improvement in pressure ulcer surface area compared to a 42% improvement in surface area in the group receiving 1.2 g/kg of protein despite the fact that the group that received the higher protein level began the study with larger surface area pressure ulcers (22.6 vs. 9.1 cm^2) (17).

The optimum dietary protein intake in patients with pressure ulcers is unknown, but may be much higher than current adult recommendation of 0.8 g/kg/day. Current recommendations for dietary intake of protein in stressed elderly patients lie between 1.2 and 1.5 g/kg/day. Yet half of chronically ill elderly persons cannot maintain nitrogen balance at this level (27). On the other hand, increasing protein intake beyond 1.5 g/kg/day may not increase protein synthesis and may cause dehydration *(28)*. The optimum protein intake for these patients has not been defined, but may lie between 1.2 and 1.5 g/kg/day.

12.5.3 Amino Acids

The association of dietary protein intake with wound healing has led to investigation of the use of specific amino acids. Glutamine is essential for the immune system function, but supplemental glutamine has not been shown to have noticeable effects on wound healing *(29)*. Arginine enhances wound collagen deposition in healthy volunteers *(30,31)* but studies on sick, wounded patients have not been done. No effect on healing of pressure ulcers has been observed with arginine supplementation *(32)*. No improvement in wound healing has been demonstrated by using high supplements of branched-chain amino acid formulations *(33)*.

12.5.4 Vitamins and Minerals

The deficiency of several vitamins has significant effects on wound healing. However, supplementation of vitamins to accelerate wound healing in the absence of a deficiency state is controversial. Vitamin C is essential for wound healing, and impaired wound healing has been observed in clinical scurvy. However, in studies of clinically impaired wound healing, 6 months of an ascorbate-free diet is required to produce a deficient state *(34)*. In animals who are vitamin C deficient, wound healing is abnormal at 7 days but completely normal at 14 days *(35)*.

There is no evidence of acceleration of wound healing by vitamin C supplementation in patients who are not vitamin C deficient *(36)*. Supertherapeutic doses of vitamin C have not been shown to accelerate wound healing *(37)*. Two clinical trials have evaluated the effect of supplemental vitamin C in the treatment of pressure ulcers. In a multi-center, blinded trial, 88 patients with pressure ulcers were randomized to either 10 or 500 mg twice daily of vitamin C. The wound closure rate, relative healing rate, and wound improvement score were not different between groups *(38)*. An earlier trial in acute surgical patients with pressure ulcers found a mean reduction in surface area at 1 month of 84% in patients treated with large doses of vitamin C compared to a reduction in surface area of 43% in the control group ($p < 0.005$) *(39)*.

The recommended daily allowance of vitamin C is 60 mg. This RDA is easily achieved from dietary sources that include citrus fruits, green vegetables, peppers, tomatoes, and potatoes.

Vitamin A deficiency results in delayed wound healing and increased susceptibility to infection *(40)*. Vitamin A has been shown to be effective in counteracting delayed healing in patients on corticosteroids *(41)*. Vitamin E deficiency does not appear to play an active role in wound healing *(42)*.

Zinc was first implicated in delayed wound healing in 1967 *(43)*. No study to date has shown improved wound healing in patients supplemented with zinc who were not zinc deficient *(44,45)*. Zinc levels have not been associated with development of pressure ulcers in patients with femoral neck fractures *(46)*. In a small study of patients with pressure ulcers, no effect on ulcer healing was seen at 12 weeks in zinc-supplemented vs. non-zinc-supplemented patients *(47)*. Indiscriminate or long-term zinc supplementation should be avoided since high serum zinc levels may inhibit healing, impair phagocytosis, and interfere with copper metabolism *(10,48,49)*. The RDA for zinc is 12 15 mg, but the intake of most elderly persons is 7–11 mg of zinc per day *(50)*, chiefly from meats and cereals.

12.6 FACTORS CONTRIBUTING TO THE NUTRITIONAL PARADOX

Traditionally, weight loss and undernutrition have been classified as due to either a relative lack of dietary protein (kwashiokor) or lack of both dietary protein and calories (marasmus) *(51)*. This classification system focused clinical attention on a lack of adequate food. Simply put, weight loss in older persons was attributed to starvation. Measures of serum proteins or anthropormorphological parameters were thought to detect early starvation. Remedial nutritional strategies were aimed at increasing voluntary food intake, prescribing hypercaloric supplements, or instituting parenteral or enteral feeding *(52)*. In general, these interventions have had only a modest effect in reversing weight loss *(53,54)*.

Recent data suggest that the major cause of weight loss in older adults is the anorexia/cachexia syndrome *(55)*. This has resulted in a new paradigm for defining the mechanism of weight loss in older persons. Simply put, underlying cachexia may be the most common cause of weight loss in long-term care settings.

Cachexia is directly related to inflammatory states, such as cancer or acquired immunodeficiency syndrome, and also occurs in other common conditions such as rheumatoid arthritis, chronic renal insufficiency, chronic obstructive pulmonary disease, ischemic cardiomyopathy, and infectious diseases *(56)*. Interleukin (IL)-1 concentrations are elevated in elderly patients with severe undernutrition of unknown etiology *(57)*, and levels of IL-1β and IL-6 can be increased in elderly persons without evidence of infection or cancer *(58)*. Tumor necrosis factor-α (TNF) levels are elevated in patients with severe undernutrition and congestive heart failure, but not in patients with congestive heart failure who do not have severe undernutrition *(59)*. Severe undernutrition occurs in both chronic infections and neoplastic disorders, suggesting that severe undernutrition develops along a common pathway and is not dependent on a specific infection or a particular neoplasm.

Measures of serum proteins are increasingly seen as acute phase reactants reflecting underlying inflammation rather than measures of starvation. Physiological stress (such as surgical operations), cortisol excess, and hypermetabolic states reduce serum albumin even in the presence of adequate protein intake. Decreases in serum albumin may reflect the presence of inflammatory cytokine production or comorbidity rather than nutritional status *(60)*. Soluble IL-2 receptor are negatively associated with albumin, prealbumin, cholesterol, transferrin, and hemoglobin. The use of albumin and cholesterol in these patients as nutritional markers could potentially lead to over diagnosis of malnutrition *(61)*. This may explain why serum albumin has not consistently been an independent predictor of pressure ulcers.

Several cytokines, particularly IL-1α, IL-1β, and IL-6, have been suggested to be elevated in subjects with pressure ulcers. Whether these levels change with healing or are predictive of healing is not known. These cytokines are known to also increase in severe undernutrition.

Serum IL-1β is elevated in patients with pressure ulcers *(62)*. Levels of IL-1α are elevated in pressure ulcers but low in acute wound fluid *(63)*. In hospitalized elderly patients suffering from bacterial pneumonia, cerebrovascular disease, or femoral bone fracture, serum IL-1 beta (but not IL-6) was higher in subjects with pressure ulcers. Albumin, hemoglobin, C-reactive protein, fibrinogen, and white cell count were also lower in subjects with pressure ulcers, despite no significant differences in age, gender, Braden scale, or underlying diseases between the two groups *(62)*.

Circulating serum levels of IL-6, IL-2, and IL-2R are higher in spinal-cord-injured patients compared to normal controls, and highest in subjects with pressure ulcers. The highest concentration of cytokines was in subjects with the slowest healing pressure ulcers *(64)*. In other studies, IL-6 serum levels were increased in patients with pressure ulcers but IL-1 and TNF were not elevated *(65)*.

Existing studies are not clear as to whether the elevation is due to the presence of a pressure ulcer or due to underlying severe undernutrition. Alternatively, the elevation of cytokine levels may be a common pathway for both conditions. The hypotheses are demonstrated in Fig. 12.1. Cytokine-mediated anorexia and weight loss are common in the population that develops pressure ulcers. The interrelationship is outlined in Table 12.4.

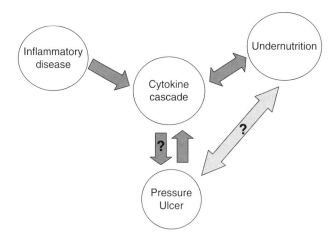

Fig. 12.1. Interaction between proinflammatory cytokines, undernutrition, and pressure ulcers. Inflammatory disease may initiate the proinflammatory cytokine cascade leading directly to development of a pressure ulcer. Alternatively, the development of a pressure ulcer may initiate the cytokine cascade leading to undernutrition through suppression of appetite and cachexia. Undernutrition may lead to development of a pressure ulcer either through cachexia and loss of body mass or through the cytokine cascade.

Table 12.4
Documented interactions among cytokines, undernutrition, and chronic wounds

---

Undernutrition
- Poor wound healing
- Increased risk of infection
- Increased incidence of pressure ulcers

Proinflammatory cytokines
- Suppress appetite
- Promote/interfere with wound healing

Chronic wounds
- Source of cytokines
- Increased association with undernutrition

12.7 INTERVENTIONS FOR INFLAMMATORY-MEDIATED CACHEXIA

This new paradigm suggests that interventions targeting weight loss in older persons must address the anorexia/cachexia continuum. Assessment of appetite can suggest anorexia as a cause of decreased food intake *(66)*. A common cause may be loss of appetite, due to dysregulation of a variety of psychological, gastrointestinal, metabolic, and nutritional factors *(67)*. Loss of appetite may initiate a vicious cycle of weight loss and increasing undernutrition.

Cytokines may regulate appetite directly through the central feeding drive. Significant interaction between the central feeding drive, neuropeptide Y, and IL-1β has been demonstrated in rats *(68,69)*. IL-1, IL-6, TNF, interferon-γ, leukemia inhibitory factor (D-factor), and prostaglandin E_2 have all been implicated in cancer-induced severe undernutrition *(70,71)*. Leptin, a central regulator of food intake and body fat mass, increases under the stress of hip operations *(72)*, but is low in undernourished men *(73)*.

The lack of effect of hypercaloric feeding in pressure ulcers may reflect that the underlying pathophysiology is cytokine-induced cachexia rather than simple starvation. Starvation is amenable to hypercaloric feeding in all but the terminally undernourished patients. Cytokine-induced cachexia is remarkably resistant to hypercaloric feeding *(74,75)*. This may explain the modest results of clinical nutrition intervention trials in older persons *(76)*.

Where possible, the underlying inflammatory condition should be sought. The importance of defining the distinction lies in developing a targeted therapeutic approach to weight loss in older persons *(77)*. Failure to distinguish among these causes of weight loss often results in frustration over the clinical response to therapeutic interventions.

Interventions to modulate cytokine activity are possible. Cytokine modulation has been postulated as a potential treatment for cachexia *(78–79)*. If a significant positive relationship exists between circulating cytokines and pressure ulcers, an opportunity for potential intervention to promote healing exists.

12.8 CONCLUSIONS

Wound nutrition is whole body nutrition. Unquestionably, providing nutritional support can prevent the effects of starvation. Death is an inevitable consequence of starvation. Whether nutrition can improve the outcome of pressure ulcers remains in dispute. Improvements in nutritional markers, such as serum protein concentrations, nitrogen balance, and weight gain, have not usually been accompanied by clinical benefits *(72,80)*.

There is no doubt that undernutrition does not have a positive effect on wound healing. However, there is no magic nutritional bullet that will accelerate wound healing. General nutritional support should be provided to persons with pressure ulcers, consistent with medical goals and patient wishes.

12.9 RECOMMENDATIONS

1. Provide optimum nutrition consistent with goals of care for each patient.
2. Optimize protein intake with a goal of 1.2–1.5 g/kg/day of protein.
3. A simple multivitamin supplement may be indicated for nutritionally compromised patients, but there are no data to support the routine use of vitamin C and zinc in patients with pressure ulcers.
4. Consider vitamin A supplements in patients on corticosteroids.
5. Consider whether a cytokine-associated inflammatory condition may be present and potentially treatable.

REFERENCES

1. Young ME. Malnutrition and wound healing. Heart Lung 1988;17:60–7.
2. Casey J, Flinn WR, Yao JST, et al. Correlation of immune and nutritional status with wound complications in patients undergoing vascular operations. Surgery 1983;93:822–7.
3. Detsky AS, Baker JP, O'Rourke K, et al. Predicting nutrition-associated complications for patients undergoing gastrointestinal surgery. JPEN 1987;11:440–6.
4. Covinsky KE, Covinsky MH, Palmer RM, Sehgal AR. Serum albumin concentration and clinical assessments of nutritional status in hospitalized older people: Different sides of different coins? J Am Geriatr Soc 2002;50:631–7.
5. Takeda T, Koyama T, Izawa Y, et al. Effects of malnutrition on development of experimental pressure sores. J Dermatol 1992;19:602–9.
6. Thomas DR, Goode PS, Tarquine PH, Allman R. Hospital acquired pressure ulcers and risk of Death. *J Amer Geriatr Soc* 1996;44:1435–40.
7. Pinchcofsky-Devin GD, Kaminski MV Jr. Correlation of pressure sores and nutritional status. J Am Geriatr Soc 1986;34:435–40.
8. Bergstrom N, Braden B: A prospective study of pressure sore risk among institutionalized elderly. J Am Geriatr Soc 1992;40:747–58.
9. Berlowitz DR, Wilking SV. Risk factors for pressure sores: a comparison of cross- sectional and cohort-derived data. J Am Geriatr Soc 1989;37:1043–50.
10. Thomas DR. The role of nutrition in prevention and healing of pressure ulcers. Clin Geriatr Med 1997;13:497–511.
11. Thomas DR. Improving the outcome of pressure ulcers with nutritional intervention: A review of the evidence. Nutrition 2001;17:121–5.
12. Bourdel-Marchasson I, Barateau M, Sourgen C, et al. Prospective audits of quality of PEM recognition and nutritional support in critically ill elderly patients. Clin Nutr 1999;18:233–40.

13. Bourdel-Marchasson I, Barateau M, Rondeau V, Dequae-Merchadou L, Salles-Montaudon N, Emeriau JP, Manciet G, Dartigues JF. A multi-center trial of the effects of oral nutritional supplementation in critically ill older inpatients. GAGE Group. Groupe Aquitain Geriatrique d'Evaluation. Nutrition 2000;16:1–5.

14. Delmi M, Rapin CH, Bengoa JM, Delmas PD, Vasey H, Bonjour JP. Dietary supplementation in elderly patients with fractured neck of the femur. Lancet 1990;335(8696):1013–6.

15. Houwing RH, Rozendaal M, Wouters-Wesseling W, Beulens JW, Buskens E, Haalboom JR. A randomised, double-blind assessment of the effect of nutritional supplementation on the prevention of pressure ulcers in hip-fracture patients. Nutrition 2003;22(4):401–5.

16. Rebecca J. Stratton, Anna-Christina Ek, Meike Engfer, Zena Moore, Paul Rigby, Robert Wolfe, Marinos Elia. Enteral nutritional support in prevention and treatment of pressure ulcers: A systematic review and meta-analysis. Ageing Res Rev 2005;4:422–50.

17. Hartgrink HH, Wille J, Konig P, et al. Pressure sores and tube feeding in patients with a fracture of the hip: a randomized clinical trial. Clin Nutr 1998;6:287–92.

18. Henderson CT, Trumbore LS, Mobarhan S, et al. Prolonged tube feeding in long-term care: Nutritional status and clinical outcomes. J Am College Clin Nutr 1992;11:309.

19. Mitchell SL, Kiely DK, Lipsitz LA. The risk factors and impact on survival of feeding tube placement in nursing home residents with severe cognitive impairment. Arch Intern Med 1997;157:327–32

20. Albina JE. Nutrition and wound healing. JPEN 1994;18:367–76.

21. Thomas DR. Issues and dilemmas in managing pressure ulcers. J Gerontol Med Sci 2001;56:M238–340.

22. Choban PS, Burge JC, Flanobaum L. Nutrition support of obese hospitalized patients. Nutr Clin Pract 1997;12:149–54.

23. Ireton-Jones CS. Evaluation of energy expenditures in obese patients. Nutr Clin Pract 1989;4:127–9.

24. Food and Nutrition Board, Institute of Medicine. Dietary Reference Intakes for Energy, Carbohydrates, Fiber, Fat, Protein and Amino Acids (Macronutrients). In: Energy: The National Academies Press, 2002:93–206.

25. Chernoff RS, Milton KY, Lipschitz DA. The effect of very high-protein liquid formula (Replete) on decubitus ulcer healing in long-term tube-fed institutionalized patients. Investigators Final Report 1990. J Am Diet Assoc 1990;90(9):A–130.

26. Breslow RA, Hallfrisch J, Guy DG, et al. The importance of dietary protein in healing pressure ulcers. J Am Geriatr Soc 1993;41:357–62.

27. Gersovitz M, Motil K, Munro HN, Scrimshaw. Human protein requirements: Assessment of the adequacy of the current Recommended Dietary Allowance for dietary protein in elderly men and women. Am J Clin Nutr 1982;35:6–14.

28. Long CL, Nelson KM, Akin JM Jr, Geiger JW, Merrick HW, Blakemore WZ. A physiologic bases for the provision of fuel mixtures in normal and stressed patients. J Trauma 1990;30:1077–86.

29. McCauley\ R, Platell C, Hall J, McCulloch R. Effects of glutamine on colonic strength anastomosis in the rat. J Parenter Enter Nutr 1991;116:821.

30. Barbul A, Lazarous S, Efron DT, et al. Arginine enhances wound healing in humans. Surgery 1990;108:331–7.

31. Kirk SJ, Regan MC, Holt D, et al. Arginine stimulates wound healing and immune function in aged humans. Surgery 1993;114:155.

32. Langkamp-Henken B. Herrlinger-Garcia KA. Stechmiller JK. Nickerson-Troy JA. Lewis B. Moffatt L. Arginine supplementation is well tolerated but does not enhance mitogen-induced lymphocyte proliferation in elderly nursing home residents with pressure ulcers. J Parenter Enter Nutr 2000;24(5):280–7.

33. McCauley C, Platell C, Hall J, McCullock R. Influence of branched chain amino acid solutions on wound healing. Aust NZ J Surg 1990;60:471.

34. Crandon JH, Lind CC, Dill DB. Experimental human scurvy. N Engl J Med 1940;223:353.

35. Levenson SM, Upjohn HL, Preston JA, et al. Effect of thermal burns on wound healing. Ann Surg 1957;146:357–68.

36. Rackett SC, Rothe MJ, Grant-Kels JM. Diet and dermatology. The role of dietary manipulation in the prevention and treatment of cutaneous disorders. J Am Acad Dermatol 1993;29:447–61.

37. Vilter RW. Nutritional aspects of ascorbic acid: Uses and abuses. West J Med 1980;133:485.

38. ter Riet G, Kessels AG, Knipschild PG. Randomized clinical trial of ascorbic acid in the treatment of pressure ulcers. J Clin Epidemiol 1995;48:1453–60.

39. Taylor TV, Rimmer S, Day B, Butcher J, Dymock IW. Ascorbic acid supplementation in the treatment of pressure sores. Lancet 1974;2:544–6.

40. Hunt TK. Vitamin A and wound healing. J AM Acad Dermatol 1986;15:817–21.

41. Ehrlich HP, Hunt TK. Effects of cortisone and vitamin A on wound healing. Ann Surg 1968;167:324.

42. Waldorf H, Fewkes J. Wound healing. Adv Dermatol 1995;10:77–96.

43. Pories WJ, Henzel WH, Rob CG, et al. Acceleration of healing with zinc sulfate. Ann Surg 1967;165:423.

44. Hallbrook T, Lanner E. Serum zinc and healing of leg ulcers. Lancet 1972;2:780.

45. Sandstead SH, Henrikson LK, Greger JL, et al. Zinc nutriture in the elderly in relation to taste acuity, immune response, and wound healing. Am J Clin Nutr 1982;36(Supp):1046.

46. Goode HF, Burns E, Walker BE. Vitamin C depletion and pressure ulcers in elderly patients with femoral neck fracture. Brit Med J 1992:305:925–7.

47. Norris JR, Reynolds RE. The effect of oral zinc sulfate therapy on decubitus ulcers. JAGS 1971;19:793.

48. Goode P, Allman R. The prevention and management of pressure ulcers. Med Clin N Amer 1989;73:1511–24.

49. Reed BR, Clark RAF. Cutaneous tissue repair: Practical implications of current knowledge: II. J Am Acad Dermatol 1985;13:919–41.

50. Gregger JL. Potential for trace mineral deficiencies and toxicities in the elderly. *In* Mineral Homeostasis in the Elderly. Ed. Bales CW, New York, Marcel Dekker, 1989, pp. 171–200.

51. Thomas DR. Distinguishing starvation from cachexia. Geriatr Clin North Am 2002;18:883–92.

52. Thomas DR. Weight loss in older adults. Rev Endocrinol Metabol 2005;6:129–36.

53. Haddad RY, Thomas DR. Enteral Nutrition and Tube Feeding: A Review of the evidence. Geriatr Clin North Am 2002;18:867–82.

54. Thomas DR. A prospective, randomized clinical study of adjunctive peripheral parenteral nutrition in adult subacute care patients. J Nutr Health Aging 2005;9:321–5.

55. Morley JE. Thomas DR. Wilson MM. Cachexia: pathophysiology and clinical relevance. Am J Clin Nutr 2006;83(4):735–43.

56. Thomas DR. Loss of skeletal muscle mass in aging: examining the relationship of starvation, sarcopenia and cachexia. Clin Nutr 2007;26(4):389–99.

57. Liso Z, Tu JH, Small CB, Schnipper SM, rosenstreich DL. Increased urine IL-1 levels in aging. Gerontology 1993;39:19–27.

58. Cederholm T, Whetline B, Hollstrom K, et al. Enhanced generation of Interleukin 1β and 6 may contribute to the cachexia of chronic disease. Am J Clin Nutr 1997;65:876–82.

59. Ikeda U, Yamamoto K, Akazawa H, et al. Plasma cytokine levels in cardiac chambers of patients with mitral stenssis with congestive heart failure. Cardiology 1996;87:476–8.

60. Friedman FJ, Campbell AJ, Caradoc-Davies. Hypoalbuminemia in the elderly is due to disease not malnutrition. Clin Exper Gerontol 1985;7:191–203.

61. Rosenthal AJ, Sanders KM, McMurtry CT, Jacobs MA, Thompson DD, Gheorghiu D, Little KL, Adler RA. Is malnutrition overdiagnosed in older hospitalized patients? Association between the soluble interleukin-2 receptor and serum markers of malnutrition. J Gerontol Series A, Biol Sci Med Sci 1998;53:M81–6.

62. Matsuyama N. Takano K. Mashiko T. Jimbo S. Shimetani N. Ohtani H. The possibility of acute inflammatory reaction affects the development of pressure ulcers in bedridden elderly patients. Rinsho Byori – Jpn J Clin Pathol 1999;47(11):1039–45.

63. Barone EJ, Yager DR, Pozez AL, Olutoye OO, Crossland MC, Diegelmann RF, Cohen IK. Interleukin-1α and collagenase activity are elevated in chronic wounds. Plast Reconstr Surg 1998;102:1023–7.

64. Segal JL, Gonzales E, Yousefi S, Jamshidipour L, Brunnemann SR. Circulating levels of IL-2R, ICAM-1, and IL-6 in spinal cord injuries. Arch Physical Med Rehab 1997;78:44–7.

65. Bonnefoy M, Coulon L, Bienvenu J, Boisson RC, Rys L. Implication of Cytokines in the aggravation of malnutrition and hypercatabolism in elderly patients with severe pressure sores. Age Ageing 1995;24:37–42.

66. Wilson MM. Thomas DR. Rubenstein LZ. Chibnall JT. Anderson S. Baxi A. Diebold MR. Morley JE. Appetite assessment: simple appetite questionnaire predicts weight loss in community-dwelling adults and nursing home residents. Am J Clin Nutr 2005;82(5):1074–81.

67. Morley JE, Thomas DR. Anorexia and aging: Pathophysiology. Nutrition 1999;15:499–503.

68. Chasse WT, Balasubramahiam A, Dayal R, et al. Hypothalamic concentration and release of neuropeptide Y into micordialyses is reduced in anorectic tumor bearing rates. Life Sci 1994;54:1869–74.

69. Leibowitz SF. Neurochemical-neuroendocrine systems in the brain controlling macronutrient intake and metabolism. Trends Neurosci 1992;12:491–7.

70. Noguchi Y, Yoshikawa T, Marsumoto A, Svaninger G, Gelin J. Are cytokines possible mediators of cancer cachexia? Jpn J Surg 1996;26:467–75.

71. Keiler U. Pathophysiology of cancer cachexia. Support Care Cancer 1993;1:290–4.

72. Straton RJ, Dewit O, Crowe R, Jennings G, Viller RN, Elia M. Plasm leptin, energy intake and hunger following total hip replacement surgery. Clin Sci 1997;93:113–7.

73. Cederholm T, Arter P, Palmviad J. Low circulation leptin level in protein-energy malnourished chronically ill elderly patients. J Intern Med 1997;242:377–82.

74. Souba WW. Drug Therapy: Nutritional Support. N Engl J Med 1997;336:41–8.

75. Atkinson S, Sieffert E, Bihari D. A prospective, randomized, double-blind, controlled clinical trial of enteral immunonutrition in the critically ill. Crit Care Med 1998;26:1164–72.

76. Milne AC, Potter J, Avenell A. Protein and energy supplementation in elderly people at risk from malnutrition. Cochrane Metabolic and Endocrine Disorders Group. Cochrane D Syst Rev. 2008;4.

77. Thomas DR. Guidelines for the use of orexigenic drugs in long-term care. Nutr Clin Pract 2006;21(1):82–7.

78. Bruerra E, Macmillan K, Hanson J, et al. A controlled trial of megestrol acetate on appetite, caloric intake, nutritional status, and other symptoms in patients with advance cancer. Cancer 1990;66:1279–82.

79. Allman RM, Goode PS, Patrick MM. Pressure ulcer risk factors among hospitalised patients with severe limitation. J Am Med Assoc 1995;273:865–70.

80. Schmoll E, Wilke H, Thole R. Megestrol acetate in cancer cachexia. Seminars Oncol 1991;1(suppl)2:32–4.

81. Heckmayr M, Gatzenneier U. Treatment of cancer weight loss in patients with advance lung cancer. Oncology 1992;49(suppl)2:32–4

82. Feliu J, gonzalez-Baron M, Berrocal A. Usefulness of megestrol acetate in cancer cachexia and anorexia. Am J Clin Oncol 1992;15:436–40.

83. Azona C, Castro L, Crespo E, et al. Megestrol acetate therapy for anorexia and weight loss in children with malignant solid tumours. Aliment Pharmacol Ther 1996;10:577–86.

84. Mantovani G, Maccio A, Bianchi A, Curreli L, Ghiani M, Santona MC, Del Giacco GS. Megestrol acetate in neoplastic anorexia/cachexia: clinical evaluation and comparison with cytokine levels in patients with head and neck carcinoma treated with neoadjuvant chemotherapy. Springer-Verlag 1995.

85. Christou NV, Meakins JL, Gordon J, et al. The delayed hypersensitivity response and host resistance in surgical patients: 20 years later. Ann Surg 1995;222:534–48.

86. Allman RM, Walker JM, Hart MK, et al. Air-fluidized beds or conventional therapy for pressure sores: A randomized trial. Ann Intern Med 1987;107:641–48.

87. Gorse GJ, Messner RL: Improved pressure sore healing with hydrocolloid dressings. Arch Dermatol 1987;123:766–71.

88. Inman KJ, Sibbald WJ, Rutledge FS Clinical utility and cost-effectiveness of an air suspension bed in the prevention of pressure ulcers. JAMA 1993;269:1139–43.

89. Anthony D, Reynolds T, Russell L. An investigation into the use of serum albumin in pressure sore prediction. J Adv Nurs 2000;32:359–65.

90. Moolten SE. Bedsores in the chronically ill patient. Arch Phys Med Rehabil 1972;53:430–38.
91. Bennett RG, Bellantoni MF, Ouslander JG: Air-fluidized bed treatment of nursing home patients with pressure sores. J Am Geriatr Soc 1989;37:235–42.
92. Brandeis GH, Morris JN, Nash DJ, et al. Epidemiology and natural history of pressure ulcers in elderly nursing home residents. J Am Med Assoc 1990;264:2905–9.
93. Trumbore LS, Miles TP, Henderson CT, et al. Hypocholesterolemia and pressure sore risk with chronic tube feeding. Clin Res 1990;38:760A.
94. Breslow RA, Hallfrisch J, Goldberg AP. Malnutrition in tubefed nursing home patients with pressure sores. J Parent Enter Nutr 1991;15:663–8.
95. Ferrell BA, Osterweil D, Christenson P: A randomized trial of low-air-loss beds for treatment of pressure ulcers. JAMA 1993;269:494–7.
96. Guralnik JM, Harris TB, White LR, et al. Occurrence and predictors of pressure sores in the National Health and Nutrition Examination survey follow-up. J Am Geriatrics Soc 1988;36:807–12.

13 Nutrition at the End of Life: Ethical Issues

Christine Seel Ritchie and Elizabeth Kvale

Key Points

- Because of the importance of food as an integral part of life, health-care providers need to be well versed in the issues surrounding nutrition and hydration in terminal illness in order to assist patients and their families in treatment decisions.
- Case law in the United States considers enteral nutrition to be medical treatment and as such recognizes a patient's right to refuse artificial nutritional support.
- The decision about feeding should be consistent with the overall goals of care: for example, curative treatment, rehabilitative treatment or palliative treatment.

Key Words: Nutrition support; terminal illness; palliative treatment

13.1 INTRODUCTION

In the final weeks of an individual's life the issues central to decisions about nutritional support shift to a cautious weighing of its burdens and benefits. Nutrition support during the final phase of life is an important and sensitive issue that must be resolved by careful discussion between patients, the patient's loved ones and the health-care team. Food is an integral part of day-to-day life. As a society, we celebrate with food and we find comfort from food. We prepare and share meals as a way of expressing love and concern. Therefore, the loss of interest in or the ability to partake in this activity often causes great distress for caregivers. Eating is so central to life in society that an illness that impairs the ability or desire to eat may compel caregivers and health-care providers to feel they must "do something" to address the situation. Health-care providers need to be well versed in the issues surrounding nutrition and hydration in terminal illness in order to assist patients and their families in treatment decisions. This chapter will focus on the benefits and

From: *Nutrition and Health: Handbook of Clinical Nutrition and Aging, Second Edition*
Edited by: C. W. Bales and C. S. Ritchie, DOI 10.1007/978-1-60327-385-5_13,
© Humana Press, a part of Springer Science+Business Media, LLC 2009

limitations of nutritional support in the final weeks to months of life. The principles presented in this chapter may not be generalizable to populations with a longer life expectancy, even if the underlying disease process is similar.

13.2 DEFINITIONS

Non-oral feeding is the provision of food by a nasogastric tube, gastrostomy tube, gastrojejunostomy tube or TPN. *Artificial hydration* is the provision of water or electrolyte solutions by any non-oral route. *Artificial nutrition* includes enteral nutrition by nasogastric tube, percutaneous endoscopic gastrostomy tube, percutaneous jejunostomy tube, gastrostomy tube or gastrojejunostomy tube.

13.3 LEGAL, RELIGIOUS AND ETHICAL PRECIDENTS FOR DECISION-MAKING

Multiple legal decisions provide guidance to physicians concerning the nutritional treatment of terminal patients. In the Karen Ann Quinlan Case of 1976, the court upheld the right to forgo life-sustaining care and clarified an important distinction between killing and letting die *(1)*. The Barber Case in 1983 involved a patient in a deep coma after cardiopulmonary arrest following surgery to close an ileostomy. The spouse testified that the patient had stated "no Karen Quinlan." However, a California lower court indicted two physicians who stopped intravenous fluids and feeding tubes for murder. The appellate court dismissed the murder charges because the charge of murder requires proof of an act of commission. Terminating artificial nutrition was considered an act of omission, not commission. The distinction between acts of omission and acts of commission is central to discussions of the ethics of terminal nutrition and hydration. Acts of omission may be unlawful if there is a known duty to act. There was no duty to act in the Barber case because of the patient's known prior wishes for no treatment *(2)*.

In 1986, the appellate court in the Bouvia Case ruled that refusal of medical support, in the instance enteral nutrition, by a competent patient is a fundamental right. The patient was a 29-year-old female with severe cerebral palsy who was bedridden, immobile, in constant pain and competent to make her own medical decisions. A feeding tube was placed against the patient's wishes. The patient then petitioned to remove the feeding tube. The lower court refused the petition stating it was a form of suicide. An appeals court later ruled in favor of the patient *(3)*. In the 1990 Nancy Cruzan case, parents of Nancy Cruzan, a woman in a persistent vegetative state, requested cessation of nutritional support. The US Supreme Court stated that patients have a right to die and that competent patients can refuse therapy. They also made it clear that artificial nutrition/hydration was no different than any other medical treatment. They ultimately found in favor of the state by concluding that the state (in this case Missouri) can set its evidentiary standard sufficiently high to require clear and convincing evidence of the patient's wishes to withdraw life support *(1)*.

Although the courts see artificial nutrition/hydration (ANH) as medical treatment, many patients and caregivers do not share this opinion; they see ANH as basic treatment that cannot be withdrawn. This tension around ANH is demonstrated by the fact that many states have separate additional statutory requirements for ANH, beyond that required for other forms of medical treatment *(4,5)*. This tension was also made manifest in public commentary surrounding the Terry Schiavo case *(6)*.

Religious and personal views often run counter to the prevailing legal/ethical position on ANH. Although the National Conference of Catholic Bishops stated that "Catholics are not obligated to use extraordinary or disproportionate means where there is no hope," nevertheless they also recommended a "presumption in favor of providing nutrition and hydration to all patients, including patients who require medically assisted nutrition and hydration, as long as this is of sufficient benefit to outweigh burdens involved to the patient." *(7,8)* Orthodox Judaism also promotes the continuation of ANH once it has been started *(9)*.

The lack of public consensus surrounding ANH requires that health-care providers pay close attention to the burden and benefit of ANH and make sure patients and their proxies understand as fully as possible the benefits, burdens and uncertainties associated with these therapies. Only then can patients truly determine what is consistent with their values and priorities.

In summary, case law in the United States confirms that enteral nutrition is considered to be medical treatment and not basic care. Patients have the right to refuse this form of medical treatment. Withdrawal of artificial nutritional support to allow a patient to die is not equivalent to euthanasia. In the former instance, the goal of discontinuing therapy is to remove burdensome interventions; in the latter, the intended result is the death of the patient.

13.4 NUTRITION AND HYDRATION IN ADVANCED ILLNESS

The scientific literature provides little guidance regarding the benefits and burden of nutrition and hydration at the end of life. A review of the literature since the publication of the initial edition of this text identified additional data highlighting the challenges of nutrition and hydration in terminal care. All of this data is limited by the observational nature of the data collected.

13.4.1 *Artificial Nutrition and Hydration in Terminal Cancer*

Unintentional weight loss and loss of appetite are cardinal symptoms in advanced cancer, and are poor prognostic indicators. Cancer cachexia occurs in 50% of patients suffering from any form of cancer at any stage *(10)*. The prevalence and degree of cachexia varies according to the pathology and stage of the malignancy. For example, cachexia occurs in 30% of patients with lymphoma and 85% of patients with gastric or pancreatic cancer *(10)*.

The impact of nutritional support in cancer cachexia remains in question. Nutritional support in cancer patients has not yet been shown to improve survival, improve tumor response, decrease toxicity, or decrease surgical complications. The only exception may be in GI, head and neck cancer patients and esophageal cancer patients who are malnourished *(11)*. These findings are somewhat equivocal, however, and balanced by studies that do not demonstrate benefit. In a recent secondary analysis of the Radiation Therapy Oncology Group 90-03 Study, a trial primarily aimed at evaluating four different radiation fractionation schedules in head and neck cancer, findings did not suggest benefit of prophylactic nutrition support (PNS). Whereas patients receiving PNS had less weight loss by the end of treatment and less grade 3–4 mucositis, by 5 years, patients receiving PNS had poorer locoregional control and poorer survival *(12)*.

Even less clinical data exist regarding nutrition support at the end of life. In one of the most well-known studies, the authors completed a prospective study of patients in a comfort care unit – the majority of whom had cancer or stroke as their terminal diagnosis. The patients were alert and competent. Food was offered and feeding was assisted but not forced. The patients were followed for thirst, hunger and dry mouth and to see if food or fluid relieved the symptoms *(13)*. Sixty-three percent of the patients studied did not experience hunger; an additional 34% had hunger initially that subsequently resolved. Similarly, 62% had no thirst or thirst only initially. Most of the patients' symptoms were easily controlled with a small amount of food or water. The authors concluded by stating that hunger and thirst were uncommon in the terminal phase despite food and water intake inadequate to sustain basal energy requirements.

13.4.2 Artificial Nutrition and Hydration in Terminal Dementia

13.4.2.1 Definitions and Prognosis of Advanced Dementia

Dementia is a progressive terminal condition caused by one of a number of conditions including Alzheimer's disease, cerebrovascular disease, congenital or acquired neurodegenerative diseases, brain tumors, AIDS and Parkinson's disease. The features of dementia include loss of higher cognitive function, loss of intelligible speech, inability to maintain oral nutrition due to loss of swallowing reflex and inability to ambulate. There were between 2.17 and 4.78 million cases of AD in the United States in 2000, of which between 44 and 57% had moderate or severe disease *(14)*. Dementia is a progressive disease that worsens in recognizable stages. The Functional Assessment Staging System (FAST) is one system used to follow the course of Alzheimer's disease thereby helping to decide how far the disease has progressed *(15)*. There are seven stages in the FAST system, with the first stage comprising essentially no symptoms and the last stage describing advanced, end-stage dementia. At stage 6, the patient needs supervision in dressing, bathing, toileting and eating and becomes dependent on the caregiver. Late deficits in this stage are incontinence and the inability to flush the toilet. Patients typically either die or are institutionalized after 3 years in this stage. This is the stage when patients with Alzheimer's disease may stop eating spontaneously, but can be encouraged to eat. At stage 7, these patients lose the ability to speak, ambulate, eat, control their muscles and smile. When patients reach this stage it is very difficult to maintain nutrition because encouragement to eat becomes less successful. Patients at stage 7 typically die within a year, and difficulty eating is a marker for the terminal phase of Alzheimer's dementia. The association of difficulty eating and terminal disease seen in Alzheimer's dementia should not be extended to other forms of dementia. For example, patients with Parkinson's disease often lose the ability to maintain adequate caloric intake at an earlier stage of their disease; in this setting, a feeding tube may be helpful *(16)*.

13.4.2.2 Addressing the Costs and Benefits of Artificial Nutrition and Hydration in Advanced Dementia

Physicians and families have difficulty discussing nutrition support for patients with advanced dementia. Nutrition may be a more emotional decision to some families than ventilator support or cardiac resuscitation. Families may articulate a concern that they are "starving" their loved one. They fear the ill person will not survive as a direct result of not eating. These issues are so difficult that sometimes physicians and families do not ever even initiate discussions of them.

Finucane attempted to address these difficult questions based on the limited clinical data available *(17)*. He reviewed primary concerns commonly cited as rationale for enteral nutrition, and included aspiration pneumonia, skin breakdown, quality of life and survival as outcomes. His review of the literature did not indicate that enteral nutrition improved any of these outcomes. However, many of the studies reviewed were not solely of patients with dementia and almost all had methodological limitations.

A potentially unintended result of tube feedings may be the subsequent use of patient restraints. Restraints can lead to distress and agitation and sedating medications to control the behavior, all negatively impacting quality of life. Gillick addressed quality of life in her research and found that advanced dementia patients who were tube fed were often deprived of taste, touch and social interaction *(18)*. Restraints were used 71% of the time in patients with dementia and feeding tubes, regardless of the type of tube. Bedfast patients with advanced dementia who receive artificial nutritional support are more likely to be restrained; the resulting immobility and incontinence (rather than nutritional factors) may explain in part the *lack* of association between tube feeding and decreased skin breakdown.

Sanders addressed survival secondary to tube feeding *(19)*. He found that in a nursing home, the patients who were fed by a gastrostomy tube and those fed by hand had the same survival rates. He found that among gastrostomy patients, patients with dementia had a much worse prognosis (54% 1-month mortality) compared to those without dementia (28% 1-month mortality). Other studies, such as that by Rimon et al., have not demonstrated that the presence of dementia incurs a higher risk for mortality among tube-fed patients *(20)*.

A recent study attempted to evaluate the level of comfort for nursing home patients with dementia who had little or no voluntary intake and for whom the decision had been made to forgo ANH. Of the 145 residents evaluated, the average number of symptoms was 2.1 at baseline, and decreased during follow-up. In patients who died within 2 weeks after baseline, the average level of discomfort decreased until they died. Pain, dehydration, restlessness and dyspnea were all associated with increased discomfort. Fluid intake in these patients did not significantly influence the level of comfort *(21)*.

13.4.2.3 Alternative Approaches to Artificial Nutrition and Hydration in Advanced Dementia

In patients with Alzheimer's dementia, skillful feeding techniques need to be employed (see Chapter 9). These include the appropriate selection of food consistencies and minimization of distractions. Additionally, adequate time for feeding and verbal cueing to chew the food and swallow must occur. The mid-day meal is when food intake is often greatest, and presents an opportunity to effectively focus feeding efforts *(22)*.

13.5 ADDRESSING TREATMENT GOALS AND DECISION-MAKING FOR TERMINAL PATIENTS

When treatment goals are discussed early following the diagnosis of a terminal condition, the patient often can decide for himself or herself what their wishes are regarding ANH at the end of life. In this way, the goals will reflect the preferences and values of the patient. A discussion about feeding wishes should also take place. Unfortunately, patients' wishes are rarely known regarding tube feeding. Friedel and Ozick described how processes of care may circumvent a meaningful informed consent process regarding gastrostomy tube placement in a small 2000 study. Of 18 patients in a New York municipal hospital scheduled for a gastrostomy tube, only one patient was deemed capable of giving informed consent. None of the other 17 patients had an Advanced Directive or Power of Attorney. The medical staff became the decision-makers and all 18 received a gastrostomy tube *(23)*.

The decision about feeding should be consistent with the overall goals of care. Treatment goals are not widely discussed because physicians feel pressed for time, feel uncomfortable discussing issues surrounding the end of life and/or were never taught how to discuss these sensitive topics. When discussing treatment goals, one must start by asking the patient what they already know. The next step is to review the patient's condition and prognosis. Then, the physician should stop to ask for questions. The physician should continue by discussing the patient's preferences and goals, then outlining appropriate treatment options – or goals as they relate to the patient's current condition: curative treatment, rehabilitative treatment or palliative treatment *(24)*. Palliative treatment focuses on improving quality of life and may include life-prolonging treatment or treatment solely aimed at comfort care. Finally, the physician should invite questions, allow time to reflect and then decide if related issues of treatment withdrawal need to be discussed (Fig. 13.1).

When a patient becomes terminal and is non-decisional, there are established guidelines as to how decisions are to be made. The advanced directive, if one exists, is the document that should be consulted first to know the patient's prior expressed wishes. If the advanced directive does not answer the specific question that needs to be addressed, then the legal guardian or the agent of the advanced directive makes the decision based on what they think the patient would have wanted or the wishes previously expressed by the patient. If there is no document and no designated decision-maker, then the first-order relative makes decisions (usually spouse, then adult children, then siblings but this may vary by state). Finally the opinion of other relatives can be considered. If none of the above exists, then the physician's judgment can be used to determine the best treatment for the patient.

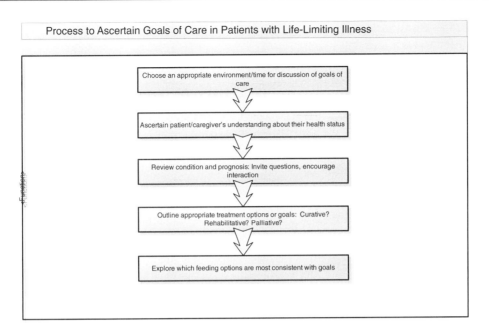

Fig. 13.1. Goal Setting Process for Nutritional Support.

13.6 PRACTICAL CONSIDERATIONS BEFORE PROVIDING ENTERAL/PARENTAL NUTRITION AND HYDRATION

Decision-making regarding the initiation of tube feeding is never easy. Practical matters may be more likely to influence the decision regarding feeding than ethical principles. In an imminently terminal patient the usual goal of care should be aggressive palliative care. This means the goal is to provide comfort care by managing the symptoms of the disease or side effects of the treatment while maintaining optimal quality of life. Thus, the goal of nutrition support in this phase of illness should be to maintain energy and strength while being attentive to potentially negative quality of life effects of coercive feeding or artificial nutrition and hydration. During this time, the physician will be most effective if they can understand the caregivers' feelings and counsel them. The caregiver may feel frustrated over the inability to find and prepare foods that are tolerated by the patient. They may also sense that the food they are offering is not providing the comfort that they were hoping for. The caregiver should be educated to understand that the loss of appetite and the inability to eat are common experiences in the terminally ill. Also, physical and emotional changes influence the ability to eat. For example, the disease itself, medications, fear or depression may make it difficult to eat. Changes in the sense of smell, diarrhea, constipation and nausea or vomiting also decrease the patient's appetite. Caregivers who push food on the patient with anorexia may inadvertently contribute to the patient's distress instead of comforting the patient. This was demonstrated by McCann et al.'s study of cancer patients *(13)*. Less is known about how dementia patients perceive symptoms related to feeding. It is hard to know if patients with dementia experience discomfort from not eating or burdens related to assisted eating because they are noncommunicative at this stage.

Thirst and hunger often appear to be diminished in the dying process. Practical options in caring for the terminally ill include eliminating most dietary restrictions and give only the amounts of foods and liquids tolerated or accepted. Finally, the patient should be assisted with meals, but not forced to eat. It can be helpful to share with caregivers that (1) withholding nutrition and hydration at the end of life can be beneficial with regard to patient comfort and (2) the injudicious use of ANH may aggravate symptoms of volume overload. Without hydration there are less oral and airway secretions, less congestion, coughing and fewer symptoms associated with ascites and edema. In addition, terminal patients cannot always cough secondary to weakness; therefore, aspiration risk increases. Peripheral edema may increase pain and predispose the patient to pressure ulcers. Finally, increased gastrointestinal fluids can cause nausea and vomiting, especially for patients with intestinal strictures or obstruction from neoplasms. During the dying process, dehydration occurs from inadequate intake and losses from GI, renal, skin and pulmonary secretions. Dehydration may lead to mental changes, which may decrease the patient's awareness of their suffering. Families are sometimes concerned about the dry mouth that occurs as a person dies. Ice chips, sips of liquid, lip moisteners, salivary substitutes, mouth swabs, hard candy and routine mouth care all help to relieve the xerostoma – dry mouth – that occurs.

13.7 SUMMARY

Caring for patients with a terminal illness such as cancer or advanced Alzheimer's disease is difficult for the family and the physician. The issues surrounding feeding are some of the hardest to resolve. This chapter defines the problems and offers guidelines.

To summarize, decisions regarding nutritional support in end-of-life care should be informed by treatment goals and patient preference. Case law regards enteral nutrition as medical treatment. With the exception of head and neck cancer and esophageal cancer, no studies have demonstrated improved survival in cancer or advanced dementia with enteral support. In advanced cancer patients, nausea and pain should be addressed and corticosteroids and progestational agents considered. In advanced dementia, emphasis should be placed on oral food intake, allowing adequate time for feeding, avoiding distractions and using verbal cueing. Every person, family and physician must decide for himself/herself to what extent to nourish a person with a terminal illness based on available information about the risks and benefits.

13.7.1 Recommendations

1. Before deciding on a specific form of nutritional support, establish treatment goals.
2. With a few exceptions, artificial nutritional support in cancer patients does not improve survival, improve tumor response, decrease toxicity or decrease surgical complications.

3. Artificial nutritional support may be appropriate in head and neck cancer patients and esophageal cancer patients who are not able to swallow properly and still have an appetite.

4. Current limited data do not demonstrate that artificial nutritional support improves survival or quality of life in advanced dementia patients.

REFERENCES

1. Annas GJ, Arnold B, Aroskar M, et al. Bioethicists' statement on the U.S. Supreme Court's Cruzan decision. N Engl J Med 1990;323(10):686–7.
2. Burck R. Feeding, withdrawing, and withholding: ethical perspectives. Nutr Clin Pract 1996;11(6):243–53.
3. Bouvia v. Superior Court (Glenchur). Wests Calif Report 1986;225:297–308.
4. Mayo TW. Living and dying in a post-Schiavo world. J Health Law 2005;38(4):587–608.
5. Sieger CE, Arnold JF, Ahronheim JC. Refusing artificial nutrition and hydration: does statutory law send the wrong message? J Am Geriatr Soc 2002;50(3):544–50.
6. Perry JE, Churchill LR, Kirshner HS. The Terri Schiavo case: legal, ethical, and medical perspectives. Ann Intern Med 2005;143(10):744–8.
7. Nutrition and hydration: moral and pastoral reflections. National Conference of Catholic Bishops Committee for Pro-life Activities. J Contemp Health Law Policy 1999;15(2):455–77.
8. Ethical and Religious Directives for Catholic Health Care Services. In: United States Catholic Conference: National Conference of Catholic Bishops; 1995; Washington, DC; 1995.
9. Dorff EN. A Jewish approach to end-stage medical care. Conserv Jud 1991;43(3):3–51.
10. Dewys WD, Begg C, Lavin PT, et al. Prognostic effect of weight loss prior to chemotherapy in cancer patients. Eastern Cooperative Oncology Group. Am J Med 1980;69(4):491–7.
11. Senesse P, Assenat E, Schneider S, et al. Nutritional support during oncologic treatment of patients with gastrointestinal cancer: Who could benefit? Cancer Treat Rev 2008.
12. Rabinovitch R, Grant B, Berkey BA, et al. Impact of nutrition support on treatment outcome in patients with locally advanced head and neck squamous cell cancer treated with definitive radiotherapy: a secondary analysis of RTOG trial 90-03. Head Neck 2006;28(4):287–96.
13. McCann RM, Hall WJ, Groth-Juncker A. Comfort care for terminally ill patients. The appropriate use of nutrition and hydration. Jama 1994;272(16):1263–6.
14. Sloane PD, Zimmerman S, Suchindran C, et al. The public health impact of Alzheimer's disease, 2000–2050: potential implication of treatment advances. Annu Rev Public Health 2002;23:213–31.
15. Reisberg B. Functional assessment staging (FAST). Psychopharmacol Bull 1988;24(4):653–9.
16. Evatt ML. Nutritional therapies in Parkinson's disease. Curr Treat Options Neurol 2007;9(3):198–204.
17. Finucane TE, Christmas C, Travis K. Tube feeding in patients with advanced dementia: a review of the evidence. Jama 1999;282(14):1365–70.
18. Gillick MR. Rethinking the role of tube feeding in patients with advanced dementia. N Engl J Med 2000;342(3):206–10.
19. Sanders DS, Carter MJ, D'Silva J, James G, Bolton RP, Bardhan KD. Survival analysis in percutaneous endoscopic gastrostomy feeding: a worse outcome in patients with dementia. Am J Gastroenterol 2000;95(6):1472–5.
20. Rimon E, Kagansky N, Levy S. Percutaneous endoscopic gastrostomy; evidence of different prognosis in various patient subgroups. Age Ageing 2005;34(4):353–7.
21. Pasman HR, Onwuteaka-Philipsen BD, Kriegsman DM, Ooms ME, Ribbe MW, van der Wal G. Discomfort in nursing home patients with severe dementia in whom artificial nutrition and hydration is forgone. Arch Intern Med 2005;165(15):1729–35.
22. Suski NS, Nielsen CC. Factors affecting food intake of women with Alzheimer's type dementia in long-term care. J Am Diet Assoc 1989;89(12):1770–3.

23. Friedel DM, Ozick LA. Rethinking the role of tube feeding in patients with advanced dementia. N Engl J Med 2000;342(23):1756.
24. Weissman DE. Establishing treatment goals, withdrawing treatments, DNR orders. In: Weissman DE, ed. Improving End-of-Life Care: A Resource Guide for Physicians. Milwaukee: The Medical College of Wisconsin, Inc.; 1998:94–100.

III COMMON CLINICAL CONDITIONS

14 Nutrition and Oral Health: A Two-Way Relationship

Kaumudi Joshipura and Thomas Dietrich

Key Points

- Overall, prevalence of edentulism among older adults has declined significantly in the past decade, increasing the ability to consume a more varied diet.
- The consumption of fermentable carbohydrates, along with increased use of xerogenic medications, place older adults at high risk for dental caries.
- Recent studies suggest an important role for calcium, vitamin D, and possibly vitamin C in reducing periodontitis risk.
- Diets high in fruits and vegetables have consistently been shown to reduce oral cancer risk.
- Tooth loss is associated with increased tendency among older adults to consume poorer quality diets, which in turn could increase cardiovascular risk.

Key Words: Edentualism; dental caries; periodontitis; oral cancer

14.1 ORAL HEALTH STATUS IN OLDER ADULTS

Oral health contributes greatly to quality of life in older adults. Poor oral health can hinder a person's ability to sustain a satisfying diet, participate in interpersonal relationships, and maintain a positive self-image (1–3). Oral health problems may lead to chronic pain, discomfort, and alterations in diet that may adversely impact systemic disease.

Dental caries and chronic periodontitis are by far the most common oral diseases in the elderly. These two diseases are the major causes of tooth loss and thus the major cause of dental morbidity. Other oral diseases or conditions are relatively rare, of lesser importance from a public health perspective and less related to nutrition, even though they may predominantly affect the elderly (e.g., denture stomatitis and other soft tissue lesions). We will therefore focus on dental caries,

From: *Nutrition and Health: Handbook of Clinical Nutrition and Aging, Second Edition*
Edited by: C. W. Bales and C. S. Ritchie, DOI 10.1007/978-1-60327-385-5_14,
© Humana Press, a part of Springer Science+Business Media, LLC 2009

chronic periodontitis, and tooth loss as the most common oral health problems in the elderly. In addition, we will also discuss oral cancer, which although rare, is often fatal and thus of high public health importance.

14.1.1 Common Oral Conditions in Older Adults

14.1.1.1 DENTAL CARIES, PERIODONTAL DISEASE AND TOOTH LOSS

Dental caries and periodontitis, both of which may result in tooth loss, are by far the most common oral diseases. The fact that in 1999–2004, 24% of seniors aged 65–74 years and 31% of those aged 75 years and older were edentulous (i.e., had lost all their natural teeth) illustrates that caries, periodontal disease and tooth loss still continue to be significant public health problems in the United States. Almost 20% of dentate seniors 65 years of age or older had at least one tooth with untreated tooth decay in 1999–2004. On average, these seniors had about 18 decayed, missing, or filled teeth. Root caries is also an important problem in older adults. In 1999–2004, 32% of adults 65–74 years and 42% of adults 75 years or older had at least one decayed or restored root surface. Moderate to severe periodontitis was 14% and 21% among seniors aged 65–74 years and 75+ years, respectively.

The good news is that data from several representative health surveys in the United States clearly demonstrate a steady decline in the prevalence of edentulism, caries, and periodontitis over the past several decades beginning in the 1960s. However, not all segments of the population have benefited equally from this trend and tremendous heterogeneity and disparities exist between socio-economic and racial/ethnic groups. For example, between 1988–1994 (NHANES III) and NHANES 1999–2004, the prevalence of edentulism declined only marginally and non-significantly from 46 to 44% for seniors aged 65 years and older below the federal poverty threshold level. However, for the group of seniors with incomes of greater than 200% of the federal poverty threshold, the prevalence of edentulism significantly declined from 24 to 17% over this decade. Similarly, caries and period-ontal health improved significantly for most segments of the population.

However, these findings and trends may not be generalized to other countries, including other industrialized countries. For example, a recent representative sur-vey of oral health in Germany found a prevalence of severe periodontal disease (defined as attachment loss 7+ mm) of 47% in 2005, an increase from the 40% prevalence found in a previous survey conducted in 1997.

In summary, dental caries and periodontitis remain highly prevalent in the elderly. Although in most industrialized countries the prevalence of caries has declined over the past decades, dental caries and periodontitis and the associated tooth loss continue to be major public health problems.

14.1.1.2 ORAL CANCER

The most serious and potentially fatal oral condition among older adults is oral cancer. By far the most common form of malignant oral cancer is squamous cell carcinoma. Squamous cell carcinoma of the oral cavity (including the gums, tongue and floor of mouth and other oral cavity), the oropharynx, hypopharynx, and larynx is frequently called 'head and neck squamous cell carcinoma' or 'head and neck cancer'.

In 2007, there were an estimated 34,360 newly diagnosed cases of oral and pharyngeal cancers (24,180 in males, 10,180 in females) and an estimated 7,550 oral and pharyngeal cancer deaths (5,180 males, 2,370 females) in the United States *(4)*. In 2000–2004, the age-standardized annual incidence of oral and pharyngeal cancers was 10.5 cases per 100,000. Incidence rates are much higher among males (15.6 per 100,000) than females (6.1 per 100,000). During the same period, the mortality from oral and pharyngeal cancer was 2.7 per 100,000 (males: 4.1/100,000, females: 1.5/100,000) *(5)*. Oral and pharyngeal cancer incidence and mortality are the highest among persons 65 years and older. In 2000–2004, the age-standardized incidence rate for this age group was 39.5/100,000 (males: 58.9/100,000, females: 25.4/100,000) and the mortality rate was 12.5/100,000 (males: 18.5/100,000, females 8.2/100,000).

There are major racial/ethnic disparities in oral cancer incidence and particularly in oral cancer mortality and survival. For example, mortality rates per 100,000 are 6.8 and 1.7 for Black males and females, compared to 3.8 and 1.5 for White males and females, respectively *(5)*. The 5-year survival rate for oral and pharyngeal cancer diagnosed between 1996 and 2002 was 60% and has only slightly improved over the past 25 years from 53% for cases diagnosed between 1975 and 1977 *(4)*. However, this improvement was limited to Whites (55% in 1975–1977 to 62% in 1996–2002), while no significant improvement in the 5-year survival rate could be observed among African Americans. Although the oral cavity and pharynx are easily accessible for inspection, only 33% of oral and pharyngeal cancers are diagnosed in a localized stage (35% among Whites, 21% among Blacks). The 5-year relative survival rate is highly dependent on stage of diagnosis. More than 80% of patients with localized disease survive 5 years, compared to 52 and 26% with regional and distant metastases, respectively. However, even if diagnosed at the same stage, Blacks have lower 5-year relative survival rates than Whites *(4)*.

14.2 IMPACT OF NUTRITIONAL STATUS ON ORAL HEALTH

14.2.1 *Plaque and Calculus Formation*

Bacteria in the mouth, or oral flora, form a complex community or biofilm that adheres to teeth and is called plaque. These bacteria ferment sugars and carbohydrates and generate acid, which can in turn dissolve minerals in tooth enamel and dentin and lead to dental caries. Furthermore, bacterial products and components elicit an inflammatory immune response in the gingival epithelium and underlying connective tissues (gingivitis) that may lead to periodontitis in susceptible individuals. Although the presence of plaque itself is not sufficient to cause either caries or periodontitis, the current understanding of the pathogenesis views bacterial plaque as a necessary cause for both diseases. Hence oral hygiene measures that aim to remove or reduce bacterial plaque are a key strategy for the prevention of both caries and periodontitis.

Plaque can be present in subjects who do not consume carbohydrates, but is more prolific and produces more acid in individuals who eat sucrose-rich food. Frequency of carbohydrate consumption, physical characteristics of food (e.g., softness and stickiness), and timing of food intake all contribute to plaque formation and composition *(6,7)*. Plaque on tooth surfaces mineralizes to form calculus or tartar, which is often covered by unmineralized biofilm *(8)*. Hence, dietary factors are important determinants of plaque quantity and quality and are therefore important in the pathogenesis of dental disease, in particular dental caries (see below).

14.2.2 Dental Caries

Dental caries are characterized by the demineralization of dental hard tissues (enamel and dentin) by acids produced by plaque bacteria. Many factors that influence the quantity and timing of bacterial acid production ultimately determine the risk of caries. Here, diet plays an important role in caries occurrence and progression. Foods containing fermentable carbohydrates result in acid production by cariogenic plaque bacteria. The production of organic acids by sugar metabolizing bacteria then leads to significant decreases in plaque pH. If plaque pH falls below 5.5 for an appreciable period of time, demineralization of dental enamel occurs. As the plaque pH varies according to the availability of fermentable carbohydrates to plaque bacteria, demineralization and remineralization processes occur in a dynamic process. Factors other than diet that affect this dynamic process include, for example, the fluoride concentration. If demineralization is not compensated by remineralization, a breakdown of the enamel surface and formation of a cavity that can extend through the dentin (the part of the tooth located under the enamel) to the pulp tissue will result.

Because of the complexity of the demineralization/remineralization processes, the effects of fermentable sugars and other carbohydrates are not just determined by their amount. Most importantly, the frequency of sugar intake (e.g., eating sweets with main meals or as snacks at multiple occasions between meals) has been clearly shown to be a major determinant of caries risk *(6)*. Furthermore, the effect of dietary intake of sugars or other carbohydrates is modified by other factors, primarily fluoride intake and oral hygiene. Fluorides (e.g., in toothpastes) have become highly abundant over the past decades and dietary factors may be less important in subjects with good oral hygiene and regular fluoride exposure *(9)*.

Artificial sweeteners such as aspartame and saccharin and sugar alcohols such as sorbitol, mannitol, and xylitol were shown to be noncariogenic in clinical trials *(10)*. Saliva also contains components that can directly attack cariogenic bacteria and contains calcium and phosphates that help remineralize tooth enamel.

14.2.3 Chronic Periodontitis

Bacterial plaque is considered a necessary cause of periodontitis, as bacterial components and products elicit an inflammatory response in the periodontal host tissues. In susceptible individuals, this inflammatory response leads to the resorption of periodontal ligament and alveolar bone. Susceptibility to periodontitis is determined by environmental and genetic host factors. For example, smoking and diabetes are established as major risk factors for periodontitis and tooth loss *(11–13)*. Genetic risk factors that increase periodontitis susceptibility have been proposed, although to date data on specific genetic factors remain equivocal *(14)*.

Diet and nutrients could affect periodontitis risk by influencing plaque quality and quantity, but may also and perhaps more importantly affect the inflammatory response and thus affect periodontitis susceptibility. It is important to note that research in the nutritional determinants of periodontitis risk is scant and most of the currently available data are from cross-sectional surveys, particularly from the third National Health and Nutritional Examination Survey (NHANES III).

In NHANES III, inverse associations of intake levels of calcium *(15)* and also dairy products *(16)* – which is highly correlated with calcium intakes in the United States – with periodontitis prevalence have been reported. Krall et al. have reported a beneficial effect of calcium and vitamin D supplementation on tooth retention *(17)* in a small cross-sectional study. Vitamin D status is associated with bone-mineral density *(18)* and vitamin D supplementation is effective in preventing bone loss and fractures *(19)*. Because osteoporosis has been proposed as a risk factor for periodontitis, we studied the association between vitamin D status and periodontitis and found lower periodontitis prevalence among subjects older than 50 years of age with higher serum concentrations of vitamin D (25-hydroxyvitamin D) *(20)*. In addition to its established effect on calcium metabolism and bone, Vitamin D also has immuno-modulatory functions by which it may reduce periodontitis suscept-ibility. This hypothesis is consistent with the finding of a strong inverse association between vitamin D status and gingivitis (a precursor of periodontitis not affected by osteoporosis) prevalence *(21)*. If vitamin D status is truly a risk factor for period-ontitis, these findings could have significant public health implications as hypovi-taminosis D is highly prevalent in the United States *(22)* and elsewhere *(23,24)*. Interestingly, vitamin D receptor polymorphisms are among the genetic factors implicated as putative risk factors for periodontitis *(14)*. However, intervention studies will be necessary to evaluate if vitamin D supplementation is effective for periodontitis prevention.

Deficiencies of ascorbate have been associated with severity of gingivitis *(25,26)*. Furthermore, NHANES III data demonstrate an inverse association between vita-min C intakes and periodontitis prevalence in the United States *(27)*. More recently, serum levels of vitamin C were shown to be strongly associated with periodontitis prevalence, which was also confirmed among never smokers *(28)*. Given our increasing understanding of the role of immune function and inflamma-tory response in periodontitis, it is likely that immune-modulating nutrients, such as some antioxidants *(28)* and omega fatty acids *(29)*. could alter the inflammatory process in periodontitis. Interestingly, Merchant et al. reported an inverse associa-tion between whole-grain intake and risk of self-reported periodontitis incidence in a large cohort of male US health professionals *(30)*. Men in the highest quintile of whole-grain intake were 23% less likely to develop periodontitis.

Recently an interesting association between obesity and periodontitis has been noted in several cross-sectional studies *(31,32)*. Given recent evidence regarding adipose tissue serving as a reservoir for inflammatory cytokines, it is possible that increasing body fat increases the likelihood of an active host inflammatory response in periodontitis *(33)*.

14.2.4 Oral Cancer

Oral cancer is generally preceded by pre-cancerous lesions, which include oral epithelial dysplasia, erythroplakia, leukoplakia, lichen planus, and submucous fibrosis (rare in Western countries). The major risk factors for oral cancer are tobacco and alcohol use. In Asian countries chewing tobacco, beetle nut, and beetle quid are major risk factors. Chewing tobacco use is also increasing in the United States. The relation between nutrition and oral cancer, and the impact of oral cancer on the patient's ability to eat and swallow are discussed below.

14.2.4.1 FRUITS AND VEGETABLES

A consistent finding across numerous studies is that a diet high in fruits and vegetables is protective against oral pre-cancer *(34–36)* and cancer *(37–39)*. A generous consumption of fruits was associated with a 20–80% reduced risk of OC even when smoking and alcohol intake and other factors including total caloric intake are taken into account *(40–42)*. Vegetables are also protective *(38,43)* although not all studies show a protective effect *(44)*. The inconsistencies may be explained by variation in specific vegetables consumed and there seems a suggestion that raw vegetables may be more important than cooked. A study evaluating specific fruits and vegetables suggested that green vegetables, salad, and apples were more protective. Tomato shows a strong and consistent inverse association for oral cancer in 12 of 15 studies *(45)*. and in one study on leukoplakia *(34)*. Raw tomatoes were more associated with reduce risk of oral cancer than cooked tomatoes *(45)*. An inverse association was also found for raw vegetables among Japanese adults *(46)*. Glutathione – an antioxidant found in fruits and vegetables – was protective only if it was derived from fruit and raw vegetables *(47)*.

14.2.4.2 ANTI-OXIDANTS AND OTHER MICRONUTRIENTS

Several nutrients found in vegetables and fruits show an inverse association with oral cancer. These include vitamin A, vitamin B_{12}, vitamin C, tocopherol (vitamin E), retinoids, carotenoids, lycopene, beta-carotene, folate, glutathione, thiamin, vitamin B_6, folic acid, niacin, and lutein have been inversely associated with oral cancer *(40,48–53)* and pre-cancer *(34,54–56)* in one or more studies. Studies that have evaluated subgroups have generally found higher beneficial effects of fruits and vegetables and their constituent micronutrients among smokers and drinkers than among abstainers *(57)*.

Retinoids and beta-carotene in controlled therapeutic doses show protective effects, with fewer new primary tumors in persons with previous oral cancers and reversals or reduction in size of premalignant lesions *(58–60)*. High doses of 13-*cis*-retinoic acid (50–100 mg/sq-m body surface area/day for a year) have been effective in the treatment of oral leukoplakia *(52)*. Sixty-seven percent of patients with this condition showed major decreases in lesion size vs. 10% among placebo group and in prevention of second primary tumors (2% had secondary tumors after a median follow-up of 32 months vs. 12% in placebo group) *(61)*. Trials using beta-carotene supplements (60 mg/day for 6 months) have shown reduced risk of oral cancers and remission of pre-cancers with an improvement of at least one grade dysplasia in 39% and no change in 61% *(62)*.

14.2.4.3 OTHER FOOD AND NUTRIENTS

A protective effect of fiber was observed for both oral submucous fibrosis and leukoplakia *(34)* and for oral cancer *(37,53)*. There is a suggestion that meat *(53,63)* desserts, maize, and saturated fats and/or butter may be risk factors *(63–65)* and that olive oil may be protective *(65)*. Nitrate, nitrite, and nitrate reductase activity in saliva *(66)* and high intake of nitrite containing meats *(37)* have been linked with increased risk. Iron is suggested to be protective against OC *(55)* and leukoplakia *(40)*.

14.3 IMPACT OF ORAL HEALTH ON NUTRITION

This section focuses on the impact of tooth loss and dentition status, oral cancer, and xerostomia on nutrition. Other aspects of oral health such as oral pain, periodontal disease, and altered taste could also have some impact on nutritional status *(67)* but will not be reviewed here.

14.3.1 Impact of Tooth Loss on Nutritional Status

A number of studies have demonstrated an association between tooth loss and dietary intake. Many studies show that edentulous individuals (people with no teeth) are more likely to eat an unhealthy diet (for example ingesting too few nutrient-dense foods and too much calorie-rich, high fat foods) compared to people with natural teeth. In studies of healthy older adults, edentulous individuals have been noted to consume fewer fruits and vegetables, lower amounts of fiber, and higher amounts of fat *(68,69)*. Joshipura et al. *(70)* observed that edentulous male health professionals consumed fewer vegetables, less fiber and carotene, more cholesterol, saturated fat, and calories than participants with 25 or more teeth after adjusting for age, smoking, exercise, and profession. Edentulous individuals are more likely to have lower intakes of micronutrients, such as calcium, iron, pantothenic acid, vitamins C, and E, than their dentate counterparts *(71)*. In summary, most of the studies relating tooth loss and nutrition suggest that people with fewer teeth are more likely to have compromised nutritional intake.

Possible changes in fruit, vegetable, and micronutrient intake after tooth loss may explain part of the recent findings suggesting associations between tooth loss and cardiovascular disease *(72)*. Therefore patients with tooth loss warrant aggressive counseling regarding methods to maintain dietary quality, such as blending or shredding fresh fruits and vegetables to preserve adequate intake *(73)*.

Although eating with dentures may be preferable to eating with no teeth, most studies suggest that the diet of denture wearers differs from the diet of people who retained their natural teeth. In Krall et al.'s *(74)* study of veterans, individuals with full dentures consumed fewer calories, thiamin, iron, folate, vitamin A, and carotene than individuals with a number of natural teeth remaining. Papas et al. *(75)*

evaluated the impact of full dentures and noted both lower intake of protein and 19 other nutrients. In a separate population, Papas et al. *(76)* reported that subjects who wore dentures consumed more refined carbohydrates, sugar, and dietary cholesterol than their dentate counterparts. The above studies could be interpreted such that the presence of dentures contributes to poorer intake across multiple nutrients compared to dentate subjects. Poor denture fit may contribute to some of these differences. However, all of these studies report cross-sectional associations, and as seen in Section 2, tooth loss resulting from caries and/or periodontitis may well be the outcome of poor nutrition rather than its determinant. Alternatively, both pathways may have a role and explain the cross-sectional association between tooth loss and diet.

Surprisingly, longitudinal studies investigating whether tooth loss leads to dietary changes (secondary to an assumed functional impairment) are scarce. Hung et al. investigated the association between self-reported incident tooth loss and concomitant dietary changes over a period of 8 years among 31,813 male US health professionals. Compared to men who did not lose teeth, men who lost five or more teeth had significant detrimental changes in dietary intakes of dietary fiber, whole fruit, dietary cholesterol and polyunsaturated fat *(77)*. Similarly, results from the Nurses Health Study showed detrimental dietary changes over a 2 year period subsequent to incident tooth loss, with a tendency for women who lost teeth to avoid hard foods such as raw carrot, fresh apple, or pear *(78)*. However, these differences were relatively small in absolute terms and their significance with respect to chronic disease risk is uncertain *(78)*.

Indirect evidence that functional impairment associated with tooth loss may be a determinant of dietary changes also comes from studies comparing different modalities of replacements for missing teeth. In a study of denture wearers in Quebec, those that wore dentures providing poor masticatory performance consumed significantly less fruits and vegetables than those with dentures that provided good masticatory performance *(79)*. Likewise in Swedish older adults, poorly fitting upper dentures were associated with decreased intake of vitamin C *(80)*. Among older Australians, women who reported poorly fitting dentures consumed greater amounts of sweets and dessert items *(81)*.

Studies have also examined dietary differences among edentulous subjects with and without dentures. Perhaps not surprisingly, edentulous subjects without dentures consumed more mashed food *(82)*. and in a study of Swedish women, edentulous women without dentures consumed more fat *(69)*. Whether the placement of dentures in an edentulous patient makes a substantial improvement in the patient's intake remains unclear. In the only available randomized controlled trial among patients with partial tooth loss, Garett et al. *(83)* found no differences in dietary intakes between patients who received either no dentures, fixed partial dentures (FPD) or removable partial dentures (RPD). Sebring et al. *(84)* studied the effect of conventional maxillary and implant-supported or conventional mandibular dentures on patients who were edentulous with no prostheses. In both groups, calorie intake decreased; percentage of calories from fat also decreased significantly over the subsequent 3 years. Lindquist *(85)* evaluated the impact of prosthetic rehabilitation, using optimized complete dentures and then tissue-integrated mandibular fixed prostheses (TIP) on 64 dissatisfied complete denture wearers. There was no change in diet after optimizing complete dentures, but there was a persistent increase in fresh fruit consumption after placement of TIP. Olivier et al. *(86)* evaluated dietary counseling in addition to prosthetic relining for the edentulous on chewing efficiency, dietary fiber intake from various sources, gastrointestinal esophageal, and colonic symptoms. Chewing ability and fiber intake from fruits and vegetables were significantly improved. However, because there was no group that did not receive dietary counseling, it was not possible to separate the effect of relining from the counseling. In summary, dietary quality may improve with the placement of dental prostheses, but the changes are not substantial. Dietary counseling at the same time prostheses are fitted may assist patients in behavioral change and in optimizing the impact of their new chewing capabilities.

The relationship between dental status and weight, weight/height, and body mass index (BMI) varies with the population studied. In Mojon et al.'s *(87)* study of nursing home residents, compromised oral functional status was associated with lower BMI (less than 21 kg/m^2) after controlling for functional dependence and age. In Hirano et al.'s *(88)* study of community-dwelling older adults, the authors reported a similar association between masticatory ability and lower body weight, after controlling for age and sex. However, in Johansson et al.'s *(68)* cross-sectional study of healthier older adults, edentulous patients actually had higher BMIs, compared to dentate subjects. In Elwood and Bate's *(89)* study of older Welsh adults, there was a trend toward higher weight and height/weight among subjects with no teeth or dentures. The differences in findings in these studies may be due to the different characteristics of the populations evaluated, with sicker older adults more likely to lose weight in response to altered dentition, and healthier older adults more likely to maintain adequate intake but alter intake to softer foods that are more calorie dense. However, as noted above, cross-sectional studies are insufficient to make causal inferences and obesity may itself be a risk factor for tooth loss secondary to caries or periodontitis *(31)*. One longitudinal study in community-dwelling older adults found that over a 1-year period of follow-up, approximately one-third of the sample had lost 4% or more of their previous total body weight; 6% of men and 11% of women lost 10% or more of their previous body weight. Edentulism remained an independent risk factor for significant weight loss (odds ratio 1.6 for 4% weight loss and 2.0 for 10% weight loss) after controlling for gender, income, advanced age, and baseline weight *(90)*.

The largest study to date to evaluate blood nutrient status in relation to dentate status is the British National Diet and Nutrition Survey *(71)*. In their cross-sectional study of 490 free-living and institutionalized older adults, the authors reported that edentulism subjects had significantly lower mean plasma levels of retinol, ascorbate, and tocopherol than dentate subjects, after controlling for age, sex, social class, and region of residence. Among dentate subjects, mean plasma vitamin C levels were positively associated with increased numbers of occlusal pairs of teeth. Another study of adults in Sweden *(68)* reported lower serum high-density lipoprotein (HDL) levels among edentulous individuals compared to those who were dentate. These results are consistent with the studies relating dietary intake to dentition status.

Once again, the vast majority of the studies mentioned are cross-sectional and have to be interpreted with caution; it is not clear if nutrition impacts tooth loss through its impact on caries and periodontal disease or if tooth loss impacts nutritional intake or both. Another important issue that needs to be considered when evaluating the evidence relating to tooth loss and diet is the possibility for residual confounding, in particular by socio-economic status and health-conscious behaviors. Although caries and/or periodontitis are the main causes of tooth loss, a decision to extract a tooth is influenced by many other factors working at the patient, provider (dentist), and community level (access to care), and confounding by socio-economic factors is a particular concern in this context (91). In summary, individuals with compromised dentition tend to have poorer dietary quality. Whether or not this association is causal, i.e., whether or not tooth loss leads to important unfavorable changes in a person's diet, is uncertain (67). Additional longitudinal studies are necessary to answer this important question.

14.3.2 Impact of Oral Cancer on Nutrition

Oral cancer (OC) has a major impact on eating and swallowing. The location or progression of the tumor itself and the side effects of treatment hamper feeding and swallowing. Side effects of radiation, primarily a result of damage to the salivary glands and reduction in saliva production, include xerostomia, dental caries, oral mucositis, and bacterial and fungal infections. Side effects of chemotherapy include mucositis, fungal infections, xerostomia, throat and mouth pain, taste changes, food aversions, nausea, and diarrhea. Other complications include aspiration, osteoradionecrosis, and trismus (92). Side effects of surgery vary according to location and extent of surgery. The oral phase of swallowing is affected by surgical resection (93). Means to counter common problems faced by OC patients are listed in Table 14.1. These interventions can improve nutritional intake and overall quality of life.

14.3.3 Impact of Xerostomia on Nutrition

With xerostomia, individuals may have inadequate lubrication and moisture in the mouth to chew food and create an adequate food bolus for swallowing. In addition, xerostomia may contribute to altered taste perception and to food sticking to the tongue or hard palate. Three studies of xerostomia have found that diet/nutrition and the quality of saliva were affected by exposure to Sjogren's syndrome (an immunologic disorder in which the body's immune system mistakenly attacks its own moisture producing glands) and xerogenic medications. Loesche et al. (94) reported that individuals with complaints of xerostomia were more likely to avoid crunchy vegetables (e.g., carrots), dry foods (e.g., bread), and sticky foods (e.g., peanut butter). Rhodus et al. (95) studied 28 patients with Sjogren's syndrome and compared them to a group of controls matched on diabetes, depression, cardiovascular disease, arthritis, age, gender, and dental health. Caloric and micronutrient intakes were significantly lower among xerostomic patients. Rhodus and Brown

Table 14.1
Problem management in oral cancer patients

Problem	Management
Xerostomia – which could lead to other problems such as caries	Sugar free mints and gums, artificial saliva, increased intake of water, or induce salivation medically by pilocarpine hydrochloride
Increased caries susceptibility	Instruction on oral hygiene and avoidance of food high in sugar, dental referral daily fluoride gel
Trismus makes chewing difficult	Recommend appropriate jaw exercises
Dysphagia	Assess swallowing ability and risk of aspiration, monitor feeding capabilities, modify food consistency as indicated, use alternative route of nutritional support if necessary (see chapter 14)
Risk of aspiration	Use airway protection techniques and use of feeding devices as indicated
Malnutrition	Obtain a dietary consultation. Consider a multivitamin/mineral supplement and/or enteral or parenteral routes for patients who cannot meet their nutritional needs by mouth

(96) also evaluated 84 older residents of an extended care facility. Energy, protein, fiber, vitamin A, C, and B_6, thiamin, riboflavin, calcium, and iron were significantly lower in the patients with xerostomia than those without. These studies suggest that xerostomia impairs optimal nutrient intake; however, these studies are hampered by their small size and cross-sectional design. Rhodus *(95)* noted that the body mass index for the xerostomic individuals with Sjogren's syndrome was significantly lower than for the control group. In their study of older extended care facility residents, they also noted a significantly lower body mass index among the xerostomic subjects *(96)*. Dormeval et al. *(97)* evaluated hospitalized older adults and noted that low unstimulated salivary flow rates were associated with low body mass index, triceps skin fold thickness, and arm circumference.

14.4 CONCLUSION

Oral conditions that affect and are affected by nutrition, including dental caries, periodontal disease, xerostomia, and oral cancer, are more common in older adults. The causal role of dietary behaviors in the pathogenesis of dental caries throughout the life has been unequivocally demonstrated. Avoidance of in-between meal snacks, use of sugar-free candy or gum, and consumption of carbohydrates with meals and water, can reduce caries incidence. Effects of nutritional factors on periodontitis risk and oral cancer risk on the one hand, and dietary effects of tooth loss on the other hand are highly plausible; however, evidence from well-designed longitudinal or intervention studies is scarce.

Nutritional modulation of immune function, as for example through the use of antioxidants, may reduce progression of periodontal disease, but intervention studies are lacking. Many epidemiologic studies have demonstrated the protective effect of fruits and vegetables and antioxidants on oral cancer risk. Studies suggest tooth loss impacts dietary quality and nutrient intake in a manner that may increase risk for several systemic diseases. Further, impaired dentition may contribute to weight change, depending on age and other population characteristics. Attention to dietary quality is particularly important among individuals with chewing disability from tooth loss or edentulism. Patients with oral cancer experience numerous complications that increase their risk for poor dietary intake. Close attention should be given to prevention of caries in patients with xerostomia, modification of food consistency in patients with dysphagia, and alternative feeding routes if nutritional needs cannot be met orally.

14.5 RECOMMENDATIONS

1. Clinicians should advise their dentate patients to restrict between-meal snacks, eat carbohydrates with meals, and limit foods that are cariogenic.
2. Consumption of fruits and vegetables appears to reduce the risk for the development of oral cancer.
3. Patients with tooth loss are at increased risk for poor/inappropriate dietary intake. Clinicians should counsel patients regarding ways to maintain good nutrition and minimize softer calorie-dense foods with low nutritional value. Pureed or shredded fruits and vegetables may serve as a means of insuring adequate intakes of these food groups. A multivitamin should be considered in this group as well.
4. Patients with oral cancer and radiation-induced xerostomia should be counseled to use sugar-free mints and gums and routinely apply fluoride to teeth to prevent dental caries.

REFERENCES

1. Swoboda J, Kiyak HA, Persson RE, et al. Predictors of oral health quality of life in older adults. Spec Care Dentist 2006;26(4):137–44.
2. Tsakos G, Steele JG, Marcenes W, Walls AW, Sheiham A. Clinical correlates of oral health-related quality of life: evidence from a national sample of British older people. Eur J Oral Sci 2006;114(5):391–5.
3. Cunha-Cruz J, Hujoel PP, Kressin NR. Oral health-related quality of life of periodontal patients. J Periodontal Res 2007;42(2):169–76.
4. Jemal A, Siegel R, Ward E, Murray T, Xu J, Thun MJ. Cancer statistics, 2007. CA Cancer J Clin 2007;57(1):43–66.
5. Ries LAG, Melbert D, Krapcho M, et al. SEER Cancer Statistics Review 1975–2004. Bethesda, MD: National Cancer Institute; 2007.
6. Gustafsson BE, Quensel CE, Lanke LS, et al. The Vipeholm dental caries study; the effect of different levels of carbohydrate intake on caries activity in 436 individuals observed for five years. Acta Odontol Scand 1954;11(3–4):232–64.
7. Zero DT. Sugars – the arch criminal? Caries Res 2004;38(3):277–85.
8. Mandel ID. Calculus update: prevalence, pathogenicity and prevention. J Am Dent Assoc 1995;126(5):573–80.
9. van Loveren C, Duggal MS. The role of diet in caries prevention. Int Dent J 2001;51(6 Suppl 1):399–406.

10. Alfin-Slater RB, Pi-Sunyer FX. Sugar and sugar substitutes. Comparisons and indications. Postgrad Med 1987;82(2):46–50, 3–6.

11. Tomar SL, Asma S. Smoking-attributable periodontitis in the United States: findings from NHANES III. National Health and Nutrition Examination Survey. J Periodontol 2000; 71(5):743–51.

12. Dietrich T, Maserejian NN, Joshipura KJ, Krall EA, Garcia RI. Tobacco use and incidence of tooth loss among US male health professionals. J Dent Res 2007;86(4):373–7.

13. Mealey BL, Oates TW. Diabetes mellitus and periodontal diseases. J Periodontol 2006; 77(8):1289–303.

14. Loos BG, John RP, Laine ML. Identification of genetic risk factors for periodontitis and possible mechanisms of action. J Clin Periodontol 2005;32 Suppl 6:159–79.

15. Nishida M, Grossi SG, Dunford RG, Ho AW, Trevisan M, Genco RJ. Calcium and the risk for periodontal disease. J Periodontol 2000;71(7):1057–66.

16. Al-Zahrani MS. Increased intake of dairy products is related to lower periodontitis prevalence. J Periodontol 2006;77(2):289–94.

17. Krall EA, Wehler C, Garcia RI, Harris SS, Dawson-Hughes B. Calcium and vitamin D supplements reduce tooth loss in the elderly. Am J Med 2001;111(6):452–6.

18. Bischoff-Ferrari HA, Dietrich T, Orav EJ, Dawson-Hughes B. Positive association between 25-hydroxy vitamin D levels and bone mineral density: a population-based study of younger and older adults. Am J Med 2004;116(9):634–9.

19. Bischoff-Ferrari HA, Willett WC, Wong JB, Giovannucci E, Dietrich T, Dawson-Hughes B. Fracture prevention with vitamin D supplementation: a meta-analysis of randomized controlled trials. Jama 2005;293(18):2257–64.

20. Dietrich T, Joshipura KJ, Dawson-Hughes B, Bischoff-Ferrari HA. Association between serum concentrations of 25-hydroxyvitamin D3 and periodontal disease in the US population. Am J Clin Nutr 2004;80(1):108–13.

21. Dietrich T, Nunn M, Dawson-Hughes B, Bischoff-Ferrari HA. Association between serum concentrations of 25-hydroxyvitamin D and gingival inflammation. Am J Clin Nutr 2005;82(3):575–80.

22. Looker AC, Dawson-Hughes B, Calvo MS, Gunter EW, Sahyoun NR. Serum 25-Hydroxyvitamin D status of adolescents and adults in two seasonal subpopulations from NHANES III. Bone 2002;30:771–7.

23. Hypponen E, Power C. Hypovitaminosis D in British adults at age 45 y: nationwide cohort study of dietary and lifestyle predictors. Am J Clin Nutr 2007;85(3):860–8.

24. Hintzpeter B, Mensink GBM, Thierfelder W, Muller MJ, Scheidt-Nave C. Vitamin D status and health correlates among German adults. Eur J Clin Nutr 2007.

25. Leggot P, Robertson, P., Rothman, D., Murray, P., Jacob R. The effect of controlled ascorbic acid depletion and supplementation on periodontal health. J Periodontol 1985;57:480–5.

26. Leggot P, Robertson, P., Jacob R., Zambon, J., Walsh, M., Armitage, G. Effects of ascorbic acid depletion and supplementation on periodontal health and subgingival microflora in humans. J Dental Res 1991;70(December):1531–6.

27. Nishida M, Grossi SG, Dunford RG, Ho AW, Trevisan M, Genco RJ. Dietary vitamin C and the risk for periodontal disease. J Periodontol 2000;71(8):1215–23.

28. Chapple IL, Milward MR, Dietrich T. The prevalence of inflammatory periodontitis is negatively associated with serum antioxidant concentrations. J Nutr 2007;137(3):657–64.

29. Kesavalu L, Vasudevan B, Raghu B, et al. Omega-3 Fatty Acid effect on alveolar bone loss in rats. J Dent Res 2006;85(7):648–52.

30. Merchant AT, Pitiphat W, Franz M, Joshipura KJ. Whole-grain and fiber intakes and periodontitis risk in men. Am J Clin Nutr 2006;83(6):1395–400.

31. Ritchie CS. Obesity and periodontal disease. Periodontol 2000 2007;44:154–63.

32. Pischon N, Heng N, Bernimoulin JP, Kleber BM, Willich SN, Pischon T. Obesity, inflammation, and periodontal disease. J Dent Res 2007;86(5):400–9.

33. Shuldiner AR, Yang R, Gong DW. Resistin, obesity and insulin resistance – the emerging role of the adipocyte as an endocrine organ. N Engl J Med 2001;345(18):1345–6.

34. Gupta PC, Hebert JR, Bhonsle RB, Sinor PN, Mehta H, Mehta FS. Dietary factors in oral leukoplakia and submucous fibrosis in a population-based case control study in Gujarat, India. Oral Dis 1998;4(3):200–6.

35. Morse DE, Pendrys DG, Katz RV, et al. Food group intake and the risk of oral epithelial dysplasia in a United States population. Cancer Causes Control 2000;11(8):713–20.

36. Maserejian NN, Giovannucci E, Rosner B, Zavras A, Joshipura K. Prospective study of fruits and vegetables and risk of oral premalignant lesions in men. Am J Epidemiol 2006;164(6):556–66.

37. Gridley G, McLaughlin JK, Block G, et al. Diet and oral and pharyngeal cancer among blacks. Nutr Cancer 1990;14(3–4):219–25.

38. Levi F, Pasche C, La Vecchia C, Lucchini F, Franceschi S, Monnier P. Food groups and risk of oral and pharyngeal cancer. Int J Cancer 1998;77(5):705–9.

39. Boeing H, Dietrich T, Hoffmann K, et al. Intake of fruits and vegetables and risk of cancer of the upper aero-digestive tract: the prospective EPIC-study. Cancer Causes Control 2006; 17(7):957–69.

40. Gupta PC, Hebert JR, Bhonsle RB, Murti PR, Mehta H, Mehta FS. Influence of dietary factors on oral precancerous lesions in a population-based case-control study in Kerala, India. Cancer 1999;85(9):1885–93.

41. Steinmetz KA, Potter JD. Vegetables, fruit, and cancer prevention: a review. J Am Diet Assoc 1996;96(10):1027–39.

42. Winn DM. Diet and nutrition in the etiology of oral cancer. Am J Clin Nutr 1995;61(2):437S–45S.

43. Day GL, Shore RE, Blot WJ, et al. Dietary factors and second primary cancers: a follow-up of oral and pharyngeal cancer patients. Nutr Cancer 1994;21(3):223–32.

44. McLaughlin JK, Gridley G, Block G, et al. Dietary factors in oral and pharyngeal cancer. J Natl Cancer Inst 1988;80(15):1237–43.

45. De Stefani E, Oreggia F, Boffetta P, Deneo-Pellegrini H, Ronco A, Mendilaharsu M. Tomatoes, tomato-rich foods, lycopene and cancer of the upper aerodigestive tract: a case-control in Uruguay. Oral Oncol 2000;36(1):47–53.

46. Takezaki T, Hirose K, Inoue M, et al. Tobacco, alcohol and dietary factors associated with the risk of oral cancer among Japanese. Jpn J Cancer Res 1996;87(6):555–62.

47. Flagg EW, Coates RJ, Jones DP, et al. Dietary glutathione intake and the risk of oral and pharyngeal cancer. Am J Epidemiol 1994;139(5):453–65.

48. Barone J, Taioli E, Hebert JR, Wynder EL. Vitamin supplement use and risk for oral and esophageal cancer. Nutr Cancer 1992;18(1):31–41.

49. Benner SE, Winn RJ, Lippman SM, et al. Regression of oral leukoplakia with alpha-tocopherol: a community clinical oncology program chemoprevention study. J Natl Cancer Inst 1993;85(1):44–7.

50. Blot WJ, Li JY, Taylor PR, et al. Nutrition intervention trials in Linxian, China: supplementation with specific vitamin/mineral combinations, cancer incidence, and disease-specific mortality in the general population. J Natl Cancer Inst 1993;85(18):1483–92.

51. Garewal HS. Beta-carotene and vitamin E in oral cancer prevention. J Cell Biochem Suppl 1993;17F:262–9.

52. Hong WK, Endicott J, Itri LM, et al. 13-cis-retinoic acid in the treatment of oral leukoplakia. N Engl J Med 1986;315(24):1501–5.

53. Zheng W, Blot WJ, Diamond EL, et al. Serum micronutrients and the subsequent risk of oral and pharyngeal cancer. Cancer Res 1993;53(4):795–8.

54. Nagao T, Ikeda N, Warnakulasuriya S, et al. Serum antioxidant micronutrients and the risk of oral leukoplakia among Japanese. Oral Oncol 2000;36(5):466–70.

55. Negri E, Franceschi S, Bosetti C, et al. Selected micronutrients and oral and pharyngeal cancer. Int J Cancer 2000;86(1):122–7.

56. Ramaswamy G, Rao VR, Kumaraswamy SV, Anantha N. Serum vitamins' status in oral leucoplakias – a preliminary study. Eur J Cancer B Oral Oncol 1996;32B(2):120–2.

57. Tavani A, Gallus S, La Vecchia C, et al. Diet and risk of oral and pharyngeal cancer. An Italian case-control study. Eur J Cancer Prev 2001;10(2):191–5.

58. Khuri FR, Lippman SM, Spitz MR, Lotan R, Hong WK. Molecular epidemiology and retinoid chemoprevention of head and neck cancer. J Natl Cancer Inst 1997;89(3):199–211.

59. Papadimitrakopoulou VA, Hong WK. Retinoids in head and neck chemoprevention. Proc Soc Exp Biol Med 1997;216(2):283–90.

60. Zain RB. Cultural and dietary risk factors of oral cancer and precancer – a brief overview. Oral Oncol 2001;37(3):205–10.

61. Hong WK, Lippman SM, Itri LM, et al. Prevention of second primary tumors with isotretinoin in squamous-cell carcinoma of the head and neck. N Engl J Med 1990;323(12):795–801.

62. Garewal HS, Katz RV, Meyskens F, et al. Beta-carotene produces sustained remissions in patients with oral leukoplakia: results of a multicenter prospective trial. Arch Otolaryngol Head Neck Surg 1999;125(12):1305–10.

63. Garrote LF, Herrero R, Reyes RM, et al. Risk factors for cancer of the oral cavity and oropharynx in Cuba. Br J Cancer 2001;85(1):46–54.

64. Fioretti F, Bosetti C, Tavani A, Franceschi S, La Vecchia C. Risk factors for oral and pharyngeal cancer in never smokers. Oral Oncol 1999;35(4):375–8.

65. Franceschi S, Favero A, Conti E, et al. Food groups, oils and butter, and cancer of the oral cavity and pharynx. Br J Cancer 1999;80(3–4):614–20.

66. Badawi AF, Hosny G, el-Hadary M, Mostafa MH. Salivary nitrate, nitrite and nitrate reductase activity in relation to risk of oral cancer in Egypt. Dis Markers 1998;14(2):91–7.

67. Ritchie CS, Joshipura K, Hung HC, Douglass CW. Nutrition as a mediator in the relation between oral and systemic disease: associations between specific measures of adult oral health and nutrition outcomes. Crit Rev Oral Biol Med 2002;13(3):291–300.

68. Johansson I, Tidehag P, Lundberg V, Hallmans G. Dental status, diet and cardiovascular risk factors in middle-aged people in northern Sweden. Community Dent Oral Epidemiol 1994;22(6):431–6.

69. Norlen P, Steen, B., Birkhed, D., Bjorn, A.L. On the relationship bewteen dietary habits, nutrients, and oral health in women at the age of retirement. Acta Odontol Scand 1993;51:277–84.

70. Joshipura KJ, Willett WC, Douglass CW. The impact of edentulousness on food and nutrient intake. J Am Dent Assoc 1996;127(4):459–67.

71. Sheiham A, Steele JG, Marcenes W, et al. The relationship among dental status, nutrient intake, and nutritional status in older people. J Dent Res 2001;80(2):408–13.

72. Joshipura K, Ritchie C, Douglass C. Strength of evidence linking oral conditions and systemic disease. Compend Contin Educ Dent Suppl 2000(30):12–23; quiz 65.

73. Joshipura K. How can tooth loss affect diet and health, and what nutritional advice would you give to a patient scheduled for extractions? J Can Dent Assoc 2005;71(6):421–2.

74. Krall E, Hayes C, Garcia R. How dentition status and masticatory function affect nutrient intake. J Am Dent Assoc 1998;129(9):1261–9.

75. Papas AS, Palmer CA, Rounds MC, Russell RM. The effects of denture status on nutrition. Spec Care Dentist 1998;18(1):17–25.

76. Papas AS, Joshi A, Giunta JL, Palmer CA. Relationships among education, dentate status, and diet in adults. Spec Care Dentist 1998;18(1):26–32.

77. Hung HC, Willett W, Ascherio A, Rosner BA, Rimm E, Joshipura KJ. Tooth loss and dietary intake. J Am Dent Assoc 2003;134(9):1185–92.

78. Hung HC, Colditz G, Joshipura KJ. The association between tooth loss and the self reported intake of selected CVD-related nutrients and foods among US women. Community Dent Oral Epidemiol 2005;33(3):167–73.

79. Laurin D, Brodeur JM, Bourdages J, Vallee R, Lachapelle D. Fibre intake in elderly individuals with poor masticatory performance. J Can Dent Assoc 1994;60(5):443–6, 9.

80. Nordstrom G. The impact of socio-medical factors and oral status on dietary intake in the eighth decade of life. Aging 1990;2(December):371–85.

81. Horwath C. Chewing difficulty and dietary intake in the elderly. J Nutr Elder 1989;9:17–24.

82. Lamy M, Mojon, Ph., Kalykakis, G., Legrand, R., Butz-Jorgensen, E. Oral status and nutrition in the institutionalized elderly. J Dentistry 1999;24:443–48.

83. Garrett N, Kapur, K., Hasse, A., Dent, R. Veterans Administration Cooperative Dental Implant Study – Comparisons between fixed partial dentures supported by blade-vent implants and removable partial dentures. PartV: Comparisons of pretreatment and posttreatment dietary intakes. J Prosthet Dent 1997;77(February):153–60.

84. Sebring N, Guckes, A., Li, S.H., McCarthy, G. Nutritional Adequacy of reported intake of edentulous subjects treated with new conventional or implant supported mandibular dentures. J Prosthet Dent 1995;74(October):358–63.

85. Lindquist L. Prosthetic rehabilitation of the edentulous mandible. Swedish Dental J 1987;48:1–39.

86. Olivier M, Laurin D, Brodeur JM, et al. Prosthetic relining and dietary counselling in elderly women. J Can Dent Assoc 1995;61(10):882–6.

87. Mojon P, Budtz-Jorgensen, E., Rapin, C. Relationship between oral health and nutrition in very old people. Age Ageing 1999;28:463–8.

88. Hirano H, Ishiyama N, Watanabe I, Nasu I. Masticatory ability in relation to oral status and general health on aging. J Nutr Health Aging 1999;3(1):48–52.

89. Elwood PC, Bates JF. Dentition and nutrition. Dent Pract Dent Rec 1972;22(11):427–9.

90. Ritchie CS, Joshipura K, Silliman RA, Miller B, Douglas CW. Oral health problems and significant weight loss among community-dwelling older adults. J Gerontol A Biol Sci Med Sci 2000;55(7):M366–71.

91. Joshipura KJ, Ritchie C. Can the relation between tooth loss and chronic disease be explained by socio-economic status? Eur J Epidemiol 2005;20(3):203–4.

92. Minasian A, Dwyer JT. Nutritional implications of dental and swallowing issues in head and neck cancer. Oncology (Williston Park) 1998;12(8):1155–62; discussion 62–9.

93. Kronenberger MB, Meyers AD. Dysphagia following head and neck cancer surgery. Dysphagia 1994;9(4):236–44.

94. Loesche W, Abrams, J., Terpenning, M., Bretz, W., Dominguez, L., Grossman, N., Hildebrandt, G., Langmore, S., Lopatin, D. Dental findings in geriatric populations with diverse medical backgrounds. Oral Surgery Oral Medicine Oral Pathology 1995;80(July):43–54.

95. Rhodus NL. Qualitative nutritional intake analysis of older adults with sjogren's syndrome. Gerodontology 1988;7:61–9.

96. Rhodus NL, Brown J. The association of xerostomia and inadequate intake in older adults. J Am Diet Assoc 1990;90(12):1688–92.

97. Dormeval V, Budtz-Jorgensen, E., Mojon, P., Bruyere, A., Rapin, C. Nutrition, general health status and oral health status in hospitalised elders. Gerodontology 1995;12:73–80.

15 Obesity in Older Adults – A Growing Problem

Dennis T. Villareal and Krupa Shah

Key Points

- The increasing prevalence of obese older adults is a major public health issue.
- Obesity causes frailty in older adults by exacerbating the age-related decline in physical function.
- Treatment plans for obese older adults should include lifestyle intervention such as weight loss, behavior modification and exercise therapy to improve physical function, quality of life and medical complications associated with obesity.
- The treatment must consider the potential adverse effects of weight loss on bone and muscle mass.

Key Words: Obesity; older adults; frailty; weight loss; exercise; behavior modification; sarcopenia

15.1 INTRODUCTORY OVERVIEW

Obesity is defined as an unhealthy excess of body fat, which increases the risk of morbidity and premature mortality. Obesity is a growing concern among adults. It not only has increased in prevalence, but has also been associated with significant morbidity and mortality. Some of its medical risks include hypertension, diabetes, hyperlipidemia, coronary artery disease, and osteoarthritis. More so in older adults, obesity exacerbates the age-related decline in physical function, impairs quality of life, and leads to frailty. The current therapeutic and management tools designed for weight loss in older persons include lifestyle intervention (diet, physical activity, and behavior modifications), pharmacotherapy, and surgery. Current evidence suggests that weight-loss therapy in obese older adults improves physical function, quality of life, and reduces medical complications. Some argue that such therapy aimed at weight loss can have potentially adverse effects on a person's muscle and bone mass.

From: *Nutrition and Health: Handbook of Clinical Nutrition and Aging, Second Edition*
Edited by: C. W. Bales and C. S. Ritchie, DOI 10.1007/978-1-60327-385-5_15,
© Humana Press, a part of Springer Science+Business Media, LLC 2009

This chapter will review the clinical issues related to obesity in older adults and provides health professionals with the appropriate weight-management guidelines on the basis of current evidence.

15.2 OBESITY: AN EPIDEMIC

Obesity continues to grow in prevalence in the United States. Data from the National Health and Nutrition Examination Survey (NHANES) indicate that approximately one-third of United States adults are obese *(1)*. In developed countries, the prevalence of obesity is increasing among older adults. The underlying reasons for the increased prevalence are an increase in the older person population, and an increase in the percentage of obesity in that population. Past studies have compared point-in-time statistics of the American older adult population and highlighted the increase in prevalence of obesity. For example, in a 10-year period between 1991 and 2000, obesity was found to grow from 14.7 to 22.9% in the 60- to 69-year age group, while obesity grew from 11.4 to 15.5% in the > 70-year age group. This represents an increase of 56 and 36% in the respective age groups *(2)*. The prevalence of obesity in older adults is likely to continue to increase, and this increase will continue to challenge our health care systems *(3)*. Furthermore, obesity poses an increasing problem for long-term care facilities *(4)*.

On a positive note, obesity is less likely to develop in the very old population (> 80-year-olds). In this age group, the prevalence rate of obesity declines precipitously. The relatively low prevalence of obesity after 80 years of age could be due to the survival advantage of being lean *(5)*. Nonetheless, more than 15% of the older American population is obese, and obesity is more common in older women than in men *(2)*. Moreover, the prevalence of obesity is not contained to the United States. It is an increasing problem of older populations throughout the world *(6)*.

15.3 PATHOPHYSIOLOGY OF OBESITY

Aging is associated with marked changes in body composition. After 30 years of age, fat-free mass (FFM), which is comprised predominantly of muscle progressively decreases, whereas fat mass increases. FFM reaches its peak during the third decade of life, while fat mass reaches its peak during the seventh decade *(7)*. Subsequently, at > 70 years of age, both indices (FFM and fat mass) decrease. Aside from quantitative changes of FFM and fat mass, aging is also associated with the redistribution of body fat and FFM. The intraabdominal fat increases with respect to aging, while the subcutaneous fat and total body fat decrease with aging *(8)*.

Body fat is accumulated when energy input exceeds energy output. Energy input does not change or even declines with aging. Energy output comprises the resting metabolic rate (accounts for ~70%), the thermal effect of food (~10%), and physical activity (20%). Aging is associated with a decrease in all major components of energy output. Resting metabolic rate decreases by 3% every decade after 20 years of age. About three-fourths of this decline can be accounted for by a loss in FFM *(9)*.

The thermal effect of food is 20% lower in older men than in younger men *(10)*. Physical activity decreases with increasing age, and it accounts for about one-half of the decrease in energy output that occurs with aging *(11)*.

As one ages, the growth hormone and testosterone production decreases, which results in a reduction in FFM and increased accumulation of fat mass *(12)*. Thyroid hormone-induced oxidative bursts are decreased with aging *(13)*. Resistance to leptin could result in a diminished ability to down-regulate appetite *(14)*. These changes in hormone levels with aging could play an important role in the pathogenesis of obesity.

15.4 MEASURING OVERWEIGHT AND OBESITY

It is difficult to accurately measure body fat mass in most clinical settings because such assessments require the use of sophisticated technologies that are not readily available. There are two measures for assessing overweight and total body fat content, which are widely used and accepted as simple methods to classify medical risk. They are body mass index (BMI) and waist circumference.

15.4.1 Body Mass Index

BMI is calculated as weight (kg)/height squared (m^2). The BMI is used to assess overweight and obesity and to monitor changes in body weight. It allows meaningful comparisons of weight status within and between populations. However, in older adults, age-related changes in body composition and loss of height caused by compression of vertebral bodies and kyphosis alter the relationship between BMI and percentage body fat. Therefore, at any given BMI value, changes in body composition tend to underestimate fatness, whereas the loss of height would tend to overestimate fatness.

Table 15.1
Overweight and obesity by BMI, waist circumference, and disease risk

			Disease risk* (relative to normal weight and waist circumference)	
	BMI (kg/m^2)	*Obesity class*	*Men < 40 inches Women < 35 inches*	*Men > 40 inches Women > 35 inches*
Underweight	< 18.5		–	–
Normal [†]	18.5–24.9		–	–
Overweight	25.0–29.9	I	Increased	High
Obesity	30.0–34.9	II	High	Very high
	35.0–39.9		Very high	Very high
Extreme obesity	> 40	III	Extremely high	Extremely high

* Disease risk for type 2 diabetes, hypertension, and CVD.

[†] Increased waist circumference can also be a marker for increased risk even in persons of normal weight.

Source: Clinical Guidelines on the Identification, Evaluation, and Treatment of Overweight and Obesity in Adults: Evidence Report. National Heart, Lung, and Blood Institute at www.nhlbi.nih.gov

15.4.2 Waist Circumference

The presence of excess fat in the abdomen out of proportion to total body fat is an independent predictor of comorbidities such as cardiovascular disease, diabetes, and hypertension *(15)*. Men with a waist circumference of > 40 inches and women with a waist circumference of > 35 inches are considered to have increased disease risk.

Table 15.1 incorporates both BMI and waist circumference in the classification of overweight and obesity and provides an indication of relative disease risk *(16)*.

15.5 HEALTH IMPLICATIONS OF OBESITY

15.5.1 Adverse Effects of Obesity

Obesity is associated with a number of health hazards. Some adverse effects include increased mortality, health complications, poor quality of life, and disability. These hazards are discussed in detail below.

15.5.1.1 MORTALITY

Obesity is associated with increased cardiovascular and overall mortality in both younger and older adults *(17,18)*. Although the *relative* risk of death associated with obesity is greater for younger adults than for older ones *(18,19)*, a high BMI increases *absolute* mortality and health risks linearly up to 75 years of age *(20)*. That is, from a clinical standpoint, the health complications associated with obesity increase linearly with increasing BMI until the age of 75. The relationship of obesity in >75 years of age with total mortality is unclear. Some previous epidemiological studies do not show that excess body weight is detrimental to mortality in advancing age *(21,22)*. However, underlying diseases that can themselves increase the risk of early mortality may cause the underestimation of the relation between obesity and mortality in older adults. Since those who are susceptible to the effects of obesity die at a younger age, the surviving group of obese older adults are said to be the "resistant" survivors. For a more comprehensive discussion of the sometimes paradoxical relationships between BMI and mortality in late life, see Chapter 9.

15.5.1.2 COMORBID DISEASE

Obesity and increased visceral fat are associated with increased morbidity and poor quality of life. Most studies evaluating obesity-related complications focus on middle-aged and younger adults. The prevalence of the medical complications associated with obesity, such as hypertension, diabetes, cardiovascular disease, and osteoarthritis, increases with age. Therefore, obesity and weight gain during middle age may contribute to medical complications, and subsequent increased health care expenditures that occur during old age *(23)*.

15.5.1.3 METABOLIC ABNORMALITIES

There is an age-related increase in the prevalence of all components of metabolic syndrome. The odds ratio for developing metabolic syndrome in those who are > 65 years relative to those who are 20–34 years of age was 5.8 in men and 4.9 in women *(24)*. Additionally, increased abdominal fat is independently associated with metabolic syndrome in adult's aged 70–79 years *(25)*. Fasting plasma glucose increases by 1–2 mg/dL and postprandial glucose by 10–20 mg/dL for each decade after 30 years. Accordingly, the prevalence of type 2 diabetes mellitus based on standard criteria is high in older persons *(26)*. The age-related increase in fat and more importantly visceral fat could be the main causative factor for the increased prevalence of diabetes mellitus and insulin resistance in the elderly.

Hypertension is extremely prevalent in the older population, affecting 65% of all persons aged >60 years *(27)*. Obesity and high blood pressure continue to be correlated, even in old age *(28)*. Obesity-related dyslipidemia (i.e., low HDL-cholesterol and high serum triglyceride concentrations) is seen in both younger and older adults. In the United States, 35–42% of white men and women who are ≥65 years of age with metabolic syndrome have low HDL-cholesterol (≤40 mg/dL in men and ≤50 mg/dL in women) and high triglyceride (≥150 mg/dL) concentrations *(24)*. Data from longitudinal studies suggest that obesity increases the risk of cardiovascular disease in older men. Elevated BMI in older men was associated with an increase in new cases of coronary artery disease, fatal and nonfatal myocardial infarction, and cardiovascular disease mortality during 12–15 years of observation *(29)*.

15.5.1.4 ARTHRITIS

Osteoarthritis (OA) is the most common type of arthritis and its prevalence increases progressively with age in both sexes in parallel with the increase in body weight and fat observed with aging. The age-related increase in prevalence of OA presumably reflects bodily changes as a result of a lifetime of being overweight, which results in chronic mechanical strain on weight-bearing joints. In a population-based study of older adults, with a mean age of 73, the relative risk of developing knee OA increased from 0.1 for a BMI lower than 20 kg/m^2 to 13.6 for a BMI of 36 kg/m^2 or higher *(30)*.

15.5.1.5 PULMONARY ABNORMALITIES

Obesity is associated with obstructive sleep apnea (OSA), obesity-hypoventilation syndrome, and pulmonary function abnormalities *(31)*. Increased fat on the chest wall decreases lung compliance, increases the work of breathing, and reduces ventilation. The prevalence of OSA increases with age. Both waist circumference and waist changes were the most powerful predictors of OSA in older obese and normal-weight men in a 30-year follow-up study *(32)*.

15.5.1.6 URINARY INCONTINENCE

The prevalence of urinary incontinence increases after the age of 65 and affects 15 to 30% of the population. Obesity contributes to the increase in prevalence of urinary incontinence in older adults, and the increase in urinary incontinence is directly associated with elevated BMI *(33)*.

15.5.1.7 CANCER

Obesity is a risk factor for several types of cancer, including breast, colon, gall-bladder, pancreas, and bladder amongst both men and women, more so in older than younger adults *(34)*. A study in older women has shown that breast cancer occurs more frequently in obese older women than all older women *(34)*.

15.5.1.8 FUNCTIONAL IMPAIRMENT AND QUALITY OF LIFE

Aging causes a progressive decline in physical function because of a continued decline in muscle mass, strength, and power and an increase in joint instability and arthritis *(35)*. These functional impairments affect activities of daily living, decrease quality of life, and lead to an increased utilization of services. Obesity has important functional implications in older adults because it worsens this age-related decline in physical function. Data from cross-sectional studies *(36–38)* and longitudinal studies *(39–41)* have consistently demonstrated a strong link between increasing BMI and worsening physical function in older persons. High BMI is associated with self-reported impairment in ADLs, limitations in mobility, decreased physical performance, and increased risk for functional decline *(38–42)*. Moreover, obesity is associated with increasing nursing home admissions *(43)*.

Although obesity is associated with increases in FFM, aging is associated with a decline in FFM (primarily skeletal muscle) and function, referred to as sarcopenia *(44)* and obesity does not appear to protect against sarcopenia. In one study *(45)*, the prevalence of sarcopenia in obese persons increased with age, suggesting that many obese persons maintain a constant fat mass while losing muscle mass. In another study *(46)*, obese older adults were found to have sarcopenia based on lower relative muscle mass and low muscle strength per muscle area [low muscle quality, Fig. 15.1] despite having more than adequate body weight which is opposite of the stereotypical frail older adult. Their functional performance, aerobic capacity, strength, balance, and walking speed were as severely reduced as the frail non-obese adults *(46)*. Thus, obesity in older adults acts synergistically with sarcopenia (sarcopenic obesity) to augment disability. Accordingly, the "sarcopenic-obese" individual has two problems that lead to frailty: (1) decreased muscle mass and strength which occur with aging, and (2) a need to carry greater weight due to excess body fat *(46,47)*. Figure 15.2 is a cross-sectional MRI image from the mid-thigh in a frail obese older adult. This figure demonstrates the excessive adipose tissue infiltration of skeletal muscle mass with obesity.

In one study *(46)*, 96% of community-living older adults with BMIs greater than 30 were frail, as determined by physical performance test scores *(48)*, peak oxygen consumption *(49)*, and self-reported ability to perform activities of daily living *(50)*. Data from another study *(51)* also demonstrated that obesity was associated with a marked increased risk of frailty (odds ratio = 3.5), determined by weakness, slowness, weight loss, low physical activity, and exhaustion. In another study *(52)* obesity was identified as one of the five modifiable risk factors that predict functional decline in both vigorous and basic activities among older women.

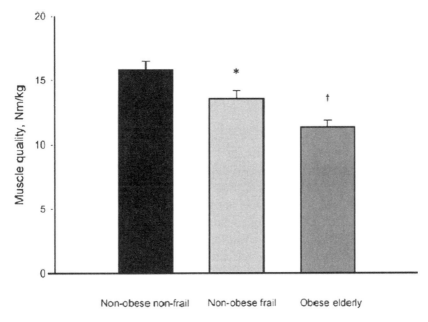

Fig. 15.1. Muscle quality (strength per muscle mass) in non-obese non-frail, non-obese frail, and obese elderly subjects (From Villareal et al. Obes Res, 2004; 12:913).

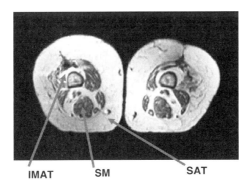

Fig. 15.2. Cross-sectional MRI image from the mid-thigh in a frail obese female participant (76 years of age). IMAT, intermuscular adipose tissue; SAT, subcutaneous adipose tissue; SM, skeletal muscle.

15.5.2 Beneficial Effects of Obesity

Increased body weight is associated with increased bone mineral density (BMD) and decreased osteoporosis and hip fracture in older men and women, whereas the converse is true for decreased body weight (53). Both body fat mass and FFM are directly correlated with BMD. Although the increase in BMD has been attributed to mechanical stress on the weight-bearing skeleton, the protective effects have also been observed in non-weight-bearing bones (54). Therefore, hormonal factors that are increased in obese persons, such as circulating estrogens, insulin, and leptin, might contribute to the osteoprotective effects of obesity, by stimulating bone formation and inhibiting bone resorption. The increase in both BMD and the extra cushioning effect of the fat surrounding crucial areas such as the hip might provide protection against hip fracture during a fall in obese older adults *(55)*.

15.6 EFFECTS OF WEIGHT LOSS IN OLDER ADULTS

15.6.1 Body Composition

Weight loss results in a decrease in both fat mass and FFM. Therefore, it is possible that weight loss in obese older persons could increase sarcopenia by worsening the age-related loss of muscle mass and in younger adults, ~75% of diet-induced weight loss is composed of fat tissue and ~25% is composed of FFM *(56)*. The relative amount of diet-induced weight loss as FFM and fat mass in older men and women is similar to that observed in younger adults *(57)*. Therefore, diet-induced weight loss does not produce a disproportionate loss of lean tissue in old persons. Despite much evidence linking high body fat to functional disability *(38,40,41)*. weight loss has not been typically instituted in obese older persons because of the fear that it will exacerbate sarcopenia. Additionally, it is a general belief among many geriatricians that some "reserve" of body fat is advantageous in the older people particularly if they are hospitalized *(58)*.

In a randomized controlled trial conducted in obese older subjects, there was no significant difference in loss of FFM after a diet-induced weight loss plus regular exercise compared with the control group who did not lose weight. These encouraging findings suggest that regular exercise can attenuate a diet-induced loss of FFM in older persons *(59)*.

15.6.2 Medical Complications

Data from young and middle-aged adults show that weight loss improves or normalizes metabolic abnormalities associated with obesity *(60)*. A recent clinical trial in obese older adults showed that moderate weight loss decreases multiple metabolic coronary heart disease risk factors simultaneously *(61)*.

15.6.3 Physical Function and Quality of Life

Moderate weight loss in conjunction with physical activity improves physical function and health-related quality of life in obese older persons. Data from studies conducted in overweight and obese older persons with or without joint disease have shown that the combination of moderate diet-induced weight loss and exercise therapy improved both subjective and objective measures of physical function and health-related quality of life and had a greater beneficial effect than did either diet or exercise interventions alone *(59,62–64)*. These findings suggest that obesity is a reversible cause of frailty and impaired quality of life in older adults.

15.6.4 Mortality

It has been observed in several population-based studies that community-dwelling older adults who lost weight, or who experienced weight variability, had an increased relative mortality risk compared with those who were weight stable *(32)*.

However, most studies did not report whether the observed weight changes were intentional or unintentional, relied on self-reported weight change, and did not distinguish between weight loss in obese and lean subjects. For obvious reasons, there are no randomized controlled trials studying the effect of weight loss on mortality.

15.6.5 Bone Mineral Density

Weight loss can have adverse effects on bone mass. Previous interventional studies conducted in young and middle-aged adults reported that weight loss causes bone loss that may be proportional to the amount of weight loss *(65–67)*. However, it is not known whether the bone loss associated with intentional weight loss increases the risk of osteoporotic fractures in obese persons.

A recent study showed that diet-induced weight loss, but not exercise-induced weight loss, is associated with reductions in BMD at weight-bearing sites, suggesting that exercise should be an important component of a weight-loss program to offset adverse effects of diet-induced weight loss on bone *(68)*. Regular exercise may potentially attenuate weight-loss-induced bone loss, and this beneficial effect may be specific for sites involved in weight-bearing exercise *(69)*. Therefore, including exercise as part of a weight-loss program is particularly important in older persons to reduce bone loss.

15.7 INTERVENTIONS AND TREATMENT

Weight loss in obese persons of any age can improve obesity-related medical complications, physical function, and quality of life. In older adults improving physical function and quality of life may be the most important goals of therapy. The current therapeutic tools and recommendations available for weight management in older persons are (1) lifestyle intervention involving diet, physical activity, and behavior modification; (2) pharmacotherapy; and (3) surgery (Table 15.2).

15.7.1 Lifestyle Intervention

Lifestyle intervention is just as effective in older as in younger subjects *(59,62–64)*. Combination of an energy-deficit diet, increased physical activity, and behavior therapy causes moderate weight loss and is associated with a lower risk of treatment-induced complications. Weight-loss therapy that minimizes muscle and bone losses is recommended for older adults who are obese and who have functional impairments or metabolic complications that can benefit from weight loss.

15.7.1.1 Diet Therapy

In order for weight loss to be successful, an energy deficit must be achieved. A low-calorie diet that reduces energy intake by 500–750 kcal/day results in a weight loss of 0.4–0.9 kg (1–2 lb)/wk and a weight loss of 8–10% by 6 mo. The diet should contain 1.0 g/kg high-quality protein/day[70], multivitamin, and mineral supplements to ensure that all daily recommended requirements are met, including 1500 mg Ca/day and 1000 IU vitamin D/day, to prevent bone loss. Very low-calorie diets (<800 kcal/day)

Table 15.2
RECOMMENDATIONS

Initial assessment

- A thorough medical history, physical examination, appropriate laboratory tests, and review of medications should be conducted to assess the patient's current health and comorbidity risks.

- Additional information such as the patient's readiness to lose weight, previous attempts at weight loss, and current lifestyle habits should be collected before initiating weight-loss therapy.

- Clinicians should help obese older adults set their personal goals and welcome participation by family members and care providers.

- Clinicians should individualize the weight-loss plan after taking into account the special needs of this population.

Diet therapy

- Advocate a modest reduction in energy intake (500–750 kcal/day) containing 1.0 g/kg high-quality protein/day, multivitamin, and mineral supplements (including 1500 mg Ca and 1000 IU vitamin D/day).

- Consider referrals to a registered dietitian for appropriate nutritional counseling and education.

Behavior therapy

- Behavior therapy should highlight both diet and exercise – the integral parts of weight loss therapy and weight maintenance. The self-monitoring of nutrient intake and better understanding of physical activity accomplishes this task.

- Consider referring to a behavioral therapist for counseling.

- Stress management, stimulus control, problem solving, contingency management, and social support should be addressed.

Exercise therapy

- Clinicians should assess the need for stress test before any physical activity.

- Advocate an exercise program that is gradual, individualized, and monitored.

- A multicomponent exercise program including stretching, aerobic activity, and strength exercises is recommended.

Additional recommendations

- Advocate a combination of energy-deficit diet, increased physical activity, and behavior therapy. Such combinations are associated with the low risk of treatment-induced complications.

- Bariatric surgery may be an option for patients who have failed multiple weight-loss attempts.

- Weight maintenance efforts should be implemented once weight-loss goals have been achieved.

should be avoided because of an increased risk of medical complications. Also, depending on the patient's cardiovascular risk status the diet therapy should be consistent with the National Cholesterol Education Program Expert Panel (Adult Treatment Panel III)'s Therapeutic Lifestyle Changes Diet *(71)*.

Referral to a registered dietitian, who has weight-management experience, is often necessary to ensure that appropriate nutritional counseling is provided. Patients should be educated on food composition, preparation, and portion control and their food preferences should be supported to improve compliance.

Successful weight loss and maintenance program should be based on sound scientific rationale. The program should be safe, nutritionally adequate, as well as practical and applicable to patient's social and ethnic background.

15.7.1.2 PHYSICAL ACTIVITY

Introducing an exercise component early in the treatment course can improve physical function and can ameliorate frailty in the older adults (59). The exercise program should be individualized according to a person's medical conditions and disability. The program should start at a low-to-moderate intensity, duration, and frequency to promote adherence and avoid musculoskeletal injuries. If possible, the program should be gradually progressed over a period of several weeks or months to a longer, more frequent, and more vigorous effort. The goals of regular exercise in obese older persons are to increase flexibility, endurance, and strength; therefore, a multicomponent exercise program that includes stretching, aerobic activity, and strength exercises is recommended. Even very old or frail persons can participate in these types of physical activities.

15.7.1.3 BEHAVIOR MODIFICATION

Clinicians should help obese older adults set personal goals, monitor progress, and use motivational strategies to improve adherence to the weight-loss program. The cognitive behavioral therapy strategies that should be considered include goal setting, self–monitoring, social support, stimulus control techniques, and problem-solving skills. Lifestyle and behavior modification can be facilitated by counseling from a dietitian, behavioral therapist, exercise specialist, or dietitian who has weight-management experience.

Changes in the diet and activity habits of older adults may be challenging. An increased burden of disease, adverse quality of life, depression, hearing and visual difficulties, and cognitive dysfunction may make it difficult to change one's lifestyle. This increase in chronic disabilities with aging reduces physical activity and exercise capacity. Common geriatric situations, such as depression, cognitive impairment, dependency on others, institutionalization, widowhood, loneliness, and isolation should be addressed, because these factors can make it more difficult to lose weight. Lifestyle-change programs should also encourage participation by family members and care providers for better compliance.

15.7.2 Pharmacotherapy

Since most clinical trials that evaluate the use of pharmacotherapy excluded older adults or included only a small number of older adults, the available data are insufficient to determine the efficacy and safety of pharmacotherapy in this population.

The use of pharmacological agents to treat obesity can add additional burden on older persons. Many obese older patients are already taking multiple medications for other diseases, which can increase the chances of nonadherence, drug–drug interactions, and errors with obesity pharmacotherapy. Moreover, potential side effects can have more serious consequences in older adults. Weight-loss drugs are seldom covered by health insurance or Medicare, which can add an additional financial burden in older patients who have a fixed income. A thorough review of all the medications should be conducted, because some may cause weight gain (e.g., antipsychotic, antidepressants, anticonvulsants, or steroids). Furthermore, weight-loss-induced clinical improvements might require changes in medications to avoid iatrogenic complications.

15.7.3 Weight-Loss Surgery

A few studies have provided information on the effectiveness and safety of bariatric surgery among older adults. Data from case series that evaluate the effect of bariatric surgery in patients who are >60 years old suggest that the relative weight loss and improvement in obesity-related medical complications are lower, whereas the perioperative morbidity and mortality are greater, in older compared to the younger patients (72). However, bariatric surgery can result in considerable weight loss and marked improvements in obesity-related physical impairment and medical complications in the older patients. The laparoscopic-adjustable gastric band is associated with fewer serious complications and a lower mortality rate; therefore the gastric band may be a better choice than the Roux-en-Y gastric bypass for older patients. However, the efficacy and safety of these procedures have not been compared in randomized trials in older adults. There should be a careful patient selection, intensive preoperative education, and expert operative and perioperative management. Surgery should be considered in selected older adults who have disabling obesity that can be ameliorated with weight loss and who meet the criteria for surgery. The preoperative evaluation should include an assessment for depression, which is common amongst older adults and could influence outcome. Postoperative management should include monitoring for nutritional and metabolic problems; particularly, vitamin B-12 deficiency, iron deficiency, and osteoporosis.

15.8 CONCLUSION

The increasing prevalence of obese older adults is a major public health issue. Decreased muscle mass with aging and the need to carry extra mass due to obesity make it particularly difficult for obese older adults to function independently and lead to the secondary complication of frailty. Treatment plans for obese older adults should include lifestyle intervention such as weight loss, behavior modification, and exercise therapy to improve physical function, quality of life, and the medical complications associated with obesity. Finally, the treatment must consider the potential adverse effects of weight loss on bone and muscle mass.

REFERENCES

1. Flegal KM, Carroll MD, Ogden CL, Johnson CL. Prevalence and trends in obesity among US adults, 1999–2000. JAMA 2002; 288(14):1723–7.
2. Mokdad AH, Bowman BA, Ford ES, Vinicor F, Marks JS, Koplan JP. The continuing epidemics of obesity and diabetes in the United States. JAMA 2001; 286(10):1195–200.

3. Arterburn DE, Crane PK, Sullivan SD. The coming epidemic of obesity in elderly Americans. J Am Geriatr Soc 2004; 52(11):1907–12.

4. Lapane KL, Resnik L. Obesity in nursing homes: an escalating problem. J Am Geriatr Soc 2005; 53(8):1386–91.

5. Wallace JI, Schwartz RS. Involuntary weight loss in elderly outpatients: recognition, etiologies, and treatment. Clin Geriatr Med 1997; 13(4):717–35.

6. Kopelman PG. Obesity as a medical problem. Nature 2000; 404(6778):635–43.

7. Gallagher D, Visser M, De Meersman RE et al. Appendicular skeletal muscle mass: effects of age, gender, and ethnicity. J Appl Physiol 1997; 83(1):229–39.

8. Beaufrere B, Morio B. Fat and protein redistribution with aging: metabolic considerations. Eur J Clin Nutr 2000; 54 Suppl 3:S48–53.

9. Tzankoff SP, Norris AH. Effect of muscle mass decrease on age-related BMR changes. J Appl Physiol 1977; 43(6):1001–6.

10. Schwartz RS, Jaeger LF, Veith RC. The thermic effect of feeding in older men: the importance of the sympathetic nervous system. Metabolism 1990; 39(7):733–7.

11. Elia M, Ritz P, Stubbs RJ. Total energy expenditure in the elderly. Eur J Clin Nutr 2000; 54 Suppl 3:S92–103.

12. Schwartz RS. Trophic factor supplementation: effect on the age-associated changes in body composition. J Gerontol A Biol Sci Med Sci 1995; 50 Spec No:151–6.

13. Mooradian AD, Habib MP, Dickerson F. Effect of simple carbohydrates, casein hydrolysate, and a lipid test meal on ethane exhalation rate. J Appl Physiol 1994; 76(3):1119–22.

14. Moller N, O'Brien P, Nair KS. Disruption of the relationship between fat content and leptin levels with aging in humans. J Clin Endocrinol Metab 1998; 83(3):931–4.

15. Kissebah AH, Krakower GR. Regional adiposity and morbidity. Physiol Rev 1994; 74(4):761–811.

16. Obesity: preventing and managing the global epidemic. Report of a WHO consultation. World Health Organ Tech Rep Ser 2000; 894:i–253.

17. Peeters A, Barendregt JJ, Willekens F, Mackenbach JP, Al Mamun A, Bonneux L. Obesity in adulthood and its consequences for life expectancy: a life-table analysis. Ann Intern Med 2003; 138(1):24–32.

18. Flegal KM, Graubard BI, Williamson DF, Gail MH. Cause-specific excess deaths associated with underweight, overweight, and obesity. JAMA 2007; 298(17):2028–37.

19. Calle EE, Thun MJ, Petrelli JM, Rodriguez C, Heath CW, Jr. Body-mass index and mortality in a prospective cohort of U.S. adults. N Engl J Med 1999; 341(15):1097–105.

20. Villareal DT, Apovian CM, Kushner RF, Klein S. Obesity in older adults: technical review and position statement of the American Society for Nutrition and NAASO, The obesity society. Am J Clin Nutr 2005; 82(5):923–34.

21. Troiano RP, Frongillo EA, Jr., Sobal J, Levitsky DA. The relationship between body weight and mortality: a quantitative analysis of combined information from existing studies. Int J Obes Relat Metab Disord 1996; 20(1):63–75.

22. Allison DB, Gallagher D, Heo M, Pi-Sunyer FX, Heymsfield SB. Body mass index and all-cause mortality among people age 70 and over: the longitudinal study of aging. Int J Obes Relat Metab Disord 1997; 21(6):424–31.

23. Daviglus ML, Liu K, Yan LL et al. Relation of body mass index in young adulthood and middle age to Medicare expenditures in older age. JAMA 2004; 292(22):2743–9.

24. Park YW, Zhu S, Palaniappan L, Heshka S, Carnethon MR, Heymsfield SB. The metabolic syndrome: prevalence and associated risk factor findings in the US population from the third national health and nutrition examination Survey, 1988–1994. Arch Intern Med 2003; 163(4):427–36.

25. Goodpaster BH, Krishnaswami S, Harris TB et al. Obesity, regional body fat distribution, and the metabolic syndrome in older men and women. Arch Intern Med 2005; 165(7):777–83.

26. Kahn SE, Schwartz RS, Porte D, Jr., Abrass IB. The glucose intolerance of aging. Implications for intervention. Hosp Pract (Off Ed) 1991; 26(4A):29–38.

27. Hajjar I, Kotchen TA. Trends in prevalence, awareness, treatment, and control of hypertension in the United States, 1988–2000. JAMA 2003; 290(2):199–206.

28. Masaki KH, Curb JD, Chiu D, Petrovitch H, Rodriguez BL. Association of body mass index with blood pressure in elderly Japanese American men. The Honolulu Heart Program. Hypertension 1997; 29(2):673–7.

29. Dey DK, Lissner L. Obesity in 70-Year-old subjects as a risk factor for 15-year coronary heart disease incidence. Obes Res 2003; 11(7):817–27.

30. Coggon D, Reading I, Croft P, McLaren M, Barrett D, Cooper C. Knee osteoarthritis and obesity. Int J Obes Relat Metab Disord 2001; 25(5):622–7.

31. Lazarus R, Sparrow D, Weiss ST. Effects of obesity and fat distribution on ventilatory function: the normative aging study. Chest 1997; 111(4):891–8.

32. Carmelli D, Swan GE, Bliwise DL. Relationship of 30-year changes in obesity to sleep-disordered breathing in the Western collaborative group study. Obes Res 2000; 8(9):632–7.

33. Brown JS, Seeley DG, Fong J, Black DM, Ensrud KE, Grady D. Urinary incontinence in older women: who is at risk? Study of Osteoporotic fractures research group. Obstet Gynecol 1996; 87(5 Pt 1):715–21.

34. Wolk A, Gridley G, Svensson M et al. A prospective study of obesity and cancer risk (Sweden). Cancer Causes Control 2001; 12(1):13–21.

35. Jordan JM, Luta G, Renner JB et al. Self-reported functional status in osteoarthritis of the knee in a rural southern community: the role of sociodemographic factors, obesity, and knee pain. Arthritis Care Res 1996; 9(4):273–8.

36. Apovian CM, Frey CM, Rogers JZ, McDermott EA, Jensen GL. Body mass index and physical function in obese older women. J Am Geriatr Soc 1996; 44(12):1487–8.

37. Himes CL. Obesity, disease, and functional limitation in later life. Demography 2000; 37(1):73–82.

38. Davison KK, Ford ES, Cogswell ME, Dietz WH. Percentage of body fat and body mass index are associated with mobility limitations in people aged 70 and older from NHANES III. J Am Geriatr Soc 2002; 50(11):1802–9.

39. Jensen GL, Friedmann JM. Obesity is associated with functional decline in community-dwelling rural older persons. J Am Geriatr Soc 2002; 50(5):918–23.

40. Launer LJ, Harris T, Rumpel C, Madans J. Body mass index, weight change, and risk of mobility disability in middle-aged and older women. The epidemiologic follow-up study of NHANES I. JAMA 1994; 271(14):1093–8.

41. Galanos AN, Pieper CF, Cornoni-Huntley JC, Bales CW, Fillenbaum GG. Nutrition and function: is there a relationship between body mass index and the functional capabilities of community-dwelling elderly? J Am Geriatr Soc 1994; 42(4):368–73.

42. Jenkins KR. Obesity's effects on the onset of functional impairment among older adults. Gerontologist 2004; 44(2):206–16.

43. Zizza CA, Herring A, Stevens J, Popkin BM. Obesity affects nursing-care facility admission among whites but not blacks. Obes Res 2002; 10(8):816–23.

44. Roubenoff R. Sarcopenia: effects on body composition and function. J Gerontol A Biol Sci Med Sci 2003; 58(11):1012–7.

45. Baumgartner RN. Body composition in healthy aging. Ann NY Acad Sci 2000; 904:437–48.

46. Villareal DT, Banks M, Siener C, Sinacore DR, Klein S. Physical frailty and body composition in obese elderly men and women. Obes Res 2004; 12(6):913–20.

47. Roubenoff R. Sarcopenic obesity: the confluence of two epidemics. Obes Res 2004; 12(6):887–8.

48. Brown M, Sinacore DR, Binder EF, Kohrt WM. Physical and performance measures for the identification of mild to moderate frailty. J Gerontol A Biol Sci Med Sci 2000; 55(6):M350–5.

49. Holloszy JO, Kohrt WM. Handbook of physiology – Aging. London: Oxford University Press, 1995.

50. Jette AM, Cleary PD. Functional disability assessment. Phys Ther 1987; 67(12):1854–9.

51. Blaum CS, Xue QL, Michelon E, Semba RD, Fried LP. The association between obesity and the frailty syndrome in older women: the women's health and aging studies. J Am Geriatr Soc 2005; 53(6):927–34.

52. Sarkisian CA, Liu H, Gutierrez PR, Seeley DG, Cummings SR, Mangione CM. Modifiable risk factors predict functional decline among older women: a prospectively validated clinical prediction tool. The study of Osteoporotic fractures research group. J Am Geriatr Soc 2000; 48(2):170–8.

53. Felson DT, Zhang Y, Hannan MT, Anderson JJ. Effects of weight and body mass index on bone mineral density in men and women: the Framingham study. J Bone Miner Res 1993; 8(5):567–73.

54. Reid IR, Cornish J, Baldock PA. Nutrition-related peptides and bone homeostasis. J Bone Miner Res 2006; 21(4):495–500.

55. Schott AM, Cormier C, Hans D et al. How hip and whole-body bone mineral density predict hip fracture in elderly women: the EPIDOS Prospective Study. Osteoporos Int 1998; 8(3):247–54.

56. Garrow JS, Summerbell CD. Meta-analysis: effect of exercise, with or without dieting, on the body composition of overweight subjects. Eur J Clin Nutr 1995; 49(1):1–10.

57. Gallagher D, Kovera AJ, Clay-Williams G et al. Weight loss in postmenopausal obesity: no adverse alterations in body composition and protein metabolism. Am J Physiol Endocrinol Metab 2000; 279(1):E124–31.

58. Inelmen EM, Sergi G, Coin A, Miotto F, Peruzza S, Enzi G. Can obesity be a risk factor in elderly people? Obes Rev 2003; 4(3):147–55.

59. Villareal DT, Banks M, Sinacore DR, Siener C, Klein S. Effect of weight loss and exercise on frailty in obese older adults. Arch Intern Med 2006; 166(8):860–6.

60. Fontana L, Villareal DT, Weiss EP et al. Calorie restriction or exercise: Effects on coronary heart disease risk factors. A randomized controlled trial. Am J Physiol Endocrinol Metab 2007.

61. Villareal DT, Miller BV, III, Banks M, Fontana L, Sinacore DR, Klein S. Effect of lifestyle intervention on metabolic coronary heart disease risk factors in obese older adults. Am J Clin Nutr 2006; 84(6):1317–23.

62. Jensen GL, Roy MA, Buchanan AE, Berg MB. Weight loss intervention for obese older women: improvements in performance and function. Obes Res 2004; 12(11):1814–20.

63. Messier SP, Loeser RF, Miller GD et al. Exercise and dietary weight loss in overweight and obese older adults with knee osteoarthritis: the arthritis, diet, and activity promotion trial. Arthritis Rheum 2004; 50(5):1501–10.

64. Miller GD, Nicklas BJ, Davis C, Loeser RF, Lenchik L, Messier SP. Intensive weight loss program improves physical function in older obese adults with knee osteoarthritis. Obesity (Silver Spring) 2006; 14(7):1219–30.

65. Avenell A, Richmond PR, Lean ME, Reid DM. Bone loss associated with a high fibre weight reduction diet in postmenopausal women. Eur J Clin Nutr 1994; 48(8):561–6.

66. Riedt CS, Cifuentes M, Stahl T, Chowdhury HA, Schussel Y, Shapses SA. Overweight postmenopausal women lose bone with moderate weight reduction and 1 g/day calcium intake. J Bone Miner Res 2005; 20(3):455–63.

67. Jensen LB, Kollerup G, Quaade F, Sorensen OH. Bone minerals changes in obese women during a moderate weight loss with and without calcium supplementation. J Bone Miner Res 2001; 16(1):141–7.

68. Villareal DT, Fontana L, Weiss EP et al. Bone mineral density response to caloric restriction-induced weight loss or exercise-induced weight loss: A randomized controlled trial. Arch Intern Med 2006; 166(22):2502–10.

69. Ryan AS, Nicklas BJ, Dennis KE. Aerobic exercise maintains regional bone mineral density during weight loss in postmenopausal women. J Appl Physiol 1998; 84(4):1305–10.

70. Campbell WW, Crim MC, Dallal GE, Young VR, Evans WJ. Increased protein requirements in elderly people: new data and retrospective reassessments. Am J Clin Nutr 1994; 60(4):501–9.

71. Executive Summary of The Third Report of The National Cholesterol Education Program (NCEP). Expert panel on detection, evaluation, and treatment of high blood cholesterol in adults (adult treatment panel III). JAMA 2001; 285(19):2486–97.

72. Sugerman HJ, DeMaria EJ, Kellum JM, Sugerman EL, Meador JG, Wolfe LG. Effects of bariatric surgery in older patients. Ann Surg 2004; 240(2):243–7.

16 Nutrition and Lifestyle Change in Older Adults with Diabetes Mellitus and Metabolic Syndrome

Barbara Stetson and Sri Prakash Mokshagundam

Key Points

- In the USA, 44% of persons with self-reported diagnosed diabetes are aged 65 years or older and 18% are over age 75 years.
- The aim of diabetes intervention is to prevent or delay the development of long-term complications of high blood glucose and related metabolic abnormalities and improve the quality of life.
- "Metabolic syndrome" refers to a cluster of abnormalities that includes hypertension, dyslipidemia, abnormal blood glucose, and abdominal obesity. Over 40% of adults over the age of 70 have the metabolic syndrome.
- Hypoglycemia is a major limiting factor in the management of diabetes. Factors that may play a role in the increased risk of hypoglycemia in older adults include poor nutritional status, cognitive dysfunction, polypharmacy, and comorbid illnesses.
- Diabetes prevalence-related comorbidities such as diabetic retinopathy, cardiovascular disease, peripheral vascular disease, and congestive heart failure may result in decreased usual activity and limit activities of daily living, including transportation, shopping for food, and ability to read food labels and restaurant menus.
- Given the high rates of depression in the diabetes population, careful assessment of depressive symptomology and its impact on dietary intake, diabetes self-care, and health outcomes is critical.

Key Words: Blood glucose; metabolic syndrome; hypoglycemia; depression; quality of life

From: *Nutrition and Health: Handbook of Clinical Nutrition and Aging, Second Edition*
Edited by: C. W. Bales and C. S. Ritchie, DOI 10.1007/978-1-60327-385-5_16,
© Humana Press, a part of Springer Science+Business Media, LLC 2009

Diabetes mellitus (diabetes) is a major health problem in the USA. The estimated number of individuals with diabetes is approximately 20.8 million, of whom 6.2 million are undiagnosed. Type 2 diabetes disproportionately affects minority populations, including African Americans, Hispanics, Native Americans, Asian Americans, and Pacific Islanders. The Pima Indians of Arizona have one of the highest rates of diabetes in the world. Risk factors for diabetes that are specific to these populations include genetic, behavioral, and lifestyle factors *(1)*.

The prevalence of obesity is rising so rapidly in so many countries that the World Health Organization has declared that there is now a global epidemic of obesity. Obesity is common in Western market economies (Europe, USA, Canada, Australia, etc.) and in Latin America and rates are increasing in sub-Saharan Africa and Asia, where rates have traditionally been low. Internationally, emergence of new cases of diabetes parallels the increases seen in Western countries and are increasing even more quickly in Asia. The risks of type 2 diabetes in these countries tend to increase at levels of body mass index generally classified as non-obese in Caucasian Westerners *(2)*. These worldwide changes are due to an accelerated prevalence of obesity, today's predominance of sedentary lifestyle, and the rapidly growing population of older adults *(3)*.

16.1 DIABETES IN OLDER US ADULTS

The graying of America is also contributing to the increasing numbers of cases of diabetes, as diabetes prevalence increases with age. In developing countries, the majority of people with diabetes are between 45 and 64 years of age. In developed countries, the majority of people with diabetes are 65 years of age or greater. In the USA, the oldest of the large baby boomer cohort are now approaching 60 years of age, and increasing numbers will soon join these ranks. The Third National Health and Nutrition Examination Survey (NHANES III) included information on type 2 diabetes and included persons 75 years of age and older. Extrapolation from this nationally representative sample indicates that in the USA 44% of persons with self-reported diagnosed diabetes are 65 years of age or older and 18% are over 75 years of age *(4)* (Fig. 16.1)

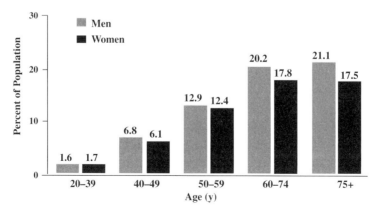

Harris, et al. *Diabetes Care.* 1998;21:518

Fig. 16.1. Estimated prevalence of diabetes in the USA: adult men and women.

16.2 HEALTH CONSEQUENCES OF DIABETES

Diabetes is a chronic disease that leads to a variety of micro and macrovascular complications that affect almost all systems in the body. While the primary abnormality in diabetes, elevated blood glucose level, remains largely asymptomatic, the consequences of sustained elevation in blood glucose are potentially devastating. Diabetes is the leading cause of blindness, chronic renal insufficiency, peripheral neuropathy, and non-traumatic limb amputations.

Type 2 diabetes exerts a tremendous economic burden, accounting for over 100 billion dollars in annual healthcare expenditures in the USA and 28% of the Medicare budget for older Americans (5). In NHANES III, among persons with type 2 diabetes over 65 years of age, 21% reported being in poor health and 35% reported having at least one hospitalization in the preceding year (6).

Cardiovascular disease (CVD) is the most frequent and costly complication of type 2 diabetes. A recent review indicates that when cardiovascular events are stratified by diabetes status, relative risk for men is twice and for women is threefold of gender-matched nondiabetics. Among all CVD events, diabetes accounted for 56% of events in men and 78% of events in women. A number of diabetes-related risk factors have been associated with CVD. Epidemiological studies have also suggested that postchallenge hyperglycemia is a risk factor for cardiovascular disease. Albuminuria in diabetics has been shown to have a CVD risk that is four to five times compared to diabetics without albuminuria, suggesting that these should be targets of preventive strategies in persons with diabetes (7).

16.3 GENERAL AIMS OF DIABETES TREATMENT

The management of diabetes requires a combination of lifestyle interventions and medications. Diabetes is often a progressive disease requiring changing therapeutic strategies. The interaction between lifestyle changes and medications must be carefully considered. Dietary intervention to maintain optimal glycemic control is a key component of management. The aims of diabetes treatment are to (1) decrease/prevent the development of long-term complications of high blood glucose and related metabolic abnormalities, (2) improve the quality of life of individuals with diabetes, (3) treat or prevent the development of symptoms of high or low blood glucose.

16.4 DIAGNOSIS AND CLASSIFICATION

16.4.1 Diagnosis

The American Diabetes Association (ADA) and the World Health Organization revised the criteria for the diagnosis of diabetes in 1997 (8, 9).and again in 2003 (10). The new diagnostic criteria emphasize fasting blood glucose levels (8). In addition to the occurrence of chronically elevated blood glucose levels, isolated postchallenge hyperglycemia is a particularly common problem among older adults who have abnormal glucose tolerance. The oral glucose tolerance test is less reproducible than the fasting plasma glucose levels and hence is used less often in routine clinical practice. The American Diabetes Association (ADA) diagnostic criteria were developed for general use and apply broadly to all age groups. No specific ADA guidelines exist for older adults.

16.4.2 Typologies of Diabetes in Older Adults

The proper classification of diabetes is important in setting goals for nutritional management of individuals with diabetes. Diabetes mellitus is broadly classified into type 1 and type 2.

16.4.2.1 TYPE 1 DIABETES

Type 1 diabetes is an autoimmune disorder resulting from cell-mediated and anti-body-mediated destruction of beta-cells of the islets (11). Insulin is required for the management of type 1 diabetes. Failure to treat with insulin results in development of an acute metabolic complication – diabetic ketoacidosis. Although type 1 diabetes most commonly occurs in the first three decades of life, it can develop at any age, even in older adults. The basic underlying mechanism of disease is autoimmune destruction of the pancreatic islets. Circulating islet cell antibodies can be demonstrated in the majority of individuals, especially in the first few years after diagnosis. In addition to new onset type 1 diabetes, older adults may have pre-existing type 1 diabetes. Type 1 diabetes, particularly of long duration, is often very "brittle" with wide fluctuations in blood glucose levels and episodes of recurrent and severe hypoglycemia.

16.4.2.2 TYPE 2 DIABETES

The majority of older adults with diabetes have type 2 diabetes, which is characterized by two defects – insulin resistance and defective insulin secretion (12). The majority of individuals with type 2 diabetes are obese. However, in the older population the proportion of subjects with type 2 diabetes who are underweight increases and could be as high as 20%. This is particularly true in the nursing home population (see section on metabolic syndrome). Type 2 diabetes results from a combination of insulin resistance, increased hepatic glucose production, and defective insulin secretion (13). Insulin resistance is generally considered the early defect in type 2 diabetes. Insulin resistance is often present in nondiabetic relatives of individuals with type 2 diabetes and in persons with impaired glucose tolerance. Several studies have also demonstrated defective insulin secretion in these at-risk individuals. Studies of Pima Indians have demonstrated that the progression from normal glucose tolerance to diabetes mellitus is associated with a progressive decline in acute insulin response to glucose (14).

The exact mechanism of insulin resistance in type 2 diabetes is unclear. A variety of genetic and environmental factors lead to decreased insulin sensitivity. Of importance to this chapter, obesity and decreased physical activity have been known to decrease insulin sensitivity. Aging is associated with a change in body composition with increase in fat mass and decrease in muscle mass (15, 16). This could be partly responsible for the increase in insulin resistance with aging. Aging is also associated with a decline in insulin secretion, particularly a blunting of the first phase insulin secretion (17). First phase insulin secretion is an important determinant of postchallenge blood glucose levels. Age-related changes in health behaviors such as increased sedentary lifestyles may also further compound these changes.

Type 2 diabetes is a progressive disorder. The progression of the clinical picture with increasing blood glucose levels, requiring increasing doses of medications, is due mainly to a progressive decline in beta-cell function. When beta-cell function is markedly reduced, exogenous insulin will be necessary to regulate blood glucose levels.

16.5 ESTABLISHING MEDICATION AND NUTRITIONAL MANAGEMENT GOALS

Once the type of diabetes is established, medication and nutritional management goals should be developed. In addition to tailoring the nutritional recommendation to assist glycemic control, consideration of other important risk factors is critical. Obesity, dyslipidemia, hypertension, and insulin resistance are important and often overlapping factors warranting consideration when planning dietary interventions for older adults with type 2 diabetes. Avoidance of hypoglycemia, particularly recurrent and/or severe hypoglycemia, is a major consideration in type 1 diabetes. Lifestyle interventions that have been recommended for the management of diabetes have positive effects on both insulin secretion and insulin resistance. Aggressive lifestyle intervention can prevent the progression of impaired glucose tolerance to diabetes and could decrease the dose and number of medications for the management of type 2 diabetes.

16.5.1 Medication Use and Glycemic Control in Older Adults with Diabetes

16.5.1.1 GOALS OF DIABETES TREATMENT IN OLDER ADULTS

The major aim of treating diabetes is to decrease the rate of micro and macrovascular disease associated with elevated blood glucose. Two landmark trials have served as the basis for current recommendations for the management of blood glucose levels in diabetes mellitus. The Diabetes Control and Complications Trial (DCCT) was conducted in adults with type 1 diabetes and compared intensive insulin treatment using multiple insulin injections or an insulin pump to conventional treatment using twice daily injections of intermediate and short-acting insulin over a follow-up period of 7 years (18, 19). The results showed significant reduction in risk of all microvascular disease endpoints in the intensively treated group. However, the DCCT did not include older adults and did not have the statistical power to analyze benefits on macrovascular risk reduction. The clear demonstration of a relationship between glycemic control, measured by reduction in hemoglobin A1c, and improved outcomes indicates that similar outcomes would be expected in older adults. A downside to tight control was indicated by the findings of higher risk of hypoglycemia in the intensively treated group. Given the burden of potential hypoglycemia and potential impact on quality of life in an older adult with a limited prognosis, the cost–benefit ratio of intensive glycemic control versus hypoglycemia risk must be carefully considered.

The United Kingdom Prospective Diabetes Study (UKPDS) was a long-term study of a variety of treatment options in adults with type 2 diabetes (19). The important findings of the UKPDS can be summarized as follows: (1) a reduction of 1% in hemoglobin A1c results in ~ 22% reduction in microvascular complications; (2) reduction in microvascular complications with reduction in hemoglobin A1c is

Table 16.1
ADA criteria for diagnosis of diabetes

	Normal (mg/dl)	Pre-diabetes (mg/dl)	Diabetes mellitus (mg/dl)
Fasting blood glucose	< 100	101–125	> 125
2-hour post glucose	< 140	141–199	≥ 200

observed irrespective of type of intervention; (3) glycemic control in type 2 diabetes mellitus worsens over time and necessitates changes in medication, irrespective of initial management approach; (4) in a subgroup of subjects treated with metformin there was a significant reduction in macrovascular disease. The DCCT did not include older adults and there were few older participants in the UKPDS. Although the UKPDS included an older population, subgroup analysis of the older age group is not available. Hence, the applicability of these studies to older adults is limited.

The ADA goal for glycemic control is to have HbA1c levels ≤ 7.0%. In addition the ADA recommends individualization of glycemic control goals with an HbA1c goal of <6% when feasible. The ADA goals are general and written to broadly apply to all persons with diabetes. However, older adult persons with diabetes require special consideration and require reassessment of goals. While the general principles of diabetes care remain the same, it is now well recognized that in managing the elderly subject additional factors need to be considered and the goals modified accordingly. The 2003 California Healthcare Foundation/American Geriatric Society (AGS) guidelines have helped clarify some of these concerns and several concepts have been subsequently incorporated in the ADA guidelines. These guidelines have emphasized the need for individualizing diabetes care, aggressively addressing cardiovascular risk factors, and stressed glycemic control in preventing microvascular complications. The AGS guidelines also recognize the importance of comorbidities that are common in the elderly subject with diabetes mellitus and have significant impact on ability to maintain strict glycemic control. These include depression, cognitive impairment, urinary incontinence, falls, pain, and polypharmacy (20).

In addition to glycemic goals, several metabolic and cardiovascular risk factors must also be considered due to the higher rates of cardiovascular disease and its substantial impact on morbidity and mortality in persons with diabetes (see Table 16.2). The high risk of cardiovascular morbidity and mortality in diabetes is due to a variety of factors. These include overall blood glucose control, glycemic fluctuation, postprandial blood glucose levels, high LDL cholesterol, low HDL cholesterol, elevated serum triglycerides, blood pressure, and altered coagulation profile. In addition, systemic inflammation, a prooxidant state, and endothelial dysfunction play a significant role. Recommendations to focus on maintaining optimal control of blood pressure in persons with diabetes are based on the positive results of such control as demonstrated in the UKPDS, Hypertension Optimal Treatment trial (HOT), and Arterial Blood Pressure Control in Diabetes (ABCD) studies (21–23). Additionally, the benefits of lowering total and LDL cholesterol have also been demonstrated in intervention studies.

Table 16.2
Factors to consider in management of diabetes in older adults[*]

1. Individualize glycemic goals. The goal, in most cases, should usually include the standard A1c target of <7%. Consider a higher goal, if appropriate, based on the following factors:
   - patient preference
   - life expectancy
   - diabetes severity
   - functional status and social support
2. Keep therapy as simple and inexpensive as possible.
3. Encourage diabetes education of the patient and primary caregivers, with the reminder that such education is covered by Medicare.
4. Treat hypertension and dyslipidemia to decrease cardiovascular risk.
5. Screen for depression and offer therapy promptly if the diagnosis is made.
6. Maintain an updated medication list and monitor regularly for adverse drug effects.
7. Screen annually for cognitive impairment and other geriatric syndromes (e.g., urinary incontinence, pain, injurious falls).

[*] Modified from Olson DE, Norris SL. Diabetes in older adults: Overview of AGS guidelines for the treatment of diabetes mellitus in geriatric populations. Geriatrics 2004; 59(April):18–25.

Reducing LDL-C has been shown to decrease cardiovascular events to 80 years of age, but studies are needed to explore the efficacy of lipid-lowering therapy in individuals older than 80 years of age. The magnitude of delay in progression of atherosclerotic disease in response to screening-guided therapy has not been well delineated. The cost-effectiveness of lipid screening in the elderly population, as well as the subgroups that would benefit most from such screening, needs to be further studied. Based on ongoing clinical trial data, the guidelines may be modified to recommend that an LDL-C no greater than 70 mg/dL is the target level for special high-risk populations. In the PROVE-IT study, a subgroup analysis showed less benefit in patients over the age of 65 compared with patients less than 65 years of age. The TNT trial excluded patients over the age of 75. Hence, the applicability of these data to the elderly population needs to be better defined. Optimal implementation of the current guidelines and the use of available agents will ultimately depend on expanding the knowledge base of health-care providers and may require far-reaching educational programs that change the way that risk-factor management is viewed by caregivers and patients alike (24–36).

Any nutritional approach to the management of diabetes mellitus must specifically address the issues related to cardiovascular risk. Cardiovascular risk reduction in diabetes mellitus is achieved through a combination of lifestyle changes and pharmacological interventions that address the multiple risk factors. A general outline of lifestyle and pharmacological approaches is shown in Table 16.3.

Taken together, these lifestyle and pharmacological interventions can play a major role in the management of cardiovascular risk reduction in persons with diabetes. The recommended goals for management of weight, blood pressure, and lipids are outlined in Table 16.4.

Table 16.3
Lifestyle and pharmacological approaches to risk-factor management

Risk factor	Lifestyle intervention	Pharmacological intervention
HbA1c	Diet and exercise	Insulin sensitizing agents, insulin secretagogues, insulin
Postprandial glucose/ glycemic excursion	Carbohydrate content of meals (amount, type, timing, personal response to CHO based on postprandial SMBG feedback)	Repaglinide (prandin), nateglinide (starlix), sitagliptin, exenatide, pramlinitide, short-acting insulins (insulin lis-pro, insulin aspart, regular insulin)
LDL cholesterol	Low cholesterol diet Exercise	Statins Bile acid binding agents Niacin
Triglyceride	Low-fat diet Exercise/weight loss	Gemfibrozil/fenofibrate Niacin (long acting) Omega 3 fatty acids
Low HDL cholesterol	Exercise Smoking cessation	Niacin Gemfibrozil/fenofibrate
High blood pressure	Low sodium diet Exercise Weight loss	Variety of antihypertensive agents (ace inhibitors preferred)
Procoagulant state	Exercise/weight loss	Aspirin
Proinflammatory state	Diet (possible) (e.g., increased proportion of less-refined CHO, increased vegetable and fruit intake, reduced saturated fat), exercise, weight loss, medication	Aspirin

16.5.1.2 GENERAL DIABETES DIETARY RECOMMENDATIONS

The general goals of nutritional recommendations for the management of diabetes mellitus include the following:

(1) Achieve and maintain blood glucose levels as outlined above.
(2) Achieve and maintain optimum lipid levels.
(3) Achieve and maintain reasonable body weight. This would include weight loss, if overweight, and weight gain, if undernourished.
(4) Prevent acute complications.
(5) Maintain overall health.

The general recommendations for macronutrient intake in diabetic diet are shown in Table 16.5. The carbohydrate composition (amount and type) of the diet has been a focus of many recommendations and subject of recent controversy. There are no well-designed studies that have compared different dietary

Table 16.4
Recommended assessment and management goals for CVD risk factors in diabetes

Parameter	Frequency	Goal
HbA1c	3–4 months	<7% (<6.5%)*
Fasting blood glucose	2–7 times a week[†]	80–120 mg/dl* (90–130 mg/dl)
Postprandial blood glucose	2–7 times a week[†]	< 140 mg/dl* (<180 mg/dl)
LDL cholesterol	Annual. 3–4 months, if abnormal	<100 mg/dl[‡]
HDL cholesterol	Annual. 3–4 months, if abnormal	> 45 mg/dl
Triglyceride	Annual. 3–4 months, if abnormal	< 200 mg/dl
Systolic blood pressure	Each visit	<130 mmHg
Diastolic blood pressure	Each visit	< 80 mmHg
Microalbuminuria	Annual	Normal or no progression
Eye examination	Annual	Normal or no progression
Neurological examination	Annual	Normal or no progression

* Recommended by AACE/ACE.
[†] No clear recommendation (need to individualize).
[‡] Optional goal of < 70 mg/dl in highest risk individuals.

approaches. In a study of a high-carbohydrate (60%) low-fat (25%) diet, compared to a low-carbohydrate (35%) high-monounsaturated fat (50%) diet *(37)*. plasma glucose, triglyceride, and VLDL cholesterol were lower in the subjects in the low-carbohydrate/high-monosaturated fat diet group. However, use of high-fat/low-carbohydrate diets could lead to more hypoglycemic episodes and ketosis.

Higher protein diets have been recommended and have been popular for weight loss. An empirical review of studies of high- versus low-protein diets found that short-term (less than 6 months), high-protein diets may help people lose more weight and body fat, due in part to increased satiety *(38)*. However, the long-term effects of high-protein diets and their efficacies in persons with diabetes have not been well tested. Higher protein content has been shown to increase risk of development and progression of diabetic nephropathy. It has been suggested that a

Table 16.5
Macronutrient content of general diabetic diet

Carbohydrates	45–65% of total caloric intake
Protein	12–20% of caloric intake
	< 30% of total caloric intake
Fat	< 200 mg /day of cholesterol
	Saturated and polyunsaturated fat
	< 10% of total caloric intake (2–40 g/day)

high-protein diet (2 g/kg of body weight) may be contraindicated for persons with poorly controlled diabetes or complications *(39)*. Low-carbohydrate diets are not recommended in the management of diabetes. Dietary carbohydrate is the major contributor to postprandial glucose concentration, and an important source of energy, water-soluble vitamins and minerals, and fiber. Thus, in agreement with the National Academy of Sciences-Food and Nutrition Board, a recommended range of carbohydrate intake is 45–65% of total calories. In addition, because the brain and central nervous system have an absolute requirement for glucose as an energy source, restricting total carbohydrate to <130 g/day is not recommended *(40)*. Monitoring carbohydrate intake, whether by carbohydrate counting, using the exchange system, or experience-based estimation remains a key component of managing diabetes. Use of complex carbohydrates is preferred. The use of lower fat content in the diet is based on the need to restrict caloric intake, improve lipid levels, and assist weight loss. While use of vitamin supplements is not generally recommended for subjects with type 2 diabetes, the ADA recommends using a multivitamin in the older adult. There has been no convincing evidence to recommend the routine use of antioxidants, vitamin E, or C in subjects with diabetes.

16.5.1.3 BALANCING DIET AND MEDICATION

The interaction of diet and medication is of particular importance in the management of diabetes. Insulin and drugs that increase insulin secretion are likely to induce hypoglycemia if meals are not taken at appropriate times. Erratic eating habits might require readjustment of medications, either dose, timing, or both. This may be of particular concern in the hospitalized older patient with diabetes. Poor eating habits might also necessitate change to medications that are less likely to cause hypoglycemia when used alone. Metformin (Glucophage) and the thiazolidenediones (pioglitazone and rosiglitazone) are least likely to cause hypoglycemia when used alone. The problem of unwanted weight gain is another issue for consideration for many overweight individuals. Insulin and the thiazolidenediones, particularly when used in combination, are most likely to result in weight gain. The deleterious effect of weight gain in overweight older adults must be evaluated in tandem with the potential benefits of improved glycemic control. Loss of appetite may occur with metformin, exenatide (a GLP-1 receptor agonist), and pramlinitide (an amylin analog) and would be of concern in the undernourished persons with diabetes. Due to concerns of lactic acidosis, metformin may not be appropriate for persons with predisposing conditions such as heart failure, liver problems, and renal insufficiency.

16.6 BODY WEIGHT AND FUNCTIONAL STATUS IN OLDER ADULTS WITH DIABETES

16.6.1 Overweight and Obesity

Overweight is not only an important risk factor for the development of diabetes; it also has a significant impact on diabetes progression and the development of complications *(41)*. Obesity is known to be a critical problem in children and young and middle-aged adults with type 2 diabetes. Only recently has the problem of obesity been systematically examined in older adults. Obesity appears to be common in older adults until the eighth decade of life and then declines in the oldest old. Data from NHANES III indicate that type 2 diabetes strongly increases in prevalence with increasing overweight in older as well as younger adults. Personal risks for diabetes were observed to be stronger for obese younger adults but still substantially elevated in older adults with odds ratio of 3.4 (95% CI, 1.1–8.3) for the most obese men over 55 years of age and 5.8 (95% CI, 4.2–7.4) for the most obese women over 55 years of age. In a study of 3-year mortality in community-dwelling older adults, unintentional weight loss and underweight BMI were associated with elevated mortality rates. Overweight or obesity and intentional weight loss were not associated with mortality. These results, which are consistent with other findings, suggest that undernutrition (e.g., low BMI, unintentional weight loss) may pose greater mortality risk in older adults than do obesity or intentional weight loss *(42)*.

16.6.2 Underweight and Malnutrition

While obesity is clearly a problem that greatly impacts diabetes in older adults, for many individuals, malnutrition may be the more pressing nutritional concern. Even with the problem of increasingly prevalent overweight, obesity and diabetes, the prevalence of obesity in older persons with diabetes is still less than that in younger persons with type 2 diabetes. This may be particularly true for the oldest old, and those who have impaired functional status. One study found that at least 21% of nursing home patients with type 2 diabetes were underweight *(43)*.

16.7 SPECIAL NUTRITION INTERVENTION SITUATIONS FOR PERSONS WITH DIABETES

16.7.1 Acute Illness, Hospitalization, Enteral and Parenteral Nutrition

Management of hyperglycemia in the hospital setting has gained increasing attention over the last few years. Hyperglycemia is common among hospitalized subjects and has been shown to be associated with higher mortality and morbidity in a variety of studies. This is particularly relevant to the older population, since they are more likely to be admitted to the hospital and have higher rates of diabetes mellitus. The American Association of Clinical Endocrinologists (AACE) and ADA have recommended glycemic targets for patients with hyperglycemia, depicted in Table 16.6 *(44, 45)*. While randomized controlled studies in the intensive care unit support the glycemic goal of < 110 mg/dl in the ICU, there are no trials that clearly justify the recommended goal blood glucose levels in the non-ICU setting *(46)*.

Table 16.6
Summary of ADA and AACE glycemic control target recommendations

Location	ADA recommendation	AACE recommendation
ICU	As close to 110 mg/dl as possible	< 110 mg/dl
General ward	As close to 90–130 mg/dl as possible; postprandial glucose < 180 mg/dl	Pre-meal < 110 mg/dl, post-meal < 180 mg/dl.

16.7.2 Hospitalization

A variety of systemic problems have been identified that affect glycemic control in the hospital setting. Barriers that may impact an individual's nutrition status and subsequently affect glycemic control include poor appetite, inability to eat, increased nutrient and calorie needs due to catabolic stress, variation in diabetes medications, and the possible need for enteral or parenteral nutrition support. Proper timing of meals and the relation to medications is important. Insulin should be administered immediately before or after a meal. Due to the wide heterogeneity in the hospital population, individualization of nutrition recommendations is a key to improving outcomes. The common practice of ordering "ADA Diet" is strongly discouraged as the ADA does not endorse any specific diet. The consistent carbohydrate meal planning system is encouraged. For this system to be effective it is important that nursing and nutrition services coordinate their services. The key areas of focus to improve inpatient glycemic control are:

(1) Establishing screening criteria for appropriate referral to a registered dietitian.
(2) Identifying nutrition-related issues in clinical pathways and patient care plans.
(3) Implementing and maintaining standardized diet orders such as consistent carbohydrate menus.
(4) Integrating blood glucose monitoring results with nutrition care plans.
(5) Using standing orders for diabetes education and diabetes MNT as appropriate.
(6) Standardizing discharge follow-up orders for MNT and diabetes education postdischarge when necessary (47, 48).

Patients requiring clear or full liquid diets should receive 200 g carbohydrate/day in equally divided amounts at meal and snack times. Liquids should not be sugar free. Patients require carbohydrate and calories, and sugar-free liquids do not meet these nutritional needs. For tube feedings, either a standard enteral formula (50% carbohydrate) or a lower carbohydrate content formula (33–40% carbohydrate) may be used. Calorie needs for most patients are in the range of 25–35 kcal/kg every 24 h. Care must be taken not to overfeed patients because this can exacerbate hyperglycemia. After surgery, food intake should be initiated as quickly as possible. Progression from clear liquids to full liquids to solid foods should be completed as rapidly as tolerated (49).

16.7.3 Enteral and Parenteral Nutrition

Enteral and parenteral nutrition might pose additional challenges in the management of patients with diabetes. While the glycemic goals for individuals receiving enteral and parenteral nutrition are the same as glycemic goals for the general population of diabetics, achievement of normoglycemia may be more difficult in patients who are acutely ill. There is evidence that poor glycemic control in subjects on parenteral or enteral nutrition is related to poor outcomes. It is estimated that up to 30% of patients who receive parenteral nutrition have diabetes. Many of these patients have no previous history of diagnosed diabetes and develop diabetes due to stress-induced increases in counterregulatory hormones and cytokines.

The relative value of high-carbohydrate versus high-fat enteral feeds for persons with diabetes has been debated (50). The most widely used commercial enteral preparations for individuals with diabetes provide, but 1 cal/ml, 40% (CHOICEdm TF; Novartis Medical Nutrition) to 34% (Glucerna; Abbott Laboratories, Inc.) carbohydrate, and 43% (CHOICEdmTF) to 49% (Glucerna) fat. They also have high-monounsaturated fatty acids (MUFA; 35% of kcal in Glucerna). MUFA has been shown to be beneficial in improving lipid profile, glycemic control, and lower insulin level *(51)*. CHOICEdmTF has a higher content of medium chain triglycerides and has no fructose. The use of insulin or oral agents in persons receiving enteral nutrition should be tailored to match the timing of feeds. Parenteral nutrition fluids are high in carbohydrate and derive only fewer calories from fat. In persons with diabetes, particularly in less severely stressed individuals, the proportion of carbohydrate may be decreased but is still very high. The usual rate of glucose infusion is 4–5 g/kg body weight and lipid infusion of 1–1/5 g/kg body weight. This requires adequate use of insulin to maintain normoglycemia *(52)*. Insulin infusion not only maintains glycemic control, but prevents protein breakdown and promotes protein synthesis.

16.7.4 Long-Term Care

Residents of long-term care facilities may face additional or unique problems. They tend to be often underweight and it is not necessary to make any caloric restrictions in these subjects. Low body weight has been associated with higher mortality and morbidity in these subjects. Restricting food choices may lead to poor overall nutritional status in these subjects and has not been shown to improve glycemic control. Hence, the use of "no concentrated sugar", "no sugar added", or "liberal diabetic diet" is discouraged *(53, 54)*.

16.8 METABOLIC SYNDROME

The term "metabolic syndrome" refers to a cluster of abnormalities that was initially described by Reaven to include hypertension, dyslipidemia, abnormal blood glucose, and abdominal obesity. Insulin resistance was considered to be a central and pathogenetic abnormality in this syndrome. Over the last 20 years a number of other clinical and laboratory features have been proposed to be components of this syndrome. Some of these features included high C-reactive protein, non-alcoholic fatty liver disease, low plasminogen activator inhibitor 1, high fibrinogen, polycystic ovarian disease, and low adiponectin levels. In 2001, the National Cholesterol Education Program-Adult Treatment Panel III (NCEP-ATP III) recommended the use of the metabolic syndrome in cardiovascular risk assessment. The NCEP-ATP III defined metabolic syndrome as shown in Table 16.7 *(55)*. A number of other organizations including the World Health Organization, International Diabetes Federation, American Association of Clinical Endocrinologists, and the American Heart Association/National Heart Lung and Blood Institute have offered different diagnostic criteria for metabolic syndrome. However, the most widely used definition in the USA is the one offered by NCEP-ATP III. The prevalence of the metabolic syndrome as defined by the NCEP-ATP III in the NHANES III increases with age. Over 40% of adults over the age of 70 have the metabolic syndrome *(56)*.

Table 16.7
ATP III criteria for diagnosis of metabolic syndrome*

Risk factor	Defining level
Abdominal obesity (waist circumference)	
Men	> 102 cm (> 40 in.)
Women	> 88 cm (> 35 in.)
Triglyceride	> 150 mg/dl
HDL cholesterol	
Men	< 40 mg/dl
Women	< 50 mg/dl
Blood pressure	< 130/85 mmHg
Fasting blood glucose	> 110 mg/dl

*Diagnosis is established if three or more risk factors are present.

The exact significance of the metabolic syndrome has been a subject of controversy. The main significance appears to be the ability of the syndrome to identify individuals who are at high risk of developing cardiovascular disease and/or diabetes. A variety of clinical studies have demonstrated an association between metabolic syndrome and risk of heart disease, stroke, and development of diabetes mellitus (57–59).

Clinical trials have not been designed specifically to test the effect of lipid and blood pressure interventions in participants with the metabolic syndrome. Diabetes may be considered as a good model of the metabolic syndrome, because 85% of diabetic subjects have the metabolic syndrome, and there are clear data that lipid and blood pressure interventions work very well in diabetic subjects. However, there are no specific studies addressing this in older adults. Many critics, including members of a joint panel of the ADA and the European Association for the Study of Diabetes, have questioned the utility of the metabolic syndrome in clinical practice. They have cited the lack of agreement in diagnostic criteria among different organizations, the absence of a clear understanding of an underlying pathophysiologic mechanism linking the multiple components of the syndrome, and the variable risk associated with individual components of the syndrome (60).

While proponents of the continued use of the metabolic syndrome in clinical practice agree with many of the criticisms, they argue that recognition of this syndrome in clinical practice is useful. The use of the metabolic syndrome may encourage healthcare providers to look for other risk factors once one risk factor is found and more importantly encourage the use of behavioral therapy such as weight loss (through a program of diet and exercise) and increased physical activity rather than always prescribing a different drug for each medical condition. The management of the metabolic syndrome is primarily aimed at reducing cardiovascular risk. In individuals with pre-diabetes, the aim would be to prevent progression to type 2 diabetes.

Several large, randomized clinical trials have demonstrated the efficacy of both lifestyle and pharmacological interventions in preventing the progression of impaired fasting glucose or impaired glucose tolerance. In the landmark Diabetes Prevention Program (DPP), progression to diabetes was reduced by 65% in the lifestyle intervention group and by 31% in the metformin group *(61)*. Over 50% of participants in the study had three or more components of the metabolic syndrome. In older participants, lifestyle intervention had an even greater impact than metformin *(62)*. Among participants aged 60 and older, lifestyle intervention reduced the risk of development of diabetes by 71%. Participants in the 60- to 85-year-old age group were the most likely to achieve weight loss (5–7% loss in body weight was achieved with dietary change and activity) and physical activity goals. No age differences were noted in reduction of caloric intake. Older participants receiving metformin, the medication intervention arm of the study, did not experience benefits to the degree as younger, heavier participants. Diabetes incidence rates over time fell with increasing age, while in the metformin group, the youngest participants showed the lowest diabetes incidence. A trend was noted that metformin had lower effectiveness relative to lifestyle change with increased participant age, despite comparable to superior medication adherence and greater weight loss in the older metformin group participants. Age differences in response to pharmacologic treatment and physiology, as well as behavior, all likely played a role in the findings. Of note, DPP participants were community dwelling, relatively healthy, and free of significant physical limitations and frailty; however, the intensive lifestyle intervention was modified to accommodate participants who developed limitations over the course of the study *(62)*.

Another large study prospectively tracked at-risk older adults and highlighted the relationship between lifestyle variables and metabolic syndrome and the feasibility and benefits of lifestyle change. Participants were 3,051 men aged 60–75 without diabetes or history of CHD, who had been followed as participants in the British Regional Heart Study and completed a 20-year follow-up. Metabolic syndrome variables and lifestyle factors, including dietary intake were also assessed. One fourth of the sample had metabolic syndrome. After adjusting for demographics and other lifestyle factors, variable associations with metabolic syndrome were as follows: BMI had the most significant, positive association; physical activity was negatively associated; total dietary fat and alcohol intake showed no significant association; no significant association for saturated or polyunsaturated fat or polysaturated:saturated (P:S) ratio. Carbohydrate intake was associated with greater odds of metabolic syndrome; participants eating a low-fat, high-carbohydrate diet had the greatest odds of having metabolic syndrome while those eating a low-carbohydrate, high-fat diet showed no increase in having metabolic syndrome. The authors cautioned against emphasizing total fat reduction in older adults given the risk of increased carbohydrate intake which could adversely impact lipids and the development of metabolic syndrome.

These large prospective studies highlight that change in lifestyle, even at older ages, influenced metabolic syndrome variables, with changes observed within 3 years of changes in behavior. Weight loss, irrespective of baseline BMI, was associated with lower risk. Increased leisure time activity was also associated with lower likelihood of metabolic syndrome *(63)*. Findings suggest that modest weight loss and exercise are appropriate for older adults at risk for type 2 diabetes, dispelling the myth that they may not be able to change and are set in their ways *(64)*.

In the DREAM trial, both rosiglitazone and metformin were shown to reduce progression to type 2 diabetes in nondiabetic adults with impaired fasting glucose and/or impaired glucose tolerance. However, rosiglitazone use was associated with weight gain *(65)*. There has been growing concern about the potential cardiovascular adverse effects of thiazolidinediones. Further, metformin is not yet approved for use in the nondiabetic population, dampening widespread use of these medications. The effect of lifestyle intervention on blood pressure, dyslipidemia, and central obesity in the setting of metabolic syndrome in individuals with normal blood glucose has not been studied. Currently, dyslipidemia and hypertension in subjects with the metabolic syndrome are managed similarly to those without metabolic syndrome.

16.9 HYPOGLYCEMIA IN OLDER ADULTS

Hypoglycemia is a major limiting factor in the management of diabetes. The incidence of hypoglycemia is relatively high in older compared to younger adults. A variety of factors may play a role in the increased risk of hypoglycemia in older adults. These include poor nutritional status, cognitive dysfunction, polypharmacy, and comorbid illnesses. Except in the severely malnourished, poor dietary intake by itself does not lead to hypoglycemia. The most common cause of hypoglycemia remains the use of blood glucose-lowering agents. Drugs that increase insulin secretion and insulin itself can cause hypoglycemia. A major finding of the DCCT, which was conducted with young, healthy adults, was that the major deleterious health consequence of tight blood glucose control in persons with type 1 diabetes is hypoglycemia. Drugs that enhance insulin sensitivity (thiazolidinediones), decrease hepatic glucose production (metformin), or decrease carbohydrate absorption (alpha-glucosidase inhibitors) have very low risk of hypoglycemia, except when used in combination with insulin or an insulin secretagogue. When medication use creates problems of consistent hypoglycemia, patients must learn how to avoid and manage hypoglycemic episodes. Older adults taking insulin who have high variability in blood glucose levels, exhibit very low average blood glucose concentrations, have had diabetes for a long duration, have a low body mass index, or who have high levels of vigorous physical activity may be at particular risk of severe hypoglycemia *(66)*.

16.9.1 Self-Monitoring and Dietary Treatment of Hypoglycemia

Frequent self-monitoring of blood glucose levels provides specific information that may serve as feedback for guiding decisions about moment-to-moment treatment needs, thus helping patients to anticipate or prevent severely low glucose levels. However, frequent blood glucose testing may be perceived as too expensive, inconvenient or painful by many persons with diabetes. Unfortunately, rather than performing frequent blood glucose testing, many individuals simply rely on their symptoms or estimates about their blood glucose levels when deciding what to eat or how vigorously to exercise or whether to operate a motor vehicle *(67)*. By increasing the frequency of blood glucose testing (at least four times per day for persons taking insulin) and making informed decisions about when to eat additional

Table 16.8
Dietary management of hypoglycemia

1. Check blood glucose level by glucose monitor.
2. If blood glucose less than 60 mg/dl or symptomatic – treat with 15 g of carbohydrate (1/2 cup juice, 1/2 cup regular soft drink, glucose gel).
3. Repeat blood sugar reading in 15 minutes after treatment and again after 60 minutes.
4. Repeat step 2 until blood glucose is > 60 mg/dl.
5. If meals are due within 60 minutes – eat meal now.
6. If meals are not due within 60 minutes follow the glucose treatment with a snack containing carbohydrate and one protein (cheese and crackers, peanut butter and crackers, skim milk and crackers, or a small sandwich).
7. If blood glucose < 40 mg/dl and/or subject is stuporous, confused, or unresponsive – give 1 amp of D50W as IV push and start D10W at 60 cc /hour. Check blood glucose every 5 minutes and repeat till blood glucose > 60 mg/dl or till awake. Give oral carbohydrate once awake.

carbohydrate (e.g., eat 15 grams of carbohydrate to raise blood glucose levels about 45 mg/dl) or to identify personal sources of vigorous physical activity contributing to low blood glucose levels, patients may learn to prevent severe hypoglycemia. Educating patients about the importance of always carrying glucose tabs or gel or fast-acting carbohydrate snacks or placing them in various locations such as the car or relative's homes may also aid in the treatment of mild to moderate hypoglycemic episodes. Recommendations for management of hypoglycemia in older adults are presented in Table 16.8.

16.9.2 Hypoglycemia Unawareness and Treatment of Hypoglycemia

As previously described, older adults with type 1 diabetes and those with type 2 diabetes who are on exogenous insulin regimens are at risk for hypoglycemia. Many individuals develop the syndrome of hypoglycemia unawareness, in which the warning symptoms that indicate that hypoglycemia is developing (e.g., tremulousness, tachycardia) are decreased or not detected. Without these warning symptoms, individuals are not able to take actions such as eating to prevent continued reductions in blood glucose levels and severe hypoglycemic episodes may result. Following episodes of hypoglycemia, counterregulatory hormone stores may not be available, and thresholds for symptoms of hypoglycemia may shift to lower glucose concentrations. Thus, patients with recurrent hypoglycemia may be particularly at risk for unawareness and for severely low hypoglycemic episodes. Failure to test blood glucose levels regularly can contribute to the problem of hypoglycemia unawareness. This cycle is particularly problematic for older adults who are highly physically active or who skip meals, do not eat sufficient quantities of food to match their insulin doses, or consume a high-fat diet which delays carbohydrate absorption and is not accounted for at the time of insulin administration.

Alcohol consumption, while not typically problematic when consumed in moderation, can pose risks for hypoglycemia in older adults taking insulin. In particular, the major risk of alcohol-related hypoglycemia is in persons in a fasting state and those who are alcohol dependent. The disinhibiting effect of alcohol poses the risk of hypoglycemia unawareness, making blood glucose monitoring essential. The potential for a delayed risk of hypoglycemia the morning after evening alcohol intake should also be emphasized *(68)*. The problem of patient hypoglycemia unawareness should be considered if the patient's HbA1c is low (e.g., <6.0), and if she/he describes inability to detect counterregulatory autonomic symptoms (e.g., tremulousness, pounding heart, anxiety, queasy stomach, sweating, flushed face) when blood glucose levels are low *(69)*. Potential barriers to blood glucose testing or adequate food consumption such as financial constraints, fear of pain, depression, or feelings of being overwhelmed by diabetes should be assessed.

Structured psychoeducational intervention and print materials to promote reduced hypoglycemia unawareness have been developed and systematically evaluated for nearly 25 years in the Blood Glucose Awareness Training (BGAT) program developed by Cox and colleagues *(67)*. The BGAT program focuses on improving the accuracy of patients' detection and interpretation of relevant blood glucose symptoms and other internal and external cues. Prospective, controlled studies including long-term follow-up indicate that training in BGAT results in improved accuracy of recognition of current blood glucose levels, improved detection of hypoglycemia in individuals with hypoglycemia unawareness, improved judgment regarding when to treat low blood glucose levels, reduced occurrence of severe hypoglycemia, improved judgment about not driving while hypoglycemic, and reduction in rate of motor vehicle violations, and better-preserved counterregulatory hormonal response during intensive insulin treatment. This intervention approach may also be appropriate for individuals who are fearful of hypoglycemia *(67, 70)*. Overall, intensive training to promote blood glucose awareness can have significant and sustained benefits and aid in more consistent dietary management of older adults who take insulin.

16.10 DIABETES LIFESTYLE CHANGE AND PHYSICAL LIMITATIONS

Diabetes-related comorbidities such diabetic retinopathy, cardiovascular disease, peripheral vascular disease, and congestive heart failure may result in decreased usual activity and limit activities of daily living (ADLs) and instrumental activities in daily living (IADLs), including limited transportation options, limited ability to shop for food, or ability to read restaurant menus. Self-monitoring may also be influenced by reduced visual acuity. Diminished fine motor skills may also impact ability to functionally conduct a finger stick, conduct the steps necessary for using the glucometer, and read the results. Due to such limitations, many older patients may require alternative choices of meters or assistance in blood glucose testing *(71)*. Comorbid health conditions increase polypharmacy, so pill boxes and mediplanners may also be helpful in simplifying medication management. Comorbid health problems that may impact self-care and dietary interventions with older adults with diabetes are presented in Table 16.9.

Table 16.9
Comorbid medical conditions influencing of diabetes dietary treatment

Comorbid diabetes-related medical conditions impacting treatment
Microvascular disease
Cardiovascular disease and related risk factors
Visual impairment
Limited mobility/frailty
Nutritional status
Obesity
Underweight/malnourished

16.11 SELF-MANAGEMENT BEHAVIORS

Few studies of diabetes self-management education (DSME) have focused on older adults. Intervention guidelines have been based on expert consensus rather than an empirical evidence base *(72)*. Some older adults may have limited knowledge and/or understanding of diabetes care. Self-management steps such as home-based blood glucose monitoring, planning for meals and adjusting food intake based on blood glucose levels, and when and how to adjust insulin or take oral diabetes agents may be influenced by cognitive functioning, physical status and personal preferences, and resources. Physical barriers to optimal diabetes dietary intake may include swallowing difficulties, poor dentition, decreased thirst or appetite, and influence of medications on taste. Psychosocial influences may include limited financial resources, difficulties with transportation, and limited social support. Given the variability in the older adult population, it has been argued that individual self-management intervention may be preferable to group-based intervention. However, several studies have demonstrated positive outcomes with group intervention with older adults with diabetes. A simplified self-care regimen is optimal *(72)*.

Provider delivery approaches to DSME to optimize care for older adults may be seen in Table 16.10.

Table 16.10
Provider delivery approaches to DSME to optimize care for older adults

Spread self-management education contact over multiple sessions.
Provide large print handouts and cues for use at home to facilitate learning and retention.
Simplify steps of the self-care regimen.
Engage caregivers and family members to enhance self-management education.
Factor in impact on day to day quality of life when considering goal of tight glycemic control.
Prioritize patients' personal preferences in overall well-being when making care-related decisions.

16.11.1 Considering Diabetes Self-Management and Dietary Guidelines Within the Context of Older Adults' Lives

The reader will note that while published guidelines are available to assist the health provider in setting general goals for metabolic control and diet and medication diabetes self-care regimens, these recommendations must of course be considered in the context of what older adults are actually doing in terms of their diabetes diet care and in the context of the demands of their day-to-day lives. The ADA guidelines, developed to be practical for the general population, recommend consideration of comorbid health problems and the specific needs of older adults, in developing dietary interventions. However, the domains of these diet-related needs are not specified and strategies for adapting the recommendations to fit the special needs of older adults with diabetes are not presented. It follows that few systematic diabetes nutrition and lifestyle change studies have included older persons (73). However, several controlled studies have been conducted in recent years and will be reviewed in sections below. Lifestyle issues influencing dietary behavior in older adults and shaping the adaptation of dietary goal and interventions to best meet the needs of older adults with diabetes will also be discussed.

16.11.2 Dietary Habits of Older Adults with Diabetes

Of the few systematic studies of dietary intake in older adults with diabetes, both the type of food consumed and the pattern of eating behaviors emerge as important influences on nutritional intake. An Australian study of adults over 65 years of age found that both diabetic and nondiabetic age-matched subjects' typical dietary habits exceeded recommended levels of dietary fat and provided inadequate fiber. Only 6% of the diabetic subjects consumed a diet with at least 50% carbohydrate and less than 30% fat (74). This suggests that recommendations to increase fiber and decrease fat may be appropriate for the majority of older adults with diabetes.

Persons with diabetes must follow a diet that incorporates healthy food choices and spacing of meals to be consistent with exogenous insulin use and physical activity, with the goal of maintaining euglycemia. Unhealthful snacking is one area that threatens optimal nutrition in older adults. A random telephone survey of 335 community-dwelling adults aged 55 and older residing in the continental USA found that the 98% of older adults reported snacking at least once each day, with evening being the most common time for snacking and nearly all snacking occurring at home (75). Taste outranked nutrition as a snack selection criteria. Fruits were popular but were chosen less often than other less healthful snacks. This suggests that the context of snacking is an important consideration influencing choices made when snacking and should be addressed when providing nutritional guidance to older patients with diabetes. Concrete suggestions for replacing highly processed, high-fat snack foods with fruits and vegetables and other nutritious snacks may assist older adults in selection of healthier snacks such as string cheese, sliced apples, or sugar-free gelatin (see reference (76) for an itemized list of practical snack foods to integrate into a diabetes diet). This may be accomplished by asking patients to generate a list of problematic meal and snack foods along with a list of healthy items that they personally deem to be healthy and tasty alternatives.

Patients may be encouraged to incorporate these alternatives into their shopping lists and keep them in the home as replacements for preferred but unhealthy items. Evening activities that are alternatives to snacking (e.g., walking, crafts) may also be encouraged.

16.12 PSYCHOSOCIAL AND BEHAVIORAL ISSUES RELATED TO SELF-CARE AND DIETARY INTAKE IN OLDER ADULTS WITH DIABETES

16.12.1 Depression

There has been a dramatic increase in the number and quality of studies of depression and diabetes in the past decade (77). A systematic review of the published literature indicates a mean prevalence rate of depression in 23.4% of persons with diabetes (versus 14.5% in controls). Further, it appears that adults with diabetes who are depressed have average HbA1c levels 0.5–1.0% higher relative to those who are nondepressed.

The depression–diabetes link is particularly salient for older adults. Data from the Epidemiological Catchment Area study of more than 18,000 adults conducted in five sites found depressive symptoms in 15% of adults over 65 years of age and lifetime rate of depression in 2% of women and 3% of men (78). The prevalence of depression in primary care geriatric clinic populations is estimated to be about 5%. In nursing homes, estimates of depression have ranged from 15 to 25% at any given point, with an incidence of 13% per year (79, 80). In older adults with diabetes, rates of depression are likely even higher. In a study of Medicare claimants, major depression was significantly more prevalent in those with diabetes and those with both diabetes and depression had more treatment visits, more time in inpatient stays and had higher medical costs, even after excluding mental health treatment-related services (81).

Prevalence and associations with depressive symptoms were examined in adults over 65 years of age with diabetes living in rural communities in the ELDER (Evaluating Long-term Diabetes Self-Management Among Elder Rural Adults) study. Nearly 16% of the sample had depressive symptomology, regardless of ethnic group. Depressive symptoms were more common in those with lower functional status, lower income, more chronic health conditions, lower education, unmarried, and in women (82).

Data from the Hispanic Established Population for the Epidemiologic Study of the Elderly (EPESE) were used to examine the prevalence of comorbid depressive symptomology and chronic medical conditions and their influence on death rates in older Mexican Americans (83). Death rates were substantially higher when a high level of depressive symptoms was comorbid with diabetes, cardiovascular disease, hypertension, stroke, and cancer. The odds of having died among persons with diabetes with high levels of depressive symptoms were three times that of diabetics without high levels of depressive symptoms. Hence, an interaction between depression and diabetes and the prevalence of other risk factors greatly increased absolute risk of mortality.

16.12.1.1 DEPRESSION AND DIETARY INTAKE IN OLDER ADULTS

In depressed older adults, indirect self-destructive behavior, such as not eating and medication nonadherence, may be more common than overt self-harming gestures such as suicide attempts, and are associated with decreased survival. Many older adults might consider depression to be a normal part of aging and may not report their symptoms to a healthcare provider. Health providers may also attribute some depressive symptoms to old age or other physical ailments or mood disturbance may be less prominent than multiple somatic complaints. Some older patients with depression may present with "failure to thrive" rather than specific complaints (84). In a large HMO study of 4,463 patients with diabetes with an average age of 63 years, major depression was associated with poor adherence to self-care reflected by poor diet, inactivity, and lower medication adherence. No differences between depressed and nondepressed participants were observed for preventive health behaviors such as screenings for retinopathy and microalbumin and foot care and blood glucose self-monitoring (85).

Given the high rates of depression in this population, careful assessment of depressive symptomology and its impact on dietary intake, related aspects of diabetes self-care, and health outcomes are critical.

16.12.1.2 DEPRESSION INTERVENTION IN DIABETES

Recidivism of depression in persons with diabetes may be relatively higher than in the general population (86). Following treatment for depression, non-remission appears to be associated with lower adherence to blood glucose monitoring, higher HbA1c levels, and higher body weight in adults with type 2 diabetes (87).

A multicenter, double-blind, placebo-controlled trial examined depression treatment in a maintenance treatment trial of adults with diabetes who had participated in a sertraline treatment trial and had achieved depression recovery. Participants aged 55 and younger and those over 55 years of age were compared with regard to differences in time to depression recurrence. Younger participant group had significant increase in time to recurrence with sertraline; however, the older group showed no intervention effect, due to a high placebo response rate. This suggests that older persons with diabetes may have unique issues with regard to optimal approaches for long-term management of depression (88).

In the Pathways Study, a randomized trial of depression focused on collaborative care in patients with diabetes. Participants received depression intervention consisting of pharmacotherapy, problem-solving treatment, or both, or routine primary care. Symptoms of depression were reduced but HbA1c did not show improvement (89). The enhanced depression care did not result in improvements in diabetes self-care behaviors including optimal dietary intake, physical activity, smoking cessation, and adherence to medications (90). Results were similar for both younger and older (over 60 years of age) study participants (91).

In summary, depression places older adults with diabetes at risk for relatively poor self-care and even increased morbidity and mortality. Risk for recidivism is high. Interventions have been found to help with mood management; however, impact on self-care behavior on metabolic control may be limited.

16.12.2 Social Support

Social isolation is an established risk factor for morbidity and mortality in numerous disease states, with the largest body of literature linking it to CVD, a common outcome of diabetes *(92)*. One avenue in which social support may impact outcomes in chronic diseases such as diabetes is through its impact on self-care behaviors. Recently widowed persons may have limited cooking skills or access to shopping. Such persons may also be depressed and withdraw from usual daily activities, including social, food-related activities such as dining out or even preparing regular meals.

The relationship between social support and mortality in older adults with diabetes was examined using data from the second cohort of the Longitudinal Study of Aging. A sample of 1,430 participants over 70 years of age with diabetes was included in analyses. Social support assessment reflected connection with individuals including relatives, friends, and neighbors as well as community support including social events, church, and senior centers. Participants with the highest levels of social support had the highest survival rate, with a 55% lower risk of death compared to those with low level of social support. Those with a medium level of support had a 41% lower risk of death. The group with the lowest social support level had the lowest survival rates. Possible mediators of the social support mortality associations were also examined. Both mental and physical factors were implicated. Participants with higher social support reported less depressed mood, less difficulty with stress, fewer bad days, and fewer limitations in activities of daily living, and better health ratings. They were also less likely to have heart disease and had a shorter duration of diabetes *(93)*.

Kuo and colleagues analyzed baseline and 2-year follow-up data from the Medicare Health Outcomes Survey to examine how physical and emotional health limitations on participation in social activities were related to future disability and death in older persons with diabetes. This focus was in contrast to a focus on financial and other non-health-related influences on social participation. Participants were 8,949 adults aged 65 and older with no ADL disability at baseline. Greater social functioning was associated with less disability and death, with an 18% less change of any ADL disability and 12% lower chance of death for each 10-point increase in social functioning, even after adjusting for demographics, comorbidities, depression, and general health. The authors concluded that disengagement from social activities due to health could be an early warning sign for later disability in ADL as well as death *(94)*.

Interestingly, studies of the impact of social support and health behavior in persons with diabetes indicate substantial gender differences. Higher levels of social support have been associated with improved glycemic control in women with type 2 diabetes. However, several studies have found that a high level of perceived support is associated with less diabetes control in men, and it may be that forms of support that are satisfactory to men may reinforce patterns of eating, drinking, and exercise that are inconsistent with optimal diabetes self-care *(95)*. In older adults who continue to live with a spouse, husband's food preference, regardless of nutritional content, is often the best predictor of family meals that are eaten *(96)*. This indicates that not only is social isolation an important influence on dietary intake, but day-to-day social support and interactions among family and friends may both positively and negatively impact dietary intake in older persons with diabetes. Thus, it is imperative to evaluate the social context of patients' food purchases and dietary intake patterns.

Diabetes education programs that include the older adult's spouse may help to promote optimal social support for healthful dietary and lifestyle changes. Substantial improvements were found for diabetes knowledge, psychosocial functioning, and metabolic control in a 6-week diabetes education program for male diabetes patients aged 65–82 years of age that included their spouses *(97)*.

These studies highlight that to provide optimal care, one must consider the social support system, provide the older adult with encouragement and reinforcement for self-care behaviors, and provide instrumental assistance and technical advice as needed. Transportation assistance to the grocery and medical appointments and assistance with food preparation, as well as financial assistance, also should be considered *(71, 98)*. Social support resources could also consider community-based options such as church groups. Church can be helpful in diabetes care as shown in study with older African-American women *(99)* or adult day centers, public health department wellness programs such as walking groups, supervised exercise programs such as the Active Older Adults programs offered at many YMCAs, or home-based nursing care or other interdisciplinary professionals to create linkages. Examples of useful resource options have been compiled in a resource guide published through the American Association of Diabetes Educators *(100)*. which is listed in the Appendix to this chapter.

16.12.3 Cognitive Dysfunction

Both cross-sectional and prospective studies associate diabetes with cognitive decline in older adults *(101)*. Cognitive function is an important consideration since impaired function can impact comprehension of health-related information, including instructions regarding self-care such as procedures for conducting blood glucose testing, making adjustments in dietary intake, and taking multiple medications *(102)*. A number of studies have indicated that older adults with both types 1 and 2 diabetes have impaired cognitive function compared with age-matched groups, and the level of impairment worsens with increases in hyperglycemia. Tightened blood glucose control, even in the absence of normoglycemia can result in cognitive improvements in older persons with diabetes *(103, 104)*.

A prospective, cohort design study of 6-year follow-up of nearly 10,000 Caucasian women over 65 years of age who were prospectively tracked in the Study of Osteoporotic Fractures, examined the association of diabetes, illness duration, and cognitive decline during follow-up. Women with diabetes had lower performance at baseline on all tests of cognitive function, with up to a twofold increased risk of cognitive impairment. A 74% increased risk of cognitive decline was noted, along with more rapid decline and greater likelihood of major cognitive decline, after controlling for major confounders. Participants who had diabetes for a duration of 15 years or more had a risk of major cognitive decline 57–114% greater than women without diabetes *(105)*.

Studies of cognitive functioning in older adults with diabetes are confounded by the influence of other comorbid medical conditions, making it challenging to distinguish the independent contributions of diabetes. Additionally, studies of cognitive function have assessed both complex and simple aspects of cognition. Despite inconsistent findings in studies comparing cognition in older adults with and without diabetes, overall it appears that diabetes is associated with impairment in more complex facets of cognitive functioning such as psychomotor efficiency and verbal memory. Influences likely impacting functioning include duration of illness, blood glucose control, and age. However, a review of studies of older adults with diabetes that examined cognitive functioning and self-management behavior found few associations. Evaluation of minor cognitive impairment and diabetes self-management task performance in a small sample of older adults also found no significant influence on performance (106, 107). At present, it remains unclear how and at what threshold cognitive impairment influences acquisition of new diabetes self-care demands.

If frequent self-monitoring and adherence to specific dietary guidelines are within the cognitive abilities of the older patient with diabetes, then attainment of tight blood glucose control may be a reasonable goal. However, intensive self-management may not always be realistic for many cognitively impaired older adults. Unfortunately, despite the potential physical benefits, intensive management and tight control may require so many day-to-day demands, that this may be difficult to practically achieve. Cognitive dysfunction can make adherence to dietary recommendations particularly difficult. For example, older adults who are cognitively impaired may not remember structured mealtimes that are coordinated with their insulin regimen. Difficulty in following a complicated meal plan, such as carbohydrate counting or using a sliding scale to match insulin units to intake, may make such treatment regimens too overwhelming to be practical. For some older individuals, a concrete, structured meal plan can help minimize ambiguity regarding their diabetes diet. See the chapter Appendix for sample resources available on-line from the American Diabetes Association. Such plans may be made in conjunction with a diabetes educator or dietitian. Helpful strategies include using cues in the environment such as regular meals, setting alarms, providing written information with large print and pictures, training videotapes, and assessing comprehension and skill by asking for demonstrations. In addition, provision of home-based caretakers or meal services may also assist the older person with cognitive impairment or significant physical disability or other barriers to obtain access to optimal nutrition that is consistent with diabetes goals.

16.12.4 Attitudes and Dietary Intake in Older Adults with Diabetes

Older patients' personal views of diabetes appear to substantially impact their levels of diabetes self-care. A study of adults over 60 years of age with type 2 diabetes found that perceptions regarding the cause of diabetes, treatment effectiveness, and seriousness of one's diabetes were all significantly associated with quality of life and negative affect. Beliefs regarding treatment effectiveness were particularly predictive of dietary intake and physical activity (108). Studies of women reflecting heterogeneous ethnic groups and socioeconomic status indicate that they conceptualize diabetes in terms of their own cognitive explanatory models, which are not necessarily congruent with the way in which health providers conceptualize diabetes *(109)*.

An interview study was conducted with adults with type 2 diabetes over 65 years of age to assess their experiences with diabetes and their health goals and practices. The mean sample age was 74 years; the majority were African American (79%) and female (57%) and had diabetes-related comorbidities. Interview synthesis indicated that this sample focused their healthcare goals outcomes from a functional and social perspective, rather than a biomedical focus. Medical professions were cited as influential (by 43% of participants); however, friends, family, and media also appeared to be critical sources of influence in shaping expectations and care goals. Participants' personal diabetes care goals predominantly centered on activities of daily living and maintaining their independence and avoiding becoming a burden to family (71%). Friends and family experiences with medical conditions and events and shaped goals (50%). Findings also suggested that these older participants did not differentiate between the relative importance of different treatment goals and the prevention of different health complications. Diet and exercise were underemphasized relative to medication management *(110)*. Effective interactions with older patients should consider their experiences and influences and focus on quality of life and patient definitions of their healthcare goals. Clearly communicating priorities in self-care tasks may facilitate adherence to medical goals and optimal care.

Many older adults may not be aware of the link between diabetes and cardiovascular disease and may benefit from enhanced education to promote awareness. A study of 1,109 adults over 45 years of age with diabetes randomly sampled from a large US mixed model managed care organization. Adults aged 65 years and older were compared to those 45–64 years of age. Older adults had longer duration of diabetes and higher rates of heart disease and cardiovascular comorbidities but were less likely to relate these health comorbidities to their diabetes *(111)*.

Collaborative provider–patient discussions regarding self-care, personal preferences and goals, and awareness of personal risk factors may enhance patient motivation and treatment engagement. Integrating patient values and beliefs about health and quality of life, considering available emotional and social support and financial resources, and addressing problem solving regarding goals and treatment adjustments may best promote long-term regimen adherence *(102)*.

16.12.5 Ethnic/Cultural Issues Influencing Self-Care and Dietary Intake

Research also suggests that ethnic minority, older adults, and traditionally "hard-to-reach" persons may have culturally unique health-related perspectives that are not effectively targeted by traditionally delivered health promotion interventions *(112)*. While being healthy appears to be important and a general awareness of what to do to stay healthy is evident, operational definitions of health in these populations are often somewhat different than that typically used in health promotion efforts. For example, focus group studies with underserved ethnic minorities found that a prevailing belief was that better health behaviors could build resistance to acute illnesses and keep them healthy, but that chronic diseases such as diabetes were due to fate and heredity and beyond their individual control. In general, participants did not appear to make the cognitive "link" between the chronic disease prevention and the importance of diet, physical activity, and weight control. Most participants expressed an interest in "doing better" but were not able to specify how such healthful changes might be made.

Qualitative evaluations of cultural influences on diabetes reveal the complexity of psychosocial influences on diabetes lifestyle change and why traditional health provider perspective-based dietary interventions with minority persons often fail. An interview-based study of 20 middle-aged Mexican American women with type 2 diabetes revealed that their personal understanding and interpretation of their diabetes was most heavily based on their family's experiences and on community influences *(109)*. From the participants' perspectives, the severity of their diabetes was indicated by being treated with insulin injections and the provider being vigilant, while treatment with oral medications and the perception that providers had a lax attitude was taken to mean that the diabetes was not severe. Having diabetes was also viewed as a confusing, silent illness, and provider provision of information was often viewed as insufficient. Participant comments revealed that many found that provider comments were predominantly focused on negative aspects of their behavior, were confrontational, and at times, petty or demeaning. Provider focus on positive gains to be made with behavior change, reinforcement for accomplishments, and avoiding pejorative terms (e.g., obese) may go a long way in engaging many patients and enhancing a more collaborative relationship.

The strong influence of family and culture on adherence to a diabetes diet and lifestyle change is also evident in focus group-based qualitative research with African-American women with type 2 diabetes. Factors influencing optimal diabetes diet and physical activity behaviors were evaluated in a study of 70 southern, predominantly rural African-American women, of whom 65% were aged 55 and older *(99)*. These women described the psychological impact of diabetes as being stronger than the physical impact, and the psychological issues reported included feelings of nervousness, fatigue, worrying, and having feelings of dietary deprivation, including craving for sweets. Participants reported considerable life stress other than diabetes, particularly having a multi-caregiver role. Family members complaining about and resistant toward healthy food preparation methods was common. Positive family support for diabetes was evident in the form of instrumental support from adult daughters or other female family members or friends to older, single, or widowed women. In addition, spirituality and religiosity emerged as a main theme in all groups. Spirituality was largely viewed as a primary source of emotional support, a positive influence on diabetes, and a contributor to quality of life. Church was described as an important source of social and emotional support and resource for coping. This study exemplifies the importance of incorporating family and the church in self-care behaviors of many southern African-American women.

These findings demonstrate why consideration of the social and cultural context of older adults' lives is critical for the development of interventions to promote diet and lifestyle change. Collaborative patient provider care and family-centered and church-based approaches may offer more appropriate avenues for efforts to promote optimal diabetes care and maximize the effective delivery of dietary interventions for many older adults with diabetes, who are frequently overrepresented in ethnic minority populations.

16.12.6 Quality of Life and Diabetes Diet in Older Adults

In considering prescribing a diabetes diet or when addressing issues related to dietary adherence, it is important to consider the impact of dietary change on the individual's quality of life. Diabetes itself poses numerous challenges and the many lifestyle demands are among the most difficult for patients. Both types 1 and 2 diabetes appear to have an impact on health-related quality of life *(113)*. Diabetes-specific quality of life issues are associated with overall well-being, and dietary restrictions and daily hassles related to diabetes care are significantly associated with treatment satisfaction as well as with general well-being. A liberalized diet and flexible insulin therapy are among the diabetes-related factors most associated with favorable quality of life in persons with type 1 diabetes *(114)*.

A Japanese study of diabetic adults over age 60 examined the burden related to having to make changes related to caloric restriction, dietary balance, regular dietary habits, restriction of favorite foods, and the amount of snacks and restrictions when eating out *(115)*. Women, relatively younger subjects, those with lower family support, chronic hyperglycemia, and taking oral diabetic agents reported greater a higher level of burden from the lifestyle changes required by the diabetes diet.

For some older adults with diabetes, functional status is an important care consideration. For individuals with limited cognitive function or multiple comorbidities, the primary goal of care may be achieving a satisfactory quality of life rather than aggressive treatment regimens. In other cases, for those who are robust, goals and care may be similar to those for younger persons with diabetes. The AGS guidelines for improving care of older persons address issues that are more common in the older diabetes population, including depression, pain, falls with injury, and declining functional status. Consistent with research, the guidelines suggest that goals be developed based on functional status and personal desires *(116)*. Of note,

Table 16.11
Psychosocial influences on diet and lifestyle change in older adults with diabetes

Dietary changes and quality of life
Social isolation / impact of social support
Depression
Hypoglycemia unawareness
Dementia / cognitive dysfunction
Competing priorities
Dislike of healthful foods or structured physical activity
Financial difficulty
Limited access to healthy foods or safe activity resources
Cultural norms that are not supportive of optimal diabetes self-care

studies of functional status and health outcomes in older adults with diabetes have found that persons with low functional level (defined by three or more limitations in IADLs or ADLs) did not benefit from aggressive intervention aimed at blood glucose control *(117)*.

As highlighted in the previous sections, psychosocial and functional influences play a substantial role in the dietary and lifestyle behaviors of older adults with diabetes. Such influences are summarized in Table 16.11.

16.13 THEORETICAL CONCEPTUALIZATIONS OF LIFESTYLE CHANGE IN OLDER ADULTS WITH TYPE 2 DIABETES

Lifestyle modification involves a complex series of behavior changes that consider the social and physical environment as well as cognitive and dispositional factors. Theories of health behavior change provide a framework for understanding such behaviors as well as providing a context for interventions to enhance lifestyle change. Two theories that have been shown to be efficacious in intervention trials aimed at promoting nutrition and lifestyle change in type 2 diabetes are Social Cognitive Theory (SCT) and the Transtheoretical Model (TTM). For an overview of these frameworks, see Chapter 3. These frameworks have been the basis of several major diabetes behavior change intervention trials, including the DCCT *(18)* and DPP *(61)*. References provided in the Appendix include theoretically based patient materials with clinical examples, including handouts and patient tip sheets.

16.14 BEHAVIOR CHANGE INTERVENTION STUDIES OF SELF-MANAGEMENT AND LIFESTYLE CHANGE IN OLDER ADULTS WITH DIABETES

Few studies of diabetes self-management education (DSME) have focused on older adults. Intervention guidelines have been based on expert consensus rather than on empirical evidence base *(72)*. It is unclear if strategies for DSME that are effective with younger patients are optimal for older adults. Consideration of issues already discussed in this chapter highlight the unique self-management issues facing many older adults with diabetes. Generational cohort influences may influence readiness for receipt of detailed self-care information. Many of the oldest of older adults may have limited knowledge or understanding of diabetes care. Self-management steps may be influenced by cognitive functioning, physical status, and personal preferences and resources. Physical barriers to optimal diabetes dietary intake may include swallowing difficulties, poor dentition, decreased thirst or appetite, poor dentition, influence of medications on taste. Psychosocial influences may include limited financial resources, difficulties with transportation, and limited social support. Providing DSME content over multiple contacts and use of memory cues such as large print handouts and cues to use at home can be helpful. A simplified self-care regimen is optimal with goals not only to maintain diabetes control but to have good quality of life *(72)*. Many clinicians advocate individual DSME for older adults, however, several studies have indicated that group education is effective. As can be seen from the few examples of controlled group diabetes education programs specifically developed for older adults, participants in intensive programs have tended to be younger, from the baby boomer generation, limiting generalizability to the larger geriatric diabetes population.

An example of one of the few theoretically based lifestyle change intervention programs developed specifically for older adults with diabetes is the "Sixty-some-thing..." study *(118)*. Principles of SCT were used to develop a 10-session self-management training program, with 102 adults over 60 years of age with type 2 diabetes. Subjects were randomly assigned to immediate or delayed intervention conditions. The intervention taught problem-solving skills and strategies for enhancing self-efficacy for overcoming personal barriers to adhering to their diabetes diet and other aspects of the diabetes regimen. The immediate intervention produced greater reductions in caloric intake and percent of calories from fat, greater weight loss, and increases in the frequency of blood glucose testing compared to delayed controls. Improvements were generally maintained at 6-month follow-up. Results from subjects receiving the delayed intervention closely approximated those for the immediate intervention subjects.

Successful lifestyle change was also achieved in the Mediterranean Lifestyle Program, a randomized intervention trial with 279 post-menopausal women with type 2 diabetes aimed at reducing CHD risk factors. The intervention component consisted of a 3-day retreat and 6 months of weekly meetings with an emphasis on a Mediterranean low saturated fat diet, moderate physical activity, stress management, smoking cessation, and group support. Behavior change was greater in participants receiving the intervention than in the usual care control group with greater changes in eating patterns, stress management, activity level, and smoking cessation. Intervention resulted in improvements in BMI, HbA1c, plasma fatty acids, and quality of life. Feasibility of this intensive approach to lifestyle change is an important issue for consideration; participants in the Mediterranean diet arm had a higher rate of attrition relative to those in the control condition *(119)*.

Chodosh and colleagues conducted a meta-analysis of self-management intervention studies that compared outcomes with usual care or control condition for diabetes, osteoarthritis, or hypertension. Twenty diabetes studies were examined for impact on HbA1c. Pooled effect size was –0.36 in favor of self-management intervention, indicating lower HbA1c in treatment groups. For fasting blood glucose outcomes, the pooled effect size for 14 studies was –0.28, equating to a decrease of 0.95 mmol/L in blood glucose level. No statistically significant differences were observed in weight change between intervention and control conditions in the 13 studies reporting this outcome. Only three studies primarily focused on diet and education, these had a pooled effect size of –0.062 for HbA1c. Post hoc analyses aimed at understanding key essential elements of the self-management intervention programs did not support any key program aspects *(123)*.

16.14.1 Diet-Specific Intervention in Older Adults with Diabetes

A randomized nutrition education intervention was developed for older adults with type 2 diabetes. Participants were age 65 and older (mean age 72) and without functional or cognitive impairment. The 10-week intervention was theoretically based utilizing SCT and meaningful learning approaches aimed at minimizing the presentation of too much information and maximizing learning by breaking information down into small pieces and successively adding upon each concept across the intervention. Strategies included limiting the content introduced at each session, meaningfully organizing the content, integrating pre-existing knowledge with new information, and modeling and in-session practice regarding decision-making related to reading food labels, food purchasing, meal planning, and using the information for diabetes self-care. Goal setting, self-monitoring, and feedback were also utilized. The intervention resulted in improved glycemic control *(120)*. greater total knowledge, positive outcome expectancies, decision-making skills, and reduced self-management barriers *(121)*.

16.14.2 *Weight Loss Intervention in Older Adults with Diabetes*

Weight loss issues that must be considered for older adults with diabetes include the impact of restrictions on quality of life and potential loss of lean muscle mass from decreased protein intake (see also Chapter 10).

In research undertaken with younger adults with diabetes, weight loss programs that combine diet, physical activity, and theoretically guided behavior change techniques have been shown to be the most effective over the short term *(1)*. Hypocaloric diets, which can improve glucose tolerance and lipid levels, may also be appropriate for older, obese persons with diabetes. Behavioral weight control interventions with persons with type 2 diabetes have found that even reductions of approximately 10% of weight loss can decrease hypertension and lipid abnormalities and improve glycemic control, with improvements related to the magnitude of weight loss *(1)*.

The Look AHEAD trial is the first large, multisite clinical trial with type 2 diabetes that compared intensive weight loss intervention to support and education. At the time of the writing of this chapter, 1 year results of the Look AHEAD trial were published *(122)*. This randomized controlled trial was conducted at multiple centers, with 5,145 adults aged 45–74 with type 2 diabetes. Group and individual meetings focused on intensive lifestyle change with weight loss and maintenance as the primary goal. Restriction of caloric intake was the primary approach with a goal of limiting total fat calories to 30%, with a maximum of 10% saturated fat and minimum of 15% from protein. Portion-controlled diets were prescribed as were structured meal plans. Home-based exercise with gradually increasing goals was utilized. Walking was encouraged although other moderate intensity activities could be chosen. A toolbox approach using algorithms and assessments of progress was used. After the initial 6-month intervention period, toolbox options included orlistat and/or advanced behavioral approaches for participants with difficulties in achieving goals. At 1 year follow-up, intervention participants had lost an average of 8.6% body weight, with greater weight loss, increased cardiovascular fitness, and improved cardiovascular risk factors relative to participants in the control condition. Use of glucose-lowering medications decreased in intervention participants and increased in control group participants. Use of antihypertensive medications increased in control group participants but remained unchanged in the intervention group. Use of lipid-lowering medications increased in both intervention and control groups, with smaller increases in intervention participants. Findings highlight that substantial weight loss is feasible in persons with type 2 diabetes. Notably, even participants on insulin lost an average of 7.6% of bodyweight from baseline. Follow-up of trial participants is planned for up to 11.5 years and will provide evidence regarding ongoing risk-factor improvements and maintenance of health gains *(122)*.

16.15 RECOMMENDATIONS FOR IMPARTING DIETARY INFORMATION

In order to impart diabetes diet information in a fashion that will lead to actual changes in behavior and maintenance of these changes, it is critical to consider the psychosocial and cultural influences that are present for each individual patient. Simple and concrete statements such as "eat less fat" or "eat less food" or "get more walking in each day" may promote learning and minimize failure. Nutrition information is best presented in sequenced manageable steps that can then be individualized to the patient's setting. Simple tip sheets and problem-solving approaches discussed in earlier sections may also be helpful. The National Diabetes Education Program (NDEP) has prepared materials adapted from those used in the DPP for use by primary care providers for middle age and older adults. These materials address motivational approaches that consider readiness for change, behavioral relapse. Materials can help to assess a patient's personal readiness for change and help in setting up a walking program. This toolkit, the NDEP GAMEPLAN (Goals, Accountability, Monitoring, Effectiveness, Prevention through a Lifestyle of Activity and Nutrition) is copyright free and contains healthcare provider information with background information and patient handouts that may be copied. For provider information, pre-diabetes risk factors are presented along with diagnostic decision trees; screening approaches are reviewed, DPP findings are summarized, and commonly asked primary care provider questions are addressed *(64)*. Information for downloading and telephone ordering is in the appendix of this chapter.

It is also important to be mindful of the range of functioning in older adults. Older adults of the World War II generation have tended to be characterized as somewhat reverential toward physicians and the healthcare system. However, baby boomers, who are now entering the realm of older adulthood, tend to differ from previous generations and tend to have high expectations of their health providers and desire additional information. This generation of "new" older adults tends to want a collaborative relationship with their healthcare provider and desires additional information including resources such as self-help publications, Internet, video, and audiotapes. They demand convenience, expect hard evidence of quality and expertise, can be skeptical of advice at face value, and are often willing to explore alternative therapies *(123)*. In order to meet the needs of the range of older adults with diabetes, it is clear that a "one size fits all" approach will not be effective. Rather issues related to culture and ethnicity and generational cohort must be considered.

16.16 RECOMMENDATIONS

1. Establish the type of diabetes and medication regimen in order to appropriately integrate dietary goals.
2. Consider the importance of cardiovascular risk, including obesity and lipids, in developing diabetes dietary goals and routine assessment.
3. Work with the patient to set a goal of achieving and maintaining reasonable body weight. For obese older adults, moderate weight loss may achieve dramatic results and exercise may greatly enhance dietary intervention. Maintenance of behavior change and weight loss is critical. For underweight adults, focus on promotion of optimal nutritional intake and functional status.
4. Educate older adults with diabetes about the rationale for diet and lifestyle change and link to health outcomes; promote self-efficacy for change.
5. Consider the risk of hypoglycemia for older adults taking insulin – particularly those with poor nutritional status, cognitive dysfunction, polypharmacy, and comorbid illness. Encourage frequent self-monitoring and dietary self-treatment and preventive strategies.
6. Assess older adults' specific dietary patterns such as food choices, quantity eaten, and unplanned snacking and the lifestyle contexts in which they occur.
7. Address psychosocial issues that may influence dietary intake, including depression, social support, cognitive status, attitudes and perceptions, and the impact of the diabetes regimen on quality of life.
8. Address and intervene within individuals' cultural context, including family influences and church, when appropriate.
9. Provide a collaborative relationship with each patient, offer resources, and provide concrete, behavioral strategies to promote behavior change.
10. When appropriate, provide self-help materials including Internet and National Diabetes Education Program and ADA and AADE resources (see Appendix).

APPENDIX

Internet Resources

- American Diabetes Association: www.ada.org
- American Association of Diabetes Educators (includes "find an educator" service locator): www.aadenet.org
- Centers for Disease Control and Prevention Section on Diabetes: http://www.cdc.gov/diabetes/
- National Institutes of Health – Institute of Diabetes and Digestive and Kidney Diseases: http://www2.niddk.nih.gov/
- NDEP GAMEPLAN (Goals, Accountability, Monitoring, Effectiveness, Prevention through a Lifestyle of Activity and Nutrition) toolkit: http://ndep.nih.gov/diabetes/pubs/GP_Toolkit.pdf800-438-5383

Books/Videos (Books from the American Diabetes Association may be ordered at 1-800-ADA-ORDER or online at www.diabetes.org/shop-for-books-and-gifts.jsp

Books on meal planning and nutrition, including structured meal suggestions and recipes are available through the ADA at "http://www.diabetes.org/shop-for-books-and-gifts/meal-planning-nutrition.jsp" (sample titles that may appeal to older adults include "Meal planning made easy", "Month of meals – All-American fare" and "Month of Meals – Old-Time Favorites").

- Pastors, JG; Arnold, MS, Daly, MS, Franz, M.,Warshaw, H.S. *Diabetes Nutrition Q&A for Health Professionals.* American Diabetes Association, Alexandria: VA, 2003.
- Anderson RM; Barr PA; Funnell MM; Arnold MS; Edwards GJ; Fitzgerald JT. *Living With Diabetes: Challenges in the African American Community.* American Diabetes Association, Alexandria: VA, 2000. (includes video vignettes)
- BJ. Anderson and RR Rubin (Eds). *Practical Psychology for Diabetes Clinicians, 2nd Ed.* Alexandria VA, American Diabetes Association, 2003.
- Mensing, C. (Ed.). *The Art and Science of Diabetes Self-Management Education: A Desk Reference for Healthcare Professionals.* Chicago, American Association of Diabetes Educators. 2006.

REFERENCES

1. Wing RR GM, Acton KJ, Birch LL, Jakicic JM, Sallis JF Jr, Smith-West D, Jeffery RW, Surwit RS. Behavioral science research in diabetes: lifestyle changes related to obesity, eating behavior, and physical activity. Diabetes Care 2001;24(1):1–2.
2. Seidell J. Obesity, insulin resistance and diabetes—a worldwide epidemic. Br J Nutr 2000;83(Suppl 1):S5–8.
3. Visscher TL SJ. The public health impact of obesity. Annu Rev Public Health 2001;22:355–75.
4. King H AR, Herman WH. Global burden of diabetes 1995–2025: prevalence, numerical estimates, and projections. Diabetes Care 1998;21(9):1414–31.
5. Ratner R. Type 2 diabetes mellitus: the grand overview. Diabetic Med 1998;15(Suppl 4):S4–7.
6. Shorr RI FL, Resnick HE, DiBari M, Johnson KC, Pahor M. Glycemic control of older adults with type 2 diabetes: Findings from the Third National Health and Nutrition Examination Survey, 1988–1994. J Am Geriatr Soc 2000;48(3):264–7.
7. Howard BV MM. Diabetes and cardiovascular disease. Curr Atheroscler Rep 2000;2(6):476–81.
8. Report of the Expert Committee on the diagnosis and classification of diabetes mellitus. Diabetes Care 1997;20:1183–97.
9. Organization WH. Definition, Diagnosis and Classification of Diabetes Mellitus and Its Complications: Report of a WHO Consultation. Part 1, Diagnosis and Classification of Diabetes Mellitus (WHO/NCD/NCS/99.2). . Geneva: World Health Org; 1999.
10. The Expert Committee on the Diagnosis and Classification of Diabetes. Follow-up report on the diagnosis of diabetes mellitus. Diabetes Care 2003;26:3160–7.
11. Falorni A KI, Sanjeevi CB, Lernmark A. Pathogenesis of insulin-dependent diabetes mellitus. Baillieres Clin Endocrinol Metab 1995;9(1):25–46.
12. Morley J. An overview of diabetes mellitus in older persons. Clin Geriatr Med 1999;15(2):211–24.
13. Dagogo-Jack S SJ. Pathophysiology of Type 2 Diabetes and Modes of Action of Therapeutic Interventions. Arch Int Med 1997;157(16):1802–17.
14. Weyer C BC, Mott DM, Pratley RE. The natural history of insulin secretory dysfunction and insulin resistance in the pathogenesis of type 2 diabetes mellitus. J Clin Invest 1999;104(6):787–94.
15. Elahi D MD. Carbohydrate metabolism in the elderly. Eur J Clin Nutr 2000;54 (Suppl 3):S112–20.
16. Beaufrere B MB. Fat and protein redistribution with aging: metabolic considerations. Eur J Clin Nutr 2000;54(Suppl 3):S48–53.
17. Chiu KC LN, Cohan P, Chuang LM Beta cell function declines with age in glucose tolerant Caucasians. Clin Endocrinol (Oxf) 2000;53(5):569–75.
18. The Diabetes Control and Complications Trial Research Group. N Engl J Med 1993 329(14):977–86.

19. Turner RC HR. Lessons from UK prospective diabetes study. Diabetes Res Clin Pract 1995;28 S151–7.

20. Brown AF MC, Saliba D, Sarkisian CA. Guidelines for improving the care of the older person with diabetes mellitus. J Am Geriatr Soc 2003;51(5 suppl):S265–80.

21. Adler AI SI, Neil HA, Yudkin JS, Matthews DR, Cull CA, Wright AD, Turner RC, Holman RR. Association of systolic blood pressure with macrovascular and microvascular complications of type2 diabetes (UKPDS 36): prospective observational study. BMJ 2000;321(7258):412–9.

22. Hansson L ZA, Carruthers SG, Dahlof B, Elmfeldt D, Julius S, Menard J, Rahn KH, Wedel H, Westerling S. Effects of intensive blood-pressure lowering and low-dose aspirin in patients with hypertension: principal results of the Hypertension Optimal Treatment (HOT) randomized trial. HOT study group. Lancet 1998;351(9118):1755–62.

23. Villarosa IP BG. The Appropriate Blood Pressure Control in Diabetes (ABCD) Trial. J Hum Hypertens 1998;12(9):653–5.

24. Goldberg RB, Mellies, M.J., Sacks, F.M., Moye, L.A., Howard, B.V., Howard, W.J., Davis, B.R., Cole, T.G., Pfeffer, M.A., Braunwald, E. The CARE Investigators. Cardiovascular events and their reduction with pravastatin in diabetic and glucose-intolerant myocardial infarction survivors with average cholesterol levels: subgroup analyses in the cholesterol and recurrent events (CARE) trial. Circulation 1998;8(98):2513–9.

25. Haffner SM. The Scandinavian Simvastatin Survival Study (4S) subgroup analysis of diabetic subjects: Implications for the prevention of coronary heart disease. Diabetes Care 1997;20(4):469–71.

26. Helmy T PA, Alameddine F, Wenger NK. Management strategies of dyslipidemia in the elderly: Review. Arch Gen Med 2005;7(4):8.

27. Sever PS DB, Poulter NR, et al. Prevention of coronary and stroke events with atorvastatin in hypertensive patients who have average or lower-than-average cholesterol concentrations, in the Anglo-Scandinavian Cardiac Outcomes Trial – Lipid Lowering Arm (ASCOT-LLA): a multi-centre randomised controlled trial. Lancet 2003;361:1149–58.

28. Shepherd J BG, Murphy MB, et al. Pravastatin in elderly individuals at risk of vascular disease (PROSPER): a randomised controlled trial. Lancet 2002;360:1623–30.

29. Cannon CP BE, McCabe CH, et al. Intensive versus moderate lipid lowering with statins after acute coronary syndromes. N Engl J Med 2004;350: 1495–504.

30. LaRosa JC GS, Waters DD, et al. Treating to New Targets (TNT) Investigators. Intensive lipid lowering with atorvastatin in patients with stable coronary disease. N Engl J Med 2005;352:1425–35.

31. Group. HPSC. MRC/BHF Heart Protection Study of cholesterol lowering with simvastatin in 20,536 high-risk individuals: a randomised placebo-controlled trial. Lancet 2002;360:7–22.

32. Downs JR CM, Weis S, et al. Primary prevention of acute coronary events with lovastatin in men and women with average cholesterol levels: results of AFCAPS/TexCAPS. Air Force/Texas Coronary Atherosclerosis Prevention Study. JAMA 1998;279:1615–22.

33. Shepherd J CS, Ford I, et al. Prevention of coronary heart disease with pravastatin in men with hypercholesterolemia. West of Scotland Coronary Prevention Study Group. N Engl J Med 1995;333:1301–7.

34. Sacks FM PM, Moye LA, et al. The effect of pravastatin on coronary events after myocardial infarction in patients with average cholesterol levels. Cholesterol and Recurrent Events Trial investigators. N Engl J Med 1996;335:1001–9.

35. Hunt D YP, Simes J, et al. Benefits of pravastatin on cardiovascular events and mortality in older patients with coronary heart disease are equal to or exceed those seen in younger patients: results from the LIPID trial. Ann Intern Med 2001;134:931–40.

36. listed Na. Randomised trial of cholesterol lowering in 4444 patients with coronary heart disease: the Scandinavian Simvastatin Survival Study (4S). Lancet 1994;344:1383–9.

37. Garg A BA, Grundy SM, Zhang ZJ, Unger RH. Comparison of a high-carbohydrate diet with a high-monosaturated fat diet in patients with non-insulin-dependent diabetes mellitus. N. Engl J Med 1988;319(13):829–34.

38. Halton TL HF. The effects of high protein diets on thermogenesis, satiety and weight loss: A critical review. J Am College Nutr 2004;23:373–85.
39. Salahudeen AK. Halting progression of renal failure: consideration beyond angiotensin II inhibition. Nephrol Dial Transplant 2002;17(11):1871–5.
40. Sheard NF CN, Brand-Miller JC, Franz MJ, Pi-Sunyer, FX, Mayer-Davis E, Kulkarni K, Geil P. Dietary Carbohydrate (Amount and Type) in the Prevention and Management of Diabetes: A statement by the American Diabetes Association. Diabetes Care 2004;27(9):2266–71.
41. Jovanovic L GB. Type 2 diabetes: the epidemic of the new millennium. Ann Clin Lab Sci 1999;29(1):33–42.
42. Locher JL, Roth, D.L., Ritchie, C.S., Cox, K., Sawyer, P., Bodner, E.V., Allman, R.M. Body mass index, weight loss, and mortality in community-dwelling older adults. J Gerontol Ser A Biol Sci Med Sci 2007;62:1389–92.
43. Mooradian AD OD, Petrawek D, Morley JE. Diabetes Mellitus in elderly nursing home patients. J Am Geriatric Soc 1988;36:391–6.
44. Garber A.J. MES, Bransome E.D. Jr., et al. American College of Endocrinology position statement on inpatient diabetes and metabolic control. Endocr Pract 2004;10:77–82.
45. Clement S. BSS, Magee M.F., Ahmann A., Smith E.P., Schafer R.G., Hirsch I.B. The American Diabetes Association Diabetes in Hospitals Writing Committee. Management of diabetes and hyperglycemia in hospitals. Diabetes Care 2004;27:553–91.
46. Inzucchi SE. Management of Hyperglycemia in the Hospital Setting. New Engl J Med 2006;355(18):1903–11.
47. Swift C.S. BJL. Nutrition therapy for the hospitalized patient with diabetes. Endocr Pract 2006;12(Suppl 3):61–7.
48. Boucher J.L. SCS, Franz M.J., Kulkarni K., Schafer R.G., Pritchett E., Clark N.G. Inpatient management of diabetes and hyperglycemia: implications for nutrition practice and the food and nutrition professional. J Am Diet Assoc 2007;107(1):105–11.
49. Association AD. Diabetes nutrition recommendations for health care institutions (Position Statement). Diabetes Care 2004;27 (Suppl. 1):S55–7.
50. Wright J. Total parenteral nutrition and enteral nutrition in diabetes. Curr Opin Clin Nutr Metab Care 2000;3:5–10.
51. Garg A. High-MUFA diets for patients with DM: a meta-analysis. Am J Clin Nutr 1998;67(Suppl 3):577S–82S.
52. Hongsermeier T BB. Evaluation of a practical technique of determining insulin requirements in diabetic patients receiving total parenteral nutrition. J Parenter Enter Nutr 1993;17:16–9.
53. Coulston A.M. MD, Reaven G.M. Dietary management of nursing home residents with non-insulin-dependent diabetes mellitus. Am J Clin Nutr 1990;51:67–71.
54. Tariq SH KE, Thomas DR, Thomson K, Philpot C, Chapel DL, Morley JE. The use of a no-concentrated-sweets diet in the management of type 2 diabetes in nursing homes. J Am Diet Assoc 2001;101:1463–6.
55. Expert Panel on Detection E, and Treatment of High Blood Cholesterol in Adults. Executive summary of the third report of the National Cholesterol Education Program (NCEP) Expert Panel on Detection, Evaluation, and Treatment of High Blood Cholesterol in Adults (Adult Treatment Panel III). JAMA 2001;285:2486–97.
56. Ford ES GW, Dietz WH. Prevalence of the metabolic syndrome among US adults: findings from the third National Health and Nutrition Examination Survey. JAMA 2002;287:356–9.
57. Stern MP WK, Gonzalez-Villalpando C, Hunt KJ, Haffner SM. Does the metabolic syndrome improve identification of individuals at risk of type 2 diabetes and/or cardiovascular disease? Diabetes Care 2004;27:2676–81.
58. Ninomiya JK LIG, Criqui MH, Whyte JL, Gamst A, Chen RS. Association of the metabolic syndrome with history of myocardial infarction and stroke in the Third National Health and Nutrition Examination Survey. Circulation 2004;109:42–6.
59. Lorenzo C OM, Williams K, Stern MP, Haffner SM. The metabolic syndrome as predictor of type 2 diabetes: the San Antonio Heart Study. Diabetes Care 2003;26:3153–9.

60. Kahn R BJ, Ferrannini E, Stern M. The metabolic syndrome: time for a critical appraisal: joint statement from the American Diabetes Association and the European Association for the Study of Diabetes. Diabetes Care 2005;28:2289–304.
61. Group. DPPR. Reduction in the incidence of type 2 diabetes with lifestyle intervention or metformin.. N Engl J Med 2002;346:393–403.
62. Crandall J, Schade, D, Ma, Y, Fujimoto, WY, Barrett-Conner, E, Fowlder, S, Dagogo-Jack, S, Andres, R. for the Diabetes Prevention Program Research Group. The influence of age on the effects of lifestyle modification and metformin in prevention of diabetes. J Gerontol 2006;61A (10):1075–81.
63. Wannamethee SG, Shaper, A.G., Whincup, P.H. Modifiable lifestyle factors and the metabolic syndrome in older men: Effects of lifestyle changes. J Am Geriatr Soc 2006;S4:1909–14.
64. Kelly JM, Marrero, D.G., Gallivan, J., Leontos, C., Perry, S. Diabetes prevention: A GAME-PLAN for success. Geriatrics 2004;59(7):26.
65. DREAM (Diabetes REduction Assessment with ramipril and rosiglitazone Medication) Trial Investigators GH, Yusuf S, Bosch J, Pogue J, Sheridan P, Dinccag N, Hanefeld M, Hoogwerf B, Laakso M, Mohan V, Shaw J, Zinman B, Holman, RR Effect of rosiglitazone on the frequency of diabetes in patients with impaired glucose tolerance or impaired fasting glucose: a randomised controlled trial. Lancet 2006;368(9541):1096–105.
66. Janssen MM SF, de Jongh RT, Casteleijn S, Deville W, Heine RJ. Biological and behavioural determinants of the frequency of mild, biochemical hypoglycaemia in patients with type 1 diabetes on multiple injection therapy. Diabetes Metab Res Rev 2000;16(3):157–63.
67. Cox DJ G-FL, Polonsky W, Schlundt D, Kovatchev B, Clarke W. Blood Glucose Awareness Training (BGAT-2): Long-term benefits. Diabetes Care 2001;24(4):637–42.
68. Meeking DR, Cavan, D.A. Alcohol ingestion and glycemic control in patients with insulin-dependent diabetes mellitus. Diabet Med 1997;14(4):279–83.
69. Bolli GB. How to ameliorate the problem of hypoglycemia in intensive as well as nonintensive treatment of type 1 diabetes. Diabetes Care 1999;22(Suppl 2):B43–52.
70. Wild D, von Maltzahn, R., Brohan, E., Christensen, T., Clauson, P., Gonder-Frederick, L. A critical review of the literature on fear of hypoglycemia in diabetes: Implications for diabetes management and patient education. Patient Educ Couns 2007;68:10–5.
71. American Association of Diabetes Educators. Special considerations for the education and management of older adults with diabetes. Diabetes Educator 2000;26(1):37–9.
72. Suhl E, Bonsignore, P. Diabetes self-management education for older adults: General principles and practical application. Diabetes Spectrum 2006;19(4):234–40.
73. Strano-Paul L, Phanumas, D. Diabetes management. Analysis of the American Diabetes Associations clinical practice recommendations. Geriatrics 2000;55(4):57–62.
74. Horwath CC, Worsley, A. Dietary habits of elderly persons with diabetes. J Am Diet Assoc 1991;91(5):553–7.
75. Cross AT, Babicz, D., Cushman, L.F. Snacking habits of senior Americans. J Nutr Elder 1995;14(2–3):27–38.
76. James-Enger K. Express lane. Diabetic cooking snack savvy. Diabetes Forecast 2002;55(6):73–5.
77. Lustman PJ, Penckofer, S.M., Clouse, R.E. Recent advances in understanding depression in adults with diabetes. Current Diabetes Report 2007;4:114–22.
78. Fombonne E. Increased rates of depression: Update of epidemiological findings and analytic problems. Acta Psychiatrica Scandinavica 1994;90:145–56.
79. Consensus Panel. Diagnosis and treatment of depression in late life. JAMA 1992;268(8):1018–24.
80. Slater SL, Katz, I.R. Prevalence of depression in the aged: Formal calculations versus clinical facts. J Am Geriatr Soc 1995;43:778–9.
81. Finkelstein EA, Bray, J.W., Chen, H., Larson, M.J., Miller, K., Tompkins, C., Keme, A., Manderscheid, R. Prevalence and costs of major depression among elderly claimants with diabetes. .Diabetes Care 2003;26:415–20.
82. Bell RA, Smith, SL, Arcury, TA, Snively, BM, Stafford, JM, Quandt, SA. Prevalence and correlates of depressive symptoms among rural older African Americans, Native Americans, and Whites with diabetes. Diabetes Care 2005;28(4):823–9.

83. Black SA, Markides, K.S. Depressive symptoms and mortality in older Mexican Americans. Ann Epidemiol 1999;9:45–52.
84. Sarkisian CA, Lachs, M.S. "Failure to thrive" in older adults. Ann Int Med 1996;124:1072–8.
85. Lin EHB, Katon, W, Von Korff, M, Rutter, C, Simon, GE, Oliver, M, Ciechanowski, P, Ludman, EJ, Bush, T, Young, B. Relationship of depression and diabetes self-care, medication adherence, and preventive care. Diabetes Care 2004;27(9):2154–60.
86. Lustman P, Griffith, L, Freedland, K, Clouse, R. The course of major depression in diabetes. Gen Hos Psychiat 1997;19:138–43.
87. Lustman P, Freedland, K, Griffith, LS, Clouse, RE. Predicting response to cognitive behavior therapy of depression in type 2 diabetes. Gen Hosp Psychiat 1998;20:302–6.
88. Williams MM, Clouse, RE, Nix, BD, Rubin, EH, Sayuk, GS, McGill, JB, Gelenberg, AJ, Ciechanowski, PS, Hirsch, IB, Lustman, PJ. Efficacy of sertraline in prevention of depression recurrence in older versus younger adults with diabetes. Diabetes Care 2007;30(4):801–6.
89. Katon W, Von Korff, M, Ciechanowski, P, Russo, J, Lin, E, Simon, G, Ludman, E, Walker, E, Bush, T, Young, G. Behavioral and clinical factors associated with depression among individuals with diabetes. Diabetes Care 2004;27(4):914–20.
90. Lin EH KW, Rutter C, Simon GE, Ludman EJ, Von Korff M, Young B, Oliver M, Ciechanowski PC, Kinder L, Walker E. Effects of enhanced depression treatment on diabetes self-care. Ann Fam Med 2006;4(1):46–53.
91. Williams JW, Katon, W, Lin, EHB, Noel, PH, Worchel, J, Cornell, J, Harple, L, Fultz, BA, Junkeler, E, Mka, VA, Unutzer, J. for the IMPACT Investigators. The effectiveness of depression care management on diabetes-related outcomes in older patients. Ann Int Med 2004;140(12):1015–25.
92. Orth-Gomer K. International epidemiological evidence for a relationship between social support and cardiovascular disease. In: S.A. Shumaker SMC, ed. Social Support and Cardiovascular Disease. New York: Plenum Press; 1994.
93. Zhang X, Norris, SL, Gregg, EW, Beckles, G. Social support and mortality among older persons with diabetes. Diabetes Educator 2007;33(2):273–81.
94. Kuo YF, Raja, MA, Peek, MK, Goodwin, JS. Health-related social disengagement in elderly diabetic patients. Association with subsequent disability and survival. Diabetes Care 2004;27(7):1630–7.
95. Kaplan RM, Hartwell, SL. Differential effects of social support and social network on physiological and social outcomes in men and women with type II diabetes mellitus. Health Psychol 1987;6:387–98.
96. Weidner G, Healy, AB, Matarazzo, J.D. Family consumption of low fat foods: Stated preference versus actual consumption. J Appl Soc Psychol 1985;15:773–9.
97. Gilden JL, Hendryz, M, Casia, C, Singh, SP. The effectiveness of diabetes education programs for older patients and their spouses. J Am Geriatr Soc 1989;37(11):1023–54.
98. Haas LB. Caring for community-dwelling older adults with diabetes: Perspectives from healthcare providers and caregivers. Diabetes Spectrum 2006;19(4):240–4.
99. Samuel-Hodge CD, Headen, SW, Skelly, AH, Ingram, AF, Keyserling, TC, Jackson, EJ, Ammerman, AS, Elasy, TA. Influences on day to day self-management of type 2 diabetes among African American women: Spirituality, the multi-caregiver role, and other social context factors. Diabetes Care 2000;23(7):928–33.
100. Brownson CA, Lovegreen, SL, Fisher, EB. Community and society support for diabetes self-management. In: Mensing C, Ed. The Art and Science of Diabetes Self-Management Education: A Desk Reference for Healthcare Professionals. Chicago: American Association of Diabetes Educators; 2006.
101. Bennett DA. Diabetes and change in cognitive function. Arch Int Med 2000;160(2):141–3.
102. Jack L, Jr., Airhihenbuwa, CO, Namageyo-Funa, A, Owens, MD, Vinicor, F. The psychosocial aspects of diabetes care. Geriatrics 2004;59(5):26–32.
103. Bent N, Rabitt, P, Metcalfe, D. Diabetes mellitus and the rate of cognitive ageing. British J Clin Psychol 2000;39(4):349–62.

104. Meneilly GS, Cheung, E, Tessier, D, Yakura, C, Tuokko, H. The effect of improved glycemic control on cognitive functions in the elderly patient with diabetes. Gerontologist 1993;48(4):M117–21.

105. Gregg EW, Yaffe, K., Cauley, JA., Rolka, DB, Blackwell, TL, Narayan, KM, Venkat, MD, Cummings, SR. for the Study of Osteoporotic Fractures Research Group. Is diabetes associated with cognitive impairment and cognitive decline among older women? Arch Int Med 2000;160(2):174–80.

106. Asimakopoulou K, Hampson, SE. Cognitive Function in type 2 diabetes: Relationship to diabetes self-management. In. Guilford, Surrey, UK: University of Surrey; 2001.

107. Asimakopoulou K, Hampson, SE. Cognitive functioning and self-management in older people with diabetes. Diabetes Spectrum 2002;15(2):116–21.

108. Hampson SE, Glasgow, RE, Foster, LS. Personal models of diabetes among older adults: Relationship to self-management and other variables. Diabetes Educator 1995;21(4):300–7.

109. Alcozer F. Secondary analysis of perceptions and meanings of type 2 diabetes among Mexican American women. Diabetes Educator 2000;26(5):785–95.

110. Huang ES, Gorawara-Bhat, R, Chin, MH. Self-reported goals of older patients with type 2 diabetes. J Am Geriatr Soc 2005;53:306–11.

111. O'Connor PJ, Desai, JR, Solberg, LI, Rush, WA, Bishop, DB. Variation in diabetes care by age: opportunities for customization. BMC Family Practice. 2003;4(16):www.biomedcentral.com/1471-2296/1474/1416.

112. White SL, Maloney, SK. Promoting healthy diets and active lives to hard-to-reach groups: market research study. Public Health Report 1990;105(3):224–31.

113. Jacobson AM. Quality of life in patients with diabetes mellitus. Semin Clin Neuropsychiatry 1997;2(1):82–93.

114. Bott U, Muhlhauser, I, Overmann, J, Berger, M. Validation of a diabetes-specific quality of life scale for patients with type 1 diabetes. Diabetes Care 1998;21(5):757–69.

115. Araki A, Izumo, Y, Inoue, J, Hattori, A, Nakamura, T, Takahashi, R, Takanashi, K, Teshima, T, Yatomi, N, Simizu, Y. Burden of dietary therapy on elderly patients with diabetes mellitus. Nippon Ronen Igakkai Zasshi 1995;32(12):904–9.

116. Olson DE, Norris, SL. Overview of AGS guidelines for the treatment of diabetes mellitus in geriatric populations. Geriatrics 2004;59(4):18–24.

117. Blaum CS, Ofstedal, MB, Langa, KM, Wray, LA. Functional status and health outcomes in older Americans with diabetes mellitus. J Am Geriatr Soc 2003;51(6):745–53.

118. Glasgow RE, Toobert, DJ, Hampson, SE, Brown, JE, Lewinsohn, PM, Donnelly, J. Improving self-care among older patients with type II diabetes: The "Sixty Something..." Study. Patient Educ Couns 1992;19(1):61–74.

119. Toobert DJ, Glasgow, RE, Stryker, LA, Barrera, M Jr., Radcliffe, JL, Wander, RC, Bagdade, JD. Biologic and quality of life outcomes from the Mediterranean lifestyle program. Diabetes Care 2003;26(8):2288–93.

120. Miller CA, Edwards, L, Kissling, G, Sanville, L. Nutrition Education improves metabolic outcomes among older adults with diabetes mellitus: Results from a randomized controlled trial. Prev Med 2002;34:252–9.

121. Miller CA, Edwards, L, Kissling, G, Sanville, L. Evaluation of a theory-based nutrition intervention for older adults with diabetes mellitus. J Am Diet Assoc 2002;102(8):1069–81.

122. Group TLAR. Reduction in weight and cardiovascular disease risk factors in individuals with type 2 diabetes. Diabetes Care 2007;30(6):1374–83.

123. Clark B. Older, Sicker, smarter, and redefining quality: The older consumer's quest for service. In: Dychtwald K, ed. Healthy Aging. Challenges and Solutions. Gaithersburg MD: Aspen Publishers, Inc.; 1999.

17 Cardiac Rehabilitation: The Nutrition Counseling Component

William E. Kraus and Julie D. Pruitt

Key Points

- Cardiac rehabilitation following a cardiac event is a comprehensive, multidisciplinary program that includes lifestyle counseling in nutrition, physical activity, stress management, and smoking cessation.
- Special concerns that can affect implementation of cardiac rehabilitation in older patients include complicated co-morbidities; functional limitations, alterations in taste, smell, and appetite; difficulties with medication use/effectiveness; and limited financial, social, and/or caregiver resources.
- Recommendations for dietary modification in cardiac rehabilitation include advice to increase the intake of fruits, vegetables, and whole grains, to substitute non-hydrogenated unsaturated fats for saturated and *trans*-fats, and to increase intake of omega-3 fatty acids.

Key Words: Cardiac event; hypertension; dyslipidemia; lipid-modified diet; sodium

17.1 CORE COMPONENTS OF CARDIAC REHABILITATION

The era of cardiac rehabilitation in the United States dates back at least 30 years, when visionaries Herman Hellerstein, Andy Wallace, and Ken Cooper, among others, envisioned that a comprehensive lifestyle approach to the rehabilitation and prevention of patients having had a cardiac event would potentially yield great benefits for the individual patient and the health-care system. Until that time, the thought of vigorous exercise in the cardiac patient soon after an event was close to anathema. Since then the concept of cardiac rehabilitation has grown into a comprehensive multidisciplinary program that includes lifestyle counseling, particularly in nutrition, physical activity, stress management, and smoking cessation, as core concepts within the program *(1)*. Despite the importance of providing

From: *Nutrition and Health: Handbook of Clinical Nutrition and Aging, Second Edition*
Edited by: C. W. Bales and C. S. Ritchie, DOI 10.1007/978-1-60327-385-5_17,
© Humana Press, a part of Springer Science+Business Media, LLC 2009

these elements to a comprehensive rehabilitation program, until recently there was little recognition of the need to find mechanisms to reimburse for many of these services *(2,3)*.

In the last 30 years the practice of cardiovascular medicine has grown and changed and with it so has, by necessity, the practice of cardiac rehabilitation. As one example, just very recently in 2006, the Center for Medicare and Medicaid Services (CMS) approved three new indications for cardiac rehabilitation reimbursement (percutaneous coronary intervention, cardiac transplantation, and valvular surgery) to accompany the previous three indications of chronic stable angina, post-bypass, and post-myocardial infarction *(4)*. More importantly and significantly, CMS recognized cardiac rehabilitation as the truly multidisciplinary program that it is beyond just exercise therapy for the cardiac patient *(4)*. And, the American Association of Cardiovascular and Pulmonary Rehabilitation, the American Heart Association, and the American College of Cardiology have combined efforts to publish the first set of performance measures for referral to and delivery of cardiac rehabilitation services *(5)*. A practical text describing the components and processes of cardiac rehabilitation is also now available *(6)*. We will present in this chapter a practical description of the nutrition component of cardiac rehabilitation with an emphasis on the dietary options available to the cardiovascular patient and some of the evidence supporting these dietary choices for the treatment and prevention of cardiovascular disease.

17.2 NUTRITION AND CARDIOVASCULAR RISK

Healthy nutrition plays an essential role in improving the cardiovascular risk profile following a cardiac event. Recognizing the impact that healthy dietary behaviors can have on recovery, the guidelines for cardiac rehabilitation have stated that "…nutritional counseling should be provided to all participants in cardiac rehabilitation" *(3)*. Research has shown that the combination of regular exercise and healthy nutrition significantly slows the progression of coronary heart disease *(1)*. Increasing fruit and vegetable intake and managing fat in the diet are also critical in the management of other heart disease risk factors such as hypertension, type 2 diabetes mellitus, and dyslipidemias of many varieties.

National guidelines specifically require assessment of and targeted intervention on the nutrition status of all cardiac rehabilitation participants *(3,5)*. The methods and tools used to achieve these requirements vary from program to program. The size of the program, additional state regulatory requirements, program resources, and other considerations will all influence the choice of tools and methods employed by each program (see Table 17.1).

One element essential for all programs to include, regardless of size or resources, is personalization of information for each participant. Personalization begins with an individualized look at the participant's nutritional status and home environment. Once an assessment is complete, each participant should have access to appropriate nutrition education. The nutrition therapy portion of a cardiac rehabilitation program should culminate with healthy personalized recommendations.

Table 17.1

Core components of nutritional counseling in cardiac rehabilitation

Evaluation – Assess caloric and nutrient intake, assess eating habits, assess target areas for intervention

Intervention – Develop individualized diet plan aimed at general heart healthy recommendations and specific risk reduction strategies; counsel participant and family; incorporate behavior change and compliance strategies

Expected Outcomes – Participant understanding of basic dietary principles, plan to address eating behavior problems, and adherence to diet

Adapted from Balady 2007

Based on the structure of the program, these services may be provided through individual consultation with program staff or through interactive classroom instruction.

17.3 SPECIAL ISSUES OF CONCERN FOR OLDER PATIENTS IN CARDIAC REHABILITATION

There are several issues that present special concerns for aged patients in the cardiac rehabilitation setting. Older patients are more likely to have more co-morbid conditions. Co-morbidities, such as arthritis, chronic obstructive coronary artery disease, and dementia hamper referral and interfere with cardiac rehabilitation goals for patients that are referred.

A hallmark of the aging process is a progressive decline in physical activity, which represents a significant cardiovascular risk factor *(7)*. Although many view frailty as a rationale to omit cardiac rehabilitation for older patients, those with limited functional capacity may actually derive the greatest benefit in the form of proportional improvements with minimal risk and improved quality of life *(8)*. Cardiac rehabilitation might actually be more beneficial in this group than in the general post-event cardiac patient in promoting return to normal functional status and establishment and maintenance of a good preventive cardiovascular program looking forward. In fact, the fourth edition of the American Association of Cardiovascular and Pulmonary Rehabilitation (AACVPR) guidelines address issues of aging and present a strong case for providing cardiac rehabilitation services to the older patient *(9)*. However, at present, it must be noted that the benefits of cardiac rehabilitation for these elderly patients remain more conceptual than proven.

Additional consideration must be given to other physiologic processes that are occurring at this stage of life, such as the likely presence of co-morbid disease states and associated treatments and altered physiologic response to food. According to Podrabsky and Remig, an average 75-year-old person has three chronic disease conditions and takes five prescription medications *(10)*. There is a high likelihood that each disease state and its treatments have its own dietary concerns/restrictions.

Some cardiovascular risk factors present particular challenges in older adults that can interact with nutritional factors. For hypertension, salt sensitivity increases the already stiff vasculature of elderly individuals and makes the use of diuretics and a first-line medication more justifiable. However, fragile volume status might make over-treatment and positional hypotensive episodes more likely, as well. Further, dietary sodium content may interact and complicate the process of finding a stable medical regimen for treatment of blood pressure, especially if the dietary sodium content is not stable. Also, alcohol use can exacerbate hypertension and alcoholism is a common geriatric problem.

With respect to dyslipidemia, while the importance of treating dyslipidemias as cardiovascular risk factors in the elder patients has sometimes been controversial, accumulating evidence suggests that treating elevated low-density lipoproteins cholesterol levels is beneficial in older subjects and that physicians should not shy away from aggressive treatment of lipids, irrespective of the age of the patient *(11)*. For example, in the PROSPER study, designed to study the efficacy of cholesterol modification with pravastatin in older adults aged 70–82, at 3.2 years the active therapy reduced the risk of a primary cardiovascular end point by 15%.

Given the well-known difficulty in compliance with statin therapy even in younger cardiovascular patients, cardiac rehabilitation can be particularly valuable for the elderly patient in encouraging compliance and reinforcing dietary approaches to cardiovascular risk factors such as hypertension and dyslipidemias. Further a propensity for depression following cardiovascular surgery, a particularly problematic issue in the elderly patient, can be addressed in the cardiac rehabilitation environment, thus encouraging compliance with risk modification through lifestyle and medical interventions.

Providing nutrition counseling to an older adult population elicits the need to discuss population-specific challenges and issues that must be considered in a comprehensive program. Implementation of the clinical recommendations that follow may be particularly difficult for this group. Many older adults have a limited or restricted ability to meet their nutritional needs. As a whole, older adults often have limited financial resources with which to purchase food, be it inexpensive canned or frozen vegetables or a loaf of whole-grain bread, much less omega-3 fatty acid-rich fish or fresh fruit. These cardiac participants may also be faced with restricted access to acquiring food, especially following a cardiac event. They may no longer drive and may be physically less able to take trips out of the house.

Dependency on others necessarily increases, leading to a lack of control on the part of the older adult. Many older adults, while still living in the community at large, are reliant upon informal caregivers. These informal caregivers, often family and friends, provide essential support and services but are untrained in how to best provide support. Including the cardiac rehabilitation participant's support persons during cardiac rehabilitation education sessions whenever possible is greatly beneficial. These caregivers are often the critical link in the participant's ability to follow through with the established nutrition goals as the caregivers may be the people who actually purchase and prepare the food. Including an element in nutrition counseling that enables caring for the caregiver to occur simultaneously provides a healthier environment for the participant and allows the caregiver to be better able to perform caregiving duties. By functioning as a team, the caregiver and cardiac rehabilitation participant can jointly work toward a healthier future, thereby returning some of the locus of control back to the participant.

Changes in taste and smell also occur with age. This may lead to poor nutrition status due to reduced enjoyment of eating, and therefore, to reduced intake of nutritious foods or to increased consumption of low-nutrient-density foods with stronger flavors such as high-fat, high-salt, and high-sugar foods. Food intolerances such as heart burn and flatulence also seem to increase with age. Incorporating nutrition solutions into personalized recommendations that not only clinically address cardiovascular disease risk but also address some of the fundamental issues faced by this population is critical for success.

17.4 DIETARY INTAKE AND THE MANAGEMENT OF HEART DISEASE

Extensive research has been conducted on many different individual nutritional components and the impact they have on coronary heart disease and the associated risk factors. Hu and Willett completed a review of 147 original investigations and reviews of major dietary factors. They identified increased intake of fruits, vegetables, and whole grains, substitution of non-hydrogenated unsaturated fats for saturated and *trans*-fats, and increased intake of omega-3 fatty acids as the major effective dietary strategies for preventing coronary heart disease *(12)*. Additionally, incorporation of low-fat dairy products on a daily basis helps to control blood pressure and may, possibly, assist in weight management, if needed *(13,14)*. Structuring cardiac rehabilitation nutrition education around these principles will support the goal of secondary prevention in participants.

17.4.1 Fruits and Vegetables

Numerous studies have demonstrated that diets rich in fruits and vegetables are correlated with prevention of coronary heart disease and associated risk factors *(12)*. In a large epidemiologic study, a significant inverse correlation between risk of coronary heart disease and intake of fruits and vegetables was noted. Every single serving/day increase in intake was correlated with a 4% decrease in risk. Individuals in the highest quintile of intake (9.15–10.15 servings/day) had a 20% lower relative risk of developing coronary heart disease than individuals in the lowest quintile of intake (2.5–2.9 servings/day). Especially beneficial were leafy green vegetables and vitamin C-rich fruits and vegetables *(15)*. In the landmark DASH trials, increased consumption of fruits and vegetables in the setting of additional healthy food behaviors led to markedly decreased blood pressure levels *(13)*. The beneficial effects of including fruits and vegetables in the diet are thought to come, in part, from the provision of potassium, fiber, phytochemicals, and the displacement of unhealthier food choices.

Counseling participants with regard to the inclusion of additional fruits and vegetables in their diets should include several key elements. First, while respecting the person's dignity, discussion of the participant's ability to procure fruits and vegetables is important. Together, explore viable options for increasing fruit and vegetable intake, such as selecting inexpensive preparations (canned or frozen), selection of in-season varieties, home-delivered farm shares, and the like. Second, emphasize the importance of serving the fruits and vegetables in a healthy and appealing manner. People often offset the benefit gained from the inclusion of fruits and vegetables in the diet by adding copious amounts of fats and sugars during food preparation. Third, starchy vegetables such as corn, potatoes, peas, and beans should be considered part of the *starches* food group and not the *vegetable* food group. Starchy vegetables, as the name implies, have carbohydrate and caloric contents similar to other starches. When developing a meal plan, treating them as such is necessary, especially when blood glucose management is a factor in meal composition. Non-starchy vegetables are very low in calories and can be consumed liberally, even in the setting of weight management. Last, encouragement in the selection of whole fruits and vegetables whether they are fresh, frozen, or canned in their own juice or light syrup over juices is important. Whole fruits and vegetables provide the extra benefit of fiber and increased satiety. In general, over-consumption of calories from fruit rarely occurs and only when a participant consumes more than the recommended amount of juice or dried fruit – less satisfying foods. Increasing the consumption of healthy fruits and vegetables is often a new experience for many cardiac rehabilitation participants. Increasing their self-efficacy through education and counseling to establish these new habits is critical in their eventual success.

17.4.2 Dietary Fats

There is undeniable evidence that diets high in saturated and *trans*-fats are significant contributors to the risk of coronary heart disease, while inclusion of healthy monounsaturated and polyunsaturated fats reduces risk. Differences in total fat intake, with the exception of difficult-to-adhere-to very low-fat diets (<10% of total calories from fat), do not significantly influence coronary heart disease risk; only substitution of saturated and *trans*-fats with monounsaturated and polyunsaturated fats significantly impact risk *(16–18)*. In the Women's Health Initiative Study, an 8.2% reduction in total fat intake (2.9% from saturated fat, 0.6% *trans*-fats, 3.3% monounsaturated fats, 1.5% polyunsaturated fats) did not result in significant impact on coronary heart disease risk. However, replacement of 5% of energy from saturated fat and 2% of energy from *trans*-fat with monounsaturated and polyunsaturated fats was associated with a 42 and 53% lower risk of coronary heart disease, respectively, in the Nurses' Health Study *(17)*. Targeted reductions and substitutions of fat subtypes are critical to improving a person's cardiac risk profile.

In addition to replacing unhealthy fats in the diet with healthy fats, inclusion of omega-3 fats in the diet can provide substantial benefits. Inclusion of approximately eight or more servings of omega-3 fatty acid-rich fish each month reduced sudden cardiac death by 50% compared to consumption of less than one serving of omega-3 fatty acid-rich fish per month *(19)*. Plant-based sources of omega-3 fatty acids have also been indicated in the secondary prevention of coronary heart disease *(19)*. It is important to encourage participants to increase omega-3 fatty acids in their diet by including two 3-ounce servings of low-contaminate (see Table 17.2) fatty fish each week and plant-based sources such as flaxseed, English walnuts, and canola oil, within the context of recent safety advisories.

Managing the balance of fat in the diet through the replacement of unhealthy saturated and *trans*-fats with healthy omega-3, monounsaturated and polyunsaturated

Table 17.2
Fish advisory (from the American Heart Association)

Some fatty fish such as shark, swordfish, king mackerel, and tilefish have higher
    mercury content. Limit your consumption of these fish to no more than 7 ounces per
    week. Consumption of fish with lower mercury contamination like tuna, red snapper,
    and orange roughly should be limited to no more than 14 ounces per week.
Women of childbearing age, pregnant woman, and children should avoid fish with high-
    mercury contamination altogether and limit lower mercury fish to not more than 12
    ounces per week.

fats can represent a significant lifestyle change for many cardiac rehabilitation participants. This will likely be a shift away from many of the comfort foods that are ingrained
in their home environment. Acknowledging the difficulty of these changes can be
helpful as the participant begins to set goals for change. Smaller, more attainable
goals lead to early success and increased self-efficacy. Work with the participant to
identify one or two specific areas for change. These targeted efforts can have substantial
returns. For example, someone who is eating high-saturated fat meat at most meals
could set a goal to prepare one vegetarian meal each week. A second goal may be
substituting canola or olive oil for shortening when cooking. Both of these actions can
considerably improve the participant's diet without an overwhelming, unnecessary, and
expensive pantry overhaul. The sheer magnitude of which could cause a total resistance
to all change.

17.4.3 Dairy Products

Dairy products may play a role in managing blood pressure and weight. Adding
low-fat dairy products such as milk and yogurt to a diet rich in fruits and vegetables
compounds the diet's ability to reduce cardiac risk. In the setting of weight maintenance, the DASH-feeding studies noted addition of two to three servings of
low-fat dairy products to the basic diet more than doubled the reduction in
blood pressure achieved by subjects using this dietary approach (13). While
these positive effects on blood pressure have been well documented, newer research
demonstrates the potential for even more risk reduction benefits when weight loss
is needed.

Recent studies support the role of calcium and low-fat dairy products in weight
and body fat mass management. Including dairy (that is, three servings of yogurt/
day) in the setting of energy restriction improved total body fat loss and trunk fat
loss by 61 and 81%, respectively (14). If energy intake is held constant and
suboptimal calcium intake is increased through the inclusion of three servings of
dairy per day, body fat mass decreases (−5.4%), especially in the trunk region
(−4.6%), while body weight remains stable (14). A recent randomized control trial
compared increased calcium through supplementation and inclusion of dairy products. While individuals taking calcium supplements experienced improvement
over the control group in weight, total body fat, and trunk fat loss, individuals in
the dairy group had substantially greater losses (14).

These smaller trials indicate that provision of dairy products in the diet is beneficial in the management of weight and body fat; however, more extensive research is needed. Virtually all cardiac rehabilitation participants can benefit from the inclusion of two to three servings of dairy products in their daily diet. Primarily, these servings should come from low-fat milk and yogurt. Cheese and butter, due to their higher fat content and lower calcium content, are considered when determining daily fat intake, rather than as dairy servings.

17.4.4 Whole Grains and Starches

In the past several years, starches have received volumes of negative press, with some fad diet creators suggesting virtual elimination of starches from the diet. However, starches are essential components of a healthy diet, as they provide certain essential nutrients and serve as a vital energy source. Starches also serve an important function in satiation. A distinction to be made during nutritional counseling is in the *type* of starch that is recommended for inclusion in the diet. While refined grains are stripped of their nutrients, whole grains provide a bounty of nutrients, phytochemicals, and fiber.

Inclusion of whole grains and cereal fiber in the diet decreases risk of disease. In the Iowa Women's Health study, Jacobs et al. found a clear inverse relationship between intake of whole grains and risk of heart disease (20). Individuals in the highest quintile of intake (3.2 servings/day) had a 30% lower relative risk of heart disease than individuals in the lowest quintile of intake (0.2 servings/day) (20). Similarly, results from the Nurse's Health Study also demonstrated a 34% lower risk of heart disease in women in the highest quintile of intake of fiber. The decreased risk was only significant for dietary fiber from cereal grains and not for fiber from fruits and vegetables (21).

Introducing or increasing whole-grain products in the diet can be structured in a stepwise manner with the goal of at least half of all starch servings in the diet provided by whole-grain products. It is productive to have the participant begin to examine the starch products currently consumed and identify areas for change. Initial increases may come from mixing whole-grain breakfast cereals with favorite refined grain cereals or purchasing whole-grain pastas. Eventually, the participant can continue to transition to additional whole-grain products as the palate and gastrointestinal tract adjust.

17.5 ADDITIONAL RELEVANT DIETARY COMPONENTS

17.5.1 Stanols and Sterols

Plant stanols and sterols are organic compounds found naturally in vegetable oils, cereals, fruits, and vegetables. Additionally, they are now added to products such as margarines and orange juice. Due to their cholesterol-like structure, these compounds interfere with the absorption of cholesterol and cholesterol-building blocks in the digestive tract. The net effect of this lowered absorption is a 6–15% reduction in LDL cholesterol (22). The Third Report of the National Cholesterol Education Program Expert Panel of Detection, Evaluation, and Treatment of High Blood Cholesterol in Adults (Adult Treatment Panel III) recommends consuming

2 g of plant stanol and sterol esters per day as a therapeutic option for managing LDL cholesterol *(22)*. To achieve an optimal intake level, inclusion of fortified products is necessary as the amount of stanols and sterols occurring naturally in foods is minimal.

17.5.2 Alcohol

While alcohol intake and a reduction in cardiac events have been widely publicized in the popular media and supported by several research studies, we do not recommend encouraging the addition of alcohol in the diet of a cardiac rehabilitation. The addictive nature and adverse consequences of over-consuming alcohol do not outweigh the potential benefits. In fact, as noted, excessive alcohol intake can be especially problematic in the older adult population. Also, it is becoming clear that many of the studies that found a potential benefit of moderate alcohol consumption perhaps overestimated the benefits by incorrectly including reformed high-volume alcoholic consumers in the abstinence group. While the proper role of alcohol in the diet and for risk reduction is still being investigated, if a participant currently consumes alcohol, the American Heart Association recommends limiting intake to not more than one drink per day for women and two drinks per day for men *(23)*.

17.5.3 Sodium

Numerous studies have shown that reducing intake of dietary sodium can have a significant effect on lowering blood pressure, a risk factor for heart disease. In the DASH – Sodium trial, subjects were maintained on sodium intakes of 3300, 2400, and 1500 mg/day in a cross-over design. Blood pressure was significantly reduced with each incremental drop in sodium intake *(24)*. While sodium intake at the lowest level showed the greatest overall lowering effect, maintaining this level of intake is difficult at best. Therefore, the American Heart Association recommends consuming no more than 2400 mg of sodium per day *(23)*. To remain under the recommended level of intake, counsel cardiac rehabilitation participants to steer clear from high-sodium seasonings, canned meats, soups and vegetables, and salty snacks and to avoid adding salt when cooking or at the table. Encourage participants to experiment with low-sodium flavor agents to maintain or add flavor to their foods.

17.6 ALTERNATIVE DIETARY PATTERNS

One of the most significant ways that cardiac risk can be reduced is through achieving and maintaining a healthy body weight. As a result, cardiac rehabilitation participants who are overweight or obese often solicit advice about initiating one of the numerous popular diets for weight loss. For this reason health professionals need to be aware of two key points emerging from research in this field. Several studies have noted that weight loss is not significantly different after 12 months in subjects following any one of the several popular diets *(25–27)*. In a randomized clinical trial comparing a low-carbohydrate, high-protein, high-fat diet to a low-calorie, high-carbohydrate, low-fat diet, greater improvement of some cardiac risk factors (triglycerides and HDL) was experienced on the low-carbohydrate diet *(25)*.

Changes in other risk factors (LDL, blood pressure, insulin sensitivity), however, were not significantly different between the groups or, as demonstrated by meta-analysis, were found to be unfavorable on a low-carbohydrate diet *(25,27)*. When discussing alternative popular diets with participants, one might examine the restrictions set forth by the popular diet which limit the consumption of the proven beneficial foods (fruits, vegetables, low-fat dairy products, and whole grains) discussed previously. The long-term health ramifications and safety of excluding or limiting these foods are, as of yet, unknown.

17.7 RECOMMENDATIONS FOR APPROPRIATE CALORIC INTAKES AND DIETARY PATTERNS

Practical implementation of a healthy diet pattern begins with laying a foundation of appropriate caloric intake. For the older adult population achieving an appropriate caloric intake may mean either a decrease or an increase in current caloric intake. In this population less strict adherence to specific calorie levels and more emphasis on healthy eating patterns are recommended. The tool presented in Table 17.3 was developed by dietitians at the Duke Center for Living and Duke Cardiac Rehabilitation program. Based on the Harris Benedict equation, it is a user-friendly tool requiring minimal calculations to approximate caloric intake (see Table 17.3). Once an appropriate calorie level is identified, determine the recommended contributions from each dietary component. A healthy cardiac diet allows for an estimated 25–30% of total calories from fat – mainly from healthy mono-unsaturated and polyunsaturated fats by limiting unhealthy saturated fat to 7% of total calories. A quick reference for total fat and saturated fat gram allowance for varying calorie levels is provided (see Table 17.4). Select the calorie level closest to the participant's calculated needs.

Table 17.3
Determining your daily calorie allowance

Step 1:	Write current weight _____ (lbs) then multiply by 10 =	_____
Step 2:	Choose one from Steps a–e below. *(It is important not to have a calorie level <1200 without evaluation by a dietitian or physician)*	
	a. If you want to tone up body, **maintain weight or lose less than 10 lbs**, then add 500	
	b. If you want to **lose 10–25 lbs**, then add 0.	_____
	c. If you want to **lose greater than 25 lbs**, subtract 500	
	d. If you **weigh 350 lbs** or more, subtract 1,000.	
	e. If you want to **gain weight**, add 1,000.	
Step 3:	Add calories in right-hand column from Steps 1 and 2.	_____**Calories**
	This is your estimated calorie needs per day	**per day**

Above calorie levels are based on a person engaging in ~30 min of exercise 3–5 days/week.

Table 17.4
Daily fat gram budget chart

Calorie needs	Maximum daily total fat gram budget	Maximum daily saturated fat gram budget
1300	40	10
1400	43	11
1600	48	12
1800	51	14
2000	58	16
2200	66	17
2400	73	19
2600	79	20
2800	87	22
3000	95	23

Table 17.5 shows two suggested meal plans that take into account the calories from fat and the recommendations for following a diet pattern rich in fruits, vegetables, low-fat dairy products, and whole grains. The first plan (Table 17.5, Chart 1) includes provisions for participants who consume beef, poultry, seafood, eggs, and cheese, while the second (Table 17.5, Chart 2) is specifically designed for lacto-ovo vegetarians (individuals who do not consume animal products with the exception of eggs, cheese, and dairy products). To advise a specific patient, begin by finding the caloric level that most closely matches the patient's recommended caloric intake on the appropriate chart.

17.8 SUMMARY

The goal of nutrition therapy for older adult cardiac rehabilitation participants is the adaptation and maintenance of healthy behaviors in order to improve a participant's cardiovascular risk profile and prevent additional cardiac events for a lifetime. Nutrition therapy begins with an assessment of current dietary practices and review of potential barriers, followed by identification of targeted areas for change and creation of an implementation plan. Including the services of a registered dietitian in a cardiac rehabilitation program may be beneficial in managing the complexities of changing dietary behaviors and providing nutrition education. Involving individuals that are part of the participant's support network and providing support and encouragement for each step a participant makes will foster continued positive changes and cardiac risk reduction.

17.9 RECOMMENDATIONS

1. Complete a thorough assessment of the nutrition status and home environment of the cardiac rehabilitation participant.

Table 17.5
Suggested meal plans for healthy eating

	Chart 1 plan for people who eat beef, poultry, seafood, eggs, and cheese									
	1300	1400	1600	1800	2000	2200	2400	2600	2800	3000
Fat grams	≤43	≤45	≤52	≤58	≤64	≤70	≤78	≤84	≤90	≤98
Starch*	4	4	5	5	6	7	7	8	9	10
Fruit	2	3	4	4	5	5	5	6	7	7
Veg	3+	4+	4+	5+	5+	6+	6+	6+	7+	7+
Dairy	2	2	2	2	2	2	3	3	3	3
M&P**	3	3	4	6	6	8	8	8	8	9

	Chart 2 plan for people who are lacto-ovo vegetarian (eat only eggs, cheese, and dairy animal products)									
	1300 VEG	1400 VEG	1600 VEG	1800 VEG	2000 VEG	2200 VEG	2400 VEG	2600 VEG	2800 VEG	3000 VEG
Fat grams	≤43	≤45	≤52	≤58	≤64	≤70	≤78	≤84	≤90	≤98
Starch*	5	5	6	6	7	8	9	10	11	12
Fruit	2	2	3	3	4	4	4	5	6	6
Veg	3+	3+	4+	5+	5+	5+	6+	6+	7+	7+
Dairy	3	3	3	3	3	3	3	3	3	3
M&P**	2	2	2	3	3	3	3	4	4	5

*Include at least half of all starch servings from whole-grain products.
**M&P = Meat (beef, poultry, seafood, pork, etc.) and other proteins such as eggs, cheese, nuts.

2. Provide education on a diet rich in fruits, vegetables, low-fat dairy products, and whole grains and low in saturated and *trans*-fats.
3. Assist participants in identifying changes they can make to substitute non-hydrogenated unsaturated fats for saturated and *trans*-fats in their diets.
4. Approximate the participant's optimal caloric intake (see Table 17.3) for supporting healthy weight maintenance goals.
5. Develop a participant-specific implementation plan that acknowledges the areas which need improvement and solutions to potential barriers.

REFERENCES

1. Ades PA. Cardiac rehabilitation and secondary prevention of coronary heart disease. N Engl J Med 2001;345(12):892–902.
2. Ades PA, Balady GJ, Berra K. Transforming exercise-based cardiac rehabilitation programs into secondary prevention centers: a national imperative. J Cardiopulm Rehabil 2001;21(5):263–72.
3. Leon AS, Franklin BA, Costa F, et al. Cardiac rehabilitation and secondary prevention of coronary heart disease: an American Heart Association scientific statement from the Council on Clinical Cardiology (Subcommittee on Exercise, Cardiac Rehabilitation, and Prevention) and the Council on Nutrition, Physical Activity, and Metabolism (Subcommittee on Physical Activity), in collaboration with the American association of Cardiovascular and Pulmonary Rehabilitation. Circulation 2005;111(3):369–76.
4. CMMS, Services CfMaM. Decision Memo for Cardiac Rehabilitation Programs. In: CAG-00089R; 2006.
5. Balady GJ, Williams MA, Ades PA, et al. Core components of cardiac rehabilitation/secondary prevention programs: 2007 update: a scientific statement from the American Heart Association Exercise, Cardiac Rehabilitation, and Prevention Committee, the Council on Clinical Cardiology; the Councils on Cardiovascular Nursing, Epidemiology and Prevention, and Nutrition, Physical Activity, and Metabolism; and the American Association of Cardiovascular and Pulmonary Rehabilitation. Circulation 2007;115(20):2675–82.
6. Kraus WE, Keteyian SJ. Cardiac Rehabilitation. Totowa, NJ: Humana Press; 2007.
7. Fleg JL, Morrell CH, Bos AG, et al. Accelerated longitudinal decline of aerobic capacity in healthy older adults. Circulation 2005;112(5):674–82.
8. Ades PA, Savage PD, Cress ME, Brochu M, Lee NM, Poehlman ET. Resistance training on physical performance in disabled older female cardiac patients. Med Sci Sports Exerc 2003;35(8):1265–70.
9. AACVPR. Special Considerations. In: Williams MA, ed. Guidelines for Cardiac Rehabilitation and Secondary Prevention Programs. Champaign, IL: Human Kinetics; 2004:135–76.
10. Podrabsky M, Remig V. Public Policy Initiative III: Meeting the Nation's New Aging Reality. ADA Times 2005;3(2).
11. Shepherd J, Blauw GJ, Murphy MB, et al. Pravastatin in elderly individuals at risk of vascular disease (PROSPER): a randomised controlled trial. Lancet 2002;360(9346):1623–30.
12. Hu FB, Bronner L, Willett WC, et al. Fish and omega-3 fatty acid intake and risk of coronary heart disease in women. Jama 2002;287(14):1815–21.
13. Appel LJ, Moore TJ, Obarzanek E, et al. A clinical trial of the effects of dietary patterns on blood pressure. DASH Collaborative Research Group. N Engl J Med 1997;336(16):1117–24.
14. Zemel MB. The role of dairy foods in weight management. J Am Coll Nutr 2005;24(6 Suppl):537S–46S.
15. Joshipura KJ, Hu FB, Manson JE, et al. The effect of fruit and vegetable intake on risk for coronary heart disease. Ann Intern Med 2001;134(12):1106–14.
16. Howard BV, Van Horn L, Hsia J, et al. Low-fat dietary pattern and risk of cardiovascular disease: the Women's Health Initiative Randomized Controlled Dietary Modification Trial. Jama 2006;295(6):655–66.

17. Hu FB, Stampfer MJ, Manson JE, et al. Trends in the incidence of coronary heart disease and changes in diet and lifestyle in women. N Engl J Med 2000;343(8):530–7.
18. Ornish D, Scherwitz LW, Billings JH, et al. Intensive lifestyle changes for reversal of coronary heart disease. Jama 1998;280(23):2001–7.
19. Kris-Etherton P, Daniels SR, Eckel RH, et al. Summary of the scientific conference on dietary fatty acids and cardiovascular health: conference summary from the nutrition committee of the American Heart Association. Circulation 2001;103(7):1034–9.
20. Jacobs DR, Jr., Meyer KA, Kushi LH, Folsom AR. Whole-grain intake may reduce the risk of ischemic heart disease death in postmenopausal women: the Iowa Women's Health Study. Am J Clin Nutr 1998;68(2):248–57.
21. Wolk A, Manson JE, Stampfer MJ, et al. Long-term intake of dietary fiber and decreased risk of coronary heart disease among women. Jama 1999;281(21):1998–2004.
22. Executive Summary of the Third Report of The National Cholesterol Education Program (NCEP) Expert Panel on Detection, Evaluation, And Treatment of High Blood Cholesterol In Adults (Adult Treatment Panel III). Jama 2001;285(19):2486–97.
23. Krauss RM, Eckel RH, Howard B, et al. AHA Dietary Guidelines: revision 2000: A statement for healthcare professionals from the Nutrition Committee of the American Heart Association. Circulation 2000;102(18):2284–99.
24. Sacks FM, Svetkey LP, Vollmer WM, et al. Effects on blood pressure of reduced dietary sodium and the Dietary Approaches to Stop Hypertension (DASH) diet. DASH-Sodium Collaborative Research Group. N Engl J Med 2001;344(1):3–10.
25. Foster GD, Wyatt HR, Hill JO, et al. A randomized trial of a low-carbohydrate diet for obesity. N Engl J Med 2003;348(21):2082–90.
26. Freedman MR, King J, Kennedy E. Popular diets: a scientific review. Obes Res 2001;9 Suppl 1:1S–40S.
27. Nordmann AJ, Nordmann A, Briel M, et al. Effects of low-carbohydrate vs low-fat diets on weight loss and cardiovascular risk factors: a meta-analysis of randomized controlled trials. Arch Intern Med 2006;166(3):285–93.

18 Chronic Heart Failure

Christopher Holley and Michael W. Rich

"Dropsy [heart failure] is usually produced when a patient remains for a long time with impurities of the body following a long illness. The flesh is consumed and becomes water. The abdomen fills with fluid; the feet and legs swell; the shoulders, clavicles, chest and thighs melt away."

–Hippocrates *(1)*

Key Points

- Heart failure is the leading cause of hospitalization in the Medicare age group. The prognosis for established heart failure in persons over age 65 is poor, with 5-year survival rates of less than 50% in both men and women.
- The pharmacotherapy of systolic heart failure is well established, with angiotensin-converting enzyme inhibitors and beta-blockers having the most proven benefit. Treatment of diastolic heart failure is an area of active research, but no therapies have been definitively shown to reduce mortality.
- Unintentional weight loss in heart failure is likely due to both increased energy utilization and decreased availability of fat, protein (amino acids), and carbohydrates despite "normal" caloric intake.
- Moderate dietary sodium restriction, such as a 2-g sodium diet, is appropriate for most patients with heart failure, and excess fluid intake should be avoided.
- Some patients will require supplementation with potassium, calcium, and/or magnesium if adequate amounts cannot be obtained from the diet. However, the importance of most vitamins and other micronutrients in the pathogenesis and treatment of chronic heart failure has not been well characterized.

Key Words: Cardiac function; systolic heart failure; diastolic heart failure; cardiac cachexia; sodium restriction; electrolytes

18.1 OVERVIEW OF HEART FAILURE

18.1.1 Background

Heart failure is a condition in which one or more abnormalities in cardiac function lead to an inability of the heart to pump sufficient blood to meet the

From: *Nutrition and Health: Handbook of Clinical Nutrition and Aging, Second Edition*
Edited by: C. W. Bales and C. S. Ritchie, DOI 10.1007/978-1-60327-385-5_18,
© Humana Press, a part of Springer Science+Business Media, LLC 2009

body's metabolic needs while maintaining normal or near-normal intracardiac pressures and blood volumes. It affects approximately 5 million Americans, and more than 500,000 new cases are diagnosed each year *(2)*. Importantly, both the incidence and the prevalence of heart failure increase progressively with advancing age *(3)*, and almost 80% of hospital admissions for heart failure occur in persons over 65 years of age *(4)*, with more than 50% occurring in persons over the age of 75 *(5)*. As a result, heart failure is the leading cause of hospitalization in the Medicare age group, and it is currently the most costly cardiovascular illness in the United States *(6)*. Moreover, it is anticipated that the rapid growth in the older adult population will result in a doubling in the number of older persons with heart failure during the next two to three decades.

18.1.2 Etiology

In the United States, chronic hypertension and coronary heart disease account for 70–80% of heart failure cases *(7, 8)*. In older women, hypertension is the most common etiology of heart failure, accounting for almost 60% of cases *(8)*. In older men, coronary heart disease and hypertension each account for 30–40% of heart failure cases *(8)*. Valvular heart disease (esp. aortic stenosis and mitral regurgitation) and nonischemic cardiomyopathies (dilated, hypertrophic, restrictive) are also common causes of heart failure in older adults. Less frequent causes include infective endocarditis, pericardial disease, thyroid disorders, and drug toxicity (e.g., anthracyclines).

18.1.3 Pathophysiology

The cardiac cycle is divided into a filling phase (diastole) and an emptying or pumping phase (systole). Impaired cardiac filling due to increased "stiffness" of the heart (e.g., from hypertension) results in increased intracardiac pressures and reduced cardiac output, leading to the syndrome of "diastolic heart failure". Conversely, damage to the heart muscle (e.g., from a myocardial infarction) results in impaired pumping action or "systolic heart failure". Although most patients with heart failure have evidence for both systolic and diastolic dysfunction, patients with significantly reduced systolic function (ejection fraction <40%) are often classified as having predominantly "systolic" heart failure, whereas patients with preserved systolic function at rest (ejection fraction >50%) are considered to have predominantly "diastolic" heart failure. Patients with heart failure and an ejection fraction of 40–49% may be viewed as having "mixed" systolic and diastolic heart failure. Recent studies indicate that about half of heart failure cases are associated with impaired systolic function, while the remainder have normal or near-normal systolic function at rest *(7, 9)*. Importantly, diastolic heart failure is more common in women than in men, in part due to the high prevalence of hypertension in women, and the proportion of patients with diastolic heart failure increases markedly with age. In the Cardiovascular Health Study, two-thirds of women over age 65 with heart failure had preserved systolic function, as compared with only 41% of men in this age group *(10)*.

Although treatment of systolic and diastolic heart failure is similar in many respects, it is important to evaluate ventricular function by echocardiography, radionuclide angiography, or cardiac catheterization in all patients with newly diagnosed heart failure because, as discussed below, there are important differences in pharmacotherapy depending on the degree of impairment in contractile function.

18.1.4 Clinical Features

The cardinal symptoms of heart failure include exertional shortness of breath and fatigue, reduced exercise tolerance, orthopnea, and lower extremity edema. Palpitations and orthostatic light-headedness are also common, but chest discomfort in the absence of ischemia is not usually present. Physical findings may include tachycardia, tachypnea, elevated jugular venous pressure, moist pulmonary rales, an S_3 or S_4 gallop, hepatomegaly, and dependent pitting edema. In patients with advanced or long-standing heart failure, there is loss of lean body mass, particularly muscle mass, which in severe cases leads to the syndrome of cardiac cachexia.

18.1.5 Prognosis

The prognosis for established heart failure in persons over age 65 is poor, with 5-year survival rates of less than 50% in both men and women (11). In addition, chronic heart failure is characterized by recurrent hospitalizations for acute exacerbations (12, 13), a marked increase in the risk of sudden death due to arrhythmia (14), and substantially impaired quality of life due to diminished activity tolerance. Although the short-term prognosis (i.e., 3–6 months) is somewhat more favorable in patients with diastolic compared to systolic heart failure, the long-term prognosis is similar (15, 16). In addition, hospitalization rates, symptom severity, and functional capacity do not differ significantly in patients with preserved versus impaired systolic function (17).

18.1.6 Treatment

Optimal treatment of chronic heart failure combines both nonpharmacological and pharmacological approaches (18). Nonpharmacological measures include patient education, dietary counseling, sodium and in some cases fluid restriction, attention to psychosocial and financial concerns, and close follow-up. Older patients with multiple comorbid conditions or complex environmental issues often benefit from a multidisciplinary approach to care delivery, involving nurses, social workers, dietitians, therapists, pharmacists, and physicians (19, 20).

The pharmacotherapy of systolic heart failure has been studied extensively over the last 25 years. Angiotensin-converting enzyme (ACE) inhibitors are the cornerstone of treatment, and available evidence indicates that these agents are at least as effective in older as in younger heart failure patients (21). Angiotensin II receptor blockers (ARBs) and the combination of hydralazine and isosorbide dinitrate are suitable alternatives in patients who are unable to tolerate ACE inhibitors (22–25). Beta-blockers have also been shown to reduce mortality and improve left ventricular function in stable heart failure patients at least up to the age of 80 (26, 27). Digoxin improves symptoms and reduces hospitalizations for heart failure but has no effect on survival, with similar effects in older and younger patients (28, 29). Diuretics are important for maintaining normal volume status and for managing acute heart failure exacerbations, but with the exception of the aldosterone

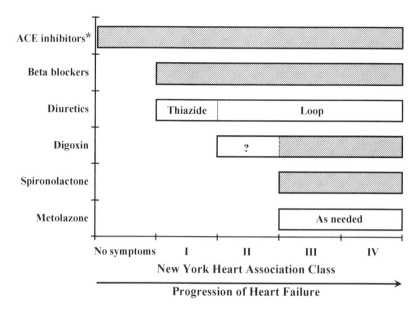

Fig. 18.1. Pharmacotherapy of left ventricular systolic dysfunction. Solid regions denote conditions for which improved outcomes have been documented in prospective randomized clinical trials.
ACE: angiotensin-converting enzyme.
* angiotensin receptor blockers (ARBs) and the combination of hydralazine and isosorbide dinitrate are acceptable alternatives or adjuncts to ACE inhibitors in selected patients.

antagonists spironolactone and eplerenone, diuretics have no discernible effect on the natural history of heart failure. Spironolactone reduces mortality in patients with advanced heart failure and is indicated in patients who remain highly symptomatic despite the above therapeutic measures *(30)*. Current pharmacotherapy of systolic heart failure is summarized in Fig. 18.1.

In contrast to systolic heart failure, treatment of diastolic heart failure has been less well studied and remains largely empiric *(17)*. Hypertension should be treated aggressively and coronary artery disease should be managed with medications and/ or revascularization as indicated. Diuretics are indicated for controlling volume overload, but over-diuresis should be avoided. The ARB candesartan has been shown to reduce hospitalizations for HF in this population, but without a significant mortality benefit *(31)*. Additional pharmacologic agents that have been shown to improve symptoms in selected patients with diastolic heart failure include nitrates, ACE inhibitors, beta-blockers, calcium channel blockers *(17, 32)*.

18.2 GENERAL NUTRITIONAL ASPECTS OF HEART FAILURE

18.2.1 Heart Failure as a Metabolic Syndrome

Heart failure is a chronic progressive disorder characterized by a host of neurohormonal, immunologic, and metabolic derangements (Table 18.1) *(33)*. In acute heart failure, activation of the sympathetic nervous system and renin-angiotensin-aldosterone axis serves to maintain cardiac output and preserve tissue perfusion. However, chronic

Table 18.1
Neurohormonal and metabolic abnormalities in chronic heart failure

Neurohormonal abnormalities
    Activation of sympathetic nervous system
        • Increased circulating norepinephrine and epinephrine
        • Sympathovagal imbalance with sympathetic dominance
    Activation of renin-angiotensin-aldosterone system
        • Increased angiotensin II levels
        • Increased aldosterone levels
    Increased atrial and brain natriuretic peptide levels
    Increased levels of endothelin-1
    Increased levels of vasopressin (anti-diuretic hormone)
    Increased cortisol levels
    Decreased levels of dehydroepiandrosterone (DHEA)*
    Increased insulin levels in noncachectic patients
    Increased growth hormone
    Normal or reduced insulin-like growth factor-1 (IGF-1)
    Thyroid dysfunction

Metabolic abnormalities
    Increased basal metabolic rate (BMR)
    Impaired peripheral blood flow (decreased nutrient delivery)
    Altered protein and fat metabolism
    Insulin resistance*

*denotes anti-anabolic effect.

activation of these systems is deleterious and perpetuates progression of the heart failure syndrome. Indeed, current therapy for heart failure focuses on antagonizing the harmful effects of these two neurohormonal pathways through the use of beta-blockers, ACE inhibitors, angiotensin receptor blockers, and aldosterone antagonists.

In addition to activation of neurohormonal systems, chronic heart failure is associated with immunological dysregulation, as evidenced by increased levels of circulating tumor necrosis factor-alpha (TNF-α), interleukins 1 and 6, soluble adhesion molecules, and certain leukocyte chemokines *(34, 35)*. Activation of these cytokines likely plays a pivotal role in the apoptosis (programmed cell death) and anorexia which are features of chronic heart failure. In addition, some cytokines may exert direct cardiotoxic effects (e.g., through increased oxygen free radical activity), thereby contributing to heart failure progression.

Many of the neurohormonal and immulogic abnormalities in chronic heart failure are also associated with important effects on metabolism. While the mechanisms underlying these effects are complex and not fully understood, the net effect is characterized by an imbalance between catabolic (tissue-wasting) and anabolic (tissue-building) factors *(36)*. Cardinal features of advanced heart failure include an increase in basal metabolic rate (BMR) *(37)*, altered protein and fat metabolism, and impaired peripheral blood flow with reduced nutrient delivery to bodily tissues. In the chronic setting of advanced heart failure, these effects lead to tissue wasting and loss of lean body mass *(38)*.

18.2.2 Cardiac Cachexia

Hippocrates's early description of "dropsy" (i.e., heart failure) cited at the beginning of this chapter provides a remarkably apt characterization of cardiac cachexia. While tissue wasting and loss of muscle mass occur early in chronic heart failure, marked tissue wasting and cachexia are hallmarks of advanced or end-stage heart failure.

Cardiac cachexia has been defined as a 6% or greater decline in lean body mass over a period of 6 months or more *(39)*. Estimates of the prevalence of cachexia in patients with heart failure vary widely, but typically range from 10 to 20%. Importantly, cachexia differs markedly from starvation (e.g., due to anorexia nervosa). Although patients with heart failure may exhibit signs of malnutrition, in cachexia there tends to be a greater loss of lean body mass, principally muscle mass but also bone mass, whereas in starvation there is preferential loss of adipose tissue in the early stages, with subsequent loss of muscle mass as malnutrition progresses *(38)*. In addition, prolonged starvation is almost invariably associated with a very low body mass index (BMI), whereas patients with cardiac cachexia may experience only modest reductions in body weight, in part due to increased extracellular fluid accumulation (edema), as well as conversion of muscle tissue to fat.

As noted above, the complex cascade of metabolic disturbances leading to cardiac cachexia is incompletely understood. However, circulating levels of TNF-α are invariably elevated in patients with cardiac cachexia; indeed, the TNF-α level is the strongest predictor of weight loss in heart failure patients *(36)*. It is currently unknown whether TNF-α plays a direct pathophysiological role in the development of cardiac cachexia or if it merely serves as a marker for the cachectic state (and the severity of heart failure), but circulating TNF-α levels are a strong independent predictor of mortality in heart failure patients. Unfortunately, early clinical trials of TNF-α blocking agents have been disappointing (see below).

Other circulating factors that may play a role in cardiac cachexia include endotoxin, leptin, and ghrelin. Circulating endotoxin may arise when extensive fluid overload causes gut edema, allowing bacterial toxins to move from the bowel to the bloodstream. This leads to increased inflammatory cytokines with systemic effects as noted above. Leptin is a peptide that acts centrally to reduce food intake and increase energy expenditure, but leptin levels have not been shown to reliably predict nutritional status or prognosis in heart failure patients *(40)*. In contrast to leptin, ghrelin is a centrally acting peptide produced in the stomach that *stimulates* food intake and can increase circulating levels of growth hormone (GH). Administration of ghrelin has shown promising results in both animal models of heart failure as well as in a small clinical trial, in which heart failure patients treated with ghrelin for 3 weeks had increased left ventricular function, exercise capacity, and lean body mass *(41, 42)*.

18.2.3 Caloric Intake, Fat, and Protein

The occurrence of weight loss in patients with advanced heart failure is somewhat paradoxical, since heart failure is often associated with reduced physical activity. Furthermore, although some heart failure patients develop anorexia, due either to heart failure itself or as a result of medications (e.g., digoxin, captopril), in the absence of anorexia caloric intake is similar in heart failure patients (including those with cachexia) and persons without heart failure *(43)*. This combination of preserved caloric intake and apparently reduced activity would be expected to result in weight gain rather than weight loss. What, then, accounts for the net loss in non-edematous body mass so frequently encountered in end-stage heart failure patients?

First, as noted above, most studies have shown that despite reduced muscle mass, there is an increase in BMR in most patients with heart failure, most likely due to increased energy requirements for respiration and the generalized catabolic state arising from neurohormonal dysregulation (esp. increased circulating catecholamines) *(37, 44)*. Second, although caloric intake is maintained, there is evidence that fat absorption is impaired in patients with heart failure, perhaps due to bowel edema *(45)*. In addition, although intestinal handling of protein appears to be preserved *(46)*, alterations in protein and carbohydrate metabolism result in impaired delivery of these nutrients to the body's tissues *(38)*. In sum, weight loss in heart failure is likely due to both increased energy utilization and decreased availability of fat, protein (amino acids), and carbohydrates despite "normal" caloric intake.

Management of cardiac cachexia is problematic, and perhaps the best approach is to focus on prevention. In this regard, preliminary data suggest that both ACE inhibitors and beta-blockers may exert favorable effects on weight loss in heart failure patients. In a post hoc analysis of the Studies of Left Ventricular Dysfunction (SOLVD) trial, patients receiving the ACE inhibitor enalapril were 19% less likely to experience weight loss of 6% or more during a mean follow-up period of 35 months compared to patients receiving placebo (p = 0.05) *(47)*. Although cachexia is an ominous sign associated with significant mortality, there is some evidence that pharmacologic treatment of HF can lead to partial reversal of the cachectic state. In a small observational study involving 13 heart failure patients with cachexia, 6 months of treatment with either carvedilol or metoprolol was associated with significant weight gain accompanied by favorable effects on plasma norepinephrine and leptin levels *(48)*. In another study of eight patients with advanced heart failure and cardiac cachexia, treatment with the combination of digoxin, the ACE inhibitor enalapril, and the loop diuretic furosemide was associated with significant clinical improvement as well as increased muscle bulk, subcutaneous fat, and serum albumin and hematocrit levels *(49)*. In contrast to these studies, two trials of etanercept, a tumor necrosis factor (TNF) blocking agent, failed to demonstrate significant benefit and were discontinued *(50)*. These findings were particularly disappointing since TNF-α may play a role in the development of cardiac cachexia.

Beyond pharmacologic approaches to prevent or treat cachexia, a few studies have examined the role of nutritional support in advanced heart failure patients. In one small randomized trial, a high caloric diet failed to result in significant changes in nutritional status or clinical outcomes of patients with advanced heart failure *(51)*. In another study, patients with moderate to severe heart failure and malnutrition received high-energy nasogastric tube feedings for 2 weeks *(52)*. Total body weight and extracellular fluid weight declined, but lean body mass increased. There was, however, no change in oxygen consumption or cardiac function *(53)*. A third study involving only six patients undergoing mitral valve surgery showed that perioperative nutritional support was associated with improved clinical status and stable cardiac function. Finally, in a study of eight patients with cardiac cachexia, infusion of branched-chain amino acids had no discernible effect on protein metabolism *(54)*. In summary, there are insufficient data at present to assess the impact of various modes of nutritional support on metabolic parameters or clinical outcomes in patients with advanced heart failure.

18.2.4 Heart Failure and Obesity

The epidemic of obesity in the United States and elsewhere has led to an increasing proportion of heart failure patients who are overweight or obese. While the preponderance of evidence indicates that patients with heart failure and increased body mass index (BMI) have better outcomes than those with cachexia or low BMI ($<18–20$ kg/m^2) *(55)*, the presence of obesity nonetheless warrants attention in heart failure patients. Obesity is independently associated with the development of heart failure as well as with hypertension, diabetes, obstructive sleep apnea, pulmonary hypertension, and other comorbidities that adversely affect the heart failure patient. Obesity may also result in misdiagnosis of heart failure, since exertional shortness of breath may be due to pulmonary hypertension or physical deconditioning, while lower extremity edema may be caused by venous insufficiency. Conversely, sedentary obese patients may not experience significant shortness of breath until heart failure is far advanced, and early diagnosis of heart failure, at a stage when treatment is more likely to be effective, can easily be missed. Finally, the excess weight carried by overweight heart failure patients results in increased cardiac work, which strains the already weakened heart. Indeed, obesity is often associated with increased heart rate and vascular resistance, effects that not only increase cardiac work but which are diametrically opposed to the actions of beta-blockers and ACE inhibitors. Thus, heart failure patients who are significantly overweight or obese should generally be counseled to gradually lose weight through a combination of increased physical activity and a modest reduction in caloric intake*(18)*. When feasible, this should be done under the guidance of a nutritionist or exercise therapist to minimize potential risks, particularly in elderly patients*(56)*(see also Chapter 15).

18.3 SPECIFIC NUTRIENTS IN HEART FAILURE

18.3.1 Water and Sodium

Activation of the renin-angiotensin-aldosterone system in patients with heart failure results in sodium and water retention. As a result, untreated heart failure is usually associated with increased total body water and total body sodium. Of note, total body sodium is generally increased even when serum sodium levels are reduced (i.e., hyponatremia). This situation occurs in patients with advanced heart failure because fluid retention is more pronounced than sodium retention, in part due to the action of vasopressin (anti-diuretic hormone). Indeed, hyponatremia in patients with heart failure is associated with more severe hemodynamic and neurohormonal disturbances and is a marker for poor prognosis *(57)*.

Diuretics are the mainstay of therapy for fluid overload in heart failure patients. Ideally, diuretic dosages are adjusted to maintain a normal state of hydration (i.e., "euvolemia"). However, over- and under-diuresis are both common, so that at any given time, a patient may be volume-overloaded, euvolemic, or relatively dehydrated, and careful assessment of volume status is thus essential in managing heart failure patients. From the practical standpoint, the simplest way to do this is by monitoring daily weights. Patients should be instructed to weigh themselves every morning without clothing, after voiding, and before eating, and weights should be recorded on a daily weight chart. An optimal or "dry" weight should be established, and variances of more than 2–3 lbs in either direction should lead to adjustments in diuretic dosage. The rationale behind this approach is that short-term variability in body weight primarily reflects changes in total body water. Note, however, that nonedematous weight may change over longer periods of time, usually decreasing but occasionally increasing if the overall nutritional status improves. Therefore, periodic reassessment of the patient's desirable weight is appropriate.

In addition to monitoring daily weights and adjusting diuretic dosages, dietary sodium restriction plays a pivotal role in maintaining normal volume status and avoiding acute heart failure exacerbations, as evidenced by the fact that several studies have shown that dietary sodium excess is a common precipitant of repetitive heart failure hospitalizations *(12, 58)*. Dietary sodium excess contributes to fluid retention, and an acute dietary sodium load (e.g., potato chips, canned soup, "fast food") may result in a sudden increase in intravascular blood volume, triggering a rise in intracardiac pressures and precipitating acute heart failure. Older patients with diastolic heart failure are particularly sensitive to salt intake and changes in blood volume and are therefore less tolerant of a salt load. While there are no clinical trial data demonstrating improved outcomes with sodium restriction, it is standard of care that heart failure patients, family members, and other caregivers should be educated about the importance of avoiding high-sodium foods and limiting daily sodium intake to no more than about 2 g *(18)*. Although some patients may find it difficult to adhere to a sodium-restricted diet, careful instruction and guidance from a dietitian is often effective in overcoming this barrier. In contrast to sodium restriction, fluid restriction is not usually required for most patients with mild to moderate heart failure unless significant renal impairment is also present. However, patients should be advised to avoid *excess* fluid intake; i.e., the oft-quoted dictum to "drink 8–10 glasses of water every day" does not apply to patients with heart failure. In addition, patients with advanced heart failure accompanied by hyponatremia may benefit from more stringent fluid restriction, e.g., 1.5 L/day total fluid intake.

18.3.2 Other Electrolytes

Apart from their effect on body water, diuretics have important effects on key electrolytes, including sodium, potassium, chloride, magnesium, and calcium. Thiazide and "loop" diuretics (furosemide, bumetanide, torsemide), as well as metolazone, promote urinary loss of sodium, potassium, chloride, and magnesium. As a result, these classes of diuretics may be associated with hyponatremia, hypokalemia, hypochloremia, and hypomagnesemia. In addition, loop diuretics increase calcium excretion and may contribute to a negative calcium balance, although hypocalcemia due to loop diuretics is uncommon. Conversely, the potassium-sparing diuretics spironolactone, eplerenone, triameterine, and amiloride, as well as the ACE inhibitors and angiotensin receptor blockers, are all associated with potassium retention and may occasionally induce significant hyperkalemia. For these reasons, serum electrolytes should be monitored periodically in patients receiving long-term diuretic therapy, especially during periods of dosage adjustment.

Diet and nutrition play an important role in managing electrolytes in heart failure patients. Unfortunately, many trials evaluating micronutrient supplementation in cardiac disease states have excluded patients with heart failure *(59)*. Despite this limitation, some general principles apply. Patients with heart failure and preserved renal function should consume a diet rich in potassium, magnesium, and calcium, but low in sodium. Most patients on chronic loop diuretic therapy will require potassium replacement, either through high-potassium foods (e.g., fresh fruits) or as potassium supplements (usually administered as potassium chloride, which also aids in chloride replacement). Diuretic-induced hyponatremia is potentially life threatening and may require hospitalization (e.g., if the serum sodium concentration falls to <120–125 meq/L). Treatment includes fluid restriction, reduction in diuretic dosage, and temporary liberalization of sodium intake. Hypomagnesemia is relatively common during long-term diuretic therapy, but may be overlooked unless serum magnesium levels are assessed. Importantly, magnesium deficiency may contribute to muscle fatigue. Treatment consists of dietary therapy and magnesium supplements. Patients with chronic heart failure often suffer bone loss (osteopenia) due to low levels of vitamin D and secondary hyperparathyroidism *(33)*. However, the value of calcium supplements, with or without vitamin D, in heart failure patients is currently unknown.

18.3.3 Other Minerals

Zinc, manganese, copper, and selenium all have anti-oxidant effects, and deficiencies of these minerals may be associated with increased lipid peroxidation and oxidative stress *(60)*. In addition, severe copper and selenium deficiency have been associated with cardiomyopathies in humans *(61, 62)*, while zinc and manganese deficiency have been associated with myocardial contractile dysfunction in laboratory animals *(63, 64)*. Diuretics appear to increase urinary zinc excretion, and clinically significant zinc deficiency is common in older heart failure patients on chronic diuretic therapy *(65)*. Conversely, serious deficiencies of manganese, copper, and selenium occur infrequently in older adults consuming a normal diet. Based on currently available data, daily intake of each of these minerals should be sufficient to meet Dietary Reference Intakes (DRIs). Although some patients with diuretic-associated zinc deficiency may benefit from zinc supplements, there are currently no data to support routine use of such supplements in older heart failure patients.

Iron is essential for the production of hemoglobin, and iron deficiency is common in older adults. Although iron deficiency has no known direct cardiotoxic effects, chronic anemia leads to an increase in cardiac work in order to preserve tissue oxygen delivery and in severe cases may lead to high-output cardiac failure. Conversely, iron overload due to multiple blood transfusions or hemochromatosis has been associated with restrictive cardiomyopathy *(66)*. Thus, iron intake should be sufficient to maintain tissue stores and prevent chronic iron deficiency anemia, but excess iron intake should be avoided.

18.3.4 Vitamins

Vitamin B_1 (thiamin) deficiency impairs oxidative metabolism and has been unequivocally linked to high-output cardiac failure *(60, 67)*. In addition, thiamin deficiency may contribute to "diuretic resistance" in patients receiving moderate to high doses of loop diuretics over a prolonged period of time *(68, 69)*. In the United States, clinically important thiamin deficiency is most commonly encountered in alcoholics and in older heart failure patients treated with loop diuretics. Of note, both digoxin and furosemide diminish uptake of thiamin by cardiac myocytes, and the effects of these drugs are additive *(70)*. Thiamin deficiency responds promptly to either oral or parenteral thiamin administration and is usually associated with substantial improvement in cardiac function and symptoms. Although chronic thiamin supplementation may be considered in selected high-risk populations (e.g., alcoholics and poorly nourished older adults treated with high-dose loop diuretics), in most cases maintaining a well-balanced diet will ensure adequate thiamin intake.

Vitamin C supplementation has been associated with improved endothelial function *(71–73)*, and some epidemiologic studies have suggested that increased intake of vitamin C correlates with reduced risk for cardiovascular disease *(74–76)*. However, there is no convincing evidence that vitamin C deficiency contributes to the development of heart failure or that vitamin C supplements are beneficial in heart failure patients *(60)*.

Vitamin E has anti-oxidant properties and reduces platelet adhesion *(77)*, and several epidemiologic studies have reported that diets high in vitamin E, alone or in combination with vitamin C, are associated with a lower incidence of coronary heart disease *(78–81)*. In contrast, several large randomized trials of vitamin E therapy failed to show significant benefit, and one meta-analysis suggested that high-dose vitamin E intake may be associated with increased mortality *(82, 83)*. Moreover, follow-up analysis of the Heart Outcomes Prevention Evaluation (HOPE) trial and its extension (HOPE-TOO) showed that vitamin E therapy was associated with increases in the incidence of heart failure and in the risk of hospitalization for heart failure *(84)*.

Deficiencies of folic acid, vitamin B_6, and vitamin B_{12} are common in older adults and contribute to age-associated increases in homocysteine levels *(85)*. In addition, elevated homocysteine is an established marker for increased risk of coronary and cerebrovascular diseases in both older and younger adults *(86–88)*, and elevated homocysteine levels have also been associated with more severe heart failure and poorer prognosis *(89)*. However, although plausible mechanisms for an adverse effect of homocysteine on myocardial function have been proposed, there is currently no convincing evidence that reduction in homocysteine levels through the use of folic acid and B-vitamin supplements lowers the risk of coronary or cerebrovascular events or improves myocardial function or outcomes in heart failure patients *(89)*.

Vitamin D is essential for maintaining normal calcium homeostasis, and marked vitamin D deficiency has been associated with decreased contractility in laboratory animals *(90)*. Vitamin D deficiency is common in older adults with or without heart failure *(91, 92)*, and although vitamin D supplementation is appropriate in these individuals, there is no evidence that such treatment alters the clinical course of patients with heart failure.

Although high-dose niacin is an effective agent for the treatment of dyslipidemia, there is no evidence that niacin deficiency contributes to the development of cardiovascular disease *(60)*. Likewise, low beta-carotene intake has been associated with increased risk for myocardial infarction *(93)*, but there is no evidence that vitamin A levels correlate with heart failure risk or that vitamin A supplements are useful in the prevention of cardiovascular disease *(60)*. Similarly, there are no established links between vitamins B_2 (riboflavin) and B_{17} (pantothenic acid) and either the development or treatment of cardiac disorders *(60)*.

18.3.5 Other Nutritional Supplements

Despite continued interest in the anti-oxidant coenzyme Q_{10} (ubiquinone), its role in the pathophysiology and treatment of heart failure remains controversial. Myocardial coenzyme Q_{10} levels are reduced in patients with heart failure, and low plasma coenzyme Q_{10} levels are associated with increased mortality *(60)*. In addition, the HMG-CoA reductase inhibitors ("statins"), which are commonly used to treat hyperlipidemia, have been associated with depletion of coenzyme Q_{10} *(94)*. Observational studies and some (but not all) small randomized trials indicate that coenzyme Q_{10} supplementation may improve LV function, symptoms, and exercise tolerance *(95–99)*. However, there are no large randomized controlled trials, and routine administration of coenzyme Q_{10} is not recommended – even in patients on statin therapy *(100)*.

Carnitine and creatine phosphate are nutritional supplements which may enhance skeletal muscle performance in some patients with heart failure *(101–103)*, but there is little evidence that oral administration improves cardiac function. In addition, there is no evidence that these agents improve long-term clinical outcomes in heart failure patients, and there are also concerns about the safety of these agents during chronic use *(60)*.

18.3.6 Multinutrient Therapy

Older heart failure patients often have multiple nutritional deficiencies, suggesting that therapeutic interventions may need to be broad based, rather than focusing on a single or even a relatively small number of micronutrients. This concept is supported by a recently reported trial, in which 30 heart failure patients over age 70 (mean age 75 years) were randomized to receive high-dose micronutrient capsules or placebo in double-blind fashion *(104)*. The capsules contained calcium, magnesium, zinc, copper, selenium, vitamin A, thiamin, riboflavin, vitamin B_6, folate, vitamin B_{12}, vitamin C, vitamin D, vitamin E, and coenzyme Q_{10}. During a 9-month follow-up period, patients receiving the micronutrient capsules demonstrated decreased left ventricular volumes, an increase in left ventricular ejection fraction (mean 5.3%), and improved quality of life scores, whereas no changes occurred in the placebo group. Although these findings require replication in a large prospective randomized trial, they provide preliminary evidence that a combination of nutritional supplements may be beneficial in older heart failure patients.

18.4 IMPACT OF HEART FAILURE MEDICATIONS AND AGE ON NUTRITIONAL PARAMETERS

18.4.1 Medication Effects

As mentioned previously, many of the agents used in the treatment of chronic heart failure may have a beneficial effect on nutritional status. Conversely, there is a risk for drug-related side effects and a potentially negative impact on the nutritional status of these patients. Diuretics directly impact fluid and electrolyte homeostasis, and diuretic-induced electrolyte abnormalities are very common. In addition, loop diuretics have been associated with thiamin deficiency, and thiazide diuretics in particular may adversely affect carbohydrate and lipid metabolism. Digoxin may be associated with nausea and anorexia, and these symptoms may occur in older patients even at therapeutic dosages. The ACE inhibitor captopril occasionally causes dysgeusia (altered taste), nausea, and anorexia, and other ACE inhibitors may be associated with similar side effects, although less frequently. Beta-blockers may also influence carbohydrate and lipid metabolism, and depressive symptoms, including reduced appetite, may occur in older patients treated with these agents. Finally, the calcium channel blockers diltiazem and especially verapamil are commonly associated with constipation in older individuals.

18.4.2 Age-Specific Nutritional Issues

Older age is associated with increased risk for a broad range of nutritional deficiencies, and this risk is potentiated by the presence of cardiovascular disease in general and by heart failure in particular. In addition, older adults are more susceptible to the adverse effects of pharmacological agents and dietary interventions on nutritional parameters, in part due to pre-existing nutritional deficiencies coupled with an increased prevalence of comorbid conditions. The latter issue may be particularly problematic, since the presence of several common comorbidities, e.g., coronary artery disease, diabetes mellitus, and renal insufficiency, may lead to serial dietary restrictions (low fat, low carbohydrate, low protein, low salt) culminating in a diet that is unpalatable and severely deficient in both calories and essential nutrients. It is therefore critically important that an appropriately detailed nutritional evaluation, including dietary history, body weight, selected laboratory tests (hemoglobin, albumin, cholesterol, electrolytes, creatinine, blood urea nitrogen), and in some cases anthropometric assessments, be incorporated into the routine management of older patients with chronic illnesses, including heart failure.

18.5 NUTRITIONAL GUIDELINES

Nutritional guidelines for managing chronic heart failure in older adults are summarized in Table 18.2. As noted previously, nutritional management of older patients begins with a nutritional assessment, ideally with the assistance of an experienced dietitian or nurse. As with other chronic illnesses, the guiding principle in making nutritional recommendations to older heart failure patients is, first and foremost, maintenance of a well-balanced diet with sufficient calories, nutrients, and fluids to meet daily requirements (see Table 18.3). In addition, the diet should be both palatable

Table 18.2
Nutritional guidelines for older adults with chronic heart failure

Component	Recommendation
Nutritional assessment – basic (all patients)	Obtain detailed dietary history; Assess body weight and habitus; Laboratory: hemoglobin, serum albumin, cholesterol, serum electrolytes (sodium, potassium, calcium, phosphorus, magnesium), creatinine, blood urea nitrogen
Supplemental (selected patients)	Anthropometric measures (e.g., skinfold thickness) Determination of lean body mass, folate and B_{12} levels, bone mineral density
General diet	Well balanced, rich in fruits and vegetables, whole grains, dairy products, lean meats
Caloric intake	Sufficient to maintain lean body mass; 1600–2000 calories/day in most cases
Protein	15–20% of total calories
Fat	25–30% of total calories
Complex carbohydrates	50–60% of total calories
Fluids	About 2 L/day
	1.5 L/day in setting of hyponatremia, severe renal failure, diuretic resistance; Avoid excess fluid intake
Electrolytes	
Sodium	2 g Na^+/day
Potassium, calcium, magnesium	Sufficient to maintain body stores and serum levels; supplement as indicated
Minerals	
Zinc, copper, manganese, selenium	Sufficient intake to meet DRIs; zinc supplements in selected patients
Iron	Sufficient to maintain body stores; avoid iron overload
Vitamins	
Thiamin (vitamin B_1)	Supplement in alcoholics, possibly patients on chronic high-dose loop diuretics
Folate, B_6, and B_{12}	Supplement if deficient (common)

(continued)

Table 18.2
(continued)

Component	Recommendation
Cholecalciferol (vitamin D$_3$)	Supplement if deficient, esp. if osteoporosis present (common)
Beta-carotene	No known relation to heart failure; maintain DRI
Riboflavin (vitamin B$_2$)	
Niacin (vitamin B$_3$)	
Ascorbic acid (vitamin C)	
Alpha-tocopherol (vitamin E)	
Dietary supplements	
Ubiquinone (coenzyme Q$_{10}$)	Unproven benefit, not recommended
Carnitine	
Creatine phosphate	

DRI: Dietary Reference Intake.

Source: Food and Nutrition Board, Institute of Medicine. Dietary Reference Intakes: The Essential Guide to Nutrient Requirements, The National Academies Press, 2006.

Table 18.3
RECOMMENDATIONS

1. Nutritional management begins with a nutritional assessment. Obtain baseline body weight and monitor at regular intervals. Recognize that caloric needs may change as body weight changes and provide increased nutritional support if rapid unintentional weight loss occurs.
2. Patients should be encouraged to choose a well-balanced diet high in fruits and vegetables as excellent sources of vitamins and electrolytes.
3. Sodium restriction to 2 g/day is usually sufficient; dietary counseling may be required to assist patients in achieving this goal.
4. Fluid restriction is not usually necessary except when hyponatremia, severe renal failure, or advanced heart failure is present; in these cases, fluid should be restricted to ~1.5 L/day. Excess fluid intake ("8–10 glasses of water per day") should be avoided.
5. Magnesium, potassium, and calcium levels should be monitored and supplemented as needed. Other nutrients of concern in high-risk patients include thiamin, folate, vitamin B$_{12}$, vitamin D, and zinc. Routine daily use of an oral vitamin/mineral supplement may be helpful in alleviating any deficits.

and "accessible", i.e., within the patient's financial means and physical capabilities. Few older patients with heart failure require a weight reduction diet, since body weight correlates inversely with mortality in heart failure patients *(55)*. Indeed, in most cases it is appropriate to prescribe a diet that will either maintain current nonedematous weight or promote a modest increase in lean body mass. Although it has been suggested that the proportion of calories derived from protein and fat should perhaps be increased in older heart failure patients *(38)*, there is little evidence to support this contention, and current recommendations are that 15–20% of total calories be derived from protein, 25–30% from fat, and the remaining 50–60% from complex carbohydrates.

Moderate dietary sodium restriction, such as a 2-g sodium diet, is appropriate for most patients with heart failure *(18)*. Patients should be instructed to avoid high-sodium foods, such as canned soups and sauces, tomato juice, most prepared lunch meats and pre-packaged frozen entrees, pickles, "fast foods", and certain ethnic foods which are high in sodium (e.g., Chinese cuisine). Moderate use of salt during cooking is acceptable, but use of salt at the table should be avoided. Dining out is potentially problematic, and patients should be advised to call ahead to see if low-sodium options are available either on the menu or by request. Patients should also be instructed about the widespread availability of alternative seasonings which contain little or no salt.

Fluid intake should be adequate to maintain hydration while avoiding volume overload. In patients with preserved renal function, about 2 L of fluid per day is appropriate. Excess fluid intake ("8–10 glasses of water a day") should be avoided, but fluid restriction is unnecessary in the absence of hyponatremia, severe renal failure, or advanced heart failure with diuretic resistance. In such cases, fluid intake should be limited to about 1.5 L/day. Alcohol should be minimized or avoided, especially in cases of alcohol-induced cardiomyopathy.

Dietary potassium, calcium, and magnesium requirements vary considerably depending on medications, renal function, and comorbid conditions (e.g., osteoporosis). As a general principle, a well-balanced diet rich in fresh fruits and vegetables, whole grain breads and cereals, and dairy products will provide sufficient amounts of potassium, calcium, and magnesium to meet normal needs. However, many older heart failure patients will require supplemental administration of one or more of these electrolytes to overcome losses through urinary excretion or as a result of other metabolic abnormalities. Since individual requirements cannot easily be predicted, periodic assessment of serum electrolyte levels is appropriate.

As discussed above, the importance of most vitamins and other micronutrients in the pathogenesis and treatment of chronic heart failure has not been well characterized, and it is therefore difficult to make specific nutritional recommendations in most cases. However, since older patients are at increased risk for multiple nutritional deficiencies, it is appropriate to maintain a high index of suspicion, particularly in frail, socially isolated, or institutionalized elders, as well as those with multiple comorbidities and those receiving multiple medications. In particular, deficiencies of folate, B_{12}, vitamin D, and zinc are common, and dietary or pharmacological supplementation is indicated when specific deficiencies are identified or suspected. In addition, since long-term administration of loop diuretics may deplete thiamin stores, thiamin replacement should be considered in such cases, particularly in the setting of increasing diuretic resistance. Finally, although evidence supporting high-dose multivitamin and mineral supplements in older heart failure patients is sparse, daily use of a non-prescription multivitamin and mineral supplement may ease concerns about multinutrient deficiencies and is unlikely to be harmful. Conversely, the use of other dietary supplements, such as coenzyme Q_{10}, carnitine, or creatine phosphate, is not currently recommended, and patients should be screened for the use of these and other neutraceuticals.

REFERENCES

1. Katz AM, Katz PB. Diseases of the heart in the works of Hippocrates. Br Heart J 1961;24:257–64.
2. Rosamond W, Flegal K, Friday G, et al. Heart disease and stroke statistics – 2007 update: a report from the American Heart Association Statistics Committee and Stroke Statistics Subcommittee. Circulation 2007;115:e69–171.
3. Kannel WB, Belanger AJ. Epidemiology of heart failure. Am Heart J 1991;121:951 7.
4. DeFrances CJ, Podgornik MN. 2004 National Hospital Discharge Survey. Advance data from vital and health statistics; no. 371. Hyattsville, MD: National Center for Health Statistics, 2006.
5. Kozak LJ, DeFrances CJ, Hall MJ. National Hospital Discharge Survey: 2004 annual summary with detailed diagnosis and procedure data. National Center for Health Statistics. Vital Health Stat 2006:13(162).
6. O'Connell JB. The economic burden of heart failure. Clin Cardiol 2000;23 (suppl):III6–10.
7. Gottdiener JS, Arnold AM, Aurigemma GP, et al. Predictors of congestive heart failure in the elderly: the Cardiovascular Health Study. J Am Coll Cardiol 2000;35:1628–37.
8. Levy D, Larson MG, Vasan RS, Kannel WB, Ho KK. The progression from hypertension to congestive heart failure. JAMA 1996;275:1557–62.
9. Vasan RS, Larson MG, Benjamin EJ, Evans JC, Reiss CK, Levy D. Congestive heart failure in subjects with normal versus reduced left ventricular ejection fraction: prevalence and mortality in a population-based cohort. J Am Coll Cardiol 1999;33:1948–55.
10. Kitzman DW, Gardin JM, Gottdiener JS, et al. Importance of heart failure with preserved systolic function in patients > 65 years of age. CHS Research Group. Cardiovascular Health Study. Am J Cardiol 2001;87:413–9.
11. Croft JB, Giles WH, Pollard RA, Keenan NL, Casper ML, Anda RF. Heart failure survival among older adults in the United States: a poor prognosis for an emerging epidemic in the Medicare population. Arch Intern Med 1999;159:505–10.
12. Vinson JM, Rich MW, Shah AS, Sperry JC. Early readmission of elderly patients with congestive heart failure. J Am Geriatr Soc 1990;38:1290–5.
13. Krumholz HM, Parent EM, Tu N, et al. Readmission after hospitalization for congestive heart failure among Medicare beneficiaries. Arch Intern Med 1997;157:99–104.
14. Ho KK, Anderson KM, Kannel WB, Grossman W, Levy D. Survival after the onset of congestive heart failure in Framingham Heart Study subjects. Circulation 1993;88:107–15.
15. Pernenkil R, Vinson JM, Shah AS, Beckham V, Wittenberg C, Rich MW. Course and prognosis in patients > 70 years of age with congestive heart failure and normal versus abnormal left ventricular ejection fraction. Am J Cardiol 1997;79:216–9.
16. Senni M, Tribouilloy CM, Rodeheffer RJ, et al. Congestive heart failure in the community: a study of all incident cases in Olmsted County, Minnesota, in 1991. Circulation 1998;98:2282–9.
17. Chinnaiyan KM, Alexander D, Maddens M, McCullough PA. Curriculum in cardiology: integrated diagnosis and management of diastolic heart failure. Am Heart J 2007;153:189–200.
18. Heart Failure Society of America. Executive summary: HFSA 2006 comprehensive heart failure practice guideline. J Card Fail 2006;12:10–38.
19. Rich MW, Beckham V, Wittenberg C, Leven CL, Freedland KE, Carney RM. A multidisciplinary intervention to prevent the readmission of elderly patients with congestive heart failure. N Engl J Med 1995;333:1190–5.
20. McAlister FA, Lawson FM, Teo KK, Armstrong PW. A systematic review of randomized trials of disease management programs in heart failure. Am J Med 2001;110:378–84.
21. Flather MD, Yusuf S, Køber L, et al. Long-term ACE-inhibitor therapy in patients with heart failure or left-ventricular dysfunction: a systematic overview of data from individual patients. Lancet 2000;355:1575–81.
22. Pitt B, Poole-Wilson PA, Segal R, et al. Effect of losartan compared with captopril on mortality in patients with symptomatic heart failure: randomized trial – the Losartan Heart Failure Survival Study ELITE II. Lancet 2000;355:1582–7.

23. Cohn JN, Tognoni G, for the Valsartan Heart Failure Trial Investigators. A randomized trial of the angiotensin-receptor blocker valsartan in chronic heart failure. N Engl J Med 2001;345: 1667–75.

24. Cohn JN, Archibald DG, Ziesche S, et al. Effect of vasodilator therapy on mortality in chronic congestive heart failure. Results of a Veterans Administration Cooperative Study. N Engl J Med 1986;314:1547–52.

25. Cohn JN, Johnson G, Ziesche S, et al. A comparison of enalapril with hydralazine-isosorbide dinitrate in the treatment of chronic congestive heart failure. N Eng J Med 1991;325:303–10.

26. CIBIS-II Investigators and Committees. The Cardiac Insufficiency Bisoprolol Study II (CIBIS II): a randomized trial. Lancet 1999;353:9–13.

27. Effect of metoprolol CR/XL in chronic heart failure: Metoprolol CR/XL Randomised Intervention Trial in Congestive Heart Failure (MERIT-HF). Lancet 1999;353:2001–7.

28. The Digitalis Investigation Group. The effect of digoxin on mortality and morbidity in patients with heart failure. N Engl J Med 1997;336:525–33.

29. Rich MW, McSherry F, Williford WO, Yusuf S, for the Digitalis Investigation Group. Effect of age on mortality, hospitalizations and response to digoxin in patients with heart failure: The DIG Study. J Am Coll Cardiol 2001;38:806–13.

30. Pitt B, Zannad F, Remme WJ, et al. The effect of spironolactone on morbidity and mortality in patients with severe heart failure. Randomized Aldactone Evaluation Study Investigators. N Engl J Med 1999;341:709–17.

31. Yusuf S, Pfeffer MA, Swedberg K, Granger CB, Held P, McMurray JJ, Michelson EL, Olofsson B, Ostergren J; CHARM Investigators and Committees. Effects of candesartan in patients with chronic heart failure and preserved left-ventricular ejection fraction: the CHARM-Preserved Trial. Lancet 2003;362:777–81.

32. Cleland JG, Tendera M, Adamus J, Freemantle N, Polonski L, Taylor J; PEP-CHF Investigators. The perindopril in elderly people with chronic heart failure (PEP-CHF) study. Eur Heart J 2006;27:2338–45.

33. Anker SD, Rauchhaus M. Heart failure as a metabolic problem. Eur J Heart Failure 1999;1: 127–31.

34. Anker SD, Rauchhaus M. Insights into the pathogenesis of chronic heart failure: immune activation and cachexia. Curr Opin Cardiol 1999;14:211–6.

35. Berry C, Clark AL. Catabolism in chronic heart failure. Eur Heart J 2000;21:521–32.

36. Anker SD, Chua TP, Ponikowski P, et al. Hormonal changes and catabolic/anabolic imbalance in chronic heart failure and their importance for cardiac cachexia. Circulation 1997;96:526–34.

37. Poehlman ET, Scheffers J, Gottlieb SS, Fisher ML, Vaitkevicius P. Increased resting metabolic rates in patients with congestive heart failure. Ann Intern Med 1994;121:860–2.

38. Freeman LM, Roubenoff R. The nutrition implications of cardiac cachexia. Nutr Rev 1994;52: 340–7.

39. Anker SD, Ponikowski P, Varney S, et al. Wasting as independent risk factor for mortality in chronic heart failure. Lancet 1997;349:1050–3.

40. Murdoch DR, Rooney E, Dargie HJ, Shapiro D, Morton JJ, McMurray JJ. Inappropriately low plasma leptin concentration in the cachexia associated with chronic heart failure. Heart 1999; 82:352–6.

41. Strassburg S, Anker SD. Metabolic and immunologic derangements in cardiac cachexia: where to from here? Heart Fail Rev 2006;11:57–64.

42. Nagaya N, Moriya J, Yasumura Y, Uematsu M, Ono F, Shimizu W, Ueno K, Kitakaze M, Miyatake K, Kangawa K. Effects of ghrelin administration on left ventricular function, exercise capacity, and muscle wasting in patients with chronic heart failure. Circulation 2004;110:3674–9.

43. Zhao SP, Zeng LH. Elevated plasma levels of tumor necrosis factor in chronic heart failure with cachexia. Int J Cardiol 1997;58:257–61.

44. Poehlman ET, Toth MJ, Fishman PS, et al. Sarcopenia in aging humans: the impact of menopause and disease. J Gerontol Biol Sci Med Sci 1995;50:73–7.

45. King D, Smith ML, Chapman TJ, Stockdale HR, Lye M. Fat malabsorption in elderly patients with cardiac cachexia. Age Ageing 1996;25:144–9.

46. King D, Smith ML, Lye M. Gastro-intestinal protein loss in elderly patients with cardiac cachexia. Age Ageing 1996;25:221–3.

47. Anker SD, Negassa A, Coats AJ, Afzal R, Poole-Wilson PA, Cohn JN, Yusuf S. Prognostic importance of weight loss in chronic heart failure and the effect of treatment with angiotensin-converting-enzyme inhibitors: an observational study. Lancet 2003;361:1077–83.

48. Hryniewicz K, Androne AS, Hudaihed A, Katz SD. Partial reversal of cachexia by beta adrenergic receptor blocker therapy in patients with chronic heart failure. J Card Fail 2003;9:464–8.

49. Adigun AQ, Ajayi AAL. The effects of enalapril-digoxin-diuretic combination therapy on nutritional and anthropometric indices in chronic congestive heart failure: preliminary findings in cardiac cachexia. Eur J Heart Failure 2001;3:359–63.

50. Mann DL, McMurray JJ, Packer M, et al. Targeted anticytokine therapy in patients with chronic heart failure: results of the Randomized Etanercept Worldwide Evaluation (RENEWAL). Circulation 2004;109:1594–602.

51. Broqvist M, Arnqvist H, Dahlstrom U, Larsson J, Nylander E, Permert J. Nutritional assessment and muscle energy metabolism in severe chronic congestive heart failure: effects of long-term dietary supplementation. Eur Heart J 1994;15:1641–50.

52. Heymsfield SB, Casper K. Congestive heart failure: clinical management by use of continuous nasoenteric feeding. Am J Clin Nutr 1989;50:539–44.

53. Paccagnella A, Calo MA, Caenaro G, et al. Cardiac cachexia: preoperative and postoperative nutritional management. J Parenteral Enteral Nutr 1994;18:409–16.

54. Morrison WL, Gibson JN, Rennie MJ. Skeletal muscle and whole body protein turnover in cardiac cachexia: influence of branched-chain amino acid administration. Eur J Clin Invest 1988;18:648–54.

55. Horwich TB, Fonarow GC, Hamilton MA, MacLellan WR, Woo MA, Tillisch JH. The relationship between obesity and mortality in patients with heart failure. J Am Coll Cardiol 2001;38:789–95.

56. Villareal DT, Apovian CM, Kushner RF, Klein S. Obesity in older adults: technical review and position statement of the American Society for Nutrition and NAASO, The Obesity Society. Obes Res 2005;13(11):1849–63.

57. Huynh BC, Rovner A, Rich MW. Long-term survival in elderly patients hospitalized for heart failure: 14-year follow-up from a prospective randomized trial. Arch Intern Med 2006;166:1892–8.

58. Ghali JK, Kadakia S, Cooper R, Ferlinz J. Precipitating factors leading to decompensation of heart failure: traits among urban blacks. Arch Intern Med 1988;148:2013–6.

59. Witte KK, Clark AL. Micronutrients and their supplementation in chronic cardiac failure. An update beyond theoretical perspectives. Heart Fail Rev 2006;11:65–74.

60. Witte KK, Clark AL, Cleland JGF. Chronic heart failure and micronutrients. J Am Coll Cardiol 2001;37:1765–74.

61. Kopp SJ, Klevay LM, Feliksik JM. Physiological and metabolic characterization of a cardiomyopathy induced by chronic copper deficiency. Am J Physiol 1983;245:H855–66.

62. Lockitch G, Taylor GP, Wong LT, et al. Cardiomyopathy associated with nonendemic selenium deficiency in a Caucasian adolescent. Am J Clin Nutr 1990;52:572–7.

63. Coudray C, Boucher F, Richard MJ, et al. Zinc deficiency, ethanol and myocardial ischemia effect lipoperoxidation in rats. Biol Trace Elem Res 1991;30:103–18.

64. Li Y, Huang TT, Carlson EJ, et al. Dilated cardiomyopathy and neonatal lethality in mutant mice lacking manganese superoxide dismutase. Nat Genet 1995;11:376–81.

65. Golik A, Cohen N, Ramot Y, et al. Type II diabetes mellitus, congestive cardiac failure and zinc metabolism. Biol Trace Elem Res 1993;39:171–5.

66. Liu P, Olivieri N. Iron overload cardiomyopathies: new insights into an old disease. Cardiovasc Drugs Ther 1994;8:101–10.

67. Djoenaidi W, Notermans SL, Dunda G. Beriberi cardiomyopathy. Eur J Clin Nutr 1992;46:227–34.

68. Seligmann H, Halkin H, Rauchfleisch S, et al. Thiamin deficiency in patients with congestive heart failure receiving long-term furosemide therapy: a pilot study. Am J Med 1991;91:151–5.

69. Shimon I, Almog S, Vered Z, et al. Improved left ventricular function after thiamin supplementation in patients with congestive heart failure receiving long-term furosemide therapy. Am J Med 1995;98:485–90.

70. Zangen A, Botzer D, Zangen R, Shainberg A. Furosemide and digoxin inhibit thiaminie uptake in cardiac cells. Eur J Pharmacol 1998;13:151–5.

71. Ting HH, Timimi FK, Haley EA, et al. Vitamin C improves endothelium-dependent vasodilation in forearm resistance vessels of humans with hypercholesterolemia. Circulation 1997; 95:2617–22.

72. Gokce N, Keaney JF, Frei B, et al. Long-term ascorbic acid administration reverses endothelial vasomotor dysfunction in patients with coronary artery disease. Circulation 1999;99:3234–40.

73. Hornig B, Arakawa N, Kohler C, Drexler H. Vitamin C improves endothelial function of conduit arteries in patients with chronic heart failure. Circulation 1998;97:363–8.

74. Khaw KT, Bingham S, Welch A, et al. Relation between plasma ascorbic acid and mortality in men and women in EPIC-Norfolk prospective study: a prospective population study. European Prospective Investigation into Cancer and Nutrition. Lancet 2001;357:657–63.

75. Gale CR, Martyn CN, Winter PD, Cooper C. Vitamin C and risk of death from stroke and coronary heart disease in cohort of elderly people. BMJ 1995;310:1563–6.

76. Enstrom JE, Kanim LE, Klein MA. Vitamin C intake and mortality among a sample of the United States population. Epidemiology 1992;3:194–202.

77. Calzada C, Bruckdorfer KR, Rice-Evans CA. The influence of antioxidant nutrients on platelet function in healthy volunteers. Atherosclerosis 1997;128:97–105.

78. Rimm EB, Stampfer MJ, Ascherio A, et al. Vitamin E consumption and risk of coronary heart disease in men. N Engl J Med 1993;328:1450–6.

79. Stampfer MJ, Hennekens CH, Manson JE, et al. Vitamin E consumption and the risk of coronary disease in women. N Engl J Med 1993;328:1444–9.

80. Losonczy KG, Harris TB, Havlik RJ. Vitamin E and vitamin C supplement use and risk of all-cause and coronary heart disease mortality in older persons: the Established Populations for Epidemiologic Studies of the Elderly. Am J Clin Nutr 1996;64:190–6.

81. Kushi LH, Folsom AR, Prineas RJ, Mink PJ, Wu Y, Bostick RM. Dietary antioxidant vitamins and death from coronary heart disease in postmenopausal women. N Engl J Med 1996;334: 1156–62.

82. Eidelman RS, Hollar D, Hebert PR, Lamas GA, Hennekens CH. Randomized trials of vitamin E in the treatment and prevention of cardiovascular disease. Arch Intern Med 2004; 164:1552–6.

83. Miller ER, Pastor-Barriuso R, Dalai D, Riemersma RA, Appel LJ, Guallar E. Meta-analysis: high-dosage vitamin E supplementation may increase all-cause mortality. Ann Intern Med 2005; 142:37–46.

84. Lonn E, Bosch J, Yusuf S, Sheridan P, Pogue J, Arnold JM, Ross C, Arnold A, Sleight P, Probstfield J, Dagenais GR. HOPE and HOPE-TOO Trial Investigators. Effects of long-term vitamin E supplementation on cardiovascular events and cancer: a randomized controlled trial. JAMA 2005;293:1338–47.

85. Selhub J, Jacques PF, Wilson PWF, et al. Vitamin status and intake as primary determinants of homocysteinemia in an elderly population. JAMA 1993;270:2693–8.

86. Eikelboom JW, Lonn E, Genest J, Hankey G, Yusuf S. Homocyst(e)ine and cardiovascular disease: a critical review of the epidemiologic evidence. Ann Intern Med 1999;131:363–75.

87. Bots ML, Launer LJ, Lindemans J, et al. Homocysteine and short-term risk of myocardial infarction and stroke in the elderly: the Rotterdam Study. Arch Intern Med 1999;159:38–44.

88. Bostom AG, Rosenberg IH, Silbershatz H, et al. Nonfasting plasma total homocysteine levels and stroke incidence in elderly persons: the Framingham Study. Ann Intern Med 1999;131:352–5.

89. Herrmann M, Taban-Shomal O, Hubner U, Bohm M, Herrmann W. A review of homocysteine and heart failure. Eur J Heart Fail 2006;8:571–6.

90. Weisshaar RE, Simpson RU. Involvement of vitamin D3 with cardiovascular function: direct and indirect effects. Am J Physiol 1987;253:E675–83.

91. MacLaughlin J, Holick MF. Aging decreases the capacity of human skin to produce vitamin D_3. J Clin Invest 1985;76:1536–8.

92. Shane E, Mancini D, Aaronson K, et al. Bone mass, vitamin D deficiency and hypoparathyroidism in congestive heart failure. Am J Med 1997;103:197–207.

93. Tavani A, Negri E, D'Avanzo B, LaVecchia C. Beta-carotene intake and risk of nonfatal acute myocardial infarction in women. Eur J Epidemiol 1997;13:631–7.

94. DePinieux G, Chariot P, Ammi-Said M, et al. Lipid-lowering drugs and mitochondrial function: effects of HMG-CoA reductase inhibitors on serum ubiquinone and blood lactate/pyruvate ratio. Br J Clin Pharmacol 1996;42:333–7.

95. Hofman-Bang C, Rehnqvist N, Swedberg K, Wiklund I, Astrom H. Coenzyme Q10 as an adjunctive in the treatment of chronic congestive heart failure. The Q10 Study Group. J Cardiac Failure 1995;1:101–7.

96. Watson PS, Scalia GM, Galbraith A, Burstow DJ, Bett N, Aroney CN. Lack of effect of coenzyme Q on left ventricular function in patients with congestive heart failure. J Am Coll Cardiol 1999;33:1549–52.

97. Khatta M, Alexander BS, Krichten CM, et al. The effect of coenzyme Q10 in patients with congestive heart failure. Ann Intern Med 2000;132:636–40.

98. Belardinelli R, Mucaj A, Lacalaprice F, et al. Coenzyme Q10 and exercise training in chronic heart failure. Eur Heart J 2006;27:2675–81.

99. Weant KA, Smith KM. The role of coenzyme Q10 in heart failure. Ann Pharmacother 2005; 39:1522–6.

100. Nawarskas JJ. HMG-CoA reductase inhibitors and coenzyme Q10. Cardiol Rev 2005;13:76–9.

101. Anand I, Chandrashekhan Y, DeGiuli F, et al. Acute and chronic effects of propionyl-L-carnitine on the hemodynamics, exercise capacity and hormones in patients with congestive heart failure. Cardiovasc Drugs Ther 1998;12:291–9.

102. The Investigators of the Study on Propionyl-L-Carnitine in Chronic Heart Failure. Study on propionyl-l-carnitine in chronic heart failure. Eur Heart J 1999;20:70–6.

103. Gordon A, Hultman E, Kaijser L. Creatine supplementation in chronic heart failure increases skeletal muscle creatine phosphate and muscle performance. Cardiovasc Res 1995;30:413–8.

104. Witte KK, Nikitin NP, Parker AC, von Haehling S, Volk HD, Anker SD, Clark AL, Cleland JG. The effect of micronutrient supplementation on quality-of-life and left ventricular function in elderly patients with chronic heart failure. Eur Heart J 2005;26:2238–44.

19 Nutrition Support in Cancer

Elizabeth Kvale, Christine Seel Ritchie,
and Lodovico Balducci

Key Points

- Older cancer patients have unique qualities related to aging that place them at nutritional risk during cancer treatment.
- Routine assessment of nutritional risk is indicated in older cancer patients.
- Comprehensive geriatric assessment of older cancer patients can assist with determination of which older patients can tolerate cancer treatment with acceptable levels of morbidity and mortality.
- Dietary counseling and nutritional intervention (especially protein supplementation) should be incorporated into the care of nutritionally compromised older cancer patients as needed.
- Cancer is a frequent indication for enteral nutrition support (when there is intact gut function) or total parenteral nutrition.
- Management of nutritional compromise should be guided by patient-centered goals of care that incorporate the existing evidence-base into decision-making.

Key Words: Cancer; aging; cachexia; inflammation

19.1 INTRODUCTION

The majority of solid tumor cancers occur in patients over the age of 65 years. Age interacts with the factors that influence disease prevalence, treatment options, prognosis, quality of life, and survivorship. The nutritional compromise that older adults may carry into illness is compounded by diminished homeostatic reserve and susceptibility to toxicities of treatment that contribute to deterioration of nutritional status during cancer treatment. Further, the clinical approach to supportive care in the older cancer patient, including nutritional support and symptom management, is influenced by concerns for the tolerance of various interventions, pharmacologic and otherwise.

From: *Nutrition and Health: Handbook of Clinical Nutrition and Aging, Second Edition*
Edited by: C. W. Bales and C. S. Ritchie, DOI 10.1007/978-1-60327-385-5_19,
© Humana Press, a part of Springer Science+Business Media, LLC 2009

In general, the literature offers little evidence to guide the management of nutritional compromise in older cancer patients. This chapter will focus on developing an understanding of the factors that contribute to nutritional risk in older cancer patients, nutritional assessment in this population, indications for nutritional support, and supportive nutrition management in cancer patients. The development of an evidence-based proactive approach to supportive nutritional care of older cancer patients is a research imperative as older patients benefit from improved survival rates in cancer treatment *(1)*.

19.2 BACKGROUND

The high prevalence of malnutrition among patients with cancer is documented in the literature. The Eastern Cooperative Oncology Group (ECOG) quantified wasting and malnutrition in cancer patients in 1980 *(2)*. In this seminal report, weight loss of greater than 10% in the previous 6 months was most common in patients with GI tumors. Twenty-six percent of patients with pancreatic cancer exceeded this benchmark, and as many as 38% of patients with measurable gastric tumor burden experienced weight loss at this level of severity. Wasting was less common among those with breast cancer (6% of patients reported >10% weight loss over 6 months), acute nonlymphocytic leukemia (4%), sarcoma (7%), prostate (10%), and favorable non-Hodgkin's lymphoma (10%). Most of these data were extracted from studies conducted in the inpatient setting, and it is likely that these patients were relatively sicker and more likely to be exposed to aggressive treatment regimens than patients in the community. Malnutrition is nearly ubiquitous in hospitalized patients with advanced cancer *(3)*. Hospitalized older cancer patients are noted to have a high prevalence of malnutrition. In one retrospective study among patients with a mean age of 74.1 years admitted to an Oncology – Acute Care for Elders (OACE) unit, 42 of 119 patients (35%) had a history of weight loss documented in their charts. The most common primary malignancy among the patients with recorded weight loss was a gastrointestinal cancer *(4)*.

Findings in an outpatient radiation oncology population (mean age of 53 years) suggest a linear relationship between cancer stage and nutritional compromise *(5)*. The prevalence of nutritional compromise across this outpatient population was high. Moderate to severe malnutrition as defined by score on a standard nutritional assessment (patient-generated subjective global assessment) was identified in greater than 40% of the head and neck cancer patients, in greater than 30% of the colorectal patients, and in greater than 20% of gastro-esophageal patients. In contrast, a study examining the prevalence of malnutrition in a general practice in the United Kingdom found that malnutrition was not significantly more common among 213 patients with a cancer diagnosis as compared to 228 patients with other chronic diseases (10% vs. 8%; $P = 0.469$). The mean age in the UK study was 68 (range 22–93) with 68% being 65 and older; thus, age did not predict malnutrition in this study *(6)*. The literature supports the conclusion that disease type and stage are important factors in the development of malnutrition in cancer. There are insufficient data to determine if the prevalence of malnutrition in cancer increases with age.

Malnutrition is associated with poor outcomes in cancer treatment. In the ECOG study cited earlier, weight loss of at least 5% in the 6 months prior to diagnosis was associated with decreased survival, reduced response rate to chemotherapy, and lower quality of life. The prognostic impact of weight loss in this study was most significant in individuals with other favorable indicators, such as preserved performance status *(2)*. Poor nutritional status predicts surgical outcomes *(7,8)*, depression *(9)*, increased risk of chemotherapy-related toxicity *(10,11)*, and quality of life *(12)* in cancer patients. Moreover, poor nutritional status is linked to health care cost. Among cancer patients discharged from one cancer center between 1993 and 1994, the presence of a discharge diagnosis of malnutrition and/or dehydration resulted in an average length of stay (ALOS) of 9.4 days, while those with a diagnosis of malnutrition alone resulted in an ALOS of 13.4 days. Patients without either diagnosis had an ALOS of 5.8 days *(13)*. There are insufficient data to determine whether older cancer patients are more vulnerable to the negative outcomes associated with poor nutritional status as compared to younger patients.

19.3 PHYSIOLOGIC BASIS FOR MALNUTRITION IN CANCER AND CONTRIBUTING FACTORS

The development of nutritional compromise in cancer patients is multifactorial. Contributing factors can be organized into factors attributed to the underlying disease, treatment-associated factors, and co-morbid conditions. "Cancer" is a spectrum of diverse illnesses that have in common the presence of an underlying neoplastic process. As noted earlier, malnutrition associated with cancer is largely determined by the type and stage of cancer. The tumor may have a direct effect on gastrointestinal function through mass effect. This may manifest as esophageal or bowel obstruction, early satiety, or anorexia. Changes in taste or smell have been noted in cancer patients *(14)*. Decreased caloric intake is observed in cancer patients and contributes to weight loss *(15,16)*. Disruption of gut physiology may also occur, with resulting diarrhea, malabsorption, or constipation. Tumor-associated symptoms may also contribute to the loss of appetite and weight loss observed in cancer patients. Poorly controlled pain, in particular, has a deleterious effect on appetite *(17)*. Fatigue and depression are highly prevalent symptoms that may also contribute to loss of appetite in cancer patients *(18)*. Screening for depression is especially important in cancer patients because this is a potentially reversible factor.

Many authors conceptualize constitutional symptoms such as fatigue, depression, and insomnia as part of the systemic inflammation and cytokine derangement identified as anorexia–cachexia syndrome *(19)*. The anorexia–cachexia syndrome is a complex metabolic/inflammatory/neuroendocrine disturbance that is incompletely understood. The adaptive physiologic response to weight loss in a healthy individual is an increase in appetite. This response is absent in persons with anorexia–cachexia syndrome and reflects a disruption of homeostatic mechanisms and metabolic imbalance. Cancer cachexia is characterized by involuntary weight loss, lean body mass wasting, and poor performance. Death is the ultimate endpoint of untreated cachexia. Metabolic disruptions observed in this syndrome include hypermetabolism, a feature which distinguishes cachexia from simple decreased caloric intake *(20)*. Altered carbohydrate metabolism is also present, with a marked increase in the hepatic production of glucose *(21)*. Protein metabolism in patients with cachexia shows both increased degradation of protein in muscle biopsies and decreased protein synthesis, with a resulting whole-body catabolism *(22)*. There is also an acute phase response, with increased total body protein turnover *(20)*. Lipid metabolism is characterized by an increase in lipid mobilization, a cytokine-mediated disruption of the clearance of triglycerides from the plasma, and a decrease in lipogenesis.

A more complete discussion of the mechanisms of cachexia is found elsewhere in this text (Chapter 11). It should be noted, however, that although cachexia is a final common pathway for many end-stage diseases, it cannot be assumed that the mediators that drive the syndrome are the same in different disease processes. Cancer cachexia is associated with an increase in pro-inflammatory cytokines such as IL-6, IL-1β, and TNF-α that may be important triggers of the acute phase response and metabolic derangement. Tumor-specific factors may also be present as important mediators of cachexia. Most notably, proteolysis-inducing factor, a glycoprotein first noted in the urine of cachectic pancreatic cancer patients *(23)*, has now been identified in many other cachexia-inducing tumors and appears to mediate cachexia via the ubiquitin–proteasome proteolytic pathway. Lipid mobilizing factor is another tumor-specific mediator involved in the increased mobilization of fat stores *(22)*.

A number of cancer treatment-related factors may contribute to the development of malnutrition in cancer patients. These include mucositis, chemotherapy-induced nausea and vomiting, diarrhea, malabsorption, anorexia, constipation, post-surgical complications, and food aversion. Mucositis, or stomatitis, is an inflammation of the oral mucosal tissue that can occur in response to chemotherapy, radiation therapy, or concomitant therapy. A recent systematic review of the literature found that the mean incidence of mucositis across 33 studies involving 6,181 patients receiving RT with or without chemotherapy was 80%. Rates of hospitalization due to mucositis for the studies that reported this outcome ($n = 700$) were 16% overall and 32% for patients receiving altered fractionation radiation therapy *(24)*. In the eight studies that reported weight loss ($n = 880$), 34% of patients reported weight loss, with 17% reporting >10% weight loss in the three studies that reported this outcome ($n = 485$). Older cancer patients are noted to have both increased susceptibility to mucositis and delayed recovery *(25)*. Protocols that include 5FU and leucovorin are especially toxic with regard to this side effect.

Chemotherapy-induced nausea and vomiting (CINV) remains an important symptom that may contribute to the development of malnutrition in older cancer patients despite significant improvements in prevention and treatment. In a recent study in the community oncology setting, 69.5% of patients received guideline-based management of acute CINV with regimens utilizing 5-hydroxytryptamine 3 (5HT3) receptor antagonists, and 77.5% received appropriate therapy for delayed CINV with regimens utilizing dexamethasone. Among 151 patients (mean age 56) enrolled prior to treatment with a new chemotherapy regimen, 36% of patients developed acute CINV and 59% developed delayed CINV during cycle 1 *(26)*.

CINV also significantly interfered with quality of life in this study. Older persons experience more CINV toxicity than younger patients and are believed to be more susceptible to the toxicity of treatment to the gastrointestinal tract because of increased proliferation of intestinal crypt cells *(27)*.

Constipation may be associated with cancer treatment and has a deleterious impact on patient appetite and quality of life. Constipation may occur either as a result of chemotherapy, or secondary to medications used in symptom management, most notably opioids and 5HT3 antagonists (which may be associated with either diarrhea or constipation). The vinca alkaloids are most likely to be associated with constipation as a side effect. Constipation and ileus are attributed to the neurotoxicity of these agents and most frequently develop within a few days of dosing and may require several weeks to resolve *(28)*. Presentation may be severe, mimicking bowel obstruction, and may require intensive nutritional support.

Malabsorption is not a frequent clinical finding in association with chemotherapy in the oncology or supportive care literature. Absorption studies, however, have found altered sugar absorption following chemotherapy that peaks at about 7 days after treatment *(29,30)*. This altered sugar absorption reflects a disruption of alimentary function that may impact absorption of other macro- and micronutrients. Altered absorption as the result of whole gut mucositis may contribute to the development of malnutrition in cancer patients, but has not been extensively studied in the clinical setting. Malabsorption in cancer patients may also result from surgical resection of gut sections *(31–33)*. Additionally, there is some evidence that the stress of surgery may contribute to an inflammatory catabolic state following surgery in cancer patients with malnutrition *(34)*.

19.4 RISK FACTORS FOR MALNUTRITION IN OLDER CANCER PATIENTS

A number of factors may place older patients at greater risk for nutritional deficits in cancer treatment than younger adults. While macronutrient absorption is well preserved in aging, a number of micronutrient deficiencies may be more common in older patients. Absorption of vitamin B_{12}, vitamin D, and calcium may be less efficient in aging, leading to deficiencies of these nutrients. Among older adults in colder climates especially, or among adults with impaired mobility or illness that may limit sun exposure, vitamin D deficiency may be present. This is of concern especially given the links between some types of cancer and vitamin D deficiency. There is a direct link between disruption of vitamin D metabolism in the older adult and calcium levels. In addition to reduced sun exposure, the kidneys become less efficient in the dihydroxylation of 25-hydroxyvitamin D to the active 1,25-dihydroxy form. Older adults may also have reduced numbers of vitamin D receptors in the gut. These changes in vitamin D metabolism result in reduced absorption of calcium. In older cancer patients who experience these deficiencies, the end result may be an increase in parathyroid hormone secretion to compensate for low calcium levels, with increased bone loss.

Low riboflavin intake, as measured by erythrocyte glutathione reductase activity coefficient, is noted in a significant minority of older Americans and may be higher in developing countries *(35)*. Estimates of vitamin B_6 deficiency may vary depending on the methodology of assessment, but in western society approximately 10% of older adults are identified as deficient using a measure of elevated transferase activity coefficient *(35)*. Vitamin B_{12} deficiency is a well-recognized problem for older adults with atrophic gastritis. Vitamin B_{12} is generally absorbed after dissociation by pepsin and hydrochloric acid from protein sources. In older adults with gut physiology resulting in a low acid environment, this absorption is impaired. Cancer treatments that result in a greater shift toward a low acid environment or impair subsequent absorption of B_{12} in the proximal small bowel may further compound this problem.

Changes in gut physiology with aging may make older adults somewhat more vulnerable to the adverse effects of cancer and cancer treatment. In general, there is an increased prevalence of motility disorders in aging. While the clinical impact on gastrointestinal function may be minimal in normal aging, this may present a significant vulnerability in cancer treatment. As noted earlier, the increased proliferation of cells lining the colonic mucosa may place older adults at greater risk for the adverse effects of cancer therapies that target dividing cells *(36)*. Additionally, aging may diminish the adaptive capacity of the gut. There is evidence from both animal studies and human studies that the ability to rebound from a period of nutritional stress is markedly diminished with aging. This underscores the importance of a proactive plan to provide aggressive nutritional support to older adults to prevent the development of periods of significant nutritional compromise.

19.5 NUTRITIONAL ASSESSMENT OF OLDER CANCER PATIENTS

There is uniform agreement in the literature that routine assessment of nutritional status is an important element in the comprehensive management of cancer patients. There is emerging recognition that comprehensive geriatric assessment (CGA) of older cancer patients can assist with determination of which older patients can tolerate cancer treatment with acceptable levels of morbidity and mortality *(37–41)*. A CGA includes nutrition assessment, and the routine institution of CGA for older patients would accomplish nutritional screening and address significant cognitive, emotional functioning, social and functional factors that contribute to nutritional risk. Unfortunately, CGA is time consuming and requires significant resources. The majority of patients over 65 probably do not require this intensive level of evaluation, however, the majority of older cancer patients are not screened at all for nutrition risk.

The assessment of older cancer patients for nutritional risk should include targeted elements of the history, physical examination, and laboratory and instrumental evaluation. One simple parameter that can be obtained from the history to screen for protein-calorie malnutrition is the presence of weight loss, usually reported as percent of usual weight lost. History of weight loss has demonstrated prognostic value in cancer populations *(2,5)*. Weight loss alone as a single indicator of nutritional state is limited by the fact that weight can change dramatically with shifts in total body water. A number of structured instruments have been developed to elicit history related to nutritional risk. Ottery's patient-generated subjective

Table 19.1
The Patient Generated Subjective Global Assessment (PG-SGA)

Global assessment
- SGA-A – Well nourished or anabolic
- SGA-B – Moderate or suspected malnutrition
- SGA-C – Severely malnourished

Nutritional Triage Recommendations (based on additive score on PG-SGA
- 0–1 No intervention at this time
- 2–3 Patient and family education by dietitian, nurse, or other clinician with pharmacologic intervention as indicated
- 4–8 Requires intervention by dietitian, in conjunction with nurse or physician
- ≥9 Indicates a critical need for improved symptom management and/or nutrient intervention options

global assessment (PG-SGA) is most reported in the literature and has been validated in a number of populations *(13,42,43)*. This instrument also has clinical utility, as patients are categorized into levels of nutritional compromise (Table 19.1) that can guide nutritional intervention. The combination of a history of weight loss and the PG-SGA in cancer patients is demonstrated to increase the sensitivity and positive predictive value of the PG-SGA significantly *(5)*, and the authors advocate this approach to assessment of nutritional risk in older cancer patients.

The physical examination offers an opportunity for the clinician to objectively assess the older patient's level of nutritional risk. Body mass index (BMI) normalizes body weight for height (BMI = weight (kg)/height (m2)) and can be a useful indicator of protein-calorie malnutrition. A BMI of less than 22 may indicate protein-calorie deficit, and a value of less than 18 is considered indicative of clinical malnutrition. Limitations on the use of BMI as a screen include the high prevalence of obesity in the United States, which may contribute to the maintenance of normal-range BMI even in the setting of advanced disease with associated weight loss *(44)*. Other physical exam findings that may indicate nutritional compromise include a loss of subcutaneous fat, muscle wasting, ankle or sacral edema, and ascites. Anthropometrics involve formal measurement of body fat depositions in assessment of nutritional state. The triceps skinfold is a frequently utilized measure of subcutaneous fat. This measure is obtained through the use of specialized calipers on the upper left arm, and through calculation of the arm muscle circumference (AMS) (AMS = arm circumference – TSF) can give an estimate of muscle maintenance or loss. The use of all anthropometrics is limited by need for specialized training in the use of these measures, and continued poor interobserver reliability even in the setting of specialized training *(45)*. The emerging methodology utilizing bioelectrical impedance vector analysis (BIVA) may offer advantages over other approaches in the assessment of body composition. Bioelectrical impedance analysis measures the opposition to flow of an alternating electrical current through body tissues. Conventional bioelectrical impedance analysis (BIA) is influenced by violations of standard assumptions of body composition including fluid status. BIVA addresses this shortcoming and offers enhanced precision in measurement of body composition to follow response to nutritional intervention *(46,47)*.

Assessment of the older cancer patient's level of nutritional risk can be augmented through assay of biochemical indicators that are sensitive to nutritional state. Serum albumin is commonly used as an indicator of nutritional state. Clinicians are familiar with this marker, and it is routinely obtained as a component of basic metabolic laboratory panels. Because albumin is synthesized by the liver, low levels (less than 3.5 g/dL) may indicate protein-calorie malnutrition. Albumin is also a negative acute phase reactant, however, and levels of serum albumin are usually reduced in the setting of stress such as acute illness or injury. Additionally, albumin levels may be affected by fluid status, liver disease, and renal disease. Despite the insensitivity of albumin as a screen for malnutrition, low levels of albumin are consistently identified in malnutrition and are associated with poor outcomes and higher mortality in a number of populations *(48–54)*. Serum albumin has a half-life of 20 days and is not responsive to acute nutritional insult or intervention.

Pre-albumin is more useful for following response to nutritional intervention than albumin, having a half-life of 1–2 days. Pre-albumin has demonstrated usefulness in predicting tolerance to surgery in some cancer populations and is predictive of prognosis following nutritional intervention in cancer patients *(8,55,56)*.

Transferrin is a hepatically synthesized protein that is responsive to nutritional status. The half-life of transferrin is 8–10 days, and it is more useful than albumin as a measure to follow response to nutritional support. Clinicians are familiar with transferrin as an iron-binding globulin, and levels of transferrin are also influenced by settings of iron deficiency or overload. Transferrin is not reliable when patients have received multiple transfusions.

Serum C-reactive protein is an acute phase reactant that is elevated in inflammatory states and reflects a catabolic balance. Elevated C-reactive protein levels are not seen in healthy aging people *(56)*. Routine early assessment in the supportive care of older oncology patients may identify individuals experiencing the inflammatory metabolic derangement that is phenotypically characterized as cachexia. Identification of persons in a catabolic and inflammatory metabolic state may help clinicians select appropriate interventions as targeted therapies for cancer cachexia become available *(57)*. Other indicators of nutritional state and metabolic balance may have utility in selected clinical or research settings. These include insulin growth factor (IGF)-1, leptin, fibronectin, specific interleukins and cytokines, delayed hypersensitivity reaction, urine creatinine, and retinol-binding protein. In males, assessment of serum testosterone is helpful, as hypogonadism has a relatively high prevalence in some cancer populations and may impair an anabolic response to nutritional support *(58–60)*.

19.6 NUTRITIONAL SUPPORT OF OLDER CANCER PATIENTS

The appropriate level of nutritional support for older cancer patients is highly individualized and determined by the patient's pretreatment status, cancer treatment plan, personal resources, and goals of care. The approach to the nutritional assessment and support of patients who elect palliative goals of care are addressed elsewhere in this text. This discussion focuses on patients who require supportive care during and after treatment. It is recommended that any older cancer patient who either screens positive for nutritional compromise or is scheduled to receive cancer treatment therapy that places them at risk for nutritional compromise receive nutritional support proactively and concurrently with cancer treatment. Therapies that place nutritionally intact older patients at risk include any radiation therapy to the oral cavity, pharynx, esophagus, stomach, or GI tract *(61)* as well as moderately to highly emetogenic chemotherapy protocols (Table 19.2). Figure 19.1 provides a general algorithm for the nutritional support of older cancer patients based on initial assessment with the SGA or other similar assessment that allows the clinician to assess the patient as nutritionally intact, compromised, or experiencing malnutrition.

19.6.1 Levels of Intervention

Nutritional support of older cancer patients must be tailored to meet the needs and goals of the patient in a proactive manner. There is evidence from a number of trials that dietary counseling and nutrition intervention increase caloric and protein intake in varied populations of cancer patients. Other outcomes that demonstrate improvement with dietary counseling and nutritional intervention in some studies include quality of life, symptom management, and patient satisfaction *(62–64)*. The essential content of dietary counseling interventions has not been evaluated in studies, but the majority of counseling intervention studies have utilized the American Dietetic Association Medical Nutrition Therapy Protocol (ADA MNT). The ADA MNT offers guidelines on the timing and frequency of dietary intervention consultation, nutrition intervention strategies, ongoing assessment, and allows for individualization of the intervention *(65)*. Nutrition supplementation, particularly protein supplementation, plays a role in the nutritional support of ill elderly patients *(66)*.

Table 19.2
Emetogenicity of common chemotherapy agents

Risk of emesis with treatment	Typical agents
High – Nearly all patients exposed	Cisplatin
	Dacarbazine
	Nitrogen mustard
Moderate – >70% of patients	Carboplatin
	Anthracyclines
	Cyclophosphamide
	Irinotecan
Low – 10–70% of patients	Mitoxandrone
	Taxanes
Minimal – <10% of patients	Hormones
	Vinca alkaloids
	Bleomycin

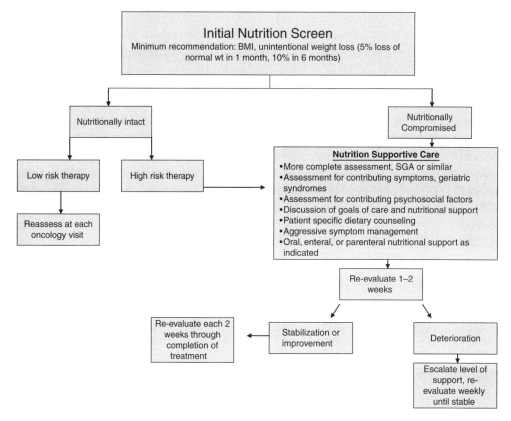

Fig. 19.1. An algorithm presenting recommendations for nutrition screening and initial evaluation and management of older cancer patients.

Similarly, calorie and protein supplementation is an important component of nutritional support for cancer patients who are not meeting energy or protein requirements. No studies were identified in the literature evaluating whether indications for supplementation in older patients are different from those for the general cancer population.

Indications for enteral nutrition in older cancer patients are the same indications that apply to other populations. Enteral nutrition is beneficial for patients with intact gut function who are unable to ingest adequate nutrients orally. There is no evidence that routine use of enteral nutrition in well-nourished cancer patients undergoing chemotherapy or surgery is helpful. Patients with cancers of the oropharyngeal region or gastrointestinal tract, however, are at particular risk for malnutrition. Head and neck cancer patients, because of the location of the tumor and also because this population has a high prevalence of co-morbid alcohol use, have a high rate of malnutrition at the time of presentation. In one study, as many as 57% of head and neck cancer patients were estimated to be nutritionally compromised at the time of presentation *(67)*. Additionally, head and neck patients are at high risk for treatment-related impairment of oral intake secondary to mucositis and dysphagia. In retrospective studies of head and neck cancer patients, placement of a percutaneous endoscopic gastrostomy (PEG) tube prior to initiation of radiation therapy was associated with less weight loss, fewer hospitalizations, and fewer treatment interruptions *(68)* and was well tolerated *(69)*. The potential benefits of prophylactic PEG placement should be weighed in the nutritional support of older head and neck cancer patients. Similarly, patients with esophageal cancer are likely to present with significant pretreatment weight loss and have a high risk of tumor or treatment-associated impairment of oral intake. In a prospective study prophylactic PEG placement prior to chemoradiation therapy reduced weight loss during therapy *(70)*. In selected populations of patients these benefits should be considered in the discussion of possible enteral support.

Enteral feeding is preferable to parenteral feeding because it is physiologically beneficial, less expensive, and better tolerated *(71)*. Cancer, however, is a frequent indication for total parenteral nutrition *(72)*. Total parenteral nutrition may support patients through periods of treatment-related gut dysfunction such as mucositis or ileus and may also be helpful for patients with disease-related gut dysfunction such as a total obstruction or high-output fistula. A number of studies have looked at the use of parenteral nutrition perioperatively for malnourished patients and were summarized in a consensus statement from the National Institutes of Health, the American Society for Parenteral and Enteral Nutrition, and the American Society for Clinical Nutrition. Based on a review of 13 prospective RCTs involving 1,250 patients, the authors identified a 10% reduction in postoperative complications for malnourished (defined by weight loss, nutritional indices, or prognostic indices) patients with gastrointestinal malignancies who received preoperative TPN for 7–10 days. No benefit was identified in malnourished patients receiving routine postoperative TPN, in fact early postoperative TPN in similar populations was associated with a 10% increase in complications. This consensus statement also suggests that nutrition support is necessary for patients unable to eat after surgery and recommends that TPN be initiated within 5–10 days in postoperative patients who are unable to eat or tolerate enteral feeding *(73)*.

19.6.2 Tailoring the Intervention

The goal of nutrition support of older cancer patients is to provide adequate energy and nutrients to maintain metabolic balance. The caloric requirements for older cancer patients can be estimated using equations, such as the Harris–Benedict equation *(74)*, developed to predict basal metabolic rate; however, it should be noted that these equations are intended for use with healthy individuals and were generally not derived from populations of older persons (Table 19.3). Generally,

Table 19.3
The Harris–Benedict equation to estimate energy expenditure

Estimates of energy expenditure

Harris–Benedict equations for basal metabolic requirements (BMR):

For males BMR $= 66.4730 + (13.7516 \times W) + (5.0033 \times H) - (6.7550 \times A)$

For females BMR $= 65.5095 + (9.563 \times W) + (1.8496 \times H) - (4.6756 \times A)$

W = weight (kg); H = height (cm); A = age (years)

estimates for cancer patients are adjusted for mild to moderate stress *(61)*. The estimate of energy expenditure in cancer patients is complicated not only by variations in predictive equations and adjustments for stress utilized *(75)* but also by method of measure of body composition *(76)*. These differences are unlikely to be of clinical significance, however. Energy expenditure can be measured utilizing indirect calorimetry, and this approach may offer advantages particularly in following the efficacy of interventions aimed at the inflammatory component of anorexia–cachexia syndrome. There is insufficient evidence, however, to support routine use of indirect calorimetry in clinical applications in older cancer patients.

19.7 CANCER ANOREXIA–CACHEXIA-SPECIFIC INTERVENTIONS

There is a burgeoning literature on interventions targeting anorexia–cachexia syndrome. This text will address only those interventions for which a sufficient literature exists to allow some assessment of their role in clinical care. Many interventions have focused on the anorexia component of the anorexia–cachexia syndrome. Corticosteroids are frequently utilized as appetite stimulants in patients with cancer based on a number of reports demonstrating improved appetite, but not weight gain *(77–79)*. Dexamethasone 3–8 mg/day or prednisone 15 mg/day are typical doses, although studies demonstrating superiority of a particular agent or dosing strategy are not found in the literature. Corticosteroids are perhaps most effective when utilized for patients with advanced disease. Positive effects on symptoms such as appetite, food intake, and sensation of well-being generally have a duration of around 4 weeks *(20)*. In older patients the utility of corticosteroids may be limited by co-morbidities such as diabetes mellitus that limit their use or by their potential to induce delirium. Additionally, well-recognized long-term side effects of glucocorticoids include immunosuppression, osteoporosis, and myopathy.

Progestational agents are frequently utilized as appetite stimulants in cancer patients and have been relatively well studied. A 2005 Cochrane database review of megestrol acetate (MA) for treatment of anorexia–cachexia syndrome summarized data from 30 trials involving 4,123 patients. The authors found a benefit for MA as compared to placebo conditions with regard to appetite improvement and weight gain. No conclusions could be drawn with regard to other important outcomes such as survival or quality of life *(80)*. Reports from the North Central Cancer Treatment group indicate that the resulting weight gain consists of primarily fat and not lean tissue. The same report indicates that doses from 480 to 800 mg/day are effective, with little benefit from higher dosing schedules *(81)*. Potential side effects of MA and related agents are significant and include the risk of deep venous thrombosis (DVT), male impotence, and menstrual bleeding in female patients.

Decision-making regarding initiating an orexigenic medication should be an intentional and thoughtful process. Neither of the two best studied agents has demonstrated improvement in survival or lean body mass. Additionally, though both of these agents stimulate appetite, they may theoretically shift patients toward a catabolic balance. The burden of side effects may limit their utility in older patients with multiple morbidities. Finally, there are no published data on the risk

Table 19.4
Appetite stimulants in cancer care

Medication	Usual dose	Desirable side effect	Undesirable side effect	Relative cost
Corticosteroids Dexamethasone Prednisone	3–8 mg q a.m. 15 mg	Improved energy level Sense of well-being Improved bone pain	Muscle atrophy Bone loss Elevated glucose	$
Megestrol acetate	480–800 mg/ day	None	DVT Hypogonadism Menstrual bleeding	$$$$
Mirtazapine	15–30 mg/ day	Antidepressant effect Sedation		$$

Abbreviations: DVT, Deep vein thrombosis; $, Reflects relative expense only

of DVT with MA administration in older adults. The authors recommend initiating an orexigenic agent only in patients for whom appetite loss is troublesome, selecting an agent based on tolerability or desirability of potential side effects and evaluating effectiveness of the intervention at a pre-determined time-point (Table 19.4). Other agents proposed or utilized as orexigenic agents but less well studied include tetra-hydrocannabinol (THC) derivatives. The limited studies in the literature suggest that these agents are not as effective in stimulating appetite or weight gain in cancer patients as MA, and side effects may limit their usefulness in an older population (82,83). The atypical antidepressant mirtazapine has been utilized based on limited evidence of appetite stimulation (84). The use of this agent is likely related to desirability of its antidepressant effect. Cyproheptadine is not effective and not well tolerated and does not have a role in the care of older cancer patients with appetite loss (85).

Current research is exploring the role of agents to address the inflammatory component of the metabolic derangement that characterizes the anorexia–cachexia syndrome. A number of small studies of varied design indicate that anti-inflammatory medications may be beneficial either alone or in combination with other agents (86–89). Thalidomide, in a randomized controlled trial conducted in 50 patients with advanced pancreatic cancer, attenuated loss of weight and lean body mass (90). The fish oil derivative, eicosapentanoic acid, generated a significant amount of interest because it targets mechanisms related to the production of proteolysis-inducing factor (91–93). Unfortunately, subsequent well-designed trials have not found evidence that fish oil agents are effective in improving appetite or maintaining lean body mass (94–96). This example stands as a cautionary tale against the over-interpretation of preliminary studies, and the use of the agents described in this paragraph should be limited to research settings until further evidence is available.

19.8 RECOMMENDATIONS

1. All older cancer patients should be evaluated for nutritional risk prior to initiation of cancer treatment. Older cancer patients who are identified as nutritionally compromised or who are receiving therapy that places them at high nutritional risk should be followed proactively for nutritional needs throughout cancer treatment.
2. Dietary counseling and nutrition intervention increases caloric and protein intake in varied populations of cancer patients and should be incorporated into the care of nutritionally compromised older cancer patients. Protein supplementation may be especially helpful in the care of older cancer patients.
3. Orexigenic agents should be initiated only in patients for whom appetite loss is troublesome. Agents should be selected based on tolerability or desirability of potential side effects, and effectiveness of the intervention should be evaluated at a pre-determined time-point.

REFERENCES

1. Brenner H. Long-term survival rates of cancer patients achieved by the end of the 20th century: a period analysis. Lancet 2002; 360(9340):1131–5.
2. Dewys WD, Begg C, Lavin PT, et al. Prognostic effect of weight loss prior to chemotherapy in cancer patients. Eastern Cooperative Oncology Group. Am J Med 1980; 69(4):491–7.
3. Nixon DW, Heymsfield SB, Cohen AE, et al. Protein-calorie undernutrition in hospitalized cancer patients. Am J Med 1980; 68(5):683–90.
4. Flood KL, Carroll MB, Le CV, Ball L, Esker DA, Carr DB. Geriatric syndromes in elderly patients admitted to an oncology-acute care for elders unit. J Clin Oncol 2006; 24(15):2298–303.
5. Ravasco P, Monteiro-Grillo I, Vidal PM, Camilo ME. Nutritional deterioration in cancer: the role of disease and diet. Clin Oncol (R Coll Radiol) 2003; 15(8):443–50.
6. Edington J, Kon P, Martyn CN. Prevalence of malnutrition in patients in general practice. Clin Nutr 1996; 15(2):60–3.
7. Tewari N, Martin-Ucar AE, Black E, et al. Nutritional status affects long term survival after lobectomy for lung cancer. Lung Cancer 2007; 57(3):389–94.
8. Geisler JP, Linnemeier GC, Thomas AJ, Manahan KJ. Nutritional assessment using prealbumin as an objective criterion to determine whom should not undergo primary radical cytoreductive surgery for ovarian cancer. Gynecol Oncol 2007; 106(1):128–31.
9. Toliusiene J, Lesauskaite V. The nutritional status of older men with advanced prostate cancer and factors affecting it. Support Care Cancer 2004; 12(10):716–9.
10. Aslani A, Smith RC, Allen BJ, Pavlakis N, Levi JA. The predictive value of body protein for chemotherapy-induced toxicity. Cancer 2000; 88(4):796–803.
11. Alexandre J, Gross-Goupil M, Falissard B, et al. Evaluation of the nutritional and inflammatory status in cancer patients for the risk assessment of severe haematological toxicity following chemotherapy. Ann Oncol 2003; 14(1):36–41.
12. Ravasco P, Monteiro-Grillo I, Vidal PM, Camilo ME. Cancer: disease and nutrition are key determinants of patients' quality of life. Support Care Cancer 2004; 12(4):246–52.
13. Ottery FD. Definition of standardized nutritional assessment and interventional pathways in oncology. Nutrition 1996; 12(1 Suppl):S15–9.
14. Nitenberg G, Raynard B. Nutritional support of the cancer patient: issues and dilemmas. Crit Rev Oncol Hematol 2000; 34(3):137–68.
15. Wigmore SJ, Plester CE, Ross JA, Fearon KC. Contribution of anorexia and hypermetabolism to weight loss in anicteric patients with pancreatic cancer. Br J Surg 1997; 84(2):196–7.
16. Staal-van den Brekel AJ, Schols AM, ten Velde GP, Buurman WA, Wouters EF. Analysis of the energy balance in lung cancer patients. Cancer Res 1994; 54(24):6430–3.

17. Cherny N. Cancer Pain: Principles of Assessment and Syndromes. In: Berger, AM, Shuster JL, Von Roenn J, eds. Palliative Care and Supportive Oncology (3rd ed.). Philadelphia, PA: Lippincott Williams and Wilkins, 2007; 3–26.

18. Portenoy R, Thaler H, Kornblith A, et al. The memorial symptom assessment scale: an instrument for the evaluation of symptom prevalence, characteristics and distress. Eur J Cancer 1994; 30A(9):1326–36.

19. Deans C, Wigmore SJ. Systemic inflammation, cachexia and prognosis in patients with cancer. Curr Opin Clin Nutr Metab Care 2005; 8(3):265–9.

20. Inui A. Cancer anorexia-cachexia syndrome: current issues in research and management. CA Cancer J Clin 2002; 52(2):72–91.

21. Tisdale MJ. Metabolic abnormalities in cachexia and anorexia. Nutrition 2000; 16(10):1013–4.

22. Skipworth RJ, Stewart GD, Dejong CH, Preston T, Fearon KC. Pathophysiology of cancer cachexia: Much more than host-tumour interaction? Clin Nutr 2007 Dec; 26(6):667–76.

23. Todorov P, Cariuk P, McDevitt T, Coles B, Fearon K, Tisdale M. Characterization of a cancer cachectic factor. Nature 1996; 379(6567):739–42.

24. Trotti A, Bellm LA, Epstein JB, et al. Mucositis incidence, severity and associated outcomes in patients with head and neck cancer receiving radiotherapy with or without chemotherapy: a systematic literature review. Radiother Oncol 2003; 66(3):253–62.

25. Balducci L, Corcoran MB. Antineoplastic chemotherapy of the older cancer patient. Hematol Oncol Clin North Am 2000; 14(1):193–212, x–xi.

26. Cohen L, de Moor CA, Eisenberg P, Ming EE, Hu H. Chemotherapy-induced nausea and vomiting: incidence and impact on patient quality of life at community oncology settings. Support Care Cancer 2007; 15(5):497–503.

27. Carreca I, Balducci L, Extermann M. Cancer in the older person. Cancer Treat Rev 2005; 31(5):380–402.

28. McDonald GB, Tirumali N. Intestinal and liver toxicity of antineoplastic drugs. West J Med 1984; 140(2):250–9.

29. Blijlevens NM, van't Land B, Donnelly JP, M'Rabet L, de Pauw BE. Measuring mucosal damage induced by cytotoxic therapy. Support Care Cancer 2004; 12(4):227–33.

30. Keefe DM, Cummins AG, Dale BM, Kotasek D, Robb TA, Sage RE. Effect of high-dose chemotherapy on intestinal permeability in humans. Clin Sci (London) 1997; 92(4):385–9.

31. Kang I, Kim YS, Kim C. Mineral deficiency in patients who have undergone gastrectomy. Nutrition 2007; 23(4):318–22.

32. Bae JM, Park JW, Yang HK, Kim JP. Nutritional status of gastric cancer patients after total gastrectomy. World J Surg 1998; 22(3):254–60; discussion 60–1.

33. Shils ME. Effects on nutrition of surgery of the liver, pancreas, and genitourinary tract. Cancer Res 1977; 37(7, Pt 2):2387–94.

34. Inoue Y, Miki C, Kusunoki M. Nutritional status and cytokine-related protein breakdown in elderly patients with gastrointestinal malignancies. J Surg Oncol 2004; 86(2):91–8.

35. Russell RM. The aging process as a modifier of metabolism. Am J Clin Nutr 2000; 72(2 Suppl):529S–32S.

36. Salles N. Basic mechanisms of the aging gastrointestinal tract. Dig Dis 2007; 25(2):112–7.

37. Hurria A, Gupta S, Zauderer M, et al. Developing a cancer-specific geriatric assessment: a feasibility study. Cancer 2005; 104(9):1998–2005.

38. Rao AV, Seo PH, Cohen HJ. Geriatric assessment and comorbidity. Semin Oncol 2004; 31(2):149–59.

39. Extermann M, Hurria A. Comprehensive geriatric assessment for older patients with cancer. J Clin Oncol 2007; 25(14):1824–31.

40. Chen CC, Kenefick AL, Tang ST, McCorkle R. Utilization of comprehensive geriatric assessment in cancer patients. Crit Rev Oncol Hematol 2004; 49(1):53–67.

41. Balducci L, Beghe C. The application of the principles of geriatrics to the management of the older person with cancer. Crit Rev Oncol Hematol 2000; 35(3):147–54.

42. Isenring E, Cross G, Daniels L, Kellett E, Koczwara B. Validity of the malnutrition screening tool as an effective predictor of nutritional risk in oncology outpatients receiving chemotherapy. Support Care Cancer 2006; 14(11):1152–6.

43. Bauer J, Capra S, Ferguson M. Use of the scored Patient-Generated Subjective Global Assessment (PG-SGA) as a nutrition assessment tool in patients with cancer. Eur J Clin Nutr 2002; 56(8):779–85.

44. Davis MP, Dreicer R, Walsh D, Lagman R, LeGrand SB. Appetite and cancer-associated anorexia: a review. J Clin Oncol 2004; 22(8):1510–7.

45. Hall JC, O'Quigley J, Giles GR, Appleton N, Stocks H. Upper limb anthropometry: the value of measurement variance studies. Am J Clin Nutr 1980; 33(8):1846–51.

46. Toso S, Piccoli A, Gusella M, et al. Bioimpedance vector pattern in cancer patients without disease versus locally advanced or disseminated disease. Nutrition 2003; 19(6):510–4.

47. Buffa R, Floris G, Marini E. Migration of the bioelectrical impedance vector in healthy elderly subjects. Nutrition 2003; 19(11–12):917–21.

48. Lohsiriwat V, Chinswangwatanakul V, Lohsiriwat S, et al. Hypoalbuminemia is a predictor of delayed postoperative bowel function and poor surgical outcomes in right-sided colon cancer patients. Asia Pac J Clin Nutr 2007; 16(2):213–7.

49. Iwata M, Kuzuya M, Kitagawa Y, Iguchi A. Prognostic value of serum albumin combined with serum C-reactive protein levels in older hospitalized patients: continuing importance of serum albumin. Aging Clin Exp Res 2006; 18(4):307–11.

50. Jiang JY, Tseng FY. Prognostic factors of anaplastic thyroid carcinoma. J Endocrinol Invest 2006; 29(1):11–7.

51. Morita T, Hyodo I, Yoshimi T, et al. Artificial hydration therapy, laboratory findings, and fluid balance in terminally ill patients with abdominal malignancies. J Pain Symptom Manage 2006; 31(2):130–9.

52. Elahi MM, McMillan DC, McArdle CS, Angerson WJ, Sattar N. Score based on hypoalbuminemia and elevated C-reactive protein predicts survival in patients with advanced gastrointestinal cancer. Nutr Cancer 2004; 48(2):171–3.

53. McMillan DC, Elahi MM, Sattar N, Angerson WJ, Johnstone J, McArdle CS. Measurement of the systemic inflammatory response predicts cancer-specific and non-cancer survival in patients with cancer. Nutr Cancer 2001; 41(1–2):64–9.

54. Brown DJ, Milroy R, Preston T, McMillan DC. The relationship between an inflammation-based prognostic score (Glasgow Prognostic Score) and changes in serum biochemical variables in patients with advanced lung and gastrointestinal cancer. J Clin Pathol 2007; 60(6):705–8.

55. Bourry J, Milano G, Caldani C, Schneider M. Assessment of nutritional proteins during the parenteral nutrition of cancer patients. Ann Clin Lab Sci 1982; 12(3):158–62.

56. Omran ML, Morley JE. Assessment of protein energy malnutrition in older persons, Part II: Laboratory evaluation. Nutrition 2000; 16(2):131–40.

57. MacDonald N. Cancer cachexia and targeting chronic inflammation: a unified approach to cancer treatment and palliative/supportive care. J Support Oncol 2007; 5(4):157–62; discussion 64–6, 83.

58. Dev R, Del Fabbro E, Bruera E. Association between megestrol acetate treatment and symptomatic adrenal insufficiency with hypogonadism in male patients with cancer. Cancer 2007; 110(6):1173–7.

59. Lackner JE, Mark I, Schatzl G, Marberger M, Kratzik C. Hypogonadism and androgen deficiency symptoms in testicular cancer survivors. Urology 2007; 69(4):754–8.

60. Brown JE, Ellis SP, Silcocks P, et al. Effect of chemotherapy on skeletal health in male survivors from testicular cancer and lymphoma. Clin Cancer Res 2006; 12(21):6480–6.

61. Mourad W, Apovian C, Still C. Nutrition Support. In: Berger A, Shuster JL, Von Roenn J, eds. Principles and Practice of Palliative Care and Supportive Oncology, 3rd ed. Philadelphia, PA: Lippincott Williams and Wilkins, 2007, pp. 741–54.

62. Ravasco P, Monteiro-Grillo I, Vidal PM, Camilo ME. Dietary counseling improves patient outcomes: a prospective, randomized, controlled trial in colorectal cancer patients undergoing radiotherapy. J Clin Oncol 2005; 23(7):1431–8.

63. Ravasco P, Monteiro-Grillo I, Marques Vidal P, Camilo ME. Impact of nutrition on outcome: a prospective randomized controlled trial in patients with head and neck cancer undergoing radiotherapy. Head Neck 2005; 27(8):659–68.

64. Isenring E, Capra S, Bauer J. Patient satisfaction is rated higher by radiation oncology outpatients receiving nutrition intervention compared with usual care. J Hum Nutr Diet 2004; 17(2):145–52.

65. ADA. Medical Nutrition Therapy Across the Continuum of Care (2nd ed.). Chicago IL: American Dietetic Association, 1998.

66. Milne AC, Potter J, Avenell A. Protein and energy supplementation in elderly people at risk from malnutrition. Cochrane Database Syst Rev 2005 April 18; (2):CD003288.

67. Lees J. Incidence of weight loss in head and neck cancer patients on commencing radiotherapy treatment at a regional oncology centre. Eur J Cancer Care (Engl) 1999; 8(3):133–6.

68. Lee JH, Machtay M, Unger LD, et al. Prophylactic gastrostomy tubes in patients undergoing intensive irradiation for cancer of the head and neck. Arch Otolaryngol Head Neck Surg 1998; 124(8):871–5.

69. Scolapio JS, Spangler PR, Romano MM, McLaughlin MP, Salassa JR. Prophylactic placement of gastrostomy feeding tubes before radiotherapy in patients with head and neck cancer: is it worthwhile? J Clin Gastroenterol 2001; 33(3):215–7.

70. Bozzetti F, Cozzaglio L, Gavazzi C, et al. Nutritional support in patients with cancer of the esophagus: impact on nutritional status, patient compliance to therapy, and survival. Tumori 1998; 84(6):681–6.

71. Schattner M. Enteral nutritional support of the patient with cancer: route and role. J Clin Gastroenterol 2003; 36(4):297–302.

72. August DA, Thorn D, Fisher RL, Welchek CM. Home parenteral nutrition for patients with inoperable malignant bowel obstruction. JPEN J Parenter Enteral Nutr 1991; 15(3):323–7.

73. Klein S, Kinney J, Jeejeebhoy K, et al. Nutrition support in clinical practice: review of published data and recommendations for future research directions. Summary of a conference sponsored by the National Institutes of Health, American Society for Parenteral and Enteral Nutrition, and American Society for Clinical Nutrition. Am J Clin Nutr 1997; 66(3):683–706.

74. Harris JA, Benedict FG. A biometric study of basal metabolism in man. Washington D.C.: Carnegie Institute of Washington, 1919.

75. Green AJ, Smith P, Whelan K. Estimating resting energy expenditure in patients requiring nutritional support: a survey of dietetic practice. Eur J Clin Nutr 2008 Jan; 62(1)150–3.

76. Korth O, Bosy-Westphal A, Zschoche P, Gluer CC, Heller M, Muller MJ. Influence of methods used in body composition analysis on the prediction of resting energy expenditure. Eur J Clin Nutr 2007; 61(5):582–9.

77. Willox JC, Corr J, Shaw J, Richardson M, Calman KC, Drennan M. Prednisolone as an appetite stimulant in patients with cancer. Br Med J (Clin Res Ed) 1984; 288(6410):27.

78. Moertel CG, Schutt AJ, Reitemeier RJ, Hahn RG. Corticosteroid therapy of preterminal gastrointestinal cancer. Cancer 1974; 33(6):1607–9.

79. Bruera E, Roca E, Cedaro L, Carraro S, Chacon R. Action of oral methylprednisolone in terminal cancer patients: a prospective randomized double-blind study. Cancer Treat Rep 1985; 69(7–8):751–4.

80. Berenstein EG, Ortiz Z. Megestrol acetate for the treatment of anorexia-cachexia syndrome. Cochrane Database Syst Rev 2005 April 18; (2):CD004310.

81. Loprinzi CL, Michalak JC, Schaid DJ, et al. Phase III evaluation of four doses of megestrol acetate as therapy for patients with cancer anorexia and/or cachexia. J Clin Oncol 1993; 11(4):762–7.

82. Strasser F, Luftner D, Possinger K, et al. Comparison of orally administered cannabis extract and delta-9-tetrahydrocannabinol in treating patients with cancer-related anorexia-cachexia syndrome: a multicenter, phase III, randomized, double-blind, placebo-controlled clinical trial from the Cannabis-In-Cachexia-Study-Group. J Clin Oncol 2006; 24(21):3394–400.

83. Jatoi A, Windschitl HE, Loprinzi CL, et al. Dronabinol versus megestrol acetate versus combination therapy for cancer-associated anorexia: a North Central Cancer Treatment Group study. J Clin Oncol 2002; 20(2):567–73.

84. Theobald DE, Kirsh KL, Holtsclaw E, Donaghy K, Passik SD. An open-label, crossover trial of mirtazapine (15 and 30 mg) in cancer patients with pain and other distressing symptoms. J Pain Symptom Manage 2002; 23(5):442–7.

85. Kardinal CG, Loprinzi CL, Schaid DJ, et al. A controlled trial of cyproheptadine in cancer patients with anorexia and/or cachexia. Cancer 1990; 65(12):2657–62.

86. Lundholm K, Gelin J, Hyltander A, et al. Anti-inflammatory treatment may prolong survival in undernourished patients with metastatic solid tumors. Cancer Res 1994; 54(21):5602–6.

87. Mantovani G, Maccio A, Madeddu C, et al. A phase II study with antioxidants, both in the diet and supplemented, pharmaconutritional support, progestagen, and anti-cyclooxygenase-2 showing efficacy and safety in patients with cancer-related anorexia/cachexia and oxidative stress. Cancer Epidemiol Biomarkers Prev 2006; 15(5):1030–4.

88. Diament MJ, Peluffo GD, Stillitani I, et al. Inhibition of tumor progression and paraneoplastic syndrome development in a murine lung adenocarcinoma by medroxyprogesterone acetate and indomethacin. Cancer Invest 2006; 24(2):126–31.

89. Cerchietti LC, Navigante AH, Peluffo GD, et al. Effects of celecoxib, medroxyprogesterone, and dietary intervention on systemic syndromes in patients with advanced lung adenocarcinoma: a pilot study. J Pain Symptom Manage 2004; 27(1):85–95.

90. Gordon JN, Trebble TM, Ellis RD, Duncan HD, Johns T, Goggin PM. Thalidomide in the treatment of cancer cachexia: a randomised placebo controlled trial. Gut 2005; 54(4):540–5.

91. Barber MD, Ross JA, Voss AC, Tisdale MJ, Fearon KC. The effect of an oral nutritional supplement enriched with fish oil on weight-loss in patients with pancreatic cancer. Br J Cancer 1999; 81:80–6.

92. Beck SA, Smith KL, Tisdale MJ. Anticachectic and antitumor effect of eicosapentaenoic acid and its effect on protein turnover. Cancer Res 1991; 51:6089–93.

93. Tisdale M. The 'cancer cachectic factor'. Supportive Care Cancer 2003; 11:73–8.

94. Bruera E, Strasser F, Palmer JL, et al. Effect of fish oil on appetite and other symptoms in patients with advanced cancer and anorexia/cachexia: a double-blind, placebo-controlled study. J Clin Oncol 2003; 21(1):129–34.

95. Burns CP, Halabi S, Clamon G, et al. Phase II study of high-dose fish oil capsules for patients with cancer-related cachexia. Cancer 2004; 101(2):370–8.

96. Laviano A, Muscaritoli M, Rossi-Fanelli F. Phase II study of high-dose fish oil capsules for patients with cancer-related cachexia: a Cancer and Leukemia Group B study. Cancer 2005; 103(3):651–2.

20 Nutrition and Chronic Obstructive Pulmonary Disease

Danielle St-Arnaud McKenzie
and Katherine Gray-Donald

Key Points

- By 2020, COPD will be the third leading cause of mortality worldwide.
- Many COPD patients are malnourished, resulting in clinical deterioration, decreased exercise capacity, and diminished survival.
- The weight loss and loss of lean mass associated with COPD reflect in part metabolic disturbances, including hypermetabolism, inflammation-induced catabolism, alterations in protein metabolism, hormonal alterations, and muscle alterations and dysfunction.
- Loss of body mass also results from low food intake caused by increased difficulty in carrying out daily food-related activities (e.g., shopping, cooking) due to fatigue, dietary problems (loss of appetite, early satiety, dyspnea associated with feeding, swallowing problems, gastroesophageal reflux), and psychological problems (solitude, depression, anxiety, attitude/beliefs).
- Hospital stay due to exacerbations and treatment with glucocorticosteroids can also contribute to reduced food consumption.
- Interventions should emphasize weight gain in patients with BMI < 25 and weight stability in those with higher BMI, positive nitrogen balance, and functional improvements in muscle strength, handgrip strength, and walking ability.
- Strategies to help meet adequate intake include eating six to seven small meals (300–500 kcal/meal) throughout the day, using energy/protein-dense foods/supplements, using affordable yet enjoyable foods, and enlisting family community support at mealtime.
- Other strategies aim to minimize the energy cost of food activities. Meal preparation time should be kept to a minimum, one-dish recipes help reduce the energy spent cleaning up after the meal, frequently used foods and kitchenware should be close at hand, and cooking more than needed on good days to ensure having good food to eat on bad days.

Key Words: Respiratory function; pulmonary disease; dyspnea; weight loss

From: *Nutrition and Health: Handbook of Clinical Nutrition and Aging, Second Edition*
Edited by: C. W. Bales and C. S. Ritchie, DOI 10.1007/978-1-60327-385-5_20,
© Humana Press, a part of Springer Science+Business Media, LLC 2009

20.1 INTRODUCTION

Chronic obstructive pulmonary disease (COPD) is prevalent and a major cause of disease and mortality. It is estimated that up to 10% of the world population over 40 years of age experience some degree of airway limitation *(1)*. The prevalence of COPD is age dependent. In Sweden, 3.5% of young adults (20–44 years) present airflow limitation. In the United States, rates of moderate COPD rise to 21% in the population aged 65–74 years and to 23% in adults aged more than 75 years *(2)*. COPD is an important cause of morbidity *(3)* and currently ranks as the fifth leading cause of mortality accounting for more than 2,750,000 deaths worldwide *(4)*. In the United States, the COPD death rate has doubled between 1970 and 2002 *(5)*. As the population ages, it is projected that by 2020, COPD will be the third leading cause of death worldwide *(6)*. The economic burden of this condition is enormous, approximately $5–25 billion per year in developed countries *(7)*.

It is well known that smoking represents the most important risk factor for COPD *(8)*. Although it is reported that only 15% of smokers have COPD, this is a misleading estimate as all smokers show decreased lung function and all will probably develop COPD unless they stop smoking *(9)*. COPD mortality is becoming more common in women than in men. Between 1980 and 2004, estimates of COPD deaths in the United States increased from 38,362 to 58,646 in men while increasing from 17,425 to 63,341 in women *(10)*. Among COPD patients receiving long-term oxygen treatment, women have a significantly greater risk of dying than men *(11)*. The differential evolution of smoking habits, increased susceptibility to smoke-related COPD, smaller stature, differential clinical course of COPD, and bias for under-diagnosis and care are suspected reasons for the changing gender profile of COPD *(12–14)*. Thus the notion that COPD is an *old man's disease* is quickly changing. As with osteoporosis, COPD may soon become an *old woman's disease (15)*. Besides smoking status, age, and gender, other risk factors for COPD include long-term exposure to air pollution and occupational dust, genes, oxidative stress, comorbidities, respiratory infections, socioeconomic status, and nutrition *(8,16)*.

COPD is a condition in which airflow obstruction, a key element, interferes with the normal function of the respiratory system. In turn, impaired lung function, depending on the severity of the disease, can curtail activities of daily living *(17)* including, food-related activities, leading to sub-optimal food intake *(18,19)*. This condition is however complex and multidimensional in nature *(20)*. Well-documented extrapulmonary effects of COPD include systemic inflammation, nutritional metabolic abnormalities, skeletal muscle dysfunction, muscle and bone loss, and weight loss *(21)*. Weight loss is a common development in COPD. Depending on the population studied and on the indicator used to determine nutritional status, between 19% and 60% of COPD patients are classified as nutritionally depleted *(22,23)*. The clinical deterioration, decreased exercise capacity, and diminished life expectancy that are associated with unintentional weight loss have been acknowledged for many years *(24–26)*. More recent findings confirm that low BMI and weight loss are important prognostic factors of mortality in patients with moderate to severe COPD *(27–30)*.

Total daily energy expenditure (TDEE) in patients with COPD is highly variable but is often increased *(31)*. This may be due to increased resting energy expenditure (REE) *(32)* and/or increased energy cost of physical activity, but this is still not clear. Proposed mechanisms for increased TDEE include a hypermetabolic state created by systemic inflammatory effects associated with COPD, glucocorticosteroid-associated catabolism, increased requirements for respiration, and greater metabolic or mechanical demands for activities. Insufficient energy intakes resulting from ambulatory dysfunction, poor appetite and breathing difficulties, as well as elevated concentrations of cytokines have also been implicated as possible reasons for weight loss. It follows that COPD-associated weight loss can result from increased energy requirements not adequately compensated by an equivalent increase in food intake, diminished food intake, or both. Evidence for the prevention or reversibility of this nutritional decline is accumulating and points to the importance of nutritional interventions in this area.

20.2 DEFINITION, PATHOPHYSIOLOGY, AND ETIOLOGY OF COPD

In 2001 the Global Initiative for Chronic Obstructive Lung disease (GOLD) redefined COPD as "a preventable and treatable disease state characterized by airflow limitation that is not fully reversible" *(8)*. COPD includes conditions associated with airflow obstruction such as chronic bronchitis and emphysema but not asthma *(33)*. Chronic bronchitis is a clinical syndrome manifested by mucus hypersecretion, a daily productive cough for at least 3 months of the year for two consecutive years, and bronchial gland hypertrophy. Emphysema, on the other hand, is a chronic lung disorder characterized by the destruction of the gas-exchanging parenchyma distal to the terminal bronchioles, leading to the collapse of the abnormally enlarged air spaces and interfering with the transfer of O_2 and CO_2 between the blood and the alveolar air *(34–36)*. COPD can remain asymptomatic for many years and a diagnosis using spirometry is usually made at the moderate to advanced stages when dyspnea is severe enough to interfere with usual activities of daily living and a major loss of lung function has occurred.

COPD pathology involves lung parenchyma, intrathoracic airways, large and small, and lung vasculature. The two prominent clinical features of COPD are increased resistance to expiratory airflow and hyperinflation of the lungs. It is characterized by chronic inflammation, tissue destruction, and airway narrowing that are responsible for the chronic airflow limitation *(34)*. The expiratory airflow resistance causes air to be progressively trapped in the lungs during expiration which accounts for most of the hyperinflation and which also reduces inspiratory capacity. With increased airway resistance and hyperinflation, the work of breathing is noticeably elevated. The exchange of O_2 and CO_2 is impaired in COPD gradually leading to CO_2 retention. As a consequence, the level of minute ventilation needed to maintain arterial carbon dioxide pressure at normal levels can be two to three times the normal rate *(37)*. Hyperinflation also affects the diaphragm, the primary muscle of inspiration, which normally assumes a dome shape that protrudes upward into the thoracic cavity. The negative pressure in a tightly curved diaphragm is necessary to move air into the lungs. With hyperinflation, the diaphragm is flattened, increasing the radius of the curvature and decreasing the pressure generated. A flattened diaphragm may not be able to produce any useful inspiratory pressure, thus adding to the work of breathing and often resulting in the recruitment of the muscles of the abdominal wall and other accessory muscles to increase ventilation *(38)*.

Recent evidence indicating that COPD is not only a lung disease is increasing. COPD is now recognized as a multi-component condition with several systemic effects that influence survival and that are the result of systemic chronic inflammation *(21,39)*. These systemic features include notable weight loss, muscle wasting and weakness, and alterations in bone metabolism.

The multidimensional nature of COPD requires a multidisciplinary approach to treatment. Carefully planned therapy should include cessation of smoking but also administration of bronchodilators and/or glucocorticosteroids, respiratory muscle training, and nutritional support aiming to maintain or improve body weight. Management and care will help alleviate some of the symptoms and improve quality of life, but will not reverse the condition.

20.3 NUTRITIONAL CONCERNS

20.3.1 Low Body Weight, Weight Loss, and Muscle Wasting in COPD

There is no consensus regarding the diagnostic criteria for malnutrition in COPD patients. Many studies have based nutritional assessment on body weight, either in terms of body mass index (BMI; kg/m^2) or percent of ideal body weight (%IBW) in their assessment. However, with rates up to 25%, prevalence of poor nutritional status is high in patients suffering from COPD *(22,40–41)*. Being underweight is a risk factor for exacerbations and hospitalization *(40,42)*. A low body weight, weight loss, and loss of fat-free mass (FFM) have consistently been shown to be independent predictors of COPD mortality *(27,29,41–46)*, whereas being overweight or obese decreases the risk of mortality *(42–43,47)*. In fact, being underweight increases the risk of dying by 4.5 times *(43)*. The 5-year survival rates for COPD patients on long-term oxygen therapy are 24%, 34%, 44%, and 59%, respectively, for patients with BMIs <20, 20–24, 25–29, and ≥30 kg/m^2 *(42)*. A BMI of at least 25 kg/m^2 confers best survival *(29)*.

In addition to body weight, weight history predicts outcomes of COPD patients. COPD patients often experience weight loss. Rates of weight loss increase with decreasing lung function and are higher in women with severe COPD compared to men (35% vs. 25%, respectively) *(27)*. Weight loss increases the odds of having a COPD exacerbation regardless of initial body weight *(40)*. Furthermore, weight loss is a predictor of mortality. A loss of 3 units in BMI in COPD patients increases the risk of all-cause mortality by 70% and doubles the risk of respiratory mortality. Even a loss of 1 unit (the equivalent of about 3 kg in a 1.7 m person) suffices to confer excess mortality *(27,29)*. However, COPD-related weight loss is reversible *(24,48)*. By contrast, weight gain in underweight and normal-weight patients with severe COPD appears to have a beneficial effect on survival *(27)*.

Using total body weight or BMI as an indicator of nutritional status is limited by the inability to distinguish between fat tissue and lean body mass. Body weight is not always sensitive to variations in body composition in COPD patients and loss of FFM is frequently found in normal-weight COPD patients *(49)*. Patients suffering from COPD frequently present with decreased lean body mass/fat-free mass that is disproportionate compared to healthy adults *(50–51)*. Thus FFM is more depleted in COPD patients than in healthy controls *(32,52)* and more in women than in men (25% vs. 11%) *(30,53)*. Loss of FFM has deleterious effects on respiratory muscle functioning, resulting in decreased strength and endurance of the respiratory muscles which is evident in their impaired capacity for strenuous exercise *(54–55)*. Moreover, poor nutritional status will weaken the immune system and may interfere with the protection of the airways contributing to undesirable clinical outcomes. FFM is associated with bone mineral density *(49)* and is a predictor of survival *(45,56)* even in subjects with mild to moderate COPD *(30)*. In fact, FFM appears to be a better predictor of mortality than BMI *(30,57)*. The COPD-associated excessive loss of FFM is often termed pulmonary cachexia *(58–59)*, and is associated with COPD-related clinical outcomes. For instance, better functional status is found with a higher lean/fat ratio in both men and women with COPD *(60)*.

Recording body weight over time can be an informative measure of nutritional status. However, since changes in body composition, particularly loss of lean body mass, may go unrecognized in COPD, other methods of accurately determining changes in body composition in COPD patients need to be adopted (see Section 20.4.4).

20.3.2 Causes of Weight Loss and Muscle Mass Depletion

Weight loss and loss of FFM reflect complex interactions between metabolic disturbances (hypermetabolism, inflammation-induced catabolism, alterations in protein metabolism, hormonal alterations, muscle dysfunction) and changes in dietary behavior leading to reduced food intake due to dyspnea, decreased appetite, fatigue, frequent hospital admissions, and treatment with glucocorticosteroids. The following sections discuss these factors.

20.3.2.1 METABOLIC DISTURBANCES

Hypermetabolism. Energy expenditure that is chronically greater than energy intake obviously leads to weight loss. In COPD patients, TDEE is frequently increased compared to healthy individuals *(31,55,61)* which when coupled with unchanged or decreased intake could account for loss of weight. TDEE is the sum of the energy expenditure of three components: resting energy expenditure (REE), physical activity, and diet-induced thermogenesis (DIT). In most healthy sedentary adults, REE is the largest component (approximately 60–70%) of TDEE. It has long been recognized that the body's usual response to a decrease in food intake is to lower REE, an adaptive mechanism necessary to sustain life *(62)*. As in other chronic diseases, this mechanism appears to fail in COPD and several studies over the years have pointed to an increase in REE as one explanation for the observed weight loss in many patients with COPD. In a well-designed study by

Schols and colleagues *(63)*, weight-losing and weight-stable COPD patients were compared. The weight-losing patients exhibited higher REE (118 ± 17% predicted) than the weight-stable patients (110 ± 11% predicted), and both groups of COPD patients had significantly higher REE than healthy controls (104 ± 6% predicted). The higher REE remained after adjusting for lean body mass and was significantly higher in the weight-losing than in the weight-stable COPD patients. Elevated REE is evident in 50–60% of COPD patients *(31–32)*. COPD patients often show a reduced physical activity level; however, increased energy expenditure from physical activity could arise from greater energy demand for the same task due to muscle inefficiency but evidence on this point is not clear *(61,64)*.

To summarize, it is well established that REE in many clinically stable patients with moderate to severe COPD is 10–20% higher than would normally be predicted from the Harris–Benedict equation *(65)*. However, it is also clear that the energy expenditure of COPD patients varies greatly, so that the energy requirements are difficult to predict. This leaves energy balance as measured by weight changes as a crude but useful measure in the clinical setting.

Catabolic processes. COPD, as many other chronic illnesses, is characterized by acute and chronic inflammation. There is elevated cytokine production that is associated with deleterious systemic effects and, in particular, the degradation of skeletal muscle mass termed "cachexia" *(66)* (see also Chapter 11). Acute exacerbations are most often triggered by an infection or some other environmental stimuli and result in hospitalization. Bacterial and viral infections are powerful catalysts for the body's immune system, resulting in a complex and coordinated interaction to rid the body of these foreign substances. The production and release of stress hormones and cytokines contribute to the resulting changes in substrate metabolism and inappropriate mobilization of muscle protein and fat. A prolonged pro-inflammatory cytokinic response has been related to unrestrained net protein catabolism (cachexia) seen in patients with cancer and other critical illnesses *(67–68)*. Tumor necrosis factor-alpha (TNF-α), a multipurpose cytokine, has been implicated in the metabolic and nutritional disturbances observed in very ill patients *(69)*. Elevated concentrations of serum TNF-α have also been found in weight-losing COPD patients, whereas weight-stable patients and healthy control subjects have much lower levels of TNF-α in their blood *(70–72)*. TNF-α is believed to be the central cytokine responsible for the loss of muscle mass in COPD patients but other cytokines may also be involved, such as C-reactive protein (CRP) and interleukins (IL-6, IL-8). Elevated CRP in patients hospitalized for acute COPD exacerbation is associated with nonrecovery of symptoms after 2 weeks and is predictor of exacerbation recurrence within 50 days *(73)*. Moreover, frequent exacerbators have lesser reduction in plasma IL-6 and IL-8 during recovery than infrequent exacerbators. Evidence for the role of chronic low-grade systemic inflammation in pulmonary cachexia is also building *(74)*. Smokers present with higher, albeit subclinical, serum concentrations of cytokines (CRP, IL-6) that persist up to 20 years after smoking cessation *(75)*. In a recent study, higher CRP and IL-6 levels were found in stable COPD patients compared to healthy controls *(76)*, but whether chronic inflammation could account for the loss of weight in stable COPD patients is not clear as cachectic and non-cachectic patients do not appear to differ in their degree of inflammatory response. Furthermore, anti-TNF-α strategies have not proven successful in reversing cachexia *(77)* which highlights the high complexity of the underlying inflammatory component of the disease and of the etiology of weight loss in COPD. Still more research is needed to clarify the relationship between the production of cytokines during the acute and chronic inflammatory processes and their contribution to elevated REE and weight loss seen in many COPD patients.

Alterations in protein and amino acid metabolism. One possible mechanism for the COPD-related loss of FFM bears on alterations in protein metabolism that increases protein turnover favoring net catabolism. In patients with severe COPD, both protein degradation and protein synthesis are enhanced which translates to a 10% increase in whole-body protein turnover compared to healthy controls *(78)*. In a recent study, subjects with Stage II–IV COPD were found to have increased urinary pseudouridine (PSU) compared to healthy subjects, indicating increased protein breakdown *(79)*. But even in stable cachectic COPD patients not treated with corticosteroids and with no increase in whole-body protein turnover, there is evidence showing an increased whole-body myofibrillar protein breakdown when compared to non-cachectic COPD patients or controls, indicating muscle protein breakdown *(80)*.

Being the building blocks for proteins, it is not surprising that amino acid metabolism is also altered in COPD *(81)*. Some studies indicate that serum gluta-mate, glutamine, and alanine may be decreased in underweight patients with COPD, particularly in those with lower FFM. However, other studies report an increase in plasma glutamate. There are also reports of metabolic disturbances with respect to branched-chain amino acids (BCAA: leucine, isoleucine, and valine), plasma BCAA being lower in underweight compared to normal-weight COPD subjects *(82)*.

Hormonal alterations. Testosterone is an anabolic hormone that drives protein accretion and muscle repair. Hypogonadism has hence been examined as a possible explanation for the COPD-related loss of FFM in men. Prevalence of low testosterone levels is about 20–40% in men with COPD which does not differ from that of the general older population *(83)*. In COPD patients, low testosterone status is linked to peripheral muscle wasting *(84)*, skeletal muscle weakness *(85)*, and reduced pulmonary function *(86)*, but not to exercise tolerance or capacity *(85,87)*. A low ratio of insulin-like growth factor to growth hormone (IGF-1/GH), evidence for GH derangement, is also reported in cachectic COPD patients *(88)*. Testosterone replacement or administration of anabolic steroids or of growth hormone results in increased body weight that is predominantly from increased lean mass and increased skeletal muscle strength *(89–93)*. Studies on its effect on exercise capacity/tolerance are conflicting *(89–90,92–93)*. The evidence so far is unconvincing and long-term testosterone replacement therapy remains controversial as it may be associated with potentially dangerous side effects *(94)*.

Muscle dysfunction. Muscle dysfunction and ambulatory limitations are common in COPD patients, resulting in detrimental effects on physical capacity and quality of life. Inefficient ventilation cannot entirely account for the overall functional limitation so other causes have been sought. This has been the focus of much research; yet, to date, it is still not clear whether muscle dysfunction is the result of myopathy or de-conditioning (i.e., disuse). There is evidence in favor of the causal role of myopathy, including morphological (shift from type I to type II fibers) and metabolic differences (oxidative activity, oxidative stress in muscles), between COPD patients and healthy subjects *(95)*. But evidence, including studies showing reduced physical activity level *(96)*, and the fact that muscles in COPD patients have the ability to respond to muscle training when training "outside the constraint of maximal ventilation" *(97)* also support the role of de-conditioning *(98)*. It is likely that both are responsible. Patients with COPD may face a downward spiral in which muscle dysfunction leads to decreased physical activity that leads to FFM loss and in turn to greater muscle dysfunction.

20.3.2.2 REDUCED FOOD INTAKE

The failure to meet increased energy/protein requirements and/or reduction in food intake leads to gradual and sometimes precipitous weight losses, depending on the extent of energy/protein imbalance. When even a small but constant reduction in energy intake is not offset by a decrease in energy expenditure, weight loss will occur. Recent studies confirm that malnourished patients have lower dietary intake than those that are well nourished *(64,99)* and that in general, COPD patients present a negative energy balance that can be attributed to insufficient food intake *(40,99)*.

Energy requirements to maintain body weight in hospitalized COPD patients is high. Energy needs have been shown to be in the range of 135–150% of REE as predicted using the Harris–Benedict equation and 1.2 g protein/kg body weight is needed to meet protein requirements *(23,100–101)*.

When malnourished patients with severe COPD strive to regain their lost weight, the necessary energy requirement is substantially increased. In an intervention study conducted in underweight lung transplant COPD candidates, an energy intake of 180% of predicted REE ($\sim$44 kcal/kg) led to a weight gain of $\geq$2 kg *(101)*. In addition, protein intake was increased to 1.8–1.9 g/kg. This increase in intake was reached using, for the most part, ordinary foods.

Reasons for low food intake in COPD are numerous but can be classified into those linked to ambulatory dysfunction (difficulties in shopping and meal preparation, fatigue), dietary problems (dyspnea, swallowing and chewing problems, loss of appetite), and psychological problems (solitude, depression, anxiety, attitude/beliefs). The presence of two or more of these problems has been associated with lower energy intake *(102)*. Furthermore, both hospitalization and COPD treatments can cause reduced food intake.

Difficulties in shopping for food, meal preparation, cooking, and fatigue. Functional limitations and fatigue in COPD affect activities of daily living, including food-related activities. A forced expiratory volume in 1 second (FEV$_1$) <50% is a predictor of a high degree of meal-related challenges *(103)*. Most patients with COPD report that feelings of breathlessness make it difficult to shop for food and transport groceries *(104)*. Chronic fatigue occurs in close to half of COPD patients *(105)*. The entire process of shopping, preparing, and eating a meal can be very tiring for some patients and, for that reason, can lead to a reduction in energy intake. Seventy percent of underweight patients with very severe COPD reported having difficulties cooking meals compared to 43% of those with normal weight *(106)*. Strategies developed by patients to cope with meal-related difficulties include cooking extra meals on "good days", eating cold meals on bad days, or using food supplements, but they often become dependent on others for these activities.

Dyspnea, dysphagia, gastroesophageal reflux. Dyspnea is a distressing manifestation in respiratory diseases that is described as unpleasant sensations of hunger for air, effort to breathe, tightness in the chest, and unsatisfactory inspiration *(107)*. Some patients with COPD experience dyspnea during chewing and swallowing, and may decrease their intake of food to avoid these unpleasant sensations. It has been suggested that the increased sense of breathlessness experienced during eating by hypoxemic but not by normoxemic patients is due to decreased arterial oxygen saturation in the former but not in the latter group of COPD patients *(108)*. In contrast, no differences in post-prandial dyspnea sensations nor in objective measures of ventilation in undernourished compared to normally nourished patients were observed with a liquid meal in a small group of patients *(109)*. As the prevalence of severe dyspnea (40%) is not linked to nutritional status in COPD patients *(22)*, other factors may be more important in the restriction of protein and energy intake.

COPD patients are also susceptible to oropharyngeal dysfunction. Dysphagia is more prevalent in COPD patients compared to healthy subjects but reported prevalence rates vary greatly. In a group of 84 male COPD patients, 85% showed some degree of dysphagia based on videofluoroscopic examination *(110)*. These patients had been referred for swallowing assessment due to suspected swallowing difficulties which could explain the high rate of dysphagia in this sample. In another study, self-reported symptoms of swallowing problems were found in 15% of a group of 100 male COPD patients compared to 4% in a control group of 50 patients without COPD or other respiratory complaints *(111)*. The presence of dysphagia in COPD patients may increase the risk of aspiration, a possible contributing factor for exacerbations though this relationship has not been clearly demonstrated. The prevalence of gastroesophageal reflux (GER) symptoms (regurgitation, heartburn) in COPD is about 20% *(111)*. GERD symptoms are associated with the degree of severity of airway obstruction. Importantly, experiencing GER symptoms once or more per week has been shown to double the rate of exacerbations compared to those with symptoms less than once per week independently of the severity of COPD *(112)*. The exact impact of oropharyngeal dysfunction on food intake in COPD patients is not known.

Loss of appetite, taste alteration, and early satiety. Appetite is an important factor influencing the motivation to eat *(113–114)*. Loss of appetite is associated with reduced food intake and weight loss in elderly individuals and is considered a clinical symptom of malnutrition *(115)*. Among COPD patients, diminished appetite is a frequent complaint *(88)*. The diminished desire to eat is related to weight loss in underweight patients. In addition, taste alterations have been reported in men with COPD *(116)*. As taste is an important determinant of food choices and consumption *(117)*, alterations in taste could also modulate the motivation to eat. Finally, early satiety is linked to lower BMI in COPD patients *(99)*. This could explain why some patients with COPD report eating smaller but more frequent meals. Loss of appetite, taste alterations, and early satiety may all contribute to reduced food consumption in individuals with COPD; the full nutritional impact of these factors on food intake in COPD remains to be determined.

Solitude, depression, and anxiety. The presence of others at mealtime is known to influence the quantity of food eaten at a meal *(118–119)*. Some COPD patients report that being alone affects their enjoyment of the meal and that being with others makes meals more appealing and more tasty *(104)*. Up to 50% of COPD patients experience anxiety and depressive symptoms *(120)*, which for some individuals may be related to decreased food consumption *(121)*.

Social norms, attitudes, and health beliefs. Early interventions to prevent weight loss in COPD would benefit from appropriately targeting dietary behaviors for the special case of COPD. According to current behavior theory, social norms are determinants of food behavior as they influence attitudes toward food and health beliefs. A "healthy diet" may be seen as one low in fat and simple carbohydrates and thus lower in total energy so that overeating with the aim of gaining or maintaining weight can be felt and seen as socially unacceptable *(122)*. COPD patients, especially female patients (25% vs. 5% in men) reported being currently on a weight loss diet *(102)*. Over 30% of women and 5% of men with COPD express a fear of gaining weight. Modifying dietary behaviors in these patients with the aim of increasing the intake of energy-dense foods requires strategies that are different from advice provided to the general population. While giving nutritional advice to patients, it is important to include those on providing or preparing the food.

Hospitalization and treatment with glucocorticoids. Some patients with COPD can maintain their weight and function reasonably well until an acute event causes a deficit in energy balance. Weight loss follows and the lost weight is not necessarily regained during the recovery period leading to a stepped decline in body weight over time *(123)*. Patients with advanced COPD have episodes of acute not fully reversible airway obstruction that increase in frequency with the severity of the disease. These exacerbations may require hospitalization where the deterioration of nutritional status (apart from that caused by the normal progression of the disease) is unfortunately a likely consequence. It is well recognized that serious nutritional deficits can develop over the course of hospitalization in a variety of clinical contexts (ICU, rehabilitation, nursing home), and this has deleterious impact on length of hospital stay and on in-hospital and post-discharge survival *(124–132)*. Failure to eat adequately may be explained by patients' dislike of the foods offered, lack of appetite, and diet prescriptions *(133)*. Failure to record patient's height and weight, failure to identify undernourished patients, lack of adequate nutritional support when problems are recognized, failure to observe and record patient's dietary intake, and prolonged use of glucose and saline IV fluids remain common institutional causes of low intake *(134)*.

Satisfaction with foods' sensory qualities (taste, temperature, appearance, and overall appeal) is related to food intake in hospital *(135)*. The most frequently cited reason for nonconsumption of the foods served is lack of taste *(136)*. The taste of food is a largely neglected nutritional factor in hospitals/institutions. Improving food taste has been shown to have a beneficial impact on nutritional status in elderly patients *(137–138)*. Adequate dietary intake is at the center of nutritional rehabilitation and weight gain, and the attending health care team is the COPD patient's first line of defense against protein and energy undernutrition during their hospital stay. Organizational factors are important determinants of the evolution of nutritional status in hospitalized patients, and improvements in this area are likely to help in preventing weight loss in hospitalized patients.

Oral or intravenous therapy with glucocorticosteroids is used in the treatment of an acute exacerbation of COPD. The anti-inflammatory and immunosuppressive properties of glucocorticosteroids are well established and are responsible for their widespread use in COPD and in other respiratory/inflammatory diseases. The catabolic effects of glucocorticosteroids on respiratory muscles are also well known. In a prospective examination of eight patients prescribed an average daily dose of 14.2 ± 8.2 mg of methylprednisolone or an equivalent for the previous 6 months, severe muscle weakness in both the respiratory and peripheral muscles developed *(139)*. The negative effect of glucocorticosteroids on protein metabolism is indicated by diminished protein formation, resulting in increased protein breakdown and reduced protein synthesis. Reduced muscle mass, thinning of skin, a reduction in the protein matrix of the bones followed by calcium loss, and a negative nitrogen balance are other well-known side effects of glucocorticosteroid treatment *(140)*.

20.4 NUTRITIONAL EVALUATION OF DEPLETION

20.4.1 Screening for Malnutrition

Given the poor nutrition prognosis for COPD patients and the possibility of frequent hospital admissions, it is important that frontline health care workers in the community and in hospitals identify individuals at risk of nutritional decline so that nutritional therapy/intervention can be initiated. Maintaining weight throughout the progression of the disease may be difficult to achieve, but weight maintenance in the early stages is easier and may be more beneficial than interventions to induce weight gain after depletion of lean body mass has occurred. Thorsdottir et al. *(141)* have developed and validated a screening tool to detect malnutrition risk in hospitalized COPD patients. This tool includes seven criteria: BMI, recent weight loss, age, dietary problems (vomiting, diarrhea, loss of appetite, problems with chewing and swallowing), recent surgery, comorbidity, and hospitalization of more than 5 days in the last 2 months. Using a 5/7 cut-off score, this screening

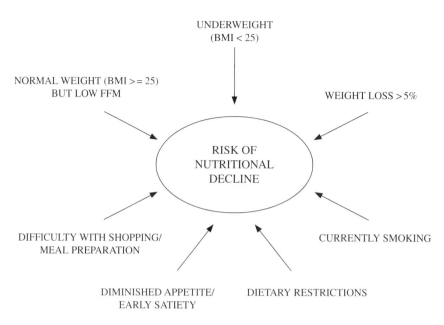

Fig. 20.1. Screening for COPD patients at risk of nutritional decline: factors that may help identify individuals at risk of becoming malnourished.

tool demonstrated excellent specificity (95%) but poor sensitivity (46%). A suggested screening process for COPD patients based on the current literature is provided in Fig. 20.1.

Reports of recent weight loss (>5% of usual body weight) can be used to identify individuals at risk. Reduced lean body mass and other factors such as diminished appetite, feelings of early satiety, difficulty obtaining and/or preparing food, dietary restrictions, and being a current smoker may identify individuals at high risk of becoming malnourished *(99)*. These individuals need to be brought to the attention of dietitians who can then conduct a nutritional assessment and begin nutritional intervention/therapy.

20.4.2 *Assessment of Nutritional Status of Patients with COPD*

Nutritional assessment is defined as a comprehensive evaluation of nutritional status, including medical history, dietary history, physical examination, anthropometric measurements, and laboratory data *(142)*. A summary of the studies determining the nutritional status of patients with COPD is shown in Table 20.1. The current desired weight range is a BMI in the range of 25–29 kg/m^2 in COPD patients *(29,42)*. BMI only requires the measurement of body weight and height. If height cannot be directly measured, it can be calculated from equations using knee height *(150–151)* or arm-span measurements *(152–153)*.

It is also desirable to measure indicators of fat-free mass as increasing fat mass may conceal FFM loss. Skinfold measurements and the bioelectrical impedance assay (BIA) are two simple, practical, and validated methods for this purpose. The skinfold thickness method involves measuring triceps, biceps, subscapular and

Table 20.1

Summary of studies assessing nutritional status of patients with COPD: criteria for malnutrition

Criteria	Study sample	% Malnourished	References
<90% UBW	38 Hospitalized patients	50	(143)
<90% IBW	Retrospective assessment of nutritional status of 77 hospitalized patients	43	(144)
<90% IBW	135 Outpatients	24	(46)
<90% IBW	779 Outpatients	24	(145)
NI comprising four parameters: Alb, Pre-Alb, TLC, %IBW	153 Rehabilitation center inpatients	19	(146)
BMI < 20 underweight and TSF < 10th percentile, MAC < 5th percentile	136 Outpatients	23	(147)
<90% IBW and <63% FFMPIBW	25 Rehabilitation center inpatients	26	(50)
NI based on anthropometric and biochemical variables	50 Hospitalized patients	60	(148)
≤85% IBW	18 Outpatients	50	(109)
BMI < 20	58 Outpatients	34	(149)
NI-based anthropometric and biochemical variables	34 Hospitalized patients	38	(141)
BMI < 20	4088 Patients on LTOT	23 M; 30 F	(42)
BMI ≤ 19.9	81 Outpatients	25	(49)
BMI < 20 and/or MAMC ≤ 15th percentile	103 Stable outpatients	23	(99)
BMI < 20	255 Outpatients on LTOT	10	(41)
BMI ≤ 21	73 Outpatients	37	(102)
BMI < 21 and FFMI ≤ 16 M; ≤15 F	412 Rehabilitation center inpatients	29 M; 27 F	(45)
BMI < 19 or BMI < 25 and weight loss <10%	42 Lung transplantation candidates	64	(106)
BMI < 20	41 Hospitalized patients	24	(40)
FFMI < 17 M, <15 F	1898 CCHS population	15 M; 16 F	(30)
BMI ≤ 21 and/or FFMI ≤ 16 M, ≤15 F	389 Outpatients	27	(22)

%UBW: percent usual body weight; %IBW: percent ideal body weight; NI: nutritional index; Alb: albumin; Pre-Alb: prealbumin; TSF: triceps skinfold; MAC: mid-arm circumference; TLC: total lymphocyte count; FFMPIBW: fat-free mass per ideal body weight; BMI: body mass index; MAMC: mid-arm muscle circumference; FFMI: fat-free mass index; LTOT: long-term oxygen therapy; CCHS: Canadian Community Health Study; F: female; M: male.

suprailiac crest skinfold thicknesses using a caliper. Sex- and age-specific equations have been developed to determine body composition from these measurements *(154)*. BIA, a technique used to assess body composition, has been shown to be a safe, convenient, and feasible alternative in the evaluation and monitoring of lean body mass in normal and critically ill subjects *(155–156)*. BIA has been used in COPD patients *(63)*. Dual-energy X-ray absorptiometry (DEXA) is the current gold standard when estimating body composition. In a recent validation study conducted in COPD patients, Lerario et al. *(157)* found that, while not perfect, both skinfold measurements and BIA are highly correlated with DEXA measurements, and both methods are considered reliable and clinically satisfactory for assessing FFM in this population.

20.4.3 Assessment of Visceral Protein Status

A more comprehensive review of the biochemical measures used in the assessment of nutritional status can be found elsewhere in this text (Chapter 4). Most patients with COPD have values within the normal range for serum albumin, transferrin, and retinol-binding protein *(145,158–160)*. A comparison of malnourished and well-nourished COPD patients revealed no differences in these indicators *(141)*.

20.4.4 Assessment of Short-Term Changes in Lean Body Mass

Depletion of lean body mass is a serious problem in a substantial portion of patients with COPD and affects their muscle strength, their ability to exercise, their daily living activities, and their survival. Short-term changes in protein metabolism can be estimated using the nitrogen balance technique *(161)*, a noninvasive and useful measure. Agreement between urinary urea nitrogen and total urea nitrogen in critically ill patients with a variety of clinical conditions has been reported *(162–163)*. Many patients with COPD are in negative nitrogen balance in hospital, indicating that protein breakdown exceeds protein synthesis *(23)*. The measurement of urinary pseudouridine standardized with urinary creatinine concentration can also be used as a marker of muscle protein depletion and of loss of FFM *(164)*.

20.5 NUTRITIONAL INTERVENTION IN MALNOURISHED STABLE PATIENTS WITH COPD

The interest in using nutritional support to improve the outcome in malnourished stable COPD patients has grown over the last two and a half decades but refeeding trials among inpatients and outpatients have had mixed results (Table 20.2). Three studies have shown that under well-controlled, experimental conditions, short-term (14–21 days) nutritional therapy will result in weight gain and improvements in respiratory muscle strength *(145,158,167–168)*. Mean weight gain in these studies ranged from 1.7 to 3.0 kg; the energy intake ranged between 1.4 and 2.2 × REE. Of the seven outpatient trials (Table 20.2), four did not show any improvements in lung function or respiratory muscle strength over periods ranging from 7 weeks to 4 months, whereas two well-controlled refeeding trials were successful in improving functional measures of physical performance *(158,171)*.

Table 20.2
Summary of refeeding trials carried out in malnourished stable COPD patients

No. of subjects, type, duration	Total intake	Comments	References
6 Patients followed in a clinical research unit for 2 weeks	3 > 1.5 × BMR 3 = 1.5×BMR	Gain of 3 kg, improvement in PImax, no change in spirometry	(145)
21 Patients, randomized and followed as outpatients for 8 weeks	~2,091 kcal, 1.7×BMR	No weight gain, no change in PImax, PEmax	(165)
25 Patients, randomized and crossover, followed as outpatients for 8 weeks	~2,350 kcal, ~1.8 × BMR	Small increase in weight, no effects on FEV1, FVC, TLC	(159)
21 Patients, randomized and followed as outpatients for 9 months	~2,118 kcal, ~82 g protein	Gain of 4 kg, improvement in respiratory muscles and grip, no change in lung function	(158)
28 Patients, randomized, placebo controlled, and followed for 13 weeks	~2,719 kcal, 2.4 × REE	Gain of 1.5 kg, no change in PImax, PEmax	(166)
10 Patients, randomized, in hospital refeeding trial, 16 days	~2,489 kcal, 2.2 × REE	Gain of 2.4 kg, improvement in PEmax, no change in FEV_1, FVC	(167)
27 Patients, randomized and followed in a clinical research unit for 3 weeks followed by 3 months as outpatients	1.7 × REE, 1.5 g protein/ kg/day	Gain of 1.7 kg, improvement in grip strength and PEmax	(168)
12 Patients followed as outpatients for 4 months	~1,670 kcal	No change in weight, FEV_1, or respiratory muscles	(169)
217 Patients participating in an intensive inpatient rehabilitation program, randomized and followed for 8 weeks	~2,400 kcal	Weight gain in both depleted and nondepleted patients. Nutrition and steroids led to a greater increase in lean body mass in depleted patients	(170)
85 Patients participating in a rehabilitation program, randomized and followed as outpatients for 7 weeks	~2,225 kcal	Small increase in weight, effect on shuttle walk test in well-nourished patients (BMI > 19 kg/m^2)	(171)

BMR: basal metabolic rate; REE: resting energy expenditure; PImax: maximal inspiratory pressure; PEmax: maximal expiratory pressure; FEV_1: forced expiratory volume in 1 second; FVC: forced vital capacity, TLC: total lung capacity.

A meta-analysis of controlled trials of >2 week supplementation in stable COPD concludes that supplementation leads to weight gain in undernourished patients, but the evidence related to the beneficial effect on lung function or quality of life shows mixed results *(172)*. The limited therapeutic effect of nutritional support in clinically stable outpatients is disappointing but may be related to the intensity and modality of the intervention and the complexity of the metabolic disturbances in COPD *(160)*. Additional therapy using other substances (ghrelin *(173)*, L-carnitine *(174)*, creatine *(175–176)*) to reverse the weight loss has shown some degree of success but is not widely used.

20.5.1 Nutritional Intervention During an Acute Exacerbation

Limited data are available evaluating the effect of oral nutritional support during an exacerbation of COPD patients. An acute exacerbation may result in a temporarily low energy and protein intake as well as an increased REE on admission to hospital *(177)*. Food intake may increase a few days after admission. Our own data indicate that an increase in oral intake to 2,370 kcal/day and 1.54 g protein/kg/day is possible during hospitalization *(23)*. However, a randomized double-blind, placebo-controlled trial, supplementing COPD patients experiencing acute exacerbation with oral liquid supplement for 9 days, found no short-term improvements in lung function or muscle strength *(178)*.

20.6 NUTRITION AND EXERCISE

While nutritional support alone in malnourished COPD patients has been met with limited success, the combination of nutritional support and an appropriate exercise program that helps improve respiratory muscle strength and increase lean mass may work better in a comprehensive rehabilitation program for COPD patients. Adding a simple but effective resistance or endurance training program may be as beneficial in COPD patients as it was found to be in healthy elderly whose walking capacity and respiratory and limb muscle strength improved after participating in an 18-month nutritional supplementation and resistance training program *(179)*. In a group of frail elderly people, the combination of resistance training and supplementation was more beneficial in improving muscle strength and mobility than supplementation alone *(180)*. Studies in COPD patients in this regard are rare. One study reports weight gain and improvement in the shuttle walk test in a supplemented group of COPD patients compared to a placebo group during a 7-week rehabilitation program *(171)*.

20.7 WHAT ARE THE NUTRITIONAL REQUIREMENTS OF COPD PATIENTS?

Optimal levels of carbohydrate, protein, and fat intake have not been defined for individuals with COPD. Recommendations for energy and protein intake have mainly come from feeding trials in malnourished, stable patients and from patients in acute care during an exacerbation of the disease.

20.7.1 Energy and Protein Intake

In malnourished, stable patients successful weight gain and nitrogen retention were achieved with energy intakes ranging from 1.4 to 2.2 × REE *(145,167–168)*. In patients hospitalized during an exacerbation, an energy intake of 1.9 × REE is achievable *(23,101)*. It is important that adequate energy intakes to cover the higher than expected energy expenditures be provided to COPD patients. Protein intakes have been high in successful refeeding studies. Intakes of ≥1.7 g/kg/day lead to good weight gain and positive nitrogen balance *(158,181)*. To achieve such a level of protein intake from food requires a fairly high energy intake. Thorsdottir *(100)* reports stable nitrogen balance during the first week of admission with intake ranging from 1.2 to 1.7 g/kg/day. Apart from determining which energy and protein requirements are needed to keep weight stable or reverse weight loss seen in COPD patients, attention has started to focus on other dietary constituents and to investigate nutritional modulation of COPD and its outcomes.

20.7.1.1 THE MACRONUTRIENT COMPOSITION OF FOOD INTAKE

The macronutrient composition of the diet for patients with COPD has received attention because an increase in dyspnea resulting from diet-induced thermogenesis is possible. A high intake of carbohydrate (≥74% of total energy) has been shown to increase carbon dioxide production and minute ventilation *(182)*; however, a more realistic intake of 53% energy from carbohydrate was shown to be well tolerated by normal-weight and underweight COPD patients. A carbohydrate-rich (60% of energy) supplement, while resulting in a higher respiratory quotient, led to faster gastric emptying and less shortness of breath than a fat-rich supplement (60% of energy) *(183)*. Although a very high intake of carbohydrate is not appropriate for this population, an intake of up to 60% of energy as carbohydrate is well tolerated. Broekhuizen et al. *(184)* showed successful clinical outcomes with a 125 ml supplement containing 60% carbohydrate, 20% protein, and 20% lipid that given three times a day contributed an extra 570 kcal/day.

20.7.1.2 AMINO ACIDS

Amino acids are implicated in intermediary metabolism. Disturbances in glutamate metabolism are found in many chronic diseases. COPD patients present low levels of plasma glutamate and glutamine *(185–186)*. However, there are no reports of glutamate supplementation studies in these patients. A more interesting area of research is branched-chain amino acids (BCAA) metabolism in COPD. Decreased levels of plasma BCAA have been reported in patients with COPD and in particular in those who were underweight *(82)*. Furthermore, diminished BCAA concentrations are related to hypermetabolism. BCAA supplementation results in increased post-prandial whole-body protein synthesis *(187)*, suggesting possible benefits in improving protein accretion in COPD patients. BCAA's are found in all protein-rich foods but especially eggs and red meat.

20.7.1.3 POLYUNSATURATED FATTY ACIDS

The potentially protective role of antioxidants and omega-3 fatty acids in lung function has recently been addressed. Clear benefits of a high intake of omega-3 fatty acids have been shown for coronary heart disease and stroke. Omega-3 fatty acids confer their health benefits through anti-inflammatory properties and their inhibitory effects in the synthesis of cytokines and mitogens *(188)*. Omega-3 fatty acids are abundantly present in mackerel, lake trout, tuna, herring, salmon, and anchovies as well as in seeds and nuts, particularly flaxseeds and walnuts. A randomized double-blind controlled trial reported beneficial effects of omega-3-rich PUFA supplements (9 g per day) for a period of 8 weeks on exercise capacity in stable COPD patients *(189)*. In another study, an omega-3-rich PUFA supplemented over a 2-year period improved the inflammatory status of COPD patients (TNF-α, interleukin-8, leukotriene B) compared to no change in another group that received an omega-6-rich PUFA supplement *(190)*. The beneficial effects of omega-3 fatty acids are still uncertain at this point. Nonetheless, they are important components of the phospholipid membrane of all cells and membrane fluidity is necessary for cells and tissues to function properly.

20.7.1.4 FRUIT AND VEGETABLE INTAKE AND ALCOHOL

The role of diet in the prevention of COPD has been explored. Consumption of fruits and vegetables is associated with decreased risk of developing COPD in smokers *(191)*. Walda et al. *(192)* also found a 10–24% diminished risk of 20-year COPD mortality with greater fruit consumption. The beneficial effect of increased fruit and vegetable intake on impaired lung function in COPD has, however, not yet been clearly demonstrated.

In contrast to increased risk of airway obstruction with heavy alcohol drinking, low to moderate alcohol intake is associated with better lung function and decreased risk of airflow limitation *(193)*. Red wine consumption may be of benefit to COPD patients. Resveratrol, a biologically active component in wine with strong antioxidant properties, may improve the inflammatory status in these patients *(194)*.

20.8 RECOMMENDATIONS

Weight loss is a common clinical feature of COPD and because the adverse effects of weight loss are well understood, adequate nutritional support and appropriate exercise program should be part of an inclusive rehabilitation program for COPD patients. Weight gain, positive nitrogen balance, and improvements in muscle strength, handgrip strength, and walking distance are possible with energy intakes ranging from 1.4 to 2.2 × REE and protein intakes of ≥1.7 g/kg/day. For obese patients, appropriate nutritious food choices along with a daily exercise program are important and weight stability with an improvement in body composition should be promoted. Increased intake of fruits and vegetables is recommended. Other food constituents, particularly antioxidants, BCAAs, and omega-3 fatty acids, may be beneficial. Every effort should be made to identify patients at risk of nutritional decline as soon as possible so that appropriate therapy can be delivered as *secondary prevention* rather than as *treatment* once serious losses have occurred.

20.8.1 *Maintaining or Improving Nutritional and Respiratory Function*

20.8.1.1 ROLE OF THE DIETITIAN

Because of the serious consequences of undernutrition in this patient group, the nutritionist should focus on the patient who is starting to lose or has lost lean body mass. COPD patients often have a poor appetite, may suffer from fatigue and breathlessness, and dietary advice should include enjoyable, energy- and protein-dense foods as eating can tire some COPD patients. The patient's usual food habits, tastes, and preferences need to be considered. It is important for the patient to try to eat what he/she enjoys. Sustaining an adequate intake at all times is a challenge. Extra food may be needed and could come in the form of energy-dense supplements spread out over the day to avoid interfering with other foods.

Many concurrent health problems may be evident but nutritional interventions in hospitals need to ensure the adequate intake of protein and energy. This can be achieved through tempting food choices (having family members bring in favorite foods), offering snacks on a very regular basis, having food available at frequent intervals without disruptions, and helping the patient and family understand the importance of nutritional therapy (despite their major preoccupation with breathing problems). At home, family support or such services as meal delivery (Meals on Wheels), home care, or community meals can help assure adequate nutritional support. Nutritious snacks, such as cheese, puddings, or commercial supplements, are very important for COPD patients, as many find it difficult to eat a lot during one meal. Thus, patients should eat frequent (six to seven) small meals (300–500 kcal/meal) of the foods they enjoy (see suggestions in Tables 20.3 and 20.4; see also Chapter 9). Strategies to help motivate the patient to eat and achieve nutritional goal also aim to reduce the energy cost of food activities so as to not tire nor discourage the patient. These include using simple recipes with a minimum of preparation time. Meals prepared, cooked, and served in one dish help minimize the

Table 20.3
Suggestions for a menu plan for a female COPD patient

Day 1	
Breakfast	1 Soft-boiled egg, 1 slice of whole-wheat toast with 3 slices of cheese, Coffee or tea with whole milk and sugar
Snack	~200 g Rice pudding with raisins
Lunch	1 Bowl of chunky vegetable soup homemade or canned and 1 turkey Sandwich (2 slices of whole-wheat bread, mustard, ~90 g sliced turkey, tomatoes, and lettuce)
Dinner	~100 g Fried salmon, 100 g boiled carrots, 40 g boiled peas, 1 small scoop (~16 g) of ice cream and ~240 g (1 cup) of applesauce
Snack	~200 g Rice pudding with raisins

This menu supplies approximately 2,190 kcal or $\geq 1.7 \times$ REE for a 75-year-old female COPD patient weighting 64 kg. Her protein intake is ~1.7 g/kg body weight. The caloric breakdown is as follows: 19% protein, 50% carbohydrate, and 31% fat.

Table 20.4
Suggestions for a menu plan for a male COPD patient

Day 2	
Breakfast	60 g Raisin cereal with 2% milk, coffee, or tea with 3.7% milk and sugar
Snack	One 5-oz can chocolate pudding
Lunch	~490 g (2 cups) beef and vegetable stew, 2 slices of bread with 2 teaspoons of butter, a few slices of tomatoes and 3 slices of cheese, 1 small scoop of ice cream (16 g), and ~244 g (1 cup) of applesauce
Snack	Tea or coffee with 2% milk
Dinner	~150 g Fried chicken, 200 g (1 cup) of rice, ~70 g (1/2 cup)
Snack	One 5-oz can chocolate pudding

This menu supplies approximately 2,470 kcal or ≥1.7 × REE for a 73-year-old male COPD patient weighing 70 kg. The protein intake is ~1.8 g/kg body weight. The energy breakdown is as follows: 20% protein, 50% carbohydrate, and 30% fat.

energy spent in cleaning up after the meal. Cooking larger meals on good days will ensure that good food is available on bad days. Placement of frequently used foods and kitchenware close at hand in the kitchen will reduce oxygen consumption and dyspnea *(195)*. These should ideally be placed on the counter or on shelves set at hip and/or shoulder height. Economic status can also present a challenge to optimal dietary intake as many COPD patients may be disabled and may have retired from work; thus affordable food suggestions are important.

High-risk individuals (Fig. 20.1) should also undergo a dietary evaluation assessing the adequacy of energy and protein intake, and acceptable foods that are nutrient dense should be offered to enhance intakes. Micronutrient requirements are likely met by using the food guide pyramid as a resource. Patients in hospitals need close monitoring to ensure that meals are not missed due to medical procedures, that snacks and meals are consumed, and that missed foods are replaced.

20.8.1.2 MAINTAINING OR IMPROVING RESPIRATORY FUNCTION

Whenever possible, in addition to nutritional interventions, patients should participate in an endurance and strength exercise program to preserve and/or improve respiratory muscle strength and build peripheral muscles. Severely depleted patients may need an exercise program adapted to their special situation. The benefits of exercise are well known *(179–180)*, and in COPD patients, they can enhance the nutritional therapy, owing to improved appetite and well-being.

It is possible that despite adherence to an exercise program and aggressive therapy some individuals cannot arrive at the desired outcome, and in such instances anabolic agents may be considered. A treatment combining nutritional supplementation with the injection of the anabolic steroid nandrolone decanoate resulted in an increase in weight and fat-free mass and improved respiratory muscle function *(170,196)*. Clearly, more work needs to be done to unravel the factors that make some patients more able to maintain nutritional adequacy, whereas others have great difficulty in this area.

In conclusion, COPD patients need to be screened to detect those who are starting to lose weight as well as those already undernourished. Dietary interventions are thought to be more useful early on in the disease process to stop the loss of lean body mass, compared to refeeding to rebuild. Waiting until patients show obvious signs of undernutrition is not in their best interest. Offering appealing foods with high energy and protein density is important and needs to be institutionalized. Disruptions in food intake for laboratory tests should be minimized. Support at home is crucial. Appealing meals need to be provided with a minimum of effort on the patient's part. Fatigue, discomfort, depression, anxiety, and reduced income from leaving a job can all lead to poor intakes. It is clear that a treatment plan for discharged patients is critical to their well-being, as breathlessness and exercise intolerance are their most distressing symptoms.

20.9 RECOMMENDATIONS

1. Routine nutritional assessment (weight and height measurements, fat-free mass assessment, and recent weight loss) for patients with COPD will help identify patients with poor status so that intervention can be initiated as early as possible.
2. Patients found to be at risk should receive nutritional surveillance, be provided adequate protein ($\geq$1.7 g/kg/day) and energy (1.4–2.2 × REE) supplements as needed, be given an appropriate exercise program, and be encouraged to include food sources rich in antioxidants and omega-3 fatty acids.
3. Interventions should emphasize weight gain in patients with BMI < 25 kg/m^2 and weight stability in those with normal body weight or overweight, positive nitrogen balance, and improvements in muscle strength, handgrip strength, and distance walked.
4. For undernourished patients at home, family and community support is crucial so that the patient can receive appealing meals with a minimum of effort in a supportive social environment. Social support is also important in the hospital setting, as is the need to minimize disruptions of food intake for medical tests.

REFERENCES

1. Chapman KR, Mannino DM, Soriano JB, Vermeire PA, Buist AS, Thun MJ, et al. Epidemiology and costs of chronic obstructive pulmonary disease. Eur Respir J 2006; 27(1):188–207.
2. Pawels RA, Rabe KF. Burden and clinical feature of chronic obstructive pulmonary disease (COPD). Lancet 2004; 364:613–20.
3. Mannino DM, Homa DM, Akimbani LJ, Ford ES, Redd SC. Chronic Obstructive Pulmonary Disease Surveillance – United States, 1971–2000. MMWR 2002; 51:1–16.
4. World Health Organization. The World Health Report: 2004: Changing history. Available from: URL: www.who.int/whr/2004/en/report04_en.pdf
5. Jemal A, Ward E, Hao Y, Thun M. Trends in the leading causes of death in the United States, 1970–2002. JAMA 2005; 294(10):1255–9.
6. Murray CJL, Lopez AD, eds. The Global Burden of Disease: A Comprehensive Assessment of Mortality and Disability from Diseases, Injuries and Risk Factors in 1990 and Projected to 2020. Cambridge (MA): Harvard University Press, 1996.
7. Tinkelman D, Nordyke RJ, Isonaka S, George D, DesFosses K, Nonikov D. The impact of chronic obstructive pulmonary disease on long-term disability costs. J Manag Care Pharm 2005; 11(1): 25–32.

8. Global Initiative for Chronic Obstructive Lung Disease. Global Strategy for the Diagnosis, Management and Prevention of COPD, 2006. Available from: URL: www.goldcopd.org

9. Rennard SI, Vestbo J. COPD, the dangerous underestimate of 15%. Lancet 2006; 367:1216–9.

10. National Center for Health Statistics. Health, United States, 2006, with chartbook on trends in the health of Americans. Hyattsville (MD), 2006.

11. Machado MC, Krishnan JA, Buist SA, Bilderback AL, Fazolo GP, Santarosa MG, et al. Sex differences in survival of oxygen-dependent patients with chronic obstructive pulmonary disease. Am J Respir Crit Care Med 2006; 174(5):524–9.

12. Costanza MC, Salamun J, Lopez AD, Morabia A. Gender differentials in the evolution of cigarette smoking habits in a general European adult population from 1993–2003. BMC Public Health [serial online] 2006; 6:130. Available from: URL: www.biomedcentral.com/1471-2458/6/130

13. Watson L, Vestbo J, Postma DS, Decramer M, Rennard S, Kiri VA, et al. Gender differences in the management and experience of Chronic Obstructive Pulmonary Disease. Respir Med 2004; 98(12):1207–13.

14. Chapman KR. Chronic obstructive pulmonary disease: are women more susceptible than men? Clin Chest Med 2004; 25:331–41.

15. Weir E. COPD death rates: projecting a female trajectory. Can Med Assoc J 2004; 170(3):334.

16. Thorn J, Björkelund C, Bengtsson C, Guo X, Lissner L, Sundh V. Low socio-economic status, smoking, mental stress and obesity predict obstructive symptoms in women, but only smoking also predicts subsequent experience of poor health. Int J Med Sci 2007; 4(1):7–12.

17. Peruzza S, Sergi G, Vianello A, Pisent C, Tiozzo F, Manzan A, et al. Chronic obstructive pulmonary disease (COPD) in elderly subjects: impact on functional status and quality of life. Respir Med 2003; 97(6):612–7.

18. Odencrants S, Ehnfors M, Grobe SJ. Living with chronic obstructive pulmonary disease: part I. Struggling with meal-related situations: experiences among persons with COPD. Scand J Caring Sci 2005; 19(3):230–9.

19. Cochrane WJ, Afolabi OA. Investigation into the nutritional status, dietary intake and smoking habits of patients with chronic obstructive pulmonary disease. J Hum Nutr Diet 2004; 17(1):3–11.

20. Agusti AG. COPD, a multicomponent disease: Implications for management. Respir Med 2005; 99(6):670–82.

21. Agusti AG. Systemic effects of chronic obstructive pulmonary disease. Proc Am Thorac Soc 2005; 2:367–70.

22. Vermeeren MA, Creutzberg EC, Schols AM, Postma DS, Pieters WR, Roldaan AC, et al., on behalf of the COSMIC Study Group. Prevalence of nutritional depletion in a large out-patient population of patients with COPD. Respir Med 2006; 100(8):1349–55.

23. Saudny-Unterberger H, Martin G, Gray-Donald K. Impact of nutritional support on functional status during an acute exacerbation of chronic obstructive pulmonary disease. Am J Respir Crit Care Med 1997; 156:794–9.

24. Schols AMWJ, Slangen, JOS, Volovics L, Wouters FM. Weight loss is a reversible factor in the prognosis of chronic obstructive pulmonary disease. Am J Respir Crit Care Med 1998; 157:1791–7.

25. Gray-Donald K, Gibbons S, Shapiro H, Macklem PT, Martin JG. Nutritional status and mortality in chronic obstructive pulmonary disease. Am J Respir Crit Care Med 1996; 153:961–6.

26. Schols AMWJ, Mostert R, Soeter PB, Wouters EFM. Body composition and exercise performance in chronic obstructive pulmonary disease. Thorax 1991; 46:695–9.

27. Prescott E, Almdal T, Mikkelsen KL, Tofteng CL, Vestbo J, Lange P. Prognostic value of weight change in chronic obstructive pulmonary disease: results from the Copenhagen City Heart Study. Eur Respir J 2002; 20(3):539–44.

28. Yohannes AM, Baldwin RC, Connolly M. Mortality predictors in disabling chronic obstructive pulmonary disease in old age. Age Ageing 2002; 31(2):137–40.

29. Marti S, Munoz X, Rios J, Morell F, Ferrer J. Body weight and comorbidity predict mortality in COPD patients treated with oxygen therapy. Eur Respir J 2006; 27(4):689–96.

30. Vestbo J, Prescott E, Almdal T, Dahl M, Nordestgaard BG, Andersen T, et al. Body mass, fat-free body mass, and prognosis in patients with chronic obstructive pulmonary disease from a random population sample: findings from the Copenhagen City Heart Study. Am J Respir Crit Care Med 2006; 173(1):79–83.

31. Slinde F, Ellegard L, Gronberg AM, Larsson S, Rossander-Hulthen L. Total energy expenditure in underweight patients with severe chronic obstructive pulmonary disease living at home. Clin Nutr 2003; 22(2):159–65.

32. Sergi G, Coin A, Marin S, Vianello A, Manzan A, Peruzza S, et al. Body composition and resting energy expenditure in elderly male patients with chronic obstructive pulmonary disease. Respir Med 2006; 100(11):1918–24.

33. International Statistical Classification of Diseases and Related Health Problems, 10th Revision, Version for 2007. Available from: URL: www.who.int/classifications/apps/icd/icd10 online

34. Shapiro SD, Ingenito EP. The pathogenesis of chronic obstructive pulmonary disease: advances in the past 100 years. Am J Respir Cell Mol Biol 2005; 32(5):367–72.

35. Hogg JC. Pathophysiology of airflow limitation in chronic obstructive pulmonary disease. Lancet 2004; 364(9435):709–21.

36. Hubmayr RD, Rodarte JR. Cellular effects and physiologic responses: lung mechanics. In: Cherniack NS, ed. Chronic Obstructive Pulmonary Disease. Philadelphia (PA): WB Saunders, 1991, pp. 79–90.

37. Rochester DF. Effects of COPD on the respiratory muscles. In: Cherniack NS, ed. Chronic Obstructive Pulmonary Disease. Philadelphia (PA): WB Saunders, 1991, pp. 134–57.

38. Tobin MJ. Respiratory muscles in disease. Clin Chest Med 1988; 9:263–86.

39. Wouters EF. Chronic obstructive pulmonary disease. 5: systemic effects of COPD. Thorax 2002; 57(12):1067–70.

40. Hallin R, Koivisto-Hursti UK, Lindberg E, Janson C. Nutritional status, dietary energy intake and the risk of exacerbations in patients with chronic obstructive pulmonary disease (COPD). Respir Med 2006; 100(3):561–7.

41. Toth S, Tkacova R, Matula P, Stubna J. Nutritional depletion in relation to mortality in patients with chronic respiratory insufficiency treated with long-term oxygen therapy. Wiener Klinische Wochenschrift 2004; 116(17–18):617–21.

42. Chailleux E, Laaban JP, Veale D. Prognostic value of nutritional depletion in patients with COPD treated by long-term oxygen therapy: data from the ANTADIR observatory. Chest 2003; 123(5):1460–6.

43. Meyer PA, Mannino DM, Redd SC, Olson DR. Characteristics of adults dying with COPD. Chest 2002; 122(6):2003–8.

44. Ringbaek TJ, Viskum K, Lange P. BMI and oral glucocorticoids as predictors of prognosis in COPD patients on long-term oxygen therapy. Chron Respir Dis 2004; 1(2):71–8.

45. Schols AM, Broekhuizen R, Weling-Scheepers CA, Wouters EF. Body composition and mortality in chronic obstructive pulmonary disease. Am J Clin Nutr 2005; 82(1):53–9.

46. Gray-Donald K, Gibbons L, Shapiro SH, Martin JG. Effect of nutritional status on exercise performance in patients with chronic obstructive pulmonary disease. Am Rev Respir Dis 1989; 140:1544–8.

47. Landbo C, Prescott E, Lange P, Vestbo J, Almdal T. Prognostic value of nutritional status in chronic obstructive pulmonary disease. Am J Respir Crit Care Med 1999; 16: 1856–61.

48. Creutzberg EC, Wouters EF, Mostert R, Weling-Scheepers CA, Schols AM. Efficacy of nutritional supplementation therapy in depleted patients with chronic obstructive pulmonary disease. Nutrition 2003; 19(2):120–7.

49. Bolton CE, Ionescu AA, Shiels KM, Pettit RJ, Edwards PH, Stone MD, Nixon LS, Evans WD, Griffiths TL, Shale DJ. Associated loss of fat-free mass and bone mineral density in chronic obstructive pulmonary disease. Am J Respir Crit Care Med 2004; 170:1286–93.

50. Schols MWJ, Soeters PB, Dingemans AM, Mostert R, Frantzen PJ, Wouters EF. Prevalence and characteristics of nutritional depletion in patients with stable COPD eligible for pulmonary rehabilitation. Am Rev Respir Dis 1993; 147:1151–6.

51. Hopkinson NS, Tennant RC, Dayer MJ, Swallow EB, Hansel TT, Moxham J, et al. A prospective study of decline in fat free mass and skeletal muscle strength in chronic obstructive pulmonary disease. Respir Res [serial online] 2007; 8:25. Available from: URL: respiratory-research.com/content/8/1/25

52. Wouters EFM. Nutrition and metabolism in COPD. Chest 2000; 117:274S–80S.

53. Kyle UG, Janssens JP, Rochat T, Raguso CA, Pichard C. Body composition in patients with hypercapnic respiratory failure. Respir Med 2006; 100:244–52.

54. Engelen MPKJ, Schols AMWJ, Baken WC, Wesseling GJ, Wouters EFM. Nutritional depletion in relation to respiratory and peripheral skeletal muscle function in out patients with COPD. Eur Respir 1994; 7:1793–7.

55. Baarends EM, Schols AMWJ, Pannemans DLE, Westerterp KR, Wouters EFM. Total free living energy expenditure in patients with severe chronic obstructive pulmonary disease. Am J Respir Crit Care Med 1997; 155:549–54.

56. Slinde F, Grönberg AM, Engström CP, Rossander-Hultén L, Larsson S. Body composition by bioelectrical impedance predicts mortality in chronic obstructive pulmonary disease patients. Respir Med 2005; 99:1004–9.

57. Soler-Cataluña JJ, Sánchez-Sánchez L, Martínez-García MA, Sánchez PR, Salcedo E, Navarro M. Mid-arm muscle area is a better predictor of mortality than body mass index in COPD. Chest 2005; 128:2108–15.

58. Kotler DP. Cachexia. Ann Intern Med 2000; 133:622–34.

59. Schols AMWJ. Pulmonary cachexia. Int J Cardiol 2002; 85:101–10.

60. Eisner MD, Blanc PD, Sidney S, Yelin EH, Lathon PV, Katz PP, et al. Body composition and functional limitation in COPD. Respir Res [serial online] 2007; 8:7. Available from: URL: respiratory-research.com/content/8/1/7

61. Slinde F, Kvarnhult K, Grönberg AM, Nordenson A, Larsson S, Hulthén1 L. Energy expenditure in underweight chronic obstructive pulmonary disease patients before and during a physiotherapy programme. Eur J Clin Nutr 2006; 60:870–6.

62. Benedict F, Miles W, Roth P, et al. Human Vitality and Efficiency Under Prolonged Restricted Diet. Carnegie Institute, Washington (DC), 1919; publication no. 280.

63. Schols AMWJ, Wouters EFM, Soeters PB, Westerterp KR. Body composition by bioelectrical-impedance analysis compared with deuterium dilution and skinfold anthropometry in patients with chronic obstructive pulmonary disease. Am J Clin Nutr 1991; 53:421–4.

64. Tang NLS, Chung ML, Elia M, Hui E, Lum CM, Luk JKH, et al. Total daily energy expenditure in wasted chronic obstructive pulmonary disease patients. Eur J Clin Nutr 2002; 5:282–7.

65. Harris JA, Benedict EG. A Biometric Study of Basal Metabolism. Carnegie Institution of Washington, Washington (DC), 1919.

66. Morley JE, Thomas DR, Wilson MMG. Cachexia: pathophysiology and clinical relevance. Am J Clin Nutr 2006; 83:735–43.

67. Tracey KJ, Cerami A. Tumor necrosis factor: a pleiotropic cytokine and therapeutic target. Annu Rev Med 1994; 45:491–503.

68. Haddad F, Zaldivar F, Cooper DM, Adams GR. IL-6-induced skeletal muscle atrophy. J Appl Physiol 2005; 98(3):911–7.

69. Van Der Poll T, Sauerwein HP. Tumor necrosis factor-α: its role in the metabolic response to sepsis. Clin Sci 1993; 84:247–56.

70. Di Francia M, Barbier D, Mege Jean L, Orehek J. Tumor necrosis factor-alpha. Levels and weight loss in chronic obstructive pulmonary disease. Am J Respir Crit Care Med 1994; 150:1453–5.

71. De Godoy I, Calhoun WJ, Donahoe M, Mancino J, Rogers RM. Elevated TNF-α production by peripheral blood monocytes of weight losing COPD patients. Am J Respir Crit Care Med 1996; 153:633–7.

72. Takabatake N, Nakamura H, Abe S, Inoue S, Hino T, Saito H, et al. The relationship between chronic hypoxemia and activation of the tumor necrosis factor-α system in patients with chronic obstructive pulmonary disease. Am J Respir Crit Care Med 2000; 1161:1179–84.

73. Perera WR, Hurst JR, Wilkinson TMA, Sapsford RJ, Müllerova H, Donaldson GC, et al. Inflammatory changes, recovery, and recurrence at COPD exacerbation. Eur Respir J 2007; 29:527–34.

74. Gan WQ, Man SFP, Senthilselvan A, Sin DD. Association between chronic obstructive pulmonary disease and systemic inflammation: a systematic review and a metaanalysis. Thorax 2004; 59:574–80.

75. Yanbaeva DG, Dentener MA, Creutzberg EC, Wesseling G, Wouters EMF. Systemic effects of smoking. Chest 2007; 131:1557–66.

76. Broekhuizen R, Grimble RF, Howell WM, Shale DJ, Creutzberg EC, Wouters EF, Schols AM. Pulmonary cachexia, systemic inflammatory profile, and the interleukin 1–511 single nucleotide polymorphism. Am J Clin Nutr 2005; 82:1059–64.

77. Rennard SI, Fogarty C, Kelsen S, Long W, Ramsdell J, Allison J, et al. The safety and efficacy of infliximab in moderate to severe chronic obstructive pulmonary disease. Am J Respir Crit Care Med 2007; 175:926–34.

78. Engelen MPKJ, Deutz NEP, Wouters EFM, Schols AMWJ. Enhanced levels of whole-body protein turnover in patients with chronic obstructive pulmonary disease. Am J Respir Crit Care Med 2000; 162:1488–92.

79. Bolton CE, Broekhuizen R, Ionescu AA, Nixon LS, Wouters EFM, Shale DJ, Schols AMWJ. Cellular protein breakdown and systemic inflammation unaffected by pulmonary rehabilitation in COPD. Thorax 2007; 62:109–14.

80. Rutten EPA, Franssen FME, Engelen MPKJ, Wouters EFM, Deutz NEP, Schols AMWJ. Greater whole-body myofibrillar protein breakdown in cachectic patients with chronic obstructive pulmonary disease. Am J Clin Nutr 2006; 83:829–34.

81. Engelen MPKJ, Schols MWJ. Altered amino acid metabolism in chronic obstructive pulmonary disease: new therapeutic perspective? Curr Opin Clin Nutr Metab Care 2003; 6:73–8.

82. Yoneda T, Yoshikawa M, Fu A, Tsukaguchi K, Okamoto Y, Takenaka H. Plasma levels of amino acids and hypermetabolism in patients with chronic obstructive pulmonary disease. Nutrition 2001; 17(2):95–9.

83. Laghi F, Antonescu-Turcu A, Collins E, Segal J, Tobin DE, Jubran A, et al. Hypogonadism in men with chronic obstructive pulmonary disease: prevalence and quality of life. Am J Respir Crit Care Med 2005; 171:728–33.

84. Debigare R, Marquis K, Cote CH, Tremblay RR, Michaud A, LeBlanc P, et al. Catabolic/anabolic balance and muscle wasting in patients with COPD. Chest 2003; 124:83–9.

85. Van Vliet M, Spruit MA, Verleden G, Kasran A, Van Herck E, Pitta F, et al. Hypogonadism, quadriceps weakness, and exercise intolerance in chronic obstructive pulmonary disease. Am J Respir Crit Care Med 2005; 172:1105–11.

86. Svartberg J, Schirmer H, Medbo A, Melbye H, Aasebo U. Reduced pulmonary function is associated with lower levels of endogenous total and free testosterone. The Tromso study. Eur J Epidemiol 2007; 22(2):107–12.

87. Laghi F, Langbein WE, Antonescu-Turcu A, Jubran A, Bammert C, Tobin MJ. Respiratory and skeletal muscles in hypogonadal men with chronic obstructive pulmonary disease. Am J Respir Crit Care Med 2004; 171:598–605.

88. Koehler F, Doehner W, Hoernig A, Witt C, Anker SD, John M. Anorexia in chronic obstructive pulmonary disease—Association to cachexia and hormonal derangement. Int J Cardiol 2007; 119:83–9.

89. Ferreira IM, Verreschi IT, Nery LE, Goldstein RS, Zamel N, Brooks D, et al. The influence of 6 months of oral anabolic steroids on body mass and respiratory muscles in undernourished COPD patients. Chest 1998; 114:19–28.

90. Yeh SS, DeGuzman B, Kramer T for the M012 Study Group. Reversal of COPD-associated weight loss using the anabolic agent oxandrolone. Chest 2002; 122:421–8.

91. Casaburi R, Bhasin S, Cosentino L, Porszasz J, Somfay A, Lewis MI, et al. Effects of testosterone and resistance training in men with chronic obstructive pulmonary. Am J Respir Crit Care Med 2004; 170:870–8.

92. Creutzberg EC, Wouters EF, Mostert R, Pluymers RJ, Schols AM. A role for anabolic steroids in the rehabilitation of patients with COPD? A double-blind, placebo-controlled, randomized trial. Chest 2003; 124:1733–42.

93. Burdet L, de Muralt B, Schutz Y, Pichard C, Fitting JW. Administration of growth hormone to underweight patients with chronic obstructive pulmonary disease. A prospective, randomized, controlled study. Am J Respir Crit Care Med 1997; 156(6):1800–6.

94. Laghi F. Low testosterone in chronic obstructive pulmonary disease, does it really matter?. Am J Respir Crit Care Med 2005; 172:1069–70.

95. Couillard A, Prefaut C. From muscle disuse to myopathy in COPD: potential contribution of oxidative stress. Eur Respir J 2005; 26:703–19.

96. Garcia-Aymerich J, Felez MA, Escarrabill J, Marrades RM, Morera J, Elosua R, et al. Physical activity and its determinants in severe chronic obstructive pulmonary disease. Med Sci Sports Exerc 2004; 36(10):1667–73.

97. Wagner PD. Skeletal muscles in chronic obstructive pulmonary disease: deconditioning, or myopathy? Respirology 2006; 11:681–686.

98. Franssen FME, Wouters EFM, Schols AMWJ. The contribution of starvation, deconditioning and aging to the observed alterations in peripheral skeletal muscle in chronic organ diseases. Clin Nutr 2002; 21(1):1–14.

99. Cochrane WJ, Afolabi OA. Investigation into the nutritional status, dietary intake and smoking habits of patients with chronic obstructive pulmonary disease. J Hum Nutr Dietet 2004; 17:3–11.

100. Thorsdottir I, Gunnarsdottir I. Energy intake must be increased among recently hospitalized patients with chronic obstructive pulmonary disease to improve nutritional status. J Am Diet Assoc 2002; 102(2):247–9.

101. Førli L, Boe J. The energy intake that is needed for weight gain in COPD candidates for lung transplantation. COPD: J Chronic Obstructive Pulmonary Disease 2005; 2:405–10.

102. Gronberg AM, Slinde F, Engstrom CP, Hulthen L, Larsson S. Dietary problems in patients with severe chronic obstructive pulmonary disease. J Hum Nutr Diet 2005; 18(6):445–52.

103. Engstrom CP, Persson LO, Larsson S, Ryden A, Sullivan M. Functional status and well being in chronic obstructive pulmonary disease with regard to clinical parameters and smoking: a descriptive and comparative study. Thorax 1996; 51(8):825–30.

104. Odencrants S, Ehnfors M, Grobe SJ. Living with chronic obstructive pulmonary disease: Part I. Struggling with meal-related situations: Experiences among persons with COPD. Scand J Caring Sci 2005; 19:230–9.

105. Theander K, Unosson M. Fatigue in patients with chronic obstructive pulmonary disease. J Adv Nurs 2004; 45(2):172–7.

106. Førli L, Moum T, Bjørtuft Ø, Vatn M, Boe J. The influence of underweight and dietary support on well-being in lung transplant candidates. Respir Med 2006; 100(7):1239–46.

107. O'Donnell DE, Banzett RB, Carrieri-Kohlman V, Casaburi C, Davenport PW, Gandevia SC, et al. Pathophysiology of dyspnea in chronic obstructive pulmonary disease: a roundtable. Proc Am Thorac Soc 2004; 4:145–68.

108. Wolkove N, Fu LY, Purohit A, Colacone A, Kreisman H. Meal-induced oxygen desaturation and dyspnea in chronic obstructive pulmonary disease. Can Respir J 1998; 5(5):361–5.

109. Gray-Donald K, Carrey Z, Martin JG. Postprandial dyspnea and malnutrition in patients with chronic obstructive pulmonary disease. Clin Invest Med 1998; 21(3):135–41.

110. Good-Fratturelli MD, Curlee RF, Holle JF. Dysphagia in VA patients with COPD referred for videofluoroscopic swallow examination. J Commun Disord 2000; 33:93–110.

111. Mokhlesi B, Morris AL, Huang CF, Curcio AJ, Barrett TA, Kamp DW. Increased prevalence of gastroesophageal symptoms in patients with COPD. Chest 2001; 119:1043–8.

112. Rascon-Aguilar IE, Pamer M, Wludyka P, Cury J, Coultas D, Lambiase LR, et al. Role of gastroesophageal reflux symptoms in exacerbations of COPD. Chest 2006; 130:1096–101.

113. Wikby K, Fagerskiold A. The willingness to eat. An investigation of appetite among elderly people. Scand J Caring Sci 2004; 18(2):120–7.

114. Kondrup J, Johansen N, Plum LM, Bak L, Larsen IH, Martinsen A, et al. Incidence of nutritional risk and causes of inadequate nutritional care in hospitals. Clin Nutr 2002; 21(6):461–8.

115. Mowé M, Bømer T. Reduced appetite. A predictor for undernutrition in aged people. J Nutr Health Aging 2002; 6(1):81–3.

116. Chapman-Novakofski K, Brewer MS, Riskowski J, Burkowski C, Winter L. Alterations in taste thresholds in men with chronic obstructive pulmonary disease. J Am Diet Assoc 1999; 99(12):1536–41.

117. Glanz K, Basil M, Maibach E, Goldberg J, Snyder D. Why Americans eat what they do: taste, nutrition, cost, convenience, and weight control concerns as influence on food consumption. J Am Diet Assoc. 1998; 98:1118–26.

118. de Castro JM, Brewer EM, Elmore DK, Orazco S. Social facilitation of the spontaneous meal size of humans occurs regardless of time, place, alcohol or snacks. Appetite 1990; 15:89–101.

119. de Castro JM, Brewer E. The amount eaten in meals by humans is a power function of the number of people present. Physiol Behav 1992; 51:121–5.

120. Mikkelsen RL, Middelboe T, Pisinger C, Stage KB. Anxiety and depression in patients with chronic obstructive pulmonary disease (COPD): a review. Nord J Psy 2004; 58(1):65–70.

121. Braun SR, Keim NL, Dixon RM, Clagnaz P, Anderegg A, Shrago ES. The prevalence and determinants of nutritional changes in chronic obstructive pulmonary disease. Chest 1984; 4:558–63.

122. Brug J, Schols A, Mesters I. Dietary change, nutrition education and chronic obstructive pulmonary disease. Patient Educ Couns 2004; 52:249–57.

123. Wilson DO, Rogers RM, Hoffman RM. Nutrition and chronic lung disease. State of the art. Am Rev Respir Dis 1985; 132:1347–65.

124. Weinsier RL, Hunker EM, Krumdieck CL, Butterworth CE. Hospital malnutrition: a prospective evaluation of general medical patients during the course of hospitalization. Am Clin Nutr 1979; 32:418–26.

125. Coats KG, Morgan SL, Bartolucci AA, Weinsier RL. Hospital-associated malnutrition: a reevaluation 12 years later. Am Diet Assoc 1993; 93:27–33.

126. McWhirter JP, Pennington CR. Incidence and recognition of malnutrition in hospital. BMJ 1994; 308:945–8.

127. Gariballa SE. Malnutrition in hospitalized elderly patients: when does it matter? Clin Nutr 2001; 20(6):487–91.

128. Paillaud E, Campillo B, Bories PN, Le Parco JC. Nutritional status in 57 elderly patients hospitalised in a rehabilitation unit: influence of causing disease. Rev Med Interne 2001; 22:238–44.

129. Pichard C, Kyle UG, Morabia A, Perrier A, Vermeulen B, Unger P. Nutritional assessment: lean body mass depletion at hospital admission is associated with an increased length of stay. Am J Clin Nutr 2004; 79:613–8.

130. Splett PL, Roth-Yousey LL, Vogelzang JL. Medical nutrition therapy for the prevention and treatment of unintentional weight loss in residential healthcare facilities. J Am Diet Assoc 2003; 103(3):352–62.

131. Incalzi RA, Landi F, Pagano F, Capparella O, Gemma A, Carbonin PU. Changes in nutritional status during the hospital stay: a predictor of long-term survival. Aging Clin Exp Res 1998; 10:490–6.

132. Sullivan DH, Walls RC. Protein-energy undernutrition and the risk of mortality within six years of hospital discharge. J Am Coll Nutr 1998; 17(6):571–8.

133. Sullivan DH, Sun S, Walls RC. Protein-energy undernutrition among elderly hospitalized patients: a prospective study. JAMA 1999; 281:2013–9.

134. Corish CA, Kennedy NP. Protein-energy undernutrition in hospital in-patients. Br Nutr 2000; 83: 575–91.

135. Paquet C, St-Arnaud-McKenzie D, Kergoat MJ, Ferland G, Dube L. Direct and indirect effects of everyday emotions on food intake of elderly patients in institutions. J Gerontol A Biol Sci Med Sci 2003; 58(2):M153–8.

136. Dupertuis YM, Kossovsky MP, Kyle UG, Raguso CA, Genton L, Pichard C. Food intake in 1707 hospitalized patients: a prospective comprehensive hospital survey. Clin Nutr 2003; 22:120–3.

137. Schiffman SS, Warwick ZS. Effect of flavor enhancement of food for the elderly on nutritional status: food intake, biochemical indices and anthropometric measures. Physiol Behav 1993; 53:395–402.

138. Mathey MF, Siebelink E, de Graaf C, Van Staveeren WA. Flavor enhancement of food improves dietary intake and nutritional status of elderly nursing home residents. J Gerontol A Biol Sci Med Sci 2001; 56(4):M200–5.

139. Decramer M, De Bock V, Dom R. Functional and histologic picture of steroid-induced myopathy in chronic obstructive pulmonary disease. Am J Respir Crit Care Med 1996; 153:1958–64.

140. Roubenoff R, Roubenoff RA, Ward LM, Stevens MB. Catabolic effects of high-dose corticosteroids persist despite therapeutic benefit in rheumatoid arthritis. Am J Clin Nutr 1990; 52:1113–7.

141. Thorsdottir I, Gunnarsdottir I, Eriksen B. Screening method evaluated by nutritional status measurements can be used to detect malnourishment in chronic obstructive pulmonary disease. J Am Diet Assoc 2001; 101(6):648–54.

142. American Society for Parenteral and Enteral Nutrition. Standards for nutrition support: hospitalized patients. Nutr Clin Prac 1995; 10:208–19.

143. Hunter AM, Carey MA, Larsh HW. The nutritional status of patients with chronic obstructive pulmonary disease. Am Rev Respir Dis 1981; 124:376–81.

144. Openbrier DR, Irwin MM, Rogers RM, Gottlieb GP, Dauber JH, Van Thiel DH, et al. Nutritional status and lung function in patients with emphysema and chronic bronchitis. Chest 1983; 1:17–22.

145. Wilson DO, Rogers RM, Sanders MH, Pennock BE, Reilly JJ. Nutritional intervention in malnourished patients with emphysema. Am Rev Respir Dis 1986; 134:672–7.

146. Schols MWJ, Mostert R, Soeters P, et al. Inventory of nutritional status in patients with COPD. Chest 1989; 96:247–9.

147. Sahebjami H, Doers JT, Render ML, Bond TL. Anthropometric and pulmonary function test profiles outpatients with stable chronic obstructive pulmonary disease. Am J Med 1993; 94:469–74.

148. Laaban JP, Kouchakji B, Dore MF, et al. Nutritional status of patients with chronic obstructive pulmonary disease and acute respiratory failure. Chest 1993; 103:1362–8.

149. Dore MF, Kouchakji B, Orvoën-frija E, Rochemaure, Laaban J.-P. Body composition in patients with chronic obstructive pulmonary disease. Comparison between bioelectrical impedance analysis and anthropometry. Rev Mal Respir 2000; 17:665–70.

150. Chumlea WC, Roche AF, Steinbaugh ML. Estimating stature from knee height for persons 60–90 years of age. J Am Geriatr Soc 1985; 33:116–20.

151. Han TS, Lean ME. Lower leg length as an index of stature in adults. Int J Obes Relat Metab Dis 1996; 20:21–7.

152. Kwok T, Whitelaw MN. The use of armspan in nutritional assessment of the elderly. J Am Geriatr Soc 1991; 39:492–6.

153. Reeves SL, Varakamin C, Henry CJK. The relationship between arm-span measurement and height with special reference to gender and ethnicity. Eur J Clin Nutr 1996; 50:398–400.

154. Durnin JVGA, Womersley J. Body fat assessed from total body density and its estimation from skinfold thickness: measurements on 481 men and women aged from 16 to 72 years. Br J Nutr 1974; 32:77–97.

155. Lukaski HC, Johnson PE, Bolonchuk W, Lykken GI. Assessment of fat free mass using bioelectrical measurements of the body. Am J Clin Nutr 1985; 41:810–7.

156. Robert S, Zarowitz BJ, Hyzy R, Eichenhorn M, Peterson EL, Popovich J. Bioelectric impedance assessment of nutritional status in critically ill patients. Am J Clin Nutr 1993; 57:840–4.

157. Lerario MC, Sachs A, Lazaretti-Castro M, Saraiva LG, Jardim JR. Body composition in patients with chronic obstructive pulmonary disease: which method to use in clinical practice? B J Nutr 2006; 96:86–92.

158. Efthimiou J, Fleming J, Gomes C, Spiro SG. The effect of supplementary oral nutrition in poorly nourished patients with chronic obstructive pulmonary disease. Am Rev Respir Dis 1988; 137:1075–82.

159. Knowles JB, Fairbarn MS, Wiggs BJ, Chan-Yan C, Pardy RL. Dietary supplementation and respiratory muscle performance in patients with COPD. Chest 1988; 93:977–83.

160. Creutzberg Eva C, Schols AMWJ, Weling-Scheepers CAPM, Buurman WA, Wouters EFM. Characterization of non response to high caloric oral nutritional therapy in depleted patients with chronic obstructive pulmonary disease. Am J Respir Crit Care Med 2000; 161:745–52.

161. Mackenzie TH, Clark N, Bistrian BR, Flatt JP, Hallowell EM, Blackburn GL. A simple method for estimating nitrogen balance in hospitalized patients: a review and supporting data for a previously proposed technique. J Am Coll Nutr 1985; 4:575–81.

162. Blackburn GL, Bistrian BR, Maini BS, Schlamm HT, Smith MF. Nutritional and metabolic assessment of the hospitalized patient. J Parent Enteral Nutr 1977; 1:11–22.

163. Milner A. Accuracy of urinary nitrogen for predicting total urinary nitrogen in thermally injured patients. J Parent Enteral Nutr 1993; 17:414–6.

164. Li Y, Wang S, Zhong N. Simultaneous determination of pseudouridine and creatinine in urine of normal children and patients with leukaemia by high performance liquid chromatography. Biomed Chromatogr 1992; 6:191–3.

165. Lewis MI, Belman MJ, Dorr-Uyemura L. Nutritional supplementation in ambulatory patients with chronic obstructive pulmonary disease. Am Rev Respir Dis 1987; 135:1062–8.

166. Otte KE, Ahlburg P, D'Amore F, Stellfeld M. Nutritional repletion in malnourished patients with emphysema. J Parent Enteral Nutr 1989; 13:152–6.

167. Whittaker JS, Ryan CG, Buckley PA, Road JD. The effects of refeeding on peripheral and respiratory muscle function in malnourished chronic obstructive pulmonary disease patients. Am Rev Respir Dis 1990; 142:283–8.

168. Rogers RM, Donahoe M, Costantino J. Physiologic effects of oral supplemental feeding in malnourished patients with chronic obstructive pulmonary disease. Am Rev Respir Dis 1992; 146:1511–7.

169. Sridhar MK, Galloway A, Lean MEJ. Study of an outpatient nutritional supplementation programme in malnourished patients with emphysematous COPD. Eur Respir J 1994; 7:720–4.

170. Schols AM, Soeters PB, Mostert R, Pluymers RJ, Wouters EF. Physiologic effects of nutritional support and anabolic steroids in patients with chronic obstructive pulmonary disease. A placebo-controlled randomized trial. Am J Respir Crit Care Med 1995; 152:1268–74.

171. Steiner MC, Barton RL, Singh SJ, Morgan MDL. Nutritional enhancement of exercise performance in chronic obstructive pulmonary disease: a randomised controlled trial. Thorax 2003; 58(9):745–51.

172. Ferreira IM, Brooks D, Lacasse Y, Goldstein RS, White J. Nutritional supplementation for stable chronic obstructive pulmonary disease. Cochrane Database Sys Rev 2005; (2):CD000998.

173. Nagaya N, Itoh T, Murakami S, Oya H, Uematsu M, Miyatake K, et al.. Treatment of cachexia with ghrelin in patients with COPD. Chest 2005; 128(3):1187–93.

174. Borghi-Silva A, Baldissera V, Sampaio LM, Pires-DiLorenzo VA, Jamami M, Demonte A, et al. L-carnitine as an ergogenic aid for patients with chronic obstructive pulmonary disease submitted to whole-body and respiratory muscle training programs. Braz J Med Biol Res 2006; 39(4):465–74.

175. Fuld JP, Kilduff LP, Neder JA, Pitsiladis Y, Lean ME, Ward SA, et al. Creatine supplementation during pulmonary rehabilitation in chronic obstructive pulmonary disease. Thorax 2005; 60(7):531–7.

176. Faager G, Söderlund K, Sköld CM, Rundgren S, Tollbäck A, Jakobsson P. Creatine supplementation and physical training in patients with COPD: a double blind, placebo-controlled study. Int J Chron Obstruct Pulmon Dis 2006; 1(4):445–53.

177. Vermeeren MAP, Schols AMWJ, Wouters EFM. Effects of an acute exacerbation on nutritional and metabolic profile of patients with COPD. Eur Respir J 1997; 10:2264–9.

178. Vermeeren MAP, Wouters EFM, Geraerts-Keeris AJW, Schols AMWJ. Nutritional support in patients with chronic obstructive pulmonary disease during hospitalization for an acute exacerbation: a randomized controlled feasibility trial. Clin Nutr 2004; 23:1184–92.

179. Bunout D, Barrera G, De la Maza P, et al. The impact of nutritional supplementation and resistance training on the health functioning of free-living Chilean elders: results of 18 months of follow-up. J Nutr 2001; 131:2441S–6S.

180. Fiatarone MA, O'Neill EF, Ryan ND, Clements KM, Solares GR, et al. Exercise training and nutritional supplementation for physical frailty in very elderly people. N Eng J Med 1994; 330:1769–75.

181. Goldstein SA, Thomashow BM, Kvetan V, Askanazi J, Kinney JM, Elwyn DH. Nitrogen and energy relationships in malnourished patients with emphysema. Am Rev Respir Dis 1988; 138:636–44.

182. Angelillo VA, Bedi S, Durfee D, Dahl J, Patterson AJ, O'Donohue WJ, Jr. Effects of low and high carbohydrate feedings in ambulatory patients with chronic obstructive pulmonary disease and chronic hypercapnia. Ann Intern Med 1985; 103:883–5.

183. Vermeeren MAP, Wouters EF, Nelissen LH, Van Lier A, Hofman Z, Schols AM. Acute effects of different nutritional supplements on symptoms and functional capacity in patients with chronic obstructive pulmonary disease. Am J Clin Nutr 2001; 73:295–301.

184. Broekhuizen1 R, Creutzberg EC, Weling-Scheepers CAPM, Wouters EFM, Schols AMWJ. Optimizing oral nutritional drink supplementation in patients with chronic obstructive pulmonary disease. B J Nutr 2005; 93:965–71.

185. Pouw EM, Schols AMWJ, Deutz NEP, Wouters EFM. Plasma, muscle amino-acid levels in relation to resting energy expenditure and inflammation in stable COPD. Am J Respir Crit Care Med 1998; 158:797–801.

186. Engelen MP, Wouters EF, Deutz NE, Menheere PP, Schols AM. Factors contributing to alterations in skeletal muscle and plasma amino acid profiles in patients with chronic obstructive pulmonary disease. Am J Clin Nutr 2000; 72:1480–7.

187. Engelen MPKJ, Rutten EPA, De Castro CLN, Wouters EFM, Schols AMWJ, Deutz NE. Supplementation of soy protein with branched-chain amino acids alters protein metabolism in healthy elderly and even more in patients with chronic obstructive pulmonary disease. Am J Clin Nutr 2007; 85:431–9.

188. Connor WE. Importance of n-3 fatty acids in health and disease. Am J Clin Nutr 2000; 71:171S–5S.

189. Broekhuizen R, Wouters EFM, Creutzberg EC, Weling-Scheepers CAPM, Schols AMWJ. Polyunsaturated fatty acids improve exercise capacity in chronic obstructive pulmonary disease. Thorax 2005; 60:376–82.

190. Matsuyama W, Mitsuyama H, Watanabe M, Oonakahara K, Higashimoto I, Osame M, et al. Effects of omega-3 polyunsaturated fatty acids on inflammatory markers in COPD. Chest 2005; 128:3817–27.

191. Watson L, Margetts B, Howarth P, Dorward M, Thompson R, Little P. The association between diet and chronic obstructive pulmonary disease in subjects selected from general practice. Eur Respir J 2002; 20:313–8.

192. Walda IC, TabakC, Smit HA, Räsänen L, Fidanza F, Menotti A, Nissinen A, Feskens EJM, Kromhout D. Diet and 20-year chronic obstructive pulmonary disease mortality in middle-aged men from three European countries. Eur J Clin Nutr 2002; 56:638–43.

193. Sisson JH, Stoner JA, Romberger DJ, Spurzem JR, Wyatt TA, Owens-Ream J, et al. Alcohol intake is associated with altered pulmonary function. Alcohol 2005 May; 36(1):19–30.

194. Culpitt SV, Rogers DF, Fenwick PS, Shah P, De Matos C, Russell REK, et al. Inhibition by red wine extract, resveratrol, of cytokine release by alveolar macrophages in COPD. Thorax 2003; 58:942–6.

195. Velloso M, Jardim JR. Study of energy expenditure during activities of daily living using and not using body position recommended by energy conservation techniques in patients with COPD. Chest 2006; 130:126–32.

196. Pape GS, Friedman M, Underwood LE, Clemmons DR. The effect of growth hormone on weight gain and pulmonary function in patients with chronic obstructive lung disease. Chest 1991; 99(6):1495–500.

21 Nutrition and Chronic Kidney Disease

Srinivasan Beddhu

Key Points

- The development and progression of chronic kidney disease (CKD) is influenced by a number of dietary factors, including salt and protein intake and energy balance (obesity).
- While the benefits of a low-protein intake in preventing the development of CKD are not firmly established, it is likely that a high-protein intake is detrimental to individuals with even mild impairment of renal function.
- In the non-CKD population, obesity is a risk factor for the development of CKD via a variety of mechanisms, including clinical or sub-clinical insulin resistance, hypertension, and possibly other metabolic derangements.
- However, the deleterious metabolic effects of obesity are neutralized by its protective nutritional effects in the moderate CKD population and the deleterious metabolic effects of obesity are outweighed by its protective nutritional effects in stage V CKD patients on dialysis.
- Specific dietary recommendations regarding protein level and intakes of sodium, phosphorus, potassium, and fluids must be carefully individualized as appropriate for the level of renal function.

Key Words: Renal disease; hemodialysis; uremia; metabolic acidosis

21.1 INTRODUCTION

Based on the National Health and Nutrition Examination Survey (NHANES), it is estimated that 13% of US adults have chronic kidney disease (CKD) and from 1988 to 1994 the prevalence of CKD has increased by 30% in 1999–2004 *(1)*. Furthermore, chronic kidney disease is a strong independent predictor of atherosclerotic events *(2)*. The annual mortality on dialysis is approximately 20% and the total Medicare cost of end-stage renal disease in 2005 is estimated to be $ 19.3 billion *(3)*. Thus, chronic kidney disease is extremely common, carrying a significant health and economic burden.

From: *Nutrition and Health: Handbook of Clinical Nutrition and Aging, Second Edition*
Edited by: C. W. Bales and C. S. Ritchie, DOI 10.1007/978-1-60327-385-5_21,
© Humana Press, a part of Springer Science+Business Media, LLC 2009

Nutritional status plays a major role in development and progression of CKD as well as strongly influences survival in established CKD. Thus, dietary interventions could potentially have a significant impact on decreasing the progression of CKD and improving the outcomes of those with established CKD. The data on (1) the impact of diet and obesity on the development and progression of CKD and (2) nutritional issues unique to advanced CKD (uremia) are reviewed in this chapter.

21.2 ROLE OF DIET AND OBESITY ON DEVELOPMENT AND PROGRESSION OF CKD

Figure 21.1 summarizes the role of diet and nutrition in the development and progression of chronic kidney disease. These are discussed below in detail.

21.2.1 Salt Intake, Hypertension and the Risk of CKD

Hypertension is a very strong risk factor for CKD. In those with pre-existing hypertension, lowering the dietary sodium intake to approximately 1 g/d is associated with better control of blood pressure *(4,5)*. In those without pre-existing hypertension, the beneficial effects of lowering salt intake have been controversial *(6)*. However, in the Dietary Approaches to Stop Hypertension (DASH)-Sodium Study, 412 participants were randomly assigned to eat either a control diet (typical of intake in the United States) or the DASH diet (which is rich in vegetables, fruits, and low-fat dairy products). Within the assigned diet, participants ate foods with high, intermediate, and low levels of sodium for 30 consecutive days each, in random order. Reducing the sodium intake from the high (approximately 3 g/d) to low (approximately 1 g/d) level reduced the systolic blood pressure by 6.7 mmHg during the control diet and by 3.0 mmHg during the DASH diet. The effects of sodium were observed in participants with and in those without hypertension. The DASH diet was associated with a significantly lower systolic blood pressure at each sodium level. As compared to the control diet with a high sodium level, the DASH diet with a low sodium level led to a mean systolic blood pressure that was 7.1 mmHg lower in participants without hypertension, and 11.5 mmHg lower in participants with hypertension. These data suggest that reducing salt intake and

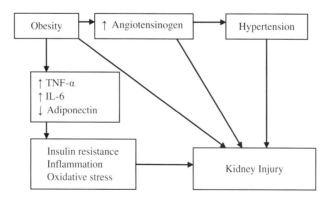

Fig. 21.1. Mechanisms of kidney damage in obesity.

increasing vegetables and fruits consumption could lower blood pressure. Nonetheless, there are no randomized controlled trials that have examined whether reduction of dietary sodium in normotensive individuals will reduce the development of CKD.

21.2.2 Protein Intake and Kidney Function

The impact of protein intake on renal function likely differs based on the level of baseline renal function. Hence, these are discussed separately.

21.2.2.1 IN THOSE WITH NORMAL RENAL FUNCTION

A high-protein, low-carbohydrate diet has been shown to produce greater weight loss in obese subjects compared to a low-fat, high-carbohydrate diet *(7)*. However, there are concerns on whether a high-protein diet will cause nephrotoxicity. In a 6 month randomized controlled trial of 65 healthy, overweight and obese individuals, dietary protein intake changed from 91.1 g/d to a 6 months intervention average of 70.4 g/d (*P*<0.05) in the low-protein group and from 91.4 to 107.8 g/d (*P*<0.05) in the high-protein group *(8)*. These resulted in changes in glomerular filtration rate (GFR) of –7.1 ml/min in the low-protein group and + 5.2 ml/min in the high-protein group (group effect: *P*<0.05). Kidney volume decreased by –6.2 cm^3 in the low-protein group and increased by + 9.1 cm^3 in the high-protein group (*P*<0.05), whereas albuminuria remained unchanged in all groups. Thus, a high-protein diet led to increased renal mass and hyperfiltration. As hyperfiltration is a recognized risk factor for progressive kidney damage, theoretically high-protein diet over the long-term might result in kidney failure. On the other hand, as discussed below, obesity per se is a risk factor for kidney damage. Therefore, the risk–benefit ratio of long-term use of high-protein diet on kidney function remains unclear. Hence, based on the currently available data *(7)*. short-term (6–12 months) use of high-protein diet in obese individuals for weight loss might be reasonable.

21.2.2.2 IN THOSE WITH CHRONIC KIDNEY DISEASE

In animal studies, restricting protein intake delays the progression of renal disease. In the Modification of Diet in Renal Diseases (MDRD) Study *(9)*. 585 patients with GFR of 25–55 ml/min/1.73 m^2 were randomly assigned to a usual-protein (1.3 g/kg) or a low-protein (0.58 g/kg) diet. Over 2.2 years of follow-up, the mean decline in the GFR did not differ significantly between the diet groups. In the same study, 255 patients with GFR of 13–24 ml/min/1.73 m^2 were randomly assigned to the low-protein (0.58 g/kg) or a very-low-protein diet (0.28 g/kg) with a keto acid-amino acid supplement. The very-low-protein group had a marginally slower decline in GFR than did the low-protein group ($P = 0.07$). In a secondary correlational analysis of the 255 patients with GFR of 13–24 ml/min/1.73 m^2, data on patients assigned to both diets were combined *(10)*. Controlled for baseline factors associated with a faster progression of renal disease, a 0.2 g/kg/d lower achieved total protein intake was associated with a 1.15 mL/min/year slower mean decline in GFR ($P = 0.011$). The authors suggested in patients with GFR less than 25 mL/min/1.73 m^2 a prescribed dietary protein intake of 0.6 g/kg/d. Nonetheless, this suggestion is based on an observational analysis of randomized controlled trial data. Moreover, protein restriction in the MDRD Study was associated with small but significant declines in various indices of nutritional status *(11)*. Thus, the role of protein restriction in slowing the progression of CKD remains controversial.

On the other hand, high-protein intake, particularly of non-dairy animal protein intake in those with mild kidney dysfunction (MDRD GFR > 55 mL/min/1.73 m^2 but <80 mL/min/1.73 m^2) in the Nurses Health Study was associated with greater decline in renal function *(12)*.

In summary, high-protein diet might be beneficial for weight loss in obese individuals with normal kidney function but should be avoided in those with even mild impairment of renal function. Therefore, kidney function should be estimated with a GFR estimating equation before high-protein diet could be prescribed particularly because mild kidney dysfunction is very common and is almost always asymptomatic.

21.2.3 Obesity, Insulin Resistance and the Risk of CKD

While the impact of obesity on cardiovascular disease, hypertension and diabetes are well established, only recently the data on the impact of obesity on kidney disease is emerging *(13–17)*. In an analysis of 320,252 adults, compared to those with normal BMI (18.5–24.9 kg/m^2), the risk of developing end-stage renal disease (ESRD) needing dialysis therapy was 1.87 (95% CI, 1.64–2.14) for those who were overweight (BMI, 25.0–29.9 kg/m^2), 3.57 (CI, 3.05–4.18) for those with class I obesity (BMI, 30.0–34.9 kg/m^2), 6.12 (CI, 4.97–7.54) for those with class II obesity (BMI, 35.0–39.9 kg/m^2), and 7.07 (CI, 5.37–9.31) for those with extreme obesity (BMI $\geq$ 40 kg/m^2) *(16)*. Furthermore, in morbidly obese individuals that underwent bariatric surgery, loss of weight is accompanied by better control of blood pressure and reduction of glomerular hyperfiltration and proteinuria *(17–19)*. These data suggest that obesity is a risk factor for kidney disease and that weight loss in obese individuals slows the rate of kidney damage.

In the above studies, higher baseline BMI remained an independent predictor for CKD/ ESRD after additional adjustments for baseline blood pressure level and the presence or absence of diabetes mellitus *(16)*. However, diabetes is only one extreme of insulin resistance. It is possible that sub-clinical insulin resistance in non-diabetic individuals might reflect a poor metabolic milieu and might still lead to kidney injury. Indeed, in non-diabetic adults, insulin resistance is associated with development of CKD *(20)*.

The potential molecular mechanisms for this phenomenon are depicted in Fig. 21.1. Adipose tissue is not a mere storage depot of fat. It is metabolically active and produces adipokines such as tumor necrosis factor-α (TNF-α), interleukin-6 (IL-6), plasminogen activator inhibitor, leptin, angiotensinogen, and adiponectin *(21–35)*. Alterations in production of these adipokines in obesity result in metabolic derangements that cause insulin resistance, dyslipidemia, hypertension, and inflammation *(21–35)*. In morbidly obese individuals that underwent bariatric surgery, loss of weight is accompanied by better control of blood pressure and reduction of glomerular hyperfiltration and proteinuria *(17–19)*. These data suggest that weight loss reduces kidney damage in obesity. A part of this effect might be mediated through hypertension and a part of this effect might be independent of hypertension. Insulin resistance, inflammation, and oxidative stress as consequences of altered production of adipokines in obesity might also result in kidney damage.

21.3 NUTRITIONAL ISSUES IN UREMIA

The previous section discussed the role of diet and obesity in the development and progression of CKD. In this section, nutritional issues that might be unique to uremia, i.e., the obesity paradox, the pathophysiology of wasting syndrome and serum phosphorus in advanced CKD are discussed.

21.3.1 Obesity Paradox in Uremia

The annual mortality in dialysis patients is approximately 20% a year. Better nutritional status as evidenced by higher body size, muscle mass, and serum albumin is associated with better survival in dialysis patients. In particular, in contrast to the associations of high body mass index (BMI) with increased mortality in the general population, high BMI is associated with better survival in dialysis patients (36–46). This phenomenon has been described as risk factor paradox or reverse epidemiology of wasting disease (40,47,48). Hence, it has been suggested that obesity is protective rather than harmful in dialysis patients (40).

Although high BMI is associated with better survival in dialysis patients, the associations of adiposity with traditional cardiovascular risk factors such as diabetes and non-traditional risk factors such as inflammation are not confined to the non-dialysis population. Indeed, previous studies have shown that in dialysis patients, adiposity and/or high BMI is associated with insulin resistance (49), diabetes (50), inflammation (51), anemia (52), coronary calcification (53,54) and carotid atherosclerosis (49). It is perplexing that if adiposity is associated with these apparent cardiovascular risk factors, why adiposity is associated with better survival in dialysis patients.

These apparently perplexing associations might be explained if (1) adiposity has dual competing effects on survival; a protective nutritional effect, and a deleterious metabolic effect resulting in insulin resistance, dyslipidemia, hypertension, and inflammation and (2) the level of kidney function modifies the relative importance of these effects (55,56). In this paradigm, the deleterious metabolic effects of obesity outweigh its protective nutritional effects in the non-CKD population, the deleterious metabolic effects of obesity are neutralized by its protective nutritional effects in the moderate CKD population and the deleterious metabolic effects of obesity are outweighed by its protective nutritional effects in stage V CKD on dialysis. In other words, the overall effects of obesity on survival vary according to the level of kidney function and there is an interaction of body size and presence or absence of CKD on survival even though the metabolic effects of adiposity are not modified by the level of kidney function.

21.3.2 Pathophysiology of Malnutrition in Uremia

Figure 21.2 summarizes the likely causal pathways for sarcopenia (low muscle mass) in uremia. Muscle mass is the net result of muscle protein synthesis and

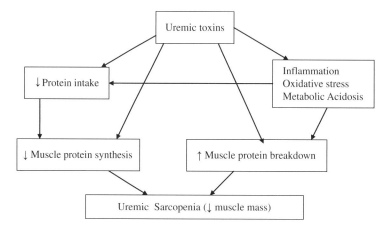

Fig. 21.2. Likely causal pathways for sarcopenia in uremia.

breakdown. Decreased protein intake results in decreased muscle protein synthesis whereas inflammation, oxidative stress, and metabolic acidosis promote muscle protein breakdown. Uremic toxins decrease muscle protein synthesis by decreasing protein intake as well as by directly inhibiting muscle protein synthetic machinery. Uremic toxins also increase muscle protein breakdown by inflammation, oxidative stress, and metabolic acidosis as well as by directly stimulating muscle protein catabolic pathways.

The relative importance of these pathways (decreased anabolism vs. increased catabolism) in causing uremic malnutrition remains controversial and the following discusses these issues.

21.3.2.1 Protein Intake and Nutritional Status

The current clinical guidelines recommend a dietary protein intake of 1.2 g/kg/d for hemodialysis patients *(57)*. These guidelines are based upon opinion and observational data. High-protein intake might result in increased serum phosphorus (discussed later) and metabolic acidosis which might be detrimental in dialysis patients. Furthermore, as discussed below, there are controversies on whether dietary interventions could impact on the nutritional status in dialysis patients.

It has been suggested that malnutrition in dialysis patients is a misnomer because uremic wasting (the state of low muscle mass, body weight and serum proteins in dialysis patients) is caused by hypercatabolism and not decreased dietary intake *(58)*. In support of this theory, there is experimental data that suggest that acidosis *(59–61)* and tumor necrosis factor-alpha (TNF-α) *(62,63)* activate ubiquitin-proteasome proteolytic system, the major pathway for protein catabolism *(64)*. Thus it has been suggested that metabolic acidosis and inflammation may be the major determinants of the "state of loss of protein" in dialysis patients *(65)*. Further, in a small study of eight peritoneal dialysis patients, increase in dialysate lactate was associated with decreased expression of skeletal muscle ubiquitin and improvement of nutritional indices in 1 month *(66)*. The other factors that are thought to contribute to this state of increased catabolism include resistance to anabolic activities of insulin and muscle breakdown induced by the hemodialysis procedure *(58,67,68)*.

Nonetheless, while increased catabolism might play a causative role in the state of low muscle mass, body mass index, and serum proteins in dialysis patients, it remains unclear whether inflammation, oxidative stress, and metabolic acidosis are the key determinants of muscle mass in uremia. Dietary intake might still be the most important determinant of muscle mass and body size in dialysis patients. For instance, after starvation for a prolonged period, malnutrition is likely in dialysis patients as in the general population. However, once malnutrition is established, it might be more difficult to reverse it in dialysis patients as compared to the general population. Thus, difficulties in reversing malnourished state with increased nutrition does not preclude that malnourished state is the result of deficient dietary intake.

21.3.2.2 Metabolic Acidosis and Nutritional Status

As discussed above, metabolic acidosis *(59–61)* activates ubiquitin–proteasome proteolytic system, the major pathway for protein catabolism *(64)* and hence, metabolic acidosis is proposed as the major cause of low muscle mass in dialysis patients. Thus, low-serum creatinine levels, which reflect lower muscle mass in prevalent hemodialysis patients is expected in those with lower serum bicarbonate concentrations. However, in an analysis of the well-dialyzed participants of the HEMO Study, the opposite association was present, i.e., those with higher serum creatinine concentrations (higher muscle mass) had lower serum bicarbonate concentrations *(69)* suggesting that acidosis was not a dominant determinant of low muscle mass.

21.3.2.3 Inflammation and Nutritional Status

TNF-α also activates the ubiquitin–proteasome proteolytic system *(63,70)*. Thus, inflammation is thought to be a major determinant of muscle wasting in dialysis patients *(58)*. In an earlier analysis of the HEMO Study data, very high levels of CRP were associated with decreased serum creatinine levels *(69)*. In a separate study, mid-thigh muscle mass estimated by computed tomography was also negatively associated with CRP *(71)*. In addition to increased catabolism, inflammation may also suppress appetite resulting in malnutrition *(47,72–74)*. Thus, it is biologically plausible that inflammation is associated with malnutrition in dialysis patients.

However, longitudinal associations between inflammation and low BMI have never been documented in dialysis patients. Furthermore, in the above longitudinal analysis of HEMO Study, each unit increase in natural logarithm of CRP was associated with a very modest 0.15 mg/dl decrease in serum creatinine *(69)*. Hence, a 10-fold increase in natural logarithm of CRP will be associated with a drop in serum creatinine of 1.5 mg% in prevalent hemodialysis patients. Thus, the association of severe inflammation with loss of muscle mass in dialysis patients is plausible but might not be sufficient to explain the high prevalence of low muscle mass in dialysis patients.

In summary, as suggested in Fig. 21.2, the effects of deficient diet and increased catabolism need not be mutually exclusive and on the other hand, they might have synergistic effects. Interventional studies are warranted to determine the relative importance of anabolic vs. catabolic processes in uremic malnutrition.

21.4 CALCIUM–PHOSPHORUS, PARATHYROID HORMONE, AND VASCULAR CALCIFICATION

With decline in GFR, phosphorus excretion in the urine is reduced and phosphorus accumulates in the blood. This results in hypocalcemia as well as hyperparathyroidism.

Higher phosphorus, calcium phosphorus product, and parathyroid hormone levels are associated with increased vascular calcification *(75)* and mortality in dialysis patients *(76)*. Serum phosphrous could be reduced by decreasing dietary phosphorus intake. However, this is problematic because foods that are high in phosphorus are also high in protein. Therefore, in order to let the patient consume reasonable amounts of protein, phosphorus binders are used along with meals. These act by binding the phosphorus in the gut resulting in non-absorbable compounds that are eliminated in stools.

Aluminum containing phosphorus binders resulted in elevated serum aluminum levels with consequent osteomalacia and dialysis dementia. Therefore, aluminum containing phosphorus binders have been largely abandoned. Currently, the two main classes of phosphorus binders are calcium based or non-calcium based. Randomized controlled trials showed that therapy with sevelamer (a non-calcium containing phosphorus binder) as opposed to calcium-containing phosphorus binders were associated with decreased arterial calcification, particularly in those who are older and have pre-existing calcification *(77,78)*. In a follow-up analysis of one of these trials of 129 incident hemodialysis patients, in sevelamer treated subjects, mortality, a pre-specified secondary end-point was lower. However, in a multi-center, randomized, open-label, parallel design trial ($n - 2,103$) of sevelamer and calcium-based binders, all-cause mortality rates and cause-specific mortality rates were not significantly different *(79)*. However, there was a significant age interaction on the treatment effect. Only in patients over 65 years of age was there a significant effect of sevelamer in lowering the mortality rate. There was a suggestion that sevelamer was associated with lower overall, but not cardiovascular-linked, mortality in older patients. Taken together with the above observational data, sevelamer might be beneficial in elderly who are likely to have arterial calcification but not in younger dialysis patients.

21.5 RECOMMENDATIONS

Table 21.1 provides a summary of diet recommendations individualized appropriately for the level of renal function. More extensive information about appropriate dieting modifications during various levels of CKD can be found at the National Kidney Foundation website (www.kidney.org).

Table 21.1
Summary of dietary recommendations based on the level of renal function

Condition	Appropriate diet
Obesity with normal kidney function (estimated MDRD GFR > 80 ml/min/ 1.73 m^2)	High protein, low carbohydrate diet (or other well-balanced, reduced calorie diet)
	No fluid restriction
Impaired kidney function (MDRD GFR 30 to ≤ 80 ml/min/1.73 m^2)	Avoid high-protein diet
	Potential benefit of protein restriction (0.6 g/kg/d)
	Salt restriction (3–4 g/d)
	No fluid restriction if urine output is normal
Stage IV CKD (MDRD GFR 15–29 ml/min/1.73 m^2)	Avoid high-protein diet
	Potential benefit of protein restriction (0.6 g /kg/d)
	Salt restriction (3–4 g/d)
	Restrict potassium and phosphorus as needed
	No fluid restriction if urine output is normal
On dialysis	Potential benefit of high-protein diet (1.2 g/kg/d)
	Salt restriction (3–4 g/d)
	Restrict potassium and phosphorus
	Restrict fluid intake to 1–1.5 L /d

In summary, diet and nutritional status exert a strong effect on both the development and the progression of CKD and is a strong determinant of the outcomes in those with established CKD. Dietary and life-style interventions could reduce the burden of CKD in the United States.

REFERENCES

1. Coresh J, Selvin E, Stevens LA, et al. Prevalence of chronic kidney disease in the United States. Jama 2007; 298(17):2038–47.
2. Beddhu S, Allen-Brady K, Cheung AK, et al. Impact of renal failure on the risk of myocardial infarction and death. Kidney Int 2002; 62(5):1776–83.
3. USRDS 2006. Annual Data Report: http://www.usrds.org
4. Melander O, von Wowern F, Frandsen E, et al. Moderate salt restriction effectively lowers blood pressure and degree of salt sensitivity is related to baseline concentration of renin and N-terminal atrial natriuretic peptide in plasma. J Hypertens 2007; 25(3):619–27.
5. Swift PA, Markandu ND, Sagnella GA, He FJ, MacGregor GA. Modest salt reduction reduces blood pressure and urine protein excretion in black hypertensives: a randomized control trial. Hypertension 2005; 46(2):308–12.
6. Fodor JG, Whitmore B, Leenen F, Larochelle P. Lifestyle modifications to prevent and control hypertension. 5. Recommendations on dietary salt. Canadian Hypertension Society, Canadian Coalition for High Blood Pressure Prevention and Control, Laboratory Centre for Disease Control at Health Canada, Heart and Stroke Foundation of Canada. CMAJ 1999; 160(9 Suppl):S29–34.
7. Gardner CD, Kiazand A, Alhassan S, et al. Comparison of the Atkins, Zone, Ornish, and LEARN diets for change in weight and related risk factors among overweight premenopausal women: the A TO Z Weight Loss Study: a randomized trial. JAMA 2007; 297(9):969–77.

8. Skov AR, Toubro S, Bulow J, Krabbe K, Parving HH, Astrup A. Changes in renal function during weight loss induced by high vs low-protein low-fat diets in overweight subjects. Int J Obes Relat Metab Disord 1999; 23(11):1170–7.

9. Klahr S, Levey AS, Beck GJ, et al. The effects of dietary protein restriction and blood-pressure control on the progression of chronic renal disease. Modification of Diet in Renal Disease Study Group. N Engl J Med 1994; 330(13):877–84.

10. Levey AS, Adler S, Caggiula AW, et al. Effects of dietary protein restriction on the progression of advanced renal disease in the Modification of Diet in Renal Disease Study. Am J Kidney Dis 1996; 27(5):652–63.

11. Kopple JD, Levey AS, Greene T, et al. Effect of dietary protein restriction on nutritional status in the Modification of Diet in Renal Disease Study. Kidney Int 1997; 52(3):778–91.

12. Knight EL, Stampfer MJ, Hankinson SE, Spiegelman D, Curhan GC. The impact of protein intake on renal function decline in women with normal renal function or mild renal insufficiency. Ann Intern Med 2003; 138(6):460–7.

13. Hallan S, de Mutsert R, Carlsen S, Dekker FW, Aasarod K, Holmen J. Obesity, smoking, and physical inactivity as risk factors for CKD: are men more vulnerable? Am J Kidney Dis 2006; 47(3):396–405.

14. Kramer H, Luke A, Bidani A, Cao G, Cooper R, McGee D. Obesity and prevalent and incident CKD: the Hypertension Detection and Follow-Up Program. Am J Kidney Dis 2005; 46(4): 587–94.

15. Gelber RP, Kurth T, Kausz AT, et al. Association between body mass index and CKD in apparently healthy men. Am J Kidney Dis 2005; 46(5):871–80.

16. Hsu CY, McCulloch CE, Iribarren C, Darbinian J, Go AS. Body mass index and risk for end-stage renal disease. Ann Intern Med 2006; 144(1):21–8.

17. Chagnac A, Weinstein T, Herman M, Hirsh J, Gafter U, Ori Y. The effects of weight loss on renal function in patients with severe obesity. J Am Soc Nephrol 2003; 14(6):1480–6.

18. Vogel JA, Franklin BA, Zalesin KC, et al. Reduction in predicted coronary heart disease risk after substantial weight reduction after bariatric surgery. Am J Cardiol 2007; 99(2):222–6.

19. Navarro-Diaz M, Serra A, Romero R, et al. Effect of drastic weight loss after bariatric surgery on renal parameters in extremely obese patients: long-term follow-up. J Am Soc Nephrol 2006; 17(12 Suppl 3):S213–7.

20. Kurella M, Lo JC, Chertow GM. Metabolic syndrome and the risk for chronic kidney disease among nondiabetic adults. J Am Soc Nephrol 2005; 16(7):2134–40.

21. Bastard JP, Jardel C, Delattre J, Hainque B, Bruckert E, Oberlin F. Evidence for a link between adipose tissue interleukin-6 content and serum C-reactive protein concentrations in obese subjects. Circulation 1999; 99: 2221–2.

22. Bhagat K, Vallance P. Inflammatory cytokines impair endothelium-dependent dilatation in human veins in vivo. Circulation 1997; 96(9):3042–7.

23. Bouloumie A, Marumo T, Lafontan M, Busse R. Leptin induces oxidative stress in human endothelial cells. Faseb J 1999; 13(10):1231–8.

24. Hotamisligil GS, Peraldi P, Budavari A, Ellis R, White MF, Spiegelman BM. IRS-1-mediated inhibition of insulin receptor tyrosine kinase activity in TNF-alpha- and obesity-induced insulin resistance. Science 1996; 271(5249):665–8.

25. Nakajima J, Mogi M, Kage T, Chino T, Harada M. Hypertriglyceridemia associated with tumor necrosis factor-alpha in hamster cheek-pouch carcinogenesis. J Dent Res 1995; 74(9):1558–63.

26. Ouchi N, Kihara S, Funahashi T, et al. Reciprocal association of C-reactive protein with adiponectin in blood stream and adipose tissue. Circulation 2003; 107(5):671–4.

27. Quehenberger P, Exner M, Sunder-Plassmann R, et al. Leptin induces endothelin-1 in endothelial cells in vitro. Circ Res 2002; 90(6):711–8.

28. Schneider JG, von Eynatten M, Schiekofer S, Nawroth PP, Dugi KA. Low plasma adiponectin levels are associated with increased hepatic lipase activity in vivo. Diabetes Care 2005; 28(9): 2181–6.

29. Stephens JM, Pekala PH. Transcriptional repression of the GLUT4 and C/EBP genes in 3T3-L1 adipocytes by tumor necrosis factor-alpha. J Biol Chem 1991; 266(32):21839–45.

30. Tham DM, Martin-McNulty B, Wang YX, et al. Angiotensin II is associated with activation of NF-kappa B-mediated genes and downregulation of PPARs. Physiol Genomics 2002; 11(1): 21–30.
31. Verma S, Li SH, Wang CH, et al. Resistin promotes endothelial cell activation: further evidence of adipokine-endothelial interaction. Circulation 2003; 108(6):736–40.
32. von Eynatten M, Schneider JG, Humpert PM, et al. Decreased plasma lipoprotein lipase in hypoadiponectinemia: an association independent of systemic inflammation and insulin resistance. Diabetes Care 2004; 27(12):2925–9.
33. Wellen KE, Hotamisligil GS. Inflammation, stress, and diabetes. J Clin Invest 2005; 115(5): 1111–9.
34. Whitehead JP, Richards AA, Hickman IJ, Macdonald GA, Prins JB. Adiponectin–a key adipokine in the metabolic syndrome. Diabetes Obes Metab 2006; 8(3):264–80.
35. Ziccardi P, Nappo F, Giugliano G, et al. Reduction of inflammatory cytokine concentrations and improvement of endothelial functions in obese women after weight loss over one year. Circulation 2002; 105(7):804–9.
36. Johansen KL, Young B, Kaysen GA, Chertow GM. Association of body size with outcomes among patients beginning dialysis. Am J Clin Nutr 2004; 80:324–32.
37. Abbott KC, Glanton CW, Trespalacios FC, et al. Body mass index, dialysis modality, and survival: analysis of the United States Renal Data System Dialysis Morbidity and Mortality Wave II Study. Kidney Int 2004; 65:597–605.
38. Aoyagi T, Naka H, Miyaji K, Hayakawa K, Ishikawa H, Hata M. Body mass index for chronic hemodialysis patients: stable hemodialysis and mortality. Int J Urol 2001; 8:S71–S5.
39. Beddhu S, Pappas LM, Ramkumar N, Samore M. Effects of body size and body composition on survival in hemodialysis patients. J Am Soc Nephrol 2003; 14:2366–72.
40. Kalantar-Zadeh K, Abbott KC, Salahudeen AK, Kilpatrick RD, Horwich TB. Survival advantages of obesity in dialysis patients. Am J Clin Nutr 2005; 81(3):543–54.
41. Kalantar-Zadeh K, Kopple JD, Kilpatrick RD, et al. Association of morbid obesity and weight change over time with cardiovascular survival in hemodialysis population. Am J Kidney Dis 2005; 46:489–500.
42. Kopple JD. Nutritional status as a predictor of morbidity and mortality in maintenance dialysis patients. ASAIO J 1997; 43:246–50.
43. Kutner NG, Zhang R. Body mass index as a predictor of continued survival in older chronic dialysis patients. Int Urol Nephrol 2001; 32:441–8.
44. Leavey SF, McCullough K, Hecking E, Goodkin D, Port FK, Young EW. Body mass index and mortality in 'healthier' as compared with 'sicker' haemodialysis patients: results from the Dialysis Outcomes and Practice Patterns Study (DOPPS). Nephrol Dial Transplant 2001; 16:2386–94.
45. Leavey SF, Strawderman RL, Jones CA, Port FK, Held PJ. Simple nutritional indicators as independent predictors of mortality in hemodialysis patients. Am J Kidney Dis 1998; 31: 997–1006.
46. Port FK, Ashby VB, Dhingra RK, Roys EC, Wolfe RA. Dialysis dose and body mass index are strongly associated with survival in hemodialysis patients. J Am Soc Nephrol 2002; 13:1061–6.
47. Kalantar-Zadeh K, Ikizler TA, Block G, Avram MM, Kopple JD. Malnutrition-inflammation complex syndrome in dialysis patients: causes and consequences. Am J Kidney Dis 2003; 42: 864–81.
48. Nishizawa Y, Shoji T, Ishimura E, Inaba M, Morii H. Paradox of risk factors for cardiovascular mortality in uremia: is a higher cholesterol level better for atherosclerosis in uremia?. Am J Kidney Dis 2001; 38:S4–S7.
49. Yamauchi T, Kuno T, Takada H, Nagura Y, Kanmatsuse K, Takahashi S. The impact of visceral fat on multiple risk factors and carotid atherosclerosis in chronic haemodialysis patients. Nephrol Dial Transplant 2003; 18:1842–7.
50. Beddhu S, Pappas LM, Ramkumar N, Samore M. Malnutrition and atherosclerosis in dialysis patients. J Am Soc Nephrol 2004; 15:733–42.
51. Axelsson J, Qureshi RA, Suliman ME, et al. Truncal fat mass as a contributor to inflammation in end-stage renal disease. Am J Clin Nutr 2004; 80:1222–9.

52. Axelsson J, Qureshi AR, Heimburger O, Lindholm B, Stenvinkel P, Barany P. Body fat mass and serum leptin levels influence epoetin sensitivity in patients with ESRD. Am J Kidney Dis 2005; 46(4):628–34.

53. Goodman WG, Goldin J, Kuizon BD, et al. Coronary-artery calcification in young adults with end-stage renal disease who are undergoing dialysis. N Engl J Med 2000; 342(20):1478–83.

54. Stompor T, Pasowicz M, Sullowicz W, et al. An association between coronary artery calcification score, lipid profile, and selected markers of chronic inflammation in ESRD patients treated with peritoneal dialysis. Am J Kidney Dis 2003; 41:203–11.

55. Kwan BC, Murtaugh MA, Beddhu S. Associations of body size with metabolic syndrome and mortality in moderate chronic kidney disease. Clin J Am Soc Nephrol 2007; 2(5):992–8.

56. Kwan BC, Beddhu S. A story half untold: adiposity, adipokines and outcomes in dialysis population. Semin Dial 2007; 20(6):493–7.

57. Clinical Practice Guidelines for Nutrition in Chronic Renal Failure. K/DOQI, National Kidney Foundation. Am J Kidney Dis 2000; 35(6 Suppl 2):S1–140.

58. Mitch WE. Malnutrition: a frequent misdiagnosis for hemodialysis patients. J Clin Invest 2002; 110:437–9.

59. Hara Y, May RC, Kelly RA, Mitch WE. Acidosis, not azotemia, stimulates branched-chain, amino acid catabolism in uremic rats. Kidney Int 1987 Dec; 32(6):808–14.

60. May RC, Kelly RA, Mitch WE. Mechanisms for defects in muscle protein metabolism in rats with chronic uremia. Influence of metabolic acidosis. J Clin Invest 1987 Apr; 79(4):1099–103.

61. May RC, Bailey JL, Mitch WE, Masud T, England BK. Glucocorticoids and acidosis stimulate protein and amino acid catabolism in vivo. Kidney Int 1996 Mar; 49(3):679–83.

62. Goodman MN. Tumor necrosis factor induces skeletal muscle protein breakdown in rats. Am J Physiol 1991 260(5):E727–30.

63. Goldberg AL, Kettelhut IC, Furuno K, Fagan JM, Baracos V. Activation of protein breakdown and prostaglandin E2 production in rat skeletal muscle in fever is signaled by a macrophage product distinct from interleukin 1 or other known monokines. J Clin Invest 1988; 81(5):1378–83.

64. Mitch WE, Goldberg AL. Mechanisms of muscle wasting. The role of the ubiquitin-proteasome pathway. N Engl J Med 1996; 335(25):1897–905.

65. Ahuja TS, Mitch WE. The evidence against malnutrition as a prominent problem for chronic dialysis patients. Semin Dial 2004 Nov–Dec; 17(6):427–31.

66. Pickering WP, Price SR, Bircher G, Marinovic AC, Mitch WE, Walls J. Nutrition in CAPD: serum bicarbonate and the ubiquitin-proteasome system in muscle. Kidney Int 2002 Apr; 61(4):1286–92.

67. Pupim LB, Flakoll PJ, Brouillette JR, Levenhagen DK, Hakim RM, Ikizler TA. Intradialytic parenteral nutrition improves protein and energy homeostasis in chronic hemodialysis patients. J Clin Invest 2002 Aug; 110((4):483–92.

68. Ikizler TA, Pupim LB, Brouillette JR, et al. Hemodialysis stimulates muscle and whole-body protein loss and alters substrate oxidation. Am J Physiol – Endocrinol Metab 2002; 282:E107–16.

69. Kaysen GA, Greene T, Daugirdas JT, et al. Longitudinal and cross-sectional effects of C-reactive protein, equilibrated normalized protein catabolic rate, and serum bicarbonate on creatinine and albumin levels in dialysis patients. Am J Kidney Dis 2003; 42(6):1200–11.

70. Goodman MN. Tumor necrosis factor induces skeletal muscle protein breakdown in rats. Am J Physiol 1991 May; 260(5 Pt 1):E727–30.

71. Kaizu Y, Ohkawa S, Odamaki M, et al. Association between inflammatory mediators and muscle mass in long-term hemodialysis patients. Am J Kidney Dis 2003; 42:295–302.

72. Kalantar-Zadeh K, Kopple JD, Block G, Humphreys MH. A malnutrition-inflammation score is correlated with morbidity and mortality in maintenance hemodialysis patients. Am J Kidney Dis 2001; 38:1251–63.

73. Stenvinkel P, Heimburger O, Paultre F, et al. Strong association between malnutrition, inflammation, and atherosclerosis in chronic renal failure. Kidney Int 1999 May; 55(5):1899–911.

74. Stenvinkel P, Heimburger O, Lindholm B, Kaysen GA, Bergstrom J. Are there two types of malnutrition in chronic renal failure? Evidence for relationships between malnutrition, inflammation and atherosclerosis (MIA syndrome). Nephrol Dial Transplant 2000; 15:953–60.

75. Chertow GM, Raggi P, Chasan-Taber S, Bommer J, Holzer H, Burke SK. Determinants of progressive vascular calcification in haemodialysis patients. Nephrol Dial Transplant 2004; 19(6):1489–96.
76. Ganesh SK, Stack AG, Levin NW, Hulbert-Shearon T, Port FK. Association of elevated serum PO(4), Ca x PO(4) product, and parathyroid hormone with cardiac mortality risk in chronic hemodialysis patients. J Am Soc Nephrol 2001; 12:2131–8.
77. Chertow GM, Burke SK, Raggi P, Treat to Goal Working Group. Sevelamer attenuates the progression of coronary and aortic calcification in hemodialysis patients. Kidney Int 2002; 62(1):245–52.
78. Block GA, Spiegel DM, Ehrlich J, et al. Effects of sevelamer and calcium on coronary artery calcification in patients new to hemodialysis. Kidney Int 2005; 68(4):1815–24.
79. Suki WN, Zabaneh R, Cangiano JL, et al. Effects of sevelamer and calcium-based phosphate binders on mortality in hemodialysis patients. Kidney Int 2007; 72(9):1130–7.

22 Nutritional and Pharmacological Aspects of Osteoporosis

David A. Ontjes and John J.B. Anderson

Key Points

- Physical activity and a healthy diet are two of the more important lifestyle contributors to bone health and the maintenance of its functions into late life, with calcium and vitamin D being the two most important dietary components.
- The changes in the skeleton over the life cycle reflect early-life gains and late-life losses that place the individual at increasing risk of fragility fractures of the bones, especially the distal radius, lumbar vertebrae and proximal femur (hip).
- Low circulating 25-hydroxyvitamin D concentration and vitamin D inadequacy are highly prevalent in elderly populations; older adults may need 700–800 IU of vitamin D and 1,000–1,200 mg of calcium as provided by the combination of their usual diet and supplements.
- Recent human studies of vitamin K supplementation suggest that this vitamin promotes more complete carboxylation of osteocalcin and other matrix Gla proteins and may be associated with a reduced risk of osteoporotic fractures.
- Available pharmacologic agents act to increase the absorption of calcium from the gastrointestinal tract, to decrease the rate of resorption of existing bone, or to increase the rate of formation of new bone. Drug therapy is usually indicated when osteoporotic fractures occur or when the measured bone mineral density is in the osteoporotic range.
- Both the incidence and prevalence of osteoporosis are increasing as populations are aging, thus preventive strategies, either primary or secondary, must be identified and implemented. A combination of good nutrition, including sufficient calcium and vitamin D and appropriate exercise, will continue to be important, even as pharmacologic agents become more widely used.

Key Words: Osteopenia; fractures; bone resorption; calcium; vitamin D; vitamin K; calcitriol; parathyroid hormone

From: *Nutrition and Health: Handbook of Clinical Nutrition and Aging, Second Edition*
Edited by: C. W. Bales and C. S. Ritchie, DOI 10.1007/978-1-60327-385-5_22,
© Humana Press, a part of Springer Science+Business Media, LLC 2009

Osteoporosis is a disease characterized by low bone mass and an increased risk of fracture. It is typically a chronic multi-factorial disease occurring in late adulthood, following menopause in women and a decade or so later in men. Osteoporosis may appear in younger individuals who may be at high risk because of hypogonadism, malabsorptive gastrointestinal disorders, and exposure to excessive glucocorticoids or other drugs capable of causing a negative calcium balance.

Both hereditary and environmental factors contribute to the multi-factorial nature of this disease. Inherited differences in bone metabolism and structure as well as lifestyle and environmental factors influence bone strength. In terms of lifestyle, both regular physical activity and a healthy diet remain two of the more important contributors to bone health and the maintenance of its functions into late life. The two most important dietary components are calcium and vitamin D. When calcium intake is not adequate, the adaptive role of the calcium regulatory system operates in an attempt to preserve calcium homeostasis. If the adaptation is not sufficient, loss of both bone mass and density will lead to fragility fractures. This review focuses on two aspects of osteoporosis: (1) dietary factors involved in prevention and (2) the role of nutrition and drugs in treatment of osteoporosis.

22.1 BONE STRUCTURE AND BONE MASS DURING THE NORMAL LIFE CYCLE

Each individual bone contains two types of bone tissue: trabecular (cancellous) and cortical (compact). Trabecular bone is the more metabolically active because it has about 8–10 times more total surface area than a similar mass of cortical bone. The mineralized surfaces are covered by bone cells, including osteoblasts that are responsible for new bone formation and osteoclasts that are involved in resorption of existing bone. Each specific bone contains both types of bone tissue, but in different proportions. For example, long bones, such as the femur, contain much more trabecular tissue at either end near the hip joint or knee joint and a much greater proportion of cortical bone in the shaft that connects the two ends. This distinction is important because most fractures occur where more metabolically active trabecular bone tissue predominates.

In early life, during the growth period of the skeleton, bone acquisition through new bone formation predominates over bone resorption. This phase typically ends by 16–18 years of age in females and 18–22 years of age in males. Once linear growth is completed, resorption of bone equals formation. In young adulthood, the net amount of bone mass remains fairly constant, though some modest gain may still occur until the mid-20s or beyond. By age 30 years, most individuals have achieved their maximum or peak bone mass that will serve as a "healthy norm" for the rest of life. Beyond the age of 30 years, a very slow loss of bone mass typically occurs with normal aging. In radiographic definitions of osteopenia and osteoporosis based on measurements of bone mass, the mean bone density of a population of 20- to 29-year-old males or females is taken as the standard of comparison for determining whether osteoporosis is present later in life.

In late life, i.e., after the menopause in women and a decade or so later in men, imbalances in the remodeling of the skeleton result in more rapid bone losses, so that reductions in both bone mineral content (BMC) and bone mineral density (BMD) occur. The loss of estrogens after the menopause and probably the later decline of androgens in men contribute to an increase in resorption, a reduction of formation, or both. The increase in resorption is triggered by increased activity of osteoclasts, and the decline in formation is directly related to decreased activity of osteoblasts. An increase in bone turnover, i.e., increased rates of resorption and formation with resorption dominating formation, accelerates the rate of bone loss. Bone turnover may be assessed indirectly by measurement of biochemical markers of bone metabolism such as the breakdown products of bone collagen or proteins derived from osteoblasts such as osteocalcin or "bone-specific" alkaline phosphatase.

The changes in the skeleton over the life cycle reflect early-life gains and late-life losses. As loss of bone mass and change in structure progress, the individual is at increasing risk of fragility fractures of bones such as the distal radius, lumbar vertebrae, and proximal femur (hip).

22.2 MEASUREMENT OF BMD IN THE DIAGNOSIS OF OSTEOPOROSIS AND PREDICTION OF FRACTURE RISK

Osteoporosis may be diagnosed based either on the presence of fragility fractures or on a demonstration of low bone mineral density. BMD is most commonly measured today using a technique called dual-energy X-radiographic analysis (DXA). Osteopenia, or too little bone, is followed by osteoporosis, or even greater bone loss, which places an individual at great risk of a fracture. Quantitative definitions of osteopenia and osteoporosis have been established for postmenopausal Caucasian women by the World Health Organization as follows: osteopenia is between 1 and 2.5 standard deviations (SDs) below the 20–29-year-old mean values for Caucasian women; and osteoporosis is more than 2.5 SDs below the 20–29-year-old means (1). These values for individuals at any adult age compared to means of healthy young adults are known as T-scores (1). According to this definition, a postmenopausal 60-year-old woman with a BMD T-score of 3.0 SDs below the 29-year-old mean at one or more skeletal measurement sites would be diagnosed as having osteoporosis even though she might not yet have evidence of osteoporotic fractures. The International Society for Clinical Densitometry recommends using a single standardized sex-specific database for calculating T-scores in postmenopausal women and elderly men. In younger patients, a Z-score expressing the standard deviation from the mean in a control population matched for age, ethnicity, and sex should be used instead of a T-score.

The WHO definition of osteoporosis does not apply to pre-menopausal women or to men. However, for these individuals as well as for postmenopausal women, a low BMD is still strongly predictive of fracture risk. In any individual the risk of fracture increases by approximately two-fold for each standard deviation of the measured T-score below young control values (2).

22.3 ETIOLOGIES AND TYPES OF OSTEOPOROSIS

No single cause predisposes an individual to osteoporosis, but rather a variety of inherited and acquired factors do so. Some of the most common clinical risk factors are summarized below. Osteoporosis results when too much bone resorption occurs, too little formation exists, or a combination of both co-exists *(3)*. The most common cause of increased bone resorption results from estrogen deficiency associated with menopause in normal women. Accelerated bone loss continues for about 10 years after menopause; then the rate of decline subsides to near the rate that exists for normal aging. Estrogen replacement in the postmenopausal period reduces the rate of resorption and stabilizes bone mass. Men with hypogonadism have accelerated bone loss similar to that of postmenopausal women *(4)*. Other conditions that cause increased bone resorption include hyperparathyroidism and hyperthyroidism.

Age-related bone loss is characterized by low rates of bone formation. This type of osteoporosis affects both men and women. Although the causation of age-related bone loss is poorly understood, it may be related in part to decreased intestinal absorption of calcium. Other factors besides advanced age may cause impaired bone formation including exposure to certain drugs, such as glucocorticoids, and immobilization or lack of mechanical stress on bone itself.

Genetic factors undoubtedly play a major role in determining both the peak bone mass of young adults and the rate of bone loss in older individuals. In population-based studies, natural variations (polymorphisms) in genes for the vitamin D receptor, the estrogen receptor, and for type I collagen matrix protein all appear to affect bone mass. The identity of other genes contributing to bone mass and strength is the subject of active ongoing research.

22.4 NON-DIETARY RISK FACTORS IMPLICATED IN ETIOLOGY OF OSTEOPOROSIS

Several non-dietary risk factors—all potentially adverse or harmful—have been identified that promote the loss of bone and the onset of fractures (see Table 22.1). Each of these factors has an associated risk, but when two or three exist together at the same time, the risk of an osteoporotic fracture can increase dramatically. For example, the small-framed older postmenopausal woman who smokes a pack of cigarettes a day, drinks two to three servings of alcohol a day, and who has little physical activity in her daily life will typically be at great risk of an early hip fracture, i.e., by age 70 years or younger. When elderly individuals suffer declines of acuity of their senses, such as vision and equilibrium, or take medications that result in the same effects, they are much more likely to fall and break their hip.

The most important of these acquired factors, in general, may be the decline in lean body mass that accompanies reduced physical activity. Declining muscle strength and tone lead to reduced physical stress on bone which is coupled to reduced bone formation and increased resorption.

Table 22.1
Risk factors: non-dietary

- Thinness with low lean body mass
- Cigarette smoking
- Excessive alcohol consumption
- Insufficient physical activity
- Drugs—over-the-counter and prescription
- Decline of sensory perceptions
- Falls
- <18.5 BMI for low lean body mass
- >Two drinks a day for men and >one drink a day for women
- Drugs like corticosteroids, anabolic steroids, phenytoin (Dilantin[R]), and others

22.5 DIETARY RISK FACTORS IMPLICATED IN ETIOLOGY OF OSTEOPOROSIS

Clearly, adequate intakes of energy, protein, essential fatty acids, and the many micronutrients are required for the support and maintenance of bone health of adults. In contrast to the benefits of a healthy diet on the skeleton, several dietary factors may have adverse effects on bone tissue (see Table 22.2). These deleterious factors are thought to operate throughout the life cycle, not just during late life. A resurgence of interest in the adequacy of calcium and vitamin D in the adult diet has resulted in new dietary recommendations.

22.5.1 Calcium and Phosphorus

The most common dietary problem associated with the development of osteoporosis has been an inadequate consumption of calcium, at least among Caucasians in western nations, including pediatric populations (5). Low calcium intakes in these nations are typically accompanied by high phosphate consumption because of the ingestion of naturally occurring phosphorus in foods, especially animal proteins, and also of phosphorus from processed foods. Phosphorus food additives are quite common in western nations because of the widespread use of processed foods—foods modified with the many applications that utilize phosphate salts of one type or another—and this source may account for almost 33% of the total phosphorus intake of consumers of processed foods (6). Cola-type soft drink beverages containing phosphoric acid also fit in this category, although technically they are not foods. The net result of consuming phosphate-rich foods may be a decreased ratio of calcium (Ca) to phosphate (P) intake (Ca:P ratio) that is worsened when individuals consume little milk, cheese, or other calcium-rich foods in their usual diets.

Intake ratios less than 0.5:1 have been established to increase parathyroid hormone secretion and, presumably, increase bone resorption (7). A constantly elevated parathyroid hormone (PTH) level leads to bone loss and a gradual decline in BMD that may lead to osteoporosis (6). The time period required to reach the osteoporotic range, according to WHO definitions, is not clear, but a long-term dietary pattern of low calcium–high phosphate may even contribute to inappropriately low bone mass in females before they reach 20 years of age, if fractures among girls and pubertal females are a valid index of low bone mass *(8)*.

Vegetarian diets may also compromise bone health through a number of possible mechanisms *(9)*. but the low calcium and vitamin D intake from a vegan dietary pattern may be largely responsible for lower bone mass among vegetarians. Calcium supplementation alone, however, has not been found in a meta-analysis to improve bone mass or density of postmenopausal osteoporotic women appreciably *(10)*. In addition, some studies have reported that calcium supplements, as opposed to calcium from foods, have little benefit for bone measurements over a year or longer *(11)*.

22.5.2 Vitamin D

The active form of vitamin D (1,25-dihydroxyvitamin D) is essential for normal absorption of dietary calcium in the gastrointestinal tract. Vitamin D deficiency leads to impaired calcium absorption and a secondary increase in the secretion of parathyroid hormone. The latter leads in turn to impaired mineralization of bone matrix and increased resorption of bone mineral with an overall reduction in bone mass and strength. Vitamin D deficiency associated with limited skin exposure to vitamin D-promoting sunlight and a low intake of vitamin D from foods, especially fortified dairy products in the United States and various deep-sea fish species in much of the world, is now considered a major risk factor for low bone mass or density and fracture. The clinical diagnosis of this condition is based on measurement of a low circulating concentration of 25-hydroxyvitamin D, which is the immediate precursor for 1,25-dihydroxyvitamin D *(12)*. Calcium intake per se has little influence on the serum 25-hydroxyvitamin D concentration *(13)*. The clinical definition of vitamin D deficiency has changed in recent years to recognize the importance of milder forms of inadequacy. Severe vitamin D deficiency, associated with classical rickets in children and osteomalacia in adults, is present when serum levels of 25-hydroxyvitamin D fall below 20 ng/ml (50 nmol/l). Milder vitamin D insufficiency occurs at serum levels ranging between 20 and 30 ng/mL. In this range rickets and osteomalacia do not occur, but intestinal absorption of calcium is still suboptimal, leading to mild secondary increases in PTH secretion and a reduced bone mass *(14)*. Only at serum 25-hydroxyvitamin D levels exceeding 30 ng/ML (75 nmol/L) is intestinal absorption of calcium optimal and excessive PTH secretion absent.

Low circulating 25-hydroxyvitamin D concentration and vitamin D inadequacy are highly prevalent in older populations, especially in Northern Europe and North America *(15)*. Vitamin D insufficiency is also very common in dark-skinned populations worldwide. A recent meta-analysis suggests that older adults, especially shut-ins, need 700–800 IU of vitamin D and 1,000–1,200 mg of calcium as a supplement to their usual diet *(16)*. In some individuals the daily requirement for vitamin D may be increased to over 1,000 IU/day due to gastrointestinal disorders causing malabsorption or to the administration of drugs causing increased vitamin D clearance.

22.5.3 Vitamin K

Vitamin K has a role in the post-translational modification of osteocalcin and possibly other matrix proteins synthesized by osteoblasts. This fat-soluble vitamin promotes insertion of a carboxyl group at the gamma position of the amino acid, glutamic acid, in the newly synthesized protein. Gamma carboxylation may occur in up to 8 or 10 sites in each osteocalcin molecule. The insertion of these carboxyl groups allows for bidentate binding of one calcium ion wherever glutamate residues exist throughout the molecule. The additional calcium-binding sites in osteocalcin are generally thought to enhance the mineralization of collagen in new bone matrix, but the specific mechanisms involved in the mineralization process remain to be elucidated. Vitamin K is now also thought to be needed for the activation of osteoblasts. A deficiency of this vitamin in the diets of older subjects may contribute to increased vertebral and hip fractures.

Recent human studies of vitamin K supplementation using naturally occurring molecules or synthetic analogs suggest that more complete carboxylation of osteocalcin and other matrix Gla proteins result from treatment, and this may be associated with a reduced risk of osteoporotic fractures. For example, one report suggests that vitamin K_2 (menaquinone-4 or MK-4) supplementation (945 mg/day) over 3 years improved both hip bone geometry and strength [17] and another suggests that menaquinone-7 (MK-7) had a longer circulating half-life than vitamin K_1 (phylloquinone) [18]. Another report, in abstract only, of older Japanese women found that a lower circulating concentration of vitamin K_2 was associated with greater risk of vertebral fracture [19]. Finally, a 2-year randomized controlled trial demonstrated that vitamin K_1 (phylloquinone) substantially increased gamma carboxylation of osteocalcin and, with supplemental calcium and vitamin D, improved BMC of the ultradistal radius [20]. Vitamin K is converted in human tissues to other forms of vitamin K.

Although only a relatively few prospective studies have been reported, the limited evidence suggests that many older adults have insufficient or deficient intakes of vitamin K from both foods and supplements that may place them at increased risk of fracture. Increasing carboxylation of osteocalcin with vitamin K is generally thought to be fracture preventive. The major food sources, dark-green vegetables, contain vitamin K_1 and they are poorly consumed by most US citizens. Food fortification of vitamin K may need to be considered to correct for the inadequate consumption of vitamin K by most in the United States. Supplementation of vitamin K may be recommended in the future.

22.5.4 Protein and Acid Load

An unusually high animal protein intake may contribute to bone loss and osteoporosis, but the mechanism has not been fully established. After metabolism of amino acids, the increased production of acids, i.e., phosphoric and sulfuric, from the degradation of phosphorus- and sulfur-containing amino acids is considered responsible for initiating bone resorption to maintain acid–base balance.

The net effect is an increased loss of calcium ions in urine, a loss that has been shown acutely after a single meal and in short-term experiments, but not in long-term studies.

22.5.5 Sodium and Potassium

High-sodium snack foods have become very popular in western nations and they contribute additional sodium to an already high intake that derives from so many foods processed with sodium or salt. Renal losses of calcium increase on high-sodium intakes because the kidneys favor sodium reabsorption at the expense of calcium ions. The net loss of calcium comes from bone, and therefore a loss of bone mass may be associated with excessive consumption of sodium-rich snack foods. Higher intakes of potassium from fruits and vegetables may help to reduce the adverse effects of sodium on calcium losses *(21)*.

22.5.6 Phytochemicals, Micronutrients, and Other Bioactive Compounds

Phytoestrogens, particularly genistein, have been demonstrated in a randomized controlled trial to result in improvement of bone density in older adults after a year of supplementation *(22)*. Further corroboration is needed before a recommendation of genistein supplementation can be advanced.

Finally, a poor diet, especially one based on limited intakes of servings of fruits and vegetables, may be deficient in other bone-essential nutrients, such as magnesium, vitamin K, zinc, anti-oxidant nutrients, and probably a dozen or more phytochemicals. Any one or a combination of these deficits may limit efficient bone formation. Adverse effects of these deficits include loss of protection against oxidative stress and poor regulation of acid–base balance, both impacting bone as a major buffering store. Several other micronutrients may have important affects on bone tissue, hence bone mass, but typically the research is limited and clinical applications remain uncertain. Whatever the mechanisms may be, bone health requires a wide variety of nutrients that are best supplied by a varied diet consisting of the recommended numbers of servings of foods each day *(21)*. Recommended intakes for energy and several key nutrients impacting on bone health are listed for the elderly in Table 22.3.

Table 22.2
Risk factors: dietary

- Poor overall diet quality
- Low calcium intake
- High phosphorus intake
- Low vitamin D status
- High animal protein and acid load
- High-sodium snacks
- Vegetarian diet

Table 22.3
Recommended daily intakes of selected nutrient variables for the elderly

| | *51–70 years* | | *>70 years* | |
	Male	*Female*	*Male*	*Female*
Energy, kcal	2,300	1,900	2,300	1,900
Protein, g	63	50	63	50
Calcium, mg[+]	1,200	1,200	1,200	1,200
Phosphorus, mg	700	700	700	700
Vitamin D, mcg[+,#]	10	10	15	15
IU[+,#]	400	400	600	600
Vitamin K, mcg[+]	120	90	120	90

[+] Intake listed as an Adequate Intake (AI). AIs represent the recommended average daily intake level based on observed or experimentally determined approximations. AIs are used when there is insufficient information to determine an RDA.

[#] While the AI for vitamin D is 400–600 IU/day, many experts recommend a slightly higher amount, 700–800 IU/day[16.]

22.6 ADAPTATION TO LOW CALCIUM INTAKES

Several adaptations to dietary intakes occur after each meal in healthy individuals; a number of these adaptations directly affect calcium homeostasis by impacting on serum PTH and 1,25-dihydroxyvitamin D concentrations. For example, low calcium intake, especially when coupled with a high phosphate consumption pattern, stimulates PTH secretion via a calcium sensor on the membranes of parathyroid gland cells. A chronically elevated PTH, in turn, stimulates the removal of calcium from bone and the loss of bone mass and density (see above).

In addition, low calcium stimulates the vitamin D regulatory mechanism by increasing the renal production of the hormonal form of vitamin D, i.e., 1,25-dihydroxyvitamin D, which leads to an increase in intestinal calcium absorption and bone utilization of calcium. The potential problem in older women, and perhaps men, is that the intestinal adaptation declines, and less calcium can be absorbed with the hormonal stimulus.

Finally, excessive, continuous PTH secretion that may develop late in life with a low Ca:high P ratio diet and low circulating 25-hydroxyvitamin D concentration contributes to continuous bone loss. It is now recognized that a persistent and continuous treatment with PTH causes significant bone loss, whereas a discontinuous treatment with PTH, i.e., once daily or once weekly, increases bone mass and density by stimulating osteoblasts to make new bone tissue. Thus, any type of low calcium intake that generates a continuous secretion and elevation of PTH in blood will have serious negative consequences on the maintenance of bone mass.

22.7 DIETARY PREVENTION AND TREATMENT—FOODS AND SUPPLEMENTS

The general principles of both primary and secondary prevention of osteoporosis are similar: (1) Increase calcium intake through foods and supplements to ≥1,000 mg/day in order to suppress PTH secretion. (2) Assure that individuals obtain sufficient amounts of vitamin D through both foods and supplements to maintain optimal calcium absorption: 800 IUs or more are now being recommended *(15)*. Sunshine exposure is also encouraged, depending on geographic latitude and time of the year. (See a full discussion of the beneficial effects of calcium and vitamin D in the following pages, under Treatment, 22.9.1.1.) (3) A healthy overall diet containing virtually all nutrients from foods, i.e., nine servings of fruits and vegetables a day, makes sense from a nutritional perspective. For the elderly, a daily supplement that contains a wide range of nutrients at recommended intake levels is both safe and inexpensive. Snack foods may also be beneficial for the elderly to help boost energy intakes to recommended amounts *(23)*. The current Dietary Reference Intakes (DRIs) should be used to guide consumers in maintaining appropriate amounts of nutrients each day *(5)*.

22.8 ADVERSE EFFECTS OF EXCESSIVE CALCIUM AND VITAMIN D SUPPLEMENTATION

Recent findings on vascular calcification, i.e., plaque mineralization and new bone formation in the medial layers of arteries, suggest that a small fraction of usual dietary calcium may be contributing to mineralization at these two sites and to heart valves in adults, particularly among the elderly *(24)*. Much remains to be learned about the signaling factors that contribute to bone formation at these sites, but current evidence supports the concern about adverse vascular flow dynamics and abnormal elasticity of these tissues, leading to suboptimal nutritional and oxygen supply to the subjacent tissues. In a recently reported study, Bolland et al. *(25)* found a link between daily use of calcium supplements (1 g elemental calcium daily, 400 mg in the morning and 600 mg in the evening versus placebo) and increased risk of cardiovascular events in healthy postmenopausal women followed for 5 years. However, the extent of cardiovascular risk from calcium supplements is still a matter of debate. A recent letter to the editor *(26)* reported that when they pooled the five trials with crude mortality data (representing 12,609 subjects) from their recent meta-analysis of calcium supplementation, there was no evidence of an increased mortality. Moreover, when Hsia et al. *(27)* examined cardiovascular disease (as a pre-specified secondary efficacy outcome) in a large randomized controlled trial of women aged 50–79 years who took 500 mg calcium with 200 IU vitamin D twice daily versus placebo, they found that supplemental calcium neither increased nor decreased coronary and cerebrovascular risk over a 7-year use period.

Calcium supplementation could also become a concern in individuals with declines in renal function; the retention of calcium may become excessive and accelerate vascular calcification. This condition may be worsened if calcium and vitamin D are bundled together in one supplement. Oral intakes of vitamin D in excess of 10,000 IU/day are associated with increases in urinary and serum calcium and may result in vitamin D intoxication when serum 25-hydroxyvitamin D levels exceed 100 ng/mL *(28)*.

22.9 DRUGS AND OTHER AGENTS USED IN PREVENTION AND TREATMENT OF OSTEOPOROSIS

The objectives of therapy in osteoporosis include prevention of further excessive bone loss, promotion of bone formation, prevention of fractures, reduction or elimination of pain, and restoration of physical function. Over the last two decades a number of anti-osteoporotic drugs have been developed. The mechanisms of action of these agents vary, as shown in Fig. 22.1. In simple terms, the agents may be characterized as acting to increase the availability of calcium from the gastrointestinal tract, to decrease the rate of resorption of existing bone, or to increase the rate of formation of new bone. Estrogens, selective estrogen receptor modulators, calcitonin, and bisphosphonates are considered to be "anti-resorptive" agents, acting to reduce rates of bone resorption by reducing osteoclast activity. Parathyroid hormone analogs in contrast are "anabolic" agents, stimulating bone formation through a primary action on osteoblasts. Most of the drugs mentioned below have been shown to improve bone density, and several have also been shown to reduce the incidence of fractures in randomized placebo-controlled clinical trials. Drugs approved by the Food and Drug Administration for the treatment of osteoporosis in the United States are generally indicated for use in patients having either documented osteoporotic fractures or bone densities at least 2 SDs below normal young controls. At the beginning of therapy it is important to obtain quantitative measurements of bone density at one or more sites of interest, for example, the lumbar spine and proximal femur. Such measurements can serve as a baseline for evaluating the effects of therapy after 1 or 2 years. Biochemical markers of bone turnover, particularly breakdown products of bone collagen, may also be used to monitor a therapeutic effect.

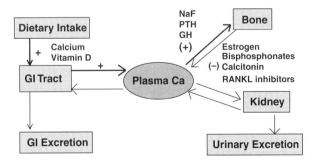

Result: Net Positive Balance

Fig. 22.1. Effects of therapeutic agents on calcium balance. Increases in dietary calcium and vitamin D result in increased gastrointestinal (GI) absorption. Anabolic agents increase bone formation [heavy arrow from plasma calcium (Ca) to bone]. Estrogen and other molecules inhibit bone resorption. NaF, sodium fluoride; PTH, parathyroid hormone; GH, growth hormone; RANK, receptor-activated nuclear factor Kappa; RANKL, RANK ligand.

22.9.1 Anti-resorptive Agents

Most of the drugs in current use are considered to be "anti-resorptive" agents, acting to reduce rates of bone resorption rather than to stimulate bone formation. When bone resorption is inhibited and bone formation continues, a modest gain in bone mass, typically 2–5%, can occur. The overall gain tends to be limited because bone formation is eventually downregulated to match the lowered rate of resorption. Anti-resorptive drugs are generally more effective in increasing the mass of trabecular bone than cortical bone because of the higher rate of resorption existing in trabecular bone.

22.9.1.1 CALCIUM AND VITAMIN D

Patients with severe vitamin D deficiency are unable to mineralize bone matrix normally and develop osteomalacia. Most individuals with only moderately limited vitamin D stores do not have osteomalacia, but they do have mild secondary hyperparathyroidism and a higher bone turnover as described above. Supplementation with calcium and vitamin D serves to increase the quantity of calcium absorbed by the gastrointestinal tract, to raise serum levels of ionized calcium, and to reduce the secretion of parathyroid hormone (29). Rates of bone resorption are thus decreased. In some clinical trials administration of calcium and vitamin D supplements to older populations in whom prevailing vitamin D stores and calcium intake are low has produced improvements in bone density (30,31).and decreased fracture rates (30,32). Other observational studies of large populations have failed to show a relationship between dietary calcium intake, calcium supplements, and fracture rates (33,34). The varying outcomes of these studies may reflect varying intakes of calcium and vitamin D in the study populations prior to entry. In spite of the uncertainty about the efficacy of calcium and vitamin D alone for treatment of osteoporosis, it is widely held that a sufficient intake of both is required as part of any anti-osteoporotic regimen. In a recently published meta-analysis, Tang et al. found support for calcium plus vitamin D supplementation in the "preventive treatment" of osteoporosis, including a 12% risk reduction in fractures of all types in the 17 trials that reported fracture as an outcome (35). Also published very recently is a study of 930 participants randomized to receive 4 years of treatment with 1,200 mg elemental calcium (as calcium carbonate) or placebo and followed for 10.8 years on average. The investigators (36) found that calcium supplementation reduced the risk of all fractures, including those due to minimal trauma, but that the beneficial effect was not sustained once treatment ended.

The requirement for dietary calcium is not precisely defined, but will depend on the age and health status of the individual. The Institute of Medicine of the National Academy of Sciences has recommended an elemental calcium intake of at least 1,000 mg/day for adults between 19 and 50 years of age, and 1,200 mg/day for adults over 50 years of age (5). To reach these levels, calcium supplements will be required in most individuals. Calcium carbonate is widely used and is the least expensive supplement. It is poorly absorbed in individuals with achlorhydria, since it is soluble only in an acidic medium. Calcium citrate is soluble at any pH and may be better absorbed in achlorhydric individuals (37).

The desirable dietary intake of vitamin D required for bone health is also poorly defined. Results vary depending upon the population studied and the degree of sun exposure. The Institute of Medicine has recommended intakes of 5 mcg/day (200 IU/day) for adults aged 31–50, 10 mcg/day (400 IU/day) for adults aged 51–70, and 15 mcg/day (600 IU/day) for those 71 and older *(5)*. The adequacy of dietary supply of vitamin D is reflected in serum levels of 25-hydroxyvitamin D. At 25-hyroxyvitamin D serum levels below 32 ng/mL (80 nmol/L) serum levels of parathyroid hormone and rates of bone resorption begin to rise. Thus, a state of vitamin D sufficiency may be defined as having serum levels of 25-hydroxyvitamin D above this threshold. A study of older adults in Boston showed that less than 50% reached the desired threshold with an average dietary intake of 5 mcg (200 IU)/day. When supplemented with vitamin D to an intake of 20 mcg (800 IU)/day, 90% reached the targeted 25-hydroxyvitamin D threshold of 80 nmol/L *(38)*. A meta-analysis of seven randomized, controlled trials of 9,820 elderly subjects showed that in five studies higher doses of vitamin D (700–800 IU/day) reduced the relative risk of both hip and non-vertebral fractures, whereas no significant reduction in fractures was seen in two studies using lower doses (400 IU/day) *(39)*. In addition to reducing fracture risk, vitamin D supplementation may improve muscle function and reduce the risk of falls in susceptible populations. In a meta-analysis of five randomized clinical trials of vitamin D in 1237 older subjects, falls were reduced by 22% in comparison to control subjects receiving calcium or placebo *(40)*. In summary, a vitamin D supplement of approximately 20 mcg (800 IU)/day is desirable in older adults with limited sun exposure. In individual patients vitamin D requirements and the response to supplementation are best assessed by monitoring serum levels of 25-hydroxyvitamin D.

22.9.1.2 ACTIVE METABOLITES OF VITAMIN D

Vitamin D is converted in the body to two active metabolites, 25-hydroxyvitamin D and 1,25-dihydroxyvitamin D (calcitriol). Calcitriol, the most active metabolite, has been evaluated for treatment of osteoporosis because of its potent effects in improving gastrointestinal calcium absorption. The results of several clinical trials have been mixed. In one of the largest trials 622 women with one or more vertebral compression fractures were randomized to receive either a calcium supplement or calcitriol (0.25 mcg twice a day). The number of new vertebral fractures was significantly reduced in the subjects receiving calcitriol during the second and third years of treatment *(41)*. In other smaller trials calcitriol failed to reduce the incidence of fractures *(42)*. Hypercalcemia and hypercalciuria are recognized complications of higher doses of calcitriol, particularly when combined with calcium supplements *(43)*. Because of the lack of conclusive evidence of benefit and the potential risk, calcitriol is not recommended as a first-line therapy in the treatment of osteoporosis.

22.9.1.3 ESTROGENS

Estrogen replacement in postmenopausal women prevents the excessive bone loss due to estrogen deficiency. Estrogens should be given together with a progestin in women who have an intact uterus to avoid the increased risk of endometrial cancer due to estrogens alone. A large number of clinical studies have shown that both orally and transdermally administered estrogens increase spine BMD in postmenopausal women (44). The Women's Health Initiative trial of estrogen replacement in postmenopausal women showed that both combined estrogen–progestin (in women with an intact uterus) and estrogen alone (in women with hysterectomies) reduced hip and vertebral fracture risk (45). Estrogens convey both benefits and risks apart from their skeletal effects. Orally administered estrogens have a beneficial effect on the serum lipid profile, but they increase the risk of venous blood clots in susceptible individuals. Long-term estrogen replacement also adversely affects the risk for breast cancer in postmenopausal women. For this reason, estrogen replacement is not considered to be the treatment of choice in most postmenopausal women if the primary objective of therapy is the prevention of fractures.

22.9.1.4 SELECTIVE ESTROGEN RECEPTOR MODULATORS (SERMS)

These synthetic analogs have the ability to act on the estrogen receptor to produce a variable spectrum of estrogen effects. SERMs are capable of replicating biological effects of natural estrogen in some target tissues, but may act as estrogen antagonists in other tissues. The basis for the observed effects may depend in part to the existence of two types of estrogen receptors existing in different tissues. In bone tissue ERβ is the more abundant whereas in reproductive tissues ERα predominates. Drugs in this class include tamoxifen, a drug used to treat breast cancer, and raloxifene, a drug approved for the treatment of osteoporosis. Raloxifene acts as an estrogen agonist with respect to bone and lipoprotein metabolism. It increases bone density and lowers serum cholesterol when given to postmenopausal women. Raloxifene does not stimulate the endometrium and like tamoxifen it acts as an estrogen antagonist in breast tissue. Thus, raloxifene is a good choice as an anti-osteoporosis drug in women at high risk of breast cancer. A large clinical trial has demonstrated that raloxifene therapy reduces vertebral fracture risk in postmenopausal women with osteoporosis (46).

22.9.1.5 ISOFLAVONES

These naturally occurring phytoestrogens are derived from plant materials and include genistein and daidzein. Isoflavones appear to act like SERMs or weak estrogen agonists in bone. Phytoestrogens in soy beans may help prevent loss of vertebral BMD in adult life and, hence, may delay the occurrence of osteoporosis (47). Only a few clinical trials in peri- and postmenopausal women have been conducted. Two different groups of investigators used doses of approximately 90 mg/day of isoflavones to obtain small increases in vertebral density after 6 months of treatment (48,49). These studies were not designed to determine whether isoflavones could reduce fracture risk. A recent study of ipriflavone, an artificial derivative of a natural molecule found in soy protein, failed to demonstrate improvement in bone density or in biochemical markers of bone metabolism in 474 postmenopausal women. In addition to having no measurable effect on bone mass, approximately 13% of women receiving the experimental compound showed a significant decrease in white blood cells *(50)*. Further research will be needed to demonstrate which isoflavones may have beneficial effects on bone and whether fracture incidence is decreased.

22.9.1.6 BISPHOSPHONATES

These useful drugs are analogs of pyrophosphate, a substance that occurs naturally in bone. Nitrogen-containing bisphosphonates are taken up by osteoclasts, resulting in reduced osteoclast activity and survival. They are among the most potent anti-resorptive drugs known. Clinical trials of several bisphosphonates, including etidronate, alendronate, risedronate, ibandronate, and zoledronic acid, indicate that bone density is increased in postmenopausal women after two or more years of treatment. Further trials with alendronate *(51)*. risedronate *(52)*. and zoledronic acid *(53)* provide strong evidence that these drugs can reduce fracture risk by 40–60% at various skeletal sites, including the spine and hip. Ibandronate has been shown to reduce the incidence of vertebral but not hip fractures *(54)*. These drugs are effective in men as well as women, making them useful for the treatment of almost all patients with osteoporosis. Bisphosphonates are effective given either orally at intervals of 1–4 weeks (alendronate, risedronate, or ibandronate) or intravenously at intervals of 1–12 months (ibandronate or zoledronic acid). Alendronate, risedronate, ibandronate, and zoledronic acid are currently approved in the United States for treatment of osteoporosis. Other drugs of this class are likely to be approved in the near future.

22.9.1.7 CALCITONIN

This natural peptide hormone is produced in small quantities by the parafollicular cells of the normal thyroid gland. Synthetic peptides having the amino acid sequence of either human or salmon calcitonin have been approved for the treatment of osteoporosis. This hormone binds directly to cell surface receptors on osteoclasts leading to reversible inhibition. The administration of synthetic human or salmon calcitonin in patients with osteoporosis causes a reduction in bone resorption and a modest increase in bone density. Clinical trials have found that intranasal calcitonin therapy reduces the occurrence of vertebral fractures but not non-vertebral fractures in postmenopausal women *(55)*. Large doses of calcitonin may have an analgesic effect through an independent action on the central nervous system.

22.9.1.8 INHIBITORS OF RANKL

The natural cytokine RANK ligand (RANKL) is a member of the TNF family of ligands and receptors. RANKL accelerates osteoclast formation and function. It is blocked by its natural antagonist, osteoprotegerin, also known as OPG, which is secreted by several types of cells, including osteoblasts. OPG thus inhibits osteoclast activation and bone resorption. OPG is detectable in the serum of adult humans and may have an important regulatory role in bone metabolism. When given as a single injection to postmenopausal women, OPG causes long-lasting inhibition of bone resorption, indicating that it is a potentially useful agent for the treatment of osteoporosis *(56)*. Denosumab, a monoclonal antibody against RANKL, is also a potent anti-resorptive agent that can increase BMD in postmenopausal women in clinical trials *(57)*.

22.9.2 Anabolic Agents

Whereas most of the currently approved therapies for osteoporosis inhibit bone resorption there has been a growing interest in therapies that can stimulate bone formation.

22.9.2.1 SODIUM FLUORIDE

Fluoride has long been known to be a potent stimulator of bone formation. When taken in higher doses of 50–100 mg/day it causes marked increases in BMD, especially in cancellous bone. Clinical trials with sodium fluoride, however, have produced conflicting results. Earlier studies indicated that although bone density was increased, bone strength was not. Treated patients showed no decrease in vertebral fractures and an increase in non-vertebral fractures, suggesting that the quality of fluoride-containing bone might be impaired *(58)*. Another group, giving a slow release formulation at a dose of 50 mg/day together with calcium citrate did find that this agent reduced new but not recurrent vertebral fractures *(59)*. Fluoride is not currently approved for use as an anti-osteoporosis drug because of continuing concerns about its effectiveness and safety.

22.9.2.2 PARATHYROID HORMONE (PTH)

This natural peptide hormone is a primary regulator of calcium homeostasis and bone metabolism in humans and other mammals. Its effects are complex, because it can activate both bone formation and bone resorption, depending on the conditions. Its primary target is the osteoblast, where it can induce proliferation and maturation, thus increasing bone collagen synthesis. These anabolic effects predominate when PTH is present intermittently and in low concentrations. At higher, more sustained concentrations the predominant effect of PTH favors increased osteoclast activity and bone resorption. These increased activities are true for both intact PTH 1-84 and for the biologically active fragment PTH 1-34. Osteoclast activation by PTH is probably mediated by cytokines such as IL-6 and RANKL. These cytokines are produced by osteoblasts under the influence of PTH and promote the recruitment and activation of osteoclasts. When recombinant human PTH 1-34 (teriparatide) is given as a subcutaneous dose of 20–40 mcg once a day, blood levels are only intermittently elevated. Clinical trials have shown that intermittent PTH administration is an effective therapy for osteoporosis, causing a marked 10–15% increase in spine BMD over a period of 24 months, with lesser increases in the femoral neck. In the largest PTH trial to date treatment reduced the risk of both vertebral and non-vertebral fractures in postmenopausal women with osteoporosis *(60)*. The drug is effective in men as well as women *(61)*.

Preliminary studies indicate that PTH can be combined or given in sequence with anti-resorptive drugs to achieve further increases in BMD *(62)*. Combined estrogen and PTH therapy appears to be more effective than estrogen therapy alone *(63)*. The concurrent administration of alendronate together with PTH 1-84 was not additive *(64)*. In contrast, the sequential administration of alendronate to postmenopausal women for 1 year following a year of therapy with PTH 1-84 resulted in a further gain of nearly 6% in BMD following a 6% increase during the first year *(65)*.

Over the short term PTH is well tolerated by most patients. Transient hypercalcemia and hypercalciuria usually return to normal within a few hours of administration. In Fisher rats given high doses of PTH for up to 2 years beginning in infancy, the risk of osteosarcoma is increased. This cancer does not appear to be a risk with therapeutic doses of PTH given to adult humans. However, the FDA recommends that therapy with PTH be limited to no more than 2 years. Because of its high cost relative to oral bisphosphonate drugs many physicians reserve PTH for use in patients who cannot tolerate bisphosphonates or who continue to have fractures or lose BMD in spite of therapy with other drugs.

22.9.2.3 GROWTH HORMONE AND INSULIN-LIKE GROWTH FACTOR-1 (IGF-1)

These natural peptides play a critical role in determining bone growth and the achievement of optimal bone mass. Growth hormone acts on target tissues to stimulate the formation and release of IGF-1. After the liver, bone is the second richest source of IGF-1 in the body. IGF-1 acts locally in bone to promote chondrocyte and osteoblast differentiation and growth. IGF-1 also circulates in the blood, where its concentration reflects the rate of secretion of growth hormone. Two studies have suggested that lower serum levels of IGF-1 are associated with a higher risk of hip and spine fractures *(66,67)*. There have been several small clinical trials determining the effects growth hormone or IGF-1 on bone metabolism and bone density, but none have been designed to examine fracture rates. In one trial involving men over age 65, treatment with recombinant human growth hormone (rhGH) caused a 1.6% increase in lumbar BMD after 6 months, but the changes were not sustained after 1 year *(68)*. A study in elderly women with recent hip fracture found that administration of IGF-1 complexed with IGF-1-binding protein 3 promoted a more rapid regain of bone mass in the contralateral hip *(69)*. Both growth hormone and IGF-1 activate the entire bone remodeling system, causing increased bone resorption as well as bone formation *(70)*. Thus far the net anabolic effects of these agents on bone have been modest. Their use in elderly individuals is also limited by side effects, including weight gain, edema, carpal tunnel syndrome, and glucose intolerance.

22.10 SUMMARY

Since both the incidence and prevalence of osteoporosis are increasing as populations are aging, greater expenditures for the care of fracture patients are almost certain to occur. Many patients will not survive the first post-fracture year. Preventive strategies, either primary or secondary, that are cheap and effective must be identified and implemented. Diet and drug therapy represent the two major approaches, but exercise programs should not be overlooked. Improving the diet with respect to calcium and vitamin D intake may be one of the most easily modifiable approaches for the prevention of osteoporosis. Walking and maintaining activities of daily living by the elderly are also critical for the prevention of osteoporosis. Any type of minimal exercise program should also yield some benefit to the retention of musculoskeletal function.

When osteoporotic fractures occur, or when bone density declines into the osteoporotic range, drug therapy is usually indicated. With the emergence of new pharmacologic agents many choices are now available. Even more choices will be available in the future. Table 22.4 summarizes the efficacy of the agents currently available, as demonstrated in published clinical trials. Choice of an appropriate agent should depend not only on demonstrated efficacy in reducing fracture risk but also on patient risk factors and individual side effects. A combination of good nutrition including sufficient calcium and appropriate exercise will continue to be important, even as anti-osteoporotic drugs become more widely used.

Table 22.4
Efficacy of anti-osteoporotic therapies in randomized clinical trials

	Increased BMD	Spine fractures	Non-spine fractures	Hip fractures
Calcium and vitamin D preparations				
Calcium monotherapy	+/−	NA	NA	NA
Vitamin D monotherapy	+/−	NA	NA	NA
Calcium + vitamin D	+	NA	+/−	+/−
Calcitriol	+/−	+/−	+/−	NA
Estrogens and SERMs				
Estrogen replacement	+	+	+	+
Raloxifene	+	+	NS	NS
Bisphosphonates				
Alendronate	+	+	+	+
Risedronate	+	+	+	+
Ibandronate	+	+	NS	NS
Zoledronic acid	±	±	±	±
Calcitonin				
Intranasal calcitonin	+	+	NS	NS
Anabolic agents				
Human PTH	+	+	+	NA
Human growth hormone	+/−	NA	NA	NA

22.11 RECOMMENDATIONS

1. Dual energy X-radiographic analysis (DXA) should be used to measure bone mineral density (BMD) in all women over age 65 and in men or younger women with known risk factors for osteoporosis. Clinical observations, such as the occurrence of fragility fractures and loss of height or back pain, are also important in making a diagnosis of osteoporosis. Serial DXA examinations done at intervals of 1 or 2 years are useful in following the response to therapy.

2. Primary and secondary prevention of osteoporosis emphasize (1) increased calcium intake to 1,000 mg/day or more and (2) intake of sufficient vitamin D to assure optimal calcium absorption (at least 800 IU/day if sun exposure is limited). The adequacy of vitamin D intake may be assessed clinically by measurement of serum 25-hydroxyvitamin D. Adequate intakes of calcium and vitamin D are essential parts of any anti-osteoporotic regime. Vitamin K may also be included in this recommendation.

3. Drug therapy is usually indicated when osteoporotic fractures occur or when BMD declines into the osteoporotic range. Most available drugs are anti-resorptive, but anabolic, bone-forming agents are also available and have proven to be effective.

   a. Estrogen replacement therapy helps prevent excessive bone loss resulting from the decline of ovarian synthesis of estrogens following the menopause, and selective estrogen receptor modulators (SERM) may also produce estrogen-like effects on bone as well. For example, raloxifene, a SERM, has been reported to reduce the risk of vertebral fractures.

   b. Bisphosphonates constitute another major class of anti-resorptive drugs that are effective in reducing osteoporotic fractures. Drugs in this class include alendronate, risedronate, ibandronate, and zoledronic acid and may be given orally or intravenously. These drugs are considered to be first-line agents in treating both women and men.

   c. Parathyroid hormone (PTH), given subcutaneously and at low doses, is the most effective anabolic agent currently available for use in patients with osteoporosis. Anti-resorptive drugs such as bisphosphonates may be useful when given sequentially after a course of therapy with PTH.

4. Maintaining physical activity is also a key component of any regimen aimed at preventing osteoporosis or treating osteoporotic patients.

5. Good nutrition from foods, including recommended amounts of calcium and vitamins D and K, should accompany the use of drugs or other therapeutic modalities. If sufficient calcium or vitamin D or vitamin K cannot be obtained from foods, supplements will be necessary.

REFERENCES

1. Kanis JA, Melton LJ, III, Christiansen C, et al. The diagnosis of osteoporosis. J Bone Miner Res 1994; 9:1137–41.
2. Marshall D, Johnell O, Wedel H. Meta-analysis of how well measures of bone mineral density predict occurrence of osteoporotic fractures. BMJ 1996;312:1254–9.
3. Raisz LG, Pathogenesis of osteoporosis: concepts, conflicts and prospects. J Clin Invest 2005, 115:3318–25.
4. Fink HA, Ensrud KE, Barrett-Conner E. et al. Association of testosterone and estradiol deficiency with osteoporosis and rapid bone loss in older men. J Clin Endocrinol Metab 2006; 91:3908–15.
5. Institute of Medicine (IOM), Food and Nutrition Board, National Academy of Sciences. Dietary Reference Intakes for Calcium, Phosphorus, Magnesium, Vitamin D, and Fluoride. Washington, DC: National Academy Press, 1997.
6. Anderson JJB, Klemmer P, Sell ML, Garner SC, Calvo MS. Phosphorus. In: Bowman B, Russell R, eds. Present Knowledge in Nutrition, 9th ed. Washington, DC: ILSI Press, 2006, pp. 383–99.

7. Kemi VE, Karkkainen MUM, Karp HJ, Laitinen KAE, Lamberg-Allardt CJE. Increased calcium intake does not completely counteract the effects of increased phosphorus intake on bone: an acute dose-response study in healthy females. Br J Nutr 2008;99:832–839.

8. Wyshak G. Teenaged girls, carbonated beverage consumption, and bone fractures. Arch Pediatr Adolesc Med 2000;154:610–13.

9. Anderson JJB. Plant-based diets and bone health: nutritional implications. Am J Clin Nutr 1999;70:539S–42S.

10. Shea B, Wells G, Cranney A, et al. Meta-analysis of therapies for postmenopausal osteoporosis. VII. Meta-analysis of calcium supplementation for the prevention of postmenopausal osteoporosis. Endocr Rev 2002;23:552–9.

11. Lambert HL, et al. Calcium supplementation and bone mineral accretion in adolescent girls: an 18-mo randomized controlled trial with 2-y follow-up. Am J Clin Nutr 2008; 87:455–62.

12. Steingrimsdottir L, Gunnarsson O, Indridason O, et al. Relationship between serum parathyroid hormone levels, vitamin D sufficiency, and calcium intake. JAMA 2005; 294:2336–41.

13. Goussous R, Song L, Dallal GE, Dawson-Hughes B. Lack of effect of calcium intake on the 25-hydroxyvitamin D response to oral vitamin D_3. J Clin Endocrinol Metab 2005; 90:707–11.

14. Holick MF. Vitamin D: a D-Lightful health perspective. Nutr Rev 2008; 66(Suppl 2):S182–S194.

15. Chapuy M-C, Preziosi P, Maamer M, et al. Prevalence of vitamin D insufficiency in an adult normal population. Osteoporosis Int 1997; 7:439–43.

16. Boonen S, Lips P, Bouillon R, et al. Need for additional calcium to reduce the risk of hip fracture with vitamin D supplementation: evidence from a comparative metaanalysis of randomized controlled trials. J Clin Endocrinol Metab 2007; 92:1415–23.

17. Knapen MH, Schurgers LJ, Vermeer C. Vitamin K_2 supplementation improves hip bone geometry and bone strength indices in postmenopausal women. Osteopor Int 2007; 18:963–72.

18. Schurgers LJ, Teunissen KJF, Hauulyak K, et al. Vitamin K-containing dietary supplements: comparison of synthetic vitamin K_1 and natto-derived menaquinone-7. Blood 2007; 109:3279–83.

19. Okano T, Tsugawa N, Kamao M, et al. Association between vitamin K and bone mineral density or bone fracture in elderly Japanese women. J Bone Miner Res 2006; 21(Suppl 1):SU290 [abstract].

20. Bolton-Smith C, McMurdo MET, Paterson CR, et al. Two year randomized controlled trial of vitamin K_1 (phylloquinone) and vitamin D_3 plus calcium on the bone health of older women. J Bone Miner Res 2007; 22:509–19.

21. New SA, Robins SP, Campbell MK, et al. Dietary influences on bone mass and bone metabolism: further evidence of a positive link between fruit and vegetable consumption and bone health? Am J Clin Nutr 2000; 71:142–51.

22. Marini H, Minutoli L, Polito F, et al. Effects of the phytoestrogen genistein on bone metabolism in osteoporotic postmenopausal women. Ann Intern Med 2007; 146:839–47.

23. Zizza C, Tayie FA, Lino M. Benefits of snacking in older Americans. J Am Diet Assoc 2007; 107:800–06.

24. Demers L. A skeleton in the atherosclerosis closet. Circulation 1995; 92:2029–32.

25. Bolland MJ, Barber PA, Doughty RN, Mason B, Horne A, Ames R, Gamble GD, Grey A, Reid IR. Vascular events in healthy older women receiving calcium supplementation: randomised controlled trial. BMJ 2008 Feb 2; 336(7638):262

26. Tang BMP, Nordin BEC. Calcium supplementation does not increase mortality. Med J Aust 2008; 188(9):547.

27. Hsia J, Heiss G, Ren H, et al. Calcium/vitamin D supplementation and cardiovascular events. Circulation 2007; 115(7):846–54.

28. Vieth R. Vitamin D supplementation, 25-hydroxyvitamin D concentrations, and safety. Am J Clin Nutr 1999; 69:842–56.

29. Riggs BL, O'Fallon WM, Muhs J, et al. Long-term effects of calcium supplementation on serum parathyroid hormone level, bone turnover and bone loss in elderly women. J Bone Miner Res 1998; 13:168–74.

30. Chapuy MC, Arlot ME, Duboeuf F, et al. Vitamin D₃ and calcium to prevent hip fractures in elderly women. New Engl J Med 1992; 327:1637–42.

31. Dawson-Hughes B, Harris SS, Krall EA, Dallal GE. Effect of calcium and vitamin D supplementation on bone density in men and women 65 years of age or older. New Engl J Med 1997; 337:670–6.

32. Recker RR, Hinders SK, Davies M, et al. Correcting calcium nutritional deficiency prevents spine fractures in elderly women. J Bone Miner Res 1996; 11:1961 6.

33. Cumming RG, Cummings SR, Nevitt MC, et al. Calcium intake and fracture risk: results from the study of osteoporotic fractures. Am J Epidemiol 1997; 145:926–34.

34. Owusu W, Willett WC, Feskanich D, et al. Calcium intake and the incidence of forearm and hip fractures among men. J Nutr 1997; 127:1782–7.

35. Tang B, Eslick G, Nowson C, et al. Use of calcium or calcium in combination with vitamin D supplementation to prevent fractures and bone loss in people aged 50 years and older: a meta-analysis. Lancet 2007; 370(9588):657–66.

36. Bischoff-Ferrari HA, Dawson-Hughes B, Walter W, et al. Effect of calcium supplementation on fracture risk: a double-blind randomized controlled trial. Am J Clin Nutr 2008; 87(6):1945–51.

37. Recker RR. Calcium absorption and achlorhydria. New Engl J Med 1985; 313:70–3.

38. Dawson-Hughes B, Harris SS. Definition of the optimal 25OH-D status for bone. In: Norman AW, Bouillon R, Thomasset M, eds. Vitamin D Endocrine System. Berkeley: University of California Riverside Press, 2000, pp. 909–14.

39. Bischoff-Ferrari HA, Willett WC, Wong JB, et al. Fracture prevention with vitamin D supplementation: a meta-analysis of randomized controlled trials. JAMA 2005; 293:2257–64.

40. Bischoff-Ferrari HA, Dawson-Hughes B, Willett WC, et al. Fall prevention by vitamin D treatment: a meta-analysis of randomized controlled trials. JAMA 2004; 291:1999–2006.

41. Tilyard MW, Spears GF, Thomson J, Dovey S. Treatment of postmenopausal osteoporosis with calcitriol or calcium. New Engl J Med 1992; 326:357–62.

42. Ott SM, Chesnut CH, III. Calcitriol treatment is not effective in postmenopausal osteoporosis. Ann Intern Med 1989; 110:267–74.

43. Aloia JF, Vaswani A, Yeh JK, et al. Calcitriol in the treatment of postmenopausal osteoporosis. Am J Med 1988; 84:401–08.

44. Bush TL, Wells HB, James MK, et al. Effects of hormone therapy on bone mineral density: results from the Postmenopausal Estrogen/Progestin (PEPI) Trial. JAMA 1996; 276:1389–96.

45. Rossouw JE, Anderson GL, Prentice RL, et al. Risks and benefits of estrogen plus progestin in healthy postmenopausal women: principal results from the Women's Health Initiative randomized controlled trial. JAMA 2002; 288:321–33.

46. Ettinger B, Black DM, Mitlak BH, et al. Reduction of vertebral fracture risk in postmenopausal women with osteoporosis treated with raloxifene. JAMA 1999; 282:637–45.

47. Anderson JJB, Garner SC. Phytoestrogens and bone. In: Adlercreutz HL, ed. Balleire's Clinical Endocrinology and Metabolism. Balliere Tyndall, London, 1998, pp. 1–16.

48. Potter SM, et al. Soy protein and isoflavones: their effects on blood lipids and bone density in postmenopausal women. Am J Clin Nutr 1998; 68(Suppl):1375S–79S.

49. Alekel DL, et al. Isoflavone-rich soy protein isolate attenuates bone loss in the lumbar spine of perimenopausal women. Am J Clin Nutr 2000; 72:844–52.

50. Alexandersen P, et al. Ipriflavone in the treatment of postmenopausal osteoporosis: a randomized controlled trial. JAMA 2001; 285:1482–8.

51. Black DM, Thompson DE, Bauer DC, et al. Fracture risk reduction with alendronate in women with osteoporosis: the Fracture Intervention Trial. J Clin Endocrinol Metab 2000; 85:4118–24.

52. Harris ST, Watts NB, Genant HB, et al. Effects of risedronate treatment on vertebral and non-vertebral fractures in women with postmenopausal osteoporosis. JAMA 1999; 282:1344–52.

53. Black DM, Delmas PD, Eastell R, et al. Once-yearly zoledronic acid for treatment of postmenopausal osteoporosis. N Engl J Med 2007; 356:1809–22.

54. Chesnut III CH, Skag A, Christiansen C, et al. Effects of oral ibandronate administered daily or intermittently on fracture risk in postmenopausal osteoporosis. J Bone Miner Res 2004; 19:1241–49.

55. Chesnut CH, Silverman S, Andriano K, et al. A randomized trial of nasal spray salmon calcitonin in postmenopausal women with established osteoporosis: the Prevent Recurrence of Osteoporosis Fractures Study. Am J Med 2000; 109:267–76.

56. Bekker PJ, Holloway D, Nakanishi A, et al. The effect of a single dose of osteoprotegerin in postmenopausal women. J Bone Miner Res 2001; 16:348–60.

57. McClung MR, Levicki EM, Cohen SB, et al. Denosumab in postmenopausal women with low bone density. New Engl J Med 2006; 354:821–31.

58. Riggs BL, Hodgson SF, O'Fallon MW, et al. Effect of fluoride treatment on the fracture rate in postmenopausal women with osteoporosis. New Engl J Med 1990;322:802–09.

59. Pak CYC, Sakhaee K, Adams-Huet B, et al. Treatment of postmenopausal osteoporosis with slow-release sodium fluoride: final report of a randomized controlled trial. Ann Intern Med 1995; 123:401–08.

60. Neer RM, et al. Effect of parathyroid hormone (1-34) on fractures and bone mineral density in postmenopausal women with osteoporosis. New Engl J Med 2001; 344:1434–41.

61. Kurland ES, Cosman F, McMahon DJ, et al. Therapy of idiopathic osteoporosis in men with parathyroid hormone: effects on bone mineral density and bone markers. J Clin Endocrinol Metab 2000; 85:3069–76.

62. Rosen CJ, Bilezekian JP. Anabolic therapy for osteoporosis. J Clin Endocrinol Metab 2001; 86:957–64.

63. Lindsay R, Nieves J, Formica C, et al. Randomised controlled study of the effect of parathyroid hormone on vertebral bone mass and fracture incidence among postmenopausal women on oestrogen with osteoporosis. Lancet 1997; 350:550–55.

64. Black DM, Greenspan SL, Ensrud KE, et al. The effects of parathyroid hormone and alendronate alone or in combination in postmenopausal osteoporosis. New Engl J Med 2003; 349:1207–15.

65. Black DM, Bilezikian JP, Ensrud KE, et al. One year of alendronate after one year of parathyroid hormone (1-84) for osteoporosis. New Engl J Med 2005; 353:555–65.

66. Sugimoto T, Nishiyama K, Kuribayshi F, Chihara K. Serum levels of IGF-1, IGFBP-2, and IGFBP-3 is osteoporotic patients with and without spine fractures. J Bone Miner Res 1997; 12:1272–79.

67. Rosen CJ. IGF-1 and osteoporosis. Clin Lab Med 2000; 20:591–602.

68. Rudman D, Feller AG, Nagrai HS, et al. Effect of human growth hormone in men over age 60. New Engl J Med 1990; 323:1–6.

69. Boonen S, Rosen C, Bouilion R, et al. Musculoskeletal effects of the recombinant human IGF-1/IGF-1 binding protein-3 complex in osteoporotic patients with proximal femoral fracture: a double-blind, placebo-controlled pilot study. J Clin Endocrinol Metab 2002; 87:1593–9.

70. Agnusdei D, Gentilelia R. GH and IGF-1 as therapeutic agents for osteoporosis. J Endocrinol Invest 2005; 28:32–6.

23 Osteoarthritis

Paola de Pablo and Timothy E. McAlindon

Key Points

- Osteoarthritis is the most common arthritis in aging adults and is the most frequent reason for joint replacement at a cost to the community of billions of dollars per year.
- There are numerous mechanisms by which micronutrients might be expected to influence the development or progression of osteoarthritis. Micronutrients that might interact with osteoarthritis include vitamins C and D and possibly vitamins E and K and selenium.
- There have been numerous positive clinical trials of glucosamine and chondroitin products for osteoarthritis. While shown to be well tolerated, recommendations regarding these compounds await further study since these results are clouded by issues of biologic plausibility, heterogeneity, publication bias, inconsistent results, and methodological problems.
- Further studies of simple and safe nutritional interventions for osteoarthritis, including clinical trials, are needed.

Key Words: Osteoarthritis; nutrition; antioxidants; vitamins; glucosamine and chondroitin products

23.1 INTRODUCTION

Osteoarthritis (OA) is the most common form of arthritis and a major cause of disability in aging adults. The economic impact of OA is important *(1–3)*. The cost associated with OA has been estimated at $15.5 billion dollars in the United States (in 1994 dollars) *(4)*. Furthermore, osteoarthritis is the most frequent reason for joint replacement at a cost to the community of billions of dollars per year.

23.1.1 Risk Factors for Osteoarthritis

The striking relationship of OA with increasing age and with heavy physical work has led to its characterization as a "degenerative" disorder. However, this paradigm is inaccurate because a variety of mechanical, metabolic, and genetic disorders may

From: *Nutrition and Health: Handbook of Clinical Nutrition and Aging, Second Edition*
Edited by: C. W. Bales and C. S. Ritchie, DOI 10.1007/978-1-60327-385-5_23,
© Humana Press, a part of Springer Science+Business Media, LLC 2009

Table 23.1
Risk factors associated with osteoarthritis

Constitutional factors	Increased age
	Female vs. male gender
	Obesity
Mechanical factors	Heavy/repetitive occupations
	Heavy physical activity
	Major joint injury
Genetic factors	Mutations in the type II collagen gene
	Acromegaly
Other factors	Muscle weakness
	Proprioceptive deficits
	Calcium crystal deposition disease
	Hemochromatosis

lead to osteoarthritis *(5)*. The contemporary view is that OA results from dynamic interaction between destructive and reparative processes within a joint. Risk factors for OA are listed in Table 23.1.

23.1.2 Pathogenesis of Osteoarthritis

Pathologically, osteoarthritis is characterized by focal damage to articular cartilage. This may range in severity from minor surface roughening to complete cartilage erosion. The process is often described as being "degenerative" or resulting from "wear and tear" and is generally noninflammatory. The cartilage damage is usually accompanied by some form of "reaction" in the surrounding bone. Osteophytes are an early bony response to cartilage damage. These consist of outgrowths of bone from the peripheral margin of the joint, which may confer some protection to an osteoarthritic joint by reducing instability. The trabeculae in the bone adjacent to an area of osteoarthritic cartilage become thickened and are susceptible to microscopic fracturing. This gives rise to the appearance of "subchondral sclerosis" on a radiograph. In severe disease, particularly where full thickness cartilage has occurred, circumscribed areas of bony necrosis may develop in the subchondral bone. These may be filled with marrow fat or with synovial fluid which has tracked from the joint space through the cartilage defect into the subchondral bone. They give rise to the radiographic appearance of "cysts." Ultimately the subchondral bone itself may be eroded, or may collapse. In some cases, a low-level synovitis develops resulting from the presence of crystal or other cartilaginous "detritus," which may contribute to damage in the joint.

Biochemically, cartilage consists of a network of collagen fibrils (predominantly type II), which constrain an interlocking mesh of proteoglycans that resist compressive forces through their affinity for water. The tissue is relatively avascular and acellular. Turnover in healthy cartilage is slow and represents a balance between collagen and proteoglycan synthesis and degradation by enzymes such as metalloproteinases. In early osteoarthritis, the chondrocytes (cartilage cells) proliferate and become metabolically active. These hypertrophic chondrocytes produce cytokines (e.g., IL-1, TNF-α), degradative enzymes (e.g., metalloproteinases), and other growth factors. Proteoglycan production is increased in early OA, but falls sharply at a later stage when the chondrocyte "fails."

23.1.3 Clinical Features of Osteoarthritis

Osteoarthritis is characterized by pain, impaired joint motion, and disability. Pain of affected joints is typically exacerbated by activity and relieved by rest. With advanced disease, pain may be noted with less activity, and may even occur at rest. Stiffness is also a common complaint, usually lasting less than 30 minutes. Osteoarthritis commonly affects the first carpometacarpal joint, proximal interphalangeal joint, distal interphalangeal joint, hip, knee, subtalar joint, first metarsophalangeal joint, and the cervical and lumbar spine.

23.1.4 Therapeutic Approach to Osteoarthritis

The impact osteoarthritis has on daily life activities in addition of its high prevalence poses a public health problem. The management of osteoarthritis is largely palliative, focusing on symptom alleviation. Current therapy of osteoarthritis consists of both non-pharmacologic (weight-loss, exercise, education programs) and pharmacologic (analgesics, non-steroidal anti-inflammatory agents [NSAIDs], intra-articular corticosteroids) approaches. However, pharmacologic therapy has shown modest, although sound, symptomatic efficacy over short-term treatment courses, but no relevant effect over the long-term use *(6)*. While overall benefit versus risk profile for NSAIDs remains favorable when used appropriately, the other available drugs for short-term relief of pain can also be associated with serious, adverse events in certain settings and patient populations. Furthermore, current therapeutic measures to treat osteoarthritis are often incompletely effective. Clearly, there is a need for safe and effective alternative therapies as well as preventive strategies. Such interventions could be addressed by nutrition.

23.1.5 Osteoarthritis and Diet

Despite a great public interest in the relationship between nutrition and arthritis, there has been relatively little focus in traditional scientific studies on this association. The biological mechanisms by which micronutrients might influence the pathophysiological processes in osteoarthritis are numerous. Such processes include oxidative damage, cartilage matrix degradation and repair, and chondrocyte function and response in adjacent bone. However, it is surprising to find that in osteoarthritis there has been relatively little focus in rigorous studies on nutrition, compared with osteoporosis, a widespread age-related skeletal disorder, where numerous studies have demonstrated associations with dietary factors. Several studies have shown apparent effects of various micronutrients on the natural history of this disorder.

Moreover, patients often ask their physicians questions about nutrition and osteoarthritis. There are numerous speculative lay publications on this subject, and health food stores offer multiple nutritional supplements represented as therapies for arthritis *(7)*. Surveys suggest that 5–8% of adults in the United States have used at least one of these supplements *(8)*. There is an important consumption of such over-the-counter nutritional remedies, with glucosamine and chondroitin ranking third among all top-selling nutritional products in the United States with sales up to $369 million per year *(9)*.

23.2 OSTEOARTHRITIS AND NUTRITIONAL SUPPLEMENTS

23.2.1 *Glucosamine and Chondroitin Sulfate*

Glucosamine and chondroitin sulfate are cartilage extracellular matrix components that have been widely marketed as nutritional supplements that have potential benefit in reducing pain and slowing the progression of osteoarthritis on the basis that they might provide a substrate for matrix synthesis and repair. However, the mechanisms of action of these components remain unclear.

23.2.1.1 GLUCOSAMINE

Glucosamine, an amino monosaccharide found in chitin, glycoproteins, and glycosaminoglycans, is also known as 2-amino-2-deoxyglucose or 2-amino-2-deoxy-beta-d-glucopyranose. D-Glucosamine is made naturally in the form of glucosamine-6-phosphate and is the biochemical precursor of all nitrogen-containing sugars *(10)*. Specifically, glucosamine-6-phosphate is synthesized from glutamine and glucose-6-phosphate *(11)*, as the first step of the hexosamine biosynthesis pathway. Glucosamine is available as a nutritional supplement in three forms, glucosamine sulfate, glucosamine hydrochloride, and N-acetyl-glucosamine.

According to our understanding of the metabolic pathways involved, glucosamine, as an amino sugar, should be rapidly degraded by the liver during "first-pass" metabolism. Early pharmacodynamic studies assessed absorption of the compounds only indirectly *(12,13)*. A pharmacokinetic study in dogs, using a refined high-performance liquid chromatographic (HPLC) assay, demonstrated that glucosamine hydrochloride has a bioavailability of about 10–12% from single or multiple doses *(14)*. Research in rats has suggested that glucosamine is substantially degraded in the lumen of the gastrointestinal tract *(15)*. To evaluate the absorption of glucosamine sulfate in humans, Biggee et al. performed a small study evaluating 10 participants with osteoarthritis, measuring serum levels of glucosamine every 15–30 min over 3 h after ingestion of the recommended 1,500 mg of oral glucosamine sulfate *(16)*. Nine of ten subjects had detectable serum glucosamine beginning to rise at 30–45 min and peaking at 90–180 min. The mean maximal serum level was 12 µmoles/L. This would provide <2% of ingested glucosamine to blood and interstitial fluid combined. Based on the very small serum levels seen in this study, the authors concluded that ingestion of standard glucosamine sulfate is unlikely to stimulate cartilage chondroitin synthesis. Persiani et al. evaluated both serum and synovial fluid levels of glucosamine in five people with knee osteoarthritis before and after administration of daily oral glucosamine for 2 weeks and found similar increase in serum synovial fluid glucosamine concentrations of 7.9 and 7.2 mol/l, respectively *(17)*.

In addition to the fact that serum levels of glucosamine are very low, the notion that exogenous glucosamine might be incorporated into the structure of hyaluronan or cartilage proteoglycans is also problematic because while glucosamine can enter the glycosaminoglycan biosynthetic pathway after its conversion to UDP-N-galactosamine, glucose is a much more abundant substrate. Recent in vitro and in vivo studies in animals have shown increases in proteoglycan synthesis by chondrocytes after addition of glucosamine to the culture medium, suggesting that instead of providing substrate for hyaluronan or cartilage proteoglycans, the mechanism of action may be an anti-inflammatory effect *(12,13)*.

A potential adverse effect of glucosamine was recently highlighted in a report from the Institute of Medicine *(18)*. Glucosamine may lead to an increase in insulin dysregulation among individuals predisposed to such problems. These concerns are based on the known ability of glucosamine to bypass the glutamine:fructose-6-phosphate amidotransferase step of hexosamine biosynthesis and desensitize glucose transport *(19)*. Insulin dysregulation of particular interest to individuals with osteoarthritis since high body mass index is a risk factor for osteoarthritis, as well as for insulin resistance, and diabetes mellitus. Glucosamine could contribute to diabetes by interfering with the normal regulation of the hexosamine biosynthesis pathway *(20)*, but a study in patients with type 2 diabetes found that glucosamine supplementation does not result in clinically significant alterations in glucose metabolism *(21)*. Preliminary studies have been reassuring; however, their interpretation has been limited by the considerable variability in measures and small numbers of participants *(22,23)*. In contrast, a recent study examining the effects of oral glucosamine on serum glucose and insulin levels at the initiation and throughout the duration of a 3-h oral glucose tolerance test was recently conducted using sera from 16 patients with osteoarthritis, but with no other diagnosed medical condition who had fasted overnight. Three participants who were found to have previously undiagnosed abnormalities of glucose tolerance demonstrated significant incremental elevations in glucose levels after ingestion of glucosamine sulphate. The other 13 participants also had incremental elevations that were not statistically significant. Glucosamine sulphate ingestion had no effect on insulin levels. These results suggest that glucosamine ingestion may affect glucose levels and consequent glucose uptake in patients who have untreated diabetes or glucose intolerance *(24)*.

While the effects of glucosamine have been well documented in animal models, less is known about its effects on glucose metabolism in humans. Indeed, much information is still needed regarding its biological effects in vivo and its effects on glucose metabolism *(25)*. The biologic effects of other constituents such as sulfate, chondroitin sulfate, and other additives need to be investigated as well. The National Institute of Health (NIH) is currently supporting a study of glucosamine supplementation in persons with obesity, given that this population may be prone to the effects of glucosamine on insulin resistance.

23.2.1.2 Chondroitin Sulfate

Chondroitin sulfate is a sulfated glycosaminoglycan composed of linear alternating units of N-acetylgalactosamine and glucuronic acid, usually attached to proteins forming proteoglycan. This compound is an important structural component of extracellular matrix and the predominant glycosaminglycan found in articular cartilage, providing much of its resistance to compression. In humans it is also found in bone, cartilage, cornea, skin, and the arterial wall. The two most common forms of chondroitin sulfate found in nutritional supplements are types A (chondroitin 4-sulfate) and C (chondroitin 6-sulfate). Chondroitin sulfate is primarily obtained from fish cartilage as well as in bovine and pork cartilage. The biologic fate of orally administered chondroitin sulfate is less clear, but some evidence exists to suggest that the compound may be absorbed following oral administration, possibly as a result of pinocytosis *(26)*. Chondroitin sulfate is able to cause an increase in RNA synthesis by chondrocytes *(27)*, which appears to correlate with an increase in the production of proteoglycans and collagens *(28–31)*. In addition, there is an evidence that chondroitin sulfate partially inhibits leukocyte elastase and may reduce the degradation of cartilage collagen and proteoglycans, which is prominent process in osteoarthritis *(32–35)*.

Jackson et al. conducted a study to delineate the pharmacokinetics of glucosamine and chondroitin sulfate *(36)*. To date, the results of this study have only been reported in abstract form. Based on typical dosages, a total of 33 subjects (11 per treatment arm) were randomized to receive glucosamine 1,500 mg, or chondroitin sulfate 1,200 mg, or the combination of both glucosamine and chondroitin sulfate for 12 weeks. Nine blood samples per subject were subsequently obtained over 36 h and plasma concentrations of glucosamine and chondroitin sulfate were measured in duplicate. Similar to the previously reported oral single-dose pharmacokinetics, plasma glucosamine levels in both the glucosamine and combination treatment groups achieved peak concentrations in 2–3 h post-dose with a terminal elimination half-life of approximately 3 h. The mean maximal plasma concentration was 211.1 ng/ml for glucosamine and 216.6 ng/ml for chondroitin sulfate. Area under the concentration–time curve (AUC) was slightly decreased in the combination group compared to glucosamine alone, while Cmax was roughly equal. Compared to oral single-dose pharmacokinetics, both AUC and Cmax for glucosamine were lower following multiple-dose ingestion. No change in plasma chondroitin sulfate from baseline was detected following multiple-dose oral administration of either chondroitin sulfate alone or combination. The authors concluded that glucosamine is substantially absorbed and rapidly eliminated following multiple-dose oral administration of glucosamine alone and in combination in humans whereas no change in basal plasma chondroitin sulfate concentration can be detected with similar administration of chondroitin sulfate alone or in combination. Compared to single-dose ingestion, a decrease in AUC and Cmax for glucosamine was observed following multiple-dose administration possibly reflecting a change in absorption, cellular utilization, or elimination with sustained oral use.

23.2.1.3 Sulfate

Cartilage proteoglycans are highly sulfated. The amount of sulfate made available to cells is an important factor in the degree of proteoglycan sulfation *(37,38)*. In vitro experiments on cultured cells suggest that increases in serum sulfate concentration enhance glycosaminoglycan synthesis *(39)*. It has also been found that the rate of sulfated glycosaminoglycan synthesis in human articular cartilage is sensitive to small deviations from physiologic sulfate concentrations *(40)*. Sulfate pools in humans are among the smallest of all species *(41)*, making them especially susceptible to physiologically relevant small changes. Sulfate balance is poorly understood and may vary on dietary factors or on dietary supplements. There are suggestions that sulphate is in itself clinically relevant *(42,43)*. A study measuring human urinary sulfate excretion after ingestion of methionine or chondroitin sulfate supplements in the setting of high or low protein diets found that more sulfate was excreted in the urine in those with high-protein diets compared to those with low-protein diets. This suggests that in the low-protein state, the body increased sulfate retention from supplements *(42)*. These observations raise the possibility that sulfate supplementation may have a beneficial role in cartilage health.

Low sulphate concentrations in blood may contribute to osteoarthritis by decreasing cartilage chondroitin sulphation. A recent study measured serum levels of sulphate during 3 h of fasting or glucose ingestion after overnight fasts to determine how much sulphate lowering may occur during this period. Sera samples of 14 patients with osteoarthritis who fasted overnight were obtained every 15–30 min during 3 h of continued fasting and during 3 h after ingestion of 75 g of glucose. Continuation of overnight fasting for 3 h resulted in a near-linear 3-h decrease in levels for all 14 patients ranging from 3 to 20% with a mean drop of 9.3%, whereas the 3-h decrease after glucose ingestion ranged from 10 to 33% with a mean drop of 18.9%. The authors concluded that a 3-h continuation of fasting caused a marked reduction in serum sulphate levels, whereas ingestion of 75 g of glucose in the absence of protein resulted in doubling the reduction, suggesting that fasting and ingestion of protein-free calories may produce periods of chondroitin under-sulphation that could affect osteoarthritis *(44)*.

23.2.2 *Efficacy for Pain and Function*

Glucosamine and chondroitin sulfate had been the subject of numerous clinical trials in Europe and Asia, all of which (until recently) had demonstrated favorable effects *(45–59)*. Also several systematic reviews and meta-analyses have examined the efficacy of glucosamine in the treatment of osteoarthritis *(60–64)*.

McAlindon et al. performed a meta-analysis and quality assessment of 15 eligible double-blind placebo-controlled clinical trials of glucosamine and chondroitin *(60)*. Most of the trials were sponsored by the product manufacturer and reported positive findings. The results highlighted methodologic problems and suggested publication bias. The aggregated effect sizes were 0.44 (95%CI 0.24–0.64) for glucosamine and 0.78 (95%CI 0.60–0.95) for chondroitin. Results were similar when pain was the outcome. As a reference, 0.2 is considered a small effect, 0.5 a moderate effect, and 0.8 a large effect *(65)*.

Towheed et al. also performed a systematic review of randomized controlled trials of glucosamine, which was recently updated *(62)*. The effect size for pain and function was 0.61 (95%CI: 0.28–0.95), consistent with a moderate effect of glucosamine. However, when the analysis was restricted to the higher quality studies, there was no benefit of glucosamine (0.19; 95%CI: –0.11 to 0.50). In the subset of trials that tested the Rotta preparation of glucosamine sulfate ($n = 10$), there was a surprisingly large effect on pain (1.31; 95%CI: 0.64–1.99) *(62)*, which is as large as the effect of a total joint replacement.

Leeb et al. conducted a meta-analysis on the effects of chondroitin sulfate, including seven double-blind, RCTs evaluating a total of 703 patients *(66)*. The estimated pooled effect sizes were 0.9 (95%CI: 0.8–1.0) for pain and 0.74 (95%CI: 0.65–0.85) for functions suggesting that chondroitin sulfate might be useful in osteoarthritis. However, the authors acknowledged major limitations in interpreting the data including small numbers of participants in the eligible trials, the fact that no study was evaluated using intent-to-treat analysis, and the evident publication bias that could lead to a relative error of 30% *(66)*.

Reichenbach et al. recently conducted a meta-analysis of randomized or quasi-randomized trials of chondroitin for osteoarthritis pain *(67)*. Twenty trials with a total of 3,846 patients with hip or knee osteoarthritis were included. There was a high degree of heterogeneity among the trials ($I^2 = 92\%$). Trials with unclear concealment of allocation, or that were not analyzed according to the intention-to-treat principle and small trials, showed larger effects in favor of chondroitin than the other trials. For nine trials, the authors had to use approximations to calculate effect sizes. After pooling the three large trials that had performed an intent-to-treat analysis, including 40% of the original sample, the effect size was –0.03 (95%CI: –0.13 to 0.07; $I^2=0\%$), corresponding to a difference of 0.6 mm on a visual analogue scale (10 cm). A meta-analysis of 12 trials showed a pooled relative risk of 0.99 (95%CI: 0.76–1.31) for any adverse event. The authors noted that trial quality was generally low, and heterogeneity among the trials made initial interpretation of results difficult. The authors concluded that large-scale, methodologically sound trials indicate that the symptomatic benefit of chondroitin is minimal or nonexistent *(67)*.

The body of evidence concerning the efficacy of glucosamine and chondroitin has been altered by the publication of recent independently funded, clinical trials some of which had completely null results *(68–71)*. The first of these enrolled 114 patients with knee osteoarthritis, naïve to glucosamine and chondroitin, and randomized them into a 2-month placebo-controlled trial of 500 mg tid of glucosamine sulfate *(71)*. They found no difference in pain outcomes between the two groups after either 30 days or 60 days of treatment.

Hughes & Carr performed a double-blind placebo-controlled randomized trial of glucosamine sulfate (1.5 g per day) in participants with severe knee OA, 23% having a Kellgren–Lawrence grade of 4 and a mean Western Ontario and McMaster Universities (WOMAC) Osteoarthritis Index score of 9.2 (SD 3.5) *(70)*. No significant between-group differences were found in the primary endpoint (pain in the affected knee, measured using a Visual Analog Scale [VAS]) or any of the secondary pain assessments in this study at 6, 12, and 24 weeks of follow-up.

Cibere et al. performed a glucosamine withdrawal trial in 137 people with knee osteoarthritis who were already using the product with at least moderate benefit *(68)*. The design was a 6-month, randomized, placebo-controlled glucosamine discontinuation trial in which enrollees were randomly assigned to placebo or to the treatment, where participants continued taking glucosamine sulfate. The primary outcome was the proportion of disease flares. Ultimately, disease flares occurred in 42% of the placebo arm and 45% of the glucosamine arm (difference –3%, 95%CI: –19 to 14). In the multivariate regression analysis, time to disease flare was not significantly different between the glucosamine and the placebo group (hazard ratio of flare $= 0.8$; $p = 0.4$). No differences were found in severity of disease flare or other secondary outcomes between placebo and glucosamine patients. Cibere et al. also analyzed samples for type II collagen degradation biomarkers as a proxy for osteoarthritis progression *(72)*. However, they found no statistically significant effect of glucosamine sulfate on type II collagen fragment levels, with the primary outcome being the ratio of C1, C2 epitope/C2C epitope in the urine and serum at baseline, 4, 12, and 24 weeks of follow-up.

Recently, Messier et al. conducted a 12-month randomized clinical trial to determine whether glucosamine hydrochloride (1.5 g) and chondroitin sulfate (1.2 g) is effective, both separately and combined with exercise, compared to placebo plus exercise program in 89 older adults with knee osteoarthritis *(73)*. The primary outcome was WOMAC function and secondary outcomes included WOMAC pain. The authors did not observe a significant difference in function and pain between the groups at 6- or 12-months follow-up *(73)*.

The Glucosamine Unum in Die Efficacy (GUIDE) Trial *(74)*, tested glucosamine sulfate among 318 patients with knee osteoarthritis, randomised to 1,500 mg u.i.d., or acetaminophen 1 g t.i.d., or placebo, for 6 months. The rescue medication was ibuprofen 400 mg. The primary endpoint was the 6-month change in the Lequesne index in the ITT population. The effect of glucosamine sulfate was significant on all parameters, e.g., Lequesne difference -1.2 (95%CI -2.3 to -0.8; WOMAC difference -4.7 (95%CI -9.1 to -0.2), and OARSI-A responders 39.6 vs. 21.2% ($p = 0.007$). Acetaminophen had more responders than placebo, but it failed to reach a significant difference on the Lequesne ($p = 0.18$) and WOMAC ($p = 0.077$) indexes. The authors concluded that glucosamine sulfate at the oral once-daily dose of 1,500 mg is an effective symptomatic medication for knee OA.

The Glucosamine/chondroitin Arthritis Intervention Trial (GAIT), sponsored by the NIH, was the largest comparative trial of these treatments *(75)*. About 1583 participants were randomized to one of the five arms: placebo, celecoxib 200 mg daily, glucosamine hydrochloride 1,500 mg daily, chondroitin sulfate 1,200 mg daily, or the combination of glucosamine hydrochloride and chondroitin sulfate. The primary outcome in this study was treatment response, defined as a 20% improvement in knee pain. The respective response rates were 60.1, 70.1, 64.0, 65.4, and 66.6%. The difference between combination treatment and placebo was reported as near statistically significant ($p = 0.09$). In a subgroup analysis of participants with a higher WOMAC score at baseline, the response rates were 54.3, 69.4, 65.7, 61.4, and 79.2%. In this analysis, the combination treatment was significantly different from placebo ($p = 0.002$). The authors concluded that glucosamine and chondroitin sulfate alone or in combination did not reduce pain effectively in the overall group. The combination therapy may be effective in treating moderate to severe knee pain due to OA. However, the study was limited by an attrition rate of at least 20% in each group, as well as high response rates in the placebo group. The analyses did not incorporate data from all time points, as study participants were also evaluated at 4, 8, 16 and 24 weeks. Also, the subgroup analysis looking at those with higher baseline pain scores appeared to be a post-hoc analysis where the placebo response rate was slightly lower and the combination treatment response rate was slightly higher. Further, the trial tested glucosamine hydrochloride and not glucosamine sulfate. Glucosamine sulfate is widely available as a dietary supplement in the United States and used as a prescription drug in Europe where it was recommended in the recent EULAR practice guidelines for knee OA in Europe *(76)*, since the results would then have provided important information that might have explained in part the study heterogeneity, as noted by Hochberg in an accompanying editorial *(77)*.

However, as noted in a recent literature review by Reginster et al. *(63)* differences in results on the efficacy of glucosamine for symptomatic osteoarthritis originate from differences in products, study design, and study populations. Most of the clinical trials with negative results were performed with glucosamine hydrochloride 500 mg three times daily, whereas most of the trials with positive results were performed with the glucosamine powder for oral solution at the dose of 1,500 mg once daily, raising the question of the importance of sulphate and its contribution to the overall effects of glucosamine, as sulphate in itself may have clinical relevance *(42,43)*. Also, the most clinically relevant results in GAIT were seen when sodium chondroitin sulphate was taken with glucosamine hydrochloride. Indeed, several of the glucosamine preparations contain other salts that could potentially influence uptake and utilization of glucosamine *(78)*. Reginster et al. concluded that glucosamine sulphate has shown positive effects on symptomatic and structural outcomes of knee OA, with the caution that these results should not be extrapolated to other glucosamine salts (e.g., hydrochloride or other preparations) in which no warranty exists about content, pharmacokinetics, and pharmacodynamics *(63)*.

Recently, Vlad et al. conducted a study to identify factors that explain heterogeneity in 15 glucosamine trials for pain in OA selected for analyses *(64)*. The summary effect size was 0.35 (95%CI: 0.14, 0.56), with $I^2 = 0.80$. Except for allocation concealment, no feature of study design explained the substantial heterogeneity. Summary effect sizes ranged from 0.05 to 0.16 in trials without industry involvement, but the range was 0.47–0.55 in trials with industry involvement. The effect size was 0.06 for trials using glucosamine hydrochloride and 0.44 for trials using glucosamine sulfate. Trials using Rottapharm products had an effect size of 0.55, compared with 0.11 for the other trials. The authors concluded that there is sufficient information to conclude that glucosamine hydrochloride lacks efficacy for pain in osteoarthritis. Among glucosamine sulfate trials, enough heterogeneity existed such that no definitive conclusion about efficacy is possible. This heterogeneity appeared to be most prominent among trials with industry involvement (with effect sizes consistently higher). Potential explanations considered by the authors for the effect of industry involvement in trials include bias due to industry involvement, inadequate allocation concealment, and different glucosamine sulfate preparations, as it is possible that the Rottapharm glucosamine sulfate preparation is more efficacious than other products *(64)*. Furthermore, findings supporting this issue were recently discussed by an expert panel *(79)*. In an accompanying editorial, Reginster conducted a meta-analysis including the three Rottapharm-supported pivotal trials of 1.5 g of glucosamine sulfate taken once a day for treatment of knee osteoarthritis evaluating a total of 624 patients, using the WOMAC as an endpoint *(80)*. The pooled effect size was 0.33 (95%CI 0.17–0.49) and there was no heterogeneity according to the I^2 calculation. The results were homogeneous, as all three studies enrolled patients with similar disease characteristics and treatment courses. The effect size was consistent across the parameters. Although the effect size is small to moderate, it is clinically relevant *(6)*.

23.2.3 Efficacy as Disease-Modifying Agents

Two large multicenter studies, sponsored by Rotta Pharmaceuticals, examined whether glucosamine might reduce rate of loss of articular cartilage *(81,82)*. These enrolled ~200 outpatients with primary knee OA into 3-year randomized controlled trials comparing glucosamine sulfate 1.5 g/day with placebo. The primary outcome in each trial was based on joint space measurements obtained from conventional, extended-view, standing antero-posterior knee radiographs, a recommended radiographic approach at that time. Both trials showed quantitatively similar benefits in the glucosamine treatment arms, with respect to the rate of loss of joint space width and symptoms. Unfortunately, the approach that was used to estimate joint space width in these randomized controlled trials has proved to be problematic, even though it was the recommended technique at the inception of the trials *(83)*. Precise measurement of this variable is contingent on highly reproducible radio-anatomic positioning of the joint and may be biased by the presence of pain. If those in the glucosamine group had less pain at their follow-up x-ray, they may have stood with the knee more fully extended, a non-physiologic position that may be associated with the femur riding up on the tibial edge, giving the appearance of a better preserved joint space. What appeared to have been a slower rate of joint space loss may have reflected between-group differences in the degree of knee extension at the follow-up radiograph.

Michel et al. reported the results of a 2-year randomized, double-blind, controlled trial of 800 mg chondroitin sulfate or placebo once daily among 300 patients with knee osteoarthritis *(84)*. The primary outcome was joint space loss over 2 years as assessed by a postero-anterior radiograph of the knee in mild flexion, a better validated technique *(85)*. Secondary outcomes included pain and function. The participants in the placebo arm exhibited significant joint space loss with a mean cumulative joint space loss of 0.14 mm ($\pm$0.61) at 2 years of follow-up compared to no change in the chondroitin arm, 0.00 mm ($\pm$0.53). In the intent-to-treat analysis, the between-group difference in mean joint space loss was 0.14 $\pm$ 0.57 mm; $p = 0.04$. In contrast, the differences in the symptom outcomes between the groups were trivial and non-significant. However, chondroitin was well tolerated, with no significant differences in rates of adverse events between the two groups. While the authors focused on the results providing evidence of structure damage modification by chondroitin, important questions remain about the internal validity of joint space width as a measure of cartilage loss, and its relevance to the clinical state of the patient with knee osteoarthritis – especially in the absence of any overt impact on symptomatic outcomes. Of note, the lack of symptomatic improvement of chondroitin sulfate in this moderate to large intervention trial further highlights the likely over-estimation of effect sizes of symptoms as an outcome reported in the two meta-analyses of this treatment.

Similarly, Reginster et al. recently conducted the multicentre Study on Osteoarthritis Progression Prevention (STOPP) assessing the effect of chondroitin sulfate on the structural progression of knee osteoarthritis, comparing orally administered chondroitin 4&6 sulfate, 800 mg and placebo over 24 months in 622 patients with knee osteoarthritis. The results of this study have been presented in abstract form at the time of this publication *(86)*. The primary outcome was the minimal joint space narrowing measured over 2 years, on digitalized radiographs (Lyon schuss view). Secondary outcomes included pain. The study groups were balanced at baseline with respect to demographic and clinical variables, including osteoarthritis severity. About 30% patients of the chondroitin sulfate group and 26% of the placebo group dropped out. The intention-to-treat analysis showed a mean (SE) joint space narrowing of 0.24 ($\pm$ 0.03) mm at 2 years in the placebo group, which was significantly reduced in the chondroitin sulfate group (0.10 ± 0.03 mm) ($p<0.01$). The per protocol analysis confirmed the results obtained in the ITT analysis. The interaction of time and treatment showed a statistically significant difference in pain VAS and WOMAC ($p<0.01$) in favour of chondroitin sulfate. The authors concluded that chondroitin sulfate significantly reduced progression of joint space narrowing compared with placebo in patients with knee osteoarthritis.

23.2.4 Omega-3 Polyunsaturated Fatty Acids

Omega 3 fatty acids are a family of polyunsaturated fatty acids (PUFAs) that have in common a carbon–carbon double bond in the ω-3 position. The human body cannot synthesize omega-3 fatty acids de novo, but can synthesize all the other necessary PUFAs from the simpler α-linolenic acid. Thus, α-linolenic acid is an essential nutrient which must be obtained from food, and the other PUFAs which can be either synthesized from α-linolenic acid within the body or obtained from dietary intake are sometimes also referred to as essential nutrients. PUFA classification is based on the position of the last double bond along the fatty acid chain. PUFAs are classified as omega-3, omega-6, or omega-9. The main dietary PUFAs are omega-3 (i.e., eicosapentenoic acid and linolenic acid) and omega-6 (i.e., arachidonic acid and linoleic acid). Omega-3 is found in fish oil and canola oil, as well as in flaxseeds, soybean, and walnuts. Omega-6 is found in sunflower oil, soybean, safflower, corn, and meat. These PUFAS are metabolized by cyclooxygenases and lipo-oxygenases into different eicosanoids. The omega-6-derived eicosanoids tend to be proinflammatory, whereas the omega-3-derived eicosanoids tend to be anti-inflammatory.

Omega-3 have a range of potentially favorable effects on chondrocytes, including decreased expression of aggrecanase, cyclo-oxygenase-2, 5-lipoxygenase, 5-lipoxygenase-activating protein, interleukin-1α, tumor necrosis factor-α, and matrix metaloproteinases-3 and -13 *(87–89)*. Overall, these results indicate that omega-3 PUFAs have anticatabolic and anti-inflammatory properties. However, a low omega-6/omega-3 ratio might also be detrimental. A dietary intervention study in rats showed that low intake of omega-6 induced cartilage surface irregularities and localized proteoglycan depletion *(90)*.

The utility of omega-3 for osteoarthritis was tested in a double blind, 24-week placebo-controlled trial of cod liver oil (10 ml containing 786 mg of eicosapentenoic acid) as an adjunct to NSAIDs in 86 patients with OA. Participants were assessed at

4-week intervals for joint pain/inflammation and disability. There was no significant benefit for the patients taking cod liver oil compared with placebo *(91)*. Further studies are needed to determine whether these are of benefit for osteoarthritis.

23.2.5 Avocado/Soybean Unsaponifiables

Avocado and soybean unsaponifiables (ASU) have anabolic, anticatabolic, and anti-inflammatory effects on chondrocytes in vitro. ASUs increase collagen synthesis *(92)*. inhibit collagenase activity *(93,94)*, increase the basal synthesis of aggrecan and reverse the IL1β- induced aggrecan synthesis inhibition *(95)*, decrease the production of matrix metalloproteinase (MMP)-3, IL-6, IL-8, and prostaglandin E2 while weakly reversing the IL1β-induced decrease in TIMP (tissue inhibiting metalloproteinase)-1 production *(92,94) (95)*. ASU decrease the production of nitric oxide (NO) and MIP-1β *(95)* while stimulating the expression of TGF-β and PAI-1 *(96)*. Further, ASUs prevent the osteoarthritic osteoblast-induced inhibition of matrix molecule production, suggesting that ASU may promote OA cartilage repair by acting on subchondral bone osteoblasts *(97)*. Piascledine (Pharmascience, Inc., Montreal, Quebec, Canada) composed of one-third avocado and two-thirds soybean unsaponifiables (ASU) *(94)* is the most frequently investigated lipid combination. In sheep with lateral meniscectomy, 900 mg once a day for 6 months reduced the loss of toluidine blue stain in cartilage and prevented subchondral sclerosis in the inner zone of the lateral tibial plateau but not focal cartilage lesions *(98)*.

Four trials were pooled in a meta-analysis that had positive results *(99)*. ASUs were evaluated on knee and hip osteoarthritis in four double-blind randomized placebo-controlled trials *(99–103)*. In two 3-month randomized controlled studies, one on knee and hip OA *(100)* and one on knee osteoarthritis *(102)*, 300 mg once a day decreased NSAID intake. There was no difference between 300 and 600 mg per day *(102)*. In a 6-month RCT on knee and hip osteoarthritis, 300 mg once a day resulted in an improved Lequesne functional index compared with placebo *(101)*. ASU had a 2-month delayed onset of action as well as residual symptomatic effects 2 months after the end of treatment. In a 2-year RCT on hip osteoarthritis, 300 mg once a day did not have an effect on joint space narrowing *(103)*, or any of the secondary endpoints (pain, function, and PGA). However, a posthoc analysis suggested that ASU might decrease joint space narrowing in patients with severe hip osteoarthritis.

Recently, Ameye and Chee conducted a best-evidence synthesis and concluded that good evidence is provided by ASU for symptom-modifying effects in knee and hip osteoarthritis but there is some evidence of absence of structure-modifying effects *(104)*. ASU seem to have symptom-modifying effects on knee and hip osteoarthritis over the medium term, however their symptom-modifying effects in the long term have not been confirmed.

23.2.6 Selenium and Iodine: Studies of Kashin–Beck Disease

Selenium, an essential micronutrient, is an integral component of iodothyronine deiodinase as well as glutathione peroxidase. Kashin–Beck disease is an osteoarthropathy of children and adolescents, which occurs in geographic areas of China in which deficiencies of both selenium and iodine are endemic. Strong epidemiologic evidence supports the environmental nature of this disease (105). Although the clinical and radiologic characteristics of Kashin–Beck disease differ from osteoarthritis, it raises the possibility that environmental factors also play a role in OA.

Selenium deficiency together with pro-oxidative products of organic matter in drinking water (mainly fulvic acid) and contamination of grain by fungi have been proposed as environmental causes for Kashin–Beck disease. The efficacy of selenium supplementation in preventing the disorder, however, is controversial. Moreno-Reyes et al. studied iodine and selenium metabolism in 11 villages in Tibet in which Kashin–Beck disease was endemic and 1 village in which is was not (106). They found iodine deficiency to be the main determinant of Kashin–Beck disease in these villages. It should be noted, however, in the three groups (1) people with disease in villages with Kashin–Beck disease, (2) people without disease in villages with Kashin–Beck Disease, and (3) people in the control group without Kashin–Beck Disease, all had selenium levels that were very low and those in the latter group had the lowest levels. In an accompanying editorial, Utiger inferred that Kashin–Beck disease probably results from a combination of deficiencies of both of these elements, and speculated that growth plate cartilage is both dependent on locally produced triiodothyronine and sensitive to oxidative damage (105). It should be noted that there is little evidence, if any, to suggest that Kashin–Beck disease has any similarities with adult-onset spontaneous osteoarthritis.

Jordan et al. evaluated the relationship between selenium and knee OA in an observational community-based population, the Johnston County Osteoarthritis Project (107), in which 940 participants submitted toenail clippings for a selenium assessment by Instrumental Neutron Activation Analysis. Radiographic knee OA was scored using a definition of Kellgren–Lawrence grade $\geq$ 2. Mean selenium levels were 0.76 parts per million (ppm) ($\pm$ 0.12). Compared with those in the lowest tertile of selenium, those in the highest tertile had an OR of 0.62 (95%CI: 0.37–1.02) for having prevalent knee OA and an OR of 0.56 (95%CI: 0.31–0.97) for having bilateral knee OA. Based on these findings, low selenium levels appear to be associated with prevalent knee OA, particularly bilateral disease.

While results from this study are provocative, there are several limitations to it. First, although the measurement of selenium via toenail clippings has been used in the past, the duration of exposure to different selenium levels cannot be ascertained using this measurement. Second, given that Kashin–Beck disease was the model from which selenium deficiency was hypothesized to be associated with OA, information on iodine status would have been of interest in this study. Admittedly, the supplementation of iodine in salt within the United States makes it less likely to find people severely deficient in iodine. However, if iodine status was predictive of OA in participants with low selenium levels as has been seen in Kashin–Beck disease, it would be important in enhancing our understanding of the role of selenium in osteoarthritis pathophysiology. Finally, it is possible that selenium concentration could be the surrogate for another unmeasured micronutrient. A randomized controlled trial of selenium supplementation (perhaps in factorial design with iodine supplementation) is needed to evaluate whether it would be effective as a DMOAD.

There is little research evaluating the efficacy of selenium in treating osteoarthritis symptoms. There is one small published clinical trial of supplemental selenium in which Hill and Bird conducted a 6-month double-blind placebo-controlled study of Selenium-ACE, a proprietary nutritional supplement in the United Kingdom, among 30 patients with either primary or secondary hip or knee OA *(108)*. The "active" treatment contained on average 144 µg of selenium as well as 450 µg, 90 mg, and 30 mg of vitamins A, C, and E, respectively. In fact, the "placebo" also contained 2.9 µg of selenium. Pain and stiffness scores remained similar for the two groups at both 3 and 6 months of follow-up. The authors concluded that their data did not support efficacy for selenium-ACE in relieving osteoarthritis symptoms.

It is unlikely that the aforementioned clinical trial will provide any insight regarding the efficacy of selenium in the treatment of symptoms in OA. With just 30 participants in the trial, it is underpowered to detect even a moderate effect of selenium. Even if investigators had found an effect of the active treatment, it would have been impossible to attribute the effects to selenium as the active treatment also contained moderate-high doses of vitamins A, C, and E. A larger randomized placebo-controlled clinical trial evaluating selenium supplementation should be conducted to evaluate its efficacy in the treatment of symptoms related to OA.

23.2.7 *Other Nutritional Products*

There appears to be an increasing number of nutritional remedies being promulgated for purported benefits in arthritis. Trials of S-adenosylmethionine also have had apparently positive results, albeit somewhat limited by adverse effects and high dropout rates *(109–114)*. A ginger-derived product has also been tested in a trial that had moderately positive results *(115)*.

23.3 OXIDATIVE DAMAGE AND OSTEOARTHRITIS

Reactive oxygen species (ROS) are molecules such as superoxide, hydrogen peroxide, and hydroxyl radical that are associated with cell damage. The use of oxygen as part of the process for generating metabolic energy produces reactive oxygen species. ROS are species with unpaired electrons, which are constantly formed in tissues via endogenous and some exogenous mechanisms *(116)*. Normally the oxygen is reduced to produce water; however, in about 1–2% of all cases, the oxygen is reduced to produce the superoxide radical (O_2^-) *(117)*, as a by-product in the electron transport chain *(118)*. Free radicals are also produced in the mitochondria and are implicated in the development or exacerbation of many common human diseases associated with aging *(116,119–121)*, including osteoarthritis *(122)*. Other endogenous sources include release by phagocytes during the oxidative burst, generated by oxidase enzymes and hypoxia-reperfusion events *(123)*. ROS can cause damage to many macromolecules including cell membranes, lipoproteins, proteins, and DNA *(124)*. Chondrocytes are important sources of ROS. Oxidative damage to cartilage is physiologically important *(125–128)*, as superoxide anions can damage collagen structure, depolymerize synovial fluid hyaluronate, and damage mitochondria, *(122,125,127,129,130)*. which may also contribute to the age-related loss of chondrocyte function *(122)*. Evidence of oxidative damage due to over-production of nitric oxide and other ROS has been demonstrated in aging and osteoarthritic cartilage *(131)* and has been correlated with the extent of cartilage damage *(132)*. Subjects with chondral or meniscal lesions also have increased levels of ROS in their synovial fluid *(133)*.

23.3.1 Antioxidant Effects

The human organism has a large multilayered antioxidant defense system *(116)*. Intracellular defense is provided primarily by antioxidant enzymes including peroxidases, catalase, and superoxide dismutase. Hyaluronic acid may also have an antioxidant role in joints *(134)*. Antioxidant micronutrients may act as reducing agents that prevent oxidative reactions often by scavenging ROS before they can damage cells, and thereby have an important function in the extracellular space where antioxidant enzymes are rare *(135)*. Antioxidants include the micronutrients ascorbate (vitamin C), alpha-tocopherol (vitamin E), beta-carotene (a vitamin A precursor), and other carotenoids. The serum concentrations of these antioxidants are primarily determined by dietary intake. The concept that antioxidant micronutrients might provide further defense against tissue injury when intracellular enzymes are overwhelmed, has led to the hypothesis that high dietary intake of these micronutrients might protect against age-related disorders. However, when ROS are produced in increased amounts like in osteoarthritis, the antioxidant capacity of cells and tissues can become insufficient to detoxify the ROS, which then contribute to cartilage degradation by inhibiting matrix synthesis, directly degrading matrix molecules, or by activating matrix metalloproteinases (MMPs) *(136)*. which are zinc-dependent proteases. In such instances, antioxidant micronutrients might provide further defense against tissue injury. Therefore, given their antioxidant properties, vitamins could have beneficial effects in osteoarthritis *(137,138)*. High dietary intake of these micronutrients may protect against age-related disorders. Since higher intake of dietary antioxidants appears beneficial with respect to outcomes such as cataract extraction and coronary artery disease *(119–121,139)* it is possible that they may confer similar benefits for osteoarthritis.

23.3.2 Antioxidant Micronutrients in Osteoarthritis

23.3.2.1 VITAMIN C

Vitamin C, also known as ascorbic acid or ascorbate, is a water-soluble antioxidant vitamin that is found naturally in citrus fruits, rose hips, blackcurrants, and strawberries, as well as in vegetables such as brussel sprouts, broccoli, peppers, cabbage, potatoes, and parsley.

Vitamin C plays several functions in the biosynthesis of cartilage molecules. Vitamin C is required for the post-translational hydroxylation of specific prolyl and lysyl residues in procollagen, through the vitamin C-dependent enzyme lysyl hydroxylase. This modification essential for stabilization of the mature collagen fibril *(119–121, 139–141)*. Vitamin C also appears to stimulate collagen biosynthesis by pathways independent of hydroxylation, perhaps through lipid peroxidation

(142). In addition, vitamin C participates in glycosaminoglycan synthesis by acting as a carrier of sulfate groups *(143)*. Therefore, relative deficiency of vitamin C may impair not only the production of cartilage but also its biomechanical quality.

Recent work on the impact of oxidative stress on cartilage has added insights into the biological mechanisms of osteoarthritis progression. Yudoh et al. studied this from the viewpoint of genomic instability and replicative senescence in human chondrocytes *(132)*. They isolated chondrocytes from articular cartilage from patients with knee osteoarthritis, looking to measure oxidative damage histologically by immunohistochemistry for nitrotyrosine (e.g., a maker of oxidative damage). They then assessed cellular replicative potential, telomere instability, and glycosaminoglycan production both under conditions of oxidative stress and in the presence of an antioxidant (ascorbic acid). Similarly, in the tissue cultures of the articular cartilage explants, they measured the presence of oxidative damage, chondrocyte telomere length, and loss of glycosaminoglycans in the presence or absence of reactive oxygen species, or ascorbic acid.

They found lower antioxidative capacity and stronger staining of nitrotyrosine in osteoarthritic regions compared with normal regions within the same cartilage explants. Oxidative damage correlated with the severity of histological damage. During continuous culture of the chondrocytes, the telomere length, replicative capacity, and glycosaminoglycan production were all decreased in the presence of oxidative stress. In contrast, treatment of cultured chondrocytes with ascorbic acid resulted in greater telomere length and replicative lifespan of the cells. Similarly, in the tissue cultures of the cartilage explants, chondrocyte telomere length and glycosaminoglycan production in the cartilage tissue subjected to oxidative stress were lower in than in the control groups, while those treated with ascorbic acid exhibited a tendency to maintain the chondrocyte telomere length and glycosaminoglycan production. These results suggest that oxidative stress induces chondrocyte telomere instability and catabolic changes in cartilage matrix structure and composition. This process may contribute to the development and/or progression of osteoarthritis.

The results of in vitro and in vivo studies support this hypothesis. Peterkovsky et al. observed decreased synthesis of cartilage collagen and proteoglycan molecules, in guinea pigs deprived of vitamin C. Moreover, addition of ascorbate to tissue cultures of adult bovine chondrocytes resulted in decreased levels of degradative enzymes and increased synthesis of type II collagen and proteoglycans *(143,144)*. Yudoh et al. recently also demonstrated greater telomere length and replicative lifespan of human chondrocytes in culture medium treated with ascorbate *(132)*. Schwartz et al. and Meacock et al. found that vitamin C supplementation in a guinea pig model of surgically induced osteoarthritis reduced the extent of joint damage *(145,146)*.

Epidemiologic data from the Framingham OA cohort also suggest that higher intake of vitamin C may reduce progression of OA *(147)*. In that study, participants had knee X-rays taken at a baseline and at follow-up approximately 8 years later. Knee osteoarthritis was classified using the Kellgren and Lawrence grading system *(148)*. Nutrient intake, including supplement-use, was calculated from dietary habits reported at the mid-point of the study using a food frequency questionnaire.

In the analyses, micronutrient intakes were ranked into sex-specific tertiles and tested to see if higher intakes of vitamin C, vitamin E, and beta-carotene, compared with a panel of non-antioxidant "control" micronutrients, were associated with reduced incidence and reduced progression of knee osteoarthritis. All analyses presented were adjusted for age, sex, body mass index, physical activity, and total energy intake. Six hundred forty participants (mean age 70.3 years) had complete assessments. There were no significant associations with vitamin C and incident radiographic knee osteoarthritis. However, with respect to *progression* of radiographic knee osteoarthritis, there was a threefold reduction in risk for those in the middle and highest tertiles compared to those in the lowest tertile of vitamin C intake (adjusted OR 0.3, 95%CI 0.1–0.6). Those in the highest tertile for vitamin C intake also had reduced risk of developing knee pain (OR 0.3, 95%CI 0.1–0.9). Reduction in risk of progression was also seen for β-carotene (OR 0.4, 95%CI 0.2–0.9) and vitamin E (OR 0.7, 95%CI 0.3–1.6) but these findings were less compelling, in that the beta-carotene association diminished substantially after adjustment for vitamin C, and the vitamin E effect was seen only in men (OR 0.07, 95%CI 0.07–0.6).

Vitamin C is a water-soluble compound with a broad spectrum of antioxidant activity due to its ability to react with numerous aqueous free-radicals and reactive oxygen species *(116)*. The extra-cellular nature of reactive oxygen species-mediated damage in joints, and the aqueous intra-articular environment, may favor a role for a water-soluble agent such as vitamin C, rather than fat-soluble molecules like β-carotene or vitamin E. In addition, it has been suggested that vitamin C may "regenerate" vitamin E at the water–lipid interface by reducing α-tocopherol radical back to α-tocopherol. Whether this occurs in vivo, however, is controversial. An alternative explanation is that the protective effects of vitamin C relate to its biochemical participation in the biosynthesis of cartilage collagen fibrils and proteoglycan molecules, rather than its antioxidant properties. No significant associations were observed for any of the micronutrients among alleged non-antioxidants.

Baker and colleagues investigated the relationship of vitamin C intake (evaluated using a food frequency questionnaire) and knee pain over a 30-month period among 324 (mostly men) participants in the Boston Osteoarthritis of the Knee study, a natural history study of knee osteoarthritis *(149)*. In this cross-sectional analysis, pain score was computed as an average of WOMAC pain scores reported at all visits. Vitamin C status was based on the average vitamin C level from all visits. Individuals with in the lowest tertile of vitamin C intake had more knee pain after adjusting for age, body mass index, and energy intake compared to those in the middle and highest tertiles of vitamin C intake with the relation being stronger in men than in women.

There are numerous reasons to expect that vitamin C might have beneficial effects in osteoarthritis, so the results of a recent study of the effects of ascorbic acid supplementation on the expression of spontaneous osteoarthritis in the Hartley guinea pig are surprising *(150)*. This rigorous investigation tested the effects of three doses of ascorbic acid on the in vivo development of histological knee osteoarthritis. The low dose represented the minimum amount needed to prevent scurvy. The medium dose was the amount present in standard laboratory guinea pig chow and resulted in plasma levels comparable with those achieved in a person consuming five fruits and vegetables daily. The high dose was the amount shown in a previous study of the guinea pig to slow the progression of surgically induced osteoarthritis *(146)*.

A positive association between the ascorbic acid supplementation and the severity of spontaneous osteoarthritis was observed with a higher dose of vitamin C being associated with greater severity of arthritis. A dose-dependent increase in all elements of the knee joint histological scores was seen across the three arms of the study. There was a significant correlation of histologic severity score with plasma ascorbate concentration ($r = 0.38$, $p = 0.01$). Of note, there was evidence of active TGF in the guinea pigs in this study, predominantly expressed in marginal osteophytes, whereas little was seen in the extracellular matrix of the articular cartilage, remote from osteophytes *(150)*. TGF has been implicated in the pathophysiology of osteoarthritis *(151–153)*, and ascorbate may function on an activator of this cytokine *(154)*. The presence of TGF in osteophytes supports a role of this cytokine in the effects of vitamin C on histologic severity.

While these findings are provocative, it remains uncertain to what extent they can be generalized to humans. The model of spontaneous osteoarthritis used in this study may not reflect the same pathology as osteoarthritis in humans. Further, it is difficult to extrapolate that the concentrations of vitamin C considered pathologic in guinea pigs are also pathologic in humans. Nonetheless, it is paradoxical that an apparently beneficial effect of dietary vitamin C was found in the Framingham cohort study *(147)* and in the Boston Osteoarthritis of the Knee study. Thus, the current knowledge predicates a need for further studies of vitamin C in humans.

One multicenter, randomized, double-blind, placebo-controlled case-crossover study of 133 patients with radiographically diagnosed hip and/or knee OA evaluated the effectiveness of 1 g of oral calcium ascorbate *(155)*. Each participant received vitamin C for 14 days and placebo for 14 days, separated by a 7-day washout period. The participants were randomized to the sequence of administration of vitamin C and placebo. The primary outcome was pain on VAS in a preselected joint. Using intent-to-treat analysis, treatment with vitamin C resulted in a greater improvement in pain compared with placebo with a mean difference of 4.6 mm ($p = 0.0078$).

These results are exciting and support the possibility that vitamin C is effective in improving symptoms related to knee and/or hip osteoarthritis. However, there are several limitations to this study preventing vitamin C from being hailed as a treatment for symptomatic osteoarthritis. First, the dosage of vitamin C used in this trial was more than 10 times that of the Recommended Dietary Allowances of 60–200 mg per day, though it has been reported that oral doses up to 3 gm daily are unlikely to cause adverse reactions. The long-term safety of such high doses of vitamin C in older patients with osteoarthritis needs further evaluation, and efficacy needs to be confirmed by longer studies. Also, this trial was relatively small and it included participants with knee and/or hip OA, likely a heterogenous population. Finally, it is unclear whether a mean difference of 4.6 mm on VAS is a meaningful difference, though vitamin C's duration of action is unclear and treatments with a prolonged duration of action potentially will have apparently diluted out effects in studies with a case-crossover design, biasing the results toward the null. Thus, the presence of even a small effect of vitamin C may still be very meaningful.

23.3.2.2 VITAMIN D

Vitamin D, also known as calciferol, is a broad term inclusive of a collection of steroid-like substances such as vitamin D_2 (ergocalciferol) and vitamin D_3 (cholecalciferol). Vitamin D is only found in animal sources and can be produced by the body with exposure to ultraviolet radiation.

Normal bone metabolism requires the presence of vitamin D. Suboptimal vitamin D levels may have adverse effects on calcium metabolism, osteoblast activity, matrix ossification, and bone density (156,157). Hence, low tissue levels of vitamin D may impair the ability of bone to respond optimally to pathophysiological processes in osteoarthritis and predispose to disease progression.

Reactive changes in the bone underlying, and adjacent to, damaged cartilage are an integral part of the osteoarthritic process (158–164). Sclerosis of the underlying bone, trabecular micro-fracturing, attrition, and cyst formation are all likely to accelerate the degenerative process as a result of adverse biomechanical changes (165,166). Other phenomena, such as osteophyte (bony spur) formation may be attempts to repair or stabilize the process (167,168). It has also been suggested that bone mineral density may influence the skeletal expression of the disease with a more erosive form occurring in individuals with "softer" bone (169). Although some cross-sectional studies have suggested a modest inverse relationship between presence of osteoarthritis and osteoporosis, recent prospective studies have suggested that individuals with lower bone mineral density are at increased risk for osteoarthritis progression (170). The idea that the nature of bony response in osteoarthritis may determine outcome has been further advanced by the demonstration that patients with bone scan abnormalities adjacent to an osteoarthritic knee have a higher rate of progression than those without such changes (171).

Animal studies suggest that vitamin D might also have direct effects on chondrocytes in osteoarthritic cartilage. They suggest that vitamin D might exert an effect on the development or progression of OA through cartilage as well as bone. Although these findings emanate from animal studies, they serve as preliminary data that these relationships may also exist in humans. During bone growth, vitamin D regulates the transition in the growth plate from cartilage to bone. It had been assumed that chondrocytes in developing bone lose their vitamin D receptors with the attainment of skeletal maturity. Corvol et al., however, found that chondrocytes isolated from mature rabbit growth plate cartilage were able to transform 25-hydroxycholecalciferol to 24,25-dihydroxy-cholecalciferol (172,173). They also observed that 24,25-dihydroxy-cholecalciferol could stimulate proteoglycan synthesis by mature chondrocytes and that it increased DNA polymerase activity in chondrocytes during cell division. Additionally, they demonstrated the presence of nuclear receptors for 24,25-dihydroxy-cholecalciferol in chondrocytes (172).

Tetlow & Woolley were able to demonstrate a regional association of vitamin D receptor expression with matrix metalloproteinase (MMP) expression in osteoarthritic human chondrocytes, a phenomenon virtually absent in normal cartilage *(174)*. In further analyses using chondrocyte culture systems, they found that $1\text{-}25(OH)_2D_3$ could up-regulate expression of matrix MMP-3, yet suppress production of MMP-9 and prostaglandin E_2. Thus, in vitro vitamin D has both enhancing and suppressive roles in the regulation of chondrocyte products. Since these could have differential effects on cartilage, and the net overall effect is unknown, Tetlow et al. (broadly) concluded that the disparate modulatory effects of $1\text{-}25(OH)_2D_3$ may be of relevance to the chondrolytic processes that occur in OA and that further research is needed.

In addition, vitamin D deficiency could also importantly affect other elements of disease in osteoarthritis, including pain and muscle weakness. Bischoff et al. investigated the in situ expression of 1,25-dihydroxyvitamin D_3 receptor in human skeletal muscle tissue *(175)*. Intraoperative periarticular muscle biopsies were obtained from 20 female patients receiving total hip arthroplasty due to osteoarthritis of the hip or an osteoporotic hip fracture or back surgery. The immunohistological distribution of the vitamin D_3 receptor was investigated using a monoclonal rat antibody to the receptor. The receptor-positive nuclei were quantified by counting 500 nuclei per biopsy. Strong intranuclear immunostaining of the vitamin D receptor was detected in human muscle cells. Biopsies of hip patients had significantly fewer receptor-positive nuclei compared to those of back surgery patients ($p = 0.002$). VDR expression was significantly correlated with age (correlation coefficient = 0.46; $p = 0.005$), but not with vitamin D levels. The data demonstrated presence of nuclear 1,25-dihydroxyvitamin D_3 receptor in human skeletal muscle.

In a subsequent study, Bischoff et al. studied whether VDR expression in vivo is related to age or vitamin D status, or whether VDR expression differs between skeletal muscle groups *(176)*. They investigated these factors and their relation to 1,25-dihydroxyvitamin D receptor (VDR) expression in muscle. Intranuclear immunostaining of the VDR was present in muscle biopsy specimens of all orthopedic patients undergoing total hip arthroplasty or spinal surgery. Older age was significantly associated with decreased VDR expression ($\beta = -2.56$; $p = 0.047$), independent of biopsy location and serum 25-hydroxyvitamin D levels.

In addition, higher BMD *(170,177)* has been found to be associated with a decrease in disease progression in persons with knee osteoarthritis. Zhang et al., in a prospective analysis found that high BMD as well as BMD gain decreased the risk of progression of radiographic knee osteoarthritis in the Framingham cohort *(170)*. Furthermore, a significant positive association between serum 25(OH)D and BMD in individuals with primary knee osteoarthritis was observed in the Framingham study, independent of sex, age, BMI, knee pain, physical activity, and disease severity *(178)*, suggesting that vitamin D supplementation may enhance BMD in persons with OA.

McAlindon et al. tested the association of vitamin D status on the incidence and progression of knee OA among the Framingham OA Cohort Study participants *(179)*. This study included both a dietary assessment and a serum assay of 25(OH)D. Dietary intake of vitamin D and serum 25(OH)D levels was unrelated to osteoarthritis incidence. Risk of progression over 8 years, however, was three- to

–fourfold higher for participants in the middle and lower tertiles of both vitamin D intake (OR for lowest vs. highest tertile 4.0, 95%CI: 1.4–11.6) and serum level (OR 2.9, 95%CI: 1.0–8.2). Low serum vitamin D level also predicted cartilage loss, assessed by loss of joint space (OR 2.3, 95%CI: 0.9–5.5) and osteophyte growth (OR 3.1, 95%CI: 1.3–7.5).

Lane et al. tested the relationship of serum 25- and 1,25-$(OH)_2$ D with the development of radiographic hip OA among older Caucasian women participating in the Study of Osteoporotic Fractures (180). They measured serum vitamin D levels in 237 subjects randomly selected from 6,051 women who had pelvic radiographs taken at both the baseline examinations and after 8 years of follow-up. They analyzed the association of vitamin D levels with the occurrence of joint space narrowing and with the development of osteophytes, and with changes in the mean joint space width and individual radiographic feature scores during the study period. The risk of incident hip OA defined as the development of definite joint space narrowing was increased for subjects who were in the middle (OR 3.21; 95%CI 1.06–9.68) and lowest (OR 3.34; 95%CI 1.13–9.86) tertiles for vitamin D compared with those in the highest tertile. Vitamin D levels were not associated with incident hip OA defined as the development of definite osteophytes or new disease. No association between serum vitamin D and changes in radiographic hip OA was found.

Recently, Felson et al. evaluated the relationship between vitamin D status and cartilage loss in osteoarthritis (181). They measured 25(OH)D levels by radioimmunoassay in participants from two longitudinal cohort studies, the Framingham Osteoarthritis Study and the Boston Osteoarthritis of the Knee Study (BOKS). The Framingham Osteoarthritis study included individuals without knee OA (87% of knees had a K/L score ≤1), while participants in BOKS had symptomatic knee osteoarthritis at baseline (21% of knees had a K/L score ≤1). In both studies, worsening was defined by radiographic tibiofemoral joint space loss. In addition, the BOKS also obtained knee MRIs at baseline and at 15 and 30 months, of which 26% had a K/L score ≤1 at baseline. The mean vitamin D level was 20 ng/ml at baseline in both studies, and about 20% of knees exhibited joint space loss during the observation periods. The investigators found no association of baseline vitamin D levels with radiographic worsening or cartilage loss measured on MRI. In the BOKS, 57% of knees with MRI showed progressive cartilage loss on follow-up MRI (≥1 cartilage plate). In this cohort, the risk of cartilage loss was lower for subjects with vitamin D deficiency (25(OH)D <20 ng/ml) compared with those with sufficient vitamin D levels (25(OH)D ≥20 ng/ml), adjusting for age, sex, BMI, and baseline cartilage score (OR 0.74, 95% CI 0.50, 1.09). However, the analyses did not distinguish between incidence of radiographic OA and progression. Also, as noted by the authors, MRIs were not acquired in a way that permits evaluation of change in cartilage volume.

Carbone et al. tested the relationship of antiresorptive drug use to structural findings and symptoms of knee osteoarthritis (182) in the Women in the Health, Aging and Body Composition Study. They found associations of use of alendronate and/or estrogen with lower structural lesion and lower pain scores (182). However, as pointed out by DeMarco (183), the original report did not account for potential influence of vitamin D on these associations. Carbone at al, therefore, re-analyzed their results to adjust for a possible effect of vitamin D supplement use *(184)*. In this study, 16% of participants used vitamin D supplements. Vitamin D supplements use was not associated with structural changes of osteoarthritis or pain severity, nor did its inclusion as a covariate in the statistical models change the formerly observed associations.

Thus, the epidemiological evidence on the relationship of vitamin D to OA, and OA progression, is conflicting. Clearly, it is a public health priority to establish whether vitamin D has disease-modifying properties for OA. Therefore, we are currently involved in conducting a randomized double-blind placebo-controlled trial evaluating longitudinal structural progression and symptom effects of vitamin D supplementation on osteoarthritis.

23.3.2.3 VITAMIN E

Vitamin E comprises eight fat-soluble compounds, tocopherols (derivatives of tocol), α-, β-, γ-, and δ-tocopherol and α-, β-, γ-, and δ-tocotrienol, produced solely by plants. Some of the richest sources of vitamin E include vegetable and nut oils, safflower, nuts, sunflower seeds, and whole grains. The most common and biologically active form is alpha-tocopherol (5,7,8 tri-methyltocol). Synthetic α-tocopherols and their esters also exist. α-Tocopherylacetate is often used commercially because vitamin E esterification protects it from oxidation. In the human body, the ester is rapidly cleaved by cellular esterases making natural vitamin E available.

Vitamin E has diverse influences on the metabolism of arachadonic acid, a pro-inflammatory fatty acid found in all cell membranes. Vitamin E blocks formation of arachidonic acid from phospholipids and inhibits lipoxygenase activity, without having much effect on cyclooxygenase *(185)*. It is, therefore, possible that vitamin E reduces the modest synovial inflammation that may accompany osteoarthritis.

In vitro effects of vitamin E on chondrocytes have been investigated. Tiku et al. showed that vitamin E reduced the catabolism of collagen by preventing the protein oxidation mediated by aldehydic down products of lipid peroxidation when chondrocytes were submitted to an oxidative burst *(186)*. Vitamin E strongly increased the sulfate incorporation while slightly reducing the glucosamine incorporation *(187)*, suggesting that it increased glycosaminoglycan (GAG) sulfation or that it increased GAG synthesis while reducing glycoprotein or glycolipid synthesis. Like vitamin C, vitamin E affected the activities of lysosomal enzymes: it decreased the activities of arylsulfatase A and acid phosphatase in cultures of human articular chondrocytes *(187)*. However, vitamin E did not affect the LPS-induced catabolism of GAGs *(188)* and did not prevent synoviocyte apoptosis induced by superoxide anions *(189)*.

Previous studies suggest that vitamin E may enhance chondrocyte growth via protection against reactive oxygen species and ultimately modulate the development of osteoarthritis *(128,190)*. In the Framingham OA Cohort, men with higher vitamin E levels were less likely to have knee OA progression compared with those with lower levels *(147)*.

Benefit from vitamin E therapy has been suggested by several small human studies of osteoarthritis *(191–194)*. In a 6 week double-blind placebo-controlled trial of 400 mg alpha-tocopherol (vitamin E) in 56 patients with osteoarthritis *(195)*, vitamin E-treated participants experienced greater improvement in every efficacy measure including pain at rest (69% better in vitamin E vs. 34% better in placebo, $p<0.05$), pain on movement (62% better on vitamin E vs. 27% on placebo, $p<0.01$), and use of analgesics (52% less on vitamin E; 24% less on placebo, $p<0.01$). The rapid response in symptoms observed in this study suggests that vitamin E does not exert a structural effect in osteoarthritis; instead, perhaps the beneficial effect results from some metabolic action such as inhibition of arachidonic acid metabolism.

Two trials concluded that vitamin E was more efficient than placebo in decreasing pain. In a small 10-day crossover trial on spondylosis, 600 mg of vitamin E per day was superior to placebo as assessed by a patient questionnaire *(196)*. One trial suggested that vitamin E was no less efficient than diclofenac, a nonsteroidal anti-inflammatory drug, in decreasing pain. In a 3-week randomized controlled trial, no significant difference was found between 544 mg of α-tocopherylacetate three times a day and 50 mg diclofenac three times a day on VAS of pain *(197)*. However, the two most recent trials failed to show any benefit over placebo on knee osteoarthritis. Vitamin E (500 IU of per day) showed no symptomatic benefit over placebo as assessed by WOMAC in a 6-month randomized controlled trial *(198)*.

Results from a 2-year double-blind, placebo-controlled trial among 136 patients with knee osteoarthritis do not support a chondroprotective effect of vitamin E. Wluka et al. tested whether vitamin E (500 IU) affects cartilage volume loss in patients with knee osteoarthritis *(199)*. The primary outcome was change in tibial cartilage volume from baseline to 2 years follow-up measured by magnetic resonance imaging. Secondary outcomes included pain, stiffness, function, and total WOMAC scores as well as the SF-36. One hundred seventeen subjects completed the study for a loss to follow-up rate of 14%. Loss of medial and lateral tibial cartilage was similar in subjects treated with vitamin E and placebo (e.g., mean loss in the medial compartment 157 vs. 187 μm^3, $p = 0.5$). There were no significant differences between the vitamin E and the placebo-treated groups in improvement of symptoms from baseline. The authors concluded that vitamin E does not appear to benefit cartilage volume loss in osteoarthritis.

However, there are limitations that should be considered in the interpretation of these results. First, this study was powered to detect a 50% reduction in the rate of cartilage loss in the treatment arm. This effect size likely was an over-estimate of any effect that could have been expected from vitamin E over a 2-year follow-up period. Second, the structural outcome measure evaluated in this study was cartilage volume assessed on MRI. This is problematic since cartilage volume uncorrected for surface area lacks construct validity *(200)*. Further, cartilage volume has not been tested for sensitivity to change, thus it is unclear whether a real change in cartilage volume within a given individual can be distinguished from measurement error. Cartilage volume needs to be comprehensively validated and evaluated for reliability before it should replace joint space narrowing on plain radiograph as the structural outcome measure recommended in OA clinical trials *(83)*. In this study, cartilage volume was the only structural outcome measured.

A recent systematic review observed that although three out of the five RCTs concluded that vitamin E decreased pain, the two longest, largest, and highest-quality trials failed to detect any symptomatic or structural effects in knee OA, suggesting that, at least for knee OA, vitamin E alone has no medium-term beneficial effect. According to the best-evidence synthesis, the authors concluded that there is no evidence of symptom-modifying efficacy for vitamin E and some evidence of inefficacy regarding structure-modifying effects *(104)*.

23.3.2.4 VITAMIN K

The primary form of vitamin K, a fat-soluble vitamin, in the diet is phylloquinone (vitamin K_1), which is concentrated in dark green leafy vegetables and vegetable oils. Although there is some endogenous production of vitamin K, a subclinical deficiency can develop by limiting dietary intake of phylloquinone. Low dietary intake of vitamin K is common, and studies evaluating biochemical measures of vitamin K status suggest that inadequate intake of vitamin K is widespread among adults in the United States and the United Kingdom *(201,202)*.

Although it is not known to have antioxidant effects, vitamin K does have bone and cartilage effects, which may be relevant for osteoarthritis. Post-translational γ-carboxylation of glutamic acid residues to form γ-carboxyglutamic acid (Gla) residues confers functionality to these "Gla" proteins *(203)*. Vitamin K is an essential co-factor for this process *(203)*. Multiple coagulation, bone, and cartilage proteins are dependent upon vitamin K because the Gla residues are required for these proteins to function appropriately. Bone and cartilage Gla proteins include growth arrest-specific protein 6 (Gas-6), and the skeletally expressed extracellular matrix proteins, osteocalcin, and matrix Gla protein (MGP) *(203–206)*.

The vitamin K-dependent γ-carboxylation of these bone and cartilage proteins is important for their normal functioning. Growth arrest-specific protein 6 (Gas-6), through its interactions with the axl tyrosine kinase receptor, prevents chondrocyte apoptosis and is involved in chondrocyte growth and development *(204)*. Low levels of vitamin K could lead to inadequate levels of functional Gas-6, contributing to increased chondrocyte apoptosis and attendant mineralization. Another Gla protein is osteocalcin, the most abundant noncollagenous protein in bone, and a potent inhibitor of hydroxyapatite mineralization. Matrix Gla protein (MGP), a protein which plays a role in chondrocyte development and maturation, is associated with mineralization in hypertrophic chondrocytes and endochondral ossification, the same process through which ostephytes form *(207,208)*. Also, MGP may inhibit mineralization via its interaction with bone morphogenetic protein-2 (BMP-2). BMP-2 is a known inducer of chondrocyte and osteoblast differentiation that signals through Smad1, together enhancing bone formation. Interference by MGP leads to diminished bone-forming capacity *(209)*, and conversely, under-carboxylated MGP could lead to increased bone-forming capacity.

Beyond being a necessary co-factor for γ-carboxylation, vitamin K compounds also exhibit anti-inflammatory properties, reducing prostaglandin E2 and interleukin-6 production and inhibiting interleukin-1 and prostaglandin E2-mediated bone resorption *(210,211)*.

The effects of inadequately functioning vitamin K-dependent proteins has been seen in warfarin (vitamin K antagonist) embryopathy, Keutel syndrome (a genetic disorder where MGP is deficient), and an MGP knock-out mouse model, all of which exhibit growth plate cartilage abnormalities *(212–214)*. These abnormalities may reflect a process similar to osteophyte formation since both cartilage plate abnormalities and osteophyte formation involve endochondral ossification. Thus, vitamin K is an important regulator of bone and cartilage mineralization and function and may play a role in osteoarthritis.

Neogi et al. investigated the potential association between vitamin K and osteoarthritis in the Framingham Osteoarthritis cohort. In their first assessment they examined the relationship between dietary vitamin K intake (evaluated using a food frequency questionnaire) and radiographic evidence of osteophytes *(215)*. They demonstrated an association between higher vitamin K intake and lower osteophyte prevalence, but the association was not significant with prevalence ratios of osteophytes from lowest to highest vitamin K intake quartiles of 1.0 (reference), 1.1, 0.8, and 0.9 (p for trend = 0.2). In a follow-up study, Neogi et al. subsequently measured vitamin K levels using plasma vitamin levels *(216)*. In this study, they showed an association between plasma phylloquinone and severity of radiographic osteoarthritis, particularly of osteophytes, in the hand and knee, after adjusting for age, sex, body mass index, femoral neck bone mineral density, total energy intake, and plasma vitamin D. The prevalence of hand and knee osteophytes in those in the highest plasma phylloquinone quartile was 40% lower than in those in the lowest quartile. No significant associations were noted for control nutrients, vitamins B1, and B2, suggesting that a healthy lifestyle does not account for these results.

If a relationship between vitamin K and osteophytes does exist, the public health benefits could potentially be enormous. However, based on these two observational studies, it is unclear whether there is an association between vitamin K and osteoarthritis. It seems reasonable to expect that plasma levels of micronutrients are more accurate measure compared with dietary intake measures, lending more credibility to the latter study supporting an association between vitamin K and osteophytes.

The data is suggestive that vitamin K deficiency is associated with features of osteoarthritis and osteoarthritis severity, thus vitamin K supplementation has the potential of being classified as a modifying osteoarthritis drug. To address this possibility, Neogi and others are currently involved in conducting a randomized double-blind placebo-controlled trial evaluating longitudinal structural and symptom effects of vitamin K supplementation on osteoarthritis.

23.4 WEIGHT LOSS

Epidemiologic data indicate that being overweight considerably increases the risk for the development of knee osteoarthritis and may also increase the susceptibility to both hip and hand OA *(217)*. Since overweight individuals do not necessarily have increased load across their hand joints, investigators have wondered whether systemic factors, such as dietary factors or other metabolic consequences of obesity, may mediate part of this relationship. Early laboratory studies in animals suggested that there is an interaction between body weight, genetic factors and diet, although attempts to demonstrate a direct effect of dietary fat intake have proved inconclusive *(218,219)*. The fact that adipose cells share a common stem cell precursor with connective tissue cells such as osteoblasts and chondrocytes has prompted investigation into the possibility that their phenotypic differentiation might be influenced by the metabolic milieu *(220)*. Indeed, fat and fatty acids can influence prostaglandin and collagen synthesis in vitro and have been associated with osteoarthritic changes in joints *(220,221)*. Preliminary evidence also suggests that leptin, an adipose tissue-derived hormone, may have anabolic effects in osteoarthritic cartilage *(222)*.

Based on these observations, weight loss is considered a priority in the management of overweight individuals with osteoarthritis. However, there have been relatively few rigorous studies testing weight loss as a therapeutic intervention to reduce symptoms, prevent disability, or delay disease progression. The Arthritis, Diet, and Activity Promotion Trial (ADAPT) examined whether long-term exercise and dietary weight loss are effective interventions for functional impairment, pain, and mobility in 316 older overweight individuals with knee osteoarthritis *(223)*. The results suggest that diet-induced weight loss and exercise-induced weight loss are independently effective but that the combination of the two is additive and more effective than either alone. Further, only the combination treatment consistently showed a significant effect.

In contrast, a more recent study by Christensen et al. *(224)* differs with the Arthritis, Diet, and Activity Promotion Trial both in design and in conclusions. On the basis that weight loss might relieve knee osteoarthritis symptoms both through biomechanical effects and influences on body fat *(225)*, they tested the effectiveness of a rapid diet-induced weight-loss intervention on overweight individuals with knee osteoarthritis, enrolling 96 persons (mostly women) with knee osteoarthritis into a comparison of a low-energy diet (LED) intervention (3.4 MJ/day ~800 kcal/day) with a control diet (5 MJ/day ~1,200 kcal/day). The low-energy diet intervention consisted of a nutrition powder taken as six daily meals that met the recommendations for a daily intake of high-quality protein such that 37% of the energy provided from the powder was from soy protein. The control intervention (i.e. "hypo-energetic diet") consisted of a traditional low-calorie high-protein diet taken in the form of ordinary foods individually chosen by participants based on recommendations from a 2-h nutritional advice session. The LED group also had weekly dietary sessions, whereas the control group was given a booklet describing weight-loss practices. The primary outcome was self-reported pain and physical function limitation measured by the WOMAC index. Changes in body weight and body composition as independent predictors of changes in knee osteoarthritis symptoms were also examined.

There were nine dropouts, mainly due to noncompliance. However, this appeared to be non-differential, so the authors performed an analysis based on completers. The LED group lost considerably more weight than the controls (11.1 vs. 4.3%) with a mean difference of 6.8% (95% CI 5.5–8.1%). The LED group also lost 2.2% more body fat (95% CI 1.5–3.0%). There were substantially greater reductions in WOMAC scores among the low-energy intervention group. The mean between-group difference for the total WOMAC index was 219.3 mm ($p = 0.005$). Oddly, this was not reflected in the Lequesne Index assessment, which detected no between-group difference. In subsidiary analyses they estimated that the "Number Needed to Treat" to obtain an improvement in WOMAC score of 50% or greater in at least one patient was 3.4. They also found that the changes in WOMAC score were best predicted by reduction of body fat, with a 9.4% significant improvement in WOMAC score for each percent of body fat reduced.

These results indicate that rapid and substantial weight loss may, by itself, translate into reduced pain and improved function in overweight patients with knee osteoarthritis. However, some caution needs to be exerted in interpreting these results. The long-term effectiveness of this short-term intervention is uncertain. The participants had obesity, and the results may not be generalizable to a less overweight population. Although the authors assert that the groups were balanced, the effect of censoring from the analysis the participants who discontinued the intervention is uncertain. The higher WOMAC scores at baseline in the LED group compared to the control group provides further evidence that the groups were not balanced at baseline. This difference makes it difficult to attribute the differences seen in the two arms at follow-up to the effect of either intervention. The greater effect of LED as measured by a greater change in WOMAC may have resulted from a stronger tendency for regression to the mean in the LED group. Indeed, the Lequesne Index was equal in both groups at baseline and this measure was not different in the two groups at 8 weeks of follow-up. Further, the LED group preferentially received more attention with weekly sessions for 8 weeks with the dietitian to encourage a high degree of compliance, whereas the hypo-energetic group only met with the dietitian for once at the beginning of the study, a difference that was not controlled for in their analyses. Finally, the study was essentially unblinded, which may also have led to between-group biases. Nevertheless, the results are interesting and underscore a need for further research into potential benefits from more extreme weight reduction interventions. For instance, preliminary results from a study of musculoskeletal complaints among morbidly obese patients undertaking gastric bypass surgery showed a 52% reduction in the number of symptomatic sites, and an approximately 50% reduction in WOMAC score, 6–12 months following the procedure *(226)*.

A meta-analysis of randomized controlled trials (RCTs) on changes in pain and function after weight loss among overweight participants with knee osteoarthritis was recently conducted *(227)*. Four RCTs including 454 participants met the inclusion criteria. Pooled effect sizes for pain and physical disability were 0.20 (95% CI 0–0.39) and 0.23 (0.04–0.42) at a weight reduction of 6.1 kg (4.7–7.6 kg). Meta-regression analysis showed that disability could be significantly improved when weight was reduced over 5.1% or at the rate of >0.24% reduction per week. Clinical efficacy on pain reduction was present, although not predictable after weight loss. Meta-regression analysis indicated that physical disability of patients with knee OA and overweight diminished after a moderate weight reduction regime. The analysis suggests that a weight loss greater than 5% should be achieved within a 20-week period (i.e., 0.25% per week).

23.5 RECOMMENDATIONS

1. Achieve or maintain ideal body weight. Even modest weight loss in overweight individuals with knee OA may reduce the risk of incident or progressive disease.
2. Diet and exercise interventions for osteoarthritis symptoms suggest that treatment with the combination is more effective than either intervention alone.
3. Encourage a generous intake of fruits and vegetables, especially those rich in vitamin C.
4. Optimize vitamin D status. For individuals not at risk of vitamin D deficiency, supplements of 600–800 IU/d are recommended. In the event that either deficiency or toxicity is suspected, measurement of serum 25-hydroxyvitamin D is appropriate. Healthy persons have levels of 25–40 ng/mL.
5. Clinical trials of glucosamine and chondroitin products suggest efficacy in treating OA symptoms. However, the question of efficacy of these treatments with respect to structural progression still remains. Definitive results are anticipated from an ongoing NIH-funded study evaluating the efficacy of these treatments.

REFERENCES

1. Gabriel SE, Crowson CS, O'Fallon WM. Costs of osteoarthritis: estimates from a geographically defined population. J Rheumatol Suppl 1995; 43:23–5.
2. Gabriel SE, Crowson CS, Campion ME, O'Fallon WM. Indirect and nonmedical costs among people with rheumatoid arthritis and osteoarthritis compared with nonarthritic controls. J Rheumatol 1997; 24(1):43–8.
3. Gabriel SE, Crowson CS, Campion ME, O'Fallon WM. Direct medical costs unique to people with arthritis. J Rheumatol 1997; 24(4):719–25.
4. Yelin E. The Economics of Osteoarthritis. New York: Oxford University Press, 1998.
5. Creamer P, Hochberg MC. Osteoarthritis. Lancet 1997; 350(9076):503–8.
6. Bjordal JM, Ljunggren AE, Klovning A, Slordal L. Non-steroidal anti-inflammatory drugs, including cyclo-oxygenase-2 inhibitors, in osteoarthritic knee pain: meta-analysis of randomised placebo controlled trials. BMJ 2004; 329(7478):1317.
7. Theodosakis J, Adderly B, Fox B. The Arthritis Cure. New York: St. Martin's Press, 1997.
8. Anonymous. U.S. nutrition industry: top 70 supplements 1997–2001. Nutrition Business Journal 2001:Chart 14.
9. Marra J. The state of dietary suplements – even slight increases in growth are better than no growth at all. Nutraceuticals World 2002:32–40.
10. Roseman S. Reflections on glycobiology. J Biol Chem 2001; 276(45):41527–42.
11. Ghosh S, Blumenthal HJ, Davidson E, Roseman S. Glucosamine metabolism. V. Enzymatic synthesis of glucosamine 6-phosphate. J Biol Chem 1960; 235:1265–73.
12. Setnikar I, Ralumbo R, Canali S, Zanolo G. Pharmacokinetics of glucosamine in man. Drug Res 1993; 43:1109–13.
13. Setnikar I, Giachetti C, Zanolo G. Absorption, distribution and excretion of radio-activity after a single I.V. or oral administration of [14C]glucosamine to the rat. Pharmatherapeutica 1984; 3:358.
14. Adebowale A, Du J, Liang Z, Leslie JL, Eddington ND. The bioavailability and pharmacokinetics of glucosamine hydrochloride and low molecular weight chondroitin sulfate after single and multiple doses to beagle dogs. Biopharm Drug Dispos 2002; 23(6):217–25.
15. Aghazadeh-Habashi A, Sattari S, Pasutto F, Jamali F. Single dose pharmacokinetics and bioavailability of glucosamine in the rat. J Pharm Pharm Sci 2002; 5(2):181–4.
16. Biggee BA, Blinn C, McAlindon T, Nuite M, Silbert J. Human serum glucosamine and sulfate levels after ingestion of glucosamine sulfate. Arthritis Rheum 2004; 50(9 Suppl):S657.

17. Persiani S, Rovati L, Foschini V, Giacovelli G, Locatelli M, Roda A. Oral bioavailability and dose-proportionality of crystalline glucosamine sulfate in man. Arthritis Rheum 2004; 50(9 Suppl):S146.
18. Academies IoMNRCotN. Glucosamine: Prototype monograph summary. In: Dietary Supplements: A Framework for Evaluating Safety. Washington DC: National Academies Press, 2005, pp. 363–4.
19. Marshall S, Yamasaki K, Okuyama R. Glucosamine induces rapid desensitization of glucose transport in isolated adipocytes by increasing GlcN-6-P levels. Biochem Biophys Res Commun 2005; 329(3):1155–61.
20. Buse MG. Hexosamines, insulin resistance, and the complications of diabetes: current status. Am J Physiol Endocrinol Metab 2006; 290(1):E1–E8.
21. Scroggie DA, Albright A, Harris MD. The effect of glucosamine-chondroitin supplementation on glycosylated hemoglobin levels in patients with type 2 diabetes mellitus: a placebo-controlled, double-blinded, randomized clinical trial. Arch Intern Med 2003; 163(13):1587–90.
22. Tannis AJ, Barban J, Conquer JA. Effect of glucosamine supplementation on fasting and non-fasting plasma glucose and serum insulin concentrations in healthy individuals. Osteoarthritis Cartilage 2004; 12(6):506–11.
23. Yu JG, Boies SM, Olefsky JM. The effect of oral glucosamine sulfate on insulin sensitivity in human subjects. Diabetes Care 2003; 26(6):1941–2.
24. Biggee BA, Blinn CM, Nuite M, Silbert JE, McAlindon TE. Effects of oral glucosamine sulphate on serum glucose and insulin during an oral glucose tolerance test of subjects with osteoarthritis. Ann Rheum Dis 2007; 66(2):260–2.
25. Biggee BA, McAlindon T. Glucosamine for osteoarthritis: Part II, biologic and metabolic controversies. Med Health R I 2004; 87(6):180–1.
26. Theodore G. Untrsuchung von 35 arhrosefallen, behandelt mit chondroitin schwefelsaure. Schweiz Rundschaue Med Praxis 1977; 66.
27. Vach J, Pesakova V, Krajickova J, Adam M. Efect of glycosaminoglycan polysulfate on the metabolism of cartilage RNA. Arzneim Forsch/Drur Res 1984; 34:607–9.
28. Ali SY. The degrdation of cartilage matrix by an intracellular protease. Biochem J 1964; 93:611.
29. Hamerman D, Smith C, Keiser HD, Craig R. Glycosaminoglycans produced by human synovial cell cultures collagen. Rel Res 1982; 2:313.
30. Lilja S, Barrach HJ. Normally sulfated and highly sulfated glycosaminoglycans affecting fibrillogenesis on type I and type II collage in vitro. Exp Pathol 1983; 23:173–81.
31. Knanfelt A. Synthesis of articular cartilage proteoglycans by isolated bovine chondrocytes. Agents Actions 1984; 14:58–62.
32. Baici A, Salgam P, Fehr K, Boni A. Inhibition of human elastase from polymorphonuclear leucocytes by gold sodium thiomalate and pentosan polysulfate (SP-54). Biochem Pharmacol 1981; 30(7):703–8.
33. Baici A. Interactions between human leucocytes elastase and chondroitin sulfate. Chem Biol Interact 1984; 51:11.
34. Marossy K. Interaction of the antitrypsin and elastase-like enzyme of the human granulocyte with glycosaminoglycans. Biochim Biophys Acta 1981; 659:351–61.
35. De Gennaro F, Piccioni PD, Caporali R, Luisetti M, Contecucco C. Effet du traitement par le sulfate de galactosaminoglucuronoglycane sur l'estase granulocytaire synovial de patients atteints d'osteoarthrose. Litera Rhumatologica 1992; 14:53–60.
36. Jackson CG, Plaas AH, Barnhill JG, Harris CL, Hua C, Clegg DO. The Multiple-dose pharmacokinetics of orally administered glucosamine and chondroitin sulfate in humans. Arthritis Rheum 2006:S1681.
37. Humphries DE, Silbert CK, Silbert JE. Glycosaminoglycan production by bovine aortic endothelial cells cultured in sulfate-depleted medium. J Biol Chem 1986; 261(20):9122–7.
38. Silbert CK, Humphries DE, Palmer ME, Silbert JE. Effects of sulfate deprivation on the production of chondroitin/dermatan sulfate by cultures of skin fibroblasts from normal and diabetic individuals. Arch Biochem Biophys 1991; 285(1):137–41.

39. Silbert JE, Sugumaran G, Cogburn JN. Sulphation of proteochondroitin and 4-methylumbelli-feryl beta-D-xyloside-chondroitin formed by mouse mastocytoma cells cultured in sulphate-deficient medium. Biochem J 1993; 296(Pt 1):119–26.

40. van der Kraan PM, Vitters EL, de Vries BJ, van den Berg WB. High susceptibility of human articular cartilage glycosaminoglycan synthesis to changes in inorganic sulfate availability. J Orthop Res 1990; 8(4):565–71.

41. Morris ME, Levy G. Serum concentration and renal excretion by normal adults of inorganic sulfate after acetaminophen, ascorbic acid, or sodium sulfate. Clin Pharmacol Ther 1983; 33(4):529–36.

42. Cordoba F, Nimni ME. Chondroitin sulfate and other sulfate containing chondroprotective agents may exhibit their effects by overcoming a deficiency of sulfur amino acids. Osteoarthritis Cartilage 2003; 11(3):228–30.

43. Hoffer LJ, Kaplan LN, Hamadeh MJ, Grigoriu AC, Baron M. Sulfate could mediate the therapeutic effect of glucosamine sulfate. Metabolism 2001; 50(7):767–70.

44. Blinn CM, Biggee BA, McAlindon TE, Nuite M, Silbert JE. Sulphate and osteoarthritis: decrease of serum sulphate levels by an additional 3-h fast and a 3-h glucose tolerance test after an overnight fast. Ann Rheum Dis 2006; 65(9):1223–5.

45. D'Ambrosio E, Casa B, Bompani R, Scali G, Scali M. Glucosamine sulphate: a controlled clinical investigation in arthrosis. Pharmatherapeutica 1981; 2(8):504–8.

46. Crolle G, D'Este E. Glucosamine sulphate for the management of arthrosis: a controlled clinical investigation. Curr Med Res Opin 1980; 7(2):104–9.

47. Drovanti A, Bignamini AA, Rovati AL. Therapeutic activity of oral glucosamine sulfate in osteoarthrosis: a placebo-controlled double-blind investigation. Clin Ther 1980; 3(4):260–72.

48. Noack W, Fsicher M, Forster KK, Rovatis LC, Senikar I. Glucosamine sulfate in osteoarthitis of the knee. Osteoarthritis Cart 1994; 2:51–9.

49. Pujalte JM, Llavore EP, Ylescupidez FR. Double-blind clinical evaluation of oral glucosamine sulphate in the basic treatment of osteoarthrosis. Curr Med Res Opin 1980; 7(2):110–4.

50. Reichelt A, Forster KK, Fischer M, Rovati LC, Setnikar I. Efficacy and safety of intramuscular glucosamine sulfate in osteoarthritis of the knee: a randomized, placebo-controlled, double-blind study. Drug Res 1994; 44:75–80.

51. Vaz AL. Double-blind clinical evaluation of the relative efficacy of ibuprofen and glucosamine sulphate in the management of osteoarthrosis of the knee in out-patients. Curr Med Res Opin 1982; 8:145–9.

52. Vajaradul Y. Double-blind clinical evaluation of intra-articular glucosamine in outpatients with gonarthrosis. Clin Ther 1981; 3(5):336–43.

53. Tapadinhas MJ, Rivera IC, Bignamini AA. Oral glucosamine sulphate in the management of arthosis: report on a multi-centre open investigation in Portugal. Pharmatherapeutica 1982; 3(3):157–68.

54. Vetter VG. Glukosamine in der therapie des degenerativen rheumatismus. Duet Med J 1965; 16:446–9.

55. L'Hirondel JL. Klinische doppelblind-studie mit oral verabreichtem chondroitinsulfat gegen placebo bei der tibiofemoralen gonarthrose (125 patienten). Litera Rhumatologica 1992; 14:77–84.

56. Kerzberg EM, Roldan EJ, Castelli G, Huberman ED. Combination of glycosaminoglycans and acetylsalicylic acid in knee osteoarthrosis. Scand J Rheumatol 1987; 16(5):377–80.

57. Mazieres B, Loyau G, Menkes CJ, et al. Chondroitin sulfate in the treatment of gonarthrosis and coxarthrosis. 5-months result of a multicenter double-blind controlled prospective study using placebo. Rev Rhum Mal Osteoartic 1992; 59(7–8):466–72.

58. Rovetta G. Galactosaminoglycuronoglycan sulfate (matrix) in therapy of tibiofibular osteoar-thritis of the knee. Drugs Exptl Clin Res 1991; 17:53–7.

59. Muller-Fassbender H, Bach GL, Haase W, Rovato LC, Setnikar I. Glucosamine sulfate com-pared to ibuprofen in osteoarthritis of the knee. Osteoarthritis and Cartilage 1994; 2:61–9.

60. McAlindon TE, LaValley MP, Gulin JP, Felson DT. Glucosamine and chondroitin for treatment of osteoarthritis: a systematic quality assessment and meta-analysis. JAMA 2000; 283(11):1469–75.

61. Richy F, Bruyere O, Ethgen O, Cucherat M, Henrotin Y, Reginster JY. Structural and symptomatic efficacy of glucosamine and chondroitin in knee osteoarthritis: a comprehensive meta-analysis. Arch Intern Med 2003; 163(13):1514–22.

62. Towheed T, Maxwell L, Anastassiades T, et al. Glucosamine therapy for treating osteoarthritis. Cochrane Database Syst Rev 2005; (2):CD002946.

63. Reginster JY, Bruyere O, Neuprez A. Current role of glucosamine in the treatment of osteoarthritis. Rheumatology (Oxford) 2007; 46(5):731–5.

64. Vlad SC, Lavalley MP, McAlindon TE, Felson DT. Glucosamine for pain in osteoarthritis: why do trial results differ? Arthritis Rheum 2007; 56(7):2267–77.

65. Cohen J. Statistical Power Analysis for the Behavioral Sciences, 2nd ed. Hillsdale, New Jersey: Lawrence Erlbaum Associates, 1988.

66. Leeb BF, Schweitzer H, Montag K, Smolen JS. A metaanalysis of chondroitin sulfate in the treatment of osteoarthritis. J Rheumatol 2000; 27(1):205–11.

67. Reichenbach S, Sterchi R, Scherer M, et al. Meta-analysis: chondroitin for osteoarthritis of the knee or hip. Ann Intern Med 2007; 146(8):580–90.

68. Cibere J, Kopec JA, Thorne A, et al. Randomized, double-blind, placebo-controlled glucosamine discontinuation trial in knee osteoarthritis. Arthritis Rheum 2004; 51(5):738–45.

69. McAlindon T, Formica M, Kabbara K, LaValley M, Lehmer M. Conducting clinical trials over the internet: feasibility study. BMJ 2003; 327(7413):484–7.

70. Hughes R, Carr A. A randomized, double-blind, placebo-controlled trial of glucosamine sulphate as an analgesic in osteoarthritis of the knee. Rheumatology (Oxford) 2002; 41(3): 279–84.

71. Rindone JP, Hiller D, Collacott E, Nordhaugen N, Arriola G. Randomized, controlled trial of glucosamine for treating osteoarthritis of the knee. West J Med 2000; 172(2):91–4.

72. Cibere J, Thorne A, Kopec JA, et al. Glucosamine sulfate and cartilage type II collagen degradation in patients with knee osteoarthritis: randomized discontinuation trial results employing biomarkers. J Rheumatol 2005; 32(5):896–902.

73. Messier SP, Mihalko S, Loeser RF, et al. Glucosamine/chondroitin combined with exercise for the treatment of knee osteoarthritis: a preliminary study. Osteoarthritis Cartilage 2007;15(11):1256–66.

74. Herrero-Beaumont G, Ivorra JA, Del Carmen Trabado M, et al. Glucosamine sulfate in the treatment of knee osteoarthritis symptoms: a randomized, double-blind, placebo-controlled study using acetaminophen as a side comparator. Arthritis Rheum 2007; 56(2):555–67.

75. Clegg DO, Reda DJ, Harris CL, et al. Glucosamine, chondroitin sulfate, and the two in combination for painful knee osteoarthritis. N Engl J Med 2006; 354(8):795–808.

76. Jordan KM, Arden NK, Doherty M, et al. EULAR Recommendations 2003: an evidence based approach to the management of knee osteoarthritis: Report of a Task Force of the Standing Committee for International Clinical Studies Including Therapeutic Trials (ESCISIT). Ann Rheum Dis 2003; 62(12):1145–55.

77. Hochberg MC. Nutritional supplements for knee osteoarthritis – still no resolution. N Engl J Med 2006; 354(8):858–60.

78. Laverty S, Sandy JD, Celeste C, Vachon P, Marier JF, Plaas AH. Synovial fluid levels and serum pharmacokinetics in a large animal model following treatment with oral glucosamine at clinically relevant doses. Arthritis Rheum 2005; 52(1):181–91.

79. Altman RD, Abramson S, Bruyere O, et al. Commentary: osteoarthritis of the knee and glucosamine. Osteoarthritis Cartilage 2006; 14(10):963–6.

80. Reginster JY. The efficacy of glucosamine sulfate in osteoarthritis: financial and nonfinancial conflict of interest. Arthritis Rheum 2007; 56(7):2105–10.

81. Reginster JY, Deroisy R, Rovati LC, et al. Long-term effects of glucosamine sulphate on osteoarthritis progression: a randomised, placebo-controlled clinical trial. Lancet 2001; 357(9252):251–6.

82. Pavelka K, Gatterova J, Olejarova M, Machacek S, Giacovelli G, Rovati LC. Glucosamine sulfate use and delay of progression of knee osteoarthritis: a 3-year, randomized, placebo-controlled, double-blind study. Arch Intern Med 2002; 162(18):2113–23.

83. Altman R, Brandt K, Hochberg M, et al. Design and conduct of clinical trials in patients with osteoarthritis: recommendations from a task force of the Osteoarthritis Research Society. Results from a workshop. Osteoarthritis Cartilage 1996; 4(4):217–43.

84. Michel BA, Stucki G, Frey D, et al. Chondroitins 4 and 6 sulfate in osteoarthritis of the knee: a randomized, controlled trial. Arthritis Rheum 2005; 52(3):779–86.

85. Vignon E. Radiographic issues in imaging the progression of hip and knee osteoarthritis. J Rheumatol Suppl 2004; 70:36–44.

86. Reginster JY, Kahan A, Vignon E. A Two-Year Prospective, Randomized, Double-Blind, Controlled Study Assessing the Effect of Chondroitin 4&6 Sulfate (CS) on the Structural Progression of Knee Osteoarthritis: STOPP (STudy on Osteoarthritis Progression Prevention). Ann Rheum Dis 2006; 65(4):L42.

87. Curtis CL, Hughes CE, Flannery CR, Little CB, Harwood JL, Caterson B. n-3 fatty acids specifically modulate catabolic factors involved in articular cartilage degradation. J Biol Chem 2000; 275(2):721–4.

88. Curtis CL, Rees SG, Cramp J, et al. Effects of n-3 fatty acids on cartilage metabolism. Proc Nutr Soc 2002; 61(3):381–9.

89. Curtis CL, Rees SG, Little CB, et al. Pathologic indicators of degradation and inflammation in human osteoarthritic cartilage are abrogated by exposure to n-3 fatty acids. Arthritis Rheum 2002; 46(6):1544–53.

90. Lippiello L. Lipid and cell metabolic changes associated with essential fatty acid enrichment of articular chondrocytes. Proc Soc Exp Biol Med 1990; 195(2):282–7.

91. Stammers T, Sibbald B, Freeling P. Efficacy of cod liver oil as an adjunct to non-steroidal anti-inflammatory drug treatment in the management of osteoarthritis in general practice. Ann Rheum Dis 1992; 51(1):128–9.

92. Mauviel A, Daireaux M, Hartmann DJ, Galera P, Loyau G, Pujol JP. Effects of unsaponifiable extracts of avocado/soy beans (PIAS) on the production of collagen by cultures of synoviocytes, articular chondrocytes and skin fibroblasts. Rev Rhum Mal Osteoartic 1989; 56(2):207–11.

93. Mauviel A, Loyau G, Pujol JP. Effect of unsaponifiable extracts of avocado and soybean (Piascledine) on the collagenolytic action of cultures of human rheumatoid synoviocytes and rabbit articular chondrocytes treated with interleukin-1. Rev Rhum Mal Osteoartic 1991; 58(4):241–5.

94. Henrotin YE, Labasse AH, Jaspar JM, et al. Effects of three avocado/soybean unsaponifiable mixtures on metalloproteinases, cytokines and prostaglandin E2 production by human articular chondrocytes. Clin Rheumatol 1998; 17(1):31–9.

95. Henrotin YE, Sanchez C, Deberg MA, et al. Avocado/soybean unsaponifiables increase aggrecan synthesis and reduce catabolic and proinflammatory mediator production by human osteoarthritic chondrocytes. J Rheumatol 2003; 30(8):1825–34.

96. Boumediene K, Felisaz N, Bogdanowicz P, Galera P, Guillou GB, Pujol JP. Avocado/soya unsaponifiables enhance the expression of transforming growth factor beta1 and beta2 in cultured articular chondrocytes. Arthritis Rheum 1999; 42(1):148–56.

97. Henrotin YE, Deberg MA, Crielaard JM, Piccardi N, Msika P, Sanchez C. Avocado/soybean unsaponifiables prevent the inhibitory effect of osteoarthritic subchondral osteoblasts on aggrecan and type II collagen synthesis by chondrocytes. J Rheumatol 2006; 33(8):1668–78.

98. Cake MA, Read RA, Guillou B, Ghosh P. Modification of articular cartilage and subchondral bone pathology in an ovine meniscectomy model of osteoarthritis by avocado and soya unsaponifiables (ASU). Osteoarthritis Cartilage 2000; 8(6):404–11.

99. Ernst E. Avocado-soybean unsaponifiables (ASU) for osteoarthritis – a systematic review. Clin Rheumatol 2003; 22(4–5):285–8.

100. Blotman F, Maheu E, Wulwik A, Caspard H, Lopez A. Efficacy and safety of avocado/soybean unsaponifiables in the treatment of symptomatic osteoarthritis of the knee and hip. A prospective, multicenter, three-month, randomized, double-blind, placebo-controlled trial. Rev Rhum Engl Ed 1997; 64(12):825–34.

101. Maheu E, Mazieres B, Valat JP, et al. Symptomatic efficacy of avocado/soybean unsaponifiables in the treatment of osteoarthritis of the knee and hip: a prospective, randomized, double-blind, placebo-controlled, multicenter clinical trial with a six-month treatment period and a two-month followup demonstrating a persistent effect. Arthritis Rheum 1998; 41(1):81–91.

102. Appelboom T, Schuermans J, Verbruggen G, Henrotin Y, Reginster JY. Symptoms modifying effect of avocado/soybean unsaponifiables (ASU) in knee osteoarthritis. A double blind, prospective, placebo-controlled study. Scand J Rheumatol 2001; 30(4):242–7.

103. Lequesne M, Maheu E, Cadet C, Dreiser RL. Structural effect of avocado/soybean unsaponifiables on joint space loss in osteoarthritis of the hip. Arthritis Rheum 2002; 47(1):50–8.

104. Ameye LG, Chee WS. Osteoarthritis and nutrition. From nutraceuticals to functional foods: a systematic review of the scientific evidence. Arthritis Res Ther 2006; 8(4):R127.

105. Utiger RD. Kashin-Beck disease – expanding the spectrum of iodine-deficiency disorders [editorial; comment]. N Engl J Med 1998; 339(16):1156–8.

106. Moreno-Reyes R, Suetens C, Mathieu F, et al. Kashin-Beck osteoarthropathy in rural Tibet in relation to selenium and iodine status [see comments]. N Engl J Med 1998; 339(16):1112–20.

107. Jordan JM, Fang F, Arab L, et al. Low selenium levels are associated with increased risk for osteoarthritis of the knee. Arthritis Rheum 2005; 52(9 Suppl):Abstract #1189.

108. Hill J, Bird HA. Failure of selenium-ace to improve osteoarthritis. Br J Rheumatol 1990; 29(3):211–3.

109. Muller-Fassbender H. Double-blind clinical trial of S-adenosylmethionine versus ibuprofen in the treatment of osteoarthritis. Am J Med 1987; 83(5A):81–3.

110. Konig B. A long-term (two years) clinical trial with S-adenosylmethionine for the treatment of osteoarthritis. Am J Med 1987; 83(5A):89–94.

111. Vetter G. Double-blind comparative clinical trial with S-adenosylmethionine and indomethacin in the treatment of osteoarthritis. Am J Med 1987; 83(5A):78–80.

112. Maccagno A, Di Giorgio EE, Caston OL, Sagasta CL. Double-blind controlled clinical trial of oral S-adenosylmethionine versus piroxicam in knee osteoarthritis. Am J Med 1987; 83(5A):72–7.

113. Glorioso S, Todesco S, Mazzi A, et al. Double-blind multicentre study of the activity of S-adenosylmethionine in hip and knee osteoarthritis. Int J Clin Pharmacol Res 1985; 5(1):39–49.

114. Najm WI, Reinsch S, Hoehler F, Tobis JS, Harvey PW. S-adenosyl methionine (SAMe) versus celecoxib for the treatment of osteoarthritis symptoms: a double-blind cross-over trial [ISRCTN36233495]. BMC Musculoskelet Disord 2004; 5(1):6.

115. Altman RD, Marcussen KC. Effects of a ginger extract on knee pain in patients with osteoarthritis. Arthritis Rheum 2001; 44(11):2531–8.

116. Frei B. Reactive oxygen species and antioxidant vitamins: mechanisms of action. Am J Med 1994; 97(Suppl 3A):5S–13S.

117. Boveris A, Oshino N, Chance B. The cellular production of hydrogen peroxide. Biochem J 1972; 128:617–30.

118. Lenaz G. The mitochondrial production of reactive oxygen species: mechanisms and implications in human pathology. IUBMB Life 2001; 52(3–5):159–64.

119. Jacques PF, Chylack LT, Taylor A. Relationships between natural antioxidants and cataract formation. In: Frei B, ed. Natural Antioxidants in Human Health and Disease. San Diego, CA: Academic Press, 1994, pp. 515–33.

120. Gaziano JM. Antioxidant vitamins and coronary artery disease risk. Am J Med 1994; 97(Suppl 3A):18S–21S.

121. Hennekens CH. Antioxidant vitamins and cancer. Am J Med 1994; 97(Suppl 3A):2S–4S.

122. Martin JA, Buckwalter JA. Aging, articular cartilage chondrocyte senescence and osteoarthritis. Biogerontology 2002; 3(5):257–64.

123. Blake DR, Unsworth J, Outhwaite JM, et al. Hypoxic-reperfusion injury in the inflamed human. Lancet 1989; 11:290–3.

124. Ames BN, Shigenaga MK, Hagen TM. Oxidants, antioxidants and the degenerative diseases of aging. Proc Natl Acad Sci USA 1993; 90:7915–22.

125. Henrotin Y, Deby-Dupont G, Deby C, De Bruyn M, Lamy M, Franchimont P. Production of active oxygen species by isolated human chondrocytes. Br J Rheumatol 1993; 32(7):562–7.

126. Henrotin Y, Deby-Dupont G, Deby C, Franchimont P, Emerit I. Active oxygen species, articular inflammation, and cartilage damage. EXS 1992; 62:308–22.

127. Rathakrishnan C, Tiku K, Raghavan A, Tiku ML. Release of oxygen radicals by articular chondrocytes: a study of luminol-dependent chemoluminescence and hydrogen peroxide secretion. J Bone Miner Res 1992; 7:1139–48.

128. Tiku ML, Allison GT, Naik K, Karry SK. Malondialdehyde oxidation of cartilage collagen by chondrocytes. Osteoarthritis Cartilage 2003; 11(3):159–66.

129. Greenwald RA, Moy WW. Inhibition of collagen gelation by action of the superoxide radical. Arthritis Rheum 1979; 22(3):251–9.

130. McCord JM. Free radicals and inflammation: protection of synovial fluid by superoxide dismutase. Science 1974; 185:529–30.

131. Loeser RF, Carlson CS, Del Carlo M, Cole A. Detection of nitrotyrosine in aging and osteoarthritic cartilage: correlation of oxidative damage with the presence of interleukin-1beta and with chondrocyte resistance to insulin-like growth factor 1. Arthritis Rheum 2002; 46(9):2349–57.

132. Yudoh K, Nguyen T, Nakamura H, Hongo-Masuko K, Kato T, Nishioka K. Potential involvement of oxidative stress in cartilage senescence and development of osteoarthritis: oxidative stress induces chondrocyte telomere instability and downregulation of chondrocyte function. Arthritis Res Ther 2005; 7(2):R380–91.

133. Haklar U, Yuksel M, Velioglu A, Turkmen M, Haklar G, Yalcin AS. Oxygen radicals and nitric oxide levels in chondral or meniscal lesions or both. Clin Orthop 2002; 403:135–42.

134. Sato H, Takahashi T, Ide H, et al. Antioxidant activity of synovial fluid, hyaluronic acid, and two subcomponents of hyaluronic acid. Synovial fluid scavenging effect is enhanced in rheumatoid arthritis patients. Arthritis Rheum 1988; 31(1):63–71.

135. Briviba K, Seis H. Non-enzymatic antioxidant defense systems. In: Frei B, ed. Natural Antioxidants in Human Health and Disease. San Diego: Academic Press, 1994, pp. 107–28.

136. Henrotin Y, Kurz B, Aigner T. Oxygen and reactive oxygen species in cartilage degradation: friends or foes? Osteoarthritis Cartilage 2005; 13(8):643–54.

137. McAlindon T, Felson DT. Nutrition: risk factors for osteoarthritis. Ann Rheum Dis 1997; 56(7):397–400.

138. Sowers M, Lachance L. Vitamins and arthritis. The roles of vitamins A, C, D, and E. Rheum Dis Clin North Am 1999; 25(2):315–32.

139. Hankinson SE, Stampfer MJ, Seddon JM, et al. Nutrient intake and cataract extraction in women: a prospective study. BMJ 1992; 305(6849):335–9.

140. Peterkofsky B. Ascorbate requirement for hydroxylation and secretion of procollagen: relationship to inhibition of collagen synthesis in scurvy. AM J Clin Nutr 1991; 54:1135S–40S.

141. Spanheimer RG, Bird TA, Peterkofsky B. Regulation of collagen synthesis and mRNA levels in articular cartilage of scorbutic guinea pigs. Arch Biochem Biophys 1986; 246:33–41.

142. Houglum KP, Brenner DA, Chijkier M. Ascorbic acid stimulation of collagen biosynthesis independent of hydroxylation. Am J Clin Nutr 1991; 54:1141S–3S.

143. Schwartz ER, Adamy L. Effect of ascorbic acid on arylsulfatase activities and sulfated proteoglycan metabolism in chondrocyte cultures. J Clin Invest 1977; 60(1):96–106.

144. Sandell LJ, Daniel LC. Effects of ascorbic acid on collagen mRNA levels in short-term chondrocyte cultures. Connect Tiss Res 1988; 17:11–22.

145. Meacock SCR, Bodmer JL, Billingham MEJ. Experimental OA in guinea pigs. J Exp Path 1990; 71:279–93.

146. Schwartz ER, Oh WH, Leveille CR. Experimentally induced osteoarthritis in guinea pigs: metabolic responses in articular cartilage to developing pathology. Arthritis Rheum 1981; 24(11):1345–55.

147. McAlindon TE, Jacques P, Zhang Y, Hannan MT, Aliabadi P, Weissman B, Rush D, Levy D, Felson DT. Do antioxidant micronutrients protect against the development and progression of knee osteoarthritis? Arthritis Rheum 1996; 39(4):648–56.

148. Kellgren J, Lawrence JS. The Epidemiology of Chronic Rheumatism: Atlas of Standard Radiographs, Vol 2. Oxford: Blackwell Scientific, 1963.

149. Baker K, Niu J, Goggins J, Clancy M, Felson D. The effects of vitamin C intake on pain in knee osteoarthritis (OA). Arthritis Rheum 2003; 48(9):S422.

150. Kraus VB, Huebner JL, Stabler T, et al. Ascorbic acid increases the severity of spontaneous knee osteoarthritis in a guinea pig model. Arthritis Rheum 2004; 50(6):1822–31.

151. Bakker AC, van de Loo FA, van Beuningen HM, et al. Overexpression of active TGF-beta-1 in the murine knee joint: evidence for synovial-layer-dependent chondro-osteophyte formation. Osteoarthritis Cartilage 2001; 9(2):128–36.

152. Scharstuhl A, Glansbeek HL, van Beuningen HM, Vitters EL, van der Kraan PM, van den Berg WB. Inhibition of endogenous TGF-beta during experimental osteoarthritis prevents osteophyte formation and impairs cartilage repair. J Immunol 2002; 169(1):507–14.

153. van Beuningen HM, Glansbeek HL, van der Kraan PM, van den Berg WB. Osteoarthritis-like changes in the murine knee joint resulting from intra-articular transforming growth factor-beta injections. Osteoarthritis Cartilage 2000; 8(1):25–33.

154. Barcellos-Hoff MH, Dix TA. Redox-mediated activation of latent transforming growth factor-beta 1. Mol Endocrinol 1996; 10(9):1077–83.

155. Jensen NH. Reduced pain from osteoarthritis in hip joint or knee joint during treatment with calcium ascorbate. A randomized, placebo-controlled cross-over trial in general practice. Ugeskr Laeger 2003; 165(25):2563–6.

156. Kiel DP. Vitamin D, calcium and bone: descriptive epidemiology. In: Rosenberg IH, ed. Nutritional Assessment of Elderly Populations: Measurement and Function. New York: Raven, 1995, pp. 277–90.

157. Parfitt AM, Gallagher JC, Heaney RP, Neer R, Whedon GD. Vitamin D and bone health in the elderly. AM J Clin Nutr 1982; 36:1014–31.

158. Radin EL, Paul IL, Tolkoff MJ. Subchondral changes in patients with early degenerative joint disease. Arthritis Rheum 1970; 13:400–5.

159. Layton MW, Goldstein SA, Goulet RW, Feldkamp LA, Kubinski DJ, Bole GG. Examination of subchondral bone architecture in experimental osteoarthritis by microscopic computed axial tomography. Arthritis Rheum 1988; 31(11):1400–5.

160. Milgram JW. Morphological alterations of the subchondral bone in advanced degenerative arthritis. Clin Orthop Rel Res 1983; 173:293–312.

161. Kellgren JH, Lawrence JS. The Epidemiology of Chronic Rheumatism: Atlas of Standard Radiographs. Oxford, UK: Blackwell Scientific, 1962.

162. Anonymous. Cartilage and bone in osteoarthrosis. Brit Med J 1976; 2:4–5.

163. Dequecker J, Mokassa L, Aerssens J. Bone density and osteoarthritis. J Rheumatol 1995; 22(Suppl 43):98–100.

164. Dedrick DK, Goldstein SA, Brandt KD, O'Connor BL, Goulet RW, Albrecht M. A longitudinal study of subchondral plate and trabecular bone in cruciate-deficient dogs with osteoarthritis followed up for 54 months. Arthritis Rheum 1993; 36:1460–7.

165. Ledingham J, Dawson S, Preston B, Milligan G, Doherty M. Radiographic progression of hospital-referred osteoarthritis of the hip. Ann Rheum Dis 1993; 52:263–7.

166. Radin EL, Rose RM. Role of subchondral bone in the initiation and progression of cartilage damage. Clin Orthop Rel Res 1986; 213:34–40.

167. Pottenger LA, Phillips FM, Draganich LF. The effect of marginal osteophytes on reduction of varus-valgus instability in osteoarthritic knees. Arthritis Rheum 1990; 33(6):853–8.

168. Perry GH, Smith MJG, Whiteside CG. Spontaneous recovery of the joint space in degenerative hip disease. Ann Rheum Dis 1972; 31:440–8.

169. Smythe SA. Osteoarthritis, insulin and bone density. J Rheumatol 1987; 14(Suppl):91–3.

170. Zhang Y, Hannan MT, Chaisson CE, et al. Bone mineral density and risk of incident and progressive radiographic knee osteoarthritis in women: the Framingham Study. J Rheumatol 2000; 27(4):1032–7.

171. Dieppe P, Cushnaghan J, Young P, Kirwan J. Prediction of the progression of joint space narrowing in osteoarthritis of the knee by bone scintigraphy. Ann Rheum Dis 1993; 52:557–63.

172. Corvol MT. Hormonal control of cartilage metabolism. Bull Schweiz Akad Med Wiss 1981–1982:205–9.

173. Corvol MT, Dumontier MF, Tsagris L, Lang F, Bourguignon J. Cartilage and vitamin D in vitro (author's transl). Ann Endocrinol (Paris) 1981; 42(4–5):482–7.

174. Tetlow LC, Woolley DE. Expression of vitamin D receptors and matrix metalloproteinases in osteoarthritic cartilage and human articular chondrocytes in vitro. Osteoarthritis Cartilage 2001; 9(5):423–31.

175. Bischoff HA, Borchers M, Gudat F, et al. In situ detection of 1,25-dihydroxyvitamin D_3 receptor in human skeletal muscle tissue. Histochem J 2001; 33(1):19–24.

176. Bischoff-Ferrari HA, Borchers M, Gudat F, Durmuller U, Stahelin HB, Dick W. Vitamin D receptor expression in human muscle tissue decreases with age. J Bone Miner Res 2004; 19(2):265–9.

177. Hart DJ, Cronin C, Daniels M, Worthy T, Doyle DV, Spector TD. The relationship of bone density and fracture to incident and progressive radiographic osteoarthritis of the knee: the Chingford study. Arthritis Rheum 2002; 46(1):92–9.

178. Bischoff-Ferrari HA, Zhang Y, Kiel DP, Felson DT. Positive association between serum 25-hydroxyvitamin D level and bone density in osteoarthritis. Arthritis Rheum 2005; 53(6):821–6.

179. McAlindon TE, Felson DT, Zhang Y, et al. Relation of dietary intake and serum levels of vitamin D to progression of osteoarthritis of the knee among participants in the Framingham Study. Ann Intern Med 1996; 125(5):353–9.

180. Lane NE, Gore LR, Cummings SR, et al. Serum vitamin D levels and incident changes of radiographic hip osteoarthritis: a longitudinal study. Study of Osteoporotic Fractures Research Group. Arthritis Rheum 1999; 42(5):854–60.

181. Felson DT, Niu J, Clancy M, et al. Low levels of vitamin D and worsening of knee osteoarthritis: results of two longitudinal studies. Arthritis Rheum 2006; 56(1):129–36.

182. Carbone LD, Nevitt MC, Wildy K, et al. The relationship of antiresorptive drug use to structural findings and symptoms of knee osteoarthritis. Arthritis Rheum 2004; 50(11):3516–25.

183. Demarco PJ, Constantinescu F. Does vitamin D supplementation contribute to the modulation of osteoarthritis by bisphosphonates? Comment on the article by Carbone et al. Arthritis Rheum 2005; 52(5):1622–3.

184. Carbone LD, Barrow KD, Nevitt MC. Reply. Arthritis Rheum 2005; 52(5):1623.

185. Panganamala RV, Cornwell DG. The effects of vitamin E on arachidonic acid metabolism. Ann NY Acad Sci 1982; 393:376–91.

186. Tiku ML, Shah R, Allison GT. Evidence linking chondrocyte lipid peroxidation to cartilage matrix protein degradation. Possible role in cartilage aging and the pathogenesis of osteoarthritis. J Biol Chem 2000; 275(26):20069–76.

187. Schwartz ER. Effect of vitamins C and E on sulfated proteoglycan metabolism and sulfatase and phosphatase activities in organ cultures of human cartilage. Calcif Tissue Int 1979; 28(3):201–8.

188. Tiku ML, Gupta S, Deshmukh DR. Aggrecan degradation in chondrocytes is mediated by reactive oxygen species and protected by antioxidants. Free Radic Res 1999; 30(5):395–405.

189. Galleron S, Borderie D, Ponteziere C, et al. Reactive oxygen species induce apoptosis of synoviocytes in vitro. Alpha-tocopherol provides no protection. Cell Biol Int 1999; 23(9):637–42.

190. Kaiki G, Tsuji H, Yonezawa T, Sekido H, Takano T, Yamashita S, Hirano N, Sano A. Osteoarthrosis induced by intra-articular hydrogen peroxide injection and running load. J Orthop Res 1990; 8(5):731–40.

191. Hirohata K, Yao S, Imura S, Harada H. Treatment of osteoarthritis of the knee joint at the state of hydroarthrosis. Kobe Med Sci 1965; 11(Suppl):65–6.

192. Doumerg C. Etude clinique experimentale de l'alpha-tocopheryle-quinone en rheumatologie et en reeducation. Therapeutique 1969; 45:676–8.

193. Machetey I, Quaknine L. Tocopherol in osteoarthritis: a controlled pilot study. J Am Ger Soc 1978; 26:328–30.

194. Scherak O, Kolarz G, Schodl C, Blankenhorn G. Hochdosierte vitamin-E-therapie bei patienten mit aktivierter arthrose. Z Rheumatol 1990; 49:369–73.

195. Blankenhorn G. Clinical efficacy of spondyvit (vitamin E) in activated arthroses. A multicenter, placebo-controlled, double-blind study. Z Orthop 1986; 124:340–3.

196. Machtey I, Ouaknine L. Tocopherol in Osteoarthritis: a controlled pilot study. J Am Geriatr Soc 1978; 26(7):328–30.

197. Scherak O, Kolarz G, Schodl C, Blankenhorn G. High dosage vitamin E therapy in patients with activated arthrosis. Z Rheumatol 1990; 49(6):369–73.

198. Brand C, Snaddon J, Bailey M, Cicuttini F. Vitamin E is ineffective for symptomatic relief of knee osteoarthritis: a six month double blind, randomised, placebo controlled study. Ann Rheum Dis 2001; 60(10):946–9.

199. Wluka AE, Stuckey S, Brand C, Cicuttini FM. Supplementary vitamin E does not affect the loss of cartilage volume in knee osteoarthritis: a 2 year double blind randomized placebo controlled study. J Rheumatol 2002; 29(12):2585–91.

200. Hunter DJ, Niu J, Zhang Y, et al. Cartilage volume must be normalized to bone surface area in order to provide satisfactory construct validity: the Framingham study. Osteoarthritis Cartilage 2004; 12(Suppl B):Abstract #M4.

201. Thane CW, Paul AA, Bates CJ, Bolton-Smith C, Prentice A, Shearer MJ. Intake and sources of phylloquinone (vitamin K1): variation with socio-demographic and lifestyle factors in a national sample of British elderly people. Br J Nutr 2002; 87(6):605–13.

202. Booth SL, Suttie JW. Dietary intake and adequacy of vitamin K. J Nutr 1998; 128(5):785–8.

203. Furie B, Bouchard BA, Furie BC. Vitamin K-dependent biosynthesis of gamma-carboxygluta-mic acid. Blood 1999; 93(6):1798–808.

204. Loeser RF, Varnum BC, Carlson CS, et al. Human chondrocyte expression of growth-arrest-specific gene 6 and the tyrosine kinase receptor axl: potential role in autocrine signaling in cartilage. Arthritis Rheum 1997; 40(8):1455–65.

205. Hale JE, Fraser JD, Price PA. The identification of matrix Gla protein in cartilage. J Biol Chem 1988; 263(12):5820–4.

206. Price PA. Gla-containing proteins of bone. Connect Tissue Res 1989; 21(1–4):51–7, discussion 7–60.

207. Newman B, Gigout LI, Sudre L, Grant ME, Wallis GA. Coordinated expression of matrix Gla protein is required during endochondral ossification for chondrocyte survival. J Cell Biol 2001; 154(3):659–66.

208. Yagami K, Suh JY, Enomoto-Iwamoto M, et al. Matrix GLA protein is a developmental regulator of chondrocyte mineralization and, when constitutively expressed, blocks endochon-dral and intramembranous ossification in the limb. J Cell Biol 1999; 147(5):1097–108.

209. Zebboudj AF, Imura M, Bostrom K. Matrix GLA protein, a regulatory protein for bone morphogenetic protein-2. J Biol Chem 2002; 277(6):4388–94.

210. Hara K, Akiyama Y, Tajima T, Shiraki M. Menatetrenone inhibits bone resorption partly through inhibition of PGE2 synthesis in vitro. J Bone Miner Res 1993; 8(5):535–42.

211. Reddi K, Henderson B, Meghji S, et al. Interleukin 6 production by lipopolysaccharide-stimulated human fibroblasts is potently inhibited by naphthoquinone (vitamin K) compounds. Cytokine 1995; 7(3):287–90.

212. Neuropathic joints. Degenerative joint disease. Arthritis Rheum 1970; 13(5):571–8.

213. Hall JG, Pauli RM, Wilson KM. Maternal and fetal sequelae of anticoagulation during pregnancy. Am J Med 1980; 68(1):122–40.

214. Luo G, Ducy P, McKee MD, et al. Spontaneous calcification of arteries and cartilage in mice lacking matrix GLA protein. Nature 1997; 386(6620):78–81.

215. Neogi T, Zhang Y, Booth S, Jacques PF, Terkeltaub R, Felson DT. Is there an association between osteophytes and vitamin K intake? Arthritis Rheum 2004; 50(9):S350.

216. Neogi T, Booth SL, Zhang YQ, et al. Low vitamin K status is associated with osteoarthritis in the hand and knee. Arthritis Rheum 2006; 54(4):1255–61.

217. Felson DT. Weight and osteoarthritis. J Rheumatol 1995; 22(suppl 43):7–9.

218. Sokoloff L, Mickelsen O. Dietary fat supplements, body weight and osteoarthritis in DBA/2JN mice. J Nutr 1965; 85:117–21.

219. Sokoloff L, Mickelsen O, Silverstein E, Jay GE, Jr, Yamamoto RS. Experimental obesity and osteoarthritis. Am J Physiol 1960; 198:765–70.

220. Aspden RM, Scheven BA, Hutchison JD. Osteoarthritis as a systemic disorder including stromal cell differentiation and lipid metabolism. Lancet 2001; 357(9262):1118–20.

221. Lippiello L, Walsh T, Fienhold M. The association of lipid abnormalities with tissue pathology in human osteoarthritic articular cartilage. Metabolism 1991; 40(6):571–6.

222. Dumond H, Presle N, Terlain B, et al. Evidence for a key role in leptin in ostoearthritis. Arthritis Rheum 2003; 48(9):S282.

223. Messier SP, Loeser RF, Miller GD, et al. Exercise and dietary weight loss in overweight and obese older adults with knee osteoarthritis: the arthritis, diet, and activity promotion trial. Arthritis Rheum 2004; 50(5):1501–10.

224. Christensen R, Astrup A, Bliddal H. Weight loss: the treatment of choice for knee osteoarthritis? A randomized trial. Osteoarthritis Cartilage 2005; 13(1):20–7.

225. Toda Y, Toda T, Takemura S, Wada T, Morimoto T, Ogawa R. Change in body fat, but not body weight or metabolic correlates of obesity, is related to symptomatic relief of obese patients with knee osteoarthritis after a weight control program. J Rheumatol 1998; 25(11):2181–6.

226. Hooper MM, Stellato TA, Hallowell PT, Moskowitz RW. Musculoskeletal findings in morbidly obese subjects before and after weight loss due to gastric bypass surgery. Arthritis Rheum 2004; 50(9 Suppl):S699.

227. Christensen R, Bartels EM, Astrup A, Bliddal H. Effect of weight reduction in obese patients diagnosed with knee osteoarthritis: a systematic review and meta-analysis. Ann Rheum Dis 2007; 66(4):433–9.

24 Post-stroke Malnutrition and Dysphagia

Candice Hudson Scharver,
Carol Smith Hammond, and Larry B. Goldstein

Key Points

- Stroke is the third leading cause of death in the USA and swallowing problems may affect half or more of stroke patients at some time during the course of their disease.
- Slightly more than half of stroke patients are malnourished. Pre-existing subclinical swallowing dysfunction may further predispose elderly stroke patients to dysphagia.
- Early dysphagia assessment is important to minimize aspiration risk and to avoid dehydration and malnutrition that can lead to further complications and impair the recovery process.
- Stroke patients with suspected dysphagia should be assessed on a timely basis with a clinical examination and appropriate instrumental tests.
- Recommendations for dietary modifications or specific therapeutic strategies to assure adequate nutritional intake, hydration and oral hygiene should be made in close consultation with a nutritionist.

Key Words: Malnutrition post-stroke; dysphagia; speech language pathologist (SLP); clinical swallow evaluation (CSE); videofluoroscopic swallow evaluation (VSE); fiberoptic endoscopic evaluation of swallow (FEES); aspiration; Frazier Water Protocol

24.1 INTRODUCTION

The American Heart Association estimates that more than 780,000 persons have strokes in the USA each year. Stroke is the third leading cause of death in the USA and stroke deaths actually rose 7.7% between 1991 and 2001 *(1)*. Malnutrition is an important preventable complication that can increase length of stay (LOS) in patients hospitalized because of acute stroke *(2)*. Dysphagia, or swallowing impairment, affects about 50% of all stroke patients and up to 75% of those with

From: *Nutrition and Health: Handbook of Clinical Nutrition and Aging, Second Edition*
Edited by: C. W. Bales and C. S. Ritchie, DOI 10.1007/978-1-60327-385-5_24,
© Humana Press, a part of Springer Science+Business Media, LLC 2009

brainstem strokes. Estimates of the prevalence of dysphagia vary due to differences in the definition of dysphagia, the method of assessment of swallowing function, the timing of the swallowing assessment after stroke, and the number and type of stroke patients studied *(3–5)*.

Pre-existing malnutrition and subclinical swallowing dysfunction can combine to enhance nutritional risk following a stroke. An older individual who has an acute stroke may compromise an already poor nutritional status due to subsequent dysphagia *(1)*. Dysphagia recovery within the first weeks after stroke varies from 43 to 86%. When dysphagia persists, it is associated with increased mortality rates and excess morbidity *(6)*.

24.2 MALNUTRITION RISK FACTORS ASSOCIATED WITH STROKE

Malnutrition and stroke often go hand in hand; malnutrition is commonly observed both before and after a stroke occurs. In a small study of 32 consecutive admissions of geriatric patients with severe stroke, Axelsson et al. *(7)* found that 56.3% were malnourished at some point during a hospital stay of more than 3 weeks. In a similar prospective study of 49 consecutively admitted stroke patients, Finestone et al. *(8)* assessed the prevalence of malnutrition during patients' stay in rehabilitation and 2–4 months following discharge. Basing the diagnosis on abnormalities for at least two of the following – body weight, sum of skinfolds, midarm muscle circumference, serum albumin and transferrin, and total lymphocyte count – the incidence of malnutrition among patients decreased from 49% on admission to 34% at 1 month, 22% at 2 months, and to 19% at 2–4 month post-discharge follow-up. Strong predictors of malnutrition upon admission were the use of tube feedings ($p = 0.043$) and dysphagia ($p = 0.032$). Diabetes mellitus and prior stroke increased the likelihood of malnutrition on admission by 58 and 71%, respectively. Malnutrition was not associated with gender, location, or type of stroke (hemorrhage versus infarct), paresis of dominant arm, socioeconomic status, or level of education. At 1 month post-rehabilitation admission, malnutrition was associated with age over 70 years ($p = 0.002$) and at follow-up was associated with recent weight loss and lack of community nursing or home nursing care. Routine interventions such as regular weighing, calorie counts, staff attention to dysphagia, dysphagia diets, and tube feeding appeared to contribute to the overall decrease in prevalence and degree of malnutrition in these patients *(8)*. In a subsequent study by this group of investigators, malnutrition ($p = 0.01$) and impaired physical functioning were the only independent predictors ($p < 0.001$) of increased length of rehabilitation stay for patients with acute stroke *(2)*. Likewise, Unosson et al. *(9)* reported that dependence in feeding at admission was associated with a progressive loss of body cell mass in stroke patients.

24.2.1 Malnutrition Due to Feeding Problems

The loss of the ability to self-feed is associated with overall disability as reflected in scales of activities of daily living *(10)*. The high degree of malnutrition reported among stroke patients may be due to a period of inadequate nutritional intake in the early post-stroke period. Foley et al. *(11)* evaluated the protein and energy intake of patients with acute stroke ($n = 91$) who received nutrition through regular diets, dysphagia diets, or enteral tube feedings at five time points within the first 3 weeks of admission. Energy intakes were higher for patients receiving enteral feedings as compared to patients on a regular ($p = 0.018$) or dysphagia diet ($p = 0.024$). Similarly, patients receiving enteral feedings had greater protein intakes than those on a regular diet ($p < 0.0001$). There was no difference in either the energy or protein intake of patients receiving a regular diet as compared to those receiving a dysphagia diet at any time point. On average, newly diagnosed, well-nourished, hospitalized patients consumed 80–91% of both their energy and protein requirements in the early post-stroke period *(11)*.

Kumlien et al. *(12)* studied stroke patients ($n = 40$) in five nursing homes in Sweden and found that 80% had some sort of dependence in eating. Extensive or total care was needed in 22.5% and all these severely dependent patients were also judged as having some cognitive impairment. Sixty percent were moderately dependent and needed limited assistance, supervision, or meal tray set up. Seven patients (17.5%) were independent and could self-feed. The number of disabilities in individual patients ranged from 1 to 7, emphasizing the eating disability complexities in stroke patients. Dysphagia was vaguely described as "a lot of" coughing and was reported in 22.5% of patients. Thirty percent were assessed as having poor food intake or poor appetite. Reasons for poor food intake included difficulties due to chewing problems (15%), difficulties handling food on the plate (17.5%), and the need for assistive feeding devices such as tools, plates, or mugs (27.5%) *(12)*.

Wade et al. *(13)* reported that 47–80% of stroke patients were affected by hemiplegia. Trunk control problems and arm and hand weakness in stroke patients contribute to positioning problems that may lead to posture deficits. Affected patients may not be able to maintain their head or body in an upright position and may have to eat one-handed, perhaps with the nondominant hand. Visual, speech, and language difficulties hinder adequate communication about needs and food preferences. Cognitive deficits limit the patient's ability to carry out the multiple sequences of activities required to eat a meal *(14)*. Depression reduces appetite and the desire to eat and has a deleterious influence upon recovery of function in activities of daily living.

The high malnutrition rate (20%) in patients admitted to the hospital with acute stroke is related to a number of associated factors (Table 24.1). Deterioration in nutritional status occurs more often in dysphagic patients and those who are dependent on feeding *(15)*. Gariballa et al. *(16)* performed an observational study of acute stroke patients and found that serum albumin concentration at admission was the only nutritional status variable that had a negative correlation with the number of infective complications ($r = -0.33$, $P < 0.0001$). This finding, however, may be confounded by the fact that albumin is altered as an acute phase response during infection.

Table 24.1
Risk factors that contribute to malnutrition in stroke patients *(6,18)*.

Distracting environment	Visual difficulties
Dehydration	Speech difficulties: aphasia, dysphonia, dysarthria
Deficits in posture	Hearing problems
Deficits in limb control	Cognitive deficits
Dependence in feeding	Depression
Poor oral hygiene, dentition	Lack of food preferences
Dysphagia	Breathing problems
Allergies	Co-morbidities

24.3 DYSPHAGIA AS A MAJOR RISK FACTOR FOR MALNUTRITION FOLLOWING STROKE

Swallowing problems may be attributed to a variety of causes. One simple definition of dysphagia describes it as a disorder of bolus flow *(17)*. Dysphagia is also described as swallowing problems that may affect the alimentary tract anywhere from the mouth to the stomach *(6,18)*.

24.3.1 Normal Swallowing and Age-Related Changes

The normal swallow consists of five stages *(19)*. including the four active stages shown in Fig. 24.1. A variety of age-related changes can affect swallow function. Systemic disease in the elderly may indirectly impair swallowing. Physiological and sensory changes such as cartilage ossification and decreased vision may affect the pre-swallow phase and food taste sensation. Lack of ability to communicate problems may further diminish intake and lead to malnutrition. Esophageal motility problems, achalasia, and presbyesophagus are associated with advanced age. Daggett et al. *(20)* performed videofluoroscopic swallow evaluation (VSE) in normal adults ($n = 98$) aged 20–94 years while the subjects were eating solids and drinking liquids. A continuous measure of airway protection was used to rate the swallows of liquid and solid food. Penetration of thin liquid into the laryngeal area above the level of the true vocal folds was observed in 7.4% of persons under 50 years of age. Normal adults over 50 years exhibited penetration on 16.8% of swallows and larger swallows were penetrated more often. Declining physiologic reserve in the elderly may predispose this group to more severe swallowing problems than that experienced by younger patients *(21,22)*. Table 24.2 lists a summary of age-related swallowing changes.

24.3.2 Oropharyngeal Dysphagia

Speech language pathologists (SLPs) scope of practice includes observation of impaired swallow function. Oropharyngeal dysphagia may be due to weakness, impaired coordination, or obstruction of the swallowing mechanism. Figure 24.2 depicts the anatomy of normal swallowing structures. The terms *penetration* and *aspiration* describe the degree of airway protection during eating and drinking.

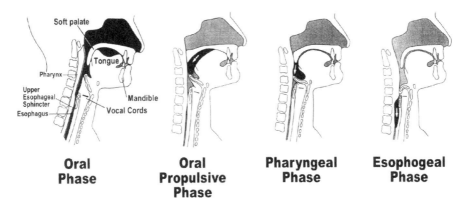

Oral Phase   **Oral Propulsive Phase**   **Pharyngeal Phase**   **Esophogeal Phase**

Fig. 24.1 The normal swallow consists of five stages. The four active stages are as follows *(19)*:
1) Anticipatory phase – affected by sight, smell, ability to hold utensils, and cognition.
2) Oral phase – affected by lip closure, buccal tone, taste, temperature, mastication.
3) Oral propulsive phase at 1–1.5 s – tongue propulsion with midline depression, nasal breathing, 1–1.5 s.
4) Pharyngeal phase, six actions at 1 s: 1 – velum elevation prevents nasal regurgitation, base of tongue retracts; 2 – anterior and superior hyoid movement and neck muscles elevate larynx; 3 – larynx closure: true folds, false folds, arytenoids anterior tilt, epiglottic base thickening and retroversion, apneic phase: breathing halted briefly; 4 – cricopharyngeal relaxation as thyrohyoid muscle elevates; 5 – tongue base drives bolus into esophagus affected by bolus viscosity; 6 – stripping action of superior, middle, and inferior pharyngeal constrictor muscles.
5) Esophageal phase at 3–20 s. Involuntary stage, slower in elderly, smooth and striated muscle, reflux prevention ensues as cricopharyngeus contracts and breathing resumes.

Penetration is defined as food or liquid entering the laryngeal area above or to the level of the vocal cords. Aspiration is defined as the passage of food or liquid below the level of the vocal cords into the proximal trachea. Another important term is *silent aspiration*. Silent aspiration occurs in a subpopulation of patients with dysphagia and is defined as the passage of bolus below the true vocal folds without cough reflex or other distress indicators. The vallecula and pyriform sinuses are spaces in the oropharynx where food and liquid normally pass through, yet often residue collects in these areas for patients with oral pharyngeal dysphagia. The vallecula is located at the base of the tongue and the pyriform sinuses are located adjacent to the larynx and above the upper esophageal sphincter (UES). Daniels et al. *(23)* compared VSE findings in acute stroke patients to normal controls and found differences in bolus timing measures, airway protection (penetration and aspiration), and residue in the valleculae and pyriform sinuses. In this exploratory study, only one of nine stroke patients had normal swallowing function at 1 month after stroke onset.

24.4 APPROACHES FOR SWALLOW ASSESSMENT: CSE, VSE, FEES

Assessments of swallowing function include both clinical and instrumental examinations, which are performed by SLP and are defined here. The clinical swallowing evaluation (CSE) is a non-instrumental assessment performed at the bedside that

Table 24.2
Age-related changes in swallowing stages *(21,22)*.

Swallow stage	Changes with age
Oral	Reduced lingual pressure
	Reduced tongue strength
	Mastication problems secondary to dentition issues
Pharyngeal	Slower swallow
	Delayed onset of airway protection (true vocal fold closure as protective measure)
	Increased apnea period with age
	Delayed upper esophageal sphincter opening
Esophageal	Contraction amplitude decreases
	Yet function remains intact for 80–90 years
	Reduced frequency of secondary peristalsis
	Increased reflux events

assesses cranial nerve and oral motor function, screens cognition and voice, assesses speech and language function, and provides patients with varied trial boluses of various volumes and consistencies of materials *(6)*. Instrumental evaluations of swallow include videofluoroscopy and endoscopic procedures. Videofluoroscopy (VSE) is a radiological study employed by the SLP to obtain a view of the oral cavity

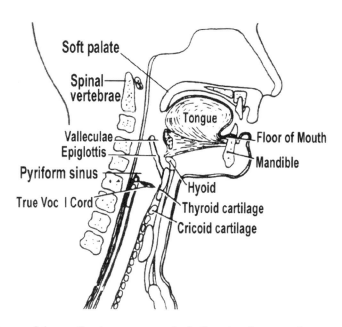

Fig. 24.2 Anatomy of the swallowing structures including the pharyngeal spaces (valleculae and pyriform sinuses often sites of bolus residue) and the true vocal folds. Residue just above the vocal folds results in laryngeal penetration of the bolus and residue below the vocal folds results in aspiration *(5)*.

and pharynx in the lateral and anterior posterior planes as patients swallow varied amounts and viscosities of liquids and solids mixed with barium. Fiberoptic endoscopic evaluation of swallow (FEES) utilizes a flexible nasendoscope passed transnasally to directly observe the pharynx and larynx before and after the swallow to detect residue and signs of aspiration. The VSE and FEES are both described as the gold standard for detection of aspiration (24).

Of 1,000 patients aged 15–104 years (mean age 72.1 ± 15.5 years) referred for VSE, 573 people (57%) aspirated and silent aspiration occurred in 296 or 52% of those who aspirated (25). Detection of silent aspiration is a critical component of the swallow examination since aspiration can lead to adverse medical outcomes including aspiration pneumonia. This suggests that VSE may be needed in addition to CSE to identify silent aspirators prior to development of further medical complications. The advantages and disadvantages of the CSE, VSE, and FEES are discussed in the following sections.

24.4.1 Clinical Swallow Evaluation

A CSE is typically performed prior to an instrumental test such as VSE in radiology or FEES. Ramsey et al. (26) reviewed the merits and limitations of available dysphagia assessment methods and found that whereas the CSE is poor at detecting silent aspiration, it is safe, relatively straightforward, and easily repeated. Although CSE provides an important way to screen for dysphagia, further refinements are necessary to improve diagnostic accuracy.

In a cohort of consecutively admitted stroke patients ($n = 128$), the independent demographic and medical status predictors of dysphagia on VSE were age >70 years, male gender, and disabling stroke with Barthel Index score <60 (17). Components of the CSE, which were independent predictors of dysphagia, were palatal weakness or asymmetry, incomplete oral clearance, and impaired pharyngeal response (cough/gurgle). Predictors of aspiration were delayed oral transit and incomplete oral clearance. The sensitivity and specificity of these predictors were not given (17). However, Mann (27) later published the manual for The Mann Assessment of Swallowing Ability (MASA) that utilized clinical, cognitive, and medical status to predict the presence of dysphagia as observed on VSE with sensitivity 73% and specificity 89%. Sensitivity was 93% and specificity was 63% to predict aspiration.

McCullough et al. (28) performed a VSE (gold standard) on 165 consecutively admitted patients with stroke (aged 35–101 years) to determine if CSE measures or a combination of measures detect aspiration. Aspiration occurred most often (40%) in the oldest age group (80–101 years) and 51% of the 43 aspirators did so "silently" (i.e., without cough or throat clear). Table 24.3 lists the CSE components and the sensitivity and specificity to predict aspiration, thus the proportion of study participants who aspirated and were positive or negative for clinical signs compared to VSE. A true positive for a sign resulted if aspiration was present on VSE when the sign or history was present on the CSE. Although time can be a factor during post-stroke recovery, the VSE was completed within 24 hours of the CSE during this study. In this study, the best measures for detecting aspiration were found to be failure of the 3-oz swallow test (coughing or choking), unilateral jaw weakness, and dysphonia. No single measure in isolation could be used to rule in or rule out aspiration (28).

Table 24.3
Clinical swallow evaluation (CSE) sensitivity and specificity*

	Sensitivity (%)	Specificity (%)
Weak jaw bilaterally against resistance	15	99
Fail 3 oz water swallow	48	95
Fail 10 ml thin liquid	38	96
Fail thick liquid	21	98
Fail 5 ml thin liquid	44	94
Breathy, weak voice	16	98
Pneumonia present	9	98
Tongue strength to palpation	64	48
Lip strength to pursing; against resistance	68	49
Weak palatal gag to tongue blade	56	51
Volitional cough (to command)	42	79
Reflexive cough after swallow	24	80
Dysphonia (hoarse voicing)	54	86
Wet, gurgly voice after swallow	22	96
Aphasia (language problems)	33	78

*Subjective clinical aspiration signs to predict aspiration on videofluorographic swallow study (VSE) (28).

New initiatives to develop objective measures to screen for the presence of dysphagia are in the exploratory stages of development. Several studies utilized pulse oximetry in addition to the CSE to determine if patients were aspirating. Ramsey et al. (29) reviewed reports by several groups in which a reduction of oxygen saturation of >2% from baseline predicted aspiration on VSE. Sensitivities ranged from 73 to 87% and specificities ranged from 39 to 87%. Ramsey (29) assessed 189 stroke patients with VSE and found no association between VSE findings of penetration and aspiration and desaturation by either >2% or 5% during swallowing and concluded that pulse oximetry alone did not have high enough sensitivity or specificity to be reliable for aspiration risk detection in acute stroke patients. Another addition to the clinical bedside evaluation that is under investigation by Smith Hammond et al. (30) is the use of airflow measures of voluntary cough. Three objective measures of voluntary cough, expulsive phase rise time, expulsive phase peak flow, and volume acceleration, were each associated with aspiration risk with areas under the curve 0.93, 0.92, and 0.86, respectively.

24.4.2 Instrumental Swallow Evaluations

The clinical swallow evaluation relies on the clinician's subjective judgments, whereas the VSE and FEES provide objective data based on the visualization of the anatomy and physiology of swallowing during deglutition. Both studies should use real foods and liquids typically consumed by the patient. The VSE and FEES instrumental swallow evaluations enable clinicians to evaluate potential therapeutic postures, bolus textures, and/or amounts that may help inform recommendations for oral nutrition. Either VSEs or FEES are recommended to identify the appropriate treatment for dysphagic patients *(24)*.

24.4.2.1 VIDEOFLUOROSCOPIC SWALLOW EVALUATION (VSE)

Videofluoroscopic swallow evaluation (VSE) provides anatomic and functional information during the swallow. The amounts and order of bolus presentation are not standardized and are at the discretion of the clinician based on the results of the CSE. Kuhlemeier et al. *(31)* studied 745 patients with acute stroke who had been referred for VSE. Aspiration was more common when patients drank from a cup than when they used a spoon ($p < 0.001$) for liquids of variable viscosity (thin, nectar, and honey thick). Aspiration was most frequent for thin liquids (e.g., coffee) compared to nectar-thick liquids and least frequent for honey-thick liquids compared to nectar-thick liquids ($p < 0.001$).

24.4.2.2 FIBEROPTIC ENDOSCOPIC EVALUATION OF SWALLOWING (FEES)

The FEES enables the SLP to directly view the pharyngeal and laryngeal areas before and after swallowing. The SLP may evaluate vocal fold function during voicing, which is valuable because abnormal voice (e.g., breathy, wet, or hoarse) is often associated with dysphagia *(28)*. Water alone could not be visualized without contrast on a x-ray; however, effects of water boluses may be observed during the FEES. Suiter and Leder *(32)* performed FEES with the 3-oz water test for 3,000 individuals of varying diagnoses. The sensitivity was 96.5% to predict aspiration during FEES and the specificity was 48.7%; 1,151 (38.4%) passed and 1,849 (61.6%) failed. Although passing the test appeared to be a good predictor of ability of patients to swallow thin liquid, failure on the 3-oz water test should not necessarily prevent a patient from taking thin liquids since the false-positive rate was 53.6%. In addition 71% of the patients were judged to tolerate an oral diet so unnecessary dietary restrictions may be avoided with instrumental assessment by an SLP.

24.4.2.3 FUNCTIONAL ORAL INTAKE SCALE (FOIS)

Typically, the functional level of oral intake or food and liquid consistency is recommended following the objective swallow evaluation. The functional oral intake scale (FOIS) was developed and is reliable, valid, and sensitive to change in stroke patients' oral intake (see Table 24.4). This seven-point ordinal scale was applied to 302 acute stroke patients at admission and after 1 and 6 months. Levels 1–3 of the scale refer to varying degrees of nonoral feeding (NPO or nothing by mouth, tube dependent, or combination of tube and oral feeding) and levels 4–7 describe oral intake without nonoral supplementation *(33)*.

Table 24.4
Functional Oral Intake Scale (FOIS) for dysphagia *(33)*.

FOIS	*Functional oral intake scale for dysphagia*
Level 1	NPO
Level 2	Tube dependent
Level 3	Tube dependent with consistent oral intake of food or liquid
Level 4	Single consistency oral diet
Level 5	Total oral, multiple consistencies, special preparation
Level 6	Total oral, multiple consistencies, specific food limits
Level 7	Total oral diet, no restrictions

24.5 SUBJECTIVE ASSESSMENT OF SWALLOWING

Stroke patients often have dysphagia symptoms with little apparent awareness, but may indicate swallowing problems when asked questions such as "Do you cough when taking medications?" Several short, easily administered scales have been used to subjectively assess and reassess patient complaints at bedside and after objective swallow assessments.

24.5.1 Dysphagia Disability Index and Reflux Symptom Index

The dysphagia disability index (DDI) is a 25-item questionnaire that probes patient opinions for functional, emotional, and physical dysphagia effects *(34)*. The reflux symptom index (RSI) is a 9-question subjective laryngopharyngeal reflux scale that rates patient complaints during the previous month for symptoms such as "cough after lying down" and "frequent throat clearing" *(35)*. These tools are simple and "user friendly" but, unfortunately, currently lack extensive reliability and validity data.

24.5.2 Patient-Centered Quality-of-Life Outcomes Tools: SWAL-CARE and SWAL-QOL

The SWAL-CARE consists of 11 SLP swallowing advice items and 4 patient satisfaction items for a total of 15 items *(36,37)*. The SWAL-QOL utilizes 44 swallowing statements, such as food sticking in the throat, choking on food or liquids, excess saliva, and gagging, rated by patients on a 5-point scale. The domains of the SWAL-CARE and SWAL-QOL scales are shown in Tables 24.5 and 24.6.

Outpatients ($n = 386$) with dysphagia on VSE, including 15.8% (61/386) with vascular disease, were sampled in the scale development and compared to 40 normal healthy adults to differentiate between those with and without dysphagia. For example, food texture dysphagia impacted scores, with 33% of dysphagic patients on pureed diets screening positive for major depression and 35% rating their health as poor. The authors stated that potentially only 5% of the normal swallowers would screen positive for major depression. SLPs, dietitians, and clinicians may thus be motivated to provide alternative maneuvers and diet modification to improve patient satisfaction and mental health with appropriate nutritional intervention.

Table 24.5
SWAL-QOL domains *(36,37)*.

SWAL-QOL – Swallowing Quality of Life – example content
Food selection – difficulty finding foods I like
Burden – dealing with my swallow problem is very difficult
Mental health – my swallow problem frustrates me
Social functioning – I do not go out to eat because of my swallow problem
Fear – I fear I may start choking when I eat food
Eating duration – it takes me forever to eat a meal

24.6 INTERVENTION FOLLOWING STROKE

Carnaby et al. *(38)* randomized acute stroke patients ($n = 306$) with clinical dysphagia to one of three treatment options: (1) usual care management by attending physician, mainly consisting of feeding supervision; (2) low-intensity swallowing therapy three times per week for 1 month with environmental modifications, advice, and dietary modifications; (3) high-intensity treatment every day for 1 month with effortful swallowing, supraglottic technique. Of those assigned to usual care, 56% returned to normal diet at 6 months as compared to 70% in the high-intensity treatment group and 64% in the low-intensity intervention group. High-intensity therapy was associated with return to normal diet ($p = 0.04$) and recovered swallow function ($p = 0.02$) by 6 months. Standard therapy and low-intensity therapy were associated with a reduction in swallowing-related medical complications for 46% of those patients in standard therapy compared to 63% allocated usual care. Chest infection due to aspiration was 47% for usual care and 26% for organized care. Patients allocated standard care swallowing therapy (36%) compared to usual care (48%) were less likely to be deceased or institutionalized at 6 months. Forty-six percent of patients allocated standard swallowing therapy (1.03–1.94) achieved prestroke swallowing functioning by 6 months ($p = 0.02$) *(39)*.

Elmstahl et al. *(39)* described nutritional effects of swallowing treatments for dysphagic stroke patients ($n = 38$). Dietary recommendations were made based on the results of VSE. Eighty-nine percent (34/38) were recommended for oral nutrition and almost 50% (17/38) required a modified diet; 11% (4/38) required parenteral nutrition or means of nutrition not through the alimentary canal but rather by injection through some other route such as subcutaneous, intramuscular, or intravenous. Swallow function as observed on VSE improved in 61% (23/

Table 24.6
SWAL-CARE domains *(36,37)*.

SWAL-CARE – Quality of care – example of content
Clinical information – weaker confidence in decisions
General advice – foods I should eat. Liquids I should drink
Patient satisfaction – clinician explained things

38) of the subjects ($p < 0.01$). Patients ($n = 23$, 61%) who improved their swallowing functions increased their albumin levels after treatment from 33.7 g/l ± 4.6 to 36.2 g/l ± 3.2 ($p < 0.01$) and also increased their total iron-binding capacity from 44.4 μmol/l ± 13.4 to 52.0 μmol/l ± 8.9 ($p < 0.01$). Patients who did not respond to therapy had decreased body weight from 75.5 kg ± 13.3 to 72.2 kg ± 11.8 ($p < 0.05$) *(39)*.

Nutritional parameters and conditions should be monitored in dysphagic stroke patients for optimal management of oropharyngeal swallowing problems.

24.6.1 Pneumonia Costs

Katzan et al. *(40)* conducted a large observational study of northeastern Ohio Medicare patients with stroke from 29 nonfederal hospitals from 1991 to 1997 to determine the effect of pneumonia on 30-day mortality. The final cohort of 11,286 stroke patients identified 5.6% (635/11,286) with pneumonia. The mean unadjusted cost of hospitalization was $21,173 (95% CI 19,421–22,925) for patients with pneumonia and $6,272 (95% CI 6,611–6,382) for patients without pneumonia. Katzan et al. *(41)* emphasized adherence to following stroke guidelines that include recommendations for assessing the ability to swallow before a patient is allowed to eat or drink due to the significant costs of possible pneumonia.

24.6.2 Pneumonia Prevention

Predictors of aspiration were investigated by Langmore et al. *(42)* in a prospective study of 189 subjects recruited from outpatient clinics, inpatient wards, and a nursing home centers. Each subject underwent CSE, esophageal clearance exams by scintigraphy, and VSE or FEES. The best predictors for aspiration were dependence for feeding and dependence for oral care (OR = 3.031, $p = 0.05$) and (OR = 2.828; $p = 0.03$).

24.6.3 Alternative Means of Nutrition

Bath et al. *(43)* assessed feeding management strategies for dysphagic stroke patients through a Cochrane review. Often stroke patients spontaneously recover swallowing function in the first 2 weeks although others with severe dysphagia may require alternative means of nutrition. Some patients are fed through a tube, inserted either in the nose and into the stomach (nasogastric tube NG) or through the skin of the abdomen into the stomach – percutaneous endoscopic gastrostomy (PEG). NG tubes are relatively easy to insert but many people find them uncomfortable and pull them out. Two controlled trials ($n = 49$) evaluated the efficacies of the two types of feeding tubes. PEG was associated with lower case fatality rates (odds ratio, OR 0.28, 95% CI 0.09–0.89) and treatment failures (OR 0.10, 95% CI 0.02–0.52) as compared to NG feeding.

The FOOD trials were randomized controlled studies designed to evaluate the effect of timing and method of enteral tube feeding for dysphagic stroke patients *(44)*. One trial ($n = 859$) determined if early initiation of enteral tube feeding versus avoidance of tube feeding improved outcomes. A second trial compared percutaneous gastrostomy (PEG) versus nasogastric tubes (NG). Patients were not tube fed or given PO for at least 7 days and were provided with parenteral fluids either intravenously or subcutaneously. In the early feeding arm, clinician choice was NG tube ($n = 367$) or PEG tubes ($n = 10$). Allocation to early tube feeding was associated with a non-significant reduction in absolute risk of death. Allocation to PEG feeding was associated with an increase in the absolute risk of death or poor outcome of 7.8% (0.0–15.5%, $p = 0.05$). There was no excess of pneumonia associated with early tube feeding. A greater risk of gastrointestinal hemorrhage occurred with early rather than avoidance of tube feeding (22 versus 11, $p = 0.04$) and with NG rather than PEG (18 versus 5, $p = 0.005$) *(44)*. The authors suggest that enteral feeding via nasogastric tube rather than PEG tube shall be offered to acute dysphagic stroke patients within the first few days of admission or within the first 2–3 weeks if necessary.

24.6.4 Dysphagia Intervention and Treatment

24.6.4.1 THICKENED LIQUIDS

Modification of diet texture and thickened liquids have been the most common treatments recommended for patients with dysphagia. In a study of various viscosity samples, healthy volunteers ($n = 8$), non-progressive brain-damaged patients (including stroke) ($n = 46$), and patients with progressive neurological diseases ($n = 46$) swallowed 3–20 ml liquid (20.4 mPa s), nectar (274.4 mPa s), and pudding (3931.2 mPa s) boluses *(45)*. During VSE, the non-progressive brain-damaged patients aspirated 21.6% of liquids, 10.5% of the nectar viscosity, and 5.3% of the pudding ($p < 0.05$). Neurodegenerative patients also improved swallowing with increased viscosity as follows: aspiration of 16.2% liquids, 8.3 % of nectar, and 2.95% pudding ($p < 0.05$).

24.6.4.2 FLUID INTAKE MAINTENANCE

The dysphagic population may consume less fluid due to increased age, physical disability, cognitive impairment, and the diminished palatability of thickened liquids. This may be a special concern for older adults, who are more prone to dehydration. Despite the typical minimum requirement of 1,700 ml of water per day, one in seven adults aged over 65 and one in four over 85 years consume less than this amount *(46)*. Often, acute stroke patients require intravenous supplementation to maintain adequate fluid intakes. Stroke patients are variable in degrees of water homeostasis. Initially, patients may be drowsy, have reduced thirst, or have infection and thus dehydrations are hyperosmolar. Dehydration may cause a rise in hematocrit and a reduction in blood pressure, potentially worsening the ischemic process. Dehydration is also an important predisposing factor in stroke recurrence. Stroke patients with high plasma osmolality levels on admission had poorer survival at 3 months *(6)*. When stroke patients were allowed to have water in addition to thickened fluids, they had a higher overall fluid intake per day than those offered only thickened liquids.

24.6.4.3 FRAZIER FREE WATER PROTOCOL

The Frazier Free Water Protocol allows for unrestricted water intake prior to and 30 minutes after a meal with aggressive oral hygiene for acute rehabilitation patients. Panther et al. *(47)* performed a retrospective chart review of dysphagic patients ($n = 234$) who were recommended thickened liquids and revealed that <1%

(2/234) developed aspiration pneumonia. Patients were assessed with VSE and were screened on water with ice chips as the usual first step in allowing water in the acute care environment. Patients treated under this protocol received medications with applesauce, pudding, yogurt or thickened liquid, and never with water if thin liquids have been aspirated on VSE. The protocol emphasizes the premise that water is relatively inert when absorbed in small amounts in the lungs *(48)*. Laryngeal aspiration is not the sole cause of aspiration pneumonia and general debilitation may contribute to dysphagia *(49)*.

A retrospective chart review of 234 dysphagic inpatients at a rehabilitation hospital compared patients who received the Frazier Water Protocol (FWP) and oral care to a control group *(50)*. There was no increase in the rate of pneumonia in those patients who received the FWP compared to pneumonia rates of 9% for the historical control group ($p < 0.097$) and 16% for concurrent controls ($p < 0.023$). FWP also increased fluid intake compared to both the historical controls ($p < 0.034$) and the concurrent controls ($p < 0.0031$).

24.6.4.4 ORAL HYGIENE EDUCATION

The Cochrane group reviewed eight eligible randomized controlled trials of post-stroke oral hygiene *(51)*. Only one trial provided stroke-specific information comparing staff attitudes and denture plaque scores to usual care. Following training, staff knowledge ($p = 0.002$) and attitudes ($p = 0.0008$) improved after 1 month and continued to improve up to 6 months. Denture plaque scores for the 67 stroke patients (dysphagia was not specified) showed that after an oral health care training program delivered to nursing home care assistants, scores were reduced at 1 ($p < 0.00001$) and 6 months ($p < 0.00001$).

24.6.4.5 DYSPHAGIA EXERCISES AND COMPENSATORY MANEUVERS

Shanahan et al. *(52)* reported elimination of aspiration in 50% (15/30) of a mixed sample of neurologically involved participants with the compensatory technique of chin tuck. Logemann et al. *(53)* examined a small cohort ($n = 5$) of lateral medullary stroke patients with unilateral pharyngeal paresis of the pyriform sinus which resulted in barium pooling. Use of a rotated head position to the weaker side (65%) resulted in an improved amount of bolus swallowed compared to head in the neutral position (33%). Anterior–posterior opening of the UES and oropharyngeal efficiency increased ($p < 0.05$) for participants with head rotated. The Shaker exercises are isometric head lift exercises designed to strengthen the suprahyoid muscles by raising the head for three lifts of 60 s each, with a 1-minute rest period in between *(54)*. The isokinetic portion of the exercise consists of 30 consecutive head lifts with constant velocity, yet without the hold, thus maximizing the strength gains with slower velocity. The exercise is used for patients with post-deglutitive aspiration, pharyngeal residue, and decreased anterior hyolaryngeal excursion resulting in decreased anteroposterior UES deglutitive opening. Twenty-seven consecutive dysphagic outpatients including 56% (15/27) with stroke were evaluated by VSE and enrolled in the Shaker exercise program. The anteroposterior upper esophageal sphincter (UES) opening ($p < 0.05$) and laryngeal anterior excursion ($p < 0.05$) improved for all 27 patients *(55)*.

During the Mendelsohn maneuver the patient is trained to "hold the swallow" and thus extend the period of superior and anterior displacement of the larynx at mid-swallow to accentuate UES opening. Surface electromyographic biofeedback (SEMG) has been used in dysphagia treatment for patients with brainstem stroke. Surface electrodes are placed submentally and provide visual feedback of muscle activity while the patient attempts the Mendelsohn maneuver (56). Crary et al. (57) used the Mendelsohn maneuver to improve oral intake in a retrospective study of 25 dysphagic post-stroke patients and 20 dysphagic head and neck cancer patients. Ninety-two percent of the stroke patients and 80% of the head and neck patients increased functional oral intake of food and/or liquid. There was a progression to total oral feeding for 55% (11/20) of the stroke patients ($p = 0.013$) and 25% (3/12) of the head and neck cancer patients. The stroke group was more likely than the cancer group to improve to normal food intake after this treatment protocol (RR = 2.2, CI -1.3 to 3.7) (57).

Neuromuscular electrical stimulation (NMES) involves applying surface electrodes across the skin to excite nerve or muscle tissue during a functional task such as swallowing (58). Further investigations with this approach are needed to determine whether NMES has greater efficacy than other swallowing treatments and is in the exploratory stage. SLPs have also been advised to monitor vital signs and screen patients with heart conditions prior to application of therapy techniques that utilize valsalva or breath hold maneuvers. Maneuvers such as the supersupraglottic and supraglottic swallow utilize prolonged voluntary closure of the glottis that may place patients at risk for cardiac arrhythmias (59).

Adequate hydration and dietary intake while avoiding aspiration pneumonia remains a goal for dysphagic stroke patients. Dysphagia diets include a progression of food and liquid textures from pureed, ground or minced, soft to modified (e.g., omitting crisp foods) until normal textures can be consumed (60). Liquid modifications range from spoon thick (like pudding), honey like (e.g., honey-thickened commercially prepared juices), nectar like (e.g., buttermilk, eggnog) to thin or regular fluids with no changes necessary. Ultimately, communication among registered dietitians, SLPs, and physicians about proper treatment remains essential for optimum nutritional management.

24.7 FUTURE RESEARCH

The number of well-designed research studies yielding positive outcomes of behavioral interventions for individuals with neurologically induced dysphagia remains limited.

Postural alterations, dietary modifications, and emerging therapy techniques that improve swallowing function in dysphagic individuals should be the focus of well-designed and controlled studies using quantifiable, relevant outcome measures. Areas of research could also include efficacy of telehealth and alternative means of service delivery consultation for dysphagia intervention, assessment of patient subjective complaints, and subsequent dietitian monitoring of nutritional parameters.

24.8 RECOMMENDATIONS

1. Prompt detection and ongoing treatment of malnutrition are recommended for the rehabilitation of stroke patients during the acute stage and at follow-up.
2. Nutrition assessment by a Registered Dietitian is recommended for every acute stroke patient.
3. VSE or FEES instrumental swallow evaluations following the clinical assessment (CSE) are recommended to identify the appropriate treatment for dysphagic patients.
4. Dysphagic patients may be prescribed dysphagia diets that are texture-modified (e.g., soft, chopped, pureed, minced, with/without thick fluids) and often high in calories (energy) and protein.
5. Adequate hydration and vigorous oral hygiene are recommended with attention to aspiration and reflux precautions as medically appropriate.
6. Other treatment options require further research and include SEMG, NMES, and positioning maneuvers, in addition to viscosity and dietary adjustments.
7. Early swallow screenings and dysphagia management by speech pathologists including diagnosis and treatment in acute stroke patients are cost-effective and may yield dramatic reductions in pneumonia rates.

Acknowledgement CHS dedicates this chapter to the memory of her mom, Betty Hudson, with appreciation for the dignity provided by Hospice during the past year and the continuing support from her husband, Jeff. CSH dedicates this chapter to the loving memory of her parents, John and Frances Smith, and her brothers and sisters for their support during their final years including Linda, Susan, John, James, Frances, Thomas, and Michael.

REFERENCES

1. Goldstein LB, Adams M, Alberts M, Appel LJ, Brass LM, Bushnell CD, Culebras A, DeGraba TJ, Gorelick PB, Guyton JR, Hart RG, Howard G, Kelly-Hayes M, Nixon JV, Sacco RL. Primary prevention of ischemic stroke: a Guideline from the American Heart Association/American Stroke Association Stroke Council. Circulation 2006; 113:873–23.
2. Finestone HM, Greene-Finestone LS, Wilson ES, Teasel RW. Prolonged length of stay and reduced functional improvement rate in malnourished stroke rehabilitation patients. Arch Phys Med Rehabil 1996; 77:340–5.
3. Mann G, Hankey GI, Camerson D. Swallowing disorders following acute stroke: prevalence and diagnostic accuracy. Cerebrovasc Dis 2000; 10:380–6.
4. Schroeder MF, Daniels SK, McClain M, Corey DM. Clinical and cognitive predictors of swallowing recovery in stroke. JRRD 2006; 43(3):301–9.
5. Hammond CS, Scharver CH, Markley LW, Kinnally, Cable M, Evanko L, Curtis D. Dysphagia evaluation, treatment, and recommendations. In: Bales CW, Ritchie CS, eds. Handbook of Clinical Nutrition and Aging. Totowa, NJ: Humana Press, Inc. 2001, pp. 547–66.
6. Kedlaya D, Brandstater ME. Swallowing, nutrition, and hydration during acute stroke care. Top Stroke Rehabil 2002 Summer;9(2):23–38.
7. Axelsson K, Asplund K, Norberg A, Eriksson S. Eating problems and nutritional status during hospital stay of patients with severe stroke. J. Am Diet Assoc 1989; 89:1092–6.
8. Finestone HM, Greene-Finestone LS, Wilson ES, Teasell RW. Malnutrition in stroke patients on the rehabilitation service and at follow-up: prevalence and predictors. Arch Phys Med Rehabil 1995; 76:310–16.
9. Unosson M, Bjurolf P, von Schonck H, Larsson J. Feeding dependence and nutritional status after acute stroke. Stroke 1994; 25:366–71.

10. Katz S, Akpom CA. A measure of primary sociobiological functions. Int J Health Serv 1976; 6:493–507.

11. Foley N, Finestone H, Woodbury M, Teasell R, Greene-Finestone L. Energy and protein intakes of acute stroke patients. J Nutr Health Aging 2006; 10(3):171–5.

12. Kumlien S, Axelsson K. Stroke patients in nursing homes: eating, feeding, nutrition and related care. J Clin Nurs 2002; 11(4):498–509.

13. Wade DT, Hewer RL. Motor loss and swallowing difficulty after stroke: frequency recovery and prognosis. Act Neurological Sand 1987; 76:50–4.

14. Steele CM, Greenwood C, Robertson EI, Seidman-Carlson R. Mealtime difficulties in a home for the elderly: not just dysphagia. Dysphagia 1997; 12:41–50.

15. Dennis M. Nutrition after stroke. Br Med Bull 2000; 56(2):466–75.

16. Gariballa SE, Parker SG, Taub N, Castleden CM. Influence of nutritional status on clinical outcome after acute stroke. Am J Clin Nutr 1998; 68:275–81.

17. Mann G, Hankey GJ. Initial clinical and demographic predictors of swallowing impairment following acute stroke. Dysphagia 2001; 16:208–15.

18. Corcoran L. Nutrition and hydration tips for stroke patients with dysphagia Nurs Times 2005 Nov 29–Dec 5;101(48)24–7.

19. Carl LL, Johnson PR, Payne J. The swallowing process and dysphagia. In: Drugs and Dysphagia How Medications Can Affect Eating and Swallowing. Austin, Texas: Pro-Ed, 2006, pp. 12–29.

20. Daggett A, Logemann J, Rademaker A, Pauloski B. Laryngeal penetration during deglutition in normal subjects of various ages. Dysphagia 2006;21(4):270–4.

21. Leslie P, Drinnan MJ, Ford GA, Wilson JA. Swallow respiratory patterns and aging: presbyphagia or dysphagia? J Gerontol Biol Sci Med Sci 2005 Mar;60(3):391–5.

22. Robbins J, Bridges AD, Taylor A. Oral, pharyngeal and esophageal motor function in aging Part 1 Oral cavity, pharynx and esophagus. 2006 doi# 10.1038/dimo39. GI Motility online 2006 (Accessed October 2007).

23. Daniels SK, Schroeder MF, McClain M, Corey DM, Rosenbeck JC, Foundas AL. Dysphagia in stroke: development of a standard method to examine swallowing recovery JRRD 2006; 43(3):347–56.

24. Smith Hammond CA, Goldstein LB. Cough and aspiration of food and liquids due to oral-pharyngeal dysphagia ACCP evidence-based clinical practice guidelines. Chest 2006; 129:154S–68S.

25. Garon GR, Engle M, Ormiston C. Silent Aspiration: results of 1,000 videofluoroscopic swallow evaluations. Neurorehab Neural Repair 1996; 10(2):121–6.

26. Ramsey D, Smithard D, Kalra L. Early assessments of dysphagia and aspiration risk in acute stroke patients. Stroke 2003; 34:1252.

27. Mann, G. MASA: The Mann Assessment of Swallowing Ability. Clifton Park, NY: Division of Thomson Learning Inc., 2002.

28. McCullough GH, Rosenbeck JC, Wertz RT, McCoy S, Mann G, McCullough K. Utility of clinical swallowing examination measures for detecting aspiration post-stroke. J Speech Lang Hearing Res 2005 Dec; 48:1280–93.

29. Ramsey DJ, Smithard DG, Kalra L. Can pulse oximetry or a bedside swallowing assessment be used to detect aspiration after stroke? Stroke 2006 Dec;37(12):2984–8.

30. Smith Hammond CA, Ying J, Horner RD, Goldstein LB, Gray L, Gonzalez-Rothi L, Bolser DC. (2006) Comparison between reflexive cough after water swallow to aerodynamic measures of voluntary cough to identify patients with stroke-related dysphagia at risk of aspiration. Poster presented at the Department of Veterans Affairs Health Services Research & Development Meeting: Managing Recovery and Health through the Continuum of Care, Arlington, VA.

31. Kuhlemeier K, Palmer J, Rosenberg D. Effect of liquid bolus consistency and delivery method on aspiration and pharyngeal retention in dysphagic patients. Dysphagia 2001; 16:119–22.

32. Suiter DM, Leder SB. Clinical utility of the 3-ounce water swallow test. Dysphagia 2008;23(3):244–50.

33. Crary MA, Carnaby Mann GD, Groher, ME. Initial psychometric assessment of a functional oral intake scale for dysphagia in stroke patients. Arch Phys Med Rehabil 2005; 86:1516–20.

34. Jacobson B, Silbergleit A, Sumlin T, Johnson A. Dysphagia Disability Index-Revised (DDI) Sept 2000. Unpublished outcome measure. Henry Ford Hospital, Detroit.

35. Belafsky PC, Postma GN, Koufman JA. Validity and reliability of the reflux symptom index (RSI). J Voice 2002; 16:274–7.

36. McHorney CA, Robbins J, Lomax K, Rosenbek JC, Chignell K, Kramer AE, Bricker DE. The SWAL-QOL and SWAL-CARE outcomes tool for oropharyngeal dysphagia in adults: III. documentation of reliability and validity. Dysphagia 2002; 17(2):97–114.

37. McHorney C, Martin-Harris B, Robbins J, Rosenbek J. Clinical validity of the SWAL-QOL and SWAL-CARE outcome tools with respect to bolus flow measures. Dysphagia 2006; 21:141–8.

38. Carnaby G, Hankey GJ, Pizzi J. Behavioral intervention for dysphagia in acute stroke: a randomized controlled trial. Lancet Neurol 2006; 5:31–7.

39. Elmstathl S, Bulow M, Ekberg O, Petersson M, Tegner H. Treatment of dysphagia improves nutritional conditions in stroke patients. Dysphagia 1999; 14:61–6.

40. Katzan IL, Cebul RD, Husak SH, Dawson NV, Baker DW. The effect of pneumonia on mortality among patients hospitalized for acute stroke. Neurology 2003; 60:620–5.

41. Katzan IL, Dawson NV, Thomas CL, Votruba ME, Cebul RD. The cost of pneumonia after acute stroke. Neurology 2007; 68:1938–43.

42. Langmore SE, Terpenning MS, Schork A, Chen Y, Murray JT, Lopatin D, Loesche WJ. Predictors of aspiration pneumonia: how important is dysphagia? Dysphagia 1998; 13:69–81.

43. Bath PMW, Bath-Hextall FJ, Smithard DG. Interventions for dysphagia in acute stroke. Cochrane Database Syst Rev 1999(4). Art. No.: CD000323. DOI:10.1002/14651858.CD000323. 2007 The Cochrane Collaboration John Wiley & Sons, Ltd.

44. Dennis M, FOOD Trial Collaboration. Effect of timing and method of enteral tube feeding for dysphagic stroke patients (FOOD): a multicentre randomised controlled trial. Lancet 2005; 365:764–72.

45. Clave P, de Kraa M, Arreola V, Girvent M, Farre R, Palomera e, Serra-Prat M. The effect of bolus viscosity on swallowing function in neurogenic dysphagia. Aliment Pharmacol Ther 2006 Nov; 1:24(9):1385–94.

46. Volkert D, Kreuel K, Stehle P. Nurtrition beyond 65- amount of usual drinking fluid and motivation to drink are interrelated in community-living independent elderly people (in German). Z Gerontol Geriatr 2004; 37:436–43.

47. Panther K. The Frazier Free Water Protocol. Perspectives Swallowing and Swallowing disorders (Dysphagia). American Speech-Language-Hearing Association Division 13 2005; 14(1): 4–9.

48. Garon BR, Engle M, Ormiston C. A randomized control study to determine the effects of unlimited oral intake of water in patients with identified aspiration. J Neurol Rehabil 1997; 11:139–48.

49. Ashford, J. Pneumonia. Factors beyond aspiration. Perspectives Swallowing and Swallowing disorders (Dysphagia). American Speech-Language-Hearing Association Division 13 2005; 14(1):10–14.

50. Bronson-Lowe C, Leising K, Bronson-Lowe D, Lanham S, Hayes S, Ronquillo A, Blake P. Effects of a free water protocol for patients with dysphagia. Dysphagia Research Society March 7, 2008 (Abstract from course syllabus printed).

51. Brady M, Furlanetto D, Hunter RV, Lewis S, Milne V. Staff-led interventions for improving oral hygiene in patients following stroke. Art. No.: CD003864. DOI: 10.1002/14651858.CD003864. pub2. Stroke 2007; 38:1115–6.

52. Shanahan TK, Logemann JA, Rademaker AW, Pauloski BR, Kahrilas PJ. Chin down posture effect on aspiration in dysphagic patients. Arch Phys Med Rehabil 1993; 74(7):736–9.

53. Logemann JA, Karilas PJ, Kobara M, Vakil NB. The benefit of head rotation on pharyngeal dysphagia. Arch Phys Med Rehabil 1989; 70(10): 767–71.

54. Easterling C, Grande B, Kern M, Sears K, Shaker R. Attaining and maintaining isometric and isokinetic goals of the shaker exercise. Dysphagia 2005; 20:133–8.

55. Shaker R, Easterling C, Kern M, Nitschke T, Massey B, Daniels S, Grande B, Kazandjian M, Dikeman K. Rehabilitation of swallowing by exercise in tube-fed patients with pharyngeal dysphagia secondary to abnormal UES opening. Gastroenterology 2002; 122:1314–21.

56. Huckabee ML, Cannito MP. Outcomes of swallowing rehabilitation in chronic brainstem dysphagia: a retrospective evaluation. Dysphagia 1999; 14:93–109.
57. Crary MA, Carnaby GD, Groher M, Helseth E. Functional benefits of dysphagia therapy using adjunctive sEMG biofeedback. Dysphagia 2004; 19:160–4.
58. Carnaby-Mann GD, Crary MA. Examining the evidence on neuromuscular electrical stimulation for swallowing. Arch Otolaryngol Head Neck Surg 2007; 133(June):564–71.
59. Chaudhuri G, Hildner CD, Brady S, Hutchins B, Aliga N, Abadilla E. Cardiovascular effects of the supraglottic and super-glottic swallowing maneuvers in stroke patients with dysphagia. Dysphagia 2002; 17(1):19–23.
60. Finestone HM, Greene-Finestone LS. Rehabilitation medicine: 2. Diagnosis of dysphagia and its nutritional management for stroke patients. CMAJ 2003 Nov 11; 169(10):1041–44.

25 Alzheimer's Disease and Other Neurodegenerative Disorders

Ling Li and Terry L. Lewis

Key Points

- Major neurodegenerative disorders affecting older adults include Alzheimer's disease (AD), Parkinson's disease, amyotrophic lateral sclerosis, and Huntington's disease. AD is the most common cause of dementia, accounting for 50–60% of all cases.
- Many AD patients experience unintentional weight loss that has negative prognostic implications, being associated with greater disease severity, a faster clinical progression rate, and increased mortality.
- Given that there are no curative therapies for neurodegenerative disorders, nutritional management may offer an opportunity to prevent or delay the onset of these devastating conditions.
- Multiple lines of evidence suggest that dietary interventions may have benefits for preventing and/or reducing the incidence of these disorders but rigorous, well-powered intervention trials are needed before clinical recommendations can be evidence-based.

Key Words: Alzheimer's disease (AD); Parkinson's disease (PD); amyotrophic lateral sclerosis (ALS); Huntington's disease (HD)

25.1 INTRODUCTION

Neurodegenerative disorders are a heterogeneous group of pathological conditions in which specific areas of the central nervous system (CNS) deteriorate progressively, resulting in cognitive or movement impairments. Major neurodegenerative disorders include Alzheimer's disease (AD), Parkinson's disease (PD), amyotrophic lateral sclerosis (ALS), and Huntington's disease (HD). Among these disorders, AD causes the most common form of dementia, accounting for 50–60% of all cases *(1)*. Although the prevalence of AD is below 1% in

From: *Nutrition and Health: Handbook of Clinical Nutrition and Aging, Second Edition*
Edited by: C. W. Bales and C. S. Ritchie, DOI 10.1007/978-1-60327-385-5_25,
© Humana Press, a part of Springer Science+Business Media, LLC 2009

individuals aged 60–64 years, it increases almost exponentially with age. In people aged 85 years or older, the prevalence of AD is between 24 and 33% in developed countries *(2)*. Currently, about 5 million people live with AD in the United States, and there are more than 20 million cases of AD worldwide. With the anticipated increase in human life expectancy, the number of people afflicted by AD is expected to double every 20 years *(2)*. Therefore, AD and other age-related neurodegenerative disorders constitute an increasing health and socio-economic problem in the world.

In recent years, much progress has been made on elucidating the etiology and pathogenic mechanisms of neurodegenerative disorders. In addition to genetic factors, environmental factors play a significant role in the development of neuro-degenerative disorders. In this chapter, we attempt to summarize the current under-standing on the pathophysiology of AD and other degenerative disorders and present recent findings on the role of nutrition in the disease process. Given that there are no satisfactory therapies available for neurodegenerative disorders, nutri-tional management may signify an effective measurement to prevent or delay the onset of these devastating disorders.

25.2 PATHOPHYSIOLOGY OF NEURODEGENERATIVE DISORDERS

25.2.1 Alzheimer's Disease

Alzheimer's disease was named after a German doctor, Alois Alzheimer, who reported the first case of AD in 1906 *(3)*. Clinically, AD is characterized by progressive cognitive impairment and changes in behavior and personality. It affects brain regions such as the entorhinal cortex, hippocampus, basal fore-brain, and amygdala that are involved in learning/memory and emotional beha-viors. Pathological hallmarks of AD include extracellular amyloid plaques and intracellular neurofibrillary tangles (Fig. 25.1). The main component of amyloid plaques is amyloid-β protein (Aβ), containing 39–43 amino acids cleaved from a large transmembrane glycoprotein, amyloid-β precursor protein (APP). Neurofibrillary tangles are intracellular fibrillar aggregates of the microtubule-associated protein tau that exhibit hyperphosphorylation and oxidative modifications.

Although the neuropathology of AD has been known for a century, our under-standing on the etiological and pathogenic mechanisms underlying AD have only occurred over the last two decades. One of prevailing hypotheses for the develop-ment of AD is the amyloid (or Aβ) cascade hypothesis *(4)* (Fig. 25.2). This hypothesis states that genetic mutations and other risk factors cause an imbalance in the metabolism of APP/Aβ. The gradual accumulation and aggregation of Aβ initiate a slow but insidious cascade that leads to synaptic alterations, microglial and astrocytic activation, modification of the normally soluble tau into oligomers and then into insoluble paired helical filaments (tangles), and progressive neuronal loss associated with multiple neurotransmitter deficiencies, culminating in cognitive failure (dementia) *(4)*. This hypothesis is based on the discovery of mutations in the genes encoding APP, presenilin-1 (PS1), and presenilin-2 (PS2) that cause the

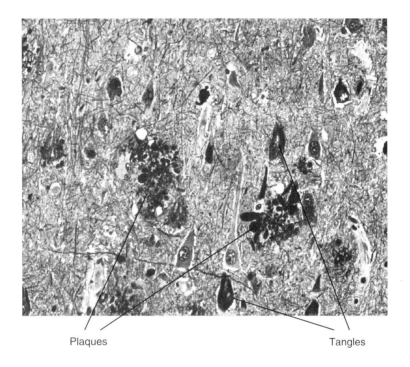

Plaques   Tangles

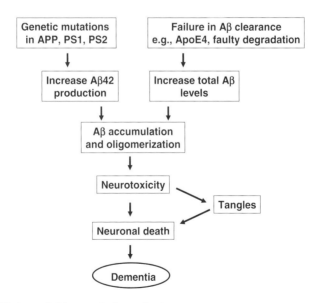

Fig. 25.2. Simplified amyloid cascade hypothesis.

early-onset (< 65 years) familial form of AD (EOFAD) *(5–7)*. This hypothesis is supported further by the fact that transgenic animals carrying EOFAD mutations in the APP gene develop some of the neuropathological and behavioral character-istics of AD *(8–10)*.

The majority of cases are the late-onset ($\geq$ 65 years) sporadic form of AD (LOAD). The cause of sporadic AD is very complex and not fully understood at present. The complexity arises from the heterogeneity of the disease involving interactions of multiple genetic and environmental factors. While aging itself is the biggest risk factor for LOAD, one of the major genetic risk factors for sporadic AD is one's genotype for apolipoprotein E (*APOE*). There are three *APOE* alleles in humans: *APOE-ε2*, *APOE-ε3*, and *APOE-ε4*, with an allele frequency of 7, 78, and 15%, respectively *(11)*. While the *APOE-ε2* allele confers some protection against AD *(12)*, the *APOE-ε4* allele is associated with an increased risk of AD *(13,14)*. One *APOE-ε4* allele increases the risk of AD by three times (heterozygotes) and two *APOE-ε4* alleles increase the risk of AD by 15 times (homozygotes) *(15)*. However, the *APOE* genotype predicts when, not whether, one is predisposed to develop AD *(16)*. In other words, the *APOE-ε4* allele decreases the age of onset. Each copy of the allele lowers the age of onset by about 10 years *(13)*. The question of how the *APOE-ε4* allele promotes AD has not been answered completely. The protein product of the *APOE* gene, apoE, plays an important role in cholesterol transport and lipid metabolism. Apo E4 is less efficient than the other apoE isoforms in recycling of membrane lipids and neuronal repair *(17)*. Apolipoprotein E has been implicated as a chaperone that modulates Aβ aggregation and deposition or clearance. In mice, apoE is necessary for Aβ deposition and plague formation *(18)*. Furthermore, expression of human Apo E2, E3, and E4 in the absence of endogenous murine apolipoprotein E leads to isoform-specific differences in amyloid load, with E4>E3>E2 *(19)*. However, about 50% of patients with LOAD do not carry the *APOE-ε4* allele, indicating that there are other genetic and/or environmental factors leading to the development of sporadic AD. Recent genetic studies have identified several other candidate genes associated with the risk of AD *(20)*. Unlike the situation with *APOE-ε4* allele, the association of these genes with AD is weak and has not been consistently replicated. Because of the heterogeneous nature of sporadic AD, it is possible that several susceptibility genes act in concert, each conferring only a minor increase in risk. In addition, the risk of AD may be modified by complex interactions between genetic factors and environmental factors such as nutrition, which will be discussed in Section 25.3.

25.2.2 Parkinson's Disease

Parkinson's disease is the second most common neurodegenerative disorder. The prevalence of PD is age related, with approximately 1% of the population being affected at 65 years, increasing to 4 to 5% in 85-year-olds *(21)*. The main clinical feature of PD is parkinsonism, a movement disorder that is characterized by tremor at rest, bradykinesia, rigidity, and postural instability *(21)*. These symptoms arise from the progressive degeneration of dopaminergic neurons leading to a profound depletion of the neurotransmitter dopamine in the striatum, a central component of the basal ganglia that is responsible for the initiation and control of movement. The pathological diagnosis of PD requires *post-mortem* findings of neuronal loss and depigmentation of the substantia nigra (SN), plus the presence of Lewy bodies (proteinaceous intracellular inclusions) in the brain stem *(22)*. The main component of Lewy bodies is the protein α-synuclein. These pathologic characteristics distinguish PD from other forms of parkinsonism *(22)*.

Like AD, the etiology of PD is complex and multifactorial. While the majority of PD cases are sporadic, PD is inherited as a Mendelian trait in a minority of cases. Studies in PD families have identified several causative genes *(23)*. Mutations in the gene *SNCA* that encodes α-synuclein (a protein that is expressed throughout the brain and has potential roles in learning, synaptic plasticity, vesicle dynamics and dopamine synthesis) cause a type of PD with autosomal-dominant inheritance and typical Lewy-body pathology. The role of aggregation of α-synuclein protein in Lewy bodies in the molecular pathogenesis of PD remains controversial *(24)*. On the other hand, mutations in the parkin gene cause autosomal recessive parkinsonism of early onset. In this form of PD, nigral degeneration is not accompanied by Lewy-body formation. Mutations in the parkin gene appear to be a common cause of PD in patients with very early onset *(23)*. The protein parkin functions as an ubiquitin–protein ligase, conjugating ubiquitin to proteins targeting them for degradation by the proteasome *(25)*. The potential importance of this cellular protein degradation pathway is also highlighted by the finding of a mutation in the gene for ubiquitin C-terminal hydrolase L1 in another small family with PD. These findings strengthen the hypothesis that proteasome dysfunction and resulting protein aggregation is central to PD. Causative mutations in other genes, *PINK1* (PTEN-induced kinase 1), *DJ-1* (an oncogene), and *LRRK2* (leucine-rich repeat kinase 2) have also been found in families with PD *(23)*. In addition, recent genome-wide association studies have identified many more susceptibility genes for PD *(26)*. However, the role of these genes in the pathogenesis of PD can only be confirmed with replication and functional proof.

Although these findings prove that there are several genetically distinct forms of PD that can be caused by mutations in single genes, there is at present no definitive evidence that any of these genes have a direct role in the etiology of the common sporadic form of PD. It is widely thought that a combination of interacting genetic and environmental causes may be responsible in the majority of PD cases. However, studies of gene–environment interactions have not yet produced any convincing results. Nevertheless, findings from genetic studies provide the rationale for developing novel therapeutic approaches to halt disease progression and have led to powerful model systems that develop PD-like pathology. It is expected that the elucidation of the molecular sequence of events leading to nigral degeneration in clearly inherited cases will provide insights on the molecular pathogenesis of the common sporadic form of PD.

25.2.3 *Amyotrophic Lateral Sclerosis*

Amyotrophic Lateral Sclerosis (ALS) results from the degeneration of motor neurons in the brain, spinal cord, and peripheral nervous system. The clinical features of ALS include weakness of the arms, legs, and face as well as difficulties with speech, swallowing, and breathing. ALS affects women and men (ratio of males to females 1.6:1), regardless of ancestry, and the risk of disease increases with age. Its clinical progression is one of the fastest of the neurodegenerative diseases, with death (often from respiratory failure) typically occurring within 3–5 years after onset. The incidence is approximately 2 per 100,000 persons per year, and the prevalence is approximately 6 per 100,000 persons *(27)*.

While the majority (∼ 90%) of ALS cases are sporadic, about 10% of ALS cases are familial forms, resulting from highly penetrant, monogenic mutations that cause disease. Mutations in the *SOD1* gene (encoding copper–zinc superoxide dismutase) account for about 2% of all patients and 20% of those with an autosomal dominant form of ALS *(27)*. In addition to *SOD1*, mutations in five other genes (encoding alsin, angiogenin, dynactin 1, senataxin, and vesicle-associated protein B) have been associated with motor neuron disease (although often not a typical ALS phenotype) in a few families *(28)*.

Little is known about the specific genes that contribute to the development of sporadic ALS. Very recently, a genome-wide association study identified 10 genetic loci that are significantly associated with sporadic ALS *(29)*. The most significant association with disease in patients as compared with controls was found for a genetic variation (single nucleotide polymorphism) near an uncharacterized gene known as *FLJ10986*. The FLJ10986 protein was found to be expressed in the spinal cord and cerebrospinal fluid of patients and of controls *(29)*. Nevertheless, the function and relevance of this gene in ALS is currently unknown.

Our understanding of the pathogenesis of ALS is also limited. Numerous mechanisms have been implicated in the selective degeneration of motor neurons in patients with sporadic ALS, including oxidative damage, excitotoxicity, apoptosis, cytoskeletal dysfunction, axonal-transport defects, inflammation, protein-processing and degradation defects, and mitochondrial dysfunction *(30)*. Most of the research has been done with the use of cellular and animal models, in particular with transgenic mice carrying human *SOD1* mutations. These mice have some features that are similar to those of patients with ALS; however, a unifying understanding of any ALS mechanism remains elusive.

25.2.4 *Huntington's Disease*

Huntington's disease (HD) is an autosomal-dominant, progressive neurodegenerative disorder. HD is characterized by selective degeneration of medium spiny neurons in the striatum, resulting in a progressive atrophy of the caudate nucleus, putamen, and globus pallidum *(31)*. In the later stage of the disease there is often atrophy in other areas, including cortical neurons. Patients with HD exhibit distinct symptoms, including chorea and dystonia, incoordination, cognitive decline, and behavioral difficulties *(31)*. Huntington's disease shows a stable prevalence in most populations of white people with about 5–7 affected individuals per 100,000. Interestingly, the rate is much lower in Asian and African populations *(32,33)*. HD is a monogenetic disorder. The underlying genetic cause of HD is a trinucleotide (CAG) repeat expansion in the gene encoding for a protein called huntingtin on chromosome 4 *(34)*. In the normal population the number of CAG repeats is maintained below 35, whereas in individuals affected by HD it ranges from 35 to more than 100, resulting in an expanded polyglutamine segment in the protein. HD can occur at any age. The age of onset is mainly determined by the number of CAG repeats (inverse correlation) in the huntingtin gene. However, other modifying genes and environmental factors have been shown to influence the age of onset for HD *(35)*. Although the function of huntingtin is still unknown, it has been hypothesized that the expanded polyglutamine segment confers a dominant toxic "gain of function" to the protein, leading to selective neuronal dysfunction and ultimately neurodegeneration *(31)*.

25.3 NUTRITIONAL FACTORS AND THE RISK OF AD AND OTHER NEURODEGENERATIVE DISORDERS

As we discussed in the previous section, causative genetic mutations only account for a small fraction of cases of neurodegenerative disorders. The cause for the majority of sporadic cases involves complex gene–environment interactions. Environmental factors (e.g., nutrition) modify the development of these disorders. To some extent, people are what they eat. The portion size, type, and content of the food that people ingest over a lifetime may affect whether or not one develops dementia in the old age. In this section, we attempt to summarize recent findings on the effect of dietary fat, cholesterol, carbohydrates, and vitamins on the development of neurodegenerative disorders.

25.3.1 *Dietary Fat and Cholesterol*

It has been long recognized that Western-type high-fat and high-cholesterol diets contribute to the prevalence of cardiovascular disease *(36)*. Recent studies indicate that high intake of saturated fat and cholesterol increase the risk of AD *(37–39)*. Kalmijn et al. reported a higher risk of incident AD with higher intake of total fat, saturated fat, and cholesterol in more than 5,000 individuals 55 years and older without dementia at baseline followed for 2 years *(37)*. Luchsinger et al. reported similar results in 980 individuals 65 years and older without dementia at baseline followed for 4 years *(38)*. This was particularly evident in people carrying the *APOE-ε*4 allele. In a population-based study, Laitinen et al. found that a moderate intake of saturated fats at midlife increases the risk of dementia and AD, especially among *APOE-ε*4 allele carriers *(39)*. These findings suggest that dietary interventions may potentially decrease the risk of dementia, particularly among genetically susceptible individuals. In transgenic mouse models of AD, we and others have also demonstrated that atherogenic diets accelerate and exacerbate AD-like neuropathology *(40–43)* and aggravate learning and memory deficits *(43)*.

In contrast, intake of unsaturated fat, in particular the omega-3 polyunsaturated fatty acids (n–3 PUFAs), is associated with a lower incidence of AD. Fish and fish oil have a high content of two important n–3 PUFAs, docosahexaenoic acid (DHA, 22 carbons long with 6 double bonds, 22:6) and its precursor eicosapentaenoic acid (EPA, 20:5). DHA can also be synthesized in the body from the n–3 fatty acid α-linolenic acid (18:3), which is present in certain vegetable oils, nuts, and seeds. However, this synthetic step is relatively inefficient *(44)*. Aside from other functions, DHA is the most prominent fatty acid in the brain and is necessary for cognitive function. DHA is especially rich in the neurons and synaptosomes of the cerebral cortex, where it occupies the number 2 position of membrane phospholipids. AD patients have a decreased level of DHA in the serum and brain compared with age-matched, nondemented subjects *(45,46)*. Increased consumption of fish (and in particular, DHA) has been reported to significantly reduce the likelihood of developing AD *(37,47,48)*. Two recent studies further support the beneficial effects of fish consumption in protection of cognitive function. The Zutphen Elderly Study *(49)* showed that fish consumers had less cognitive decline over a 5-year period than did fish nonconsumers in men aged 70–89 years. A linear relation was found between the estimated intake of DHA and EPA (DHA + EPA) and the prevention of cognitive decline. A DHA + EPA intake of about 380 mg/day seemed to prevent cognitive decline. This amount of DHA + EPA would be found in 20 g of Chinook salmon or in 100 g of cod. Two to three meals of fish per week would supply approximately 380 mg EPA + DHA/day. The Minneapolis study of 2,251 white men and women also showed that the n–3 PUFAs retarded the decline in cognition over time *(50)*. In addition, several studies in animal models have demonstrated that dietary supplement of DHA protects against the development of AD-like neuropathology and memory impairment *(51–54)*. Although the underlying mechanisms by which n–3 PUFAs exert beneficial effects on cognitive function are not fully understood, several mechanisms have been suggested including the anti-inflammatory properties of n–3 PUFAs *(55)*. The sum of these data provides the rationale for a clinical trial of fish, fish oil, or both in elderly patients prone to the development of AD. It should be noted that Morris et al. *(56)* linked dietary intake of fish and omega-3 fatty acids with slower rates of cognitive decline in a prospective cohort study of adults aged >65 years but the effect seemed to be associated with fats (saturated, polyunsaturated, trans) other than the omega-3 fatty acid component of the diet. This might imply unique benefits from fish consumption that are not available from omega-3 fatty acid supplements alone. In fact, the American Heart Association has recommended that all adults consume two fish meals per week to reduce the risk of cardiovascular disease *(57)*.

For other neurodegenerative disorders, the information about dietary fat associations is limited. For PD, several epidemiological studies have shown inconsistent results on the association between dietary fat intake and the risk of PD *(58 61)*. Most of these studies were retrospective case–control studies focusing mainly on intake of total fat and saturated fat. Two recent prospective population-based studies indicate that higher intake of unsaturated fat is associated with a lower risk of PD *(62,63)*. In accordance with these findings, a high intake of PUFAs is associated with a decreased risk of developing ALS *(64)* and treatment with EPA is beneficial for symptom management in patients with HD *(65)*. Clearly, further clinical studies are required to fully explore the effects of n–3 PUFAs on any type of neurodegenerative disease.

25.3.2 Dietary Carbohydrates and Glucose Tolerance

Another common concern about the traditional Western diet focuses on a high intake of refined sugars, mainly sucrose and high fructose corn syrup. Compelling evidence indicates that excess consumption of sweet foods, particularly sugar-sweetened beverages, plays an important role in the epidemic of obesity around the world *(66)*. In the United States, the percentage of children who are overweight has doubled, and the percentage of teenagers who are overweight has tripled in the last two decades *(67,68)*. Overweight children are at an increased risk to become obese adults *(69)*. Even moderate obesity can contribute to chronic metabolic abnormalities leading to type 2 diabetes mellitus (T2DM) *(70)* characterized by glucose intolerance and hyperinsulinemia.

Recently, numerous epidemiological studies suggest that T2DM is associated with an increased risk of AD *(71)*, independent of the risk for vascular dementia *(72,73)*. The mechanisms by which T2DM may impact AD, however, are not well understood. Several lines of evidence indicate that insulin itself and metabolic abnormalities pertinent to diabetes may affect the generation and degradation of $A\beta$ *(74)*. It has been noted that AD patients have a preference for high-carbohydrate foods *(75–77)* and that this shift of food preference is associated with poorer memory *(77)*. However, a recent study with a cohort of elderly subjects in New York City did not observe a positive association between glycemic load and the risk of AD *(78)*. Recently, we conducted a study in a transgenic mouse model of AD, in which we provided sucrose-sweetened water to the mice fed a regular low-fat diet *(79)*. Our results show that intake of sucrose-sweetened water (approximates five cans of 12 oz sugar-sweetened beverages for a 2,000 calorie diet in humans) induced insulin resistance and exacerbated AD-like memory impairment and brain amyloid pathology *(79)*. Our findings are of tremendous importance given that the consumption of sugar-sweetened beverages has increased dramatically in the past few decades and will most likely remain high in modern societies. Controlling the consumption of sugar-sweetened beverages may be an effective way to curtail the risk of developing AD.

In findings similar to those from studies of AD, dietary intake of carbohydrates and midlife adiposity have been associated with an increased risk of PD *(62,80)*. While there is interest in the potential benefit of a low-carbohydrate/high-fat ketogenic diet regimen for managing symptoms of PD *(81)*, there are also concerns about the long-term effects of consuming this high-fat and potentially unhealthy diet over time. Although the information on the effect of dietary carbohydrates on the development of ALS and HD is very limited, some evidence indicates that insulin resistance/aberration of carbohydrate metabolism is associated with ALS *(82,83)* and that there is an increased frequency of diabetes mellitus in patients with HD *(84)*.

25.3.3 *Vitamins and Antioxidants*

25.3.3.1 VITAMINS B_6 AND B_{12}, FOLATE, AND HOMOCYSTEINE

Elevated homocysteine levels in the plasma are a known risk factor for cardiovascular disease and stroke *(85,86)* and may be related to increased risk of AD as a neurotoxin or as a result of vascular changes associated with homocysteine *(87)*. Homocysteine is an intermediate to methionine and cysteine. Folate and vitamin B_{12} are required to convert homocysteine to methionine. Vitamin B_6 is required to convert homocysteine to cysteine *(87)*. Thus, deficiencies of vitamin B_6, vitamin B_{12}, and folate could lead to increased concentrations of homocysteine by preventing or slowing its conversion to methionine and cysteine, respectively.

Interestingly, in the Framingham study, homocysteine concentrations over 14 μmol/l doubled the risk of developing AD, while vitamin B_6, vitamin B_{12}, and/or folate showed no association with risk of AD (88). In contrast, the Nun study showed that decreased serum folate levels were associated with increased neocortical atrophy at autopsy in participants with significant numbers of plaques and tangles (89), whereas levels of for vitamins B_6 or B_{12} were not associated with neocortical changes (89).

Other studies specifically designed to measure the interactive effects among vitamins B_6 and B_{12}, folate, and homocysteine have produced conflicting results. Low concentrations of folate and vitamin B_{12} have been shown to be associated with a high (90) to even a doubled risk (91) of developing AD. Treatment with vitamin B_{12} improved cognitive function in patients with high serum homocysteine concentrations (92), whereas vitamin B_{12} supplement had no effect on cognition in demented patients (93) or had no association with AD prevention or progression (94,95). Higher intake of folate has been related to a lower risk of AD, while intakes of vitamin B_6 and B_{12} have no effect on the risk of AD (96). Cochrane reviews for folate, vitamins B_{12}, and B_6 have not supported a cognitive benefit for these nutrients (97–99). The discrepancies among these studies may be attributed to different patient populations, age of subjects as well as the dosage of supplements administered. Nevertheless, experimental studies in cultured cells and in animal models have shown that homocysteine displays multiple aspects of neurotoxicity (100) and that folic acid deficiency and homocysteine impairs DNA repair in hippocampal neurons and increases $A\beta$ toxicity (101).

In summary, elevated plasma homocysteine concentrations have been associated with an increased risk of developing dementia and AD. However, it is not clear if an elevation in total homocysteine concentration is a "risk factor" with a direct pathophysiological role in the development of the disease or merely a "risk marker" reflecting an underlying process such as oxidative stress responsible for both the high homocysteine concentrations and the development of AD. Given that vitamin therapy with folate, B_{12}, and B_6 lowers plasma homocysteine levels and could significantly reduce the risk of stroke and dementia permitting healthy brain aging, clinical trials should be conducted to evaluate the role of these vitamins and other homocysteine-lowering treatments in the primary prevention of AD, as well as their role in preserving cognition among persons with mild cognitive impairment and early dementia.

High plasma homocysteine levels have also been reported in patients with PD (102,103). Higher dietary intake of vitamin B_6 but not folate and B_{12} has been shown to be associated with a significantly decreased risk of PD in a prospective population-based Rotterdam study (104). Interestingly, this association is only in smokers, suggesting that the antioxidant properties of vitamin B_6 rather than its involvement in homocysteine metabolism may be responsible for its anti-PD effects (104). For ALS, no studies in human patients are available. Experimental studies in cultured neurons and in animal models suggest that homocysteine plays an important role in the pathogenesis of ALS and that therapy with antioxidants and vitamin supplement may slow the neurodegenerative process in human ALS (105,106). For

HD, a recent study has shown hyperhomocysteinemia in HD patients compared to controls *(107)*. Therefore, dietary vitamin supplements that lower homocysteine levels could provide beneficial effects on HD.

25.3.3.2 ANTIOXIDANT NUTRIENTS

Oxidative stress increases with aging and is associated with the manifestation and progression of age-related neurodegenerative diseases *(108)*. It is as of yet unknown if the production of oxidative species is a primary cause or a secondary result of the disease progression. The production and deposition of Aβ early in AD cause an increase in oxidative stress by reducing redox-active copper (Cu(II)) and iron (Fe(III)) concentrations in the brain, resulting in a subsequent increase in reactive oxygen species, oxidative stress, and neuronal damage *(109)*. The idea that reactive oxygen species may be causative has led to research involving antioxidants as possibly being involved in the prevention of the progression or even protecting against AD.

The majority of antioxidant research has revolved around the naturally occurring vitamins found in the diet. These include vitamin E (α-tocopherol), vitamin C (ascorbic acid), and carotenes. It has been shown that vitamin E suppresses signaling involved in the inflammatory cascade and reduces lipid peroxidation and the resultant oxidative stress produced by Aβ *(110)*. Vitamin C could have an affect on the production of neuronal signaling molecules norepinephrine and epinephrine. Vitamin C also blocks the reduction of nitrites preventing the formation of DNA-mutating nitrosamines. Like vitamin E, carotenes may have a protective effect by preventing lipid peroxidation *(110,111)*.

Human studies to date are conflicting concerning intake of antioxidants in relation to dementia and AD. The benefit of dietary supplementation with vitamin C and E have been shown to decrease the risk of vascular dementia, increase cognitive function in non-AD subjects *(112,113)*, and decrease the development of AD *(114)*. Vitamin E supplement alone *(115)*, vitamins E and C from dietary sources *(116,117)*, and addition of β-carotene to the diet *(118)* have been shown to decrease the incidence of AD. However, other studies have found no relation between intake of vitamin C, vitamin E, and carotenes and the risk of AD *(119,120)* The discrepancies may be attributed to the different ages and level of dementia of the study participants during testing. The studies that found beneficial effects were generally in younger patients with impaired cognition. The studies that found no benefits were generally in a population of older patients in whom there are a natural increase of oxidative species and a decrease in dietary intake, both of which would reduce the pool of available antioxidants.

A meta-analysis of eight observational studies on the effect of vitamin C, vitamin E, and β-carotene intake on the risk of PD shows that dietary intake of vitamin E protects against PD *(121)*. This protective effect was found with both moderate intake and high intake of vitamin E, although the possible benefit associated with high intake of vitamin E was not significant. These studies did not suggest any protective effects associated with vitamin C or β-carotene *(121)*. In a transgenic mouse model of ALS, dietary supplementation with vitamin E delays onset of clinical disease and slows progression *(122)*. However, no clear evidence for a beneficial effect of vitamin E administration *(123)* or a variety of antioxidant combinations *(124)* has been obtained in humans. In HD patients, while treatment with vitamin E had no effect on neurological and neuropsychiatric symptoms in the treatment group overall, post-hoc analysis revealed a significant selective therapeutic effect on neurological symptoms for patients early in the course of the disorder *(125)*, suggesting that antioxidant therapy may slow the rate of motor decline early in the course of HD.

In addition to vitamin E, vitamin C, and carotenes, other dietary antioxidants such as flavonoids and polyphenols have also been shown to protect against neurodegenerative disorders *(126)*. Detailed discussion on these other dietary antioxidants is beyond the scope of this chapter.

25.3.4 Dietary Patterns

Some of the conflicting findings on dietary effects discussed in the previous sections may partly result from the fact that most studies focus on individual dietary constituents and not the overall diet as a whole. Dietary pattern analysis has recently received growing attention in relation to many diseases because people do not consume foods or nutrients in isolation but rather as a variety of components of their daily diet. Several studies have investigated the effect of composite dietary patterns on the risk for AD. One such dietary pattern is the Mediterranean diet (MeDi). The MeDi is characterized by high intake of vegetables, legumes, fruits, and cereals; high intake of unsaturated fatty acids (mostly in the form of olive oil) and low intake of saturated fatty acids; a moderately high intake of fish; a low-to-moderate intake of dairy products (mostly cheese or yogurt); a low intake of meat and poultry; and a regular but moderate amount of ethanol, primarily in the form of wine and generally during meals *(127)*. Apparently, this diet includes many of the components reported as potentially beneficial for AD and cognitive performance. Interestingly, while individual food groups were not significantly associated with risk for AD, Scarmeas et al. found the composite Mediterranean dietary pattern to be associated with a reduction in AD risk *(128)*. This is consistent with other studies that show benefit from the composite Mediterranean diet on cognitive performance *(129)* as well as protection from death from any cause *(130)*. Recently, it has also been demonstrated that the adherence to MeDi reduces the mortality in AD *(131)*. Whether the Mediterranean dietary pattern is beneficial to other neurodegenerative disorders has not been investigated. While studies of dietary patterns more closely approximate the way individuals consume nutrients in combinations, there are limitations to this approach of diet analysis. Since there is no acknowledged "gold standard" analytical approach for dietary pattern analysis, different conclusions can be derived from the same dietary data depending upon the technique employed *(132)* and it is impossible to make comparisons across studies *(133)*. Thus while this approach is promising for its ability to develop public health messages in terms of food group servings, further refinement of the technique is needed before conclusions appropriate to drive consumer recommendations can be made *(132)*.

25.3.5 Caloric Restriction as a Preventative Strategy

As with the quality of the diet, the quantity of the diet influences health and disease. The ability of caloric restriction (CR) during adult life to lengthen life span has been observed consistently in many different species of mammals *(134)*. Findings from studies of human populations and animal models indicate that reduced food intake may protect against AD. A prospective epidemiological study of a large cohort in New York City provided evidence that individuals with a low calorie intake have a reduced risk of developing AD *(38)*, whereas excessive calorie intake is associated with increased risk of AD *(78)*. CR has been shown to reduce the development of amyloid pathology in the brain of transgenic mouse models of AD *(135,136)* suggesting that CR can suppress a key pathogenic process in AD. Recently, in a mouse model of AD with both amyloid plagues and neurofibrillary tangles, CR and intermittent fasting (IF) dietary regimens have been shown to ameliorate age-related deficits in cognitive function *(137)*. The underlying mechanisms, however, may or may not be related to Aβ and tau pathologies.

Other studies have also shown that CR and IF diets are neuroprotective and improve functional outcome in animal models of PD and HD *(138)*. The animal studies suggest that CR and IF may benefit the brain by reducing levels of oxidative stress and by enhancing cellular stress resistance mechanisms. Interestingly, no benefit of dietary restriction has been observed on disease onset or progression in animal models of ALS *(139)*. It is hypothesized that one reason that motor neurons might be selectively vulnerable to low-energy diets is that they are unable to engage neuroprotective responses to energetic stress response involving the protein chaperones, such as, heat-shock protein-70 *(139)*.

Whether we can translate findings in animal models to humans awaits randomized clinical trials. Most previous human studies have compared overweight (excess caloric intake) vs. normal weight people. It is not known whether it is beneficial for a normal weight individual to be on a calorie-restricted diet. Also, the actual values for optimum calorie intake and meal frequency may vary considerably among individuals because of factors such as their activity level, age, and sex. More studies are needed before any recommendations can be made on CR or IF dietary regimens.

25.3.6 Weight Loss in Neurodegenerative Disorders

Weight loss in AD has been well documented through out the history of the disease. Dr. Alois Alzheimer included weight loss as one of his findings in the original case in 1906 *(3)*. The National Institute of Neurological and Communicative Disorders and Stroke and the Alzheimer's Disease and Related Disorders Association Work Group include weight loss as a clinical feature consistent with the diagnosis of AD *(140)*. Various studies have shown that AD patients were more likely to lose weight compared to controls *(141–143)*. Weight loss in AD has negative prognostic implications, which is associated with greater disease severity, a faster clinical progression rate, and increased mortality. In contrast, modest weight gain is associated with a slower progression rate and reduced mortality *(144)*. However, the questions about when and why many patients with AD lose weight have not been answered.

Recent studies indicate that weight loss may occur several years prior to the clinical diagnosis in AD. In the Honolulu-Asia Aging Study, Stewart et al. tracked the progression of weight loss and development of dementia in a large population sample of men over a period of 32 years *(145)*. They found that on average, the men who became demented weighed neither more nor less during middle age than those who did not develop dementia. However, during the 6 years prior to diagnosis, the men who developed dementia lost a significantly greater amount of weight (an average of 0.8 lb per year) compared with the men who did not develop dementia evaluated at the same follow-up visit. Fifty-seven percent of the men with incident dementia had a weight loss of 11 lb or more during the 6-year period prior to their dementia diagnosis compared with 35% of those who did not develop dementia. Weight loss accelerated in the 3-year period prior to diagnosis *(145)*. Consistently, Johnson et al. also showed that accelerated weight loss precedes diagnosis in AD *(146)*. In this study, participants ($n = 449$) were older adults (65–95 years) who were enrolled as control subjects without dementia and followed longitudinally (6 years on average). Of the 449 participants, 125 developed AD. Participants without dementia lost about 0.6 lb per year, those who developed AD, doubled that rate of weight loss (1.2 lb per year) during the year prior to the detection of AD. As a group, participants who eventually developed AD weighed less (about 8 lb) at study enrollment (i.e., when they did not have dementia) than participants who remained without dementia. These findings suggest that weight loss may accelerate before the diagnosis of AD and may be a preclinical indicator of AD *(146)*.

The mechanisms responsible for weight loss prior to a diagnosis of AD are not well understood. The years just prior to a diagnosis of AD are often marked by a period of mild cognitive impairment (MCI), a transitional stage between normal cognitive aging and dementia in which memory problems start to become apparent. The current data suggest that weight loss begins concurrently with the MCI stage of AD or perhaps even earlier in many patients. MCI is associated with many behavioral, metabolic, and neuro-anatomical changes. It has been shown that patients with MCI have a higher prevalence of apathy, irritability, anxiety, and depression that may affect appetite or interest in meal preparation *(147)*. Patients with MCI have smaller hippocampi and faster rates of hippocampal atrophy than normal controls *(148)*, and medial temporal lobe atrophy is correlated with lower body weight in AD *(149)*. The presence of *APOE-ε4* allele may be associated with weight loss in AD. Women with AD carrying the *APOE-ε4* allele were more likely to lose weight than women without the *APOE-ε4* allele *(150)*. It has also been found that the presence of *APOE-ε4* allele may cause increased cortisol levels in the cerebrospinal fluid of patients with AD compared to elderly normal controls *(151)*. Changes in the concentration of leptin, an appetite-suppressing hormone secreted by adipocytes, may also contribute to weight loss in AD. Therefore, these studies suggest that there may be complex relationships between *APOE-ε4* allele, susceptibility to AD, weight loss, hormonal dysregulation, hippocampal atrophy, and cognitive impairment. Ultimately, the underlying disease process is presumably responsible for the weight loss in AD, but the details are unclear. It is possible that weight loss is a consequence of the pathologic features affecting various brain structures required to maintain body-weight homeostasis. These changes may alter behavior, appetite, and energy expenditures that lead to lower body weight. Alternatively, weight loss may be part of a generalized stress response or downregulation of energy needs associated with reduced brain function *(152)*. Future multiple-year MCI and AD longitudinal studies and clinical trials will help to elucidate the relationship between weight loss, brain metabolism, neuropathology, and cognitive decline.

Weight loss has also been observed in patients diagnosed with PD. As in AD, several studies have shown that weight loss precedes the clinical diagnosis. Logroscino et al. reported that a body mass index (BMI) decline occurred about 5 years prior to clinical diagnosis of PD in a prospective study among 10,812 men in the Harvard Alumni Health Study *(153)*. Consistent with these data, another prospective study showed that weight loss may precede the diagnosis of PD by 2–4 years *(154)*.

The mechanisms for weight loss in PD may result from subclinical effects of the disease, such as changes in dietary habits, or functional impairments secondary to rigidity and bradykinesia. Difficulty swallowing, a common clinical manifestation of PD, could also play an important role in decreasing food intake. Alternatively, it may be that some of the early nonmotor symptoms, such as constipation and sleep disorders, which can precede the diagnosis by many years *(154)* cause the weight loss. Degeneration of dopaminergic neurons in PD may induce an altered pattern of food intake. Since dopamine acts as a potent inhibitor of feeding and patients with PD usually have increased energy intake *(59)*, the decline in BMI cannot be explained by reduced energy intake *(154)*. However, motor symptoms associated with PD could increase energy requirements. Thus, it is possible that even in the presence of higher energy intake, there may be a negative energy balance, leading to a decrease in BMI. In addition, hippocampus atrophy, which has been associated with weight loss in AD *(149)*, has been observed in nondemented PD patients *(155)* and, therefore, may also be associated with weight loss in PD patients.

In ALS, denervation and a rapid reduction in physical activities usually cause muscle atrophy in ALS patients, which is consistent with a reduction in fat-free mass (FFM) *(156)*. It is known that FFM is the primary determinant of resting energy expenditure (REE), which represents $\sim 60\%$ of total daily energy expenditure *(156)*. Thus, given that ALS patients have decreased weights and FFM values compared to the control group, one would expect overall hypometabolism or normometabolism in ALS patients. However, several studies have reported a state of hypermetabolism in ALS *(156,157)*. The reason for this seemingly paradoxical phenomenon is currently unknown. Nevertheless, in the late stage of ALS, there is a progressive decreases in body fat, lean body mass, muscle power, and nitrogen balance and an increase in resting energy expenditure *(156)*. As ALS patients experience a chronically deficient intake of energy, it has been recommended to augment energy intake rather than increase the consumption of high-protein nutritional supplements in ALS patients *(156)*.

In HD, studies in both animal models and humans have shown that weight loss is a feature associated with the disease onset *(158,159)*. However, it is unknown whether weight loss is a component of HD manifestation that can be detected at an early stage of the disease or merely a consequence of increased energy expenditure through involuntary movements. Gaba et al. compared 24-h energy expenditure (EE) and energy intake in persons with early midstage HD with those of matched control subjects and found that 24-h EE was 11% higher in the HD subjects than in the control subjects *(160)*. In this study, HD subjects were able to maintain positive energy balance when offered adequate amounts of food in a controlled setting *(160)*. In a transgenic mouse model of HD, van der Burg et al. recently found that weight loss is not caused by decreased caloric intake or increased locomotor activity, but is associated with increased metabolic rate and changes in several factors regulating metabolism, such as reduced levels of hypothalamic peptides and altered uncoupling proteins *(158)*.

25.4 CLOSING REMARKS

With the advance of modern technologies, genetic factors contributing to neuro-degenerative disorders continue to be unveiled. Although familial forms of diseases only represent a small fraction of cases, deciphering the cause of these rare inherited cases helps to understand the development of the more common sporadic form of these disorders. Establishment of animal models recapitulating symptoms of human diseases through genetic manipulations has provided unprecedented tools to study gene–gene and gene–environment interactions in the molecular pathogenesis of neurodegenerative disorders and to test potential nutritional or pharmaceutical interventions. However, data from animal experiments should not be extrapolated directly to humans. Randomized clinical trials are required before any dietary or medical recommendations can or should be made. Unfortunately, while numerous observational human studies have been conducted, there have been no large rando-mized trials to assess the impact of dietary interventions on the progression of AD and other neurodegenerative disorders. Nevertheless, multiple lines of evidence suggest that dietary interventions have benefits for preventing and/or reducing the incidence of these disorders. As study continues in the future, new knowledge will increase our understanding of the effects of nutrition on neurodegenerative dis-eases. For the time being, any impact that nutrition may have on the pathogenesis of neurodegenerative disorders could have large implications for improving quality and quantity of life, as well as reducing the burden on the health care system. Certainly, more rigorous studies are needed to establish recommendations to main-tain optimal nutrition for successful aging.

25.5 RECOMMENDATIONS

1. Because recent studies indicate that high intakes of saturated fat and cholesterol increase the risk of AD, individuals with a family history of AD or other risk factors should be counseled to decrease saturated fat intake.
2. Two to three meals of fish per week (supplying approximately 380 mg EPA + DHA/ day) may reduce AD risk.
3. A Mediterranean-type diet has been shown in epidemiologic studies to be associated with lower rates of dementia.

REFERENCES

1. Blennow K, de Leon MJ, Zetterberg H. Alzheimer's disease. Lancet 2006; 368(9533):387–403.
2. Ferri CP, Prince M, Brayne C, et al. Global prevalence of dementia: a Delphi consensus study. Lancet 2005; 366(9503):2112–7.
3. Moller HJ, Graeber MB. The case described by Alois Alzheimer in 1911. Historical and conceptual perspectives based on the clinical record and neurohistological sections. Eur Arch Psychiatry Clin Neurosci 1998; 248(3):111–22.
4. Hardy J, Selkoe DJ. The amyloid hypothesis of Alzheimer's disease: progress and problems on the road to therapeutics. Science 2002; 297(5580):353–6.
5. Goate A, Chartier-Harlin MC, Mullan M, et al. Segregation of a missense mutation in the amyloid precursor protein gene with familial Alzheimer's disease. Nature 1991; 349(6311):704–6.
6. Sherrington R, Rogaev EI, Liang Y, et al. Cloning of a gene bearing missense mutations in early-onset familial Alzheimer's disease. Nature 1995; 375(6534):754–60.
7. Levy-Lahad E, Wasco W, Poorkaj P, et al. Candidate gene for the chromosome 1 familial Alzheimer's disease locus. Science 1995; 269(5226):973–7.
8. Games D, Adams D, Alessandrini R, et al. Alzheimer-type neuropathology in transgenic mice overexpressing V717F beta-amyloid precursor protein. Nature 1995; 373(6514):523–7.
9. Hsiao K, Chapman P, Nilsen S, et al. Correlative memory deficits, A beta elevation, and amyloid plaques in transgenic mice. Science 1996; 274(5284):99–102.
10. Jankowsky JL, Fadale DJ, Anderson J, et al. Mutant presenilins specifically elevate the levels of the 42 residue beta-amyloid peptide in vivo: evidence for augmentation of a 42-specific gamma secretase. Hum Mol Genet 2004; 13(2):159–70.
11. Strittmatter WJ, Roses AD. Apolipoprotein E and Alzheimer's disease. Annu Rev Neurosci 1996; 19:53–77.
12. Corder EH, Saunders AM, Risch NJ, et al. Protective effect of apolipoprotein E type 2 allele for late onset Alzheimer disease. Nat Genet 1994; 7(2):180–4.
13. Corder EH, Saunders AM, Strittmatter WJ, et al. Gene dose of apolipoprotein E type 4 allele and the risk of Alzheimer's disease in late onset families. Science 1993; 261(5123):921–3.
14. Poirier J, Davignon J, Bouthillier D, Kogan S, Bertrand P, Gauthier S. Apolipoprotein E polymorphism and Alzheimer's disease. Lancet 1993; 342(8873):697–9.
15. Farrer LA, Cupples LA, Haines JL, et al. Effects of age, sex, and ethnicity on the association between apolipoprotein E genotype and Alzheimer disease. A meta-analysis. APOE and Alzheimer Disease Meta Analysis Consortium. JAMA 1997; 278(16):1349–56.
16. Meyer MR, Tschanz JT, Norton MC, et al. APOE genotype predicts when – not whether – one is predisposed to develop Alzheimer disease. Nat Genet 1998; 19(4):321–2.
17. Poirier J. Apolipoprotein E in animal models of CNS injury and in Alzheimer's disease. Trends Neurosci 1994; 17(12):525–30.
18. Bales KR, Verina T, Dodel RC, et al. Lack of apolipoprotein E dramatically reduces amyloid beta-peptide deposition. Nat Genet 1997; 17(3):263–4.
19. Holtzman DM, Bales KR, Tenkova T, et al. Apolipoprotein E isoform-dependent amyloid deposition and neuritic degeneration in a mouse model of Alzheimer's disease. Proc Natl Acad Sci USA 2000; 97(6):2892–7.
20. Tanzi RE, Bertram L. New frontiers in Alzheimer's disease genetics. Neuron 2001; 32(2):181–4.
21. Fahn S. Description of Parkinson's disease as a clinical syndrome. Ann NY Acad Sci 2003; 991:1–14.
22. Hughes AJ, Daniel SE, Ben-Shlomo Y, Lees AJ. The accuracy of diagnosis of parkinsonian syndromes in a specialist movement disorder service. Brain 2002; 125(Pt 4):861–70.
23. Farrer MJ. Genetics of Parkinson disease: paradigm shifts and future prospects. Nat Rev Genet 2006; 7(4):306–18.
24. Chen L, Feany MB. Alpha-synuclein phosphorylation controls neurotoxicity and inclusion formation in a Drosophila model of Parkinson disease. Nat Neurosci 2005; 8(5):657–63.
25. Shimura H, Hattori N, Kubo S, et al. Familial Parkinson disease gene product, parkin, is a ubiquitin-protein ligase. Nat Genet 2000; 25(3):302–5.

26. Maraganore DM, de Andrade M, Lesnick TG, et al. High-resolution whole-genome association study of Parkinson disease. Am J Hum Genet 2005; 77(5):685–93.

27. Mitchell JD, Borasio GD. Amyotrophic lateral sclerosis. Lancet 2007; 369(9578):2031–41.

28. Pasinelli P, Brown RH. Molecular biology of amyotrophic lateral sclerosis: insights from genetics. Nat Rev Neurosci 2006; 7(9):710–23.

29. Dunckley T, Huentelman MJ, Craig DW, et al. Whole-genome analysis of sporadic amyotrophic lateral sclerosis. N Engl J Med 2007; 357(8):775–88.

30. Bruijn LI, Miller TM, Cleveland DW. Unraveling the mechanisms involved in motor neuron degeneration in ALS. Annu Rev Neurosci 2004; 27:723–49.

31. Walker FO. Huntington's disease. Lancet 2007; 369(9557):218–28.

32. Takano H, Cancel G, Ikeuchi T, et al. Close associations between prevalences of dominantly inherited spinocerebellar ataxias with CAG-repeat expansions and frequencies of large normal CAG alleles in Japanese and Caucasian populations. Am J Hum Genet 1998; 63(4):1060–6.

33. Wright HH, Still CN, Abramson RK. Huntington's disease in black kindreds in South Carolina. Arch Neurol 1981; 38(7):412–4.

34. Gusella JF, Wexler NS, Conneally PM, et al. A polymorphic DNA marker genetically linked to Huntington's disease. Nature 1983; 306(5940):234–8.

35. Wexler NS, Lorimer J, Porter J, et al. Venezuelan kindreds reveal that genetic and environmental factors modulate Huntington's disease age of onset. Proc Natl Acad Sci USA 2004; 101(10):3498–503.

36. Kuller LH. Nutrition, lipids, and cardiovascular disease. Nutr Rev 2006; 64(2 Pt 2):S15–26.

37. Kalmijn S, Launer LJ, Ott A, Witteman JC, Hofman A, Breteler MM. Dietary fat intake and the risk of incident dementia in the Rotterdam Study. Ann Neurol 1997; 42(5):776–82.

38. Luchsinger JA, Tang MX, Shea S, Mayeux R. Caloric intake and the risk of Alzheimer disease. Arch Neurol 2002; 59(8):1258–63.

39. Laitinen MH, Ngandu T, Rovio S, et al. Fat intake at midlife and risk of dementia and Alzheimer's disease: a population-based study. Dement Geriatr Cogn Disord 2006; 22(1):99–107.

40. Refolo LM, Malester B, LaFrancois J, et al. Hypercholesterolemia accelerates the Alzheimer's amyloid pathology in a transgenic mouse model. Neurobiol Dis 2000; 7(4):321–31.

41. Shie FS, Jin LW, Cook DG, Leverenz JB, LeBoeuf RC. Diet-induced hypercholesterolemia enhances brain A beta accumulation in transgenic mice. Neuroreport 2002; 13(4):455–9.

42. Levin-Allerhand JA, Lominska CE, Smith JD. Increased amyloid-levels in APPSWE transgenic mice treated chronically with a physiological high-fat high-cholesterol diet. J Nutr Health Aging 2002; 6(5):315–9.

43. Li L, Cao D, Garber DW, Kim H, Fukuchi K. Association of aortic atherosclerosis with cerebral beta-amyloidosis and learning deficits in a mouse model of Alzheimer's disease. Am J Pathol 2003; 163(6):2155–64.

44. Connor WE, Connor SL. The importance of fish and docosahexaenoic acid in Alzheimer disease. Am J Clin Nutr 2007; 85(4):929–30.

45. Tully AM, Roche HM, Doyle R, et al. Low serum cholesteryl ester-docosahexaenoic acid levels in Alzheimer's disease: a case-control study. Br J Nutr 2003; 89(4):483–9.

46. Soderberg M, Edlund C, Kristensson K, Dallner G. Fatty acid composition of brain phospholipids in aging and in Alzheimer's disease. Lipids 1991; 26(6):421–5.

47. Morris MC, Evans DA, Bienias JL, et al. Consumption of fish and n-3 fatty acids and risk of incident Alzheimer disease. Arch Neurol 2003; 60(7):940–6.

48. Schaefer EJ, Bongard V, Beiser AS, et al. Plasma phosphatidylcholine docosahexaenoic acid content and risk of dementia and Alzheimer disease: the Framingham Heart Study. Arch Neurol 2006; 63(11):1545–50.

49. van Gelder BM, Tijhuis M, Kalmijn S, Kromhout D. Fish consumption, n-3 fatty acids, and subsequent 5-y cognitive decline in elderly men: the Zutphen Elderly Study. Am J Clin Nutr 2007; 85(4):1142–7.

50. Beydoun MA, Kaufman JS, Satia JA, Rosamond W, Folsom AR. Plasma n-3 fatty acids and the risk of cognitive decline in older adults: the Atherosclerosis Risk in Communities Study. Am J Clin Nutr 2007; 85(4):1103–11.

51. Hashimoto M, Tanabe Y, Fujii Y, Kikuta T, Shibata H, Shido O. Chronic administration of docosahexaenoic acid ameliorates the impairment of spatial cognition learning ability in amyloid beta-infused rats. J Nutr 2005; 135(3):549–55.

52. Calon F, Lim GP, Yang F, et al. Docosahexaenoic acid protects from dendritic pathology in an Alzheimer's disease mouse model. Neuron 2004; 43(5):633–45.

53. Lim GP, Calon F, Morihara T, et al. A diet enriched with the omega-3 fatty acid docosahexaenoic acid reduces amyloid burden in an aged Alzheimer mouse model. J Neurosci 2005; 25(12): 3032–40.

54. Green KN, Martinez-Coria H, Khashwji H, et al. Dietary docosahexaenoic acid and docosapentaenoic acid ameliorate amyloid-beta and tau pathology via a mechanism involving presenilin 1 levels. J Neurosci 2007; 27(16):4385–95.

55. Blok WL, Katan MB, van der Meer JW. Modulation of inflammation and cytokine production by dietary (n-3) fatty acids. J Nutr 1996; 126(6):1515–33.

56. Morris MC, Evans DA, Tangney CC, Bienias JL, Wilson RS. Fish consumption and cognitive decline with age in a large community study. Arch Neurol 2005; 62(12):1849–53.

57. Lichtenstein AH, Appel LJ, Brands M, et al. Diet and lifestyle recommendations revision 2006: a scientific statement from the American Heart Association Nutrition Committee. Circulation 2006; 114(1):82–96.

58. Anderson C, Checkoway H, Franklin GM, Beresford S, Smith-Weller T, Swanson PD. Dietary factors in Parkinson's disease: the role of food groups and specific foods. Mov Disord 1999; 14(1):21–7.

59. Logroscino G, Marder K, Cote L, Tang MX, Shea S, Mayeux R. Dietary lipids and antioxidants in Parkinson's disease: a population-based, case-control study. Ann Neurol 1996; 39(1):89–94.

60. Chen H, Zhang SM, Hernan MA, Willett WC, Ascherio A. Dietary intakes of fat and risk of Parkinson's disease. Am J Epidemiol 2003; 157(11):1007–14.

61. Powers KM, Smith-Weller T, Franklin GM, Longstreth WT, Jr., Swanson PD, Checkoway H. Parkinson's disease risks associated with dietary iron, manganese, and other nutrient intakes. Neurology 2003; 60(11):1761–6.

62. Abbott RD, Ross GW, White LR, et al. Environmental, life-style, and physical precursors of clinical Parkinson's disease: recent findings from the Honolulu-Asia Aging Study. J Neurol 2003; 250(Suppl 3):III30–9.

63. de Lau LM, Bornebroek M, Witteman JC, Hofman A, Koudstaal PJ, Breteler MM. Dietary fatty acids and the risk of Parkinson disease: the Rotterdam study. Neurology 2005; 64(12): 2040–5.

64. Veldink JH, Kalmijn S, Groeneveld GJ, et al. Intake of polyunsaturated fatty acids and vitamin E reduces the risk of developing amyotrophic lateral sclerosis. J Neurol Neurosurg Psychiatry 2007; 78(4):367–71.

65. Puri BK, Leavitt BR, Hayden MR, et al. Ethyl-EPA in Huntington disease: a double-blind, randomized, placebo-controlled trial. Neurology 2005; 65(2):286–92.

66. Bray GA, Nielsen SJ, Popkin BM. Consumption of high-fructose corn syrup in beverages may play a role in the epidemic of obesity. Am J Clin Nutr 2004; 79(4):537–43.

67. Hedley AA, Ogden CL, Johnson CL, Carroll MD, Curtin LR, Flegal KM. Prevalence of overweight and obesity among US children, adolescents, and adults, 1999–2002. JAMA 2004; 291(23):2847–50.

68. Ogden CL, Flegal KM, Carroll MD, Johnson CL. Prevalence and trends in overweight among US children and adolescents, 1999–2000. JAMA 2002; 288(14):1728–32.

69. Whitaker RC, Wright JA, Pepe MS, Seidel KD, Dietz WH. Predicting obesity in young adulthood from childhood and parental obesity. N Engl J Med 1997; 337(13):869–73.

70. Grundy SM. Multifactorial causation of obesity: implications for prevention. Am J Clin Nutr 1998; 67(3 Suppl):563S–72S.

71. Haan MN. Therapy Insight: type 2 diabetes mellitus and the risk of late-onset Alzheimer's disease. Nat Clin Pract Neurol 2006; 2(3):159–66.

72. Leibson CL, Rocca WA, Hanson VA, et al. Risk of dementia among persons with diabetes mellitus: a population-based cohort study. Am J Epidemiol 1997; 145(4):301–8.

73. Stolk RP, Breteler MM, Ott A, et al. Insulin and cognitive function in an elderly population: the Rotterdam Study. Diabetes Care 1997; 20(5):792–5.
74. Craft S, Watson GS. Insulin and neurodegenerative disease: shared and specific mechanisms. Lancet Neurol 2004; 3(3):169–78.
75. Mungas D, Cooper JK, Weiler PG, Gietzen D, Franzi C, Bernick C. Dietary preference for sweet foods in patients with dementia. J Am Geriatr Soc 1990; 38(9):999–1007.
76. Greenwood CE, Tam C, Chan M, Young KW, Binns MA, van Reekum R. Behavioral disturbances, not cognitive deterioration, are associated with altered food selection in seniors with Alzheimer's disease. J Gerontol A Biol Sci Med Sci 2005; 60(4):499–505.
77. Young KW, Greenwood CE, van Reekum R, Binns MA. A randomized, crossover trial of high-carbohydrate foods in nursing home residents with Alzheimer's disease: associations among intervention response, body mass index, and behavioral and cognitive function. J Gerontol A Biol Sci Med Sci 2005; 60(8):1039–45.
78. Luchsinger JA, Tang MX, Mayeux R. Glycemic load and risk of Alzheimer's disease. J Nutr Health Aging 2007; 11(3):238–41.
79. Cao D, Lu H, Lewis TL, Li L. Intake of sucrose-sweetened water induces insulin resistance and exacerbates memory deficits and amyloidosis in a transgenic mouse model of Alzheimer's disease. J Biol Chem 2007;282(50):36275–82.
80. Abbott RD, Ross GW, White LR, et al. Midlife adiposity and the future risk of Parkinson's disease. Neurology 2002; 59(7):1051–7.
81. Vanitallie TB, Nonas C, Di Rocco A, Boyar K, Hyams K, Heymsfield SB. Treatment of Parkinson disease with diet-induced hyperketonemia: a feasibility study. Neurology 2005; 64(4):728–30.
82. Reyes ET, Perurena OH, Festoff BW, Jorgensen R, Moore WV. Insulin resistance in amyotrophic lateral sclerosis. J Neurol Sci 1984; 63(3):317–24.
83. Hubbard RW, Will AD, Peterson GW, Sanchez A, Gillan WW, Tan SA. Elevated plasma glucagon in amyotrophic lateral sclerosis. Neurology 1992; 42(8):1532–4.
84. Podolsky S, Leopold NA, Sax DS. Increased frequency of diabetes mellitus in patients with Huntington's chorea. Lancet 1972; 1(7765):1356–8.
85. Eikelboom JW, Lonn E, Genest J, Jr, Hankey G, Yusuf S. Homocyst(e)ine and cardiovascular disease: a critical review of the epidemiologic evidence. Ann Intern Med 1999; 131(5):363–75.
86. Toole JF, Malinow MR, Chambless LE, et al. Lowering homocysteine in patients with ischemic stroke to prevent recurrent stroke, myocardial infarction, and death: the Vitamin Intervention for Stroke Prevention (VISP) randomized controlled trial. JAMA 2004; 291(5):565–75.
87. Morris MS. Homocysteine and Alzheimer's disease. Lancet Neurol 2003; 2(7):425–8.
88. Seshadri S, Beiser A, Selhub J, et al. Plasma homocysteine as a risk factor for dementia and Alzheimer's disease. N Engl J Med 2002; 346(7):476–83.
89. Snowdon DA, Tully CL, Smith CD, Riley KP, Markesbery WR. Serum folate and the severity of atrophy of the neocortex in Alzheimer disease: findings from the Nun study. Am J Clin Nutr 2000; 71(4):993–8.
90. Clarke R, Smith AD, Jobst KA, Refsum H, Sutton L, Ueland PM. Folate, vitamin B_{12}, and serum total homocysteine levels in confirmed Alzheimer disease. Arch Neurol 1998; 55(11): 1449–55.
91. Wang HX, Wahlin A, Basun H, Fastbom J, Winblad B, Fratiglioni L. Vitamin B(12) and folate in relation to the development of Alzheimer's disease. Neurology 2001; 56(9):1188–94.
92. Nilsson K, Gustafson L, Hultberg B. Improvement of cognitive functions after cobalamin/folate supplementation in elderly patients with dementia and elevated plasma homocysteine. Int J Geriatr Psychiatry 2001; 16(6):609–14.
93. Eastley R, Wilcock GK, Bucks RS. Vitamin B_{12} deficiency in dementia and cognitive impairment: the effects of treatment on neuropsychological function. Int J Geriatr Psychiatry 2000; 15(3):226–33.
94. Kwok T, Tang C, Woo J, Lai WK, Law LK, Pang CP. Randomized trial of the effect of supplementation on the cognitive function of older people with subnormal cobalamin levels. Int J Geriatr Psychiatry 1998; 13(9):611–6.

95. Crystal HA, Ortof E, Frishman WH, Gruber A, Hershman D, Aronson M. Serum vitamin B$_{12}$ levels and incidence of dementia in a healthy elderly population: a report from the Bronx Longitudinal Aging Study. J Am Geriatr Soc 1994; 42(9):933–6.

96. Luchsinger JA, Tang MX, Miller J, Green R, Mayeux R. Relation of higher folate intake to lower risk of Alzheimer disease in the elderly. Arch Neurol 2007; 64(1):86–92.

97. Malouf M, Grimley EJ, Areosa SA. Folic acid with or without vitamin B$_{12}$ for cognition and dementia. Cochrane Database Syst Rev 2003(4):CD004514.

98. Malouf R, Grimley EJ, Areosa SA. Vitamin B$_{12}$ for cognition. Cochrane Database Syst Rev 2003(3):CD004326.

99. Malouf R, Grimley Evans J. The effect of vitamin B$_6$ on cognition. Cochrane Database Syst Rev 2003(4):CD004393.

100. Ho PI, Ortiz D, Rogers E, Shea TB. Multiple aspects of homocysteine neurotoxicity: glutamate excitotoxicity, kinase hyperactivation and DNA damage. J Neurosci Res 2002; 70(5):694–702.

101. Kruman II, Kumaravel TS, Lohani A, et al. Folic acid deficiency and homocysteine impair DNA repair in hippocampal neurons and sensitize them to amyloid toxicity in experimental models of Alzheimer's disease. J Neurosci 2002; 22(5):1752–62.

102. Kuhn W, Roebroek R, Blom H, et al. Elevated plasma levels of homocysteine in Parkinson's disease. Eur Neurol 1998; 40(4):225–7.

103. Muller T, Werne B, Fowler B, Kuhn W. Nigral endothelial dysfunction, homocysteine, and Parkinson's disease. Lancet 1999; 354(9173):126–7.

104. de Lau LM, Koudstaal PJ, Witteman JC, Hofman A, Breteler MM. Dietary folate, vitamin B$_{12}$, and vitamin B$_6$ and the risk of Parkinson disease. Neurology 2006; 67(2):315–8.

105. Sung JJ, Kim HJ, Choi-Kwon S, Lee J, Kim M, Lee KW. Homocysteine induces oxidative cytotoxicity in Cu,Zn-superoxide dismutase mutant motor neuronal cell. Neuroreport 2002; 13(4):377–81.

106. Chung YH, Hong JJ, Shin CM, Joo KM, Kim MJ, Cha CI. Immunohistochemical study on the distribution of homocysteine in the central nervous system of transgenic mice expressing a human Cu/Zn SOD mutation. Brain Res 2003; 967(1–2):226–34.

107. Zoccolella S, Martino D, Defazio G, Lamberti P, Livrea P. Hyperhomocysteinemia in movement disorders: current evidence and hypotheses. Curr Vasc Pharmacol 2006; 4(3):237–43.

108. Reynolds A, Laurie C, Lee Mosley R, Gendelman HE. Oxidative stress and the pathogenesis of neurodegenerative disorders. Int Rev Neurobiol 2007; 82:297–325.

109. Huang X, Moir RD, Tanzi RE, Bush AI, Rogers JT. Redox-active metals, oxidative stress, and Alzheimer's disease pathology. Ann NY Acad Sci 2004; 1012:153–63.

110. Butterfield DA, Castegna A, Drake J, Scapagnini G, Calabrese V. Vitamin E and neurodegenerative disorders associated with oxidative stress. Nutr Neurosci 2002; 5(4):229–39.

111. Pitchumoni SS, Doraiswamy PM. Current status of antioxidant therapy for Alzheimer's disease. J Am Geriatr Soc 1998; 46(12):1566–72.

112. Masaki KH, Losonczy KG, Izmirlian G, et al. Association of vitamin E and C supplement use with cognitive function and dementia in elderly men. Neurology 2000; 54(6):1265–72.

113. Grodstein F, Chen J, Willett WC. High-dose antioxidant supplements and cognitive function in community-dwelling elderly women. Am J Clin Nutr 2003; 77(4):975–84.

114. Zandi PP, Anthony JC, Khachaturian AS, et al. Reduced risk of Alzheimer disease in users of antioxidant vitamin supplements: the Cache County Study. Arch Neurol 2004; 61(1):82–8.

115. Morris MC, Evans DA, Bienias JL, Tangney CC, Wilson RS. Vitamin E and cognitive decline in older persons. Arch Neurol 2002; 59(7):1125–32.

116. Engelhart MJ, Geerlings MI, Ruitenberg A, et al. Dietary intake of antioxidants and risk of Alzheimer disease. JAMA 2002; 287(24):3223–9.

117. Morris MC, Evans DA, Bienias JL, et al. Dietary intake of antioxidant nutrients and the risk of incident Alzheimer disease in a biracial community study. JAMA 2002; 287(24):3230–7.

118. Fukao A, Tsubono Y, Kawamura M, et al. The independent association of smoking and drinking with serum beta-carotene levels among males in Miyagi, Japan. Int J Epidemiol 1996; 25(2):300–6.

119. Luchsinger JA, Tang MX, Shea S, Mayeux R. Antioxidant vitamin intake and risk of Alzheimer disease. Arch Neurol 2003; 60(2):203–8.
120. Sano M, Ernesto C, Thomas RG, et al. A controlled trial of selegiline, alpha-tocopherol, or both as treatment for Alzheimer's disease. The Alzheimer's disease cooperative study. N Engl J Med 1997; 336(17):1216–22.
121. Etminan M, Gill SS, Samii A. Intake of vitamin E, vitamin C, and carotenoids and the risk of Parkinson's disease: a meta-analysis. Lancet Neurol 2005; 4(6):362–5.
122. Gurney ME, Cutting FB, Zhai P, et al. Benefit of vitamin E, riluzole, and gabapentin in a transgenic model of familial amyotrophic lateral sclerosis. Ann Neurol 1996; 39(2):147–57.
123. Desnuelle C, Dib M, Garrel C, Favier A. A double-blind, placebo-controlled randomized clinical trial of alpha-tocopherol (vitamin E) in the treatment of amyotrophic lateral sclerosis. ALS riluzole-tocopherol Study Group. Amyotroph Lateral Scler Other Motor Neuron Disord 2001; 2(1):9–18.
124. Vyth A, Timmer JG, Bossuyt PM, Louwerse ES, de Jong JM. Survival in patients with amyotrophic lateral sclerosis, treated with an array of antioxidants. J Neurol Sci 1996; 139(Suppl):99–103.
125. Peyser CE, Folstein M, Chase GA, et al. Trial of d-alpha-tocopherol in Huntington's disease. Am J Psychiatry 1995; 152(12):1771–5.
126. Esposito E, Rotilio D, Di Matteo V, Di Giulio C, Cacchio M, Algeri S. A review of specific dietary antioxidants and the effects on biochemical mechanisms related to neurodegenerative processes. Neurobiol Aging 2002; 23(5):719–35.
127. Trichopoulou A, Costacou T, Bamia C, Trichopoulos D. Adherence to a Mediterranean diet and survival in a Greek population. N Engl J Med 2003; 348(26):2599–608.
128. Scarmeas N, Stern Y, Tang MX, Mayeux R, Luchsinger JA. Mediterranean diet and risk for Alzheimer's disease. Ann Neurol 2006; 59(6):912–21.
129. Panza F, Solfrizzi V, Colacicco AM, et al. Mediterranean diet and cognitive decline. Public Health Nutr 2004; 7(7):959–63.
130. Knoops KT, de Groot LC, Kromhout D, et al. Mediterranean diet, lifestyle factors, and 10-year mortality in elderly European men and women: the HALE project. JAMA 2004; 292(12):1433–9.
131. Scarmeas N, Luchsinger JA, Mayeux R, Stern Y. Mediterranean diet and Alzheimer disease mortality. Neurology 2007; 69(11):1084–93.
132. Bailey RL, Gutschall MD, Mitchell DC, Miller CK, Lawrence FR, Smiciklas-Wright H. Comparative strategies for using cluster analysis to assess dietary patterns. J Am Diet Assoc 2006; 106(8):1194–200.
133. Kant AK. Dietary patterns and health outcomes. J Am Diet Assoc 2004; 104(4):615–35.
134. Weindruch R, Sohal RS. Seminars in medicine of the Beth Israel Deaconess Medical Center. Caloric intake and aging. N Engl J Med 1997; 337(14):986–94.
135. Patel NV, Gordon MN, Connor KE, et al. Caloric restriction attenuates Abeta-deposition in Alzheimer transgenic models. Neurobiol Aging 2005; 26(7):995–1000.
136. Wang J, Ho L, Qin W, et al. Caloric restriction attenuates beta-amyloid neuropathology in a mouse model of Alzheimer's disease. Faseb J 2005; 19(6):659–61.
137. Halagappa VK, Guo Z, Pearson M, et al. Intermittent fasting and caloric restriction ameliorate age-related behavioral deficits in the triple-transgenic mouse model of Alzheimer's disease. Neurobiol Dis 2007; 26(1):212–20.
138. Mattson MP. Energy intake, meal frequency, and health: a neurobiological perspective. Annu Rev Nutr 2005; 25:237–60.
139. Mattson MP, Cutler RG, Camandola S. Energy intake and amyotrophic lateral sclerosis. Neuromolecular Med 2007; 9(1):17–20.
140. McKhann G, Drachman D, Folstein M, Katzman R, Price D, Stadlan EM. Clinical diagnosis of Alzheimer's disease: report of the NINCDS-ADRDA Work Group under the auspices of Department of Health and Human Services Task Force on Alzheimer's Disease. Neurology 1984; 34(7):939–44.
141. Gillette-Guyonnet S, Nourhashemi F, Andrieu S, et al. Weight loss in Alzheimer disease. Am J Clin Nutr 2000; 71(2):637S–42S.

142. Wang SY, Fukagawa N, Hossain M, Ooi WL. Longitudinal weight changes, length of survival, and energy requirements of long-term care residents with dementia. J Am Geriatr Soc 1997; 45(10):1189–95.

143. White H, Pieper C, Schmader K, Fillenbaum G. Weight change in Alzheimer's disease. J Am Geriatr Soc 1996; 44(3):265–72.

144. White H, Pieper C, Schmader K. The association of weight change in Alzheimer's disease with severity of disease and mortality: a longitudinal analysis. J Am Geriatr Soc 1998; 46(10):1223–7.

145. Stewart R, Masaki K, Xue QL, et al. A 32-year prospective study of change in body weight and incident dementia: the Honolulu-Asia Aging Study. Arch Neurol 2005; 62(1):55–60.

146. Johnson DK, Wilkins CH, Morris JC. Accelerated weight loss may precede diagnosis in Alzheimer disease. Arch Neurol 2006; 63(9):1312–7.

147. Feldman H, Scheltens P, Scarpini E, et al. Behavioral symptoms in mild cognitive impairment. Neurology 2004; 62(7):1199–201.

148. Jack CR, Jr., Petersen RC, Xu Y, et al. Rates of hippocampal atrophy correlate with change in clinical status in aging and AD. Neurology 2000; 55(4):484–89.

149. Grundman M, Corey-Bloom J, Jernigan T, Archibald S, Thal LJ. Low body weight in Alzheimer's disease is associated with mesial temporal cortex atrophy. Neurology 1996; 46(6):1585–91.

150. Vanhanen M, Kivipelto M, Koivisto K, et al. APOE-epsilon4 is associated with weight loss in women with AD: a population-based study. Neurology 2001; 56(5):655–9.

151. Peskind ER, Wilkinson CW, Petrie EC, Schellenberg GD, Raskind MA. Increased CSF cortisol in AD is a function of APOE genotype. Neurology 2001; 56(8):1094–8.

152. Grundman M. Weight loss in the elderly may be a sign of impending dementia. Arch Neurol 2005; 62(1):20–2.

153. Logroscino G, Sesso HD, Paffenbarger RS, Jr., Lee IM. Body mass index and risk of Parkinson's disease: a prospective cohort study. Am J Epidemiol 2007; 166(10):1186–90.

154. Chen H, Zhang SM, Hernan MA, Willett WC, Ascherio A. Weight loss in Parkinson's disease. Ann Neurol 2003; 53(5):676–9.

155. Camicioli R, Moore MM, Kinney A, Corbridge E, Glassberg K, Kaye JA. Parkinson's disease is associated with hippocampal atrophy. Mov Disord 2003; 18(7):784–90.

156. Kasarskis EJ, Berryman S, Vanderleest JG, Schneider AR, McClain CJ. Nutritional status of patients with amyotrophic lateral sclerosis: relation to the proximity of death. Am J Clin Nutr 1996; 63(1):130–7.

157. Desport JC, Preux PM, Magy L, et al. Factors correlated with hypermetabolism in patients with amyotrophic lateral sclerosis. Am J Clin Nutr 2001; 74(3):328–34.

158. van der Burg JM, Bacos K, Wood NI, et al. Increased metabolism in the R6/2 mouse model of Huntington's disease. Neurobiol Dis 2008;29(1):41–51.

159. Djousse L, Knowlton B, Cupples LA, Marder K, Shoulson I, Myers RH. Weight loss in early stage of Huntington's disease. Neurology 2002; 59(9):1325–30.

160. Gaba AM, Zhang K, Marder K, Moskowitz CB, Werner P, Boozer CN. Energy balance in early-stage Huntington disease. Am J Clin Nutr 2005; 81(6):1335–41.

26 Nutrition and Late-Life Depression

Martha E. Payne

Key Points

- Depression is a significant problem for older adults.
- Dietary factors related to either vascular risk or brain health may be important for depression.
- Obesity may promote depression.
- Inadequate omega-3 fatty acid consumption or levels may be related to depression.
- Folate is important for vascular and brain health but its role in depression is unclear.
- Brain lesions and their potential dietary etiology may be significant for depression.

Key Words: Depression; brain lesions; serotonin; folate; omega-3; vascular

26.1 INTRODUCTION

Mental disorders such as depression may be amenable to nutrition interventions as are physical illnesses. However, much less research has investigated this topic. Late-life depression, or depression in individuals 60 and over, is a type of psychiatric illness known as a mood disorder. Its most characteristic symptoms are sadness, feeling blue or down, and being uninterested in typical activities. Low mood is a persistent phenomenon as opposed to the daily "ups and downs" that most individuals experience. Often this dysphoria is not associated with a negative event or situation in the person's life. However, mood disturbance is not the only effect of depression. Depressed individuals also have groups of symptoms known as vegetative and ideational. Vegetative symptoms include appetite and energy disturbances. Ideational symptoms encompass feelings of guilt and hopelessness, as well as suicidal thoughts.

26.1.1 Diagnosis of Depression

A diagnosis of major depression requires that five or more of the following symptoms have endured for at least 2 weeks: (1) feeling depressed, sad, or blue,

From: *Nutrition and Health: Handbook of Clinical Nutrition and Aging, Second Edition*
Edited by: C. W. Bales and C. S. Ritchie, DOI 10.1007/978-1-60327-385-5_26,
© Humana Press, a part of Springer Science+Business Media, LLC 2009

(2) loss of interest or pleasure, (3) increased or decreased sleep, (4) increased or decreased appetite with weight change, (5) feeling agitated, restless, or slowed down, (6) feelings of worthlessness or excessive guilt, (7) low energy, (8) difficulty concentrating, and (9) feeling that life is not worth living *(1)*. Major depression may be diagnosed by a general practitioner, psychiatrist, psychologist, or other health professional. However, in approximately half of cases, depression is undetected and untreated *(2)*.

26.1.2 Impact of Depression

Depression is a serious and common mental disorder. The World Health Organization has determined that depression is the fourth leading cause of disease burden and the leading cause of years lived with disability *(3)*. Depression is also responsible for the majority of the 870,000 suicides per year *(4)*. Prevalence of depression in the older adult community typically ranges from 2.7 to 10.1% *(5,6)*. Depression is even more prevalent among those in poor health and those in nursing homes *(3)*. Late-life depression increases health-care costs by 50%, and this difference is not accounted for by mental health-care expenditures *(7)*. Part of this may relate to the adverse effects of depression upon medical health and mortality rates. Depression increases one's risk of heart disease and hip fractures *(8)*. Beyond the medical impact, depression can lead to impaired psychosocial and cognitive functioning and other disabilities which can have a tremendous impact upon the emotional and socioeconomic well-being of relatives, caregivers, and the community. Lost productivity, financial costs, and diminished quality of life for family members typically follow. The stigma of mental illness may lead to humiliation, isolation, and unemployment.

26.1.3 Etiology of Late-Life Depression

Prevention of late-life depression is obviously preferable to alternative approaches. Numerous factors affect one's risk of depression, including socioeconomic status, gender, social support, genetic factors, stressful life events, comorbid medical conditions, and ischemic brain lesions. Nutrition may be another critical factor in the etiology of depression. Although less studied than some other risk factors for depression, it is also more amenable to modification. Diet may influence depression risk by promoting vascular illnesses which lead to depression, by causing neuronal changes or other intermediary factors that promote depression, or potentially by directly promoting depression (see Fig. 26.1). Likely, all three pathways occur and there may be overlap between them.

Vascular risk factors and vascular diseases, including hypertension, atherosclerosis, heart disease, cerebrovascular disease, stroke, diabetes mellitus, and ischemic brain lesions are more common in those with late-life depression than in younger depressed subjects *(9–11)*. Individuals with depression are at greater risk for heart disease and diabetes, while these medical patient populations are at greater risk for depression *(12–14)*. In addition, patients with both a depression and an accompanying vascular medical disease are at greater risk for poor outcomes, including death. For example, after a myocardial infarction, patients with comorbid depression are four times more likely to die within 18 months, as compared to patients without depression *(15)*. The presence of vascular risk factors and, in particular, ischemic brain lesions has led to the vascular depression hypothesis of late-life

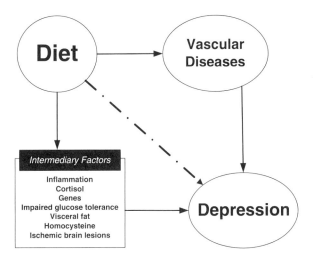

Fig. 26.1. A dietary mechanism for late-life depression. Diet may promote (or prevent) depression by influencing one's risk of vascular diseases, including atherosclerosis and diabetes. These vascular diseases are known to promote late-life depression. In addition, diet may promote (or prevent) depression by altering neuronal health, neurotransmitter levels, or other intermediary factors. There is also the potential for direct effects of diet upon depression.
Figure adapted from Payne et al. *(26)*.

depression which states that cerebrovascular disease may precipitate, predispose, or perpetuate geriatric depression *(16,17)*. In this scenario damage occurs to the cerebral vasculature, particularly the small cerebral vessels, leading to brain ischemia and neuronal cell death. The resulting ischemic brain lesions may promote depression if they impact mood regulation pathways. See Fig. 26.2 for an example of brain lesions which are seen with late-life depression.

Nutritional factors that are most likely to be related to the etiology of depression are those which have vascular effects – a large number of candidates to be sure – and

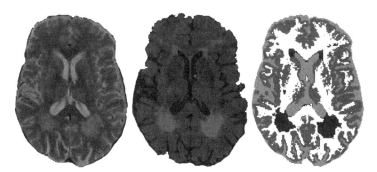

Fig. 26.2. Brain lesions seen on magnetic resonance imaging (MRI). Lesions, believed to be of ischemic origin, occur with aging but particularly with late-life depression. These lesions are believed to damage mood regulation pathways which can lead to depression. Lesions are shown here on a proton density image *(left)*, fluid-attenuated inversion recovery (FLAIR) image *(center)*, and tissue classification image *(right)*. Lesions are black on tissue classification image. Images courtesy of the Neuropsychiatric Imaging Research Laboratory (NIRL) at Duke University Medical Center.

those which affect neuronal health. Omega-3 fatty acids and folate have been studied more than other nutrients in relation to depression and are known to promote both vascular and nervous system health. As mentioned above, cerebro-vascular disease may damage the mood-regulating regions of the brain leading to depression. In terms of brain health, neurotransmitter metabolism, myelin production, and membrane fluidity are critical. Serotonin and other neurotransmitter systems appear to be dysregulated with depression. Nutrient balance can greatly impact all of these nervous system functions as well as vascular health.

26.2 ENERGY BALANCE

Weight changes are known to occur with depression but it is not clear if energy imbalance is related etiologically to depression. Weight loss is common among depressed individuals, and depression is considered the number one explanation for weight loss in the older adults *(18)*. Weight loss is of particular concern since older adults are already at increased nutritional risk. Underweight individuals are more likely to report depressive symptoms *(19)*. Unfortunately, the majority of studies which have examined weight loss and depression have been cross-sectional so it is usually not possible to determine which came first. At the other end of the spectrum, with weight gain and obesity, is another subset of depressed subjects.

26.2.1 Obesity

Obesity has been associated with depression and may be etiologically related via several mechanisms: vascular risk, weight-loss dieting behaviors, and physical impairment, all of which have been associated with risk of depression *(20–22)*. The Alameda County Study demonstrated that baseline obesity, as defined by body mass index (BMI) $\geq$ 30, increased the risk of depression during 5 year follow-up of 1,886 adults over age 50 *(23)*. Baseline depression did not affect later risk of obesity. In addition to obesity, intra-abdominal fat (IF) in particular has been linked with depression. This type of central adiposity is considered a more potent vascular risk factor than is obesity in general. Two small studies of depressed subjects measured IF with computed tomography (CT) scan. Depressed women were found to have greater IF than controls *(24)*. Interestingly, most of the depressed subjects were within 10% of ideal body weight. Depression may be associated with abnormalities of IF apart from its association to obesity. The second study followed older adult subjects for 1–2 years and found that depressed subjects accumulated more IF than non-depressed individuals *(25)*.

26.3 DIETARY FAT

Types of fat in the diet may be an important determinant of depression. Dietary fats are divided into four general categories: saturated, trans-unsaturated, mono-unsaturated, and polyunsaturated, with the latter subdivided into omega-6 and omega-3 fats. Saturated fat, a known cardiovascular risk factor, has been found to be higher among older adult depressed subjects than in comparison subjects *(26)*. Omega-3 polyunsaturated fats have been studied for their potential protective role in depression. Total omega-3 fats as well as the ratio of omega-3 to omega-6 fats are believed to be important determinants of health. A decline in both of these measures over the past 100 years has been blamed for an increase in many conditions, including depression *(27)*.

26.3.1 Omega-3 Fatty Acids

Omega-3s are found in fatty cold-water fish, including salmon and mackerel, as well as vegetable seeds and oils, including flaxseed, walnuts, and canola oil. Docosahexaenoic acid (DHA), found in seafood and fortified foods, is the omega-3 fatty acid considered most critical to brain health. DHA makes up 3% of dry brain weight and is higher in gray matter than white matter *(28)*. There are many potential mechanisms to explain a protective relationship between omega-3s and late-life depression (see Table 26.1). Both brain and vascular effects are likely critical. Individuals with depression exhibit many of the same characteristics as do those with omega-3 fatty acid deficiency, including altered neurotransmission, decreased cerebral blood flow, increased pro-inflammatory cytokines, and neuronal atrophy *(28)*. Rats who were fed an omega-3 deficient diet had decreased brain levels of DHA and increased depressive behaviors *(29)*. Rats and pigs who were given omega-3 supplements during development exhibited higher levels of serotonin, a neurotransmitter believed to be dysregulated and insufficiently active during depression *(30,31)*.

Population studies of fish consumption were the first to implicate omega-3 fats as protective nutrients for depression. An ecological study which compared nations on depression and fish consumption found a negative correlation between the two variables ($r = -0.8$, $p < 0.005$) *(32)*. A 60-fold difference in depression prevalence across countries existed, and this variation was similar to differences in coronary disease mortality, which may indicate that depression and heart disease are related to similar dietary factors. Two Finnish population studies examined depression and fish intake cross-sectionally, but few older adult subjects were included. One study ($n = 1,767$) found that frequent fish consumption (2 or more servings per week) was

Table 26.1
Potential mechanisms for a role of omega-3 fatty acids in depression

Brain	Vascular
Serotonin metabolism	Decrease triglycerides
Membrane fluidity	Decrease blood pressure
Cellular regulation	Decrease inflammation
Neurite growth	Decrease cardiac events
Neuron size	Protection from oxidative stress
Nerve growth factor	Decrease platelet aggregation
Glial cells	
Protection from oxidative stress	
Glucose metabolism	

Taken from Sinclair et al. *(28)*.

associated with a decreased risk of depressive symptoms as measured by the Beck Depression Inventory (odds ratio = 0.63) as compared to infrequent consumption, while controlling for age, sex, marital status, education, employment status, disability, region, income, general health, smoking, alcohol and coffee consumption, and physical activity *(33)*. The second study which examined the Northern Finland 1966 Birth Cohort (*n* = 5,689) determined that among women only rare fish consumers (1 or fewer servings per month) were at increased risk of depression (odds ratio = 2.6) compared with more frequent consumers, after controlling for body mass index, serum cholesterol, and socioeconomic status *(34)*. Another logistic regression model which controlled for smoking, ethanol intake, physical inactivity, and marital status yielded similar results.

Omega-3 fatty acids are believed to be the critical nutrient to explain these seafood and depression findings. Therefore, dietary omega-3s have been studied in relationship to depression, and several of these projects have included or focused on older adults. Surprisingly, most of these studies failed to find a significant relationship between omega-3 intake and depression in multivariable analyses. A study performed in the Netherlands (*n* = 332) found that older adults (aged 70–90) who consumed 156 mg or greater omega-3 fatty acids per day were less likely to report depressive symptoms than those who consumed less than 59 mg per day (odds ratio = 0.46) *(35)*. However, cross-sectional studies in Australia, Finland, and the United Kingdom showed no association between dietary omega-3s and either depression or depressive symptoms (*n* = 755, 29,133, and 2,982, respectively) *(36–38)*.

Biomarkers of omega-3 fatty acids have also been investigated in relationship to depression although, unfortunately, most studies have not measured dietary intake nor have most included older adult subjects. A small clinical study found lower omega-3 levels in the phospholipids and cholesterol esters of those with depression than in control subjects *(27)*. Along with other measures this result was considered indicative of activation of the inflammatory response system, which might lead to increases in lipid peroxidation, neuronal membrane damage, and disturbed serotonin metabolism. In Crete a study of healthy adults showed an inverse relationship between adipose tissue DHA, a long-term indicator of omega-3 intake, and depression *(39)*. A positive association has been found between the omega-6 to omega-3 ratio and depression, including in a study of older adult heart patients *(27,40)*.

In addition to associating low omega-3 levels with depression, studies have found them to be associated with depression biomarkers and depression severity. Low plasma levels of DHA predicted low levels of 5-hydroxyindoleacetic acid (5-HIAA), a metabolite of serotonin *(41)*. Low levels of 5-HIAA are associated with not only depression but also suicidal behavior *(42,43)*. A study of major depression patients found that omega-3 levels were negatively correlated with Hamilton depression scores *(44)*. In addition, the omega-6 to omega-3 ratio was positively associated with depression scores. Furthermore, these correlations were not explained by dietary intake *(27,44)*.

In summary, there is evidence that omega-3 fatty acid metabolism is altered in people with depression but the origin of this dysregulation is unclear. Longitudinal research is needed to determine if altered fatty acid metabolism leads to depression, is the result of depression, or is caused by a third factor. The role of dietary omega-3s may or may not be central to lowered physiological levels of omega-3s in depressed individuals. However, it is possible that supplementation may be used to correct such imbalances. There is evidence that fish consumption may be protective but, given the negative findings for dietary omega-3s, other nutrients in fish may explain such a relationship. To date, studies of omega-3 fats, fish consumption, and depression have been cross-sectional. It is imperative to study these relationships in a longitudinal manner, particularly in older adult populations, in order to ascertain any potentially protective role for these nutrients and foods in late-life depression.

26.4 THE B VITAMINS

Micronutrients, particularly the B vitamins, may be important to the etiology and progression of late-life depression. Folate (vitamin B_9) and cobalamin (vitamin B_{12}) have been the most extensively studied of the B vitamins. Both vitamins have roles in promoting brain health through neurotransmitter synthesis, myelin formation, and energy metabolism.

26.4.1 Folate

Folate occurs naturally in green vegetables, orange juice, strawberries, starchy vegetables, eggs, beans, and whole grains, among other foods. The synthetic form (folic acid) is found in dietary supplements and fortified refined grain products. As there are few body stores of folate, a steady supply is needed in the diet. As shown in Table 26.2, folate may influence mental health through its many functions in the brain and throughout the body. Low serum folate levels were reported as early as the 1960s when a study of psychiatric inpatients found that half had low folate levels (45). Temporality is unclear not only because of the cross-sectional nature of study but because many of the low folate subjects were alcoholics or had used folate-diminishing medications (45). Another inpatient study found that depressed subjects had lower folate levels than either non-depressed psychiatric patients or non-psychiatric patients (46). Folate levels were also inversely associated with Hamilton Depression scores. Other clinical studies have consistently associated

Table 26.2
Potential mechanisms for a role of folate in depression

Brain	*Systemic*
Serotonin synthesis	Methylation
Neurotransmitter release	Membrane phospholipids
Methylation	Homocysteine (to methionine)
Neurotransmitters	
Membrane phospholipids	
Formation of S-adenosyl methionine	
(sole methyl donor of nervous system)	

Taken from Paul et al. (80).

low folate status with depression and depression severity *(47–51)*. Unfortunately, studies have failed to control for comorbid disease and other depression risk factors. Researchers have speculated that folate deficiency is secondary to depression *(47,52)* perhaps because of increased utilization during a depressive episode *(49)*.

Population and community studies of older adult populations have examined the relationship between folate status and late-life depression in a cross-sectional manner. Two studies found a correlation between depressive symptoms and low blood folate levels *(53,54)* and five failed to find a significant association between folate status and depression *(55–59)*. The only longitudinal study to examine this question was conducted with middle-aged men and found that low dietary intake of folate was associated with later diagnosis of depression *(60)*.

A couple of factors should be considered when interpreting the conflicting literature on folate, namely, homocysteine and genetic polymorphisms. Homocysteine is a known to be damaging to both cardiovascular and neurological health and can become elevated in the absence of adequate folate. A cross-sectional study of older adults found a positive association between homocysteine levels and Beck depression inventory scores, even after controlling for folate levels and other covariates *(61)*. Low folate levels may have been a marker for or confounded with elevated homocysteine in previous studies. To further complicate matters, genes for folate metabolism have been associated with depression. A 677 C→T polymorphism (cytosine replaced by thymine at nucleotide 677) for methylenetetrahydrofolate reductase (MTHFR) is associated with low folate levels. Individuals who are homozygous for MTHFR (677 C→T) are more likely to have current or prior depression *(58,62,63)*. Disparate results from epidemiological studies may have resulted from genetic variability in the population samples but this cannot be confirmed at present.

26.4.2 Vitamin B_{12}

Cobalamin or vitamin B_{12} is found in foods of animal origin and has been investigated for its role in mental disorders including depression and dementia. Vitamin B_{12} functions in folate metabolism, serotonin synthesis, myelin sheath formation, and as a coenzyme for fatty acid and amino acid oxidation *(57)*. Cobalamin deficiency can lead to permanent neurological damage even in the absence of anemia or macrocytosis, the classic signs of B_{12} deficiency *(64)*.

Epidemiological studies have had conflicting results with regards to depression and cobalamin status. A Baltimore study of geriatric women with physical disabilities ($n = 700$) showed that those with B_{12} deficiency were more likely to be severely depressed than were non-deficient participants (odds ratio = 2.05), after controlling for sociodemographics and health status *(57)*. The prevalence of depression (31.7%) was high, as predicted by the presence of functional disabilities. A study of vitamin status among older individuals living in Rotterdam ($n = 806$) also found that B_{12} deficiency was associated with depression, after controlling for cardiovascular disease and functional disability *(59)*. In contrast, the European SENECA project population study, the Hordaland Health Study (Norway), and an Australian community study failed to find a significant association between cobalamin levels and depression *(53,55–58)*. What may be critical is the distinction between cobalamin deficiency and cobalamin status. The two studies examining B_{12} deficiency found an association with depression, while those looking at B_{12} levels did not.

26.5 BRAIN LESIONS

Since ischemic brain lesions are considered to be important etiologically for late-life depression, the nutritional correlates of brain lesions may be critical for depression as well. These brain lesions are seen on magnetic resonance imaging (MRI) and computed tomography (CT) scans (see Fig. 26.2 for MRI example) and occur with both aging and depression (65). However, individuals with late-life depression are more likely to have brain lesions than are older non-depressed subjects (16,66,67). Lesions indicate damage to gray and white matter in the brain (68) and have been linked both cross-sectionally and longitudinally with cognitive impairment and dementia (69–71), stroke (72). functional impairment (73). as well as late-life depression (74). Lesions are known to be associated with both the persistence and the worsening of depressive symptoms (74,75). Postmortem studies have shown that these brain lesions result from ischemic damage (76).

Given their ischemic etiology, nutrients related to vascular risk may be important. A handful of studies have examined this issue. The Austrian Stroke Prevention Study found that plasma concentrations of the antioxidants α-tocopherol and lycopene were negatively associated with brain lesions (77). Moderate alcohol consumption has been negatively associated with white matter lesions in the Cardiovascular Health Study, after controlling for covariates (78). High-fat dairy products have been found to be positively associated with brain lesions in an older depression sample (79). This area of research is promising for the potential to prevent brain damage and its sequelae, but longitudinal studies are needed.

26.6 RECOMMENDATIONS

Depression is a major problem for older adults. Nutrition may be one avenue by which depression can be prevented. Folate, omega-3 fats, and obesity, as well as other nutritional factors which affect vascular and brain health, may be related to late-life depression. Longitudinal studies are needed in order to examine the etiological relationships among nutritional status, diet, and late-life depression. However, since older individuals with depression are known to be at increased risk for heart disease and other vascular illnesses, it would be prudent to recommend that depressed individuals modify their diet as would be recommended for anyone at risk for cardiovascular disease.

REFERENCES

1. American Psychiatric Association: Diagnostic and Statistical Manual of Mental Disorders, 4th ed. Washington, DC: American Psychiatric Association, 1994.
2. Kessler RC, Berglund P, Demler O, et al. The epidemiology of major depressive disorder: results from the National Comorbidity Survey Replication (NCS-R). JAMA 2003; 289(23):3095–105.
3. WHO. The World Health Report 2001. Mental Health: New Understanding, New Hope. Geneva, Switzerland: World Health Organization, 2001.

4. Murray CJL, Lopez, AD, eds. Global burden of disease: a comprehensive assessment of mortality and disability from diseases, injuries and risk factors in 1990 and projected to 2020. Boston, MA: Harvard University Press, 1996.

5. Steffens DC, Skoog I, Norton MC, et al. Prevalence of depression and its treatment in an elderly population: the Cache County study. Arch Gen Psychiatry 2000; 57(6):601–7.

6. Roberts RE, Kaplan GA, Shema SJ, Strawbridge WJ. Prevalence and correlates of depression in an aging cohort: the Alameda County Study. J Gerontol B Psychol Sci Soc Sci 1997; 52(5):S252–8.

7. Unutzer J, Patrick DL, Simon G, et al. Depressive symptoms and the cost of health services in HMO patients aged 65 years and older. A 4-year prospective study. JAMA 1997; 277(20):1618–23.

8. Mussolino ME. Depression and hip fracture risk: the NHANES I epidemiologic follow-up study. Public Health Rep 2005; 120(1):71–5.

9. Baldwin RC, Tomenson B. Depression in later life. A comparison of symptoms and risk factors in early and late onset cases. Br J Psychiatry 1995; 167(5):649–52.

10. Conway C, Steffens D. Geriatric depression: further evidence for the 'vascular depression' hypothesis. Curr Opin Psychiatry 1999; 12:463–70.

11. Krishnan KR, McDonald WM. Arteriosclerotic depression. Med Hypotheses 1995; 44(2):111–5.

12. Gonzalez MB, Snyderman TB, Colket JT, et al. Depression in patients with coronary artery disease. Depression 1996; 4(2):57–62.

13. Goodnick PJ, Henry JH, Buki VM. Treatment of depression in patients with diabetes mellitus. J Clin Psychiatry 1995; 56(4):128–36.

14. Aromaa A, Raitasalo R, Reunanen A, et al. Depression and cardiovascular diseases. Acta Psychiatr Scand Suppl 1994; 377:77–82.

15. Frasure-Smith N, Lesperance F, Talajic M. Depression and 18-month prognosis after myocardial infarction. Circulation 1995; 91(4):999–1005.

16. Krishnan KR, Hays JC, Blazer DG. MRI-defined vascular depression. Am J Psychiatry 1997; 154(4):497–501.

17. Alexopoulos GS, Meyers BS, Young RC, Campbell S, Silbersweig D, Charlson M. 'Vascular depression' hypothesis. Arch Gen Psychiatry 1997; 54(10):915–22.

18. Morley JE, Mooradian AD, Silver AJ, Heber D, Alfin-Slater RB. Nutrition in the elderly. Ann Intern Med 1988; 109(11):890–904.

19. Kolasa KM, Mitchell JP, Jobe AC. Food behaviors of southern rural community-living elderly. Arch Fam Med 1995; 4(10):844–8.

20. Jorm AF, Korten AE, Christensen H, Jacomb PA, Rodgers B, Parslow RA. Association of obesity with anxiety, depression and emotional well-being: a community survey. Aust N Z J Public Health 2003; 27(4):434–40.

21. Musante GJ, Costanzo PR, Friedman KE. The comorbidity of depression and eating dysregulation processes in a diet-seeking obese population: a matter of gender specificity. Int J Eat Disord 1998; 23(1):65–75.

22. Ross CE. Overweight and depression. J Health Soc Behav 1994; 35(1):63–79.

23. Roberts RE, Deleger S, Strawbridge WJ, Kaplan GA. Prospective association between obesity and depression: evidence from the Alameda County Study. Int J Obes Relat Metab Disord 2003; 27(4):514–21.

24. Thakore JH, Richards PJ, Reznek RH, Martin A, Dinan TG. Increased intra-abdominal fat deposition in patients with major depressive illness as measured by computed tomography. Biol Psychiatry 1997; 41(11):1140–2.

25. Weber-Hamann B, Werner M, Hentschel F, et al. Metabolic changes in elderly patients with major depression: evidence for increased accumulation of visceral fat at follow-up. Psychoneuroendocrinology 2006;31(3):347–54.

26. Payne ME, Hybels CF, Bales CW, Steffens DC. Vascular nutritional correlates of late-life depression. Am J Geriatr Psychiatry 2006; 14:787–95.

27. Maes M, Christophe A, Delanghe J, Altamura C, Neels H, Meltzer HY. Lowered omega3 polyunsaturated fatty acids in serum phospholipids and cholesteryl esters of depressed patients. Psychiatry Res 1999; 85(3):275–91.

28. Sinclair AJ, Begg D, Mathai M, Weisinger RS. Omega 3 fatty acids and the brain: review of studies in depression. Asia Pac J Clin Nutr 2007; 16(Suppl 1):391–7.

29. DeMar JC, Jr., Ma K, Bell JM, Igarashi M, Greenstein D, Rapoport SI. One generation of n-3 polyunsaturated fatty acid deprivation increases depression and aggression test scores in rats. J Lipid Res 2006; 47(1):172–80.

30. de la Presa Owens S, Innis SM. Docosahexaenoic and arachidonic acid prevent a decrease in dopaminergic and serotoninergic neurotransmitters in frontal cortex caused by a linoleic and alpha-linolenic acid deficient diet in formula-fed piglets. J Nutr 1999; 129(11):2088–93.

31. Innis SM, de La Presa Owens S. Dietary fatty acid composition in pregnancy alters neurite membrane fatty acids and dopamine in newborn rat brain. J Nutr 2001; 131(1):118–22.

32. Hibbeln JR. Fish consumption and major depression. Lancet 1998; 351(9110):1213.

33. Tanskanen A, Hibbeln JR, Hintikka J, Haatainen K, Honkalampi K, Viinamaki H. Fish consumption, depression, and suicidality in a general population. Arch Gen Psychiatry 2001; 58(5):512–3.

34. Timonen M, Horrobin D, Jokelainen J, Laitinen J, Herva A, Rasanen P. Fish consumption and depression: the Northern Finland 1966 birth cohort study. J Affect Disord 2004; 82(3):447–52.

35. Kamphuis MH, Geerlings MI, Tijhuis MA, Kalmijn S, Grobbee DE, Kromhout D. Depression and cardiovascular mortality: a role for n-3 fatty acids? Am J Clin Nutr 2006; 84(6):1513–7.

36. Appleton KM, Peters TJ, Hayward RC, et al. Depressed mood and n-3 polyunsaturated fatty acid intake from fish: non-linear or confounded association? Soc Psychiatry Psychiatr Epidemiol 2007; 42(2):100–4.

37. Hakkarainen R, Partonen T, Haukka J, Virtamo J, Albanes D, Lonnqvist J. Food and nutrient intake in relation to mental wellbeing. Nutr J 2004; 3(1):14.

38. Jacka EN, Pasco JA, Henry MJ, Kotowicz MA, Nicholson GC, Berk M. Dietary omega-3 fatty acids and depression in a community sample. Nutr Neurosci 2004; 7(2):101–6.

39. Mamalakis G, Kalogeropoulos N, Andrikopoulos N, et al. Depression and long chain n-3 fatty acids in adipose tissue in adults from Crete. Eur J Clin Nutr 2006;60(7):882–8.

40. Schins A, Crijns HJ, Brummer RJ, et al. Altered omega-3 polyunsaturated fatty acid status in depressed post-myocardial infarction patients. Acta Psychiatr Scand 2007; 115(1):35–40.

41. Hibbeln JR, Umhau JC, George DT, Salem N, Jr. Do plasma polyunsaturates predict hostility and depression? World Rev Nutr Diet 1997; 82:175–86.

42. Lidberg L, Tuck JR, Asberg M, Scalia-Tomba GP, Bertilsson L. Homicide, suicide and CSF 5-HIAA. Acta Psychiatr Scand 1985; 71(3):230–6.

43. Samuelsson M, Jokinen J, Nordstrom AL, Nordstrom P. CSF 5-HIAA, suicide intent and hopelessness in the prediction of early suicide in male high-risk suicide attempters. Acta Psychiatr Scand 2006; 113(1):44–7.

44. Adams PB, Lawson S, Sanigorski A, Sinclair AJ. Arachidonic acid to eicosapentaenoic acid ratio in blood correlates positively with clinical symptoms of depression. Lipids 1996; 31(Suppl):S157–61.

45. Hunter R, Jones M, Jones TG, Matthews DM. Serum B_{12} and folate concentrations in mental patients. Brit J Psychiat 1967; 113:1291–95.

46. Ghadirian AM, Ananth J, Engelsmann F. Folic acid deficiency and depression. Psychosomatics 1980; 21(11):926–9.

47. Abou-Saleh MT, Coppen A. The biology of folate in depression: implications for nutritional hypotheses of the psychoses. J Psychiatr Res 1986; 20(2):91–101.

48. Bell IR, Edman JS, Marby DW, et al. Vitamin B_{12} and folate status in acute geropsychiatric inpatients: affective and cognitive characteristics of a vitamin nondeficient population. Biol Psychiatry 1990; 27(2):125–37.

49. Levitt AJ, Joffe RT. Folate, B_{12}, and life course of depressive illness. Biol Psychiatry 1989; 25(7):867–72.

50. Botez MI, Young SN, Bachevalier J, Gauthier S. Effect of folic acid and vitamin B_{12} deficiencies on 5-hydroxyindoleacetic acid in human cerebrospinal fluid. Ann Neurol 1982; 12(5):479–84.

51. Bottiglieri T, Hyland K, Laundy M, et al. Enhancement of recovery from psychiatric illness by methylfolate. Lancet 1990; 336(8730):1579–80.

52. Morris MS, Fava M, Jacques PF, Selhub J, Rosenberg IH. Depression and folate status in the US Population. Psychother Psychosom 2003; 72(2):80–7.

53. Sachdev PS, Parslow RA, Lux O, et al. Relationship of homocysteine, folic acid and vitamin B_{12} with depression in a middle-aged community sample. Psychol Med 2005; 35(4):529–38.

54. Ramos MI, Allen LH, Haan MN, Green R, Miller JW. Plasma folate concentrations are associated with depressive symptoms in elderly Latina women despite folic acid fortification. Am J Clin Nutr 2004; 80(4):1024–8.

55. Eussen SJ, Ferry M, Hininger I, Haller J, Matthys C, Dirren H. Five year changes in mental health and associations with vitamin B_{12}/folate status of elderly Europeans. J Nutr Health Aging 2002; 6(1):43–50.

56. Lindeman RD, Romero LJ, Koehler KM, et al. Serum vitamin B_{12}, C and folate concentrations in the New Mexico elder health survey: correlations with cognitive and affective functions. J Am Coll Nutr 2000; 19(1):68–76.

57. Penninx BW, Guralnik JM, Ferrucci L, Fried LP, Allen RH, Stabler SP. Vitamin B(12) deficiency and depression in physically disabled older women: epidemiologic evidence from the Women's Health and Aging Study. Am J Psychiatry 2000; 157(5):715–21.

58. Bjelland I, Tell GS, Vollset SE, Refsum H, Ueland PM. Folate, vitamin B_{12}, homocysteine, and the MTHFR 677C->T polymorphism in anxiety and depression: the Hordaland Homocysteine Study. Arch Gen Psychiatry 2003; 60(6):618–26.

59. Tiemeier H, van Tuijl HR, Hofman A, Meijer J, Kiliaan AJ, Breteler MM. Vitamin B_{12}, folate, and homocysteine in depression: the Rotterdam Study. Am J Psychiatry 2002; 159(12):2099–101.

60. Tolmunen T, Hintikka J, Ruusunen A, et al. Dietary folate and the risk of depression in Finnish middle-aged men. A prospective follow-up study. Psychother Psychosom 2004; 73(6):334–9.

61. Almeida OP, Lautenschlager N, Flicker L, et al. Association between homocysteine, depression, and cognitive function in community-dwelling older women from Australia. J Am Geriatr Soc 2004; 52(2):327–8.

62. Lewis SJ, Lawlor DA, Davey Smith G, et al. The thermolabile variant of MTHFR is associated with depression in the British Women's Heart and Health Study and a meta-analysis. Mol Psychiatry 2006;11(4):352–60.

63. Kelly CB, McDonnell AP, Johnston TG, et al. The MTHFR C677T polymorphism is associated with depressive episodes in patients from Northern Ireland. J Psychopharmacol 2004; 18(4):567–71.

64. Lindenbaum J, Healton EB, Savage DG, et al. Neuropsychiatric disorders caused by cobalamin deficiency in the absence of anemia or macrocytosis. N Engl J Med 1988; 318(26):1720–8.

65. Uehara T, Tabuchi M, Mori E. Risk factors for silent cerebral infarcts in subcortical white matter and basal ganglia. Stroke 1999; 30(2):378–82.

66. Firbank MJ, Lloyd AJ, Ferrier N, O'Brien JT. A volumetric study of MRI signal hyperintensities in late-life depression. Am J Geriatr Psychiatry 2004; 12(6):606–12.

67. Taylor WD, MacFall JR, Payne ME, et al. Greater MRI lesion volumes in elderly depressed subjects than in control subjects. Psychiatry Res 2005; 139(1):1–7.

68. Taylor WD, Payne ME, Krishnan KR, et al. Evidence of white matter tract disruption in MRI hyperintensities. Biol Psychiatry 2001; 50(3):179–83.

69. de Groot JC, de Leeuw FE, Oudkerk M, et al. Cerebral white matter lesions and cognitive function: the Rotterdam Scan Study. Ann Neurol 2000; 47(2):145–51.

70. van den Heuvel DM, Ten Dam VH, de Craen AJ, et al. Increase in periventricular white matter hyperintensities parallels decline in mental processing speed in a non-demented elderly population. J Neurol Neurosurg Psychiatry 2006; 77(2):149–53.

71. Vermeer SE, Prins ND, den Heijer T, Hofman A, Koudstaal PJ, Breteler MM. Silent brain infarcts and the risk of dementia and cognitive decline. N Engl J Med 2003; 348(13):1215–22.

72. Vermeer SE, Hollander M, van Dijk EJ, Hofman A, Koudstaal PJ, Breteler MM. Silent brain infarcts and white matter lesions increase stroke risk in the general population: the Rotterdam Scan Study. Stroke 2003; 34(5):1126–9.

73. Steffens DC, Bosworth HB, Provenzale JM, MacFall JR. Subcortical white matter lesions and functional impairment in geriatric depression. Depress Anxiety 2002; 15(1):23–8.

74. Steffens DC, Krishnan KR, Crump C, Burke GL. Cerebrovascular disease and evolution of depressive symptoms in the cardiovascular health study. Stroke 2002; 33(6):1636–44.

75. Steffens DC, Helms MJ, Krishnan KR, Burke GL. Cerebrovascular disease and depression symptoms in the cardiovascular health study. Stroke 1999; 30(10):2159–66.

76. Thomas AJ, O'Brien JT, Davis S, et al. Ischemic basis for deep white matter hyperintensities in major depression: a neuropathological study. Arch Gen Psychiatry 2002; 59(9):785–92.

77. Schmidt R, Hayn M, Fazekas F, Kapeller P, Esterbauer H. Magnetic resonance imaging white matter hyperintensities in clinically normal elderly individuals. Correlations with plasma concentrations of naturally occurring antioxidants. Stroke 1996; 27(11):2043–7.

78. Mukamal KG, Longstreth WT, Jr., Mittleman MA, Crum RM, Siscovick DS. Alcohol consumption and subclinical findings on magnetic resonance imaging of the brain in older adults. Stroke 2001; 32:1939–46.

79. Payne ME, Haines PS, Chambless LE, Anderson JJB, Steffens DC. Food group intake and brain lesions in late-life vascular depression. Int Psychogeriatr 2007; 19(2):295–305.

80. Paul RT, McDonnell AP, Kelly CB. Folic acid: neurochemistry, metabolism and relationship to depression. Hum Psychopharmacol 2004; 19(7):477–88.

IV NEW FRONTIERS IN PREVENTIVE NUTRITION

27

Providing Food and Nutrition Choices for Home and Community Long Term Living

Dian O. Weddle and Nancy S. Wellman

Key Points

- Food and good nutrition are key factors in helping older adults remain healthy and independent in the community by reducing chronic disease risk and disability, promoting health and supporting increased mental and physical functioning.
- The current long term care system is shifting from a nursing facility focus to one that provides an array of services in home and community-based settings.
- A new long term care system is emerging led in part by two Federal agencies, the Centers for Medicare and Medicaid and the Administration on Aging.
- A philosophical and operational shift needs to be made from a provider/service driven model to one where consumers, families and caregivers are empowered to make their own long term care decisions.

Key Words: Home and community-based long term care system; nutrition and food services; service rebalancing; Centers for Medicare and Medicaid Services; U.S. Administration on Aging; Aging and Disability Resource Center (ADRC)

27.1 INTRODUCTION

Food and good nutrition are key factors in helping older adults remain healthy and independent in the community *(1,2)*. Together, they reduce chronic disease risk and disability, promote health and support increased mental and physical functioning and active engagement with life *(1,3–6)*. Food and good nutrition play preventive roles being integral to health promotion, risk reduction and disease treatment *(1,7)*. Nutrition therapy is cost effective; it prevents malnutrition, in the form of undernutrition and obesity, and is central to management of common chronic diseases. Older adults are in need of a system that coordinates provision of food and

From: *Nutrition and Health: Handbook of Clinical Nutrition and Aging, Second Edition*
Edited by: C. W. Bales and C. S. Ritchie, DOI 10.1007/978-1-60327-385-5_27,
© Humana Press, a part of Springer Science+Business Media, LLC 2009

nutrition services and programs necessary to maintain well-being and independence to support home and community long term living *(1)*. The current long term care system is shifting from a nursing facility focus to one that provides an array of services in home and community-based settings. Nutrition, nursing, social service and healthcare practitioners can facilitate a system promoting food and nutrition services as critical elements for long term living.

27.2 NUTRITION AND CHRONIC DISEASE MANAGEMENT

The top nine chronic health conditions in older persons, (heart disease, hypertension, stroke, diabetes, cancer, arthritis, emphysema, asthma, chronic bronchitis) all have dietary and nutritional implications *(3,4,8,9)*. They affect the ability to remain independent in the community and are greatly exacerbated by malnutrition either as undernutrition or as obesity *(3)*. Heart disease, stroke, cancer and diabetes are among the most common and costly *(4)*. When obesity is a co-morbidity, they account for almost all Medicare spending over the last 15 years *(10)*. Food and nutrition interventions help control the high costs of disease treatment and hospitalizations through risk reduction, delayed disease onset and symptom management. Nutrition therapy improves the efficacy and effectiveness of the associated medical, pharmaceutical and rehabilitative treatments *(1)*.

27.3 NUTRITIONAL STATUS, MALNUTRITION
AND FUNCTIONALITY

Nutritional status is closely associated with an older person's functionality and ability to remain independent. Thus, a goal of improving nutritional status through an adequate diet is to prevent obesity or underweight, both considered malnutrition, and have the greatest impact on health *(3)*. Obesity and underweight impede functionality and independent living *(4,11)*. Malnourished older adults have limited muscle strength, more exhaustion and reduced physical activity placing them at risk for falls and hip fractures *(11)*. These reduce quality of life, threaten independence and increase healthcare costs.

Obesity, the most common nutritional disorder, is a risk factor for age-related chronic diseases described earlier *(3,10)* (also see Chapter 15). Obesity in the Medicare population grew from 11.7% in 1987 to 22.5% in 2002. Healthcare spending for this population grew from 9.4% of the federal budget in 1987 to 24.8% in 2002 *(10)*. Obese older adults are more likely to become disabled and report difficulties with ADLs and IADLs that impact their functional independence *(12)*.

Undernutrition in older adults often has multi-factors which must be addressed for a successful resolution. Several chronic diseases (e.g., cancer, COPD, and Alzheimer's) are risk factors themselves because the unintended, undetected weight loss contributes to frailty. Research shows independent relationships among the quantity and quality of diets and frailty *(13,14)* (also see Chapter 9). Without nutrition therapy and other long term care services to support recovery, individuals are at greater risk for premature nursing home placement. Unaddressed, undernutrition and frailty result in costly long term care services. Malnutrition, unintentional weight loss and dehydration are serious risk factors for development of pressure ulcers. Depending on progression, pressure ulcers need aggressive nutrition therapy, medication management, nursing care and sufficient time to heal *(15)* (also see Chapter 12).

A nutrition screening, assessment and intervention process detects and prevents malnutrition *(3)*. Routine screening and targeted intervention is encouraged as a cost-effective way to promote health and manage disease in older adults by the Nutrition Screening Initiative *(16)*. Nutrition screening, assessment and targeted interventions help keep older adults in their communities and reduce the costs associated with medications, hospital care and nursing home stays *(17)*.

27.4 CAREGIVERS AND NUTRITIONAL STATUS

Informal unpaid family caregivers are important in improving or maintaining the nutritional status of their care recipients. Informal caregivers provide the majority of care for underserved populations, including those in rural areas, those suffering from dementia and those receiving hospice care. Caregivers prepare meals, assist with eating, and, when necessary, administer and monitor home enteral nutrition *(1)*.

Caregivers may not have the skills and information needed to encourage eating, modify food texture or evaluate the appropriateness of nutritional supplements *(18)*. Nutrition education or individualized nutrition counseling for specific diseases and conditions (e.g., Alzheimer's disease, hip fracture recovery) is needed *(19)*. The responsibility of providing nutrition care and sustenance that impact the health and life of the care recipient contributes to caregiver burden. Informal caregivers must also be concerned with their own nutritional status. The stress of caregiving places them at malnutrition risk. They may skip meals, consume unhealthy foods and neglect management of their own chronic diseases or conditions *(20)*.

27.5 THE CHANGING LONG TERM CARE SYSTEM

A new long term care system is emerging led in part by two Federal agencies, the Centers for Medicare and Medicaid Services (CMS, www.cms.hhs.gov) and the U.S. Administration on Aging (AoA, www.aoa.gov). CMS administers Medicaid and Medicaid Home and Community Waiver Programs for limited income populations including older persons and those with disabilities. The AoA administers the Older Americans Act (OAA). The OAA authorizes the National Aging Network (Aging Network) and the largest home and community-based comprehensive coordinated service system for adults 60+ in the USA. The Aging Network consists of 56 State Units on Aging (SUAs, www.nasua.org), 655 Area Agencies on Aging (AAA, www.eldercare.gov), 236 Tribal & Native Organizations plus thousands of service providers, adult care centers, caregivers and volunteers. The Aging Network provider system delivers the array of nutrition, supportive, health promotion and disease prevention and caregiver support services authorized under the OAA.

The Medicaid program is the largest source of public funding for long term care in the United States. Of the Medicaid monies spent on older adults, 69% goes for long term care. Facility-based long term care accounts for 70% of Medicaid spending *(21)*. CMS, as the predominant funder of long term care, and AoA, with its Aging Network of HCBC agencies and service providers, are collaborating to better serve the different long term care needs and better manage the associated high costs of providing care to the current heterogeneous, multi-racial and ethnic population of adults aged 60 + . Further, the large numbers of those aged 45–64 who turn 65 during the next two decades will require long term care. These efforts are rebalancing the present long term care system away from long term placement in nursing home facilities to one providing an array of home and community-based care options *(21,22)*.

Rebalancing requires a different way of thinking about the delivery of and payment for home and community-based care services. Key long term care components integral to rebalancing include 1. provision of alternatives to nursing home placement (referred to as nursing home diversion). 2. Promotion of consumer-directed choice in services provided. 3. Provision of a single access point consumers use for information and referral. CMS and AoA have been providing grants and operational support to states as they rebalance and expand home- and community-based care services. Table 27.1 describes three ongoing demonstrations: Money Follows the Person Rebalancing Demonstration for nursing home diversion *(23)*. Cash and Counseling for consumer choice *(24)*; and Aging and Disability Resource Center (ADRC) Grants for improved information, access and referral *(25)*.

These rebalancing efforts are showing signs of success. A study of New Jersey's nursing home transition program followed all 1,354 residents discharged from nursing facilities during 2000 for 1 year post-discharge. Of these, 628 were in the nursing home diversion program and were transitioning to the community. Of all 1,354 persons, 72.6% remained in the community 1-year post-discharge; 17% had a long term readmission (>90 days); and 18.6% died. Within the diversion group, 75% remained in the community, 8.1% had a long term readmission and 16.9% died. Significant predictors of a long term readmission 1 year post-discharge were being male, single, not satisfied with living situation, living with others, and falling within 8–10 weeks post-discharge. The authors conclude that predisposing factors, not need, influenced readmission. They suggest individuals needing the most monitoring during transition from nursing home to home setting be identified and provided with risk-specific preventive measures *(26)*.

Another study examined Health and Retirement Study data to determine whether or not state spending for home- and community-based care made a difference in the risk of a first long term nursing home admission among older adults. Findings indicated that residing in a state with higher home- and community-based care spending resulted in a significantly lower risk of nursing home admission among older persons who were childless. The authors conclude that doubling the amount of state spending on home- and community-based care for childless older persons would reduce the risk of nursing home admission by 35% *(27)*.

Grants to establish ADRCs have been awarded to 43 states and territories since the program's inception in 2003. Grantees serve persons aged 60 + and at least one

Table 27.1

Long term care rebalancing demonstration projects

Project	Purpose	Current activity
Money Follows the Person Rebalancing Demonstration. (www.hcbs.org)	1. Increase home and community-based care & decrease facility-based care; 2. allow money to follow the person by removing barriers so individuals receive services in setting of choice; 3. Assure continuity of HCBS for eligible individuals who transition from institutional to community settings	In 2007, CMS awarded 31 states $1.436 billion in grants to transition 37,731 individuals out of institutional settings over a 5-year period (23).
Cash and Counseling. (www.cashandcounseling. org)	1. Promote consumer choice by allowing elderly and disabled Medicaid recipients who receive personal care services to direct their own care, purchase services/products needed and select the provider	To date, 15 states have developed or are in the process of developing Cash and Counseling Projects (24).
Aging and Disability Resource Center Demonstration Grant Initiative (ADRCs) (www.adrc-tae.org)	1. Include older adults and at least one other younger persons with disabilities in at least one community of all income groups; 2. Develop comprehensive "one-stop", single point of entry for older adults and individuals with disabilities who are seeking information, access and referral to home- and community-based services.	Since the ADRC program was launched in 2003, 43 states and territories have received 3-year competitive grants (25).

population of younger persons with disabilities in at least one community of all income levels. An interim report reveals that establishment of the single point of long term care information, access and referral has resulted in ADRCs establishing partnerships with home- and community-based long term care providers and institutions to better inform consumers about options. These providers include home- and community-based care intake agencies, hospitals and discharge planners, doctors' offices and rehabilitation nursing homes. These types of organizations and the services provided accounted for 55% of all contacts/requests made to the ADRCs. This suggests that ADRCs have begun to be known as trusted sources for long term care information *(25)*. Grantees also report priority efforts to establish working partnerships with their colleague populations of younger persons with disabilities and to include all those with vested community interests in the planning process *(25)*.

A comparison of the three original Cash and Counseling demonstration projects conducted in Arkansas, Florida and New Jersey indicated that there were commonalities across projects as well as variation in types of supportive services offered and with project implementation *(28)*. The Demonstration project showed that frail older adults and persons with disabilities are able to assume responsibility for self-direction and make choices about the care they want. It also demonstrated that it is feasible to pay a family member or friend to provide personal assistance. The model can be adapted across states and offered as a service choice. Although it is not for everyone, Cash and Counseling participants were quite receptive *(29)*.

27.6 OPPORTUNITIES AND CHALLENGES WITHIN THE HOME- AND COMMUNITY-BASED LONG TERM CARE SYSTEM

Dietitians, nutritionists, nurses, social service and health-care practitioners can make a difference in meeting the food and nutrition long term living needs of older persons. But, to be most effective, they must understand how home- and community-based care services are delivered and financed.

The inclusion of the full array of food and nutrition services is frequently missing in the home- and community-based long term care system. Food and nutrition are viewed as two separate non-intersecting parallel systems: 1. food as part of a social and supportive services system and 2. nutrition as part of a medical problem-oriented treatment. These systems must be joined and coordinated for food and nutrition to fulfill their critical role in promoting health and independence for community-residing older persons.

Medicaid/Medicaid Waiver programs are state-administered program and each state sets its own eligibility and service guidelines. The Medicaid Waiver Program provides home- and community-based services to frail community residing recipients who are nursing home eligible. Services vary as individual states determine what will be provided. Although it differs by state, some do offer nutrition services including home delivered meals, risk assessment and nutrition counseling *(30)*.

The OAA is also a state-administered program. While it is available to individuals aged 60+, it targets those who are poor and of minority status, including Native Americans, rural residing and those with limited English speaking ability.

The OAA supports bottom-up, state and community-driven planning. AAAs develop local area plans that are integrated with their state plan. Tribal organizations develop programs specific to needs and cultural customs.

At both the state and the local levels, Aging Network agencies and providers juggle a multitude of funding streams and client eligibility requirements as they develop strategic and operational plans to provide the mix of home- and community-based care services to meet the needs of a consumer base with different levels of frailty. Eligibility requirements and funding streams vary across home- and community-based care programs. Multiple funding streams are generally used to pay for home- and community-based care. Many states provide home- and community-based care through the OAA Aging Network structure; although the majority of funding for these agencies may come from the Medicaid Waiver Program for services to persons who are nursing home eligible and may be frailer than those served with OAA funds. In addition, funds may come from other state or local programs and private support funding (31). It is a state and community decision as to what home- and community-based services will be offered and to whom. As a result, there is great variability across the nation with respect to the availability and variety of food and nutrition services provided.

A philosophical and operational shift must be made from a provider-/service-driven model to one where consumers, families and caregivers are empowered to make their own long term care decisions. Service providers need to ensure that their services are more client, caregiver and family centered. As consumers begin to choose from a menu of long term care options and service providers, they will select services based on self identified and prioritized needs. Providers must understand this. Providers may conduct focus groups and develop products and social marketing plans that explain to consumers how their services meet needs and add value.

The OAA state and local planning approach is very grassroots oriented. Armed with convincing data and an understanding of the system, nutrition, nursing, social service and health-care practitioners have a real opportunity to advocate for inclusion of appropriate food and nutrition services. The ADRCs are becoming the "one-stop-shop" for home- and community-based long term care. Within their states and communities, practitioners should work with their Aging Network colleagues and the diverse partners involved with the ADRC development and implementation process to build awareness of the importance of food and nutrition to health. Practitioners must then direct consumers to accurate information and appropriate referral mechanisms.

27.7 THE OLDER AMERICANS ACT AND REBALANCED LONG TERM CARE

The OAA, the cornerstone of cost effective, comprehensive, coordinated, high-quality home- and community-based services, promotes rebalanced long term care. Reauthorized in October 2006, the OAA authorizes the Assistant Secretary for Aging, SUAs, and AAAs to promote the development and implementation of a comprehensive State system of long term care that enables older adults to receive client-centered services at home in the community (32). The OAA includes provisions to 1. empower consumers to make informed decisions about choices for long term living; 2. build prevention into community living through evidence-based health promotion and disease prevention programs designed for older adults; and 3. target high-risk, nursing home eligible, non-Medicaid or private pay individuals and delay institutionalization through choices for home and community care to meet individualized needs and preferences *(32)*.

The OAA promotes a more client centered, cost effective home- and community-based model. These models increase consumer satisfaction and save money by lower use of high cost emergency rooms and institutional care *(21)*. The South Carolina Lieutenant Governor's Office on Aging documented the value of OAA services, including congregate and home-delivered meals, in avoiding expensive inpatient hospital services. OAA clients had a significantly reduced rate of emergency department visits due to ambulatory care sensitive conditions including dehydration, pneumonia, etc., when compared with a matched sample of Medicaid clients. Ambulatory care sensitive conditions are typically preventable or controllable and should not normally require inpatient or emergency care. Additionally, the Office on Aging compared intensity of OAA services between clients receiving three or more meals per week (higher intensity) with those receiving fewer meals (lower intensity). Those receiving three or more meals per week had significantly lower rates of ambulatory care sensitive condition-related inpatient admissions than the lower intensity comparison group. With adjustments for race, gender, age and service durations, congregate clients who ate more meals per week had significantly fewer inpatient admissions and emergency department visits than those eating fewer meals. Home delivered meal clients receiving more meals per week also had significantly fewer inpatient admissions than the comparison group *(33)*. At the time of the study, South Carolina's average cost of a year's worth of home delivered meals was $1,107 (FY 2004, FY 2005) compared to the $25,000–$37,000 average cost of a year's stay in a nursing home (FY 2005) *(33)*.

The OAA and its evidence-based Nutrition Program have the experience, network and programs in place for a flexible, client- and family-centered home- and community-based long term care system. OAA services are highly rated *(34)*. Older adults and families turn to it for top quality, accurate information; appropriate, safe services and competent providers to serve long term living needs.

27.8 THE OAA NUTRITION PROGRAM AS A SERVICE MODEL FOR FOOD AND NUTRITION CHOICES

The OAA Nutrition Program can serve as a service model for preventive food and nutrition choices in home- and community-based long term living *(35)*. It is popular, grassroots and consumer oriented. A key foundation service, the Nutrition Program has a history of documented, substantial contributions to the health and social well-being of its participants *(35,36)*. It is well integrated into home and community settings through coordination with community partners. Its care-planning process includes nutrition screening, nutrition assessment, nutrition education, nutrition counseling with follow-up when appropriate *(32,35,36)*.

The OAA Nutrition Program goals are to reduce hunger and food insecurity; to promote socialization of older adults and to promote the health and well-being of older adults by assisting them to gain access to nutrition and other disease prevention and health promotion services to delay onset of adverse health conditions resulting from poor nutritional health or sedentary behavior *(32)*. These goals are met and implemented through the array and consumer directed choices of services funded through the OAA (Tables 27.2 and 27.3). There is coordination with other federally funded food and nutrition programs to improve access to services designed to reduce hunger and food insecurity and improve disease management (Table 27.4).

The OAA Nutrition Program is at the front line of primary, secondary and tertiary prevention by providing clients with 1. safe nutritionally adequate meals, and nutrition and fitness guidance for healthy lifestyles; and 2. nutrition risk screening, assessment, education and counseling to effectively manage chronic disease *(1,7)*. For example, diabetes coupled with inactivity results in a more than fourfold increased risk of future nursing home admissions *(37)*. As states' priorities differ, provision of Nutrition Program services is not standard across the country. Each state, with local community involvement, decides which OAA Nutrition Program services it will offer.

Federal regulations require that The OAA Nutrition Program adhere to the latest edition of the *Dietary Guidelines for Americans* (*DGA*) and use the newest nutrient requirement knowledge and guidance *(32)*. The *DGA* makes specific quantitative recommendations regarding nutrients and food components and emphasizes a healthy weight and physical activity *(5)*. The Dietary Reference Intakes (DRIs) emphasize that older adults have specific nutrient requirements due to the aging process. The DRIs help prevent nutritional deficiency, reduce the risk of chronic diseases and improve health over the long run *(38)*.

The OAA Nutrition Program employs a continuous quality improvement process to maintain adherence to the latest scientific evidence and highest performance

Table 27.2

OAA Nutrition Program services

1) Meals to improve food & nutrient intakes and reduce hunger and food insecurity.
2) Nutrition screening and assessment as needed to identify nutrition related risk.
3) Congregate dining site participation to promote active social engagement.
4) Nutrition education to motivate dietary & behavioral change through information about timely nutrition topics.
5) Nutrition counseling to enable chronic disease management.
6) Referrals & coordination to connect consumers & caregivers, for example:
   - community partners for health promotion/disease prevention services,
   - in-home services,
   - food and nutrition assistance programs,
   - facility based discharge planners for post discharge meals,
   - Medicaid Waiver home and community based services to delay nursing home placement,
   - Medicare Medical Nutrition Therapy Diabetes Self Management Programs,
   - Medicare Medical Nutrition Therapy Pre-dialysis Renal Disease.

Table 27.3

OAA Food and Nutrition choices for consumers and caregivers

1) Community dining options including congregate sites and restaurant vouchers when appropriate;
2) Culturally appropriate meals, entrée choices, soup and salad bars;
3) Home-delivered meals including therapeutic meals – renal diets, texture modifications, hot or frozen meals, daily or weekly deliveries, meals for older caregivers;
4) In-depth individualized nutrition counseling for disease management – heart disease, diabetes, hypertension, cancer;
5) Disease-specific group counseling sessions – diabetes, osteoporosis, obesity, unintended weight loss; and
6) Interactive nutrition education sessions on healthy eating, food labels, food safety, and physical activity tailored to older adults and caregivers.

Table 27.4

Federally funded food and nutrition programs

US DHHS	USDA
• Older Americans Act	• Food Stamp Program
• Indian Health Service	• Food Stamp Nutrition Education Program
○ Clinical nutrition services	• Food Distribution Program on Indian Reservations
• Medicaid HCBS Waiver Program (1915 b&c)	• Commodity Supplemental Food Program
○ 38 states offer meals	• The Emergency Food Assistance Food Program
• Medicare	• Child & Adult Care Food Program
○ Medical Nutrition Therapy	• Senior Farmers Market Nutrition Program
○ Preventive services	

standards. Mechanisms measure customer satisfaction, analysis of dietary intake, assurance of nutritional quality and safe food, assurances for families and adherence to consistency of standards. It also provides appropriate training and guidance on nutritional aspects to case managers as they assess for need for services, training for homemakers and personal care assistants in appropriate shopping and meal preparation and modification, and for family primary caregivers to help them provide adequate nourishment for their care recipient and themselves.

The *Second National Pilot Survey of Older Americans Act Title III Service Recipients* shows that the Nutrition Program successfully targets the vulnerable and frail including the underserved, those of minority status, those residing in rural areas and those with limited access to food. The Home Delivered Nutrition Program serves the frailest and most functionally impaired. This important social community link delays institutionalization *(34)*. Participants age 75 and older comprise 73% of home delivered and 62% of congregate Nutrition Program clients.

Over half of all participants live alone, including 61% of the homebound. About 70% of the homebound have difficulty with one or more ADLs and 29% have three or more ADL limitations. The latter qualifies them as needing nursing home level of care – a care level that indicates significant frailty *(34)*. The mid-day meal provides half or more of the day's total food intake for 66% of home delivered and 56% of congregate Nutrition Program participants. The congregate Nutrition Program helps participants remain independent and engaged though meals, culturally appropriate nutrition education and physical activity and social interaction *(34)*.

27.9 OAA NUTRITION PROGRAM AND FRAIL OLDER ADULTS

Consumers are increasingly diverse with respect to severity of impairments, information and referral needs, mix of nutrition therapies needed, and health/medical/social services needed *(1)*. The Aging Network will continue to target and serve an increasingly frailer, impaired and more underserved population *(39)*. Thus, the OAA Nutrition Program providers, including registered dietitians, tackle the double risk threats to consumers' loss of independence and lessened quality of life. The double risks are caused by 1. inadequate food and nutrient intakes and 2. poor dietary compliance for disease management *(37)*. Without OAA services, about 29% of participants receiving home-delivered meals were impaired to the extent that they were eligible for more costly nursing home placement *(22)*. If these individuals had been in nursing homes, regulations would require they be carefully monitored for malnutrition risk, evaluated for declines in nutrition status, and have a documented care plan to prevent deterioration. Registered dietitians and health practitioners should alert Aging Network colleagues to the potential dangers of ignoring malnutrition. Standard procedures should include care planning and treatment or referral guidelines to prevent a problem. Continuous quality improvement outcome monitors and staff training guidelines should be established.

27.10 PROVIDING FOOD AND NUTRITION CHOICES FOR HOME AND COMMUNITY LONG TERM LIVING: RECOMMENDATIONS

Older adults willingly make nutrition-related lifestyle changes when information is relevant to their needs and they understand how to make the changes *(40)*. Even with serious impairments and chronic conditions, older adults prefer to be at home rather than in a nursing facility *(22)*.

Prioritized nutrition services contribute to health and independence and reduce early nursing home admissions. OAA Nutrition Program components provide choice and can be integrated into the home- and community-based long term care system as illustrated by the following recommendations *(41)*.

1. Use the Aging and Disability Resource Centers (ADRCs) to empower consumers to make informed nutrition-related choices for long term living:

    • Provide One-Stop-Shopping to reduce nutrition risk and promote healthy eating through consumer-tested informational brochures, Internet materials, evidence-based interventions and information about available nutrition services in homes and the community (e.g., home delivered meals, congregate dining sites, Medicare-covered diabetes treatment);

- Include two to three key nutrition questions on the Intake and Referral form to prioritize service referrals and reduce malnutrition risk, use the information for appropriate interventions such as healthy meals, nutrition counseling, family caregivers support;
- Reduce nutrition risk, food insecurity and hunger through information and referrals to programs that increase access to food (e.g., Food Stamp Program, food banks, Senior Farmers' Market Nutrition Program);
- Ensure that nutrition needs are addressed as part of Medicaid Waiver or State Funded home- and community-based service options;
- Inform consumers about possible private pay options for services.

2. Build prevention into community living through evidence-based nutrition related Health Promotion and Disease Prevention programs designed for older adults:

- Regularly offer nutrition screening, nutrition education and nutrition counseling;
- Provide information and referrals for consumers and families about local evidence-based Health Promotion and Disease Prevention programs;
- Partner with Health Promotion and Disease Prevention programs to increase accessibility in senior centers and congregate dining sites; and
- Offer choices of appropriate Health Promotion and Disease Prevention programs to homebound consumers.

3. Meet the nutrition needs of persons at high risk of nursing home placement to delay institutionalization through choices in home- and community-based care:

- Train case managers and other gatekeeper assessors to screen for nutrition risk and identify the need for referral for nutrition assessment, diagnosis, treatment, care planning and monitoring.
- Provide choices in home-delivered meals including special diets, texture modifications, hot or frozen meal choices and daily or weekly deliveries.
- Provide choices in hospital and nursing home discharge planning for comprehensive nutrition services including meals, individualized nutrition counseling for disease management with follow-up indicators of care effectiveness.

The direct roles of food and nutrition in promoting health, reducing risks and managing chronic diseases have been established. Science documents the role nutrition has in maintaining functional independence and protecting quality of life. Registered dietitians, nutrition, nursing, social service and health-care practitioners can work with Aging Network colleagues to assure inclusion of food and nutrition choices that support long term living for older adults.

Acknowledgments: Recognition and thanks goes to Jean Lloyd, MS, RD, National Nutritionist, U.S. Administration on Aging for suggestions on an earlier version. This article was adapted from the draft paper, *The Older Americans Act Nutrition Program: Providing Consumers and Caregivers with Food and Nutrition Choices for Healthy, Independent Long Term Living: A Challenge Brief,* presented at the 4th State Units on Aging Nutritionists/Administrators Conference. Baltimore, MD. August 29, 2006. This project was supported in part by grant 90AM2768 from the Administration on Aging, U.S. Department of Health and Human Services, Washington, DC.

REFERENCES

1. American Dietetic Association Position Statement: Nutrition across the spectrum of aging. J Am Diet Assoc. 2005; 105:616–33.
2. Rowe JW, Kahn RL. Successful Aging. New York, NY: Pantheon Books, 1998.
3. Institute of Medicine, Committee on Nutrition Services for Medicare Beneficiaries. The Role of Nutrition in Maintaining Health in the Nation's Elderly: Evaluating Coverage of Nutrition Services for the Medicare Population. Washington, DC: National Academy Press, 2000.
4. Federal Interagency Forum on Aging-Related Statistics. Older Americans Update 2006: Key Indicators of Wellbeing. Washington, DC: US Government Printing Office, May 2006.
5. US Department of Health and Human Services and US Department of Agriculture. Dietary Guidelines for Americans, 2005, 6th ed. Washington, DC: US Government Printing Office, January 2005 (Accessed December 17, 2007, at www.health.gov/dietaryguidelines/)
6. Centers for Disease Control and Prevention. Health Information for Older Adults. (Accessed December 17, 2007, at www.cdc.gov/aging/info.htm)
7. Wellman NS. Prevention, prevention, prevention: nutrition for successful aging. J Am Diet Assoc. 2007; 107:741–3.
8. Bales CW, Fischer JG, Orenduff MC. Nutritional interventions for age-related chronic disease. Generations 2004; 28:54–60.
9. Bales CW, Ritchie CS, eds. Handbook of Clinical Nutrition and Aging. Totowa, NJ: Humana Press, 2004.
10. Thorpe KE, Howard DH. The rise in spending among Medicare beneficiaries: the role of chronic disease prevalence and changes in treatment intensity. Health Affairs 2006; w378–88.
11. Sharkey JR. The influence of nutritional health on physical function: a critical relationship for homebound older adults. Generations 2004; 28:34–8.
12. Reynolds SL, Saito Y, Crimmins EM. The impact of obesity on active life expectancy in older American men and women. Gerontologist 2005; 45:438–44.
13. Bartali B, Frongillo EA, Bandinelli S, et al. Low nutrient intake is an essential component of frailty in older persons. J Gerontol A Biol Sci Med Sci 2006; 61:589–93.
14. Bischoff HA, Staehelin HB, Willet WC. The effect of undernutrition in the development of frailty in older persons. J Gerontol A Biol Sci Med Sci. 2006; 61:589–93.
15. Niedert KC, Dorner B, Consultant Dietitians in Health Care Facilities DPG. Nutrition Care of the Older Adult: A Handbook for Dietetics Professionals Working throughout the Continuum of Care. Chicago, IL: American Dietetic Association, 2004.
16. Nutrition Screening Initiative. Nutrition Intervention Manual for Professionals Caring for Older Adults, April 1992. Available at: (Accessed September 25,2007, at eatright.org/ada/files/NSI/OlderAmericansComplete1.pdf)
17. Wellman NS, Kamp BF. Add life to years. In: Kaufman MK, ed. Nutrition in Promoting the Public's Health—Strategies, Principles and Practice. Sudbury, MA: Jones & Bartlett Pub, Inc, 2007.
18. Silver HJ, Wellman NS. Family caregiver training is needed to improve outcomes for older adults using home care technologies. J Am Diet Assoc 2002; 102:831–6.
19. Silver HJ, Wellman NS. Nutrition education may reduce burden in family caregivers. J Nutr Educ Behav. 2002; 34:S53–8.
20. Silver HJ. The nutrition-related needs of family caregivers. Generations 2004; 28(3):61–4.
21. Testimony of Dennis Smith, Director, CMSO, USDHHS, before the House Committee on Small Business, US House of Representatives. July 10, 2006. (Accessed September 20, 2007, at www.cms.hhs.gov/apps/media/press/testimony.asp?Counter = 1893)
22. Remarks of Josefina G Carbonell, Assistant Secretary for Aging, USDHHS, at Choices for Independence: A National Leadership Summit. December 5, 2006. (Accessed September 20, 2007, at www.aoa.gov/summit/main_site/summithome.aspx)
23. Centers for Medicare and Medicaid Services. Money Follows the Person (MFP) Rebalancing Demonstration. (Accessed September 20, 2007, at www.cms.hhs.gov/DeficitReductionAct/Downloads/MFP.FactSheet.pdf)

24. Cash & Counseling. (Accessed December 17, 2007, at www.cashandcounseling.org/about/index_html)

25. The Lewin Group. The Aging and Disability Resource Center (ADRC) Demonstration Grant Initiative: Interim Outcomes Report. Falls Church, VA, November 2006. (Accessed September 20, 2007, at www.adrc_tae.org/documents/InterimReport.pdf)

26. Howell S, Silverberg M, Quinn MV, Lucas A. Determinants of remaining in the community after discharge: results from New Jersey's nursing home transition program. Gerontologist 2007; 47:535–47.

27. Muramatsu N, Yin Y, Campbell RT, Hoyem RL, Jacob MA, Ross CO. Risk of nursing home admission among older Americans: does states' spending on home-and community-bases services matter? J Gerontol B Soc Sci 2007; 62B:S153–9.

28. Phillips B, Schneider B. Commonalities and variations in the cash and counseling programs across the three demonstration states. HSR 2007; 42(Part 2):397–413. (Special Issue).

29. Mahoney KJ, Fishman NW, Doty P, Squillance MR. The future of cash and counseling: the framers view. HSR 2007; 42(Part 2):550–65. (Special Issue).

30. National Resource Center on Nutrition, Physical Activity & Aging. Creative Solutions: Using Medicaid Waiver Funding for Nutrition Services. Miami, FL: Florida International University, 2007. (Accessed September 20, 2007, at www.nutritionandaging.fiu.edu/creative_solutions/hcbs.asp)

31. Payne M, Applebaum R, Molea M, Ross DE. Funding services from the bottom up: an overview of senior services levy programs in Ohio. Gerontologist 2007; 47:555–8.

32. US Administration on Aging. Older Americans Act. (Accessed September 20, 2007 at www.aoa.gov/oaa2006/Main_Site/oaa/HR%206197%20(as%20psd%20Sen).pdf)

33. Presentation by Bruce Bondo, MPA, Deputy Director for Policy and Planning and Jeanette Brodie, MPH, Program Information Manager, South Carolina Lieutenant Governor's Office on Aging, at the Administration on Aging Performance Outcome Measures Project Training Conference. Washington, DC, December 7, 2006.

34. Highlights from the Pilot Study: Second National Survey of Older Americans Act Title III Service Recipients. (Accessed September 20, 2007, at https://www.gpra.net/nationalsurvey/files/2ndhighlights.pdf)

35. Millen BE, Ohls JC, Ponza M, McCool AC. The Elderly Nutrition Program: an effective national framework for preventive nutrition interventions. J Am Diet Assoc 2002; 102:234–40.

36. Mathematica Policy Research, Inc. Serving Elders at Risk, Older Americans Act Nutrition Programs: National Evaluation of the Elderly Nutrition Program 1993–1995, Vol. I: Title III Evaluation Findings. Washington, DC: USDHHS, 1996.

37. Valiyeva E. Lifestyle-related risk factors and risk of future nursing home admission. Arch Intern Med 2006; 166:985–90.

38. Institute of Medicine Food and Nutrition Board. Dietary Reference Intakes: The Essential Guide to Nutrient Requirements. Washington, DC: National Academy Press, 2006.

39. US Administration on Aging. AoA Strategic Plan 2007–2012. (Accessed September 20, 2007, at www.aoa.gov/about/strategic/strategic.asp)

40. McCamey MA, Hawthorne NA, Reddy S, Lombardo M, Cress ME, Johnson MA. A statewide educational intervention to improve older Americans' nutrition and physical activity. Fam Econ Nutr Rev 2003; 15:47–57.

41. National Resource Center on Nutrition, Physical Activity & Aging. Older Americans: Making Food and Nutrition *Choices* for a Healthier Future: The Older Americans Act Nutrition Program in the US Administration on Aging's *Choices for Independence*. Miami, FL: Florida International University, April 2007. (Accessed September 20, 2007, at www.nutritionandaging.fiu.edu/Center_Initiatives/Fd%20Choices%20Paper%20April%202007.pdf)

28 Dietary Supplements: Current Knowledge and Future Frontiers

Rebecca B. Costello, Maureen Leser, and Paul M. Coates

Key Points

- The evidence base for the use of dietary supplements is well documented for some, such as the essential nutrients, but it is scant for others, such as some designer supplements and botanicals.
- Popular dietary supplements consumed by older Americans that have national public health recommendations and established guidelines for use include omega-3 fatty acids; the B vitamins, folic acid, vitamins B_6, and B_{12}; vitamins D and E; and the minerals, calcium, potassium, and magnesium.
- Dietary supplements not endorsed by national public health recommendations but with a considerable evidence base upon which randomized clinical trials can be planned or guidelines for use might be formulated include coenzyme Q10, creatine, and the botanicals black cohosh, French pine bark, *Ginkgo biloba,* and saw palmetto.
- A clinical deficiency of vitamins or minerals, other than iron and possibly vitamin D, is uncommon in the United States, except for certain high-risk groups.
- In situations when recommended nutrients cannot be obtained by food alone or retained by the body because of impaired absorption or other physiologic limitations, dietary supplements can provide benefit. When taken, multivitamin preparations should include 100% of the daily value for vitamins B_{12}, B_6, D, and folic acid.

Key Words: Dietary supplements; nutrients; botanicals; DSHEA; needs assessment; evidence-based guidelines; health claims

The findings and views reported in this chapter represent those of the contributing authors and not necessarily those of the National Institutes of Health and are not intended to constitute an "authoritative statement" under the Food and Drug Administration rules and regulations.

From: *Nutrition and Health: Handbook of Clinical Nutrition and Aging, Second Edition*
Edited by: C. W. Bales and C. S. Ritchie, DOI 10.1007/978-1-60327-385-5_28,
© Humana Press, a part of Springer Science+Business Media, LLC 2009

28.1 INTRODUCTION

Nutritionists and health professionals continually stress that the establishment of healthy dietary patterns is the core to maintaining health and well-being. However, a large majority of the American public believes that dietary supplements as nutrients, herbs, and other food components may help manage or lessen the impact of age-associated disorders and conditions. The evidence base for the use of dietary supplements is scant for some and well documented for others. The need for vitamin D and calcium in the prevention of osteoporosis due to bone mineral loss is well established (1–5). B vitamins, such as folic acid, vitamins B_6, and B_{12}, play a role in preventing blood vessel diseases and maintaining normal neurologic function (6–11). It should be noted that clinical deficiency of vitamins or minerals, other than iron and now possibly vitamin D (12), is uncommon in the United States, except for certain high-risk groups. Of rising concern are the increased blood levels of folic acid in response to folic acid fortification in the United States (13).

Evidence is emerging from a few well-designed randomized clinical trials to suggest that some herbal supplements may also have beneficial effects. Herbs are sold in many forms, including fresh, dried, liquid, or solid extracts, and as tablets, capsules, powders, and tea bags. However, with many herbal preparations, it is not fully understood how they work, nor is the active component always known. While approximately 30% of all drugs used today are derived from plants (14–16), very few of the herbal supplement products on the market have been vigorously tested in clinical studies for efficacy and toxicity. The lack of solid scientific evidence makes it extremely difficult to quantify true risks and benefits of supplements in many cases. Also, the risk of interactions between supplements and other medications among the elderly may be higher because they take more prescription medicines on the average than do younger adults. This review focuses on the rationale and evidence base for the use of dietary supplements for health maintenance and the risk factors associated with chronic disease. Needs, assessment tips, and resource references are also provided through the use of tables and figures.

28.1.1 Overview of Dietary Supplement Use

Total sales of dietary supplements in 2007 were estimated to be 22 billion dollars in the United States. Between 1990 and 1997, the prevalence of high-dose vitamin and herbal use increased by 130 and 380%, respectively (17); however, growth appears to have slowed in recent years. Currently the annual growth rate in herbal sales is less than 6%. In contrast, the annual growth in functional foods has enjoyed about a 9% per year growth rate (18). During this same time period, sales of prescription drugs have sharply declined. For 2006–2008, the growth forecast places specialty supplements at the top at a 7–9% growth rate followed by sports nutritionals at 4–6%. And to reach these goals, consumers are switching from single-nutrient supplements to combination and condition-specific formulae (19). One of the most high-potential marketing practices is to cross-promote supplements and foods with the sales of prescription medications for specific conditions. This practice can be seen in grocery stores, drug store chains, and even in health and food magazines. For example, vitamins, multivitamins, and minerals are among the items most frequently purchased with a heart-related prescription drug in a drugstore or grocery store *(20)*. According to consumer usage model estimates, 4.2% of U.S. adults or 9.8 million people are committed or "heavy" users of supplements and 30.6% or 72.2 million are "regular" users *(18)*.

Supplement use has been shown to increase with age. Use is higher among women than among men and in non-Hispanic whites. It is consistent with more healthy lifestyles and less overweight, and is correlated with higher family income and level of education. Findings from the Third National Health and Nutrition Examination Survey, 1988–1994 (NHANES III), suggest that 40% of Americans use dietary supplements *(21)*. Approximately 56% of middle-age and older adults consume at least one supplement on a regular basis. Use of supplements among the elderly was highest (55%) in women 80 years and above, compared to 42.3% in men of the same age group *(21)*. The demographics of any supplement use are consistent with the use of the most commonly consumed dietary supplements: multivitamin/multimineral, vitamin E, vitamin C, calcium, antacid calcium, and B-vitamin complex *(22)*. A smaller survey of 2,743 U.S. English-speaking residents conducted by the FDA, the 2002 Health and Diet Survey, identified any dietary supplement use in 73% of participants, 85% of whom used a multivitamin/multimineral supplement, 77% used a specialized vitamin/mineral supplement, and 42% used herbal or botanical supplements *(23)*. On the other hand, when users of complementary and alternative (CAM) practices were queried, only 18.9% noted use of herbal products in the last 12 months and of those who consumed herbs, 57.3% used the herbs to treat a specific health condition, but most often not in accordance with evidence-based guidelines *(24)*. It should be noted that cross-comparisons of surveys on dietary supplement use do not lend themselves to pooling as the collection instruments or formats are different (phone survey, 24-h recall, food frequency questionnaire), subjects may be queried for different time periods of use (over the last week versus last month versus last year), and respondents may be confused about what types of supplements belong to within the category being questioned.

28.1.2 Regulation of Dietary Supplements

The Dietary Supplement Health and Education Act (DSHEA) (Public Law 103-417) laid the foundation for the current regulatory framework for dietary supplements. This law amended the Federal Food, Drug, and Cosmetic Act of 1938 "to establish standards with respect to dietary supplements." Dietary supplements have been defined by DSHEA to include a product (other than tobacco) intended to supplement the diet that bears or contains one or more of the following dietary ingredients: a vitamin; a mineral; an herb or other botanical; an amino acid; a dietary substance for use by man to supplement the diet by increasing the total dietary intake; or a concentrate, metabolite, constituent, extract, or combination of any of the ingredients described above. Under DSHEA, FDA regulates safety, manufacturing, and product information, such as claims in product labels, package inserts, and accompanying literature. The FDA cannot require testing of dietary supplements *prior* to marketing (except for new dietary ingredients); however, while manufacturers are prohibited from selling dangerous products, DSHEA gives the FDA permission to remove a product from the marketplace only when the FDA proves that the product is dangerous to the health of Americans. If in the labeling or marketing of a dietary supplement product a claim is made that the product can diagnose, treat, cure, or prevent disease, such as "cures cancer," the product is said to be an unapproved new drug and thus is being sold illegally.

In July 2007, the FDA announced a final rule establishing regulations to require current good manufacturing practices (CGMPs) for dietary supplements and finished products. The regulations establish the CGMPs needed to ensure quality throughout the manufacturing, packaging, labeling, and storing of dietary supplements. The final rule includes requirements for setting quality control procedures, designing and constructing manufacturing plants, and testing ingredients and the finished product. It also includes requirements for recordkeeping and handling consumer product complaints. Manufacturers are required to evaluate the identity, purity, strength, and composition of their dietary supplements. If dietary supplements contain contaminants or do not contain the dietary ingredient they are represented to contain, FDA would consider those products to be adulterated or misbranded. The aim of the final rule is to prevent inclusion of the wrong ingredients, too much or too little of a dietary ingredient, contamination by substances such as natural toxins, bacteria, pesticides, glass, lead, and other heavy metals, as well as improper packaging and labeling (25).

The Dietary Supplement and Nonprescription Drug Consumer Protection Act (the "AER bill") (Public Law 109-462), a bill requiring manufacturers of dietary supplements and over-the-counter (OTC) products to submit serious adverse event reports (SAERs) to the FDA, was passed by Congress in December 2006. This act, which amends the Federal Food, Drug, and Cosmetic Act, will become effective within 1 year of the date it was signed into law (i.e., December 2007). Once effective, companies will be required to include contact information on their products' labels for consumers to use in reporting adverse events. Companies must further notify the FDA of any serious adverse event reports within 15 business days of receiving such reports. Under this act, a "serious adverse event" would be defined as any adverse event resulting in death, a life-threatening experience, inpatient hospitalization, a persistent or significant disability or incapacity, or a congenital anomaly or birth defect, as well as any adverse event requiring a medical or surgical intervention to prevent one of the aforementioned conditions, based on reasonable medical judgment.

28.1.3 Health Claims for Foods and Supplements

The Nutrition Labeling and Education Act of 1990 (Public Law 101–535) allows labeling of conventional foods and dietary supplements with health claims. These are statements describing relationships between a food or food component and disease, such as cardiovascular disease, or health-related condition, such as hypertension, which is considered a surrogate marker for disease. Health claims are limited to claims about disease risk reduction and cannot be used to suggest cure, mitigation, treatment, or prevention of disease. To qualify as a health claim, FDA may use several ways to exercise its oversight in determining which health claims may be used on a label or in labeling for food. Readers are referred to the FDA for details (www.cfsan.fda.gov/~dms/flguid.html). In some cases biomarkers are evaluated for predicting risk of disease, and health claims may be requested and approved based on evidence that a nutrient or dietary supplement influences a biomarker in a way associated with reduced disease risk. When evidence suggests potential benefit but the strength of evidence is not considered strong, a qualified health claim may be allowed. Qualified health claims must be accompanied by language: "FDA has determined that this evidence is limited and not conclusive." Presently FDA has authorized 14 health-disease-related claims for foods and/or dietary supplements, to include 7 for cardiovascular, 3 for cancer, and 1 for osteoporosis-related conditions. A total of 16 qualified health claims have been issued: 7 for cardiovascular, 5 for cancer, 1 for cognitive function and 1 for diabetes-related conditions. In addition, structure–function claims can be made on labels of conventional foods and dietary supplements to describe the role of substances intended to affect the normal structure or function in humans. FDA need not review or authorize structure–function claims prior to use. Nutrient content claims and dietary guidance statements can also be utilized to provide additional health information about a particular nutrient or food component, but these typically do not relate to dietary supplements.

28.1.4 Evaluating the Quality of the Literature on Studies Using Dietary Supplements

In most cases the initial scientific suggestion of benefit for a bioactive food component, botanical compound, or dietary supplement has come from epidemiologic comparisons of different cohorts, with interpretation limited by both different endogenous rates of disease and varied levels of consumption of the substance of interest. Early hypothesis-generating studies are typically supported by subsequent case-control studies and, in some cases, larger prospective cohort studies. For example, for cardiovascular disease, surrogate endpoints in the smaller prospective studies include effects on low-density lipoprotein cholesterol (LDL-C), platelet function, endothelial function, and immune/inflammatory activity (26). In some cases larger randomized controlled trials have been conducted using clinical outcome endpoints. Studies of herbal medicines have suffered from poorly characterized interventions and often studies of short duration. Independent examination has revealed great inconsistency between product labeling and actual compound concentration (27–29). In this setting there are significant safety concerns with potential adulteration of supplements with active prescription compounds, contamination of preparations, herb–herb interactions, and herb–drug interactions (30–32). Research into the safety and efficacy of botanical compounds has been hampered by this lack of standardization (33). To improve the design and reporting of randomized controlled trials using herbal medicine interventions, members of the CONSORT (Consolidated Standards of Reporting Trials) group have developed a checklist to assist investigators (34).

The process of systematic review has become well established for the evaluation of specific medical treatments and diagnostic tests. In fact, the U.S. Preventive Services Task Force (USPSTF) has recently refined its methods of evidence review and assessment to create more useable documents to serve clinicians' needs (35).

However, systematic reviews of nutrition topics and dietary supplements pose unique challenges unlike those of conventional drugs. One challenge is that of defining energy balance and using control subjects that are energetically equivalent to test subjects. Among observational studies, it is critical to account for potential confounding factors, such as background diet, food preparation method, weight, health status, medications record, and disease-specific risk factors *(36)*. In fact, it may be argued that evaluating a diet–health relationship is precisely the circumstance in which systematic review techniques can be most appropriate and effective because they are transparent and objective, and the search and review strategies could be exactly reproduced by others *(37)*. An additional value of the evidence-based review approach is that a systematic review can readily be updated to include emerging relevant data. The FDA is currently in the process of adopting an evidence-based review system for the scientific evaluation of health claims for foods and food components *(38)*.

28.2 RECOMMENDED DIETARY INTAKES FOR THE ELDERLY

Nutrients and other dietary components are essential for normal growth, development, and good health, including prevention of chronic and degenerative diseases. In 1989 the Food and Nutrition Board of the Institute of Medicine, National Academy of Sciences, replaced Recommended Dietary Allowances, used since 1941 to recommend nutrient needs, with a comprehensive set of Dietary Reference Intakes (DRIs) that have broader and more useful applications *(39–41)*. DRIs are recommended nutrient intakes that promote functional endpoints associated with good health and limit adverse health effects associated with deficient or excessive intakes. In addition to categories for younger adults, infants, and children, the DRIs include demographic categories for adults 51–70 years and 71 + years as indicated by new evidence indicating micronutrient needs change with aging. Ample evidence from studies in older people supports a need for such guidance. Aging is associated with physiologic changes that have significant nutrition impact. Aging and loss of sex steroids alter bone remodeling and accelerate bone loss, increasing the need for calcium and vitamin D. Less vitamin D is synthesized in skin in older adults. Achlorhydria (decreased gastric acidity) impairs intestinal absorption of vitamin B_{12}, calcium, and folic acid. At the same time, increased frequency of acute and chronic illness such as cardiovascular disease, diabetes, and pressure ulcers and more prevalent use of prescription and over-the-counter medications may alter nutrient metabolism and needs. Barriers such as poor dentition, decreased dietary variety, decreased mobility, and difficulties in acquiring and preparing foods present additional challenges for meeting nutrition needs in older adults.

A modified food pyramid for adults over 70 years has been developed by The Jean Mayer USDA Human Nutrition Research Center on Aging at Tufts University (www.hnrc.tufts.edu). In addition to suggesting numbers of servings of each food group consistent with USDA's MyPyramid, it includes guidelines for water equivalents and advises adults in this age group to consult with a health care provider about their potential need for calcium, vitamin D, and vitamin B_{12}

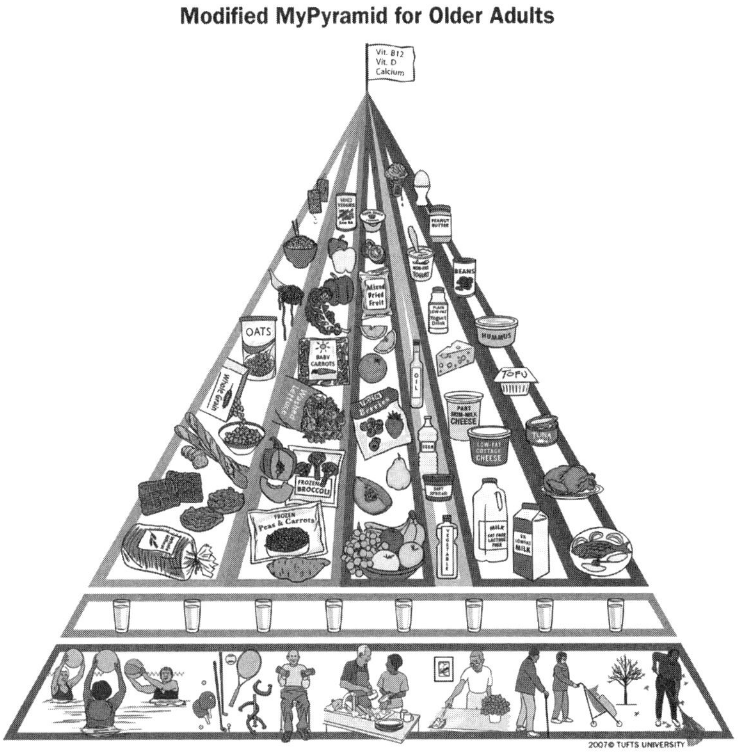

Modified MyPyramid for Older Adults

Fig. 28.1. Food pyramid for 70+ adults. Reprinted with permission from Tufts University, 1999.

supplements. (see Fig. 28.1). The interactive MyPyramid also provides individualized dietary recommendations for various age and gender categories and activity levels, including older adults, and can be accessed online at www.mypyramid.gov.

Dietary assessment surveys examine intake of various demographic groups. In the case of older adults, research suggests that many consume diets inconsistent with science-based recommendations and may benefit from dietary modifications and/or dietary supplementation. Results of a 6-month dietary assessment of adults aged 65–93 years living in rural North Carolina suggested many did not consume recommended servings of grains, fruits, vegetables, and dairy as suggested by the Food Guide Pyramid but did consume excess discretionary calories *(42)*. Similarly, diets of over 90% of the 447 older Puerto Rican and Dominican adults whose intake was assessed via validated food frequency questionnaire did not meet estimated requirements for vitamin E from food alone. Their plasma concentrations were associated with intake patterns supporting the concern that alpha-tocopherol intake was insufficient in this group *(43)*. Mean daily servings of fruits, grains, and dairy products among community-dwelling women aged 50–69 years from a rural Midwestern community were below target levels recommended in the 2000 Dietary Guidelines for Americans *(44)*. In another study, lack of dietary variety predicted inadequate macronutrient and micronutrient intakes in community-dwelling older adults surveyed for the 1994–1996 Continuing Survey of Food

Intakes by Individuals. Findings support the need for nutrition counseling for older adults and perhaps a role for dietary supplements *(45)*. In fact, when nutrient intake adequacy of vitamin/mineral supplement users was compared to nonusers aged 51 years and older, 80% or more of supplement users met the EAR for vitamins A, B_6, B_{12}, C, E, folate, iron, and zinc while less than 50% of both users and nonusers met the EAR for folate, vitamin E, and magnesium from food alone. This analysis of adults from the Continuing Survey of Food Intakes by Individuals and Diet and Health Knowledge Survey in 1994 and 1996 suggested that dietary supplements partially compensated for inadequate micronutrient intake from food *(46)*.

The American Dietetic Association (ADA) takes the position, consistent with most health care organizations, that the best way to meet the DRIs is to wisely choose a wide variety of foods *(47)*. Foods provide many compounds essential to health and adopting eating habits that provide recommended amounts of nutrients and other healthful substances may be the only step needed to meet current dietary recommendations. However, there are times when recommended nutrients cannot be obtained by food alone or retained by the body because of impaired absorption or other physiologic limitations. In these instances, dietary supplements or fortified foods can contribute needed vitamins and minerals and play a valuable role in maintaining health.

Among the popular dietary supplements, primarily vitamins and minerals, consumed by older Americans, many have national public health recommendations and established guidelines for use based on considerable evidence. A select number of these supplements are summarized in Table 28.1.

28.2.1 B Vitamins and Cognitive Function

For example, there is increasing but contradictory information that B vitamins, such as folic acid, vitamins B_6, and B_{12}, play a role in preventing blood vessel diseases and in maintaining normal neurologic function *(6,8,48–55)*. High plasma homocysteine concentrations are associated with impaired cognitive abilities and increased risk of cardiovascular disease *(56–58)*, and supplemental folic acid, vitamin B_{12}, and vitamin B_6 can decrease elevated homocysteine levels *(59)*. Data collected from participants of the Framingham Study showed that elevated homocysteine levels (>14 mmol/L) doubled the chance that a participant would develop Alzheimer's disease (AD); each 5 mmol/L elevation was shown to increase the risk of AD by 40% *(53)*. A cohort of 816 dementia-free older adults with a mean age of 74 was followed for newly diagnosed dementia and AD. Over a 4-year period subjects with elevated homocysteine and lower folate levels at baseline were more likely to develop dementia and AD; in subjects with plasma homocysteine >15 μmol/L the hazard ratio for dementia was 2.08 *(60)*. Whether hyperhomocysteinemia is a risk marker or risk factor for dementia is yet to be determined. Four small, short-duration studies found no effects of folic acid, with or without vitamin B_{12} on cognitive decline in older adults *(57)*. More recently, the FACIT trial examined the effect of 3-year folic acid supplementation on cognitive function in adults aged 50–70 years as assessed by memory testing. They found that an 800 μg folic acid supplement taken daily for 3 years significantly improved cognitive function

Table 28.1
Nutrients with intake recommendations for the elderly with public health guidelines

Nutrient	Specific needs/requirements for older adults	IOM/DRI values (for elderly age ranges)	Public health guideline
Folic acid	Intake of supplemental folic acid should not exceed 1,000 micrograms (µg)/day to prevent folic acid from triggering symptoms of vitamin B_{12} deficiency If 50 years of age or older, have B_{12} status checked before taking a folic acid supplement	RDA: 400 mcg UL: 1,000 mcg	*FDA-qualified health claims177:* "It is known that diets low in saturated fat and cholesterol may reduce the risk of heart disease. The scientific evidence about whether folic acid (folate), vitamin B_6, and vitamin B_{12} may also reduce the risk of heart disease and other vascular diseases is suggestive, but not conclusive. Studies in the general population have generally found that these vitamins lower homocysteine, an amino acid found in the blood. It is not known whether elevated levels of homocysteine may cause vascular disease or whether high homocysteine levels are caused by other factors. Studies that will directly evaluate whether reducing homocysteine may also reduce the risk of vascular disease are not yet complete" "As part of a well-balanced diet that is low in saturated fat and cholesterol, Folic Acid, Vitamin B_6 and Vitamin B_{12} may reduce the risk of vascular disease. FDA evaluated the above claim and found that, while it is known that diets low in saturated fat and cholesterol reduce the risk of heart disease and other vascular diseases, the evidence in support of the above claim is inconclusive" AHA: "Available evidence is inadequate to recommend folate and other B-vitamin supplements as a means to reduce CVD risk at this time" *(178)* ACCF: Cannot recommend at this time folic acid supplementation for vascular disease if homocysteine is not elevated *(179)*

(continued)

Table 28.1
(continued)

Nutrient	Specific needs/requirements for older adults	IOM/DRI values (for elderly age ranges)	Public health guideline
Vitamin B_{12}	Supplementation is recommended for adults 50 years of age because of the high incidence of atrophic gastritis, which limits absorption of vitamin B_{12} from foods Older adults taking proton pump inhibitors or metformin for diabetes are at risk of impaired B_{12} absorption	RDA: 2.4 mcg UL: not established	IOM: Because 10–30% of older people may be unable to absorb naturally occurring vitamin B_{12}, it is advisable for those older than 50 years to meet their RDA mainly by consuming foods fortified with vitamin B_{12} or a vitamin B_{12}-containing supplement *(180)* *FDA-qualified health claims(177)*: "It is known that diets low in saturated fat and cholesterol may reduce the risk of heart disease. The scientific evidence about whether folic acid (folate), vitamin B_6, and vitamin B_{12} may also reduce the risk of heart disease and other vascular diseases is suggestive, but not conclusive. Studies in the general population have generally found that these vitamins lower homocysteine, an amino acid found in the blood. It is not known whether elevated levels of homocysteine may cause vascular disease or whether high homocysteine levels are caused by other factors. Studies that will directly evaluate whether reducing homocysteine may also reduce the risk of vascular disease are not yet complete" "As part of a well-balanced diet that is low in saturated fat and cholesterol, Folic Acid, Vitamin B_6 and Vitamin B_{12} may reduce the risk of vascular disease. FDA evaluated the above claim and found that, while it is known that diets low in saturated fat and cholesterol reduce the risk of heart disease and other vascular diseases, the evidence in support of the above claim is inconclusive"

(continued)

Table 28.1
(continued)

Nutrient	Specific needs/requirements for older adults	IOM/DRI values (for elderly age ranges)	Public health guideline
			AHA: "Available evidence is inadequate to recommend folate and other B-vitamin supplements as a means to reduce CVD risk at this time" (178)
Vitamin B_6	No specific guidelines for B_6 intakes	RDA: men: 1.7 mg, women: 1.5 mg UL: 100 mg	*FDA-qualified health claims* (177): "It is known that diets low in saturated fat and cholesterol may reduce the risk of heart disease. The scientific evidence about whether folic acid (folate), vitamin B_6, and vitamin B_{12} may also reduce the risk of heart disease and other vascular diseases is suggestive, but not conclusive. Studies in the general population have generally found that these vitamins lower homocysteine, an amino acid found in the blood. It is not known whether elevated levels of homocysteine may cause vascular disease or whether high homocysteine levels are caused by other factors. Studies that will directly evaluate whether reducing homocysteine may also reduce the risk of vascular disease are not yet complete" "As part of a well-balanced diet that is low in saturated fat and cholesterol, Folic Acid, Vitamin B_6 and Vitamin B_{12} may reduce the risk of vascular disease. FDA evaluated the above claim and found that, while it is known that diets low in saturated fat and cholesterol reduce the risk of heart disease and other vascular diseases, the evidence in support of the above claim is inconclusive" AHA: "Available evidence is inadequate to recommend folate and other B-vitamin supplements as a means to reduce CVD risk at this time" (178)

(continued)

Table 28.1
(continued)

Nutrient	Specific needs/requirements for older adults	IOM/DRI values (for elderly age ranges)	Public health guideline
Magnesium	Elderly are at a higher risk of low magnesium intake due to restrictive diets	RDA men: 420 mg RDA women: 320 mg UL: 350 mg from dietary supplements only	Dietary Guidelines for Americans, 2005, notes that intake levels of magnesium "may be low enough to be of concern" for adults, children, and adolescents (181) ACCF: Possibly useful for cardiovascular indications. Consider magnesium supplementation for those at risk (179)
Potassium	Elderly are at higher risk of hyperkalemia due to increased incidence of impaired kidney function Potassium supplements, other than the small amount included in a multivitamin, should only be taken under the specific guidance and instruction of a health care provider	RDA: 4,700 mg UL: not established Maximum potassium intake should be 4,700 in those with impaired urinary excretion of potassium	Dietary Guidelines for Americans, 2005, notes for individuals with hypertension, blacks, and middle-aged and older adults: "Aim to consume no more than 1,500 mg of sodium per day and meet the potassium recommendation (4,700 mg/day) with food" (181) FDA Health Claims (2000) (97): "Epidemiological and animal studies indicate that the risk of stroke-related deaths is inversely related to potassium intake over the entire range of blood pressures, and the relationship appears to be dose dependent. The combination of a low-sodium, high potassium intake is associated with the lowest blood pressure levels and the lowest frequency of stroke in individuals and populations. Although the effects of reducing sodium intake and increasing potassium intake would vary and may be small in some individuals, the estimated reduction in stroke-related mortality for the population is large" "Vegetables and fruits are also good sources of potassium. A diet containing approximately 75 mEq (i.e., approximately 3.5 g of elemental potassium) daily may contribute to reduced risk of stroke, which is especially

(continued)

Table 28.1
(continued)

Nutrient	Specific needs/requirements for older adults	IOM/DRI values (for elderly age ranges)	Public health guideline
			common among blacks and older people of all races. Potassium supplements are neither necessary nor recommended for the general population
			Diets containing foods that are a good source of potassium and that are low in sodium may reduce the risk of high blood pressure and stroke." JNC 7 recommends moving toward diets low in sodium and high in potassium
			IOM: Adequate intake (AI) for potassium is set at 4.7 g (120 mmol)/day for all adults. This level of dietary intake (i.e., from foods) should maintain lower blood pressure levels, reduce the adverse effects of sodium chloride intake on blood pressure, reduce the risk of recurrent kidney stones, and possibly decrease bone loss (182)
			AHA (2006): Increase potassium by eating 8–10 servings of fruits and vegetables each day. High-potassium intake is associated with reduced blood pressure in people with or without high blood pressure (178)
Selenium	Selenomethionine is generally considered to be the best absorbed and utilized form of selenium	RDA: 55 mcg UL:400 mcg	FDA qualified health claims(183): Claim 1: "Selenium may reduce the risk of certain cancers. Some scientific evidence suggests that consumption of selenium may reduce the risk of certain forms of cancer. However, FDA has determined that this evidence is limited and not conclusive" Claim 2: "Selenium may produce anticarcinogenic effects in the body. Some scientific evidence suggests that

(continued)

Table 28.1
(continued)

Nutrient	Specific needs/requirements for older adults	IOM/DRI values (for elderly age ranges)	Public health guideline
			consumption of selenium may produce anticarcinogenic effects in the body. However, FDA has determined that this evidence is limited and not conclusive" American Cancer Society states there is no evidence at this time that supplements can reduce cancer risk, and some evidence exists that indicates that high-dose supplements can increase cancer risk. High-dose selenium supplements are not recommended, as there is only a narrow margin between safe and toxic doses (136) AHA notes that "antioxidant vitamin supplements or other supplements such as selenium to prevent CVD are not recommended" (178)
Omega-3 fatty acids	Older adults with cardiovascular disease should consider omega-3 fatty acids and the potential need to modify diet to increase EPA and DHA intake while decreasing n-6 fat intake	DRIs not determined	*FDA-qualified health claims (184):* "The scientific evidence about whether omega-3 fatty acids may reduce the risk of coronary heart disease (CHD) is suggestive, but not conclusive. Studies in the general population have looked at diets containing fish and it is not known whether diets or omega-3 fatty acids in fish may have a possible effect on a reduced risk of CHD. It is not known what effect omega-3 fatty acids may or may not have on risk of CHD in the general population" "Consumption of omega-3 fatty acids may reduce the risk of coronary heart disease. FDA evaluated the data and determined that, although there is scientific evidence supporting the claim, the evidence is not conclusive" AHA: "On the basis of the available data, the AHA recommends that patients without documented CHD eat a variety of fish, preferably oily fish at least twice a week.

(continued)

Table 28.1
(continued)

Nutrient	Specific needs/requirements for older adults	IOM/DRI values (for elderly age ranges)	Public health guideline
			Patients with documented CHD are advised to consume ≈ 1 g of EPA + DHA per day; 2–4 g per day for hypertriglyceridemia" (178)
			ACCF: Recommend omega-3 supplements 1–2 g/day if insufficient omega-3 intake from fish for individuals at risk (179)
Calcium	Increased intakes recommend for prevention of osteoporosis	AI 51 + years: 1,200 mg UL: 2,500 mg	Dietary Guidelines for Americans, 2005, lists calcium in its list of nutrients for which dietary intakes "may be low enough to be of concern" for adults, children, and adolescents (181)
			FDA health claim (185):
			"Adequate calcium intake throughout life is linked to reduced risk of osteoporosis through the mechanism of optimizing peak bone mass during adolescence and early adulthood and decreasing bone loss later in life"
			FDA-qualified health claim (186):
			"Some evidence suggests that calcium supplements may reduce the risk of colon/rectal cancer; however, FDA has determined that this evidence is limited and not conclusive"
			"Very limited and preliminary evidence suggests that calcium supplements may reduce the risk of colon/rectal polyps. FDA concludes that there is little scientific evidence to support this claim"

(continued)

Table 28.1
(continued)

Nutrient	Specific needs/requirements for older adults	IOM/DRI values (for elderly age ranges)	Public health guideline
Vitamin D	Supplementation recommended for individuals at risk of low intakes and/or sun exposure	AI 51–70 years: 10 mcg (400 IU) AI 71+ years: 15 mcg (600 IU) UL: 50 mcg (2,000 IU)	Dietary Guidelines for Americans, 2005, states that "older adults, people with dark skin, and people exposed to insufficient ultraviolet band radiation (i.e. sunlight) consume extra vitamin D from vitamin D fortified foods and/or supplements" (181)
			The FDA is considering a proposal to amend the existing health claim for calcium to allow claims of a reduced risk of osteoporosis with the consumption of both calcium and vitamin D
			American Cancer Society states that more research is needed to define optimal blood and intake levels for cancer risk reduction, but recommended intake is likely to fall between 200 and 2,000 IU, depending on age and other factors that modify vitamin D status (136)

AHA, American Heart Association; ACCF, American College of Cardiology Foundation; AI, adequate intake; DHA, docosahexaenoic acid; EPA, eicosapentaenoic acid; FDA, Food and Drug Administration; CHD, coronary heart disease; DRIs, Dietary Reference Intakes; JNC 7, Seventh Report of the Joint National Committee on Prevention, Detection, Evaluation, and Treatment of High Blood Pressure; IOM, Institute of Medicine; RDA, Recommended Dietary Allowances; UL, upper level of intake.

(a secondary endpoint) as compared to placebo *(8)*. The Chicago Health and Aging project investigated the association between cognitive change and fruit and vegetable intake over the course of 6 years and found that high vegetable but not fruit consumption may be associated with slower rate of cognitive decline in older age *(61)*. However, researchers examining this same group of 3,718 residents 65 years and older for associations between folate and vitamin B_{12} intake and cognitive decline unexpectedly found that the rate of cognitive decline among persons in the top fifth of total folate intake (median intake 742 µg/day) was more than twice that of those in the lowest fifth of folate intake (median intake 186 µg/day) while high total vitamin B_{12} intake was associated with slower cognitive decline only among the oldest participants *(62)*. A 2-year trial that provided a daily supplement containing 100 µg folic acid, 500 µg vitamin B_{12}, and 10 mg vitamin B_6 to 276 healthy subjects aged 65 years and older did not find significant differences in cognition test scores between those receiving the supplement versus a placebo *(55)*. Additional studies are needed to examine these apparently contradictory results and determine whether benefit from supplemental folic acid may be more effective when given at younger ages or for longer periods of time. Few trials have investigated the effect of vitamin B_6 supplementation on cognition. Two small studies of short duration that examined the effect of vitamin B_6 on cognitive decline in elderly men and women showed no effects. A Cochrane database review failed to find studies confirming cognitive benefit of vitamin B_6 on cognition in older adults and called for randomized controlled trials to explore the possible associations between vitamin B_6 supplementation and cognitive impairment *(57)*. Dementia and poor memory are among the symptoms of vitamin B_{12} deficiency *(63)*. A review of 1,000 community-dwelling adults aged 75 years or older indicated that those with vitamin B_{12} or holotranscobalamin levels in the bottom versus top quartiles had a twofold risk and a threefold risk, respectively, of cognitive impairment. This study adds to evidence linking deficient B_{12} levels with cognitive impairment but additional trials are needed to assess clinical significance of B_{12} supplementation *(64)*.

28.2.2 B Vitamins and Cardiovascular Health

To examine the effects of B-vitamin supplementation on cardiovascular events in patients with vascular disease, 5,522 patients 55 years or older with vascular disease or diabetes received daily treatment with either placebo or combination pill providing 2.5 mg folic acid, 40 mg vitamin B_6, and 1 mg vitamin B_{12} for 5 years *(51)*. While mean homocysteine levels decreased by 2.4 µmol/L in the treatment group versus a 0.8 µmol/L increase in the placebo group, active treatment did not significantly decrease risk of death from cardiovascular causes. However, folic acid supplementation may positively influence vascular function. Fifty-six patients with coronary artery disease were randomized to receive a low dose (400 µg) or a high dose (5 mg) of folic acid per day for 7 weeks before undergoing coronary artery bypass grafting. Low-dose supplements improved markers of vascular function, with no further improvement from higher doses. Benefit appeared to be related to vascular tissue levels of 5-methyltetrahydrofolate *(65)*. To examine the effect of 5-methyltetrahydrofolate on coronary circulation 14 patients with ischemic heart disease were enrolled in a double-blind, placebo-controlled crossover trial to examine the effect of 30 mg folic acid on myocardial blood flow. This study suggested that high-dose oral folic acid could acutely lower blood pressure and enhance coronary dilation in patients with coronary artery disease *(11)*.

Additional studies are needed, and are underway, to further explore the complex relationships between B vitamins, cognitive function, and cardiovascular disease. The Institute of Medicine advises adults aged 50 years and older to obtain most of their vitamin B_{12} from supplements or fortified food. The RDA for vitamin B_{12} is 2.4 µg. This level of supplementation is available in multivitamin preparations that provide 100% of the DV. Vitamin B_{12} does not have an upper intake level (UL) because higher intakes are not associated with adverse health events. Older persons may obtain cardiovascular and cognitive benefit from consuming 400 µg folic acid per day from a dietary supplement or from fortified foods. This level of supplementation is available in basic multivitamin preparations that provide 100% of the DV. The UL for folic acid is 1,000 µg (or 1 mg) because of the potential for folic acid to trigger vitamin B_{12} deficiency. This is particularly important for older persons because at least 30% may have difficulty absorbing vitamin B_{12} from food and are therefore at higher risk of developing a vitamin B_{12} deficiency. There is no evidence suggesting specific benefit of supplemental vitamin B_6 but taking a daily multivitamin providing 100% of the DV for B_6 (2.0 mg/day) is considered safe and acceptable. The upper intake level (UL) for vitamin B_6 is 100 mg/day.

A prime example of the complex interrelationships between diet, dietary supplements, and health is recent concern over increased blood levels of folic acid in response to folic acid fortification in the United States and Canada. Folic acid fortification of enriched uncooked cereal grains, mandated since 1998 in the United States, has been credited with achieving its goal of reducing the number of births complicated by neural tube defects. However, over this same time frame the downward trend in colorectal incidence in the United States and Canada reversed, leading some researchers to examine the association between these phenomena *(13)*. Folate is important to DNA, is a known cofactor in one-carbon transfer reactions, and is considered a modulator of colorectal cancer development *(66)*. Epidemiologic studies suggest an inverse relationship between folate intake and incidence of colorectal cancer *(67)*. However, animal studies suggest that the timing and dose of folic acid supplementation may influence its effect on development or inhibition of colorectal cancer *(66)*. In normal epithelial tissue, folate deficiency may promote neoplastic transformation while moderate folate supplementation may suppress tumor development. The opposite may be true in established tumors or microscopic neoplasms, as animal studies suggest that folate deficiency may inhibit while supplementation may promote progression of established colorectal neoplasms *(66)*. Researchers continue to investigate molecular alterations associated with folic acid supplementation and methylenetetrahydrofolate reductase (MTHFR) on the development of colorectal adenoma *(68)*.

The Prostate, Lung, Colorectal, and Ovarian (PLCO) Cancer Screening Trial recently reported a potential harmful effect of high folate intake on the risk of breast cancer. Women who ingested ≥400 µg/day of supplemental folic acid had a 20% greater risk of developing breast cancer than those who did not consume supplemental folic acid. In this study greater total folate intake, mainly influenced by folic acid supplementation, increased breast cancer risk by 32% *(69,70)*. This data contributes evidence suggesting dual modulatory effects of folate on cancer development.

Fruits and vegetables are good sources of folate but older adults are at greater risk of folate insufficiency because of lower dietary intake, health conditions, and medications that may impair folate absorption *(71)*. A study examining folate intake and serum folate levels in adults aged 60 years and older utilized three 24-h dietary recalls to assess dietary folate intake from natural food sources and fortified foods and a questionnaire to assess folate intake from dietary supplements. Dietary assessment suggested an adequate fruit and vegetable intake. Mean dietary folate intake from natural and fortified food sources was within the recommended range but 21% of participants consumed <320 µg dietary folate equivalents (DFE) per day and 2% consumed more than 1,000 µg/day, amounts consistent with inadequate and excess folate intake. Participants with lower total dietary folate intakes consumed less folate from natural sources and fortified foods and had a lower energy intake than participants with higher folate intakes. Mean intake of folate naturally occurring in food was 214 µg/day and provided 46% of folate intake (i.e., DFE). Ready-to-eat cereals provided 22% of folate intake, and folate from supplements and fortified foods other than ready-to-eat cereals contributed 32% of folate intake. In this study taking a dietary supplement and eating foods fortified with folic acid were important contributions to an adequate folate intake. Using dietary supplements did increase the percentage of participants consuming more than the upper folate intake level from 2 to 34% but decreased the proportion at risk of folate inadequacy from 21 to 12.5%. In subjects completing biochemical assessment there was a low risk of folate deficiency even though 12.5% consumed less than 320 µg DFE per day. Seniors with a high total DFE intake had serum folate levels more than twice as high as those with a low DFE intake. Folate intake greater than the upper intake level increases risk of vitamin B_{12} insufficiency or deficiency because folate can mask vitamin B_{12} deficiency. However, B_{12} intake was adequate in subjects examined because both fortified foods and dietary supplements consumed contained vitamin B_{12} as well as folic acid. Dietary assessment of older adults can identify those who may benefit from dietary supplements and/or fortified foods to meet their daily folate needs and who may be at risk of excess folate intake if they consume folate in dietary supplements *(71)*.

28.2.3 *Vitamin E and Cancer Prevention*

There are various forms of vitamin E. Alpha-tocopherol is considered the most active form. Unlike other vitamins, the form of á-tocopherol made in the laboratory and found in many supplements (all Rac-α-tocopherol) is not identical to the natural form (RRR-á-tocopherol) and is not quite as active as the natural form. The RDA of 15 mg equals 22 IU of natural and 33 IU of synthetic vitamin E. Studies suggest that daily doses of vitamin E ranging from 200 to 1,200 IU are safe when taken up to 4 months *(72–74)*.

Epidemiologic and experimental work suggest multiple therapeutic potentials for antioxidant properties of vitamin E. Unfortunately clinical trials investigating this potential have been generally negative despite the apparent relevance of oxidative stress to atherosclerosis and cancer. Inconsistent study results make it difficult to reach a firm conclusion on the effect of vitamin E and antioxidants on cancer risk. The NIH State-of-the Science Conference Statement on Multivitamin/Mineral Supplements and Chronic Disease Prevention, published in 2006 *(75)*, states, "Most of the studies we examined do not provide strong evidence for beneficial health-related effects of supplements taken singly, in pairs, or in combinations of three or more."

Vitamin E did not significantly impact primary prevention of cancer in the Women's Health Study, which randomized 39,876 female health professionals to receive 600 IU vitamin E or a placebo on every other day over 10 years *(76)*. However, the follow-up phase of the Alpha-Tocopherol Beta-Carotene (ATBC) study, which randomly assigned over 29,000 50- to 69-year-old male smokers to 50 mg α-tocopherol, 20 mg β-carotene, both, or a placebo for 5–8 years, demonstrated a 32% decrease in the incidence and a 41% decrease in mortality from prostate cancer among subjects who received α-tocopherol supplements *(77)*. Post-intervention follow-up of this study continued to indicate preventive effects of α-tocopherol on prostate cancer, but researchers believe this association requires confirmation in other trials *(78)*. Interestingly, researchers also found that men in the higher quintiles of baseline serum alpha-tocopherol had lower risks of total and cause-specific mortality than those in the lowest quintile, suggesting health benefits of higher circulating concentrations of alpha-tocopherol within the normal range *(79)*. A pooled analysis of eight prospective studies from North America and Europe examining correlations between intakes of vitamin antioxidants, multivitamins, and lung cancer concluded that regardless of smoking habits and lung cancer cell type, available data do not support an association between intakes of vitamins A, C, E, folate, and lung cancer risk *(80)*.

28.2.4 Vitamin E and Cardiovascular Health

Research has identified a role of reactive oxygen species in promoting atherogenesis. As an antioxidant with inflammatory properties vitamin E may mediate free radical reactions that oxidize LDL cholesterol and initiate or promote atherosclerosis *(81,82)*. However, results of clinical trials have not supported a role for vitamin E in the primary or secondary prevention of heart disease.

More recently, the Women's Health Study indicated that taking 600 IU vitamin E every other day for approximately 10 years did not decrease cardiovascular mortality in healthy women, but there were decreased cardiovascular deaths (primarily sudden death in women aged 65 years or older *(83)*.. In further analysis of the same data those women randomized to vitamin E demonstrated an overall 21% reduced risk of venous thromboembolism (VTE). The observed risk reduction was 44% among those with prothrombotic mutations or a personal history of VTE. The authors also noted that overall vitamin E was associated with lower bleeding risk than those observed for low-dose aspirin *(84)*. The HOPE trial, which examined the effect of vitamin E supplementation on major cardiovascular events in people with vascular disease or diabetes who were 55 years or older, demonstrated neither benefit nor risk from vitamin E supplementation. A 7-year follow-up of this trial, HOPE-2, suggested vitamin E supplementation failed to prevent cancer or major cardiovascular events *(85)*.

A number of meta-analyses have evaluated a range of CVD outcomes for varying forms and doses of vitamin E in a variety of patient populations *(86–90)*. Most of these studies concluded that vitamin E provided no benefit and no increased risk. However, one controversial meta-analysis did suggest that high-dosage vitamin E supplements ($\geq$ 400 IU/day) may increase all-cause mortality *(88)*. This meta-analysis has been criticized for its statistical design and inclusion of mixed treatments including some with β-carotene and overweighting the vitamin E group for diseases and mortality risk factors *(91)*. Lacking from most vitamin E intervention trials are plasma measures of vitamin E and/or determination of a subject's oxidative stress status.

However, some researchers argue that it may be too optimistic to expect a single-vitamin supplement to overcome effects of poor dietary habits and sedentary lifestyle and their contribution to known risk factors such as hypertension and hypercholesterolemia on the incidence of heart disease *(82)*.

The weakness of evidence supporting treatment role for vitamin E for cardiovascular disease and cancer led the FDA to reject health claims for vitamin E. At this time there is insufficient evidence to recommend vitamin E supplements for older adults, but multivitamin supplements providing 100% of the DV for vitamin E for adults (10 mg) are considered safe. However, as part of a multifactorial approach to heart disease treatment vitamin E supplements may provide value by decreasing biomarkers of heart disease. This may be particularly important for older adults, as indicated by a small study of 102 apparently healthy community-dwelling adults aged 80 years and older who had plasma vitamin E, beta-carotene, vitamin C, and products of lipid peroxidation measured. During 7.4 months of follow-up subjects whose plasma vitamin E levels were in the highest quartile had a risk of cardiovascular events one-sixth that of those with plasma vitamin E levels in the lowest quartile *(92)*. Additional studies are needed to further explore the benefit of vitamin E in treating and/or preventing heart disease but older persons at high risk of developing cardiovascular disease or who have existing cardiovascular disease should discuss the benefit of taking a supplement of vitamin E, not to exceed the UL of 1,000 IU/day, with their cardiologist or primary physician. This level of supplementation is not available in most multivitamin preparations and usually requires a specific vitamin E supplement.

28.2.5 Potassium and Cardiovascular Health

The Framingham Heart Study estimated the 20-year risk of developing hypertension as >90% for men and women not yet hypertensive by ages 55–65 years. The significance of this risk to health, according to a meta-analysis of 61 studies including almost 1,000,000 adults, is that each 20 mmHg increase in systolic blood pressure (SBP) doubles the risk of a fatal coronary event. Thus, preventing and/or treating hypertension via non-pharmacologic and pharmacologic therapies is of utmost importance. The Institute of Medicine, American Heart Association, and Seventh Report of the Joint National Committee on Prevention, Detection,

Evaluation, and Treatment of High Blood Pressure (JNC 7 Report) recommend increased intake of foods high in potassium to reach an intake goal of 4,700 mg/day. Combining an overall healthy diet with weight loss, lower salt intake, and higher potassium intake can prevent and treat hypertension, according to a new American Heart Association scientific statement, published in *Hypertension: Journal of the American Heart Association*. This report *(93)* recommends increasing potassium intake by eating 8–10 servings of fruits and vegetables each day. "High potassium intake is associated with reduced blood pressure in people with or without high blood pressure. Potassium intake reduces blood pressure more in blacks than in whites. The recommended potassium intake level is 4,700 mg/day. However, this amount may be too high for people with impaired kidney function or severe congestive heart failure."

The DASH study (Dietary Approaches to Stop Hypertension), a controlled-feeding study, suggested that high blood pressure could be significantly lowered by a diet that emphasizes fruits, vegetables, and low-fat dairy foods. Such a diet will be high in magnesium, potassium, and calcium, and low in sodium and fat *(94, 95)*.

The PREMIER study, which investigated the effects of lifestyle modifications based on established recommendations (lower salt intake, weight loss, and physical activity) alone and with the addition of the DASH dietary pattern, suggested this intake pattern enhances blood pressure lowering effects of established recommendations when used for 6 months *(96)*. The FDA *(97)* has approved the following health claim statements: "Epidemiological and animal studies indicate that the risk of stroke-related deaths is inversely related to potassium intake over the entire range of blood pressures, and the relationship appears to be dose dependent. The combination of a low-sodium, high-potassium intake is associated with the lowest blood pressure levels and the lowest frequency of stroke in individuals and populations. Although the effects of reducing sodium intake and increasing potassium intake would vary and may be small in some individuals, the estimated reduction in stroke-related mortality for the population is large" and "vegetables and fruits are also good sources of potassium. A diet containing approximately 75 mEq (i.e., approximately 3,500 mg elemental potassium) daily may contribute to reduced risk of stroke, which is especially common among blacks and older people of all races. Potassium supplements are neither necessary nor recommended for the general population" *(97)*. Potassium supplements other than small doses that may be included in a multivitamin/mineral preparation should only be taken under the specific guidance and instruction of a physician.

28.2.6 Selenium and Cancer Prevention

Selenium is incorporated in approximately 25 selenoproteins. Two selenoproteins, thioredoxin reductase (TrxR) and glutathione peroxidase (GPx), are antioxidant enzymes that provide protection against free radicals. Recent evidence points to a role for selenium compounds as well as selenoproteins in cancer prevention through modulation of free radical metabolism, among other biologic mechanisms *(98)*. Some studies indicate incidence of some cancers is lower among people with higher blood levels or intake of selenium *(99–101)*. It was hoped that the Selenium and Vitamin E Cancer Prevention Trial (SELECT) would clarify the role of selenium in prostate cancer but the study was halted early in October 2008 as the data showed no benefit that selenium and vitamin E supplements, taken alone or together prevented prostate cancer [NCI press release]. FDA had approved a qualified health claim in 2003 recognizing the anticarcinogenic potential of selenium *(102)*. [NCI.www.cancer.gov/newscenter/pressreleases/SELECTresults 2008, accessed 11/ 24/08.]. There is insufficient evidence to support supplementing a diet that already provides recommended amounts of selenium, and the NHANES III cross-sectional data hints to high levels of serum selenium being positively associated with the prevalence of diabetes *(103)*. Additionally, long-term supplementation (an average of 7 years) with 200 µg/day of selenium in participants of the Nutritional Prevention for Cancer trial demonstrated on secondary data analysis that selenium supplementation did not seem to prevent cardiovascular disease *(104)* and may have increased the risk for type 2 diabetes mellitus *(105)*. It is recommended that older adults unable or unwilling to increase dietary selenium intake may benefit from a multivitamin/mineral supplement providing 100% of the RDA for selenium, 55 µg. The UL for selenium is 400 µg.

28.2.7 *Calcium, Vitamin D, Osteoporosis, and Fractures*

28.2.7.1 CALCIUM

More than 99% of total body calcium is stored in bones and teeth to support their structure. Consuming adequate calcium and vitamin D throughout infancy, child-hood, and adolescence and engaging in weight-bearing exercise will maximize bone strength and bone density to help prevent osteoporosis and fractures later in life. In 1993 the FDA authorized a health claim for food labels on calcium and osteoporo-sis in response to scientific evidence indicating that an inadequate calcium intake contributes to low peak bone mass and is a risk factor for osteoporosis *(106)*. The claim states that "adequate calcium intake throughout life is linked to reduced risk of osteoporosis through the mechanism of optimizing peak bone mass during adolescence and early adulthood" and decreasing bone loss later in life.

Over 1.5 million fractures per year are attributed to osteoporosis. These fractures contribute 12–18 billion dollars to direct health care cost each year *(107)*. The U.S. Surgeon General issued a comprehensive report on bone health and treatment that recommends a treatment plan including calcium and vitamin D supplementation *(108)*.

However, data on the efficacy of calcium for preventing bone fractures in healthy postmenopausal women remain equivocal. The Women's Health Initiative followed over 36,000 postmenopausal women aged 50–79 years and found that calcium with vitamin D supplementation (1,000 mg calcium as calcium carbonate with 400 IU vitamin D_3 daily versus placebo) resulted in a small but significant improvement in hip bone density but did not significantly reduce hip fractures *(3)*.

The National Institutes of Health Consensus Development Conference on Opti-mal Calcium Intake suggested 1,500 mg of calcium per day for postmenopausal women who are not on estrogen therapy and for all men and women over 65 years *(109)*. However, the Institute of Medicine determined there are insufficient data to increase the current AI to this level. Researchers continue to examine this issue but sufficient valid data have been assembled to advise adults aged 50 years and older, particularly those unable to consume milk or dairy foods and/or those who have difficulty consuming adequate calories, to consult with their physician about their need for calcium supplements. However, in one of the first studies examining the relationship between vascular events and calcium intake, Bolland et al. *(110)* found an association between daily use of calcium supplements and increased risk of cardiovascular events. Their work involved secondary analysis of a randomized clinical trial designed to assess effects of calcium supplementation on bone density and fracture incidence in healthy postmenopausal women. Subjects received either 1 g elemental calcium daily, 400 mg in the morning and 600 mg in the evening, or placebo for 5 years. Using standardized definitions of vascular events, authors found a statistically significant increase in myocardial infarction in the calcium group, though not in stroke. The number needed to treat for 5 years to cause one myocardial infarction was 44, one stroke 56, and one cardiovascular event 29 while the number needed to treat to prevent one symptomatic fracture was 50. This study suggests the need for more work examining this issue, especially in view of the possibility that positive effects of calcium supplements on bone mineral density may be offset by negative vascular effects *(110)*. The UL for calcium, or maximum level associated with no adverse health effects, is 2,500 mg consumed in foods and supplements.

28.2.7.2 VITAMIN D

Dietary surveys that include metabolic measures indicate that physiologic changes that occur with aging limit conversion of vitamin D to its most active form. In addition, recent evidence suggests that vitamin D inadequacy is pandemic but that vitamin D deficiency can be prevented by sensible sun exposure and adequate supplementation. Some researchers believe intake of approximately 1,000 IU vitamin D_3 per day may be needed to help maintain blood levels of 25-hydroxyvitamin D in the recommended range of 30–60 ng/mL *(111)*. Intestinal calcium transport increased by 45–60% in women when 25-hydroxyvitamin D levels were increased from an average of 20–32 ng/mL, and evidence is mounting that a level of 30 ng/mL correlates with sufficient vitamin D *(112,113)*. Emerging evidence supports a role for vitamin D in bone health and suggests a role for vitamin D supplements for adults aged 50 years and over. An evidence report by the Agency for Healthcare Quality and Research (AHRQ) in 2006 *(114)* and NIH Conference entitled "Vitamin D and Health in the 21st Century: An Update" held in September 2007 noted that the effects of vitamin D alone, independent of calcium intake, could not be evaluated in most trials. Vitamin D_3 at 700 IU daily with calcium supplementation compared to placebo was noted to have a small beneficial effect on bone mineral density and reduces the risk of fractures and falls, although benefit may be confined to specific subgroups. The AHRQ report also noted that it is difficult to define specific blood levels of markers for vitamin D status that indicate optimal levels for bone health. One challenge is that current methods, which measures serum 25(OH) D as the marker, yield highly inconsistent results. Older evidence related to toxicity from excess intakes of vitamin D established an UL of 50 mg/day (or 2,000 IU). However, one question addressed by a new evidence report was, "Does intake of vitamin D above current reference intakes lead to toxicity?" This report concluded that vitamin D intake above current reference intakes was generally well tolerated from 22 randomized trials reviewed, although the most relevant trials were

not adequately designed to assess long-term risks. The increased risk of hypercalcemia and hypercalciuria was not clinically significant and the only significant adverse
event identified was an increase in renal stones in the WHI trial in women 50–79 years
who took 400 IU vitamin D_3 and 100 mg calcium/day (3). Results of the NIH
Conference may help practitioners determine the need for vitamin D supplementation
for their older patients, the recommended dose of vitamin D for bone health in older
adults, and the risk of toxicity. At the same time, the Canadian Consensus Conference
on Osteoporosis, 2006 Update (115) notes, "Although it might not be sufficient as
the sole therapy for osteoporosis, routine supplementation with calcium (1,000 mg/
day) and vitamin D_3 (800 IU/day) is still recommended as mandatory adjunct therapy
to the main pharmacological interventions (antiresorptive and anabolic drugs)."

The Institute of Medicine, Committee on Dietary Reference Intakes for Vitamin
D and Calcium (2008–2010) will undertake a study to assess current relevant data
and update as appropriate the DRIs for vitamin D and calcium. The review will
include consideration of chronic and non-chronic disease indicators. The study will
also incorporate, as appropriate, systematic evidence-based reviews of the literature
and an assessment of potential indicators of adequacy and of excess intake. Indicators for adequacy and excess will be selected based on the strength and quality of
the evidence and the demonstrated public health significance, taking into consideration sources of uncertainty in the evidence.

28.2.8 Magnesium and Reduction of Cardiovascular Risk

Magnesium is involved in electrolyte metabolism, helps regulate blood sugar, is
involved in energy metabolism and protein synthesis, and promotes normal blood
pressure (116). Aging is associated with magnesium depletion. Decreases in dietary
intake, coupled with decreases in intestinal magnesium absorption and increases in
urinary magnesium losses, place older adults at risk of magnesium deficiency. Older
adults are also more likely to take medication that promotes urinary loss of
magnesium such as some diuretics. On the other hand, some older adults are
more likely to take magnesium-containing laxatives than younger adults, a habit
that could result in excess magnesium intake and possible toxicity. This may be
especially significant if there is pre-existing renal insufficiency, which can impair the
kidney's ability to excrete excess magnesium.

In a multicenter controlled-feeding study of dietary patterns to lower blood
pressure, the nutrient intake goal for magnesium was 500 mg/day (117). The
Dietary Approaches to Stop Hypertension (DASH) intervention study suggested
high blood pressure could be significantly lowered by a diet that emphasizes fruits,
vegetables, and low-fat dairy foods. Such a diet will be high in magnesium, potassium, and calcium and low in sodium and fat (94,95).

A review of over 28,000 women enrolled in the Women's Health Study indicated
those in the highest quintile of magnesium intake had a decreased risk for hypertension compared with those in the lowest quintile (118). A recent review of nutritional
effects on blood pressure indicated that high levels of magnesium as well as potassium, calcium, and soy seem to have some beneficial effect on hypertension but
more research is needed before specific recommendations can be made (119). An
alternate view was presented by a Cochrane review of the effect of combined

calcium, potassium, and/or magnesium supplementation for management of primary hypertension in adults, which was unable to find robust evidence to support a role for these supplements in the treatment of hypertension but noted that the methodological quality of trials reviewed was not well reported *(120)*.

Diabetes, which is seen with increased frequency in older adults, can result in increased magnesium excretion when poorly controlled and can contribute to decreased levels of magnesium.

Observational data is confusing and mixed regarding the association of magnesium intakes and risk of developing type 2 diabetes mellitus. A recent meta-analysis of eight prospective primary prevention cohort studies of 9,702 men and 15,365 women aged 35–65 years with recorded intakes of magnesium by food frequency questionnaires showed a significant inverse association between magnesium intake and risk of type 2 diabetes mellitus *(121)* as was also seen in the analysis of the NHANES III data *(122)*. However, magnesium intake was not related to diabetes risk in the European Prospective Investigation into Cancer and Nutrition (EPIC)-Potsdam study *(121)*.

To determine the effect of magnesium restriction on glucose, cholesterol, and electrolyte metabolism, 13 postmenopausal women were fed a magnesium deficient diet for 78 days, then given a daily supplement providing 200 mg magnesium as gluconate for an additional 58 days while continuing to consume the magnesium-depleted diet. During the magnesium depletion period five women experienced changes in heart rhythm, with four of the women necessitating early entry into the magnesium repletion phase of the study to correct the rhythm abnormalities. Metabolic changes observed during the magnesium depletion period include impaired glucose homeostasis, altered cholesterol metabolism, and increased urinary excretion of sodium and potassium *(123)*. The magnesium depleted diet provided 101 mg magnesium/day, 33% of the current RDA. This study suggested that consuming 101 mg dietary magnesium per day is inadequate and may compromise cardiovascular health. These findings also question results of a pooled analysis of metabolic studies that suggested neutral magnesium balance could be maintained on a diet providing 165 mg/day in healthy subjects *(124)*. Contradictory results of these studies indicate the need for additional research into this issue.

While research indicates a role of magnesium in the pathophysiology of blood pressure and risk of developing type 2 diabetes, there is insufficient evidence to recommend magnesium supplementation for treating cardiovascular risk factors. However, the Joint National Committee on Prevention, Detection, Evaluation, and Treatment of High Blood Pressure states that diets that provide plenty of magnesium are positive lifestyle modifications for individuals with hypertension. General multivitamin/mineral supplements often provide approximately 25% of the daily recommended need for magnesium, an amount considered safe. Thus, those needing oral supplemental magnesium will need to take a separate magnesium supplement if recommended by a medical doctor. Currently it is thought that all forms of magnesium supplements are absorbed equally well; however, some forms of oral supplements may result in diarrhea. The RDA for magnesium for adult women (>31 years) is 320 mg. The RDA for magnesium for adult men (>31 years) is 420 mg. The UL for magnesium for older adults, 350 mg/day, represents intake from drugs and supplements only and does not include intake from food and water.

28.2.9 Omega-3 Fatty Acid and Cardiovascular Health

Essential omega-6 lipids promote development of pro-inflammatory eicosanoids while essential omega-3 lipids promote development of anti-inflammatory eicosanoids. The anti-inflammatory effects of omega-3 fats may partially explain their cardioprotective benefit because incorporation of these fats in cell membranes promotes vasodilation, stimulates antiarrhythmic effects, and promotes vascular patency (125,126).

Several large diet studies support cardiovascular benefits of omega-3 fats. The Diet and Infarction Trial (DART) randomized 2,033 men approximately 42 days after their first myocardial infarction (MI) to receive or not receive advice to consume oily fish twice per week. After 2 years of follow-up and repeat dietary consultation, the "fish advice" group experienced a 20% reduction in total mortality as compared to those who were not advised to consume more fish (127). Estimated daily intake of EPA and DHA in the "fish advice" group was 900 mg. In a follow-up study, DART-2 examined the effect of giving advice to eat fish or providing fish oil capsules to men with angina on mortality. This study did not result in statistically significant mortality differences and its implementation has been widely criticized. Many researchers do not consider its results reliable (128,129). By far the largest secondary prevention trial was the GISSI-Prevenzione that enrolled over 11,000 persons in Italy who had survived a MI for a median of 16 days. Researchers examined the effect of 850 mg EPA and DHA/day and 300 mg vitamin E/day in a factorial design. Treatment with EPA and DHA resulted in significant reductions in total mortality (20%), predominantly by a reduction in sudden cardiac death (45%) (130). Finally, the Japan EPA Lipid Intervention Study (JELIS), a combined primary and secondary prevention trial, randomized over 18,000 patients with hyperlipidemia to receive 1.8 g/day EPA or be a control. At baseline 20% of the participants had a history of coronary heart disease and all patients with hyperlipidemia were treated with simvastatin or pravastatin. After 4.6 years the primary endpoint of "major coronary events" was significantly reduced by 19% in the EPA group (131). A 2006 meta-analysis of 18 trials, most of limited size and duration, reported smaller, non-significant reductions in total mortality (13%) and total CVD event (5%) (132).

Many countries, including Canada, Sweden, United Kingdom, Australia, and Japan, as well as the World Health Organization and North Atlantic Treaty Organization have made formal population-based dietary recommendations for omega-3 fatty acids. The American Heart Association/American College of Cardiology and European Society for Cardiology recommend an intake of 1 g/day of the two marine omega-3 fatty acids, eicosapentaenoic acid (EPA) and docosahexaenoic acid (DHA), for secondary prevention, cardiovascular prevention, treatment of post-myocardial infarction, and prevention of sudden cardiac death (133).

Overall, these studies confirm recommendations of the American Heart Association for all adults to increase their dietary intake of DHA and EPA by consuming at least 2 servings/week of natural DHA and EPA food sources, such as salmon, trout, swordfish, and mackerel. Including natural sources of α-linolenic acid omega-3 fatty acids in their diet, such as ground flax seed and walnuts, may also be helpful but the conversion of α-linolenic acid to EPA/DHA may be limited. Consuming 1 g/day of an 85% EPA + DHA ethyl ester concentrate or 1.8 g/day EPA as ethyl ester may provide similar protection. Older adults are encouraged to discuss these options with a qualified health care provider, who could consider pros and cons of various treatment options, individual needs, and potential interactions such as use of blood thinners (e.g., warfarin).

The National Heart, Lung, and Blood Institute of the NIH established dietary recommendations based on evidence associating dietary components with cardioprotective levels of total cholesterol, HDL cholesterol, LDL cholesterol, and triglycerides. These recommendations, published as Therapeutic Lifestyle Changes (TLC) in the most recent National Cholesterol Education Program Adult Treatment Panel III document, include intake goals for total fat, saturated fat, dietary cholesterol, plant stanols, and soluble fiber. However, the TLC diet in ATP III does not include intake recommendations for omega-3 fatty acids *(134)*.

Omega-3 fat fatty acids lower triglyceride levels and the FDA has approved a prescription form of omega-3 fatty acids for treatment of very high triglyceride levels. The prescription product contains 0.84 g of EPA and DHA in every 1 g capsule. Approximately 2–4 g of omega-3 fatty acids per day is required for triglyceride lowering. EPA/DHA capsules are available without prescription, but the American Heart Association advises people to utilize EPA/DHA therapy under a physician's care *(135)*.

28.2.10 Omega-3 Fatty Acids and Cancer Prevention

The American Cancer Society states that research has not yet demonstrated whether the possible benefits of fish consumption may be duplicated by taking omega-3 fatty acids or fish oil supplements *(136)*. Omega-3 fatty acids have been shown to suppress biosynthesis of arachidonic acid-derived eicosanoids, which influence angiogenesis, apoptosis, cell proliferation, inflammation, and immune cell function *(137)*. Laboratory data also suggest that omega-3 fatty acids may protect against cancer by affecting transcription factor activity, gene expression, and signal transduction pathways; estrogen metabolism; generation of free radicals; and, possibly, insulin sensitivity and cell membrane fluidity *(137)*. Preclinical work suggests that nutrition intervention with omega-3 fatty acids may improve cell sensitivity to chemotherapy *(138)*. Available epidemiologic data on omega-3 fatty acids and the risk of cancer have been inconsistent. A review of 65 studies examining the effects of omega-3 fats on cancer risk conducted by the Southern California Evidenced-Based Practice Center and published in the Journal of the American Medical Association did not provide evidence for a significant association between omega-3 fatty acids and cancer incidence and concluded that dietary supplementation with omega-3 fatty acids is unlikely to prevent cancer *(139)*. However, it is not clear whether omega-3 fatty acids reduce, increase, or have no effect on incidence of total cancer or site-specific cancers, or on cancer mortality. Additional studies are indicated to explore biochemical potential of omega-3 fats on cancer development. At this time there are no completed or ongoing randomized clinical trials in the United States for the primary prevention of cancer in a general population. Therefore, there is insufficient evidence to recommend omega-3 fats for cancer prevention.

28.2.11 Multivitamin/Mineral Supplements for Health Maintenance

Aging is associated with physiologic changes that diminish organ function, impair utilization of many nutrients, and result in dysregulation of the immune system.

Superimposed on these changes are functional deficits and sociologic barriers that contribute to deficient eating patterns in many older adults. Together, these factors may influence risk of chronic disease such as cardiovascular disease, increase the risk of infections, and contribute to osteoporosis in older adults. Multivitamin/mineral preparations formulated at 100% of the daily value (DV) may be an effective way to safely maintain desirable blood and tissue micronutrient levels and help promote good health.

An 8-week, double-blind, placebo-controlled clinical trial among 80 adults ranging in age from 50 to 87 years indicated that a multivitamin supplement providing 100% of the daily value was able to decrease the prevalence of suboptimal plasma levels of several vitamins and improve micronutrient status beyond what was achieved through a diet rich in fortified foods *(140,141)*. Approximately 900 community-dwelling men and women 65 years and older were randomized to receive a multivitamin/mineral supplement or placebo daily for 12 months. Memory and verbal fluency were assessed at baseline and after 12 months. No benefit was evident in those under 75 years but there was weak evidence for a beneficial effect in those aged 75 years and over and those at increased risk of micronutrient deficiency *(142)*.

An NIH Consensus Conference on use of multivitamin and mineral supplements (MVMs) reported that MVMs are used by individuals who practice healthier lifestyles, making interpretation of observational studies on the overall relationship between MVM use and general health outcomes difficult to interpret. They reported that there is insufficient knowledge about the actual amount of total nutrients that Americans consume from diet and supplements. Public assurance of the safety and quality of MVMs is inadequate because manufacturers are not required to report adverse events and the FDA has no regulatory authority to require labeling changes or to help inform the public of these issues and concerns. The conference report recommended that the FDA's purview over these products be authorized and implemented. The report concluded that present evidence is insufficient to recommend either for or against the use of MVMs by the American public to prevent chronic disease *(75)*. However, evidence also suggests that older adults may not be heeding current dietary messages, increasing the likelihood that they may benefit from multivitamin supplements. Targeting nutrition messages through age-adjusted tools, such as the Tufts modified food pyramid, and individual nutrition assessment to determine the need for dietary supplements for adults over 70 years may be especially helpful. Elders must be cautioned against indiscriminate use of supplements. When taken, multivitamin preparations chosen should include 100% of the DV for vitamins $B_{12,}$ B_6, D, and folic acid, because these are key nutrients for the elderly and surveys suggest inadequate dietary intake among older persons.

28.3 DIETARY SUPPLEMENTS WITH ACCUMULATING EVIDENCE FOR RISK REDUCTION FOR CHRONIC DISEASE

Among popular dietary supplements consumed by Americans there are many not endorsed by national public health recommendations that have a considerable evidence base upon which large randomized trials can be launched or guidelines for use can be formulated by professional societies. A select number of these supplements are reviewed in Table 28.2 and several of the more promising supplements are discussed in this section.

Table 28.2
Summary of select dietary supplement ingredients purported for conditions and disorders of aging

Dietary ingredient(s)	Purported or popular use	Existing evidence base	Current guidance on use
Alpha-lipoic acid	Improves glucose utilization in type 2 diabetes mellitus	Small placebo-controlled trials (oral and IV) RCTs and meta-analysis (IV studies) (187)	Data is inconclusive. Benefit largely limited to studies of IV use IV and oral lipoic acid are approved for treatment of diabetic neuropathy in Germany
	Treats or prevents peripheral nephropathy	Cochrane Review in process NIH RCTS currently underway for glycemic control and for diabetic neuropathy (see www.clinicaltrials.gov)	
	Treats symptoms of dementia	Cochrane Review (2004) noted lack of RCTs to perform review for dementia (188)	Until RCTs become available alpha-lipoic acid cannot be recommended to treat dementia
Antioxidants	Reduction of risk for cancer	Observational studies RCTs Meta-analyses (87,88,189)	FDA (2003) approves qualified health claim for antioxidant vitamins E and/or vitamin C and reduced risk of certain kinds of cancer (194)

(continued)

Table 28.2
(continued)

Dietary ingredient(s)	Purported or popular use	Existing evidence base	Current guidance on use
		AHRQ Evidence Report Cancer (October 2003) concluded there is scant evidence that vitamin C or vitamin E beneficially affect survival or prevent formation of new tumors except for prostate cancer and vitamin E. Isolated findings of benefit which require confirmation (190)	American Cancer Society states there is no evidence at this time that supplements can reduce cancer risk, and some evidence exists that indicates that high-dose supplements can increase cancer risk (136)
		AHRQ Evidence Report (May 2006): Present evidence is insufficient to recommend either for or against the use of vitamin or mineral antioxidants to prevent chronic disease (89).	USPSTF (2003) concludes that the evidence is insufficient to recommend for or against the use of supplements of vitamins A, C, or E; multivitamins with folic acids, or antioxidant combinations for the prevention of cancer or cardiovascular. Rating: I Recommendation (195)

(continued)

Table 28.2
(continued)

Dietary ingredient(s)	Purported or popular use	Existing evidence base	Current guidance on use
	Reduction of risk for CVD	Observational studies RCTs Meta-analyses (88,191) AHRQ Evidence Report CVD (July 2003): Scant evidence that vitamin C or E beneficially affects survival (192)	AHA cannot recommend the use of antioxidant vitamin supplements to prevent CVD (178)
	Slowing progression of age-related macular degeneration	AHRQ Evidence Report (May 2006): Present evidence is insufficient to recommend either for or against the use of vitamin or mineral antioxidants to prevent chronic disease, but may prevent advanced age-related macular degeneration in high-risk individuals (89) Cochrane Review (2006) identified eight RCTs on review and found that supplementation with antioxidants and zinc may be of modest benefit in people with AMD (193)	American Diabetes Association (2007) states routine supplementation with antioxidants is not advised because of lack of evidence of efficacy and concern related to long-term safety (evidence level A) (196)

(continued)

Table 28.2
(continued)

Dietary ingredient(s)	Purported or popular use	Existing evidence base	Current guidance on use
Ayurvedic remedies (such as cinnamon, *Gymnema sylvestre, Coccinia indica,* holy basil, and fenugreek)	Reduces the risk of diabetes, lowers blood glucose levels	CCTs RCTs	Long history of traditional use
		AHRQ *Systematic Review* (2001) While there was great heterogeneity among the small and underpowered studies there is sufficient data for several herbs; *Coccinia indica,* holy basil, fenugreek and *Gymnema sylvestre* and herbal formulas (Ayush-82 and D-400) to warrant further studies (*197*)	American Diabetes Associations notes that there is insufficient evidence to demonstrate efficacy of individual herbs and supplements in diabetes management (*196*)
Black cohosh root (*Actaer racemosa*)	Reduces peri- and postmenopausal symptoms	RCTs Cochrane Systematic Review (*198*)	Black cohosh was used in North American Indian medicine for malaise, gynecological disorders, kidney disorders, malaria, rheumatitis, and sore throat

(continued)

Table 28.2
(continued)

Dietary ingredient(s)	Purported or popular use	Existing evidence base	Current guidance on use
		"In general, the study of botanicals as treatments for hot flashes is still in its infancy. There are major methodological problems associated with studying products that are not standardized" (75) Ongoing NIH studies for menopausal symptoms – see www.clinicaltrials.gov	AHPA notes restriction of use for pregnant and nursing women (199) Fresh or dried rhizome with attached roots is an approved nonprescription drug for oral use in German Commission E monographs (200) American College of Obstetricians and Gynecologists stated may be helpful in short term (6 months or less) for women with vasomotor symptoms of menopause (201) Society of Obstetricians and Gynecologists of Canada (SOGC) may be recommended for the reduction of mild vasomotor symptoms. Evidence level IB (202)

(continued)

Table 28.2
(continued)

Dietary ingredient(s)	Purported or popular use	Existing evidence base	Current guidance on use
			USP categorizes black cohosh as a class 2 herb, meaning that it is safe to use with an appropriate cautionary statement on the packaging (203)
			Suggest monitoring liver enzymes with long-term use
Carnitine (L-carnitine, propionyl-L-carnitine and acetyl-L-carnitine)	Reduces symptoms of intermittent claudication	RCTs	ACC/AHA Practice Guidelines (2005) notes that effectiveness of propionyl-L-carnitine as a therapy to improve walking distance in patients with intermittent claudication is not well established (level of evidence: B) (207)
	Reduces symptoms of intermittent claudication	RCTs	
	Protects against cognitive decline and dementia	Meta-analysis (204) Cochrane Systematic Review (April 2003): No evidence of benefit of acetyl-L-carnitine in the areas of cognition, severity of dementia, functional ability, or clinical global impression. RCTs (2- to 52-week studies) (205)	

(continued)

Table 28.2
(continued)

Dietary ingredient(s)	Purported or popular use	Existing evidence base	Current guidance on use
	Reduces elevated lipids and symptoms of neuropathy associated with HIV and AIDS		
	Reduces side effects of end-stage renal disease and hemodialysis	Small studies	
		The 2004 NIH Conference on carnitine concluded that studies of the benefits of carnitine supplementation in individuals who are carnitine-sufficient have been mostly inconclusive. "Further research is warranted to determine whether carnitine, taken as a dietary supplement, is beneficial in patients with heart failure, peripheral vasculopathies, neuropathies, neurodegenerative disorders, diabetes, obesity and thyroid disorders, HIV-related disorders, male infertility, and/or cancer-related cacheixa" (206)	

(continued)

Table 28.2
(continued)

Dietary ingredient(s)	Purported or popular use	Existing evidence base	Current guidance on use
		NIH trials currently ongoing for type I diabetes, peripheral vascular disease, HIV-infections, and hemodialysis (see www.clinicaltrials.gov)	
Cinnamon (*Cinnamomum verum*)	Reduces blood glucose levels and/or elevated cholesterol levels	Small RCTs	An ayurvedic remedy with a long history of traditional use for diabetes
			AHPA notes that long-term use of therapeutic doses or during pregnancy is not recommended (*199*)
			Commission E approved the internal use of cinnamon for loss of appetite and dyspeptic complains. Clinical efficacy not established for purported uses (*200*)
			American Diabetes Associations notes that there is insufficient evidence to demonstrate efficacy of individual herbs and supplements in diabetes management (*196*)

(continued)

Table 28.2
(continued)

Dietary ingredient(s)	Purported or popular use	Existing evidence base	Current guidance on use
Chromium (as chromium chloride, picolinate, nicotinate, high chromium yeast, or chromium citrate)	Reduces blood glucose levels	RCTs Meta-analyses (36,208)	Data is inconclusive and efficacy is marginal
	Supports lipid lowering	RCTs Eight NIH trials currently ongoing for type II diabetes, cardiovascular disease, and obesity (see www.nih.clinicaltrials.gov)	FDA Qualified Health Claim (2005) that states "One small study suggests that chromium picolinate may reduce the risk of insulin resistance, and therefore possibly may reduce the risk of type 2 diabetes. FDA concludes, however, that the existence of such a relationship between chromium picolinate and either insulin resistance or type 2 diabetes is highly uncertain" (209) American Diabetes Association (2007) supplementation in individuals with diabetes or obesity has not been clearly demonstrated and therefore cannot be recommended (level of evidence: E) (196)

(continued)

Table 28.2
(continued)

Dietary ingredient(s)	Purported or popular use	Existing evidence base	Current guidance on use
Coenzyme Q10 (ubiquinone)	Reduces risk and or symptoms of CVD and congestive heart failure (CHF)	RCTs Meta-analysis (147,150) AHRQ Evidence Review (2003) concludes "the value of CoQ10 in patients with CVD is still an open question. with neither convincing evidence supporting nor refuting evidence of benefit or harm" (192)	Large number of studies, data is mixed. The Japanese government has approved CoQ10 for treatment of CHF. ACCF: notes mortality benefit for CoQ10 is not yet established (179)
	Reduces risk for diabetes	Small RCTs	
	Reduces risk of Parkinson's disease	RCTs: Suggestive of slowing disease progression, but data are inconclusive	American Academy of Neurology notes there is no evidence of neuroprotection for coenzyme Q10 (204)
	Prevents or treat cancer	AHRQ Systematic Review (2003) evidence does not support beneficial effect to help prevent or treat cancer (192)	National Collaborating Center for Chronic Conditions (2006) – CoQ10 should not be used as a neuroprotective therapy for people with Parkinson's disease except in the context of clinical trials (211)

(continued)

Table 28.2
(continued)

Dietary ingredient(s)	Purported or popular use	Existing evidence base	Current guidance on use
Creatine	Reduces progression of degenerative neurological conditions (Huntington and Parkinson's diseases)	Small RCTs NIH sponsoring two major multicenter phase III RCTs to determine if creatine can slow the progression of Parkinson or Huntington's disease (see www.clinicaltrials.gov)	Studies are lacking, efficacy cannot be confirmed
Echinacea (*E. angustifolia, E. pallida, E. purpurea*)	Prevents and reduces symptoms of cold and flu	CCTs RCTs Meta-analyses (*204,212,213*) Cochrane Systematic Review (2006): Efficacy is marginal. There is some evidence that preparations based on aerial parts of *E. purpurea* might be effective for the early treatment of colds in adults (*214*)	Long history of traditional use to treat or prevent colds, flu, and other infections AHPA lists as an herb that can be consumed safely when used appropriately (*199*) Approved by Commission E in Germany as a nonprescription drug (*E. purpurea* herb and *E. pallida* root (*200*)

(continued)

Table 28.2
(continued)

Dietary ingredient(s)	Purported or popular use	Existing evidence base	Current guidance on use
Garlic (*Allium sativum L.*)	Reduction of risk for CVD	RCTs Meta-analyses (*215–217*) AHRQ Evidence Report (October 2000): 37 RCTs demonstrated various garlic preparations led to modest, favorable short-term effects of garlic on lipids and 10 small RCTs showed promising effects on antithrombotic factors. Effects on clinical outcomes not established; effects on blood glucose levels or insulin sensitivity and blood pressure were none to minimal and are not conclusive (*218*) Cochrane Systematic Review for garlic and peripheral arterial occlusive disease (1997) noted that one small trial of short duration found no statistically significant effect of garlic on walking distance (*219*)	Long history of traditional use as food and medicinal APHA notes concerns for this herb are based on therapeutic use (use during nursing) and dosage and may not be relevant to its consumption as a spice (*199*) German Commission E approves use of fresh or carefully dried bulb as a nonprescription drug (*200*) American Heart Association notes that garlic has "no major role" in lipid lowering (total and LDL cholesterol) (*221*)

(continued)

Table 28.2
(continued)

Dietary ingredient(s)	Purported or popular use	Existing evidence base	Current guidance on use
	Reduce risk of cancer	Meta-analysis (214) AHRQ Evidence Report (October 2000) notes scant data to suggest but not prove that dietary garlic consumption is associated with decrease odds of laryngeal gastric, colorectal, and endometrial cancer and adenomatous colorectal polyps (218)	American Cancer Society states there is no evidence that phytochemicals taken as supplements are as beneficial as the vegetables, fruits, beans, and grains from which they are extracted (136)
Ginkgo biloba (Ginkgo leaf extract)	Improves cognitive function	RCTs Meta-analysis (157,222)	AHPA recognizes as an herb that can be safely consumed when used appropriately (214)
		Cochrane Systematic Review (2007): At this time there is no convincing evidence that Ginkgo biloba is efficacious for dementia and cognitive impairment (223)	Standardized dry extracts (35-67:1) with less than 5 ppm ginkolic acids are approved drugs by German Commission E (200)
		Ongoing NIH multicenter study to evaluate the effectiveness of ginkgo in preventing dementia and Alzheimer's disease in older adults (see www.clinicaltrials.gov)	Society of Obstetricians and Gynecologists of Canada (2006) notes available data does not support use in healthy adults with normal cognition (202)

(continued)

Table 28.2
(continued)

Dietary ingredient(s)	Purported or popular use	Existing evidence base	Current guidance on use
	Relief from pain due to intermittent claudication and symptoms of peripheral vascular disease	RCTs Meta-analyses (224–227)	ACC/AHA Practice Guidelines (2005) notes the effectiveness of *Ginkgo biloba* to improve walking distance for patients with intermittent claudication is marginal and not well established (level of evidence: B) (207)
Glucosamine HCl or sulfate	Helps maintain healthy joints (protects joints and tendons and decreases inflammation)	RCTs Meta-analyses (228,229) Cochrane Systematic Review (April 2005) evaluated 20 studies with 2,570 patients. Pooled result from studies using a non-Rotta brand preparation or adequate allocation concealment failed to show benefit in pain and by the Western Ontario	FDA (October 2004) denied a health claim for glucosamine, glucosamine sulfate, and glucosamine and chondroitin sulfate and a reduced risk of osteoarthritis, joint degeneration, and cartilage deterioration (232)

(continued)

Table 28.2
(continued)

Dietary ingredient(s)	Purported or popular use	Existing evidence base	Current guidance on use
		MacMaster (WOMAC) function. Studies evaluating the Rotta brand preparation showed that glucosamine was superior to placebo in the treatment of pain and functional impairment (230) NIH funded GAIT trial demonstrated that "this rigorous, large-scale study showed that the combination of glucosamine and chondroitin sulfate appeared to help people with moderate-to-severe pain from knee osteoarthritis, but not those with mild pain," "Because of the small size of the moderate-to-severe pain subgroup, the findings in this group for glucosamine plus chondroitin sulfate should be considered preliminary and need to be confirmed in a study designed for this purpose" (231)	

(continued)

Table 28.2
(continued)

Dietary ingredient(s)	Purported or popular use	Existing evidence base	Current guidance on use
Grape seed extract (GSE) (*Vitis vinifera*)	Reduces symptoms and discomfort due to chronic venous insufficiency	Small controlled trials	Considered GRAS in the United States European regulations vary from country to country but GSE is commonly allowed in pharmacopeias as a nonprescription herbal drug (233) If marketed in Canada, it is a Natural Health Product GSE have shown some beneficial antioxidant effects in preliminary clinical trials. However, few trials have looked at specific disease or conditions, and little scientific evidence is available
Green tea (*Camellia sinensis*)	Reduces risk of CVD	Observational studies RCTs Meta-analysis (234)	Use of tea in China dates back to 2700 BC Regulatory status is considered a food in many countries and is listed as GRAS in the United States

(continued)

Table 28.2
(continued)

Dietary ingredient(s)	Purported or popular use	Existing evidence base	Current guidance on use
			AHPA notes no cautions for green tea (*199*)
			In May 2006 FDA denied health claim for green tea and CVD as the majority of studies were deemed insufficient for drawing scientific conclusions due to poor study design and reporting (*238*)
	Reduces risk of cancer	Observational studies Meta-analyses (*235–237*)	FDA approved a qualified health claim for green tea and breast and prostate cancer (*239*)
			Claim 1:
			"Two studies do not show that drinking green tea reduces the risk of breast cancer in women, but one weaker, more limited study suggests that drinking green tea may reduce this risk. Based on these studies, FDA concludes that it is highly unlikely that green tea reduces the risk of breast cancer" or

(continued)

Table 28.2
(continued)

Dietary ingredient(s)	Purported or popular use	Existing evidence base	Current guidance on use
			Claim 2:
			"One weak and limited study does not show that drinking green tea reduces the risk of prostate cancer, but another weak and limited study suggests that drinking green tea may reduce this risk. Based on these studies, FDA concludes that it is highly unlikely that green tea reduces the risk of prostate cancer"
			American Cancer Society notes that at present tea has not been proven to reduce cancer risk in humans (136)
			USP issued cautionary statement that in rare cases extracts from green tea have been reported to adversely affect the liver (203)
Gymnema or Gurmar (*Gymnema sylvestre*)	Reduces blood sugar levels	Small clinical trials	Long history of traditional use as an ayurvedic remedy for diabetes. Current evidence is limited

(continued)

Table 28.2
(continued)

Dietary ingredient(s)	Purported or popular use	Existing evidence base	Current guidance on use
			American Diabetes Associations notes that there is insufficient evidence to demonstrate efficacy of individual herbs and supplements in diabetes management (196)
Horse chestnut seed extract (*Aesculus hippocastanum L.*)	Reduces swelling and discomfort due to chronic venous insufficiency	RCTs Meta-analysis (240) Cochrane Systematic Review (2006) shown to be efficacious and safe with short-term use (2–16 weeks) based on review of 17 RCTs (241)	For centuries, horse chestnut seeds, leaves, bark, and flower have been used for a variety of conditions and diseases Is an approved drug in the German Commission E Monographs for over-the-counter use and prescription (parenteral) use (200)
Lutein	Prevents or slow development of age-related macular degeneration	Observational RCTs Meta-analysis (242) NIH-funded AREDS II study will evaluate lutein in combination with other nutrients for AMD (see www.nih.clinicaltrials.gov)	Certain lutein products have GRAS status in the United States (233)

(continued)

Table 28.2
(continued)

Dietary ingredient(s)	Purported or popular use	Existing evidence base	Current guidance on use
Lycopene	Reduces risk of cancer (prostate, lung, colorectal, gastric, breast, cervical, ovarian, endometrial, or pancreatic)	Observational studies Small clinical studies	FDA denied a health claim for lycopene and cancer risk reduction due to lack of credible evidence supporting a relationship between lycopene consumption, either as a food ingredient, a component of food, or as a dietary supplement (243)
Melatonin (N-acetyl-5-methoxytryptamine)	Reduces symptoms of jet lag	RCTs Cochrane Systematic Review (April 2002): Melatonin found effective in preventing or reducing jet lag and timing is important. It should be recommended to adult travelers flying across five or more time zones, particularly in an easterly direction. and especially if they have previously experienced jet lag (244)	Melatonin is a hormone produced by the pineal gland in the brain. It is manufactured synthetically and used as a supplement FDA has classified melatonin as an orphan drug for use in circadian rhythm sleep disorders in blind children and adults with minimal light or no light perception

(continued)

Table 28.2
(continued)

Dietary ingredient(s)	Purported or popular use	Existing evidence base	Current guidance on use
		AHRQ Evidence Report (November 2004) notes melatonin is not effective in alleviating the sleep disturbance aspect of jet lag and shift-work disorder (245)	
	Aids in sleep disorders	RCTs	
		Meta-analysis (246,247)	
		AHRQ Evidence Report (November 2004): Evidence suggests melatonin is not effective in treatment of most primary and secondary sleep disorders with short-term use. Some evidence to suggest melatonin is effective in treating delayed sleep phase syndrome short-term use (245)	
Milk thistle (Silybum marianum)	Reduces risk of liver damage from alcohol, viral infections, and toxins	RCTs	Milk thistle extracts have been used as traditional herbal remedies for almost 2,000 years and extracts are still widely used to protect the liver against toxins. Current evidence suggests that milk
		Meta-analysis (248)	
		AHRQ Evidence Report (2000) could not establish efficacy due to poor trial design and reporting (249)	

(continued)

Table 28.2
(continued)

Dietary ingredient(s)	Purported or popular use	Existing evidence base	Current guidance on use
		Cochrane Systematic Review (2005): Present evidence is insufficient due to poor quality of RCTs reviewed to support or refute use of milk thistle for alcoholic and/or hepatitis B or C virus liver diseases (250)	thistle extracts have an important hepatoprotective as well as anticancer, antidiabetic, and cardioprotective effect; however, high quality studies are limited (251) APHA states herb can be safely consumed when used appropriately (199) Approved as a nonprescription drug by the German Commission E for crude and standardize milk thistle preparations (200)
Pine bark extract	Reduces symptoms and discomfort due to chronic venous insufficiency	RCTs	In many countries, Pycnogenol® is used as a food supplement and has GRAS status in the United States
	Decrease risk of venous thrombosis and thrombophlebitis on long-distance travel	UPBEAT Study is a phase II RCT to evaluate the cardiovascular effects (blood pressure, glycemic control and lipoprotein profile) of pine bark extract in 130 healthy subjects (see www.clinicaltrials.gov)	In Greece, Switzerland, Colombia, and Venezuela, it is a nonprescription herbal drug (233)

(continued)

Table 28.2
(continued)

Dietary ingredient(s)	Purported or popular use	Existing evidence base	Current guidance on use
Policosanols	Enhances lipid lowering	RCTs Meta-analysis (252)	Efficacy data from newer studies outside of Cuba failed to report any benefit
Phosphatidylserine	Reduces risk of cognitive dysfunction	RCTs Phase IV RCT ongoing in Israel to test the efficacy of phosphatidylserine-omega3 compared to placebo in elderly with age-associated memory impairment (see www.clinicaltrials.gov)	FDA (2004) approved qualified health claim for dietary supplements and soy-derived phosphatidylserine for reduced risk of dementia based on very limited and preliminary scientific research (253)
Saw palmetto (*Serenoa repens*)	Relief from symptoms of BPH	RCTs Meta-analysis (254,255) Cochrane Systematic Review (2002): a review of 21 RCTs demonstrated that saw palmetto provides mild-to-moderate improvement in urinary symptoms and flow measures comparable to that of finasteride (standard treatment) with fewer adverse treatment events (256)	APHA lists as herb that can be safely consumed when use appropriately (199) Dried fruit and other galenical preparations or lipophilic extracts are approved by German Commission E as nonprescription drugs (200)

(continued)

Table 28.2
(continued)

Dietary ingredient(s)	Purported or popular use	Existing evidence base	Current guidance on use
		Ongoing NIH-funded trial to reduce risk of prostate cancer with saw palmetto (see www.clinicaltrials.gov)	American Urological Association guidelines states that phytotherapeutic agents and other dietary supplements cannot be recommended for treatment of BPH at this time (257)
			European Association of Urology (2004) Further studies meeting the criteria proposed by WHO-BPH conference (12-month duration, RCT, placebo-controlled) are required before plant extracts can be recommended for the treatment of lower urinary tract symptoms (258)
SAMe (S-adenosyl-L-methionine)	Relieves symptoms of osteoarthritis	RCTs	Orally, SAMe has been used for depression, anxiety,
	Relieves symptoms of depression	Meta-analysis (259) AHRQ Evidence Report and meta-analyses (August 2002) found SAMe more effective than placebo for relief of	heart disease, fibromyalgia, osteoarthritis, bursitis, tendonitis, chronic lower back pain, dementia,

(continued)

Table 28.2
(continued)

Dietary ingredient(s)	Purported or popular use	Existing evidence base	Current guidance on use
		symptoms of depression (28 studies), pain of osteoarthritis (10 studies), pruruitis of cholestasis of pregnancy (8 studies), and intrahepatic cholestasis (6 studies). Treatment with SAMe was found equivalent to standard therapy for depression and osteoarthritis but not for liver disease (260) Meta-analysis (259)	Alzheimer's disease, slowing the aging process, chronic fatigue syndrome improving intellectual performance, liver disease, and Parkinson's disease (166)
	Reducing inflammation and damage from liver conditions	Cochrane Systematic Review (2006): Present evidence (9 RCTs, 434 patients) is insufficient due to poor quality of RCTs reviewed to support or refute use of SAMe for patients with alcoholic liver diseases (262) NIH-sponsored clinical trials with SAMe ongoing for major depression and depression in Parkinson's disease (see www.clinicaltrials.gov)	

(continued)

Table 28.2
(continued)

Dietary ingredient(s)	Purported or popular use	Existing evidence base	Current guidance on use
Soy protein and/OR	Reduces risk of CVD	Observational RCTs Meta-analyses *(263–265)* AHRQ Evidence Report (August 2005) of 68 RCTs on soy and cardiovascular health showed soy to have a small effect on lipids as there was great heterogeneity across studies for soy formulations and doses *(266)* Trials of soy are mixed; the majority of studies do not show benefit. Their effectiveness and long-term safety need to be studied in rigorous clinical trials in diverse populations of women *(75)*	FDA approved health claim for soy protein in 1999 for a reduced risk of heart disease from the consumption of four daily servings of foods containing > 6.25 g soy protein or a total daily intake of 25 g/day *(185)* AHA states there is no meaningful benefit of soy protein consumption with regard to HDL, cholesterol, triglycerides, or lipoprotein(a) aside from replacing animal and dairy products that contain saturated fat and cholesterol *(178)*

(continued)

Table 28.2
(continued)

Dietary ingredient(s)	Purported or popular use	Existing evidence base	Current guidance on use
Soy Isoflavones	Reduces risk of CVD Reduces postmenopausal symptoms	Meta-analyses (267–270) Ongoing NIH phase II, III RCT of soy isoflavones on atherosclerosis progression, cognition, bone mineral density, breast tissue density changes in postmenopausal women (WISH Study, see www.clinicaltrials.gov)	
St. Johns wort (SJW) (*Hypericum perforatum*)	Reduces symptoms associated with depression	RCTs Meta-analyses (271–274) Cochrane Systematic Review (2005): Efficacy is mixed based on level of depression – most studies have evaluated individuals with major depression of mild-to-moderate intensity. In patients who meet criteria for major depression, several recent placebo-controlled trials suggest that the tested	St. John's wort has been used for centuries to treat mental disorders and nerve pain AHPA lists caution with use based on earlier in vitro research and the German Commission E monograph. SJW may interact with pharmaceutical MAO inhibitors (199)

(continued)

Table 28.2
(continued)

Dietary ingredient(s)	Purported or popular use	Existing evidence base	Current guidance on use
		hypericum extracts have minimal beneficial effects while other trials suggest that hypericum and standard antidepressants have similar beneficial effects (275)	Approved by German Commission E as a nonprescription drug for internal and external use (200) There is evidence that St. John's wort may be of benefit in mild or moderate depression; health care professionals should not prescribe or advise its use by patients because of uncertainty about appropriate doses, variation in preparations, and potential serious interactions with other drugs (276)
	Reduces anxiety and mood disturbances associated with menopause	Ongoing clinical trials for generalized social anxiety disorder and St. John's wort combined with kava kava in the treatment of major depressive disorder (see www.clinicaltrials.gov)	

AHA, American Heart Association; AHPA, American Herbal Products Association; CVD, cardiovascular disease; ACCF, American College of Cardiology Foundation; CHF, congestive heart failure; FDA, Food and Drug Administration; GRAS, generally recognized as safe;RCT, randomized controlled trial; USP, U.S. Pharmacopoeia; USPSTF, U.S. Preventative Services Task Force.

28.3.1 Black Cohosh (Cimicifuga racemosa)

Black cohosh has been used in Germany since the mid-1950s to manage menopausal symptoms. The pharmacologically active ingredients are prepared from the rhizome and root. The mechanism of action of black cohosh has not been defined. Previously it was thought that black cohosh might exert its beneficial effects through estrogen receptors to moderate menopausal symptoms and hot flashes, but now black cohosh extracts have been shown to possess serotonergic activity *(143)* Much of the published evidence for the safety and efficacy of black cohosh is based on the utilization of a commercial extract, approved in Germany, known as "Remifemin." It has been used in clinical studies lasting up to 12 months. The quality of black cohosh products in American markets is variable. Although some of the data are contradictory, the majority of research findings suggest that black cohosh extract has a modest effect on menopausal hot flashes, night sweats, anxiety, and insomnia. The Herbal Alternatives for Menopause Trial (HALT) *(144)* included 351 pre- and postmenopausal women aged 45–55 years. This 12-month, five-arm trial included the following treatment groups: (1) black cohosh (160 mg, 70% ethanol extract); (2) multibotanical (including 200 mg black cohosh); (3) multibotanical plus soy counseling; (4) hormone therapy (conjugated equine estrogen 0.625 daily); and (5) placebo. Hormone therapy was found to significantly reduce vasomotor symptoms compared with the other four groups, and there were no differences detected between the other four treatment arms at any point in the trial. In general, women earlier in menopause have a better response to black cohosh than those later in menopause. Because of the inclusion of a placebo and an estrogen replacement group in HALT, a 12-month follow-up, and a 92% trial completion rate, HALT provides strong evidence that black cohosh lacks efficacy in this population for the treatment of menopausal symptoms. The National Institutes of Health has provided support for 10 clinical studies to evaluate pharmacokinetics, safety, and efficacy of black cohosh for menopausal-related symptoms. Additional long-term randomized controlled trials are currently ongoing; the University of Chicago, Illinois, and Columbia University are evaluating black cohosh for treating menopausal symptoms, and a third 12-week phase IV study at the University of Pennsylvania will evaluate black cohosh as an alternative therapy for treating menopause-related anxiety. In addition to black cohosh, the UIC phase II study is also evaluating in parallel the efficacy of red clover. It will assess the safety of chronic dosing by evaluating uterine (endometrial biopsies), breast (mammography), and hematology parameters (CBC and chemistry lab values) at baseline and 1 year *(145)*.

The safety profile of black cohosh has recently been reviewed. Due to case reports of liver toxicity, health authorities from several countries require a cautionary label on black cohosh products. However, no toxicity has been observed during phase I or ongoing phase II studies (NIH/ODS Black Cohosh Safety Workshop, in press).

28.3.2 Coenzyme Q10 (Ubiquinone) and Cardiovascular Health

Coenzyme Q10 (CoQ10) is involved in oxidative phosphorylation and the generation of ATP. In addition, CoQ10 acts as a free radical scavenger and membrane stabilizer. Japanese scientists first reported therapeutic properties of CoQ10 in the 1960s. Some evidence suggests that CoQ10 might improve the efficiency of energy production in heart tissue and thus assist the heart during times of physical and/or oxidative stress. Most of the consistent evidence for a positive outcome with regard to CoQ10 therapy is in congestive heart failure patients. There have been over 40 controlled trials of the clinical effects of CoQ10 on cardiovascular disease. Early studies showed benefit in subjective (quality of life) and objective (increased LVEF, stroke index, decrease in hospitalizations) parameters while more recent studies have been less supportive. A recent systematic review (146) evaluated nine randomized trials of CoQ10 in heart failure and concluded that there were non-significant trends toward increased ejection fraction and reduced mortality; however, there were insufficient numbers of patients for meaningful results. An updated meta-analysis by Sander et al. (147) which included 11 clinical trials evaluating CoQ10 in doses ranging from 60 to 200 mg/day for 1–6 months found a net improvement in ejection fraction, with a more profound effect being seen in patients who were not receiving angiotensin-converting enzyme inhibitors. A subgroup analysis suggested, counter to current thinking, that subjects with worse stages of congestive heart failure did not fare better than those at an earlier stage of disease. This may suggest that less diseased hearts may possess more salvageable myocardial tissue.

Reduced plasma or serum levels of CoQ10 have been documented in observational studies and a number of randomized controlled clinical trials in patients on statin therapies. The largest trial included 1,049 patients with noted reductions in plasma CoQ10 levels of 38 and 27% after treatment with atorvastatin (148). The decrease in blood CoQ10 levels with statin treatment may be related to reduced synthesis of CoQ10 as well as a decrease in circulating levels of CoQ10, as CoQ10 is carried in the blood by LDL cholesterol. Supplementation with CoQ10 (100 mg/day for 30 days) was shown to decrease muscle pain by 40% in patients with myopathic symptoms associated with statin treatment (149). A recent systematic review on the role of coenzyme Q10 in statin-associated myopathy confirmed that statin treatment reduced circulating levels of CoQ10 and that supplementation can raise circulating levels of CoQ10. Data on the effect of CoQ10 on supplementation on myopathic symptoms are scarce and contradictory. Therefore these authors concluded that there was insufficient evidence to prove the etiologic role of CoQ10 deficiency in statin-associated myopathy and that large, well-designed clinical trials are required to address this issue (150). At this time routine use of CoQ10 is questionable in statin-treated patients but it may be an alternative for patients on statins with myalgia who cannot be satisfactorily treated with other agents.

In general, CoQ10 appears to be safe. No significant side effects have been found, even in studies that lasted a year. CoQ10 chemically resembles vitamin K. Since vitamin K counters the anticoagulant effect of warfarin, case reports associate CoQ10 therapy with decreased INR in patients on warfarin therapy; however, 100 mg CoQ10 daily had no effect on the INR in patients on warfarin in a randomized, double-blind, placebo-controlled, crossover trial (151). Typical doses range from 100 to 200 mg divided two to three times daily. However, caution is still advised if patients take CoQ10 and warfarin as CoQ10 may decrease the effectiveness of warfarin.

28.3.3 Coenzyme Q10 (Ubiquinone) and Cognitive and Neurological Health

There is also great interest in investigating the potential usefulness of CoQ10 in treating neurodegenerative diseases. CoQ10 has shown some promise for slowing the progression of Parkinson's disease. In a large NIH double-blind, placebo-controlled clinical trial, 80 subjects with early Parkinson's disease were given either CoQ10 300, 600, or 1,200 mg daily or placebo and followed for 16 months. The results suggested that CoQ10, especially at the highest dose, might have slowed disease progression (152). Shults and colleagues further investigated the safety of doses up to 3,000 mg/day of CoQ10, which was found to be well tolerated (153). In a more recent trial conducted in Germany, 131 patients with mid-stage Parkinson's disease (on stable medications for Parkinson's disease) were randomized to 300 mg of CoQ10 daily compared to a placebo for 3 months. The investigators failed to document significant differences between the groups although both the treatment and the placebo groups demonstrated significant improvement in Unified Parkinson's Disease Rating Scale (UPDRS) scores. The investigators concluded that the study did not support the hypothesis that restoring the impaired energy metabolism of dopaminergic neurons leads to beneficial effects in patients being treated for mid-stage Parkinson's disease (154).

Overall, the small neuroprotection trials performed with coenzyme Q10 in Parkinson's disease so far have been encouraging, but further evidence is required before it can be recommended routinely.

Coenzyme Q10 has also been studied in patients with Huntington's disease, an inherited disorder affecting 30,000 Americans. In addition, another 35,000 people exhibit some symptoms and as many as 75,000 people carry the abnormal gene that will cause them to develop the disease (155). The 30-month NIH trial entitled "Coenzyme Q10 and Remacemide Evaluation in Huntington's Disease or CARE-HD" failed to provide conclusive evidence of a slowing of disease progression and benefit by CoQ10 (156) in 347 subjects in the early stages of disease. However it was an important study as it is the first to show a hopeful trend toward slowing of disease with a particular therapy and provides clues to work with in future studies.

28.3.4 Ginkgo Biloba and Cognitive Function

Ginkgo biloba is one of the most popular medicines in Germany and France where physicians prescribe it for memory lapses, dizziness, anxiety, headaches, tinnitus, and other problems. Ginkgo leaf and its extracts, or GBE, contain several constituents including flavonoids, terpenoids, and organic acids. Each of these chemicals should exist in a specific amount in clinical quality GBE, representing a standard for assessing ginkgo products. Although many of ginkgo's constituents have intrinsic pharmacological effects individually, there is some evidence that the constituents work synergistically to produce more potent pharmacological effects than any individual constituent. The pharmacological actions that are most clinically relevant for ginkgo include anti-ischemic, anti-edema, anti-hypoxic, free radical scavenging, and hematologic effects and inhibition of platelet aggregation and adhesion. It is not known to what extent any of these effects contribute to the clinical effectiveness of ginkgo in dementias. Results from two independent meta-analyses (157,158) concluded that *G. biloba* extract in doses of 120–240 mg/day for treatment periods ranging from 4 weeks to 12 months significantly improved objective measures of cognitive function compared to placebo. Ginkgo has the potential to cause bleeding, especially in combination with anticoagulants *(159)*; however this was not shown to be the case in a study of ginkgo and coenzyme Q10 in 24 patients on stable warfarin as the mean dosage of warfarin did not change during either 4-week treatment period *(151)*. Similarly, in a double-blind, crossover randomized trial, ginkgo combined with aspirin did not alter bleeding times or other noted coagulation parameters beyond that of aspirin alone *(160)*. The Ginkgo Evaluation of Memory Study (GEM) to evaluate the safety and efficacy of G. Biloba (EGb761), 240 mg daily, in 3,000 subjects over the age of 75 to determine if ginkgo prevents dementia or Alzheimer's disease was recently published [JAMA, 2008]. The primary outcome of interest was incidence of all-cause dementia; secondary outcomes included rate of cognitive and functional decline, the incidence of cardiovascular and cerebrovascular events, and mortality. Study participants were followed for an average of approximately six years. Ginkgo showed no overall effect for reducing all types of dementia or Alzheimer's disease in this elderly cohort. The GEM study did not find significant adverse effects from ginkgo *(161)*. It is interesting to note that 27.4% of the GEMS cohort reported use of some type of nonvitamin/nonmineral dietary supplement at entry into study *(162)*. Over 9% of these participants were taking ginkgo and were unwilling to give up their current ginkgo supplements or would not accept assignment to placebo *(163)*. Complete ascertainment of intake of dietary supplements is crucial for later data analysis and interpretation of study outcomes. Similarly, The GuidAge study, a 5-year double-blind randomized trial of the efficacy of 240 mg/day EGb 761 for the prevention of Alzheimer disease in patients over 70 years with a memory complaint, is the largest study to be carried out in Europe on the prevention of Alzheimer disease. Final results should be available in 2010 *(164)*.

28.3.5 *French Pine Bark (Pinus pinaster Extract) and Chronic Venous Insufficiency*

Unlike many botanical supplements, one formulation of French pine bark, Pycnogenol[®], has received much attention in the popular press touting a record of 36 double-blind, placebo-controlled trials and is one of the fastest growing supplements on the U.S. market with sales up 25% in 2006 *(165)*. The strongest evidence for pycnogenol relates to improving heart health and treating chronic venous insufficiency and has also been evaluated for reducing venous complications from diabetes.

Pycnogenol is an extract from the bark of the French maritime pine tree. It consists of a mixture of bioflavonoids and is shown to have potent antioxidant properties. Pycnogenol's active constituents include the bioflavonoid monomers catechin, epicatechin, and taxifolin and phenolic acids. In chronic venous insufficiency procyanidins in pycnogenol reduce capillary permeability, which contributes to edema and microbleeding, by cross-linking capillary wall proteins such as collagen and elastin. Pycnogenol might also help prevent capillary permeability due to the antioxidant effects of several of its constituents *(166)*.

The UPBEAT Study (Understanding Pine Bark Extract as an Alternative Treatment) will test the efficacy of pine bark extract (Flavangenol[®]) in lowering blood pressure and improving glycemic control and plasma lipoprotein profiles. Changes in body weight, antioxidative capacity, anti-inflammatory markers, blood coagulation factors, and liver function tests will also be assessed. One hundred and thirty participants with mild or moderately elevated CVD risk factors will be enrolled into a phase II randomized double-blind, placebo-controlled study. The study is being conducted at Stanford University and is expected to complete in August 2008 *(167)*.

28.3.6 Saw Palmetto (Serenoa repens) and Prostate Health

Saw palmetto has been gaining in popularity for many years as an alternative therapy for treating the symptoms associated with benign prostatic hyperplasia (BPH). The ripe fruit of saw palmetto is used in several forms, including ground and dried fruit or whole berries. It is available as liquid extract, tablets and capsules and as an infusion or a tea. Modern saw palmetto preparations contain lipids extracted from the powdered berries. The primary ingredients include saturated and unsaturated fatty acids, as well as free and conjugated plant sterols *(168)*. Many saw palmetto products are standardized based on the fatty acid content. Saw palmetto has been shown to have anti-androgenic, anti-proliferative, and anti-inflammatory properties that seem to be responsible for improving BPH. Saw palmetto does not seem to affect overall prostate size, but shrinks the inner prostatic epithelium. Saw palmetto might slow prostate cell proliferation by inhibiting fibroblast growth factor and epidermal growth factor and stimulating apoptosis *(166)*. The fruit of saw palmetto has been shown in vitro to inhibit the 5-alpha-reductase and aromatase, to prevent the conversion of testosterone to dihydrotestosterone, which may be significant in the development of BPH *(169)*. In a number of double-blind controlled studies, some lasting up to 1 year, saw palmetto was found to improve urinary flow rate and most other measures of prostate disease compared to placebo. However, in a recent (2006) well-designed clinical trial, 225 men with moderate-to-severe BPH found no improvement with 320 mg saw palmetto daily for 1 year versus placebo *(170)*. However, at this time there is not enough evidence to support the use of saw palmetto for reducing the size of an enlarged prostate. Saw palmetto is well tolerated, is essentially nontoxic, and has no known drug interactions. An advantage of saw palmetto, unlike the standard therapy of Proscar®, is that it does not appear to affect PSA readings allowing clinicians to more accurately screen for prostate cancer.

28.3.7 Combination Supplements with Demonstrated Efficacy

The Age-Related Eye Disease Study (AREDS) was designed as a study of the clinical course of age-related macular degeneration (AMD) and lens opacities, as well as a randomized controlled trial of high-dose antioxidants and zinc to reduce progression of these diseases. The nutrients evaluated included 500 mg of vitamin C; 400 IU of vitamin E (synthetic DL-form of alpha-tocopherol acetate); 15 mg of β-carotene; 80 mg of zinc as zinc oxide; and 2 mg of copper as copper oxide. AREDS enrolled 4,757 participants, 55–80 years of age between 1992 and 1998. Participants were randomized to one of four treatments: (1) zinc alone; (2) antioxidants alone; (3) a combination of antioxidants and zinc; or (4) a placebo. The benefits of these nutrients were seen only in people who began the study at high risk for developing advanced AMD, those with intermediate AMD, and those with advanced AMD in one eye only. In this group, those taking "antioxidants plus zinc" had the lowest risk of developing advanced stages of AMD and its accompanying visual loss. Those in the "zinc alone" or "antioxidant alone" groups also reduced their risk of developing advanced AMD, but at more moderate rates compared to the "antioxidants plus zinc" group. Those in the placebo group had the highest risk of developing advanced AMD. The combination of high-dose antioxidant vitamins and zinc provided a moderate reduction of the risk (34%) of developing advanced AMD over a median of 6.3 years of follow-up in persons at high risk *(171)*. However, these nutrient supplements had no significant effect on development or progression of age-related cataract. The AREDS formula is the first demonstrated treatment for people at high risk for developing AMD. The findings of a protective effect of zinc alone demonstrated in AREDS were confirmed in the Blue Mountains Eye Study, an Australian population-based cohort study of 3,654 participants followed for 10 years for incident early, late, and the development of any AMD. However, the findings of a protective effect shown in AREDS and the Rotterdam Study of a combination of high doses of zinc, B-carotene, and vitamins C and E on AMD could not be confirmed in Blue Mountains Eye Study *(172)*.

Because of the success of AREDS, the NIH has launched a follow-on phase III efficacy study, AREDS II *(173)*. AREDS II is a multicenter randomized trial of 4,000 participants designed to assess the effects of oral supplementation of high doses of lutein (10 mg/day) and zeaxanthin (2 mg/day) and omega-3 long-chain polyunsaturated fatty acids (1 g/day) as docosahexaenoic acid (DHA) and eicosapentaenoic acid (EPA) for the treatment of AMD and cataract. The main study objective is to determine if these nutrients will decrease an individual's risk of progression to advanced AMD. As in AREDS, this study may help people at high risk for advanced AMD maintain useful vision for a longer time.

The cost-effectiveness of vitamin therapy (antioxidants plus zinc) for AMD was evaluated using a 50-year projection model. Individuals older than 50 years with AMD using U.S. cost and prevalence data were included in the model. Incidence of early AMD was based on published studies. Post-incident disease progression was included with unpublished data from the AREDS study. Compared with no vitamin therapy, vitamin therapy yielded a cost-effectiveness ratio of $21,387 per quality-adjusted life years considering extent of disease progression, years and severity of visual impairment, cost of ophthalmic care and nursing home services. The percentage of patients with AMD who ever developed visual impairment in the better-seeing eye was lowered from 7.0 to 5.6% *(174)*. The authors concluded, despite some assumptions made in formulating the model, vitamin therapy compares favorably with other medical therapies to prevent visual impairment from AMD and to improve health more generally and support the use of vitamin therapy in patients 50 years or older diagnosed with AMD.

28.4 GUIDELINES FOR SAFE USE OF DIETARY SUPPLEMENTS

Dietary supplements influence drug therapy within and outside of acute care facilities. Recognizing widespread consumer use of dietary supplements and their potential effect on medical status and treatment, the Joint Commission has included vitamins, herbals, and nutraceuticals in their scope of medications addressed by safety standards. Health care facilities are expected to establish standardized processes to compare current medication orders to medications usually taken, thus eliminating medication errors such as omissions, drug interactions, duplications, and dosing discrepancies in care settings as diverse as admission, surgery, discharge, and all transitions of care. In the case of dietary supplements, this requires health care staff to inquire about use of vitamins, minerals, herbals, botanicals, and nutraceuticals. Registered dietitians (RDs) are very familiar with most vitamin and mineral supplements and many herbal, botanical, and nutraceutical products, many of which incorporate vitamins and minerals. However, admission assessment interviews are most often conducted by registered nurses (RNs), who are often unfamiliar with dietary supplements. Shared accountability encourages processes that utilize input from all disciplines.

Complicating safety assessment of dietary supplements is the absence of standardization regulations for dietary supplements. As stated on the Office of Dietary Supplements Web site, "Standardization is a process manufacturers may use to ensure batch-to-batch consistency of their products. Standardization may involve markers that could help identify a consistent product and may provide a measure of quality control. Because no legal or regulatory definition for standardization exists for dietary supplements, there is no association between standardization statements on dietary supplement packages and quality." The FDA, to address this concern, published a rule in June 2007 regarding good manufacturing practices for dietary supplements.

Some hospitals have banned patient use of dietary supplements because of inconsistent manufacturing standards and concern that admission assessment of such substances will not be accurately completed *(175)*. Directors of pharmacies in acute care settings were surveyed in 2004 to determine institutional policies and practices related to the use of dietary supplements. Responses were received from 25% of pharmacy directors surveyed. Of these, 62% reported having policies in place but 30% of those reported a policy preventing use of dietary supplements *(176)*. Pharmacy directors reported concerns about the consistency of dietary supplement formulation and lack of FDA review of supplements.

28.4.1 Safeguarding Against Potential Interactions with Dietary Supplements

Herbal products, unlike most conventional drugs, provide a complex mixture of bioactive entities, which may or may not provide therapeutic activity. The active ingredient is frequently not known and complete characterization of all the chemical constituents is lacking as described in Table 28.3. As with conventional drugs, many herbal products are therapeutic at one dose and toxic at another. Concurrent use of herbs may mimic, magnify, or oppose the effect of drugs. The importance of unrecognized interactions between herbs and conventional drugs is particularly relevant in cardiology because many cardiovascular drugs have a narrow therapeutic window. Additional references for herb–drug interactions are listed in the Appendix and the reader is urged to consult these for more detailed information. An herbal product can affect clinical laboratory test results by direct assay interference, most commonly with immunoassays; exert physiologic effects through either toxicity or enzyme induction; or produce interference by contaminants.

Table 28.3
Distinctions between herbal versus manufactured drugs

Criteria	Drug	Herbal
Time frame of use	Typically a short tradition of use	Typically a long tradition of use
Ingredient characteristics	Active ingredient known	Active ingredient often not known, marker compounds used as a standard
	Pure compound available	Pure compound not available, often as mixtures
	Composition constant	Composition variable – due to season, temperature, and harvest conditions
Mechanism of action	Mechanism often known	Mechanism often unknown
Safety	Toxicology often known	Toxicology often not studied but at least evaluated
	Frequently a narrow therapeutic window	Generally a wide therapeutic window
	Adverse effects frequent	Adverse effects rare or unknown and not well characterized
Regulatory oversight	Require FDA approval for marketing	Does not require FDA approval for marketing

28.4.2 Clinical Needs Assessment for Dietary Supplements

Given the prevalence of inadequate dietary intake of many vitamins and minerals among older adults as well as the potential for inappropriate supplement use in this group, health providers should routinely screen elderly clients for the need for dietary supplementation and risk of excessive supplement intake. The following questions can identify conditions, symptoms, or situations that suggest the need to consider counseling on the use of fortified foods and/or dietary supplements in addition to advising older adults to make appropriate dietary changes.

1. Has the older client been diagnosed with physical conditions or does he/she display symptoms that may indicate a need for fortified foods and/or dietary supplements?

   - Osteoporosis: calcium, vitamin D, magnesium
   - Alcohol abuse: "B" vitamins, magnesium
   - Gastrointestinal abnormalities including diarrhea, fat malabsorption, and atrophic gastritis: vitamin B_{12}, fat-soluble vitamins (A, D, E, K)
   - Renal insufficiency: vitamin D
   - Cardiovascular disease: vitamin E, folic acid
   - Nutritional anemia: iron, vitamin B_{12}, folic acid
   - Weight loss, anorexia, nausea, or other symptom indicative of an inadequate intake: general multivitamin supplement

2. Does the older client have intake behaviors that place him/her at high risk for nutritional deficiency or excess?

- Eats fewer than two complete meals per day
- Drinks one or more (women) or two or more (men) alcoholic beverages per day
- Eats less than two servings of vegetables per day
- Eats less than two servings of fruits per day
- Takes multiple dietary supplements per day
- Takes doses of vitamin and mineral supplements that exceed the RDA unless recommended by a physician (pay particular attention to intakes greater than the UL)
- Is unaware of doses of supplements taken

3. Do body composition changes or physical limitations suggest potential for nutrient deficiencies?

- BMI < 21
- Has unintentionally lost or gained 10 lbs in the last 6 months
- Has difficulty chewing or swallowing
- Has physical disabilities that limit ability to shop for and/or prepare food

4. Are there lifestyle habits or conditions that may impair normal intake?

- Is housebound
- Has clinical evidence of depressive illness
- Needs assistance with self-care
- Demonstrates mental/cognitive impairment

5. Does the older client have disease or medical condition that may contraindicate use of dietary supplements?

- Do medical problem(s) such as impaired renal function limit excretion of supplements such as magnesium?
- Are medications taken that may adversely interact with dietary supplements?

On a daily basis clinicians are faced with interpreting the evidence, established and emerging, in order to provide the best care for their patients. Decision making occurs on many levels. Level I decisions are influenced by one's personal experience with the disease and capacity to deal with risk as "Would you have this done for yourself or for someone else in your immediate family?" A clinician making level II recommendations for his/her patient will also be influenced by prior experience, but the strength of the scientific evidence may play greater role. Lastly, a level III recommendation may be viewed as an across-the-board public health recommendation for a population, and this recommendation must be based even more on rigorous assessment of the scientific evidence. As the data on dietary supplements become more robust and evidence accrues clinicians will feel more confident in the recommendations on the use of dietary supplements for health maintenance and risk reduction both at the individual and at the group levels.

Acknowledgments The authors thank Abby Ershow, Sc.D., Lora Wilder, Sc.D., R.D., and Joseph Betz, Ph.D, for their valuable comments on the manuscript and Susan Pilch, Ph.D., M.L.S, for her expertise and technical editing.

APPENDIX: REFERENCE SOURCES FOR INFORMATION ON DIETARY SUPPLEMENTS AND HERB–DRUG INTERACTIONS

Monographs

- American Herbal Pharmacopeia and Therapeutic Compendium (AHP)

  Monographs are available through their Web site by to writing the American Herbal Pharmacopeia PO Box 5159, Santa Cruz, and CA 95063. Phone: 831-461-6318, e-mail: ahpadmin@got.net. Web address: http://www.herbal-ahp.org.
- The Complete German Commission E Monographs

  Blumenthal M, Goldberg A, Brinkmann J, eds. Herbal Medicine-Expanded Commission E Monographs. American Botanical Council, Austin, TX, 1999.

  The American Botanical Council, PO Box 201660 Austin, TX 78720. Phone: 512-33 1-8868 and Web address: www.herbalgram.org.
- WHO Monographs on Selected Medicinal Plants

  The World Health Organization (WHO) monographs are published in "WHO Monographs on Selected Plants," vol. 1, by WHO, Geneva, Switzerland, 1999, ISBN 92-4154517-8. The WHO Web site is http://www.who.org.

Books and Publications

- Barrett M. The Handbook of Clinically Tested Herbal Remedies. Haworth Herbal Press, 2004. 2-volume set. ISBN: 0-789-02723-2.
- Coates PM, Blackman MR, Cragg GM, Levine M, Moss J, and White JD, eds. Encyclopedia of Dietary Supplements. New York, NY: Marcel Dekker, 2005. ISBN: 0-824-75504-9.
- Fugh-Berman A. The 5-Minute Herb and Dietary Supplement Consult. Philadelphia, PA: Lippincott Williams & Wilkins, 2003, 475 pp. ISBN: 0-683-30273-6.
- Institute of Medicine of the National Academies. Dietary DRI Reference Intakes: The Essential Guide to Nutrient Requirements. Washington, DC: National Academies Press, 2006. ISBN: 0-309-10091-7, Web site: www.iom.edu/CMS/3788/29985/37065.aspx

  Description: A reference volume which reviews each nutrient and recommendations for daily intake.
- Newall CA, Anderson LA, Philpson JD. Herbal Medicines—A Guide for Health-Care Professionals. London, UK: The Pharmaceutical Press, 1996.
- Schulz V, Hansel R, Tyler VE. Rational phytotherapy: A Reference Guide for Physicians and Pharmacists, 5th ed. Berlin: Springer-Verlag, 2004.

Databases

- **Consumer Labs**

  www.consumerlabs.com

  Consumer labs evaluates commercially available dietary supplement products for composition, potency, purity, bioavailability, and consistency of products. The Natural Pharmacist database offers consumer-oriented information. Products that meet their criteria can receive a ConsumerLab seal of approval (annual subscription fee of $29.95).

- **MedlinePlus**

  MedlinePlus Health Information, National Library of Medicine, National Institutes of Health.

  Herbal Information page contains up-to-date, quality health care information on herbs and herbal medicine. Dietary Supplements page provides links, including the latest research, on dietary supplements.

  Herbal Information: http://www.nlm.nih.gov/medlineplus/herbalmedicine.html

  Dietary Supplements: http://www.nlm.nih.gov/medlineplus/dietarysupplements.html

- **Natural Medicines Comprehensive Database**

  www.naturaldatabase.com

  Created by the publishers of the Pharmacist's Letter. You can search by DS or commercial product name. Monographs include extensive information about common uses, evidence of efficacy and safety, mechanisms, interactions, and dosage. It is extensively referenced and updated daily. Also CME, listserv, and interactions information available (individual subscriber $92/year).

- **Natural Standard**

  www.naturalstandard.com

  This is an independent collaboration of international clinicians and researchers who created a database which can be searched by CAM subject or by medical condition. Quality of evidence is graded for each supplement (individual subscriber $99/year).

- **International Bibliographic Information on Dietary Supplements (IBIDS)**

  ods.od.nih.gov/Health_Information/IBIDS.aspx. IBIDS is produced by the Office of Dietary Supplements, NIH, along with the Food and Nutrition Information Center, National Agricultural Library, and USDA. The IBIDS database provides access to bibliographic citations and abstracts from published, international, and scientific literature on dietary supplements. IBIDS contains over 750,000 citations on the topic of dietary supplements from four major database sources: biomedical-related articles from MEDLINE, botanical and agricultural science from AGRICOLA, worldwide agricultural literature through AGRIS, and selected nutrition journals from CAB Abstracts and CAB Health. IBIDS is easy to search and available free of charge through the Internet.

REFERENCES

1. Dawson-Hughes B, Harris SS, Krall EA, Dallal GE, Falconer G, Green CL. Rates of bone loss in postmenopausal women randomly assigned to one of two dosages of vitamin D. Am J Clin Nutr 1995; 61:1140–5.
2. Chapuy MC, Arlot ME, Duboeuf F, et al. Vitamin D3 and calcium to prevent hip fractures in the elderly women. N Engl J Med 1992; 327(23):1637–42.
3. Jackson RD, LaCroix AZ, Gass M, et al. Calcium plus vitamin D supplementation and the risk of fractures. N Engl J Med 2006; 354(7):669–83.
4. Grant AM, Avenell A, Campbell MK, et al. Oral vitamin D3 and calcium for secondary prevention of low-trauma fractures in elderly people (Randomised Evaluation of Calcium Or vitamin D, RECORD): a randomised placebo-controlled trial. Lancet 2005; 365(9471):1621–8.
5. O'Brien. Combined calcium and vitamin D supplementation reduces bone loss and fracture incidence in older men and women. Nutr Rev 1998; 56:148–58.
6. Brouwer IA, van Dusseldorp M, Thomas CM, et al. Low-dose folic acid supplementation decreases plasma homocysteine concentrations: a randomized trial. Am J Clin Nutr 1999; 69:99–104.

7. Bostom AG, Rosenberg IH, Silbershatz H, et al. Nonfasting plasma total homocysteine levels and stroke incidence in elderly persons: the Framingham Study. Ann Intern Med 1999; 131(5):352–5.

8. Durga J, van Boxtel MPJ, Schouten EG, et al. Effect of 3-year folic acid supplementation on cognitive function in older adults in the FACIT trial: a randomised, double blind, controlled trial. Lancet 2007; 369(9557):208–16.

9. Picciano MF, Yetley EA, Coates PM. Folate and health. In: CAB Rev Perspect Agric Vet Sci Nutr Nat Resour 2007:No. 018.

10. Refsum H, Ueland PM, Nygard O, Vollset SE. Homocysteine and cardiovascular disease. Annu Rev Med 1998; 49:31–62.

11. Tawakol A, Migrino RQ, Aziz KS, et al. High-dose folic acid acutely improves coronary vasodilator function in patients with coronary artery disease. J Am Coll Cardiol 2005; 45:158–64.

12. Holick MF. Vitamin D deficiency. N Engl J Med 2007; 357(3):266–81.

13. Mason JB, Dickstein A, Jacques PF, et al. A temporal association between folic acid fortification and an increase in colorectal cancer rates may be illuminating important biological principles: a hypothesis. Cancer Epidemiol Biomarkers Prev 2007; 16(7):1325–9.

14. Oubre AY, Carlson TJ, King SR, Reaven GM. From plant to patient: an ethnomedical approach to the identification of new drugs for the treatment of NIDDM. Diabetologia 1997; 40:614–7.

15. Schmidt BM, Ribnicky DM, Lipsky PE, Raskin I. Revisiting the ancient concept of botanical therapeutics. Nat Chem Biol 2007; 3(7):360–6.

16. Cragg GM, Newman DJ. Plants as a source of anti-cancer agents. J Ethnopharmacol 2005; 100(1–2):72–9.

17. Eisenberg DM, Davis RB, Ettner SL, et al. Trends in alternative medicine use in the United States, 1990–1997: Results of a follow-up national survey. JAMA 1998; 280:1569–75.

18. Nutrition Business Journal. NBJ's Supplement Business Report 2006: an analysis of markets, trends, competition and strategy in the U.S. dietary supplement industry. Boulder, CO: New Hope Natural Media, 2006.

19. Sloane E. Why people use vitamin and mineral supplements. Nutr Today 2007; 42(2):55–61.

20. HealthFocus International. The 2005 HealthFocus Trend Report: A National Study of Public Attitudes and Actions Toward Shopping and Eating. St. Petersburg, FL: HealthFocus International, 2005.

21. Ervin RB, Wright JD, Kennedy-Stephenson J. Use of dietary supplements in the United States, 1988–94. Vital Health Stat 11 1999; 244(i–iii):1–14.

22. Radimer K, Bindewald B, Hughes J, Ervin B, Swanson C, Picciano MF. Dietary supplement use by US adults: data from the National Health and Nutrition Examination Survey, 1999–2000. Am J Epidemiol 2004; 160(4):339–49.

23. Timbo BB, Ross MP, McCarthy PV, Lin CTJ. Dietary supplements in a national survey: prevalence of use and reports of adverse events. J Am Diet Assoc 2006; 106(12):1966–74.

24. Bardia A, Nisly NL, Zimmerman BM, Gryzlak BM, Wallace RB. Use of herbs among adults based on evidence-based indications: findings from the National Health Interview Survey. Mayo Clin Proc 2007; 82(5):561–6.

25. Current Good Manufacturing Practice in Manufacturing, Packaging, Labeling, or Holding Operations for Dietary Supplements; Final Rule. Fed Regis 72(121):34751–958. Washington, DC: Food and Drug Administration, 2007. (Accessed August 3, 2007, at www.cfsan.fda.gov/~lrd/fr07625a.html)

26. Krucoff MW, Costello RB, Mark D, Vogel JHK. Complementary and alternative medical therapy in cardiovascular care. In: Fuster V, O'Rourke RA, Walsh R, et al., eds. Hurst's the Heart. New York: McGraw Hill, 2007.

27. Foreman J. St. John's Wort: Less than Meets the Eye. Globe Analysis Shows Popular Herbal Antidepressant Varies Widely in Content, Quality. Boston Globe 2002 January 10, Sect. C1.

28. Harkey MR, Henderson GL, Gershwin ME, Stern JS, Hackman RM. Variability in commercial ginseng products: an analysis of 25 preparations. Am J Clin Nutr 2001; 73(6):1101–6.

29. Draves AH, Walker SE. Analysis of the hypericin and pseudohypericin content of commercially available St John's Wort preparations. Can J Clin Pharmacol 2003; 10(3):114–8.

30. Ernst E. Adulteration of Chinese herbal medicines with synthetic drugs: a systematic review. J Intern Med 2002; 252(2):107–13.

31. Izzo AA, Ernst E. Interactions between herbal medicines and prescribed drugs: a systematic review. Drugs 2001; 61:2163–75.

32. Hu Z, Yang X, Ho PC, et al. Herb-drug interactions: a literature review. Drugs 2005; 65:1239–82.

33. Wolsko PM, Solondz DK, Phillips RS, Schachter SC, Eisenberg DM. Lack of herbal supplement characterization in published randomized controlled trials. Am J Med 2005; 118(10):1087–93.

34. Gagnier JJ, Boon H, Rochon P, et al. Reporting randomized, controlled trials of herbal interventions: an elaborated CONSORT statement. Ann Intern Med 2006; 144(5):364–7.

35. Barton MB, Miller T, Wolff T, et al. How to read the new recommendation statement: methods update from the U.S. Preventive Services Task Force. Ann Intern Med 2007; 147(2):123–7.

36. Balk EM, Tatsioni A, Lichtenstein AH, Lau J, Pittas AG. Effect of chromium supplementation on glucose metabolism and lipids: a systematic review of randomized controlled trials. Diabetes Care 2007; 30(8):2154–63.

37. Coates PM. Evidence-based reviews in support of health policy decisions. J Natl Cancer Inst 2007; 99(14):1059.

38. Evidence-Based Review System for the Scientific Evaluation of Health Claims. Draft Guidance. Washington, DC: Office of Nutrition, Labeling and Dietary Supplements, Center for Food Safety and Applied Nutrition, Food and Drug Administration, 2007. (Accessed August 3, 2007, at www.cfsan.fda.gov/~dms/hclmgui5.html)

39. Murphy SP, Poos MI. Dietary Reference Intakes: summary of applications in dietary assessment. Public Health Nutr 2002; 5(6A):843–9.

40. Barr SI. Applications of dietary reference intakes in dietary assessment and planning. Appl Physiol Nutr Metab 2006; 31(1):66–73.

41. Barr SI, Murphy SP, Poos MI. Interpreting and using the dietary references intakes in dietary assessment of individuals and groups. J Am Diet Assoc 2002; 102(6):780–8.

42. Vitolins MZ, Tooze JA, Golden SL, et al. Older adults in the rural South are not meeting healthful eating guidelines. J Am Diet Assoc 2007; 107(2):265–72.e3.

43. Gao X, Martin A, Lin H, Bermudez OI, Tucker KL. Alpha-Tocopherol intake and plasma concentration of Hispanic and non-Hispanic white elders is associated with dietary intake pattern. J Nutr 2006; 136(10):2574–9.

44. Boeckner LS, Pullen CH, Walker SN, Oberdorfer MK, Hageman PA. Eating behaviors and health history of rural midlife to older women in the midwestern United States. J Am Diet Assoc 2007; 107(2):306–10.

45. Roberts SB, Hajduk CL, Howarth NC, Russell R, McCrory MA. Dietary variety predicts low body mass index and inadequate macronutrient and micronutrient intakes in community-dwelling older adults. J Gerontol A Biol Sci Med Sci 2005; 60(5):613–21.

46. Sebastian RS, Cleveland LE, Goldman JD, Moshfegh AJ. Older adults who use vitamin/mineral supplements differ from nonusers in nutrient intake adequacy and dietary attitudes. J Am Diet Assoc 2007; 107(8):1322–32.

47. American Dietetic Association. Position of the American Dietetic Association: Food fortification and dietary supplements. J Am Diet Assoc 2001; 101:115–25.

48. Bronstrup A, Hages M, Pietrzik K. Lowering of homocysteine concentrations in elderly men and women. Int J Vitam Nutr Res 1999; 69(3):187–93.

49. Selhub J, Jacques PF, Wilson PF, Rush D, Rosenberg IH. Vitamin status and intake as primary determinants of homocysteinemia in an elderly population. JAMA 1993; 270:2693–8.

50. Boushey CJ, Beresford SAA, Omenn GS, Motulsky AG. A quantitative assessment of plasma homocysteine as a risk factor for vascular disease. JAMA 1995; 274(13):1049–57.

51. Lonn E, Yusuf S, Arnold MJ, et al. Homocysteine lowering with folic acid and B vitamins in vascular disease. N Engl J Med 2006; 354(15):1567–77.

52. Bønaa KH, Njolstad I, Ueland PM, et al. Homocysteine lowering and cardiovascular events after acute myocardial infarction. N Engl J Med 2006; 354(15):1578–88.

53. Seshadri S, Beiser A, Selhub J, et al. Plasma homocysteine as a risk factor for dementia and alzheimer's disease. N Engl J Med 2002; 346:476–83.

54. Tucker KL, Qiao N, Scott T, Rosenberg I, Spiro A, III. High homocysteine and low B vitamins predict cognitive decline in aging men: the Veterans Affairs Normative Aging Study. Am J Clin Nutr 2005; 82(3):627–35.
55. McMahon JA, Green TJ, Skeaff CM, Knight RG, Mann JI, Williams SM. A controlled trial of homocysteine lowering and cognitive performance. N Engl J Med 2006; 354(26):2764–72.
56. Gillette Guyonnet S, Abellan Van Kan G, Andrieu S, et al. IANA task force on nutrition and cognitive decline with aging. J Nutr Health Aging 2007; 11:132–52.
57. Malouf R, Grimley Evans J. The effect of vitamin B_6 on cognition. Cochrane Database of Systematic Reviews 2003; Issue 4. Art. No.: CD004393. DOI: 10.1002/14651858.CD004393.
58. Doncheva N, Penkov A, Velcheva A, Boev M, Popov B, Niagolov Y. Study of homocysteine concentration in coronary heart disease patients and comparison of two determination methods. Ann Nutr Metab 2007; 51(1):82–7.
59. Peeters AC, van Aken BE, Blom HJ, Reitsma PH, den Heijer M. The effect of homocysteine reduction by B-vitamin supplementation on inflammatory markers. Clin Chem Lab Med 2007; 45(1):54–8.
60. Ravaglia G, Forti P, Maioli F, et al. Homocysteine and folate as risk factors for dementia and Alzheimer disease. Am J Clin Nutr 2005; 82(3):636–43.
61. Morris MC, Evans DA, Tangney CC, Bienias JL, Wilson RS. Associations of vegetable and fruit consumption with age-related cognitive change. Neurology 2006; 67(8):1370–6.
62. Morris MC, Evans DA, Bienias JL, et al. Dietary folate and vitamin B_{12} intake and cognitive decline among community-dwelling older persons. Arch Neurol 2005; 62(4):641–5.
63. Bottiglieri T. Folate, vitamin B_{12}, and neuropsychiatric disorders. Nutr Rev 1996; 54(12):382–90.
64. Hin H, Clarke R, Sherliker P, et al. Clinical relevance of low serum vitamin B_{12} concentrations in older people: the Banbury B_{12} study. Age Ageing 2006; 35(4):416–22.
65. Shirodaria C, Antoniades C, Lee J, et al. Global improvement of vascular function and redox state with low-dose folic acid: implications for folate therapy in patients with coronary artery disease. Circulation 2007; 115(17):2262–70.
66. Kim YI. Folate, colorectal carcinogenesis, and DNA methylation: lessons from animal studies. Environ Mol Mutagen 2004; 44(1):10–25.
67. Giovannucci E. Epidemiologic studies of folate and colorectal neoplasia: a review. J Nutr 2002; 132(8 Suppl):2350S–5S.
68. van den Donk M, Pellis L, Crott JW, et al. Folic acid and vitamin B-12 supplementation does not favorably influence uracil incorporation and promoter methylation in rectal mucosa DNA of subjects with previous colorectal adenomas. J Nutr 2007; 137(9):2114–20.
69. Stolzenberg-Solomon RZ, Chang SC, Leitzmann MF, et al. Folate intake, alcohol use, and postmenopausal breast cancer risk in the Prostate, Lung, Colorectal, and Ovarian Cancer Screening Trial. Am J Clin Nutr 2006; 83(4):895–904.
70. Kim YI. Does a high folate intake increase the risk of breast cancer? Nutr Rev 2006; 64(10 Pt 1):468–75.
71. Mulligan JE, Greene GW, Caldwell M. Sources of folate and serum folate levels in older adults. J Am Diet Assoc 2007; 107:495–99.
72. Dereska NH, McLemore EC, Tessier DJ, Bash DS, Brophy CM. Short-term, moderate dosage Vitamin E supplementation may have no effect on platelet aggregation, coagulation profile, and bleeding time in healthy individuals. J Surg Res 2006; 132(1):121–9.
73. Meydani SN, Meydani M, Blumberg JB, et al. Assessment of the safety of supplementation with different amounts of vitamin E in older adults. Am J Clin Nutr 1998; 68:311–8.
74. Morinobu T, Ban R, Yoshikawa S, Murata T, Tamai H. The safety of high-dose vitamin E supplementation in healthy Japanese male adults. J Nutr Sci Vitaminol (Tokyo) 2002; 48(1):6–9.
75. National Institutes of Health. National Institutes of Health State-of-the-Science Conference statement: management of menopause-related symptoms. Ann Intern Med 2005; 142:1003–13.
76. Buring JE. Aspirin prevents stroke but not MI in women: vitamin E has no effect on CV disease or cancer. Cleve Clin J Med 2006; 73(9):863–70.

77. Heinonen OP, Albanes D, Virtamo J, et al. Prostate cancer and supplementation with alpha-tocopherol and beta-carotene: incidence and mortality in a controlled trial. J Natl Cancer Inst 1998; 18:440–6.

78. Virtamo J, Pietinen P, Huttunen JK, et al. Incidence of cancer and mortality following alpha-tocopherol and beta-carotene supplementation: a postintervention follow-up. JAMA 2003; 290(4):476–85.

79. Wright ME, Lawson KA, Weinstein SJ, et al. Higher baseline serum concentrations of vitamin E are associated with lower total and cause-specific mortality in the Alpha-Tocopherol, Beta-Carotene Cancer Prevention Study. Am J Clin Nutr 2006; 84(5):1200–7.

80. Cho E, Hunter DJ, Spiegelman D, et al. Intakes of vitamins A, C and E and folate and multivitamins and lung cancer: a pooled analysis of 8 prospective studies. Int J Cancer 2006; 118(4):970–8.

81. Blumberg JB. An update: vitamin E supplementation and heart disease. Nutr Clin Care 2002; 5(2):50–5.

82. Traber MG. Heart disease and single-vitamin supplementation. Am J Clin Nutr 2007; 85(1):293S–9S.

83. Lee IM, Cook NR, Gaziano JM, et al. Vitamin E in the primary prevention of cardiovascular disease and cancer: the Women's Health Study: a randomized controlled trial. JAMA 2005; 294(1):56–65.

84. Glynn RJ, Ridker PM, Goldhaber SZ, Zee RY, Buring JE. Effects of random allocation to vitamin E supplementation on the occurrence of venous thromboembolism: report from the Women's Health Study. Circulation 2007; 116(13):1497–503.

85. Lonn E, Bosch J, Yusuf S, et al. Effects of long-term vitamin E supplementation on cardiovascular events and cancer: a randomized controlled trial. JAMA 2005; 293(11):1338–47.

86. Vivekananthan DP, Penn MS, Sapp SK, Hsu A, Topol EJ. Use of antioxidant vitamins for the prevention of cardiovascular disease: meta-analysis of randomised trials. Lancet 2003; 361(9374):2017–23.

87. Bjelakovic G, Nikolova D, Gluud LL, Simonetti RG, Gluud C. Mortality in randomized trials of antioxidant supplements for primary and secondary prevention: systematic review and meta-analysis. JAMA 2007; 297(8):82–57.

88. Miller III ER, Pastor-Barriuso R, Dalal D, Riemersma RA, Appel LJ, Guallar E. Meta-analysis: high-dosage vitamin E supplementation may increase all-cause mortality. Ann Intern Med 2005; 142(1):37–46.

89. Huang HY, Caballero B, Chang S, et al. Multivitamin/Mineral Supplements and Prevention of Chronic Disease. (Prepared by The Johns Hopkins University Evidence-based Practice Center, under Contract No. 290-02-0018). Rockville, MD: Agency for Healthcare Research and Quality, 2006.

90. Shekelle PG, Morton SC, Jungvig LK, et al. Effect of supplemental vitamin E for the prevention and treatment of cardiovascular disease. J Gen Intern Med 2004; 19(4):380–9.

91. Hathcock JN, Azzi A, Blumberg J, et al. Vitamins E and C are safe across a broad range of intakes. Am J Clin Nutr 2005; 81(4):736–45.

92. Mezzetti A, Zuliani G, Romano F, et al. Vitamin E and lipid peroxide plasma levels predict the risk of cardiovascular events in a group of healthy very old people. J Am Geriatr Soc 2001; 49(5):533–7.

93. Appel LJ, Brands MW, Daniels SR, Karanja N, Elmer PJ, Sacks FM. Dietary approaches to prevent and treat hypertension: a scientific statement from the American Heart Association. Hypertension 2006; 47(2):296–308.

94. Sacks FM, Obarzanek E, Windhauser MM, et al. Rationale and design of the Dietary Approaches to Stop Hypertension trial (DASH). A multicenter controlled-feeding study of dietary patterns to lower blood pressure. Ann Epidemiol 1995; 5(2):108–18.

95. Sacks FM, Appel LJ, Moore TJ, et al. A dietary approach to prevent hypertension: a review of the Dietary Approaches to Stop Hypertension (DASH) Study. Clin Cardiol 1999; 22(7 Suppl):III6–10.

96. Lien LF, Brown AJ, Ard JD, et al. Effects of PREMIER lifestyle modifications on participants with and without the metabolic syndrome. Hypertension 2007:HYPERTENSIONAHA.107. 089458.

97. Health Claim Notification for Potassium Containing Foods. Washington, DC: Office of Nutritional Products, Labeling and Dietary Supplements, Center for Food Safety and Applied Nutrition, Food and Drug Administration, 2000. (Accessed August 3, 2007, at www.cfsan.fda.gov/~dms/hclm-k.html)

98. Piekutowski K, Makarewicz R, Zachara BA. The antioxidative role of selenium in pathogenesis of cancer of the female reproductive system. Neoplasma 2007; 54:374–8.

99. Li H, Stampfer MJ, Giovannucci EL, et al. A prospective study of plasma selenium levels and prostate cancer risk. J Natl Cancer Inst 2004; 96(9):696–703.

100. Chen Y, Hall M, Graziano JH, et al. A prospective study of blood selenium levels and the risk of arsenic-related premalignant skin lesions. Cancer Epidemiol Biomarkers Prev 2007; 16(2):207–13.

101. Brinkman M, Reulen RC, Kellen E, Buntinx F, Zeeers M. Are men with low selenium levels at increased risk of prostate cancer? Eur J Cancer 2006; 42:2463–71.

102. Greenwald P, Anderson D, Nelson SA, Taylor PR. Clinical trials of vitamin and mineral supplements for cancer prevention. Am J Clin Nutr 2007; 85(1):314S–7.

103. Bleys J, Navas-Acien A, Guallar E. Serum selenium and diabetes in U.S. adults. Diabetes Care 2007; 30(4):829–34.

104. Stranges S, Marshall JR, Trevisan M, et al. Effects of selenium supplementation on cardiovascular disease incidence and mortality: secondary analyses in a randomized clinical trial. Am J Epidemiol 2006; 163:694–9.

105. Stranges S, Marshall JR, Natarajan R, et al. Effects of long-term selenium supplementation on the incidence of type 2 diabetes: a randomized trial. Ann Intern Med 2007; 147:217–23.

106. Health Claims: Calcium and Osteoporosis. 21 CFR 101.72. Washington, DC: Food and Drug Administration, 1993. (Accessed August 3, 2007, at www.cfsan.fda.gov/~lrd/cf101-72.html)

107. Gass M, Dawson-Hughes B. Preventing osteoporosis-related fractures: an overview. Am J Med 2006; 119(4):S3–S11.

108. Bone Health and Osteoporosis: A Surgeon General's Report: Fact Sheets. Washington DC: Office of the Surgeon General, Department of Health and Human Services, 2004. (Accessed August 3, 2007, at www.surgeongeneral.gov/library/bonehealth/factsheet.html)

109. Optimal Calcium Intake. NIH Consensus Statement Online 1994 June 6–8; 12(4):1–31 Bethesda, MD: NIH Consensus Development Program, 1994. (Accessed September 21, 2007, at consensus.nih.gov/1994/1994OptimalCalcium097html.htm)

110. Bolland MJ, Barber PA, Doughty RN, et al. Vascular events in healthy older women receiving calcium supplementation: randomised controlled trial. BMJ 2008; 336(7638):262–6.

111. Holick MF. The role of vitamin D for bone health and fracture prevention. Curr Osteoporos Rep 2006; 4(3):96–102.

112. Heaney RP, Dowell MS, Hale CA, Bendich A. Calcium absorption varies within the reference range for serum 25-hydroxyvitamin D. J Am Coll Nutr 2003; 22(2):142–6.

113. Dawson-Hughes B, Heaney RP, Holick MF, Lips P, Meunier PJ, Vieth R. Estimates of optimal vitamin D status. Osteoporos Int 2005; 16(7):713–6.

114. Cranney A, Horsley T, O'Donnell S, et al. Effectiveness and Safety of Vitamin D in Relation to Bone Health. (Prepared by University of Ottawa Evidence-based Practice Center, under Contract No. 290-02-0021). Rockville, MD: Agency for Healthcare Research and Quality, 2007.

115. Brown JP, Fortier M, Frame H, et al. Canadian consensus conference on osteoporosis, 2006 update. J Obstet Gynaecol Can 2006; 28(2 Suppl 1):S95–S112.

116. Institute of Medicine, Food and Nutrition Board. Dietary Reference Intakes for Calcium, Phosphorus, Magnesium, Vitamin D, and Fluoride. Washington, DC: National Academy Press, 1997.

117. Your Guide to Lowering Your Blood Pressure with DASH, NIH Publication No. 06-4082. Bethesda, MD: National Heart, Lung, and Blood Institute, 2006. (Accessed August 3, 2007, at www.nhlbi.nih.gov/health/public/heart/hbp/dash/new_dash.pdf)

118. Song Y, Sesso HD, Manson JE, Cook NR, Buring JE, Liu S. Dietary magnesium intake and risk of incident hypertension among middle-aged and older US women in a 10-year follow-up study. Am J Cardiol 2006; 98(12):1616–21.

119. Myers VH, Champagne CM. Nutritional effects on blood pressure. Curr Opin Lipidol 2007; 18(1):20–4.

120. Beyer FR, Dickinson HO, Nicolson DJ, Ford GA, Mason J. Combined calcium, magnesium and potassium supplementation for the management of primary hypertension in adults. Cochrane Database Syst Rev 2006; Issue 3. Art. No.: CD004805. DOI: 10.1002/14651858. CD004805.pub2.

121. Schulze MB, Schulz M, Heidemann C, Schienkiewitz A, Hoffman K, Boeing H. Fiber and magnesium intake and incidence of type 2 diabetes: a prospective study and meta-analysis. Arch Intern Med 2007; 167:956–65.

122. Ford ES, Li C, McGuire LC, Mokdad AH, Liu S. Intake of dietary magnesium and the prevalence of the metabolic syndrome among U.S. adults. Obesity (Silver Spring) 2007; 15(5):1139–46.

123. Nielsen FH, Milne DB, Klevay LM, Gallagher S, Johnson L. Dietary magnesium deficiency induces heart rhythm changes, impairs glucose tolerance, and decreases serum cholesterol in post menopausal women. J Am Coll Nutr 2007; 26:121–32.

124. Hunt CD, Johnson L. Magnesium requirements: new estimations for men and women by cross-sectional statistical analyses of metabolic magnesium balance data. Am J Clin Nutr 2006; 84:843–52.

125. Calder PC. Polyunsaturated fatty acids and inflammation. Prostaglandins Leukot Essent Fatty Acids 2006; 75(3):197–202.

126. de Lorgeril M. Essential polyunsaturated fatty acids, inflammation, atherosclerosis and cardio-vascular diseases. Subcell Biochem 2007; 42:283–97.

127. Burr ML, Fehily AM, Gibert JF, et al. Effects of changes in fat, fish, and fibre intakes on death and myocardial reinfarction: diet and reinfarction trial (DART). Lancet 1989; 2(8666):757–61.

128. Burr ML, Ashfield-Watt PAL, Dunstan FDJ, et al. Lack of benefit of dietary advice to men with angina: results of a controlled trial. Eur J Clinl Nutr 2003; 57(2):193–200.

129. Ness AR, Ashfield-Watt PAL, Whiting JM, Smith GD, Hughes J, Burr ML. The long-term effect of dietary advice on the diet of men with angina: the diet and angina randomized trial. J Hum Nutr Diet 2004; 17(2):117–9.

130. Gruppo Italiano per lo Studio della Sopravvivenza nell'Infarto miocardico. Dietary supplementation with n-3 polyunsaturated fatty acids and vitamin E after myocardial infarction: results of the GISSI-Prevenzione trial. Gruppo Italiano per lo Studio della Sopravvivenza nell'Infarto miocardico. Lancet 1999; 354:447–55.

131. Yokoyama M, Origasa H, Matsuzaki M, et al. Effects of eicosapentaenoic acid on major coronary events in hypercholesterolaemic patients (JELIS): a randomised open-label, blinded endpoint analysis. Lancet 2007; 369:1090–8.

132. Hooper L, Thompson RL, Harrison RA, et al. Risks and benefits of omega 3 fats for mortality, cardiovascular disease, and cancer: systematic review. BMJ 2006 332:752–60.

133. Smith Jr, SC, Allen J, Blair SN, et al. AHA/ACC guidelines for secondary prevention for patients with coronary and other atherosclerotic vascular disease: 2006 update: endorsed by the National Heart, Lung, and Blood Institute. Circulation 2006; 113(19):2363–72.

134. Introduction to the TLC Diet. Bethesda, MD: National Heart, Lung, and Blood Institute, 2006. (Accessed August 3, 2007, at www.nhlbi.nih.gov/cgi-bin/chd/step2intro.cgi)

135. McKenney JM, Sica D. Role of prescription omega-3 fatty acids in the treatment of hypertrigly-ceridemia. Pharmacotherapy 2007; 5:715–28.

136. Kushi LH, Byers T, Doyle C, et al. American Cancer Society Guidelines on Nutrition and Physical Activity for cancer prevention: reducing the risk of cancer with healthy food choices and physical activity. CA Cancer J Clin 2006; 56:254–81.

137. Larsson SC, Kumlin M, Ingelman-Sundberg J, Wolk A. Dietary long-chain n-3 fatty acids for the prevention of cancer: a review of potential mechanisms. Am J Clin Nutr 2004; 79(6):935–45.

138. Pardini RS. Nutritional intervention with omega-3 fatty acids enhances tumor response to anti-neoplastic agents. Chem Biol Interact 2006; 162(2):89–105.

139. MacLean CH, Newberry SJ, Mojica WA, et al. Effects of omega-3 fatty acids on cancer risk: a systematic review. JAMA 2006; 295(4):403–15.

140. McKay DL, Perrone G, Rasmussen H, Dallal G, Blumberg JB. Multivitamin/mineral supplementation improves plasma B-vitamin status and homocysteine concentration in healthy older adults consuming a folate-fortified diet. J Nutr 2000; 130:3090–6.

141. McKay DL, Perrone G, Rasmussen H, et al. The effects of a multivitamin/mineral supplement on micronutrient status, antioxidant capacity and cytokine production in health older adults consuming a fortified diet. J Am Coll Nutr 2000; 19:613–21.

142. McNeill G, Avenell A, Campbell M, et al. Effect of multivitamin and multimineral supplementation on cognitive function in men and women aged 65 years and over: a randomised controlled trial. Nutr J 2007; 6(1):10.

143. Burdette JE, Liu J, Chen SN, et al. Black cohosh acts as a mixed competitive ligand and partial agonist of the serotonin receptor. J Agric Food Chem 2003; 51(19):5661–70.

144. Newton KM, Reed SD, LaCroix AZ, Grothaus LC, Ehrlich K, Guiltinan J. Treatment of vasomotor symptoms of menopause with black cohosh, multibotanicals, soy, hormone therapy, or placebo: a randomized trial. Ann Intern Med 2006; 145(12):869–79.

145. ClinicalTrials.gov: Use of Black Cohosh and Red Clover for the Relief of Menopausal Symptoms. Bethesda, MD: National Library of Medicine, National Institutes of Health, 2007. (Accessed August 3, 2007, at clinicaltrials.gov/ct/show/NCT00066144?order3)

146. Rosenfeldt F, Hilton D, Pepe S, Krum H. Systematic review of effect of coenzyme Q10 in physical exercise, hypertension and heart failure. Biofactors 2003; 18(1–4):91–100.

147. Sander S, Coleman CI, Patel AA, Kluger J, White CM. The impact of coenzyme Q10 on systolic function in patients with chronic heart failure. J Card Fail 2006; 12(6):464–72.

148. Davidson M, McKenney J, Stein E, et al. Comparison of one-year efficacy and safety of atorvastatin versus lovastatin in primary hypercholesterolemia. Am J Cardiol 1997; 79(11):1475–81.

149. Caso G, Kelly P, McNurlan MA, Lawson WE. Effect of coenzyme Q10 on myopathic symptoms in patients treated with statins. Am J Cardiol 2007; 99:1409–12.

150. Marcoff L, Thompson PD. The role of coenzyme Q10 in statin-associated myopathy: a systematic review. J Am Coll Cardiol 2007; 49:2231–7.

151. Engelsen J, Nielsen JD, Winther K. Effect of coenzyme Q10 and Ginkgo biloba on warfarin dosage in stable, long-term warfarin treated outpatients. A randomised, double blind, placebo-crossover trial. Thromb Haemost 2002; 87(6):1075–6.

152. Shults CW, Oakes D, Kieburtz K, et al. Effects of coenzyme Q10 in early Parkinson disease: evidence of slowing of the functional decline. Arch Neurol 2002; 59(10):1541–50.

153. Shults CW, Flint Beal M, Song D, Fontaine D. Pilot trial of high dosages of coenzyme Q10 in patients with Parkinson's disease. Exp Neurol 2004; 188(2):491–4.

154. Storch A, Jost WH, Vieregge P, et al. Randomized, double-blind, placebo-controlled trial on symptomatic effects of coenzyme Q(10) in Parkinson disease. Arch Neurol 2007; 64(7):938–44.

155. Learning About Huntington's Disease. 2007. (Accessed December 10, 2007, at www.genome.gov/10001215)

156. Huntington Study Group. A randomized, placebo-controlled trial of coenzyme Q10 and remacemide in Huntington's disease. Neurology 2001; 57(3):397–404.

157. Oken BS, Storzbach DM, Kaye JA. The efficacy of ginkgo biloba on cognitive function in Alzheimer's disease. Arch Neurol 1998; 55:1409–15.

158. Ernst E, Pittler MH. Ginkgo biloba for dementia: a systematic review of double-blind placebo controlled trials. Clin Drug Invest 1999; 17:301–8.

159. Fugh-Berman A. Herb-drug interactions. Lancet 2000; 355:134–8.

160. Wolf HR. Does Ginkgo biloba special extract EGb 761 provide additional effects on coagulation and bleeding when added to acetylsalicylic acid 500 mg daily? Drugs R D 2006; 7(3):163–72.

161. DeKosky ST, Williamson JD, Fitzpatrick AL, et al; Ginkgo Evaluation of Memory (GEM) Study Investigators. Ginkgo biloba for prevention of dementia: a randomized controlled trial. JAMA. 2008; 300(19):2253–62.

162. Nahin RL, Fitzpatrick AL, Williamson JD, et al. Use of herbal medicine and other dietary supplements in community-dwelling older people: baseline data from the ginkgo evaluation of memory study. J Am Geriatr Soc 2006; 54:1725–35.

163. Fitzpatrick AL, Fried LP, Williamson J, et al. Recruitment of the elderly into a pharmacologic prevention trial: the Ginkgo Evaluation of Memory study experience. Contemp Clin Trials 2006; 27(6):541–53.
164. Vellas B, Andrieu S, Ousset PJ, Ouzid M, Mathiex-Fortunet H, GuidAge Study Group. The GuidAge study: methodological issues. A 5-year double-blind randomized trial of the efficacy of EGb 761 for prevention of Alzheimer disease in patients over 70 with a memory complaint. Neurology 2006; 67(9 Suppl 3):S6–11.
165. Health: it's all in the bark. 2006. (Accessed December 15, 2007, at www.newsweek.com/id/44416)
166. Natural Medicines Comprehensive Database. Therapeutic Research Center, 1999 – (Accessed December 11, 2007, at www.naturaldatabase.com. Subscription required.)
167. ClinicalTrials.gov: Understanding Pine Bark Extract as an Alternative Treatment (UPBEAT) Study. Bethesda, MD: National Library of Medicine, National Institutes of Health, 2007. (Accessed August 3, 2007, at www.clinicaltrials.gov/ct/show/NCT00425945?order=2)
168. Schulz V, Hänsel R, Tyler VE. Rational Phytotherapy: A Physicians' Guide to Herbal Medicine, 4th ed. Berlin: Springer Verlag, 2001.
169. Koch E, Biber A. Pharmacological effects of sabal and urtica extracts as basis for a rational drug therapy in benign prostate hyperplasia Urologe – Ausgabe B 1994; 34(2):90–5.
170. Bent S, Kane C, Shinohara K, et al. Saw palmetto for benign prostatic hyperplasia. N Engl J Med 2006; 354(6):557–66.
171. Age-Related Eye Disease Study Research Group. A randomized, placebo-controlled, clinical trial of high-dose supplementation with vitamins C and E, beta carotene, and zinc for age-related macular degeneration and vision loss: AREDS report no. 8. Arch Ophthalmol 2001; 119(10):1417–36.
172. Tan JSL, Wang JJ, Flood V, Rochtchina E, Smith W, Mitchell P. Dietary antioxidants and the long-term incidence of age-related macular degeneration: the Blue Mountains Eye Study. Am Acad Ophthalmol 2007(July 28):[epub ahead of print].
173. ClinicalTrials.gov: Age-Related Eye Disease Study 2 (AREDS2). Bethesda, MD: National Library of Medicine, National Institutes of Health, 2007. (Accessed August 3, 2007, at www. clinicaltrials.gov/ct2/show/NCT00345176?term=areds&rank=1)
174. Rein DB, Saaddine JB, Wittenborn JS, et al. Cost-effectiveness of vitamin therapy for age-related macular degeneration. Ophthalmology 2007; 114:1319–26.
175. Cohen MH, Hrbek A, Davis RB, Schachter SC, Eisenberg DM. Emerging credentialing practices, malpractice liability policies, and guidelines governing complementary and alternative medical practices and dietary supplement recommendations: a descriptive study of 19 integrative health care centers in the United States. Arch Intern Med 2005; 165(3):289–95.
176. Bazzie KL, Witmer DR, Pinto B, Bush C, Clark J, Deffenbaugh Jr, J. National survey of dietary supplement policies in acute care facilities. Am J Health Syst Pharm 2006; 63(1):65–70.
177. Qualified Health Claims Subject to Enforcement Discretion: B Vitamins & Vascular Disease. Washington, DC: Office of Nutritional Products, Labeling, and Dietary Supplements, Center for Food Safety and Applied Nutrition, Food and Drug Administration, 2000. (Accessed February 2, 2007, at www.cfsan.fda.gov/~dms/qhc-sum.html#bvitamins)
178. Lichtenstein AH, Appel LJ, Brands M, et al. Diet and lifestyle recommendations revision 2006: a scientific statement from the American Heart Association Nutrition Committee. Circulation 2006; 114:82–96.
179. Vogel JHK, Bolling SF, Costello RB, et al. Integrating complementary medicine into cardiovascular medicine. A report of the American College of Cardiology Foundation Task Force on Clinical Expert Consensus Documents (Writing Committee to Develop an Expert Consensus Document on Complementary and Integrative Medicine). J Am Coll Cardiol 2005; 46(1):184–221.
180. Institute of Medicine, Food and Nutrition Board. Dietary Reference Intakes for Thiamin, Riboflavin, Niacin, Vitamin B$_6$, Folate, Vitamin B$_{12}$, Pantothenic Acid, Biotin, and Choline. Washington, DC: National Academy Press, 1998.
181. U.S. Department of Health and Human Services and U.S. Department of Agriculture. Dietary Guidelines for Americans, 6th ed. Washington, DC: U.S. Government Printing Office, 2005.
182. Institute of Medicine, Food and Nutrition Board. Dietary Reference Intakes: Water, Potassium, Sodium, Chloride, and Sulfate. Washington, DC: National Academy Press, 2004.

183. Qualified Health Claims Subject to Enforcement Discretion: Selenium & Cancer. Washington, DC: Office of Nutritional Products, Labeling, and Dietary Supplements, Center for Food Safety and Applied Nutrition, Food and Drug Administration, 2003. (Accessed February 2, 2007, at www.cfsan.fda.gov/~dms/qhc-sum.html#selenium)

184. Qualified Health Claims Subject to Enforcement Discretion: Omega-3 Fatty Acids & Coronary Heart Disease. Washington, DC: Office of Nutritional Products, Labeling, and Dietary Supplements, Center for Food Safety and Applied Nutrition, Food and Drug Administration, 2004. (Accessed February 2, 2007, at www.cfsan.fda.gov/~dms/qhc-sum.html#omega3)

185. A Food Labeling Guide – Appendix C. Washington, DC: Center for Food Safety and Applied Nutrition, Food and Drug Administration, 2000. (Accessed August 3, 2007, at www.cfsan.fda.gov/~dms/flg-6c.html)

186. Qualified Health Claims Subject to Enforcement Discretion: Calcium and Colon/Rectal Cancer & Calcium and Recurrent Colon/Rectal Polyps. Washington, DC: Office of Nutritional Products, Labeling, and Dietary Supplements, Center for Food Safety and Applied Nutrition, Food and Drug Administration, 2005. (Accessed August 3, 2007, at www.cfsan.fda.gov/~dms/qhcca2.html)

187. Ziegler D, Nowak H, Kempler P, Vargha P, Low PA. Treatment of symptomatic diabetic polyneuropathy with the antioxidant alpha-lipoic acid: a meta-analysis. Diabet Med 2004; 21(2):114–21.

188. Sauer J, Tabet N, Howard R. Alpha lipoic acid for dementia. Cochrane Database Syst Rev 2004; Issue 1. Art. No.: CD004244. DOI: 10.1002/14651858.CD004244.pub2.

189. Bjelakovic G, Nagorni A, Nikolova D, Simonetti RG, Bjelakovic M, Gluud C. Meta-analysis: antioxidant supplements for primary and secondary prevention of colorectal adenoma. Aliment Pharmacol Ther 2006; 24(2):281–91.

190. Hardy M, Coulter I, Shekelle P, et al. Effect of the Supplemental Use of Antioxidants Vitamin C, Vitamin E, and Coenzyme Q10 for the Prevention and Treatment of Cancer. (Prepared by Southern California-RAND Evidence-based Practice Center, under Contract No. 290-97-0001). Rockville, MD: Agency for Healthcare Research and Quality, 2003.

191. Bleys J, Miller III ER, Pastor-Barriuso R, Appel LJ, Guallar E. Vitamin-mineral supplementation and the progression of atherosclerosis: a meta-analysis of randomized controlled trials. Am J Clin Nutr 2006; 84:880–7.

192. Shekelle P, Morton S, Hardy M, et al. Effect of Supplemental Antioxidants Vitamin C, Vitamin E, and Coenzyme Q10 for the Prevention and Treatment of Cardiovascular Disease. (Prepared by Southern California-RAND Evidence-based Practice Center, under Contract No. 290-97-0001). Rockville, MD: Agency for Healthcare Research and Quality, 2003.

193. Evans JR. Antioxidant vitamin and mineral supplements for slowing the progression of age-related macular degeneration. Cochrane Database Syst Rev 2006; Issue 2. Art. No.: CD000254. DOI: 10.1002/14651858.CD000254.pub2.

194. Qualified Health Claims Subject to Enforcement Discretion: Antioxidant Vitamins & Cancer. Washington, DC: Office of Nutritional Products, Labeling, and Dietary Supplements, Center for Food Safety and Applied Nutrition, Food and Drug Administration, 2003. (Accessed February 2, 2007, at www.cfsan.fda.gov/~dms/qhc-sum.html#antioxidant)

195. U.S. Preventive Services Task Force. Routine vitamin supplementation to prevent cancer and cardiovascular disease: recommendations and rationale. Ann Intern Med 2003; 139:51–5.

196. American Diabetes Association. Nutrition Recommendations and Interventions for Diabetes: a position statement of the American Diabetes Association. Diabetes Care 2007; 30(Suppl 1):S48–65.

197. Hardy M, Coulter I, Venuturupalli S, et al. Ayurvedic Interventions for Diabetes Mellitus: A Systematic Review. (Prepared by Southern California Evidence-Based Practice Center/RAND, under Contract No. 290-97-0001). Rockville, MD: Agency for Healthcare Quality and Research, 2001.

198. Lethaby AE, Brown J, Marjoribanks J, Kronenberg F, Roberts H, Eden J. Phytoestrogens for vasomotor menopausal symptoms. Cochrane Database of Systematic Reviews 2007; Issue 4. Art. No.: CD001395. DOI: 10.1002/14651858.CD001395.pub3.

199. McGuffin M, Hobbs C, Upton R, Goldberg A, eds. American Herbal Products Association's Botanical Safety Handbook Boca Raton: CRC Press, 1997.

200. Blumenthal M, Busse WR, Goldberg A, et al., eds. Complete German Commission E Monographs: Therapeutic Guide to Herbal Medicines. Austin, TX: American Botanical Council, 1998.

201. American College of Obstetricians and Gynecologists Committee on Practice Bulletins – Gynecology. ACOG Practice Bulletin. Clinical Management Guidelines for Obstetrician-Gynecologists. Use of botanicals for management of menopausal symptoms. Obstet Gynecol 2001; 97(6):S1–11.

202. Belisle S, Blake J, Basson R, et al. Canadian consensus conference on menopause, 2006 update. J Obstet Gynaecol Can 2006; 28(2 Suppl 1):S7–S94.

203. United States Pharmacopeial Convention, Committee of Revision. The United States Pharmacopeia. Rockville, MD: United States Pharmacopeial Convention, Inc., 2007.

204. Montgomery SA, Thal LJ, Amrein R. Meta-analysis of double blind randomized controlled clinical trials of acetyl-L-carnitine versus placebo in the treatment of mild cognitive impairment and mild Alzheimer's disease. Int Clin Psycholpharmacol 2003; 18(2):61–71.

205. Hudson S, Tabet N. Acetyl-l-carnitine for dementia. Cochrane Database Syst Rev 2003; Issue 2. Art. No.: CD003158. DOI: 10.1002/14651858.CD003158.

206. Alesci S, Manoli I, Costello R, et al. Carnitine. The science behind a conditionally essential nutrient. Proceedings of a conference. March 25–26, 2004. Bethesda, Maryland, USA. Ann NY Acad Sci 2004; 1033(ix–xi):1–197.

207. Hirsch AT, Haskal ZJ, Hertzer NR, et al. ACC/AHA 2005 Practice Guidelines for the management of patients with peripheral arterial disease (lower extremity, renal, mesenteric, and abdominal aortic): a collaborative report from the American Association for Vascular Surgery/Society for Vascular Surgery, Society for Cardiovascular Angiography and Interventions, Society for Vascular Medicine and Biology, Society of Interventional Radiology, and the ACC/AHA Task Force on Practice Guidelines (Writing Committee to Develop Guidelines for the Management of Patients With Peripheral Arterial Disease): endorsed by the American Association of Cardiovascular and Pulmonary Rehabilitation; National Heart, Lung, and Blood Institute; Society for Vascular Nursing; TransAtlantic Inter-Society Consensus; and Vascular Disease Foundation. Circulation 2006; 113(11):e463–654.

208. Althuis MD, Jordan NE, Ludington EA, Wittes JT. Dietary chromium supplements and glucose and insulin response: the results of a meta-analysis. Am J Clin Nutr 2002; 76(1):148–55.

209. Qualified Health Claims Subject to Enforcement Discretion: Chromium Picolinate & Diabetes. Washington, DC: Office of Nutritional Products, Labeling, and Dietary Supplements, Center for Food Safety and Applied Nutrition, Food and Drug Administration, 2005. (Accessed February 2, 2007, at www.cfsan.fda.gov/~dms/qhc-sum.html#chromium)

210. Suchowersky O, Gronseth G, Perlmutter J, Reich S, Zesiewicz T, Weiner WJ. Practice parameter: neuroprotective strategies and alternative therapies for Parkinson disease (an evidence-based review). Report of the Quality Standards Subcommittee of the American Academy of Neurology. Neurology 2006; 66:976–82.

211. Parkinson's Disease: National Clinical Guideline for Diagnosis and Management in Primary and Secondary Care. London: National Collaborating Centre for Chronic Conditions, National Institute for Health and Clinical Excellence, 2006. (Accessed August 3, 2007, at www.nice.org.uk/cg035)

212. Schoop R, Klein P, Suter A, Johnston SL. Echinacea in the prevention of induced rhinovirus colds: a meta-analysis. Clin Ther 2006; 28(2):174–83.

213. Shah SA, Sander S, White CM, Rinaldi M, Coleman CI. Evaluation of echinacea for the prevention and treatment of the common cold: a meta-analysis. Lancet Infect Dis 2007; 7(7):473–80.

214. Linde K, Barrett B, Wolkart K, Bauer R, Melchart D. Echinacea for preventing and treating the common cold. Cochrane Database Syst Rev 2006; Issue 1. Art. No.: CD000530. DOI: 10.1002/14651858.CD000530.pub2.

215. Stevinson C, Pittler MH, Ernst E. Garlic for treating hypercholeresterolemia. A meta-analysis of randomized clinical trials. Ann Intern Med 2000; 133:420–9.

216. Silagy CA, Neil HA. A meta-analysis of the effect of garlic on blood pressure. J Hypertens 1994; 12(4):463–8.

217. Warshafsky S, Kamer RS, Sivak SL. Effect of garlic on total serum cholesterol. A meta-analysis. Ann Intern Med 1993; 119(7 Pt 1):599–605.

218. Mulrow C, Lawrence V, Ackermann R, et al. Garlic: Effects on Cardiovascular Risks and Disease, Protective Effects Against Cancer, and Clinical Adverse Effects. (Prepared by San Antonio Evidence-based Practice Center, under Contract No. 290-97-0012). Rockville, MD: Agency for Healthcare Quality and Research, 2000.

219. Jepson RG, Kleijnen J, Leng GC. Garlic for peripheral arterial occlusive disease. Cochrane Database Syst Rev 1997; Issue 2. Art. No.: CD000095. DOI: 10.1002/14651858.CD000095.

220. Fleischauer AT, Poole C, Arab L. Garlic consumption and cancer prevention: meta-analyses of colorectal and stomach cancers. Am J Clin Nutr 2000; 72(4):1047–52.

221. Fletcher B, Berra K, Ades P, et al. Managing abnormal blood lipids: a collaborative approach. Circulation 2005; 112(20):3184–209.

222. Hopfenmuller W. Evidence for a therapeutic effect of Ginkgo biloba special extract. Meta-analysis of 11 clinical studies in patients with cerebrovascular insufficiency in old age. Arznei-mittelforschung 1994; 44(9):1005–13.

223. Birks J, Grimley EV, Evans J. Ginkgo biloba for cognitive impairment and dementia. Cochrane Database Syst Rev 2007; Issue 2. Art. No.: CD003120. DOI: 10.1002/14651858.CD003120.pub2.

224. Horsch S, Walther C. Ginkgo biloba special extract EGb 761 in the treatment of peripheral arterial occlusive disease (PAOD) – a review based on randomized, controlled studies. Int J Clin Pharmcol Ther 2004; 42(2):63–72.

225. Pittler MH, Ernst E. Ginkgo biloba extract for the treatment of intermittent claudication: a meta-analysis of randomized trials. Am J Med 2000; 108(4):276–81.

226. Moher D, Pham B, Ausejo M, Saenz A, Hood S, Barber GG. Pharmacological management of intermittent claudication: a meta-analysis of randomised trials. Drugs 2000; 59(5):1057–70.

227. Schneider B. Ginkgo biloba extract in peripheral arterial diseases. Meta-analysis of controlled clinical studies. Arzneimittelforschung 1992; 42(4):428–36.

228. McAlindon TE, LaValley MP, Gulin JP, Felson DT. Glucosamine and chondroitin for treatment of osteoarthritis: a systematic quality assessment and meta-analysis. JAMA 2000; 283(11):1469–75.

229. Richy F, Bruyere O, Ethgen O, Cucherat M, Henrotin Y, Reginster JY. Structural and symptomatic efficacy of glucosamine and chondroitin in knee osteoarthritis: a comprehensive meta-analysis. Arch Intern Med 2003; 163(13):1514–22.

230. Towheed TE, Maxwell L, Anastassiades TP, et al. Glucosamine therapy for treating osteoarthritis. Cochrane Database Syst Rev 2005; Issue 2. Art. No.: CD002946. DOI: 10.1002/14651858.CD002946.pub2.

231. Efficacy of Glucosamine and Chondroitin Sulfate May Depend on Level of Osteoarthritis Pain: Press Release, February 22, 2006. (Accessed December 10, 2007, at nccam.nih.gov/news/2006/022206.htm)

232. Letter Regarding the Relationship Between the Consumption of Glucosamine and/or Chondroitin Sulfate and a Reduced Risk of: Osteoarthritis; Osteoarthritis-related Joint Pain, Joint Tenderness, and Joint Swelling; Joint Degeneration; and Cartilage Deterioration (Docket No. 2004P-0059). Washington, DC: Office of Nutritional Products, Labeling, and Dietary Supplements, Center for Food Safety and Applied Nutrition, Food and Drug Administration, 2004. (Accessed October 5, 2007, at www.cfsan.fda.gov/~dms/qhcosteo.html)

233. Coates PM, Blackman MR, Cragg GM, Levine M, Moss J, White JD, eds. Encyclopedia of Dietary Supplements. New York, NY: Marcel Dekker, 2007.

234. Taubert D, Roesen R, Schomig E. Effect of cocoa and tea intake on blood pressure: a meta-analysis. Arch Intern Med 2007; 167(7):626–34.

235. Seely D, Mills EJ, Wu P, Verma S, Guyatt GH. The effects of green tea consumption on incidence of breast cancer and recurrence of breast cancer: a systematic review and meta-analysis. Integr Cancer Ther 2005; 4(2):144–55.

236. Sun CL, Yuan JM, Koh WP, Yu MC. Green tea, black tea and breast cancer risk: a meta-analysis of epidemiological studies. Carcinogenesis 2006; 27(7):1310–5.

237. Sun CL, Yuan JM, Koh WP, Yu MC. Green tea, black tea and colorectal cancer risk: a meta-analysis of epidemiologic studies. Carcinogenesis 2006; 27(7):1301–9.

238. Qualified Health Claims: Letter of Denial – Green Tea and Reduced Risk of Cardiovascular Disease (Docket No. 2005Q-0297). Washington, DC: Office of Nutritional Products, Labeling,

and Dietary Supplements, Center for Food Safety and Applied Nutrition, Food and Drug Administration, 2006. (Accessed September 25, 2006, at www.cfsan.fda.gov/~dms/qhcgtea2. html)

239. Qualified Health Claims Subject to Enforcement Discretion: Green Tea & Cancer. Washington, DC: Office of Nutritional Products, Labeling, and Dietary Supplements, Center for Food Safety and Applied Nutrition, Food and Drug Administration, 2005. (Accessed February 2, 2007, at www.cfsan.fda.gov/~dms/qhc-sum.html#gtea)

240. Siebert U, Brach M, Sroczynski G, Berla K. Efficacy, routine effectiveness, and safety of horse-chestnut seed extract in the treatment of chronic venous insufficiency. A meta-analysis of randomized controlled trials and large observational studies. Int Angiol 2002; 21(4):305–15.

241. Pittler MH, Ernst E. Horse chestnut seed extract for chronic venous insufficiency. Cochrane Database Syst Rev 2006; Issue 1. Art. No.: CD003230. DOI: 10.1002/14651858.CD003230.pub3.

242. Chong EW, Wong TY, Kreis AJ, Simpson JA, Guymer RH. Dietary antioxidants and primary prevention of age related macular degeneration: systematic review and meta-analysis. BMJ 2007; 335(7623):755.

243. Kavanaugh CJ, Trumbo PR, Ellwood KC. The U.S. Food and Drug Administration's evidence-based review for qualified health claims: tomatoes, lycopene, and cancer. J Natl Cancer Inst 2007; 99(14):1074–85.

244. Herxheimer A, Petrie KJ. Melatonin for the prevention and treatment of jet lag. Cochrane Database Syst Rev 2002; Issue 2. Art. No.: CD001520. DOI: 10.1002/14651858.CD001520.

245. Buscemi N, Vandermeer B, Pandya R, et al. Melatonin for Treatment of Sleep Disorders. (Prepared by University of Alberta Evidence-based Practice Center, under Contract No. 290-02-0023). Rockville, MD: Agency for Healthcare Quality and Research, 2004.

246. Brzezinski A, Vangel MG, Wurtman RJ, et al. Effects of exogenous melatonin on sleep: a meta-analysis. Sleep Med Rev 2005; 9:41–50.

247. Buscemi N, Vandermeer B, Hooton N, et al. Efficacy and safety of exogenous melatonin for secondary sleep disorders and sleep disorders accompanying sleep restriction: meta-analysis. BMJ 2006; 332:385–93.

248. Jacobs BP, Dennehy C, Ramirez G, Sapp J, Lawrence VA. Milk thistle for the treatment of liver disease: a systematic review and meta-analysis. Am J Med 2002; 113(6):506–15.

249. Mulrow C, Lawrence V, Jacobs B, et al. Milk Thistle: Effects on Liver Disease and Cirrhosis and Clinical Adverse Effects. (Prepared by San Antonio Evidence-based Practice Center, under Contract No. 290-97-0012). Rockville, MD: Agency for Healthcare Quality and Research, 2000.

250. Rambaldi A, Jacobs BP, Iaquinto G, Gluud C. Milk thistle for alcoholic and/or hepatitis B or C virus liver diseases. Cochrane Database Syst Rev 2005; Issue 2. Art. No.: CD003620. DOI: 10. 1002/14651858.CD003620.pub2.

251. Tamayo C, Diamond S. Review of clinical trials evaluating safety and efficacy of milk thistle (Silybum marianum [L.] Gaertn.). Integr Cancer Ther 2007; 6(2):146–57.

252. Chen JT, Wesley R, Shamburek RD, Pucino F, Csako G. Meta-analysis of natural therapies for hyperlipidemia: plant sterols and stanols versus policosanol. Pharmacotherapy 2005; 25(2):171–83.

253. Qualified Health Claims Subject to Enforcement Discretion: Phosphatidylserine & Cognitive Dysfunction and Dementia. Washington, DC: Office of Nutritional Products, Labeling, and Dietary Supplements, Center for Food Safety and Applied Nutrition, Food and Drug Administration, 2003. (Accessed February 2, 2007, at www.cfsan.fda.gov/~dms/qhc-sum. html#phosphat)

254. Boyle P, Robertson C, Lowe F, Roehrborn C. Updated meta-analysis of clinical trials of Serenoa repens extract in the treatment of symptomatic benign prostatic hyperplasia. BJU Int 2004; 93(6):751–6.

255. Wilt TJ, Ishani A, Rutks I, MacDonald R. Phytotherapy for benign prostatic hyperplasia. Public Health Nutr 2000; 3(4A):459–72.

256. Wilt T, Ishani A, Stark G, MacDonald R, Mulrow C, Lau J. Serenoa repens for benign prostatic hyperplasia. Cochrane Database Syst Rev 2000; Issue 3. Art. No.: CD001423. DOI: 10.1002/14651858.CD001423:CD001423.

257. Chapter 1: Diagnosis and treatment recommendations. American Urological Association, 2003. (Accessed August 3, 2007, at www.auanet.org/guidelines/main_reports/bph_management/chapt_1_appendix.pdf)

258. Berges R. The impact of treatment on lower urinary tract symptoms suggestive of benign prostatic hyperplasia (LUTS/BPH) progression. Eur Urol Suppl 2004; 3(4):12–7.

259. Bressa GM. S-adenosyl-l-methionine (SAMe) as antidepressant: meta-analysis of clinical studies. Acta Neurol Scand Suppl 1994; 154:7–14.

260. Hardy M, Coulter I, Morton SCPD, et al. S-Adenosyl-L-Methionine for Treatment of Depression, Osteoarthritis, and Liver Disease. (Prepared by Southern California Evidence-Based Practice Center, under Contract No. 290-97-0001). Rockville, MD: Agency for Healthcare Research and Quality, 2006.

261. Frezza M. A meta-analysis of therapeutic trials with ademetionine in the treatment of intrahepatic cholestasis. Ann Ital Med Int 1993; 8(Suppl):48S–51S.

262. Rambaldi A, Gluud C. S-adenosyl-L-methionine for alcoholic liver diseases. Cochrane Database Syst Rev 2006; Issue 2. Art. No.: CD002235. DOI: 10.1002/14651858.CD002235.pub2.

263. Reynolds K, Chin A, Lees KA, Nguyen A, Bujnowski D, He J. A meta-analysis of the effect of soy protein supplementation on serum lipids. Am J Cardiol 2006; 98(5):633–40.

264. Anderson JW, Johnstone BM, Cook-Newell ME. Meta-analysis of the effects of soy protein intake on serum lipids. N Eng J Med 1995; 333:276–82.

265. Zhan S, Ho SC. Meta-analysis of the effects of soy protein containing isoflavones on the lipid profile. Am J Clin Nutr 2005; 81(2):397–408.

266. Balk E, Chung M, Chew P, et al. Effects of Soy on Health Outcomes. (Prepared by Tufts-New England Medical Center Evidence-based Practice Center under Contract No. 290-02-0022). Rockville, MD: Agency for Healthcare Research and Quality, 2005.

267. Taku K, Umegaki K, Sato Y, Taki Y, Endoh K, Watanabe S. Soy isoflavones lower serum total and LDL cholesterol in humans: a meta-analysis of 11 randomized controlled trials. Am J Clin Nutr 2007; 85(4):1148–56.

268. Zhuo XG, Melby MK, Watanabe S. Soy isoflavone intake lowers serum LDL cholesterol: a meta-analysis of 8 randomized controlled trials in humans. J Nutr 2004; 134(9):2395–400.

269. Weggemans RM, Trautwein EA. Relation between soy-associated isoflavones and LDL and HDL cholesterol concentrations in humans: a meta-analysis. Eur J Clin Nutr 2003; 57(8):940–6.

270. Nelson HD, Vesco KK, Haney E, et al. Nonhormonal therapies for menopausal hot flashes: systematic review and meta-analysis. JAMA 2006; 295(17):2057–71.

271. Kasper S, Dienel A. Cluster analysis of symptoms during antidepressant treatment with Hypericum extract in mildly to moderately depressed out-patients. A meta-analysis of data from three randomized, placebo-controlled trials. Psychopharmacology (Berlin) 2002; 164(3):301–8.

272. Kim HL, Streltzer J, Goebert D. St. John's wort for depression: a meta-analysis of well-defined clinical trials. J Nerv Ment Dis 1999; 187(9):532–8.

273. Röder C, Schaefer M, Leucht S. Meta-analysis of effectiveness and tolerability of treatment of mild to moderate depression with St. John's Wort. Fortschr Neurol Psychiatr 2004; 72(6):330–43.

274. Whiskey E, Werneke U, Taylor D. A systematic review and meta-analysis of Hypericum perforatum in depression: a comprehensive clinical review. Int Clin Psychopharmacol 2001; 16(5):239–52.

275. Linde K, Mulrow CD, Berner M, Egger M. St. John's Wort for depression. Cochrane Database Syst Rev 2005; Issue 2. Art. No.: CD000448. DOI: 10.1002/14651858.CD000448.pub2.

276. Depression: management of depression in primary and secondary care, NICE Clinical Guideline 23. London: National Collaborating Centre for Mental Health, National Institute for Health and Clinical Excellence, 2004. (Accessed August 3, 2007, at guidance.nice.org.uk/CG23).`

29

Minimizing the Impact of Complex Emergencies on Nutrition and Geriatric Health: Planning for Prevention is Key

Connie Watkins Bales and Nina Tumosa

Key Points

- Complex emergencies (CEs) can occur anywhere and are defined as crisis situations that greatly elevate the risk to nutrition and overall health (morbidity and mortality) of older individuals in the affected area.
- In urban areas with high population densities and heavy reliance on power-driven devices for day-to-day survival, CEs can precipitate a rapid deterioration of basic services that threatens nutritionally and medically vulnerable older adults.
- The major underlying threats to nutritional status for older adults during CEs are food insecurity, inadequate social support, and lack of access to health services.
- The most effective strategy for coping with CEs is to have detailed, individualized pre-event preparations. When a CE occurs, the immediate relief efforts focus on establishing access to food, safe water, and essential medical services.

Key Words: Disaster relief; food insecurity; humanitarian crisis; undernutrition

29.1 INTRODUCTION AND DEFINITIONS

The most common issues impacting on the nutritional well-being of elderly persons are comprehensively addressed in the preceeding 28 chapters of this edition of the *Handbook of Clinical Nutrition and Aging*. This chapter focuses on a different type of concern, one that can overshadow all other threats to health when a serious disaster strikes. That subject is the welfare of aged persons when catastrophic events pose a direct (or indirect) threat to nutrition and health *(1,2)*. While there is a large body of literature on the health impact of natural and man-made disasters (e.g., droughts, floods, military conflicts) and associated long-term food shortages in the third world, surprisingly little information is available about the short and intermediate-term

From: *Nutrition and Health: Handbook of Clinical Nutrition and Aging, Second Edition*
Edited by: C. W. Bales and C. S. Ritchie, DOI 10.1007/978-1-60327-385-5_29,
© Humana Press, a part of Springer Science+Business Media, LLC 2009

consequences of emergency situations in developed countries. In these situations, high population densities and heavy reliance on power-driven devices for day-to-day survival (e.g., electrical power for mass transit, elevators to reach living quarters, medical devices, and refrigeration of foods and medicines) can accelerate the speed with which a catastrophic, health-threatening situation develops. In 2005, the plight of the elderly evacuees from New Orleans (pre-storm population approaching 485,000) following Hurricane Katrina provided a dramatic demonstration of how essential services can rapidly deteriorate in a well-developed, highly populated urban environment following a major disaster and place older individuals in eminent mortal danger.

29.1.1 Definitions

In order to lay the foundation for this discussion, we begin with some definitions (See Table 29.1). While terms like "disaster relief" and "humanitarian crisis" may be

Table 29.1
Glossary of terms

Complex emergencies	Any of a number of crisis situations that greatly elevate the health risk of individuals in the affected area; examples are natural disasters like floods and earthquakes; urban health emergencies like fires, epidemics, and blackouts; and terrorist acts like massive bombings or poisonings of food or water supplies. Resolution of these emergencies requires collaboration between multiple groups.
Acute protein/calorie malnutrition (PCM)	PCM or "wasting" is associated with recent rapid weight loss, i.e., as in emergency situations (as opposed to chronic malnutrition).
Chronic energy deficiency (CED)	An intake of energy that is below the minimum requirement for a period of several months or years. In order to achieve energy steady state, the energy expenditure must drop to match the low intake, ultimately leading to underweight and low levels of physical activity.
Nutritional rehabilitation	Restoration of weight and healthy nutrition through the provision of appropriate foods based on established protocols.
Food rations	A shelf-stable pre-packaged dry ration that meets minimum daily intake recommendations for calories and other nutrients. Used to temporarily meet critical nutritional needs when food supply is inadequate. Examples: Meals Ready to Eat or MREs (1,250 kcal) are often distributed in complex emergencies in the United States; General food rations or GFRs (2,100 kcal) are distributed in many countries in sub-Saharan Africa.

(continued)

Table 29.1
(continued)

Complementary food ration	A complementary ration to the general food ration is sometimes provided. Typically, it consists of fresh fruit and vegetables, condiments, tea, etc. It is especially appropriate when the population of concern is completely reliant on food assistance.
"Wet" feeding	Food rations prepared and cooked on-site as opposed to rations that are taken home for preparation in the household (dry rations).
Fortification of foods	Typically, fortified foods have had supplemental vitamins and/or minerals added.
Hunger	The uneasy or painful sensation caused by lack of food.
Malnutrition	The medical condition caused by an improper or insufficient diet that can refer to undernutrition resulting from inadequate consumption, poor absorption, or excessive loss of nutrients. Malnutrition results from an inappropriate amount or quality of nutrient intake over a long period of time.
Food insecurity	The inability to obtain nutritionally adequate and safe food; or the inability to obtain it in socially acceptable ways
Food insufficiency	Inadequate amount of food intake due to a lack of food.
Epidemics and pandemics	An epidemic is a disease outbreak that affects numbers of the population in excess of what would normally be expected in a defined community, geographical area, or season. A pandemic refers to this type of disease outbreak that is occurring over a wide geographic area and affecting an exceptionally high proportion of the population.

Source: Borrel, A. Addressing the nutritional needs of older people in emergency situations in Africa: Ideas for action. HelpAge International Africa Regional Development Centre, Westlands, Nairobi, 2001.

more familiar, the most broadly acceptable term for these threatening situations is "complex emergency" *(2)*. Complex emergencies (CEs) can occur anywhere and are defined as any of a number of crisis situations that greatly elevate the risk to nutrition and overall health of individuals in the affected area. Examples include natural disasters like floods and earthquakes, urban health emergencies like fires, epidemics and blackouts, and terrorist acts like massive bombings or poisonings of food or water supplies (see Table 29.2). CEs were originally associated with wars, genocide, and political strife, where innocent civilians were forced to endure loss of access to shelter, food, appropriate clothing, and timely medical care. Such emergencies have traditionally been associated with populations in developing nations, not those in the so-called developed countries. However, with increasing

Table 29.2

Examples of complex emergencies that impact nutrition and health

Natural disasters and extreme weather

    Hurricanes, tornadoes, floods, tsunamis, tidal waves

    Earthquakes, mudslides

    Ice storms and blizzards

    Heat (#1 killer of the elderly)

Unintentional and intentional man-made and population-related emergencies

    Fires and structural collapse

    Terrorism and bio-terrorism

        Explosions and implosions

        Intentional contamination of food and/or water supplies

    Epidemic or pandemic infectious disease cut break

globalization of the world's societies and economies and news coverage documenting world events, it has become clear that CEs can and do occur in both developed and developing world locations.

Nutritional risk is commonly elevated in CEs and is most likely to occur when the crisis is protracted or recurrent. Table 29.1 includes definitions for factors related to inadequate food intake (e.g., food insecurity, hunger), the resulting nutritional problems (e.g., malnutrition, acute protein/calorie malnutrition), and terms used to discuss interventions for undernutrition (e.g., food rations, nutritional rehabilitation).

29.1.2 Food Insecurity and Federal Food Assistance Programs

Even in the absence of a crisis, older persons are well recognized to be at greater risk than the remainder of the adult population for food insecurity and hunger. Some of the many factors that contribute to increased nutritional vulnerability of older adults are listed in Table 29.3. In 2001, food insecurity and hunger affected at least 1.4 million households in the United States that contained older members *(3)*. People in 20% of those households also experienced hunger, in addition to food insecurity. Most of these older persons are suffering from food insecurity due to lack of income or due to their place of residence. Residents of the South are more apt to experience food insecurity, as are residents of cities and all elders who live alone *(3)*.

Table 29.3

Risk factors for food insecurity in older persons

• Household composition	• Lacking access to nutritionally adequate diets
• Poverty	
• Functional impairments	• Depression
• Social isolation	• Reduction in taste, smell, sight, touch
• Reduced ability to regulate energy intake	• Poor health
	• Poor dentition

Source: Magkos et al. *(41)*.

Recognizing the day-to-day nutritional vulnerability of its poor and elderly citizens, the U.S. government has a number of programs in place to provide assistance to elders at risk for food insecurity and hunger. Mandated by the Older American's Act, the Elderly Nutrition Program (ENP) provides a minimum of one-third of the daily calories required by recipients through daily meals and nutrition services to people aged 60 or older in group settings, such as senior centers and churches, or in the home, through home-delivered meals. The ENP provides an average of 1 million meals per day to older Americans. These meals are targeted toward highly vulnerable elderly populations, including the very old, people living alone, people below or near the poverty line, minority populations, and individuals with significant health conditions or physical or mental impairments. On an average the meals generously meet the RDA requirements, supplying more than 33% of the Recommended Dietary Allowances (RDAs) for key nutrients, thus significantly increasing the dietary intakes of ENP participants. The meals are also "nutrient dense", that is, they provide high ratios of key nutrients per calories. The most recent evaluation of the ENP program occurred in 1996 and was conducted by Mathematica Policy Research, Inc. (www.mathematica-mpr.com/nutrition/enp.asp). The resulting report clearly confirms that the ENP program recipients are at nutritional risk. It was found that between 80 and 90% of participants had incomes below 200% of the poverty level (twice the rate for the overall elderly population in the United States). More than twice as many Title III participants lived alone, compared with the overall elderly population. Approximately, two-thirds of the participants were either overweight or underweight, placing them at increased risk for nutrition and health problems. Title III home-delivered participants had more than twice as many physical impairments, compared with the overall elderly population. Although (and perhaps because) the success of the ENP program is well recognized, 41% of Title III ENP service providers have waiting lists for home-delivered meals, suggesting a significant unmet need for these meals. It would appear that even in times of relative calm and prosperity for most Americans, there are elderly citizens who are persistently in a state of nutritional crisis.

29.2 COMPLEX EMERGENCIES THAT THREATEN HEALTH

When nutritionally and medically vulnerable older persons encounter a complex emergency, there is an increase in morbidity and mortality rates. This is due to both short-term insufficient nutrition and the resulting long-term increased mental stress and disability, decreased resistance to infection, and exacerbation of chronic diseases (4), all of which make obtaining proper nutrition more difficult in a cyclic pattern. Many different types of CEs produce similar challenges. The consequences of a shortage of edible food and/or potable water, regardless of the type of emergency that produced that shortage, are multifold and can lead to increased physical and mental harm to older people (5). Reduced access to essential medical care heightens the immediate risk. A more extensive listing of the immediate impact of various complex emergencies and the resulting nutritional and health consequences is shown in Table 29.4.

Table 29.4
Immediate impact and nutritional/health consequences of complex emergencies

Immediate impact	Consequence
Loss of access to safe, adequate water supply	Dehydration; increased risk of delirium: inability to administer medications or keep sterile medical materials
Loss of safe, adequate food supply	Acute protein calorie malnutrition
Lack of access to special foods, nutritional products	Acute undernutrition due to loss of availability of pureed foods, tube feeding formulas, thickened liquids, other special foods
Loss of access to life-sustaining medical care, e.g., insulin injections, dialysis, respiratory support	Deteriorating medical condition, renal toxicity, hyperglycemia, etc.
Emotional trauma	Increased confusion; exacerbated dementia symptoms; poor food intake even if food is available
Loss of basic utilities	Extremes of heat/cold; inability to preserve foods and medications; inability to prepare foods
Damage to or loss of housing	Functional limitations, dysmobility or secondary injuries due to lack of lighting, safe environment

The likelihood of having to provide care for older persons during a CE is greater than one might think at first. As previously noted, Table 29.2 provides a list of common CEs that have the potential to cause nutrition-related health risks. The impact of these crises on the nutritional state and overall health of older adults is discussed in more detail in the following sections.

29.2.1 Natural Disasters and Extreme Weather

29.2.1.1 Hurricanes, Tornadoes, and Floods

The 2005 hurricane season in the United States, most notably Hurricanes Rita and Katrina, left no doubt that older persons continue to be disproportionately affected by hurricanes *(6,7)* just as they were with Hurricane Andrew in 1992 *(8)*. Older Floridians who were affected by Hurricane Charley in 2004 found that the hurricane not only disrupted their quality of life but also disrupted their medical care *(9)*. Persons with pre-existing conditions such as diabetes mellitus, heart disease, and physical disabilities were especially affected. Approximately one-third of the older residents in the area had a worsening of their conditions post-hurricane, including a lack of access to prescription medicine and loss of routine medical care for pre-existing conditions. Medically related deaths were linked to the loss of power (resulting in loss of access to oxygen) and to exacerbation of cardiac disease. Hurricane Iniki in Hawaii and the Great Hanshin-Awaji Earthquake in Japan were associated with an increase in the rate of diabetes mellitus-associated deaths for a year following the disaster *(10,11)*.

In a study of residents in the high-impact area of Hurricane Andrew, one-third of persons had high levels of PTSD *(12)*, which was attributed to variables such as property damage, exposure to life-threatening situations, and injury.

Tornadoes, while typically more limited in the size of the area affected than a hurricane, are often even more physically destructive. Although no research has been published on their specific effects on physical and mental health, it is well recognized that tornadoes can lead to many of the same dangers noted for hurricanes; the disruption of home care services and meal delivery to homebound elderly persons are of concern. The situation can become life threatening not only to the older persons who are critically dependent on these services but also to their dedicated care providers who often risk much to ensure the delivery of food and medical care to their clients (personal communication from Area Agency on Aging of Southwestern Illinois grantees to NT).

Floods are a relatively common disaster and are often associated with earthquakes or hurricanes. Besides trauma and drowning, the most common conditions associated with floods are an increase in gastrointestinal symptoms. Increased preventable conditions following the crisis include gastroenteritis *(13)*, acute respiratory infections including asthma *(10)*, and increased post-traumatic stress which can persist for years after the event *(14)*.

29.2.1.2 EARTHQUAKES

In the aftermath of an earthquake, as with the other natural disasters already mentioned, access to basic life-sustaining nutrients and hydration as well as to basic and specialized medical care may be partially or completely disrupted. Due to the magnitude and scope of the destruction that occur with a major earthquake, the restoration of infrastructure to fully support the inhabitants of the region may take months or even years to be accomplished. Earthquakes result in a three-fold increase in deaths from myocardial infarction, a doubling of the frequency of strokes, increased blood pressure levels, and increased coagulability of blood *(15,16)*. Increased rates of cardiac arrests occurring after loss of power *(17)* and deaths due to increased incidence of coronary heart disease *(18)* and myocardial infarctions *(19,20)* are also reported. Deterioration of mental health occurs and post-traumatic stress is also prevalent *(21,22)*. Emotional stress can persist for months *(23,21)*. In particular, the displacement of elderly persons from their places of residence and their social and medical supports can have a dramatic negative effect on health and quality of life (See Fig. 29.1). Displacement following a CE has been linked with a significant increase in mortality rates *(15,16)*. The confusion of the displacement, as well as loss of access to appropriate diet and medications, prevents older individuals from monitoring and treating their medical conditions. Inappropriate diet has been directly linked to decreased glycemic control and increased mortality in diabetic patients following an earthquake *(11)*.

Fig. 29.1 In emergency situations, evacuations displace elderly persons from their social and medical support systems and can negatively impact health and quality of life.

29.2.1.3 EXTREME HEAT AND COLD

The type of naturally occurring CE that is most threatening for older persons in terms of numbers affected each year comes during periods of temperature extremes, especially heat waves, claiming about 400 lives annually in the United States alone, more than the deaths caused by all other disasters combined. At greatest risk are poor persons who live in inner cities, those with chronic illnesses, and those home-bound. Heat disasters are often aggravated by power outages, which prevent people from keeping cool, bathing properly, and storing food at proper temperatures *(24)*. In the 1993 heat wave in Philadelphia, there was a 26% increase in total mortality, with a 98% increase in cardiovascular deaths, particularly in those persons over

65 years of age *(25)*. In France, during the period 1971–2003, there were six major heat waves, resulting in thousands of deaths; the mortality ratios increased with age after 55 years and in the over age 75 years cohort; the death rate was higher for women than for men *(26)*.

Although little research has been published about the health effects of ice storms and blizzards, the loss of power leaves older persons stranded at home, increasing the risk for ingestion of inadequate calories and inappropriately prepared food and/ or spoiled food. The risk of exposure combined with the risk of house fires or carbon monoxide poisoning due to use of unsafe heating devices pose serious threats at a time when emergency services may not available due to the extreme weather conditions.

29.2.2 *Unintentional and Intentional Man-Made Disasters*

29.2.2.1 FIRES

Fires increase the extent of cardio-respiratory problems, which results in exacerbation of chronic diseases *(27)*. People who already suffer from mental health problems or medically unexplained physical symptoms *(28)* and gastrointestinal morbidity *(29)* can develop an exacerbation of these problems *(16,29)* once they become a victim of a fire. Even when no injuries result, fires almost certainly force displacement of their victims, adversely affecting quality of life and manifestation of chronic diseases.

29.2.2.2 EPIDEMICS, PANDEMICS, AND UNINTENTIONAL FOOD BORNE ILLNESS OUTBREAKS

A serious infectious global pandemic is one of the most threatening of all complex emergencies, and calls back memories of the most devastating infectious disease outbreak on record, the Great Flu Epidemic of 1918–1919, which killed an estimated 20–40 million people worldwide. The spread of this epidemic was linked to the trans-global transportation of soldiers during World War I. Today, world travel and the importation of foods and other products are very common. Thus, in the event of a serious epidemic in one country, there is a high likelihood of quick transmission to others. The outbreak of SARS, a severe acute respiratory illness caused by a coronavirus, was first reported in Asia in February 2003 and spread to more than two dozen countries in North America, South America, Europe, and Asia (sickening 8,098 and killing 774) before the global outbreak was contained (http://www.cdc.gov/NCIDOD/SARS/factsheet.htm). In recognition of the severe strain that a major disease outbreak can place on health systems, the World Health Organization (WHO) advocates for an "integrated global alert and response system for epidemics and other public health emergencies" that allows for "a collective approach to the prevention, detection, and timely response" for these emergencies (http://www.who.int/csr/en/). The WHO is currently coordinating the global response to human cases of H5N1 avian influenza (bird flu) with regards to the threat of a future influenza pandemic.

A widespread illness or intoxication from a food source could also threaten nutritional and overall health. While these outbreaks are typically limited in scope and short lived, the potential for more widespread and dangerous effects exists due to the centralized nature of the US food distribution chain and the clustering of very large populations into a small geographical area. (See more on this topic in Section 29.2.2.3.)

29.2.2.3 TERRORISM AND BIO-TERRORISM

While other complex emergencies produce far more damage and deaths each year than are caused by terrorism, the destruction of the Twin Towers in New York City and a portion of the Pentagon in Washington DC on September 11, 2001, focused the attention of Americans upon the potentially devastating effects of an intentional man-made disaster. The development of the Department of Homeland Security was a tangible product of the national response to implied threats of bio-terrorism.

A terrorist attack such as one causing explosions and collapse of buildings would result in the interruption of basic living functions in a manner similar to previously discussed emergencies like earthquakes, tornadoes, or fires. Disruptions to necessities of daily living and loss of power and access to medical care would be major concerns. A bioterrorist attack would have very different potential consequences for the well-being of the elderly, potentially causing widespread illness and/or hunger and dehydration. The propagation of an illness over a wide geographical area could be lethal for a substantial number of older adults, who are typically among the most medically vulnerable. During the anthrax attacks in 2001, all emergent cases involved adults over 50 years old, with the one fatal case affecting a 94-year-old woman *(30)*. Intentional contamination of food or water supplies with a toxin or infectious agent also has the potential to cause an outbreak of poisonings or illness over a wide geographical area. In this situation, the outbreak could be slow and/or diffuse and the cause difficult to ascertain, delaying the recognition and treatment of the problem. For example, in 2006, bagged spinach contaminated (unintentionally) by *Escherichia coli* infected over 200 Americans (killing three) in 26 states before the strain was isolated and eradicated. Similarly, intentional waterborne diseases or toxins would be difficult to detect and could impact a vulnerable population more severely than a healthy population, due to delayed recognition and reporting of the contamination *(31)*. In the case of deliberate food/water contamination, nutritional health is affected directly (by reducing the availability of safe food and water) as well as indirectly (by the symptoms of illness and the reduced access to an over-burdened medical care system). In fact, the deliberate poisoning of food has already occurred in the United States, when in 1984 members of the Rajneesh religious cult contaminated salad bars in The Dalles, Oregon, with *Salmonella typhimurium*. Though it was only a trial run for a more extensive attack that was planned to disrupt local elections later that year, the contamination caused 751 people to develop salmonellosis in a 2-week period. Other isolated examples of intentional food contaminations have also been reported in the United States and Canada *(32)*.

Coping with complex emergencies due to terrorism is for the most part a new challenge, at least in the United States. Despite considerable effort to prepare for these scenarios, our experience in dealing with the aftermath is limited, yet, unfortunately, our experience is likely to grow in the future. Experts warn that a major terrorist attack on the United States is very likely (29–50%) to occur within the next 10 years (CFR Online Debate).

29.2.2.4 Summary

Heat, cold, hurricanes, tornadoes, floods, fires, illness, terrorism, and other disasters endanger health and claim elderly lives. Sometimes the effects are immediate, but more often an increase in morbidity and mortality occurs progressively after the disaster as survivors experience a continued decrease in the quality of life and increased nutritional risk due to displacement and a loss of basic resources. These events result in increased disability, which further impairs the ability of older persons to maintain access to safe food and water and sustain proper nutrition and hydration, and so the spiral continues downward. Recovery from food insecurity and poor nutrition is more difficult for persons who are poor, socially isolated, cognitively impaired, and/or old. The more risk factors people possess, the faster their decline.

29.3 MINIMIZING NUTRITIONAL AND HEALTH RISKS DUE TO CEs

All of the disasters described in this chapter threaten nutritional and metabolic health because they disrupt access to food, water, and vital medical treatment *(33)*. Older persons with pre-existing chronic conditions are particularly vulnerable to these disruptions. Preparation for and resolution of the aftermath of these emergencies require collaboration between multiple stakeholders and takes time. There are no easy fixes to CEs.

29.3.1 Conceptual and Programmatic Overview

The underlying causes of malnutrition in older adults during CEs are (1) insufficient household food security, (2) inadequate social and care environments, and (3) poor public health and inadequate health services *(2)*. The basis for current governmental and humanitarian responses to nutritional crises builds on lessons learned in the earliest organized relief efforts (circa 1940–1950). During the 1970s, guidelines began to be published following experiences with relief efforts in places like Biafra and Ethiopia *(2)*. In the subsequent decades, the experiences of various crises have progressively shaped what are, today, the characteristic challenges, and avenues of support available to older adults who are caught in CE situations in any given country. With increasing recognition that the elderly are uniquely vulnerable to CEs, efforts are underway to develop specific recommendations and resources for this population group. Table 29.5 lists some of the resources available, along with web links. HelpAge International (www.helpage.org) is a global network of more than 70 not-for-profit organizations in 50 countries who are working for improvements in the lives of older people. This group has published a manual of guidelines for best practice during disasters and humanitarian crises (See Table 29.5). The Sphere Project Minimum Standards in Disaster Response project (http://www. sphereproject.org/content/view/27/84) advocates for the use of community-based systems to implement the

Table 29. 5
Resources (Web sites and links to publications and bulletins)

(Title or description, followed by web link)

American Red Cross
www.redcross.org/services/disaster/0,1082,0_217_,00.html

Federal Emergency Management Agency (FEMA)
www.fema.gov/areyouready/
Food and Water in an Emergency
http://www.fema.gov/pdf/library/f&web.pdf

Food and Nutrition Service (FNS) of the USDA
http://www.fns.usda.gov/disasters/disaster.htm

HelpAge International
Addressing the Nutritional Needs of Older People in Emergency Situations
http://www.helpage.org/Resources/Manuals

National Recommendations for Disaster Food Handling
foodsafety.ifas.ufl.edu/HTML/tn001.htm

Hunger Issue Brief: Hunger and Food Insecurity Among the Elderly
http://www.centeronhunger.org/pdf/Elderly.pdf

Sphere Project
Minimum Standards in Disaster Response:
http://www.sphereproject.org/content/view/27/84

Food Safety Risks
http://www.ific.org/publications/other/consumersguideom.cfm

care of older individuals in these circumstances. In the United States, a number of national organizations, including the Federal Emergency Management Agency (FEMA), The American Red Cross, and various branches of the military take responsibility for rescue and relief efforts following a major CE but the contribution of the private sector to the relief effort is traditionally also a substantial one. This type of broad-based support is necessary but makes it more difficult to consistently implement age-related guidelines for relief efforts once they are in the field. Coordinating the advance preparation efforts for CEs, however, is a more tangible goal.

29.3.2 *Emergency Preparedness in Structured Living Communities*

As is true for almost all health issues, the best way to address the nutritional and related health risks that accompany CEs is to take preventive measures. In the case of nursing homes and assisted living facilities, many states require that these institutions have a substantial reserve food and water supply and that they have a well-delineated disaster and evacuation plan. The specifics of these requirements vary on a state-by-state basis. However, attention to the development of specialized evacuation plans (individualized for resident needs) is more focused since the nursing home-related deaths recorded during the hurricanes on the gulf coast in 2005. Two such incidents included the drowning of 34 nursing home residents in St. Bernard Parish, Louisiana, due to a failure to comply with evacuation orders during Hurricane Katrina, and the bus accident in which 24 Houston, Texas, nursing home residents being evacuated from Hurricane Rita died in a fire that was sparked by mechanical problems and fed by the explosions of the passengers' oxygen tanks.

Beyond the obvious need for institutions and organizations like long-term care and hospice agencies to have detailed plans for evacuations and emergency conditions, there is also a need to identify "at risk" older adults living in the community. This would involve developing registries of "vulnerable populations" of elders based on degree of factors like contact need, predominant special impairment, and predominant life-support supply need, if any. By doing so, vulnerable elders could be easily identified in the event of a disaster and better supplied with assistance. Such registries are currently implemented in some instances (examples are available in California, www.aging.ca.gov, and Florida, www.broward.org/atrisk), but a more systematic approach has yet to be employed. These registries will most likely need to be local in origin and maintenance in order that control of sensitive health data would remain confidential. However, it would be preferable for the structure of the databases to be developed in a uniform format in order to facilitate the sharing of important data across local and regional entities. Once successful programs and examples are created, their implementation by all interested parties should then be straightforward.

29.3.3 Emergency Preparedness at Home

Emergencies require flexibility and the ability to survive changes in regular routines. This flexibility can be easier to achieve if people have a few necessary and familiar objects with them to assist with performing certain everyday chores, such as eating properly, taking medications, and changing into clean clothes. In order to assist people in getting prepared for the disruptions that inevitably occur during an emergency, the FEMA and The American Red Cross recommend that every family have an emergency preparedness kit that contains food, water, clothing, medical supplies, flashlight, and other supplies that will aid their survival for 3–5 days. By the time recommended objects are placed in a backpack, the entire kit weighs between 45 and 50 pounds. This is clearly too much weight for an older person to handle safely.

29.3.3.1 HELPING OLDER ADULTS TO BE PREPARED: A QUALITY IMPROVEMENT STUDY OF EMERGENCY KITS FOR ELDERS

The Health Resources and Services Administration (HRSA) provided funding to the Gateway Geriatric Education Center of Missouri and Illinois (grant number D31HP70122) for train-the-trainer programming to teach 150 health-care professionals in the spring of 2006 how to create an emergency preparedness kit that was light, compact and specific for older adults. This kit consisted of a small satchel, a flashlight, a photo album (to store copies of prescriptions, insurance cards, evacuation plans, contact phone numbers, and family pictures), a pill box and a pamphlet

Table 29.6
Senior-specific emergency kit contents

• Bottled water	• Emergency contact information
• Family pictures	• FEMA's "Are You Ready?" booklet
• Cash (at least $5)	• Three days to 1 week supply of medicines
• Pet evacuation plan	• Extra pair of glasses and hearing aids and extra batteries
• Identification bracelet	
• Flashlight and batteries	• List of medications and written prescriptions for those medications

introducing the FEMA Web site. The trainees were then taught what other materials should be added to the kit to make it appropriate for a particular individual (Table 29.6). Upon completion of this training each of the 150 trainees received two complete kits, one to use as an example during their subsequent training sessions of other health-care providers and the other to be given to a disadvantaged older person whom they deemed at risk during an emergency. Each participant provided an e-mail address in order to be contacted 1 year following their training to determine the outcomes of their training.

One year after training, the 150 trainees were contacted by e-mail. Twenty-three of the e-mail addresses were no longer valid. Of the remaining 127 trainees, 67 filled out and returned the survey within 2 weeks (53% response rate). An additional 18 surveys were returned after a second e-mail blast (85/127, for a final response rate of 67%).

The survey asked if, as a result of their training, had the trainees:

1. Given the extra kit to an older adult?
2. Determined if that kit had been used during an emergency?
3. Used their own emergency kits for training, and if not, why?
4. Used their own emergency kits during an emergency?

Responses to the quality improvement survey are summarized in Table 29.7. The majority of the trainees (94%) had given the extra kit to an older person and many

Table 29.7
Responses to quality improvement survey

Question: Did you?	No. of yes responses (%)	No. of no responses (%)
1. Did you give the extra kit to an older adult?	120 (94)	7 (6)
2. Was that kit had been used during an emergency?	59 (46)	68 (54)
3. Did you use your own emergency kits to train others?	19 (15)	108 (85)
4. Did you use your own emergency kits during an emergency?	23 (18)	104 (82)

of the respondents indicated that the person was either an older relative or a neighbor. However, few respondents (15%) had provided any training to other health-care providers on how to create these kits. Barriers cited included lack of money to purchase kit contents, lack of commitment or permission from supervisors, lack of time to provide the training, and lack of time for their colleagues to receive training.

The percentage of older adults that were reported to have used their emergency kits by the time of the end point survey was higher than expected (46%), especially given that only 18% of the (younger) trainees reported using their kits. However, a review of the disruptive weather patterns in the 11 counties in eastern Missouri and southwestern Illinois where the trainees (and therefore, presumably of the older adults receiving the extra kits) lived, indicated that three area-wide power outages had occurred between August 2006 and January 2007. All of these three power failures lasted 1–3 weeks, with the rural areas in southwestern Illinois being the last to get power restored each time. Each of these power failures affected at least a half million citizens each time. Numerous cooling or heating stations were set up for older adults, thereby allowing them to evacuate from their homes during the days in August and to receive warm meals during the November and January power failures. Multiple public service announcements encouraged people to evacuate their homes completely until power was restored, so many older adults either moved in with relatives who did have power or went to hotels. Under those conditions, it is reasonable to expect older persons to take their emergency kits with them. Many of the health-care provider trainees reported that they had gone to work daily. A brief second query to 10 trainees who had used their kits and 10 trainees who had not used their kits indicated that both sets had gone to work daily and returned home at night, even if they had no power at home. (These health-care providers worked in facilities with working generators.) Several of those that took their kits with them indicated that the kits provided them with some measure of safety while traveling icy roads in November and January. Those that had not used their kits indicated no perceived change in their normal safety.

This quality improvement study shows that emergency kits for older adults are used during an emergency. Community-dwelling older adults appear to be more vulnerable to weather emergencies than are the health-care providers who care for them, as evidenced by the differences in usage rates of the kits by both groups through three lengthy power outages. Upon review of the barriers that prevented trainees from providing training to other health-care providers, it is possible that it would have been more appropriate to provide train-the-trainer programs to older adults rather than to health-care providers. Peer-to-peer training might have had the added advantage of motivating trainers to find community funding to make kits for distribution because of a greater perceived personal need for the kits.

29.3.4 Intervention Strategies: Providing Aid During and Following a CE

Because every emergency event presents a unique challenge, this section offers general information about coping with the major nutritional concerns, namely shortages of food and water and overall loss of access to social support and health-related resources. Optimal public health and nutrition relief includes a broad range of interventions and needs to utilize strong programmatic interconnections to meet the aforementioned needs.

29.3.4.1 COPING WITH FOOD AND WATER SHORTAGES

In the immediate aftermath of a CE, the supplies of food and water may be extremely limited. In this event, food can be more safely rationed than water. A general guideline is that the minimum adult ration be one well-balanced meal per day, with the utilization of vitamin/mineral supplements, protein drinks, "power bars", or other fortified foods as meal extenders if available. However, water should not be rationed due to the very rapid effects of dehydration. Individuals are advised to drink what is needed today and search for more water on a daily basis. Indicators of dehydration in the elderly differ from those in younger individuals; increased thirst, reduced skin turgor are not reliable markers. Better indicators include tongue dryness, longitudinal tongue furrows, dry mucous membranes of the nose and mouth, eyes that appear sunken, upper body weakness, speech difficulty, and confusion *(34)*.

When there is a loss of power to the home, perishable foods are to be consumed first, followed by foods from the freezer. Frozen foods should be safe to eat for at least 2 days following the power loss. At this point, nonperishable, staple foods would be the only safe source of nutrients.

As conditions stabilize, food aid will begin to become available. The recommended actions to be facilitated for older adults include (1) achieve/improve access to food aid (rations, supplemental feeding programs, etc.); (2) ensure that the rations are easy to prepare and consume; and (3) assure that the rations being used meet the nutritional requirements of older adults *(35)*. The USDA's Food and Nutrition Service (FNS) coordinates with State, local, and voluntary organizations to provide food for shelters and also distributes food packages and authorizes states to issue emergency food stamp benefits to individuals. As part of the National Response Plan, FNS supplies food to disaster relief organizations such as The Red Cross and the Salvation Army for mass feeding or household distribution. These organizations, along with other private donors, support the supply of water and food rations to affected areas.

There are several concerns related to the access and appropriateness of food aid for elderly individuals (again, see resources listed in Table 29.5). Access to the aid is a concern because disabilities and medical problems may prevent elderly individuals from reaching the distribution centers. Another concern is the composition of the food rations, which may not be appropriate in consistency for persons who have dentures or who lack teeth and that may not be adequate in nutritional composition. Food rations vary in composition; not all are developed for the primary purpose of post-CE relief. In the United States, the Meal, Ready-to-Eat (MRE), although first developed for use in the space program and now widely used by the armed forces, is one form of ration that is commonly distributed to civilians who need food following CEs. Having been designed for soldiers in a high activity situation, the MREs are much higher in sodium (5,500 g) and fat (136 g) than is optimal, especially for older adults *(36)*. Likewise, the texture, packaging, and preparation of MREs were not developed with the intention of use by older adults.

In an effort to supplement the nutritional needs of elderly citizens and to meet federal recommendations for increased emergency preparedness, the Administration on Aging (AoA) sought and received special funding to provide shelf stable meals that could be delivered to participants of the home-delivered-meal programs. These meals, which have a shelf life of approximately 16 months, are delivered with instructions to consume them during emergencies when regular home-delivered meal service is disrupted. The program is new so, to date, no evaluations have been done to determine what becomes of those meals (e.g., are they saved for emergencies or eaten to supplement other meals). No policy has been created to determine liability for any sickness caused by consumption of meals that are beyond their expiration date *(personal communication from Area Agency on Aging of Southwestern Illinois and the MidEast Area Agency on Aging to NT)*.

Obtaining adequate food and water is only one step on the road to recovery where elderly persons are vulnerable to food insufficiency. Once food is obtained it must then be stored properly, prepared properly, and then ingested without health risk. In each of these steps, older persons are also at increased risk, compared to the rest of the population. This is because these older persons have additional risk factors for poor nutrition such as functional impairments, social isolation, reduced ability to regulate energy intake, greater susceptibility to depression, decreased ability to taste and smell, poor dentition, and poor health. All of these items (listed in Table 29.3) can lead to malnutrition, if not starvation, in older persons.

29.3.4.2 RE-ESTABLISHMENT OF BASIC SERVICES AND ACCESS TO MEDICAL CARE

Following a CE, the speed with which basic services such as heating/cooling, shelter, and water supply can be restored will be a major factor in the recovery of older persons. Past experience has shown that cold, loss of mobility, access to services, and psychological stress and trauma are some of the most important factors contributing to undernutrition in older people following a CE *(37,38)*. In particular, the loss of social networks and support systems increases the vulnerability of these individuals *(2)* and needs to be corrected as soon as possible to prevent further deterioration as the days following the event go by. The best approach is to utilize programming strategies that address the needs of older adults without undermining their independence and discouraging their ability to support themselves *(39,2)*.

The restoration of medical facilities and the provision of transportation to appropriate medical facilities in unaffected areas are not under the control of the individual clinician or caregiver. These efforts are usually dependent on the local police and military forces who take charge post-CE. Additionally, medical facilities will vary in their ability to handle the CE, depending on the type of emergency. For example, the response to a CE such as a hurricane (which would probably slow down access to the facility) would be very different than that required for an infectious disease epidemic (when admissions might very quickly exceed capacity) *(40)*. The challenge for the clinician on the front line is to stabilize the older patient until access to more formal support can be restored. Thus, the aforementioned preparedness efforts are key in preventing the acceleration of medical conditions from chronic to life threatening. The availability of medical records and prescription medicines, as recommended for the evacuation kits of older adults, can play a critical role in this regard.

29.4 RECOMMENDATIONS

In summary, the long list of complicated and threatening CEs that can affect the nutritional status and overall medical welfare of older adults underscores the fact that ALL older adults and their care givers, as well as administrators of structured living facilities, should plan for and be physically and psychologically prepared for the event of a serious CE.

1. Home-dwelling elders should be prepared for a CE by stocking a 2-week safety supply of food, water, and medications, having a carry-away disaster pack with medicines and other essential supplies, and having a delineated evacuation plan.
2. Administrators/Medical Directors should ensure that nursing homes and assisted living facilities are prepared with food and water supplies and an alternate source of power and have detailed, individualized evacuation plans for each resident. Ideally, a multidisciplinary team should utilize age-specific guidelines to design and implement a CE-preparedness plan.
3. In the future, there is a need for conceptual advances in understanding the causes of undernutrition in older adults during a CE and the development of better advance preparations and response mechanisms.

Acknowledgments The authors thank Caroline Friedman for researching the historic and current events cited here.

REFERENCES

1. Toole MJ, Waldman RJ. The public health aspects of complex emergencies and refugee situations. Annu Rev Public Health 1997; 18:283–312.
2. Young H, Borrel A, Holland D, Salama P. Public nutrition in complex emergencies. Lancet 2004; 364(9448):1899–909.
3. Nord, M. Food security rates are high in elderly households. Food Rev 2002; 25(2): 19–24. www. crs.usda.gov/publications/FoodReview/Sept2002/frvol25i2d.pdf)
4. Wolfe, WS, Olson, CM, Kendall A, Frongillo EA. Hunger and food insecurity in the elderly. J Aging Health 1998; 10(3):327–50.
5. Bruemmer B. Food biosecurity. J Am Diet Assoc 2003; 103(6):687–91.
6. Centers for Disease Control and Prevention (CDC). Morbidity surveillance after Hurricane Katrina – Arkansas, Louisiana, Mississippi, and Texas, September 2005. MMWR Morb Mortal Wkly Rep 2006 Jul 7; 55(26):727–31.
7. Centers for Disease Control and Prevention (CDC). Public health response to Hurricanes Katrina and Rita – Louisiana, 2005. MMWR Morb Mortal Wkly Rep 2006 Jan 20; 55(2):29–30.
8. Combs DL, Parrish RG, McNabb SJ, Davis JH. Deaths related to Hurricane Andrew in Florida and Louisiana, 1992. Int J Epidemiol 1996 Jun; 25(3):537–44.
9. Centers for Disease Control and Prevention (CDC). Rapid assessment of the needs and health status of older adults after Hurricane Charley – Charlotte, DeSoto, and Hardee Counties, Florida, August 27–31, 2004. MMWR Morb Mortal Wkly Rep Sep 17; 53(36):837–40.

10. Hendrickson LA, Vogt RL, Goebert D, Pon E. Morbidity on Kauai before and after Hurricane Iniki. Prev Med 1997; 26(5 Pt 1): 711–6.

11. Kirizuka K, Nishizaki H, Kohriyama K, Nukata O, Arioka Y, Motobuchi M, et al. Influences of The Great Hanshin-Awaji Earthquake on glycemic control in diabetic patients. Diabetes Res Clin Pract 1997; 36(3):193–6.

12. Anthony JL, Lonigan CJ, Vernberg EM, Greca AM, Silverman WK, Prinstein MJ. Multisample cross-validation of a model of childhood posttraumatic stress disorder symptomatology. J Trauma Stress 2005 Dec; 18(6):667–76.

13. Reacher M, McKenzie K, Lane C, Nichols T, Kedge I, Iversen A, Hepple P, Walter T, Laxton C, Simpson J, Lewes Flood Action Recovery Team. Health impacts of flooding in Lewes: a comparison of reported gastrointestinal and other illness and mental health in flooded and non-flooded households. Commun Dis Public Health 2004 Mar; 7(1):39–46.

14. Verger P, Rotily M, Baruffol E, Boulanger N, Vial M, Sydor G, Pirard P, Bard D. Evaluation of the psychological consequences of environmental catastrophes: a feasibility study based on the 1992 floods in the Vaucluse. Sante 1999 Sep–Oct; 9(5):313–8.

15. Osaki Y, Minowa M. Factors associated with earthquake deaths in the great Hahshin-Awaji earthquake, 1995. Am J Epidemiol 2001; 153(2):153–6.

16. Dirkzwager AJ, Greivink L, van der Velden PG, Yzermans CJ. Risk factors for psychological and physical health problems after a man-made disaster. Prospective Study Br J Psychiatry 2006; 189:144–9.

17. Freese J, Richmond NJ, Silverman RA, Braun J, Kaufman BJ, Clair J. Impact of a citywide blackout on an urban emergency medical services system. Prehospital Disaster Med 2006 Nov–Dec; 21(6):372–8.

18. Kario K, Ohashi T. Increased coronary heart disease mortality after the Hanshin-Awaji earthquake among the older community on Awaji Island. Tsumna Medical Association. J Am Geriatr Soc 1997; 45(5):610–3.

19. Dobson AJ, Alexander HM, Malcolm JA, Steele PL, Miles TA. Heart attacks and the Newcastle earthquake. Med J Aust 1991; 155(11–12):757–61.

20. Ogawa K, Tsuji I, Shiono K, Hisamichi S. Increased acute myocardial infarction mortality following the 1995 Great Hanshin-Awaji earthquake in Japan. Int J Epidemiol 2000; 29(3):449–55.

21. Caldera t, Palma L, Penayo U, Kullgren G. Psychological impact of the hurricane Mitch in Nicaragua in a one-year perspective. Soc Psychiatry Epidemiol. 2001; 36(3):108–14.

22. Che CC, Yeh TL, Yand YK, Chen SJ, Lee IH, Fu LS, Yeh CY, Hsu HC, Tsai WL, Cheng SH, Chen LY, Si YC. Psychiatric morbidity and post-traumatic symptoms among survivors in the early stage following the 1999 earthquake in Taiwan. Psychiatry Res 2001; 105(1–2):13–22.

23. Lima BR, Chavez H, Samaniego N, Pompei MS, Pai S, Santacruz H, Lozano J. Disaster severity and emotional disturbance: Implications for primary mental health care in developing countries. Acta Psychiatr Scans 1989; 79(1):74–82.

24. Centers for Disease Control and Prevention (CDC). Heat-related deaths – Philadelphia and United States, 1993–1994. MMWR Morb Mortal Wkly Rep 1994 Jul 1; 43(25):453–5

25. Wainwright SH, Buchanan SD, Mainzer HM, Parrish RG, Sinks TH. Cardiovascular mortality – the hidden peril of heat waves. Prehospital Disaster Med 1999 Oct–Dec; 14(4):222–31.

26. Rey G, Jougla E, Fouillet A, Pavillon G, Bessemoulin P, Frayssinet P, Frayssinet P, Clavel J, et al. The impact of major heat waves on all-cause and cause-specific mortality in France from 1971 to 2003. Int Arch Occup Environ Health 2007 Jul; 80(7):615–26.

27. Mott JA, Mannino DM, Alverson CJ, Kiyu A, Hashim J, Lee T, Falter K, Redd SC. Cardiorespiratory hospitalizations associated with smoke exposure during the 1997, Southeast Asian forest fires. Int J Hyg Environ Health 2005; 208(1–2):75–85.

28. Den Ouden DJ, Dirkzwager AJ, Yzermans CJ. Health problems presented in general practice by survivors before and after a fireworks disaster: associations with mental health care. Scand J Prim Health Care 2005; 23(3):137–41.

29. Yzermans CJ, Donker GA, Kerssens JJ, Dirkzwager AJ, Soeteman RJ, ten Veen PH. Health problems of victims before and after disaster: a longitudinal study in general practice. Int J Epidemiol 2005; 34(4):820–6.

30. Atlas RM. Bioterrorism: from threat to reality. Annu Rev Microbiol 2002; 56:167–85. Epub 2002 Jan 30.

31. Meinhardt PL. Water and bioterrorism: preparing for the potential threat to U.S. water supplies and public health. Annu Rev Public Health 2005; 26:213–37.

32. Sobel J, Khan AS, Swerdlow DL. Threat of a biological terrorist attack on the US food supply: the CDC perspective. Lancet 2002 Mar 9; 359(9309):879–80.

33. Noji EK. Public health issues in disasters. Cut Care Med 2005 Jan; 33(1 Suppl):529–33.

34. Gross CR, Lindquist RD, Woolley AC, Granieri R, Allard, K., Allard K, Webster B. Clinical indicators of dehydration severity in elderly patients. J Emerg Med 1992 May–June; 10(3):267–74.

35. Borrel A. Addressing the Nutritional Needs of Older People in Emergency Situations in Africa: Ideas for Action. Westlands, Nairobi, HelpAge International Africa Regional Development Centre, 2001.

36. Deuster PA, Singh A. Nutritional Considerations for Military Deployment. Military Preventative Medicine: Mobilization and Deployment. P. W. Kelley, Office of the Surgeon General, Department of the Army, 2003. 1:317–340.

37. Pieterse S, Ismail S. Nutritional risk factors for older refugees. Disasters 2003 Mar; 27(1):16–36.

38. Vespa J, Watson F. Who is nutritionally vulnerable in Bosnia-Hercegovina? BMJ 1995 Sep 9; 311(7006):652–4.

39. Peachey, K. Ageism: a factor in the nutritional vulnerability of older people? Disasters 1999 Dec; 23(4):350–8.

40. Petrosillo N, Puro V, Caro A, Di Caro A, Ippolito, G. The initial hospital response to an epidemic. Arch Med Res 2005 Nov–Dec; 36(6):706–12.

41. Magkos F, Arvaniti F, Piperkou I, Katsigaraki S, Stamatelopoulos K, Sitara M, et al. Identifying nutritionally vulnerable groups in case of emergencies: experience from the Athens 1999 earthquake. Int J Food Sci Nutr 2004; 55(7):527–36.

Index

A

Achalasia, 123
Acts of omission and acts of commission, 236–237
Acute protein/calorie malnutrition (PCM), 636
Age-Related Eye Disease Study (AREDS), 614–615
Age-related macular degeneration (AMD), 614–615
   Age-Related Eye Disease Study (AREDS), 113
   dietary and plasma vitamin E, epidemiologic studies of, 107
   and dietary vitamin C, role of, 106–107
   dietary vitamin E, role of, 107
   linoleic acid, intake of, 108
   lutein and zeaxanthin, 107–108
   omega-3 fatty acid intake, 108–109
   prevalence of, 99–100
   retina, zinc concentration in, 109–111
Age-related maculopathy (ARM)
   fish intake and, 109
Aging
   Aging and Disability Resource Center (ADRC), 542–545, 549
   aging eye and diseases
      visual impairment and blindness in US population, 99–100
      cataracts and AMD, 100–101
      cataract and dietary vitamin C studies of, 102–103
      dietary vitamin E, protective effect of, 102–103
      lutein and zeaxanthin, relationship between, 104–105
      omega-3 fatty acids, 105–106
   Aging Older Americans Act (OAA) nutrition services, 12
   annual percent growth of elderly population in developed and developing countries, 34
   behavioral theory for nutrition interventions, 28–29
   Cardiovascular disease (CVD), 281
      cardiac cachexia, 338
      management of, 339
      TNF-α role in, 338–340

chronic heart failure
   neurohormonal and metabolic abnormalities in, 337
   nutritional guidelines for older adults with, 346–347
   recommendations for clinicians, 347
chronic kidney disease (CKD), 146, 403–404
   dietary recommendations, on level of renal function, 410–411
   hypertension and salt intake, 404–405
   protein intake, impact of, 405–406
   obesity, impact of, 406–407
   phosphorus binders, use of, 410
chronic obstructive pulmonary disease, 374–375
cross-cultural examinations and, 40
dementia
   artificial nutrition and hydration, 238
   costs and benefits, 239
   food consistencies and minimization of distractions, 239–240
diarrhea, 126–127
and esophageal sensitivity, 122
gastroesophageal reflux disease (GERD) and heartburn, 123–124
insulin secretion, 282–283
and micronutrient needs, 558
oral health in, 248
   chronic periodontitis, 250–252
   dental caries, 250
   oral cancer, 252–253
   plaque and calculus formation, 249–250
   oral and pharyngeal cancer incidence and mortality, 249
osteoarthritis (OA), 439
osteoporosis, 72, 418, 433–435, 575
physical function, 268, 273
pro-inflammatory cytokines, 211–212
public health and technology, 19–20
sarcopenia, 184–200
sensory impairments and
   age-related macular degeneration (AMD), 90–91
   clinicians recommendations, for patients with nutritional disorders, 95–96
   in medicated older individuals, 87

Aging (*Cont.*)
    odor sensations, 88–89
    intra-individual variation, 92–93
    sensory losses, 90–92
    taste alterations, medications and medical
        conditions associated with, 86–87
    taste, sense of, 78–79
    presbyopia, 90
  swallowing changes, 484
  tooth loss
    and BMI index, 255
    British National Diet and Nutrition
        Survey, 255
    denture usage and, 254–255
    and dietary changes, 254
    Nurses Health Study, 254
    recommendations, 258
    study of adults in Sweden, 255–256
  total body fat, 264–265
  weight loss and
    algorithm for care providers, 162
    and anorexia/cachexia syndrome, 227
    body composition, 270
    causes for, 208
    MEALS-ON-WHEELS mnemonic for
        treatable causes of, 66
    mortality impact of body weight and,
        159–160
  See also Anorexia; Cachexia; Dietary
      supplements
Albumin level, in cancer patient, 362
Alpha-Tocopherol Beta-Carotene (ATBC)
    study, 572
Alzheimer's disease, 90, 167–168, 238,
    560, 606
  ambloid cascade hypothesis, 501
  apolipoprotein E (*APOE*) genes, 502
  late-onset sporadic form of, 502
  pathophysiology of, 500
  and related disorders association work
    group, 511
  weight loss in, 511–512
  See also Dementia
American Association of Cardiovascular and
    Pulmonary Rehabilitation (AACVPR),
    320–322
American Association of Clinical
    Endocrinologists (AACE), 289
  glycemic control target recommendations
    summary of, 289
American College of Cardiology, 320
American College of Sports Medicine
    Recommend Resistance Exercise, 197
American Diabetes Association (ADA), 560

American Diabetic Association Medical
    Nutrition Therapy Protocol (ADA
    MNT), 363
  diagnostic criteria, 281–282, 284
American Geriatric Society (AGS), 284
  guidelines for, 306
American Heart Association, 320
  fish advisory, 325
American Red Cross, 646–647
Amino acids and wound healing, 226
Amyloid (Aβ) cascade hypothesis, 500–501
Amyotrophic lateral sclerosis, 503–504, 506,
    508, 513
Anemia, 71, 133–134
Angiotensin-converting-enzyme (ACE), 146
  angiotensin II receptor blockers (ARBs)
    inhibitors, 335–336
  inhibitors, 335–336, 345
Anorexia, 208
  anorexia/cachexia syndrome, 357
    interventions on, 366–367
Appetite Hunger and Sensory Perception
    (AHSP) questionnaire, 67
  weight loss in older persons, 68
Arthritis, Diet, and Activity Promotion Trial
    (ADAPT), 465
Artificial hydration and nutrition, 236
Auditory system in old age, 91–92
Austrian Stroke Prevention Study, 531
Average length of stay (ALOS), 357

B

Basal metabolic rate (BMR), 337–339
Benign prostatic hyperplasia (BPH), 614
Bioelectrical impedance analysis (BIA), 361–362
Blood Glucose Awareness Training (BGAT)
    program, 296
Blue Mountains Eye Study (BMES), 109
Body mass index (BMI), 43–44, 159, 161,
    361, 376, 407
Bone structure and mass, 418–419
Boston Osteoarthritis of Knee Study (BOKS), 460
Brain lesions, and depression, 525, 531

C

Cachexia, 159, 207, 357–358
  anorexia/cachexia syndrome, 208, 210–211
  causes and mechanisms, 209–211
  and clinical recommendations, 214–215
  consequences, tumor necrosis factor and
    interleukin, levels of, 211
  cytokine modulation, 229
  inflammatory-mediated, 229

interventions for, 214
nutritional markers for, 70
starvation and, 212–214
weight loss and undernutrition, 208
Calcitonin, 431
Calcium, 146
hypercalcemia, 148
parathyroid hormone (PTH) secretion, 147
hypocalcemia, causes of, 147–148
Caloric restriction (CR) dietary regimen, 511
Cancer, malnutrition in
contributing factors, 357–359
physiologic basis, 357
risk factors, for older patients, 359–360
studies and data on, 356–357
See also Nutrition
Cardiovascular disease (CVD), 281
cardiac cachexia, 338
management of, 339
TNF-α role in, 338–340
cardiac rehabilitation, 325–326
alcohol and sodium intake, 327
caloric intakes, recommendations
for, 328–329
Center for Medicare and Medicaid
Services (CMS), 320
co-morbidities, 321–323
core components of, 319–320
dietary fats, 324–325
healthy body weight, achieving and
maintaining, 327–328
individual nutritional components, diets
rich in fruits and vegetables,
323–324
meal plans for healthy eating, 330–331
nutritional counseling, core components
of, 321
nutrition and cardiovascular risk, 320–321
nutrition therapy, goal of, 330
stanols and sterols, 326–327
cardiovascular health study cohort, 193
clinical recommendations, 329–331
diabetes risk factors, recommended assessment
and management goals for, 287
dietary intake and management
dairy products, 325–326
fats, 324–325
fruits and vegetables, 323–324
whole grains and starches, 326
Carnitine and creatine phosphate, 344
Cataract, eye disease
and age-related macular degeneration
(AMD), 99–100
physiological basis of, 100–101

dietary vitamin C studies, 102–103
and AMD risk, role of, 106–107
dietary vitamin E, protective effect of, 103–104
LINXIAN trial, 112
lutein and zeaxanthin, relationship
between, 104–105
omega-3 fatty acids, 105–106
Roche European-American Anticataract
Trial (REACT), 113
serum evaluation of, 103
Center for Medicare and Medicaid Services
(CMS), 12, 320, 541
Chemotherapy-induced nausea and
vomiting, 358–359
Child and Adult Care Food Program, 13
China
alcohol consumption in adult, 48–49
cigarette smokers, 47–48
cross-cultural examinations, 40
dietary patterns, changes in, 41–42
disease patterns and prevalence, changes
in, 42–43
hypertension, 44
obesity, 43–44
type 2 diabetes (T2D), 44–45
elderly population
alcohol consumption, 48
Alzheimer's disease (AD), prevalence
of, 52
with chronic health problems by age
cohort, 52
consumption of alcohol in, 49
death causes of, 51–52
Family Law, 53
nutritional status and health behaviors, 47
organized physical activity, 58
poor health-related life risk factors, 52–53
prevalence of smoking in, 47–48
prolonged life and population structure
changes of, 45–47
sources of income for urban and rural, 54
WHO framework convention of tobacco
control, 55
food price indices, 58
physical activity behaviors, 49
cause of death in, 51
levels of, 51
participating in exercise, 50
population pyramids for, 46
size and proportion of elderly population
in, 46
social services and health care system
graying population, medical and
economic impact of, 55

China (*Cont.*)
    older adults, global behavior change in, 56–57
    public health responses, 55–56
    under-and over-nutrition, soaring global food costs and dual challenge of, 57–58
Chondroitin sulfate, 443–444
Chromium and glucose tolerance factor (GTF), 153
Chronic bronchitis, 375
Chronic heart failure
    neurohormonal and metabolic abnormalities in, 337
    nutritional guidelines for older adults with, 346–347
    recommendations for clinicians, 347
Chronic kidney disease (CKD), 146, 403–404
    dietary recommendations, on level of renal function, 410–411
    diet, impact of
        hypertension and salt intake, 404–405
        protein intake, impact of, 405–406
    obesity, impact of, 406–407
    phosphorus binders, use of, 410
    See also Uremia
Chronic obstructive pulmonary disease, 374–375
    definition, by GOLD, 375
    exercise program for, 388
    fat-free mass (FFM) loss and, 375–376
    low body weight, weight loss, and muscle wasting, 376–377
    metabolic disturbances and, 376
        catabolic processes, 378–379
        hormonal alterations, 379
        hypermetabolism, 377–378
        muscle dysfunction, 379
    nutritional evaluation of depletion
        lean body mass, assessment of, 386
        malnutrition, screening for, 383–384
        nutritional assessment of patients, 384–386
        visceral protein status, assessment of, 386
    nutritional intervention, in malnourished stable patients, 386–388
    during acute exacerbation, 388
    nutritional requirements in
        alcohol intake, 390
        amino acids, 389
        energy and protein intake, 389
        fruits and vegetables intake, 390
        macronutrient composition of diet, 389
        polyunsaturated fatty acids, 389–390

    pathophysiology and etiology of, 375–376
    protein and amino acid metabolism, alterations in, 379
    recommendations for, 390
        maintaining/improving respiratory function, 392
        menu plan, 391–392
        nutritionist, role of, 391
    reduced food intake in, 380
        anxiety and depression, 382
        dysphagia and gastro-esophageal reflux, 381
        dyspnea, 381
        glucocorticosteroids treatment, 383
        hospitalization, 382–383
        loss of appetite, 381–382
        meal-related difficulties, 380–381
        social norms and health beliefs, 382
        taste alteration and early satiety, 382
    TDEE in, 375
Clinical swallowing evaluation (CSE), 483–487
Colorectal cancer, 570
Complex emergencies (CEs), 635–652
    earthquakes, 641
    epidemics and pandemics, 643
    examples of, 638
    extreme heat and cold, 642–643
    federal food assistance programs, 638
    fires, 643
    food borne illness outbreaks, 643
    and food insecurity, 638–639
    hurricanes, tornadoes, and floods, 640–641
    immediate impact and nutritional/health consequences of, 640
    intervention strategies, 649–650
        basic services and medical facilities, restoration of, 651–652
        clinical recommendations, 652
        food and water shortages, coping with, 650–651
    minimizing of nutritional and health risk, 645
        conceptual and programmatic overview, 645–646
        emergency preparedness at home, 647–649
        emergency preparedness in structured living communities, 646–647
    terrorism and bio-terrorism, 644–645
Comprehensive geriatric assessment (CGA), 360
Constipation, and cancer, 359
Consumer information processing model, 21
COPD, *see* Chronic obstructive pulmonary disease
Coronary heart disease (CHD), 37
Corticosteroids, and cancer patients, 366

Cytokine
    modulation and cachexia, 229
    undernutritus and chronic wounds, 229

D

Daily calorie allowance, determining of, 328
Daily fat gram budget chart, 329
Dehydration, 139–140
Dementia
    artificial nutrition and hydration, 238
        costs and benefits, 239
        food consistencies and minimization of
            distractions, 239–240
Denosumab, 432
Dental caries, 250
Department of Homeland Security, 644
Depression, 523
    brain lesions and cobalamin, 531
    B vitamins and
        cobalamin, 530–531
        folate, 529–530
    and diabetes
        attitudes and dietary intake, 303–304
        cognitive dysfunction, 302–303
        data from epidemiological catchment area
            study, 299
        and dietary intake, 300
        ethnic/cultural issues, 304–306
        quality of life, 306–307
        social support, 301–302
        weight loss and, 309–310
    diagnosis of, 523–524
    dietary fats and, 526–527
        omega-3 fatty acids, 527–529
    dietary mechanism for, 525
    factors in etiology of, 524–526
    impact of, 524
    obesity and, 526
    and weight loss, 526
DETERMINE checklist, 67
Diabetes mellitus, 280
    adult men and women, estimated prevalence
        of diabetes in USA, 280
    Diabetes Control and Complications Trial
        (DCCT), 283–284
    Diabetes Prevention Program (DPP), 293
    Diabetes self-management education
        (DSME), 297, 307
        dietary habits of older adults with, 298–299
        provider delivery approaches to, 297
        self-management and dietary guidelines
            for, 298
    diabetic diet, macronutrient content of, 287

in, older US adults
    diet and medication, interaction of, 288
    nutritional recommendations, general
        goals of, 286–288
    treatment, goals, 283–286
    type 1 diabetes, 282
    type 2 diabetes, 282–283
in older adults, factors to consider in
    management of, 285
Dietary Ancillary Study of Eye Disease
    Case-Control Study, 108–109
Dietary Approaches to Stop Hypertension
    (DASH), 323, 574, 577
    DASH-feeding studies, 325
    diet plan, 25–26
    sodium trial, 327
    trials, 323
Dietary Guidelines for Americans (DGA), 547
Dietary supplements, 554
    dietary folate equivalents (DFE), 571
    Dietary Supplement and Nonprescription
        Drug Consumer Protection Act, 556
    Dietary Supplement Health and Education
        Act, 555
    with evidence base and promising use, 581
        alpha-lipoic acid, 582
        antioxidants, 582–583
        ayurvedic remedies, 585–586
        black cohosh root, 585, 610
        carnitine, 587–588
        chromium, 590
        cinnamon, 589
        coenzyme Q_{10}, 591, 610–612
        creatine, 592
        Echinacea, 592
        garlic, 593–594
        Ginkgo biloba, 594, 612–613
        glucosamine HCL/sulfate, 595–596
        grape seed extract, 597
        green tea, 597
        Gymnema, 599
        horse chestnut seed extract, 600
        lutein, 600
        lycopene, 601
        melatonin, 601–602
        milk thistle extracts, 602–603
        phosphatidylserine, 604
        pine bark extract, 603
        policosanols, 604
        St. Johns wort (SJW), 608
        SAMe (S-adenosyl-L-methionine), 605–606
        saw palmetto, 604, 614
        soy protein, 607
    evidence-based review system for, 557–558

Dietary supplements (*Cont.*)
  guidelines for safe use of, 615–616
    clinical needs assessment, 617–618
    safeguarding against potential
        interactions, 616–617
  health claims for, 556–557
  herbal *vs.* manufactured drugs, distinctions
      between, 617
  and older adults, 558–560
    B vitamins, 560, 569–571
    calcium, 567, 575–576
    magnesium, 564, 577–578
    multivitamin/mineral supplements,
        580–581
    omega-3 fatty acid, 566–567, 579–580
    potassium, 564–565, 573–574
    selenium, 565–566, 574–575
    vitamin D, 568, 576–577
    vitamin E, 571–573
  regulation of, 555–556
  use of, 554–555
Dilutional hyponatremia, 143
DREAM trial, 294
Dronabinol, 167–168
Duke Cardiac Rehabilitation program, 328
Dysgeusia, 86
Dysphagia, 122–123, 381, 479–480
  dysphagia disability index (DDI), 488
  intervention and treatment of
    exercises and compensatory maneuvers,
        492–493
    fluid intake maintenance, 491
    Frazier Free Water Protocol, 491–492
    oral hygiene education, 492
    thickened liquids, 491
Dyspnea, 381

E

Eastern Cooperative Oncology Group (ECOG),
    report on cancer patients, 356
Eating process
  biological and psychosocial
      phenomenon, 13–14
  healthy, social ecological model of, 3–4
  intrapersonal food choices, 5–7
  under-eating, 9
Edentulism prevalence, 248
Elderly Nutrition Program (ENP), 639
Emergency Food Assistance Program, 12
Emphysema, 375
Epidemiological Catchment Area study, data
    from, 299
European Association for Study of Diabetes, 292

Evaluating Long-Term Diabetes
    Self-Management Among Elder Rural
    Adults (ELDER), 299
Eye disease
  antioxidants, human studies on dietary intake
    cataract and dietary vitamin C studies
        of, 102–103
    dietary vitamin E, protective effect of,
        102–103
    lutein and zeaxanthin, relationship
        between, 104–105
    omega-3 fatty acids, 105–106
  Blue Mountains Eye Study (BMES), 109
  dietary intakes, specific recommendations
      for, 114–116
  eye disease case–control study, 107–108
  human eye, organization and
      terminology, 100
  nutrient supplements effect of
    age-related macular degeneration
        (AMD), 113–114
    cataract, 111–113

F

Fat-free mass (FFM), 376–377
Fecal impaction, 127
Federal Emergency Management Agency
    (FEMA), 646–648
Federal Food, Drug, and Cosmetic Act, 555–556
Fiberoptic endoscopic evaluation of swallow
    (FEES), 485, 487
First Health and Nutrition Examination
    Survey, 107
Flavangenol, 613
Folate
  deficiencies and
    dementia and macrocyticanemia, 73
    gastritis and, 125–126
    anemia, 133
    oral cancer, 252
    cardiovascular disease and stroke, 507–508
    depression, 529–530
    fortification, 570
Food
  complementary food ration, 637
  and fluid intakes
    clinical recommendations, 177
    commercially available nutritional
        supplements, 170–174
    dining environment and nutritional value,
        162–163
    feeding support and assistance, 163, 165
    partial/total nutritional support, 175–177

therapeutic interventions, 161
use of appetite stimulants, 165–168
food choice questionnaire, 5
FOOD (Feed or Ordinary Diet) trials in, 175
food stamps nutrition education program, 12
insecurity, 638
pyramid, for adults over 70 years, 558–559
rations, 636
Frailty, 71–72, 159
Framingham Osteoarthritis Study, 460
Frazier Free Water Protocol, 491–492
Functional Acuity Contrast Test (FACT)
    Chart, 94
Functional oral intake scale (FOIS), 487–488

G

Gastroesophasal reflux (GER), 381
Gastrointestinal (GI) complaints, 121–122
anemia, 133–134
Barrett's esophagus, 123
bleeding, 131–132
constipation, 130
diarrhea, 126–127
dysphagia, 122–123
fecal impaction, 127
gastritis and peptic ulcer disease, 125–126
gastroesophageal reflux disease (GERD) and
    heartburn, 123–125
hepatitis, 132–133
inflammatory bowel disease (IBD), 128
involuntary passage of, 129–130
ischemic colitis, 127–128
lactose intolerance, 129
LES pressure, 124–125
microscopic colitis, 128
small bowel bacterial overgrowth (SBBO), 128
Genistein, 424
Glaucoma and lowering intraocular pressure
    (IOP), 91
Global initiative for chronic obstructive lung
    disease (GOLD), 375
Glucosamine, 442–443
glucosamine/chondroitin arthritis
    intervention trial (GAIT), 447
Glucosamine Unum in Die Efficacy
    (GUIDE) trial, 447
poor health-related life risk factors in, 52–53
Growth hormone (GH) circulating levels, 338

H

Health
aging and body composition study, 186, 193
6-m walk and chair-stand tests, 195

Health Belief Model (HBM), 21
health-care providers, 235–236
Health Outcomes Survey, 301
Health Resources and Services
    Administration (HRSA), 647
Healthy Eating Index (HEI), 27
promotion interventions in, older adults,
    eating behavior, 6–7
Health behavioral theories, chronic diseases
cardiovascular disease prevention, 26–27
change model stages of, 21
clinical recommendations, 30
community-based diabetes education
    program, 25
Consumer Information Processing Model, 21
diabetes mellitus, 23–25
Dietary Approaches to Stop Hypertension
    (DASH), 25–26
dietary interventions in older adults,
    24, 27–29
practical applications of, 28–29
ecological perspective, 20
Health Belief Model (HBM), 21
Healthy Eating Index (HEI), 27
heart failure, health belief model, 26
hyperlipidemia, 26
meaningful learning theory of, 21
Social Cognitive Theory (SCT), 22
stage-based dietary counseling strategies, 22
stages of change model, 21
theory of meaningful learning, 21
Heart failure, 333–334
cardiac cycle, pathophysiology, 334–335
cardinal symptoms of, 335
diuretics therapy for, 341
medication and age, 345
neurohormonal and metabolic
    abnormalities, 337
nutrients in
anti-oxidant coenzyme Q_{10}
    (ubiquinone), 344
minerals, 342–343
vitamins, 343–344
water and sodium, 340–341
nutritional aspects
caloric intake, 339–340
cardiac cachexia, 338
metabolic syndrome, 336–338
obesity in, 340
nutritional guidelines for, 346–348
optimal treatment of, 335–336
pharmacotherapy of left ventricular systolic
    dysfunction, 336
prognosis for, 335

Heart Outcomes Prevention Evaluation (HOPE)
    trial, 343
Helicobacter pylori infestation of stomach, 123
HelpAge International, 645–646
Hepatitis, 132–133
Herbal Alternatives for Menopause Trial
    (HALT), 610
Hexosamine biosynthesis pathway, and
    glucosamine, 443
Hispanic Established Population for
    Epidemiologic Study of Elderly (EPESE)
    data from, 299
HMO study, 300
Home-and community-based long-term care
    system, 542–544
    opportunities and challenges in, 544–545
    See also Long-term care system
Homocysteine, 507–508
Honolulu-Asia Aging Study, 512
    See also Alzheimer's disease
Huntington's disease, 504–505, 509–510, 513–514
Hyperinflation, in COPD, 375–376
Hypoglycemia
    dietary management of, 295
    self-monitoring and dietary treatment,
        294–295
    unawareness and treatment, 295–296
Hyponatremia, 141–142

I

Ibandronate, 431
Inflammatory bowel disease (IBD), 128
Institute of Medicine of National Academy
    of Sciences, 428
Instrumental Activities of Daily Living
    (IADLs), 296
Insulin-like growth factor-1 (IGF-1), 433
Interleukin (IL)-1 concentrations in older
    adults, 227
International Society for Clinical
    Densitometry, 419
Intra-abdominal fat (IF), and depression, 526
Ischemic colitis, 127–128
Isoflavones, 430–431

J

The Jean Mayer USDA Human Nutrition
    Research Center on Aging, 558
Johnston County Osteoarthritis Project, 452

K

Kashin–Beck disease, 451–452

L

Lactose intolerance, 129
Late-life depression, *see* Depression
Lean tissue mass, 70
Legal decisions and nutritional treatment of
    terminal patients, 236–237
Long-term care system, 541–544
    rebalancing demonstration projects of, 543
    See also Older Americans Act (OAA)
Look AHEAD trial, 309
Low-energy diet (LED) intervention, 465
Lutein and zeaxanthin serum levels, 107–108

M

Magnesium, 148
    hypomagnesemia and hypermagnesemia, 149
Malnutrition, definition of, 637
Malnutrition Universal Screening Tool (MUST)
    for elderly inpatients and outpatients, 67
Mann Assessment of Swallowing Ability
    (MASA), 485
Medicaid Home-and Community-Based Care
    Service Waiver Programs (HCBC), 12
Medicaid Nutrition Services, 12
Medicaid program, 542
Medicaid Waiver Program, 544–545
Mediterranean diet (MeDi), 510
Mediterranean Lifestyle Trial, 25
Megestrol acetate (MA), and cancer
    patients, 366–367
Mendelsohn maneuver, 493
Metabolic syndrome, 291–294
    ATP III criteria for diagnosis of, 292
Methylenetetrahydrofolate reductase
    (MTHFR), 530, 570
Microscopic colitis, 128
Mild cognitive impairment (MCI), 512
Minerals
    food sources and recommended intakes
        for, 138
    studies of dietary intake and nutritional
        status of, 150
    trace minerals
        chromium, 153
        copper, 151
        iron, 152
        manganese deficiency in, 153
        molybdenum, 153–154
        selenium, 152–153
        zinc, 150–151
Mini Nutritional Assessment (MNA) scale, in
    older person, 68–70
Mirtazapine, 367

Monounsaturated fatty acids (MUFA), 291
Mucositis, 358
Multivitamin and mineral supplements
    (MVMs), 581
Muscle loss, 184

N

Nandrolone anabolic drugs, 168
Nasogastric (NG) feeding in, 175
National Aging Network (Aging Network),
    541, 545, 549
National Cholesterol Education Program-Adult
    Treatment Panel III (NCEP-ATP III), 291
National Cholesterol Education Program
    Expert Panel of Detection, Evaluation,
    and Treatment of High Blood Cholesterol
    in Adults (Adult Treatment Panel III),
    third report, 326–327
National Diabetes Education Program
    (NDEP), 309
National Health and Nutrition Examination
    Survey (NHANES), 403
Third National Health and Nutrition
    Examination Survey (NHANES III), 109,
    193, 280–281
National Institute of Neurological and
    Communicative Disorders and Stroke,
    511
NDEP GAMEPLAN (Goals, Accountability,
    Monitoring, Effectiveness, Prevention
    through Lifestyle of Activity and
    Nutrition), tool kit, 309
Neurodegenerative disorders, 499–500, 612
    Alzheimer's disease, 500–502
    amyotrophic lateral sclerosis, 503–504
    clinical recommendations, 514
    Huntington's disease, 504–505
    nutritional factors, effect of, 505
        antioxidant nutrients, 509–510
        caloric restriction (CR), 511
        carbohydrates and glucose tolerance,
            506–507
        dietary fat and cholesterol, 505–506
        dietary pattern analysis, 510
        vitamins, 507–509
    Parkinson's disease, 502–503
    weight loss in, 511–514
Neuromuscular electrical stimulation
    (NMES), 493
New Mexico Elder Health Survey, 185
Non-medicated older persons, sensory (and
    cognitive) tests, 94–95
Non-oral feeding, 236

Nutrition
    artificial nutrition and hydration in terminal
        cancer, 237–238
        dementia, 238–240
    assessment strategies
        body mass as indicator of, 66
        dehydration, clinical diagnosis
            of, 71
        delayed-type cutaneous hypersensitivity
            in, 71
        dietary intake, 70
        disability, Katz basic ADL and Lawton's
            IADLs, 72–73
        Mini Nutritional Assessment (MNA)
            scale, 68–70
        nutritional markers, 70–71
        osteoporosis, screened for, 72
        questionnaires, 67
        Simplified Nutrition Assessment
            Questionnaire (SNAQ), 67–70
        skinfold thickness, 70
        true nutritional status of, 65
        vitamin and trace element deficiency, 73
        weight loss, cause of, 66
    and chronic disease management, 540
    and food services, for older adults,
        539–541
        See also Home-and community-based
            long-term care system
    frailty, 157–158
        assessment, 71–72
        body weight and weight loss, mortality
            impact of, 159–160
        definition of, 158
        food and dining environment, improving
            esthetics of, 162–163
        functional feeding, 163–165
        orexigenic agents, 165–168
        partial and total nutrition support,
            175–177
        protein/calorie supplements,
            168–175
        therapeutic interventions, 161–162
        undernutrition causes of, 160–161
    impact of
        oral cancer, 256
        tooth loss on, 254–256
        xerostomia, 256–257
    nutritional support for older cancer
        patients, 362–363
        clinical recommendations, 368
        intervention, tailoring of, 365–366
        levels of intervention, 363–365
        nutritional assessment, 360–362

Nutrition (*Cont.*)
  Nutrition Labeling and Education Act of
      1990, 556
  and rehabilitation, 636
  undernutrition, causes of, 160–161
  and weight loss, 227–229

O

Obesity, 263
  adverse effects of
    cancer, 268
    comorbid disease, 266
    metabolic syndrome, 266–267
    mortality, 266
    obstructive sleep apnea (OSA), 267
    osteoarthritis (OA), 267
    urinary incontinence, 267
  beneficial effects of, bone mineral density
      (BMD), 269–270
  fat-free mass (FFM) and, 264–265
  functional impairment and quality of life,
      268–269
  interventions and treatment
    behavior modification, 273
    diet therapy, 271–273
    lifestyle intervention, 271
    pharmacotherapy, 273–274
    physical activity, 273
    weight-loss surgery, 274
  overweight measurement and
    body mass index (BMI), 265
    waist circumference, 266
  prevalence, 264
Odor sensations, 88
  perceptual olfactory
    causes of, 89–90
    losses in, 89
  trigeminal activation, 88–89
Older adults
  chronic diseases, nutritional therapies
      in, 19
  comparisons in senses, 92–95
  constipation, 130
  dehydration clinical diagnosis of, 71
  delayed-type cutaneous hypersensitivity in, 71
  detection thresholds (DT), 79
  diarrhea
    drug-induced diarrhea, 127
    infectious gastroenteritis, 126
    involuntary passage of, 129–130
  dietary intake, assessment of, 70
  gastritis and peptic ulcer disease, 125
    type A and type B gastritis, 126

gastrointestinal (GI) complaints,
      121–122
  medication-related alterations in, 86–87
  Mini Nutritional Assessment (MNA)
      scale, 68–70
  nutritional screening tools for, 67
  oral health in, 247
    oral cancer, 248–249
    tooth loss and dental caries, 248
  perceptual olfactory losses, 89
  psychosocial influences on diet and
      lifestyle, 306
  sensory
    function, assessment of, 93–94
    impairments, food, 78–79
  sensory losses
    auditory system, function of, 91–92
    somatosensory system, 92
    vision, 90–91
  Simplified Nutrition Assessment
      Questionnaire (SNAQ) for, 67
  skinfold thickness, 70
  suprathreshold taste studies, 86
  taste, sense of, 78–79
  true nutritional status of, 65
  vitamin and trace element deficiency, mild
      degrees of, 73
  weight loss, cause of, 66
Older adults, eating behavior
  community factors, 10–11
    intervention strategies, 11
  ecological approach and, 13–14
  healthy eating, 3, 5
    social ecological model of, 4
  institutional factors, 8–9
    intervention strategies, 10
  intrapersonal factors
    food choices, 5–6
    intervention strategies for, 6–7
    social support and networks with, 7–8
    strategies for, 8
  public policy factors, 11–13
    strategies for improve nutritional well-
        being, 13
Older Americans Act (OAA), 541,
      544–545, 639
  OAA Nutrition Program
    and frail older adults, 549
    as service model for food and nutrition
        choices, 546–549
  and rebalanced long-term care, 545–546
Older US adults
  Cardiovascular disease (CVD), 281
  diabetes, 280

attitudes and, 303–304
classification of, 282
cognitive dysfunction, 302–303
depression intervention in, 300
diagnosis and, 281–282
dietary intake in, 300
enteral and parenteral nutrition, 290–291
ethnic minority, 304–306
health consequences of, 281
hospitalization, glycemic control systemic
    problems, 290
hyperglycemia, 289
lifestyle change and physical limitations,
    296–297
lifestyle interventions and medications,
    combination of, 281
long-term care facilities, 291
malnutrition and, 289
medication use and glycemic control in,
    283–288
overweight and, 288–289
quality of life and, 306–307
social isolation, 301–302
studies of depression and, 299
Diabetes Self-Management Education
    (DSME), 297–299
hypoglycemia
    self-monitoring and dietary treatment of,
        294–295
    unawareness and, 295–296
imparting dietary information, 310
metabolic syndrome, 291–294
type 1 diabetes, 282
type 2 diabetes, 282–283
    self-management and lifestyle change,
        307–310
Olfactory receptors (ORs), 88
olfactory receptor neurons (ORNs), 88
    bulb, glomeruli in, 88
    epithelium, 89
Omega-3 fatty acids, 390
intake, 108–109
See also Age-related macular
    degeneration (AMD)
Oral health in older adults, 248
impact of nutritional status on
    chronic periodontitis, 250–252
    dental caries, 250
    oral cancer, 252–253
    plaque and calculus formation, 249–250
oral and pharyngeal cancer incidence and
    mortality, 249
Orexigenic agents, 366–368
Oropharyngeal dysphagia, 482–483

Osteoarthritis (OA), 439
antioxidant micronutrients in
    vitamin C, 454–458
    vitamin D, 458–461
    vitamin E, 461–463
    vitamin K, 463–464
clinical features of, 441
and diet, 441–442
glucosamine and chondroitin sulfate
    as disease-modifying agents, 449–450
    efficacy for pain and function, 445–449
and nutritional supplements
    avocado and soybean unsaponifiables
        (ASU), 451
    chondroitin sulfate, 443–444
    glucosamine, 442–443
    omega-3 polyunsaturated fatty acids,
        450–451
    S-adenosylmethionine, 453
    selenium and iodine, 451–453
    sulfate, 444–445
oxidative damage and, 453–454
    antioxidant effects, 454
pathogenesis of, 440–441
risk factors for, 439–440
therapeutic measures to, 441
and weight loss, 464–466
Osteocalcin, 463
Osteopenia, 419
Osteoporosis, 72, 418, 433–435, 575
and anabolic agents
    growth hormone, 433
    parathyroid hormone (PTH), 432–433
    sodium fluoride, 432
anti-resorptive agents and
    active metabolites of vitamin D, 429
    bisphosphonates, 431
    calcitonin, 431
    calcium and vitamin D, 428–429
    estrogens, 429–430
    isoflavones, 430–431
    RANKL, inhibitors of, 431–432
    selective estrogen receptor modulators
        (SERMS), 430
BMD measurement in, 419
causes and types of, 420
dietary risk factors with, 421
    calcium and phosphorus, 421–422
    key nutrients for elderly, 425
    phytoestrogens and vegetarian diet, 424
    protein and acid load, 423–424
    sodium and potassium, 424
    vitamin D, 422
    vitamin K, 423

Osteoporosis (*Cont.*)
  low calcium intake and, 425
  non-dietary risk factors and, 420–421
  prevention of, 425–426
Osteoprotegerin (OPG), 431
Osteoradionecrosis, 256
Ottery's patient-generated subjective global
    assessment (PG-SGA), 360–361
Oxandrolone anabolic drugs, 168

P

Parathyroid hormone (PTH), 421,
    432–433
Parenteral nutrition, for cancer patients, 365
Parkinson's disease, 90, 238, 502–503,
    506, 508, 513
  See also Dementia
Percutaneous endoscopic gastrostomy
    (PEG), 365
Periodontitis, 248
  chronic periodontitis, 250–252
  vitamin D status and, 251
Peripheral neuropathies, 92
Phosphorus fortification, 421
Phylloquinone, 423
Piascledine, 451
Plaque and calculus formation, 249–250
Policy for Nutrition of Older
    Adults, 11–12
Postural hypotension, 71
Potassium
  hyperkalemia, 146
  hypokalemia, 144
    causes of, 145
Presbyopia, 90
Pressure ulcers, 219
  clinical recommendations, 230
  and nutrition, 220
    amino acids, 226
    daily caloric requirements, 225–226
    in healing, 224
    markers, 220–221
    in prevention of, 221–224
    vitamins and minerals, 226–227
Pro-inflammatory cytokines
  age-related changes in, 211–212
  undernutrition, 229
PROSPER study, 322
Prostate, Lung, Colorectal, and Ovarian
    (PLCO) Cancer Screening Trial,
    570–571
Protein-based nutritional supplements in older
    adults, 196

PROVE-IT study, 285
Pulmonary cachexia, 377

R

Radiation therapy oncology group 90-03
    study, 237–238
Raloxifene, 430
Reflux symptom index (RSI), 488
Resting energy expenditure (REE), 375,
    377–378, 380, 513
Resveratrol, 390
Retina
  age-related macular degeneration
    (AMD), 101
  docosahexaenoic acid (DHA), 108
  lens role of, 100–101
  retinal pigment epithelium (RPE)
    pigment of, 101
  zinc concentration in, 109–111
Risk-factor management, lifestyle and
    pharmacological approaches, 286
Roche European-American Anticataract Trial
    (REACT), 113
Rosenbaum Pocket Vision Screening
    Chart, 94
Rotterdam Study, 508

S

Sarcopenia
  Activities Of Daily Living (ADLs) and
    Instrumental Activities Of Daily
    Living (IADLs), 186–193
  age-related reduction in skeletal muscle
    size, 184
  classification and prevalence, 185
  clinical recommendations, 200–201
  definition of, 183–184
  intramyocellular and extramyocellular lipids,
    194–195
  morbidity in, 193–194
  mortality risk in, 194
  prevention and treatment, 195–197, 200
  sarcopenic obesity, 159
  skeletal muscle strength, reduction in, 186
  in uremia, 407–408
SCALES Nutritional tool, 67
SCREEN II (Seniors in Community, Risk
    Evaluation for Eating and Nutrition), 67
Selective estrogen receptor modulators
    (SERMS), 430, 435
Senile miosis, 90
Senior Farmers' Market Nutrition Program, 13

Sensory impairments, age-related
  clinicians recommendations, for patients with
      nutritional disorders, 95–96
  age-related macular degeneration
      (AMD), 90–91
  in medicated older individuals, 87
  odor sensations, 88–89
  in older adults
      intra-individual variation, 92–93
      sensory losses, 90–92
      taste alterations, medications and medical
          conditions associated with, 86–87
      taste, sense of, 78–79
  presbyopia, 90
Serum IL-1β in patients with pressure
      ulcers, 228
Sevelamer, 410
Shaker exercises, 492
Silent aspiration, 483, 485
Simplified Nutrition Assessment Questionnaire
      (SNAQ), 67–68
Sjogren's syndrome, 256–257
Skeletal muscle
  age-related changes in composition, 194–195
  high-protein diets, long-term effect of, 195
  inflammatory cytokines and, 196
  mass values, 185
  mechanisms of muscle wasting, 213
  and resistance exercise, 198–199
  testosterone replacement therapy and, 200
Small bowel bacterial overgrowth (SBBO), 128
Smell sense, 88
  older adults, perceptual olfactory
      causes of, 89–90
      losses in, 89
Social Cognitive Theory (SCT), 22
Sodium, 140, 147–148
  hypercalcemia, 148
  hyponatremia with
      contracted ECFV, 141–142
      expanded and normal ECFV, 143
Somatosensory system in old age, 92
Speech language pathologists (SLPs), 482
Sphere Project Minimum Standards in Disaster
      Response Project, 645–646
Stages of change model, 21
Stomatitis, see Mucositis
Stroke, 479–480
  with dysphagia, intervention after, 488–490
      dysphagia intervention and treatment,
          491–493
      feeding strategies, 490–491
      pneumonia prevention, 490
  malnutrition in, 480

  dysphagia and, 482–483
  factors for malnutrition, 480–482
  See also Dysphagia; Swallow
Studies of Left Ventricular Dysfunction
      (SOLVD) trial, 339
Study to Help Improve Early Evaluation and
      Management of Risk Factors Leading to
      Diabetes (SHIELD), 56
Subjective Global Assessment in-Hospital
      Tool, 67
Surface electromyographic biofeedback
      (SEMG), 493
Swallow
  age-related swallowing changes, 484
  assessments, approaches for, 483–485
      clinical swallow evaluation (CSE), 485–486
      fiberoptic endoscopic evaluation of
          swallowing (FEES), 487
      functional oral intake scale (FOIS), 487–488
      videofluoroscopic swallow evaluation
          (VSE), 487
  normal, 482–483
  subjective assessments, 488
      dysphagia disability index (DDI), 488
      patient-centered quality-of-life outcomes
          tools, 488
      reflux symptom index (RSI), 488
      SWAL-CARE scales domains, 488–489
      SWAL-QOL scales domains, 488–489
Symbol Digit Modalities Test (SDMT), 94
Syndrome of inappropriate ADH (SIADH), 143
Systolic heart failure, left ventricular,
      pharmacotherapy of, 336

T

Tactile spatial sensitivity thresholds, 94
Tamoxifen, 430
Taste sense, 78–79
  alterations, and medical conditions, 86–87
  detection thresholds (DT) in, 79, 86
      comparison of taste detection and
          recognition thresholds for older and
          young subjects for broad range of
          stimuli, 80–85
  receptors, intravascular taste, 87
  suprathreshold taste studies, 86
Terminal patients
  decision-making, 240
  goal setting process, 241
  legal decisions and nutritional treatment,
      236–237
  practical considerations, 241–242
  recommendations for clinicians, 242–243

Thalidomide, 367
Theory of meaningful learning, 21
Therapeutic lifestyle changes (TLC), 580
Tooth loss
    and BMI index, 255
    British National Diet and Nutrition
        Survey, 255
    denture usage and, 254–255
    and dietary changes, 254
    Nurses Health Study for, 254
    recommendations, 258
    study of adults in Sweden, 255–256
Total daily energy expenditure, 375, 377
Transferrin, 362
Type 2 diabetes mellitus (T2DM), 507, 578
    prevalence of, 44–45

U

Ubiquitin–proteasome proteolytic pathway, 358
Unhappy Triad, 158, 209
    See also Cachexia
Unified Parkinson's Disease Rating Scale
        (UPDRS), 612
United Kingdom Prospective Diabetes Study
        (UKPDS), 283–284
United States Department of Agriculture
        (USDA) Programs, 12
    USDA Commodity Supplemental Food
        Program, 13
Uremia
    malnutrition in, 407–408
    nutritional status in
        inflammation and, 409–410
        metabolic acidosis, 409
        protein intake, 408–409
    obesity paradox, 407
U.S. Administration on Aging (AoA), 541
U.S. Preventive Services Task Force
        (USPSTF), 557

V

Vascular calcification, 410, 426
Venous thromboembolism (VTE), 572

Videofluoroscopic Swallow Evaluation (VSE),
        482, 484–485, 487
Vision in old age, 90–91
Vitamins
    deficiency and wound healing, 226
    folate, *see* Folate
    osteoarthritis
        vitamin C, 454–458
        vitamin E and, 461–463
    vitamin B$_{12}$ deficiency, and cancer
        patients, 360
    vitamin D
        and cancer patients, 359
        and osteoarthritis, 458–461

W

Wasting disease, 208
Water loss, dehydration and hypernatremia,
        139–140
Weight loss and older persons
    algorithm for care providers, 162
    and anorexia/cachexia syndrome, 227
    body composition, 270
    causes for, 208
    MEALS-ON-WHEELS mnemonic for
        treatable causes of, 66
    mortality impact of body weight and,
        159–160
Western Ontario and McMaster Universities
        (WOMAC), 446–447, 465–466
Whole-body skeletal muscle
        thresholds, 185
Women in Health, Aging and Body
        Composition Study, 460
Wound healing and nutrition, 219–220

X

Xerostoma, 242
Xerostomia, 124, 256–257

Z

Zinc and wound healing, 226–227

About the Editors

Dr. Adrianne Bendich is a Clinical Director, Medical Affairs at GlaxoSmithKline (GSK) Consumer Healthcare where she is responsible for leading the innovation and medical programs in support of many well-known brands including TUMS and Os-Cal. Dr. Bendich had primary responsibility for GSK's support for the Women's Health Initiative (WHI) intervention study. Prior to joining GSK, Dr. Bendich was at Roche Vitamins Inc. and was involved with the groundbreaking clinical studies showing that folic acid-containing multivitamins significantly reduced major classes of birth defects. Dr. Bendich has co-authored over 100 major clinical research studies in the area of preventive nutrition. Dr. Bendich is recognized as a leading authority on antioxidants, nutrition and immunity and pregnancy outcomes, vitamin safety and the cost-effectiveness of vitamin/mineral supplementation.

Dr. Bendich is the editor of nine books including "Preventive Nutrition: The Comprehensive Guide For Health Professionals" co-edited with Dr. Richard Deckelbaum, and is Series Editor of "Nutrition and Health" for Humana Press with 29 published volumes including "Probiotics in Pediatric Medicine" edited by Dr. Sonia Michail and Dr. Philip Sherman; "Handbook of Nutrition and Pregnancy" edited by Dr. Carol Lammi-Keefe, Dr. Sarah Couch and Dr. Elliot Philipson; "Nutrition and Rheumatic Disease" edited by Dr. Laura Coleman; "Nutrition and Kidney Disease" edited by Dr. Laura Byham-Grey, Dr. Jerrilynn Burrowes and Dr. Glenn Chertow; "Nutrition and Health in Developing Countries" edited by Dr. Richard Semba and Dr. Martin Bloem; "Calcium in Human Health" edited by Dr. Robert Heaney and Dr. Connie Weaver and "Nutrition and Bone Health" edited by Dr. Michael Holick and Dr. Bess Dawson-Hughes.

Dr. Bendich served as an Associate Editor for "Nutrition" the International Journal; served on the Editorial Board of the Journal of Women's Health and Gender-based Medicine, and was a member of the Board of Directors of the American College of Nutrition.

Dr. Bendich was the recipient of the Roche Research Award, is a *Tribute to Women and Industry* Awardee and was a recipient of the Burroughs Wellcome Visiting Professorship in Basic Medical Sciences, 2000-2001. In 2008, Dr. Bendich was given the Council for Responsible Nutrition (CRN) Apple Award in recognition of her many contributions to the scientific understanding of dietary supplements. Dr. Bendich holds academic appointments as Adjunct Professor in the Department of Preventive Medicine and Community Health at UMDNJ and has an adjunct appointment at the Institute of Nutrition, Columbia University P&S, and is an Adjunct Research Professor, Rutgers University, Newark Campus. She is listed in Who's Who in American Women.

Connie Watkins Bales, PhD, RD, FACN is Associate Research Professor in the Division of Geriatrics, Department of Medicine, and Senior Fellow in the Center for the Study of Aging and Human Development at Duke University Medical Center. She also serves as Associate Director for Education/Evaluation of the Geriatrics Research, Education, and Clinical Center at the Durham VA Medical Center. Dr. Bales received her doctorate in Nutritional Sciences at the University of Tennessee-Knoxville and held an appointment in the Graduate Nutrition Division at the University of Texas-Austin before joining the faculty at Duke. Her research endeavors over the past two decades have focused on a variety of topics in the field of nutrition and she has published broadly on nutrition and aging-related topics, including nutritional frailty, dietary determinants of aging, energy balance, and metabolic syndrome. Dr. Bales also edits the *Journal of Nutrition for the Elderly*.

Christine S. Ritchie, MD, MSPH, FACP is Associate Professor of Medicine, Director of the University of Alabama at Birmingham (UAB) Center for Palliative Care, Director of the Palliative and Supportive Care Section, Division of Gerontology, Geriatrics and Palliative Care, and Director of the UAB Geriatric Education Center. She oversees the palliative care academic programs at the VA and UAB and is a GRECC investigator. A graduate of Davidson College and the University of North Carolina (UNC)-Chapel Hill Medical School, she received her internal medicine and geriatrics training at UAB. In 1998, after serving on the faculty at UAB and Boston University, she joined the faculty at the University of Louisville where she began the palliative care and geriatrics program at the University of Louisville (UofL)-affiliated Louisville VA Medical Center. For this work, she received the VA National Mark Wolcott Award for Excellence in Clinical Care. She subsequently returned to UAB to direct the UAB Center for Palliative Care and the Palliative and Supportive Care Section of the Division of Gerontology, Geriatrics and Palliative Care. She engages in clinical care and research related to advanced illness, nutrition, and healthcare delivery-related issues, and provides nutrition, palliative care and geriatrics teaching to faculty, trainees, and students.